HIV AND THE NEW VIRUSES

HIV AND THE NEW VIRUSES

Second Edition

Angus Dalgleish

Division of Oncology,
Department of Cellular and Molecular Sciences,
St George's Hospital Medical School,
Cranmer Terrace,
London SW17 0RE, UK

Robin Weiss

Windeyer Institute of Medical Sciences,
University College London,
46 Cleveland Street,
London W1P 6DB, UK

ACADEMIC PRESS
Harcourt Brace & Company, Publishers
San Diego London Boston New York Sydney Tokyo Toronto

ACADEMIC PRESS
525 B Street, Suite 1900, San Diego,
California 92101–4495, USA
http://www.apnet.com

ACADEMIC PRESS
24–28 Oval Road
LONDON NW1 7DX
http://www.hbuk.co.uk/ap/

A catalogue record for this book is available from the British Library

Library of Congress Card Number: 98-83211

ISBN 0–12–200741–7

Typeset by J&L Composition Ltd, Filey, North Yorkshire
Printed in Great Britain by MPG Books Ltd, Bodmin, Cornwall
99 00 01 02 03 04 MP 9 8 7 6 5 4 3 2 1

CONTENTS

LIST OF CONTRIBUTORS

Alicia Alonso Centre for HIV Research, ICAPB, University of Edinburgh, Edinburgh EH9 3JN, UK

Harold Baum Division of Life Sciences, King's College Hospital, Campden Hill Road, London W8 7AH, UK

Neil Berry Division of Retrovirology, NIBSC, Blanche Lane, Potters Bar EN6 3QG, UK

Andrew J Leigh Brown Centre for HIV Research, Institute of Cell Animal and Population Biology, University of Edinburgh, West Mains Road, Edinburgh EH9 3JN, UK

William F Carman Institute of Virology, University of Glasgow, Church Street, Glasgow G11 5JR, UK

Mario Clerici Cattedra di Immunologia, Centro Ricerche LITA, H.L. Sacco, Universita di Milano, Via Venezian, 1, 21033 Milano, Italy

Alberto Clivio Dipartimento di Biologia e Genetica per le Scienze Mediche, Centro Ricerche LITA, H.L. Sacco, Universita di Milano, Via Venezian, 1, 21033 Milano, Italy

Angus G Dalgleish Division of Oncology, Department of Cellular and Molecular Sciences, St George's Hospital Medical School, Cranmer Terrace, London SW17 0RE, UK

Klaus-Michael Debatin University Children's Hospital, Prittwitzstrasse 43, D-89075 ULM, Germany

Barbara Ensoli Laboratory of Virology, Istituto Superiore di Sanità, Viale Regina Elena, 299, 00161 Rome, Italy

Daniela Fenoglio Unit of Retroviral Immunology, Department of Immunology, San Martino Hospital, University of Genoa and Advanced Biotechnology Center, Largo Benzi, 10, 16132 Genoa, Italy

Jeremy A Garson Department of Virology, University College London Medical School, Windeyer Building, 46 Cleveland Street, London W1P 6DB, UK

Antoine Gessain Unité d'Oncologie Virale, Départment du SIDA et des Rétrovirus, Institut Pasteur, 28 Rue du Dr Roux, 75724 Paris Cedex 15, France

Frances Gotch Department of Immunology, Chelsea and Westminster Hospital, 369 Fulham Palace Road, Chelsea, London SW10 9NH, UK

Marie-Lise Gougeon Départment SIDA et Rétrovirus, Unité d'Oncologie Virale, Institut Pasteur, 28 Rue du Dr Roux, 75724 Paris, Cedex 15, France

Elizabeth Hounsell Department of Biochemsitry and Molecular Biology, University College London, Gower Street, London WC1E 6BT, UK

Paul Kellam Department of Virology, Institute of Cancer Research, Royal Cancer Hospital, Chester Beatty Laboratories, 237 Fulham Road, London SW3 6JB, UK

Jonathan R Kerr Department of Virology, University College London Medical School, Windeyer Building, 46 Cleveland Street, London W1P 6DB, UK

Giuseppina Li Pira Unit of Retroviral Immunology, Department of Immunology, San Martino Hospital, University of Genoa and Advanced Biotechnology Center, Largo Benzi, 10, 16132 Genoa, Italy

Paolo Lusso Unit of Human Virology, DIBIT, San Raffaele Scientific Institute, Milano 20132, Italy

Renaud Mahieux Laboratory of Gene Expression and Receptor Biology, NCI/NIH, Building 41, Bethesda, MD 20892, USA

Fabrizio Manca Unit of Retroviral Immunology, Department of Immunology, San Martino Hospital, University of Genoa and Advanced Biotechnology Center, Largo Benzi, 10, 16132 Genoa, Italy

Dean L Mann Division of Immunogenetics, University of Maryland Medical System, 22 South Greene Street, P2F01E, University Center, Baltimore, MD 21201-595, USA

Stephen McAdam Department of Immunology, Chelsea and Westminster Hospital, 369 Fulham Palace Road, Chelsea, London SW10 9NH, UK

Myra McClure Division of Medicine, Jefferies Research Trust, Imperial College of Medicine at St Mary's, Praed Street, London W2 1NY, UK

Paola Monini Laboratory of Virology, Istituto Superiore di Sanità, Viale Regina Elena, 299, 00161 Rome, Italy

Michael Norcross Division of Haematological Products, Center for Biologics Evaluation and Research, FDA, Bethesda, MD20892, USA

Clive Patience Biotransplant Incorporated, Building 75, 3rd Avenue, Charlestown Navy Yard, MA 02129, USA

B. Matija Peterlin Howard Hughes Medical Institute, Mount Zion Cancer Center, University of California, San Francisco, CA 94143-0703, USA

Andrew Phillips Department of Primary Care and Population Sciences, Royal Free Hospital School of Medicine, Rowland Hill Street, London NW3 2PF, UK

Lisa Rosenblum Division of Medicine, Jefferies Research Trust, Imperial College of Medicine at St Mary's, Praed Street, London W2 1NY, UK

Hanneke Schuitemaker Department of Clinical Viro-Immunology, Central Laboratory of the Netherlands Red Cross Blood Transfusion Service, Plesmanlaan 125, 1006 AD, Amsterdam, The Netherlands

Cecilia Sgadari Laboratory of Virology, Istituto Superiore di Sanità, Viale Regina Elena, 299, 00161 Rome, Italy

Gene M Shearer Experimental Immunology Branch, NCI, NIH, Bethesda, MD 20892, USA

Peter Simmonds Department of Medical Microbiology, University of Edinburgh, Medical School, Teviot Place, Edinburgh EH8 9AG, UK

Donald B Smith Department of Medical Microbiology, University of Edinburgh, Medical School, Teviot Place, Edinburgh EH8 9AG, UK

Simon J Talbot Department of Medical Microbiology, Edinburgh University, Medical School, Teviot Place, Edinburgh EH8 9AG, UK

Richard Tedder UCL Medical School, Department of Virology, The Windeyer Building, 46 Cleveland Street, London W1P 6DB, UK

Christian Trautwein Abteilung Gastroenterologie und Hepatologie, Medischinische Hochsschule Hannover, Hannover, Germany

Mark A Wainberg McGill University AIDS Centre, Lady Davis Institute–Jewish General Hospital, 3755 Chemin Cote Ste-Catherine, Montreal, Quebec, Canada H3T 1E2

Ian V D Weller Department of Sexually Transmitted Diseases, University College London Medical School, Mortimer Market Centre off Capper Street, London W1CE 6AU, UK

Denise Whitby Institute of Cancer Research, Chester Beatty Research Institute, 237 Fulham Road, London SW3 6JB, UK

Mark K Williams Unit of Human Virology, DIBIT, San Raffaele Scientific Institute, Milano 20132, Italy

PREFACE TO THE FIRST EDITION

A decade ago mankind seemed to be on the brink of conquering viral disease. Smallpox had finally been eradicated, safe and efficacious vaccines were becoming available for most of the childhood infections and there was a general air of confidence that virally induced cancers would similarly yield to preventive measures. The recognition of acquired immune deficiency syndrome (AIDS) in 1981 shattered our complacency. Nevertheless, there has been remarkable progress in our understanding of AIDS since the causative agent, human immunodeficiency virus (HIV) was first isolated in 1983. The first six chapters in the volume chart the course of the AIDS epidemic, the molecular and cellular biology and pathology of HIV infection, and the prospects and problems in harnessing this knowledge for therapeutic and preventive means.

During the past decade other human viruses came to light for the first time, even though they may have an ancient provenance in human populations. The human T-cell leukaemia viruses (HTLV-I, HTLV-II) were discovered in 1980 and 1982 representing the first human retroviral pathogens, new human papilloma virus genomes (HPV-16, HPV-18) were identified in 1983 associated with cervical cancer, human herpes virus type 6 was recognized in 1987 and most recently in 1989, hepatitis C virus (HCV) has been identified among the non-A, non-B hepatitis infections. Some of these viruses (HTLV, HHV6) have been isolated by the classic method of propagation in culture, while others (HPV, HCV) were identified by molecular cloning.

The chapters contained in this volume were written to enable scientists and clinicians interested in human viral infection to obtain topical reviews that critically evaluate relevant data and concepts from the rapidly burgeoning literature. The chapters are contributed by leading investigators who have been enjoined to review a wide field without stifling their personal views. The subject matter is necessarily selective. We have, for example, omitted discussion of human papovaviruses or of the interesting animal lentiviruses related to HIV that have come to light since the discovery of HIV.

We are most grateful to the authors for their contributions and for ensuring that they are up to date including 1990 references. We thank Sue King at Academic Press for her understanding and persistence in producing this volume.

Angus G. Dalgleish
Robin A. Weiss

PREFACE TO THE SECOND EDITION

A decade ago we agreed to edit a book of authoritative and easy-to-read reviews on the major aspects of the human immunodeficiency virus (HIV), the causative agent of acquired immune deficiency syndrome, which itself had been recognized for less than a decade at the time of editing. It became apparent while planning the book that there would be much reference to the first human retrovirus, human T cell lymphotrophic virus type 1 (HTLV-1) and its relative HTLV-2, and we therefore commissioned a comprehensive review of these viruses. It became apparent that HIV and HTLV-1 were not the only new viruses to be discovered in the 1980s, and we broadened the remit to include reviews of herpesvirus 6 and the new hepatitis viruses.

In the last decade there has been a tremendous amount of information on HIV, its molecular structure, and the way it interacts with the host. The reviews presented in this volume are timely and summarize the major issues that have emerged since the first book. In order to accommodate this amount of information, the scope of the book is much broader than its predecessor.

It has been brought to our attention that the first book was referred to as much for the chapters on the new viruses as for those on HIV and AIDS. Indeed, it appeared pertinent to address the fact that viruses other than HIV were being discovered for the first time. In the past ten years a number of viruses have been discovered and we have taken the opportunity to include comprehensive reviews not only of the new viruses such as human herpesvirus 8 (HHV-8), the causative agent of Kaposi's sarcoma, but also of new aspects about other viruses only recently discovered at the time of publication of the first edition. This book also contains new information about older viruses such as hepatitis B virus, which can appear enigmatic to those not intricately acquainted with its molecular structure and pathogenesis, and also serves as an introduction to the new hepatitis viruses.

The last decade has seen the discovery and characterization of new neurological diseases and the identification of new associated viruses. We have therefore included a review on this topic as well as on the increasing documentation and association of some clinical conditions associated with endogenous retroviruses.

Although this is not a comprehensive textbook on virology, the issues covered here are of major interest to clinicians and scientists alike who are interested in the emergence of these new infectious agents. We are not aware of any other source of such comprehensive reviews of the major issues in HIV and the new viruses.

We thank Tessa Picknett, Duncan Fatz, Emma White and Lilian Leung at Academic Press for working patiently with us on this volume.

Angus G. Dalgleish
Robin A. Weiss

PREFACE TO THE FIRST EDITION

Chapter 1

HIV-1:
Control of gene expression by the viral regulatory proteins Tat and Rev

Alicia Alonso and B. Matija Peterlin

INTRODUCTION

Expression of the human immunodeficiency virus type 1 (HIV-1) in infected cells results from a complex interplay between cellular proteins and viral components (DNA, RNA and proteins). Prominent among virally encoded proteins are Tat and Rev, which are unique to HIV-1 and related lentiviruses. After extensive mutagenesis of the viral genome, both proteins were found to be essential for viral gene expression, replication and cytopathology (Dayton et al, 1986; Feinberg et al, 1986; Fisher et al, 1986; Sodroski et al, 1986). Although both proteins mediate their functions through viral RNA elements, Tat via the transactivation response element (TAR) and Rev via the Rev response element (RRE), the control points at which Tat and Rev act and their mechanisms of action are fundamentally different. Tat modifies the nascent transcription complex allowing for efficient elongation of transcription, which increases levels of full-length viral transcripts. Rev acts post-transcriptionally to export incompletely spliced viral mRNAs, which code for structural proteins, from the nucleus to the cytoplasm. Thus, both proteins are required to activate fully the quiescent provirus. Since they are indispensable for viral replication, they are also attractive candidates for therapeutic intervention. Therefore, unraveling their mechanism of action is of great biological and clinical interest. This chapter reviews the roles of these regulatory proteins in the viral life cycle, as well as structures and functions of Tat and Rev and their target sequences TAR and RRE. Their interactions with cellular proteins and forms of therapeutic intervention are also addressed.

Tat, Rev AND THE PROVIRAL LIFE CYCLE

Integration of the viral DNA into the host genome generates a provirus, which is a necessary step in the life cycle of all retroviruses. Once this event has occurred, the HIV-1 provirus behaves as a cellular gene, with the promoter and transcriptional start site located in the 5′ long terminal repeat (LTR), and a termination/polyadenylation site located in the 3′ LTR (Figure 1.1A; see also Figure 1.4 and The promoter region). Viral replication depends on the activation of the LTR and on the expression, from the 9 kb

HIV and the New Viruses Second Edition
ISBN 0-12-200741-7

(A)

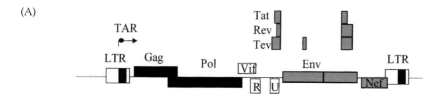

(B)

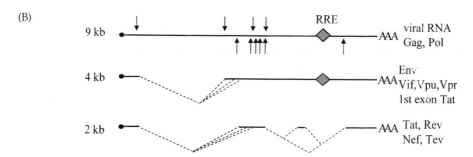

Figure 1.1 HIV-1 genomic organization and mRNA expression
(A) The 9 kb proviral genome is shown schematically flanked by cellular sequences (dotted lines). The 5' LTR sequences contain the promoter. Transcription, indicated by the arrow, starts within R (solid black). Positioned within nucleotides 1 to 60 is TAR, indicated by a solid dot. Coding sequences for the viral proteins are depicted by rectangles. The solid line within *env* separates gp120 from gp41, R= *vpr*, U= *vpu*. Spliced exons of *tat*, *rev* and *tev* are indicated above. The 3' LTR contains termination signals.
(B) The three mRNA species present during active infection are depicted. TAR is present in all of them (solid dot); a diamond indicates the presence of Rev. Splice donor sites are indicated above the line representing the full-length transcript; splice acceptor sites are indicated below. Their positions are approximate. Sizes of the unspliced, singly spliced and multiply spliced mRNA are indicated on the left, and on the right the proteins they encode. Dotted lines represent different splicing events.

provirus, of nine different open reading frames (ORFs). Of these, three encode structural proteins, and six encode regulatory and accessory proteins.

Viral mRNAs

In T cells or macrophages which support replication of HIV-1, three sizes of viral mRNAs can be detected by northern blot analysis performed with the 3' LTR polyA probe. They are 9 kb, 4 kb and 2 kb in length respectively (Figure 1.1B). The 9 kb transcripts, which represent the genomic mRNA, are homogeneous in size. They are unspliced and serve two purposes: they are the viral RNA which is packaged into new virions; and they translate the structural Gag proteins and the enzymatic Pol polyprotein, encoding reverse transcriptase, protease and integrase. In contrast, the 4 kb and 2 kb transcripts are heterogeneous in size and their existence depends on the choice of alternative splicing signals located along the 9 kb transcript (Figure 1.1B). The 4 kb transcripts are singly spliced and code for Env, Vpr, Vpu, Vif and the first exon of Tat. The 2 kb transcripts are multiply spliced, and code for Tat, Rev, Nef and Tev. To date, 15 alternatively spliced transcripts have been described (Kim et al, 1989; Robert-Guroff et al, 1990; Schwartz et al, 1990a,b; Neumann et al, 1994).

Temporal analyses of these viral mRNAs, in T cells that had undergone a single round of HIV-1 replication, determined that their expression is tightly regulated. The appearance of the 2 kb species precedes that of the 4 kb and 9 kb species: they are thus referred to as early and late transcripts respectively (Feinberg et al, 1986; Kim et al, 1989). Since Tat and Rev were translated from early transcripts, they were considered likely candidates for controlling late events in viral replication. Moreover, mutations in either of these two proteins abrogated viral replication (Dayton et al, 1986; Feinberg et al, 1986; Fisher et al, 1986; Sodroski et al, 1986). Both proteins control coordinately the expression and localization of viral transcripts.

Tat phenotype

Analyses of the total viral RNA content in cells infected with viruses containing mutations in Tat demonstrated low abundance of all three mRNA species. This defect was corrected when cells were complemented with the wild-type Tat (Feinberg et al, 1991; Adams et al, 1994). By performing steady-state RNA and nuclear run-on analyses, where the changes in the density of promoter proximal and promoter distal transcripts were monitored, it was determined that Tat increased rates of transcription from the HIV-1 LTR (hence its name *transactivator of transcription*). Tat allows RNA polymerase II (RNAPII) to traverse the length of the viral genome, thus dramatically increasing levels of elongated transcripts (Kao et al, 1987; Laspia et al, 1989, 1990; Feinberg et al, 1991). This effect of Tat was dependent on a viral *cis*-acting RNA sequence, called the transactivation response element (TAR) which is located precisely at the start of all viral transcripts. TAR extends from nucleotides +1 to +60 and forms a stable stem-loop structure (Rosen et al, 1985; Kao et al, 1987; Feng and Holland, 1988; Hauber and Cullen, 1988; Jakobovits et al, 1988; Berkhout and Jeang, 1989; Garcia et al, 1989; Selby et al, 1989; Roy et al, 1990a,b).

In the absence of Tat, the predominant viral transcripts present in infected or transfected cells are non-polyadenylated RNAs of about 60 bp, which contain TAR. They are referred to as short transcripts. Since TAR is resistant to nuclease degradation, short transcripts are probably the result of nuclease degradation of incompletely elongated transcripts which are constantly being initiated at the HIV-1 LTR by the inducer of short transcripts (IST) (see The promoter region; Kao et al, 1987; Selby et al, 1989; Toohey and Jones, 1989). When Tat is present, it binds to TAR as nascent RNA (Berkhout et al, 1989), represses the synthesis of short transcripts (Pendergrast and Hernandez, 1997), and modifies the transcription complex allowing complete elongation of viral transcripts. These events are discussed in detail in Tat and the transcription complex.

Rev phenotype

When viral transcripts from cells infected with viruses bearing mutations in *rev* were analyzed, a significant difference was apparent when nuclear and cytoplasmic RNAs were fractionated and compared. All three mRNA species were present in the nucleus, but only 2 kb transcripts were observed in the cytoplasm (Emerman et al, 1989; Felber et al, 1989; Hammarskjold et al, 1989; Malim et al, 1989). The appearance of 9 kb and 4 kb transcripts in the cytoplasm was restored only after the coexpression of Rev in the same cells (Knight et al, 1987). Thus, Rev allows for the movement of the unspliced and singly spliced mRNA species from the nucleus to the cytoplasm. Rev derives its name, the *regulator of expression of virion proteins*, because these late transcripts encode the structural and enzymatic proteins needed for virion assembly.

Splicing versus nuclear export

Some groups have hypothesized that Rev also inhibits aspects of pre-mRNA splicing, which leads indirectly to enhanced pre-mRNA export (Hammarskjold et al, 1989; Kjems and Sharp, 1993). However, it has now been demonstrated that Rev interacts directly with the nuclear export machinery (see Rev and the nuclear transport of mRNAs). In contrast to cellular transcripts, which do not exit the nucleus until all introns have been removed, the two late transcripts of HIV-1 contain at least one complete unexcised intron, i.e. the second major HIV-1 intron which codes for Env (Figure 1.1B). Like Tat, the activity of Rev is dependent on the presence of a *cis*-acting RNA element. This sequence, the Rev response element, is between 240 and 351 nucleotides long. It is located precisely in the second major intron of HIV-1, just 3′ to the junction between the gp120 and gp41 subunits of *env* (Rosen et al, 1988; Felber et al, 1989; Hadzpoulou-Cladaras et al, 1989; Malim et al, 1989b; Mann et al, 1994). The RRE sequence is therefore present only in unspliced and singly spliced mRNA species. How interactions between Rev and the RRE lead to nuclear export of the late viral transcripts is discussed on pp. 13–18.

The proviral life cycle

Tat and Rev control the switch from latency to active replication of the provirus in the following manner (Figure 1.2). After integration, the virus remains in a transcriptionally latent state, characterized by the presence of short TAR transcripts in the cytoplasm (Feinberg et al, 1991; Adams et al, 1994). Cellular activation signals overcome this latency by increasing rates of initiation of HIV-1 transcription from the LTR (see Tat and the transcription complex; Jones and Peterlin, 1994). All introns from these primary transcripts are removed in the nucleus by the splicing machinery. Thus, multiply spliced 2 kb transcripts encoding Tat and Rev accumulate (Kim et al, 1989). These early transcripts are exported rapidly to the cytoplasm, where Tat and Rev are translated and imported into the nucleus (Figure 1.2A) (Malim et al, 1989a; Perkins et al, 1989; Ruben et al, 1989; Siomi et al, 1990). Tat then interacts with TAR and the transcription complex and increases tremendously the amount of full-length polyadenylated transcripts in the nucleus (Feinberg et al, 1991; Adams et al, 1994). Because of Rev and its interaction with the nuclear export pathway, these unspliced and singly spliced transcripts are now translocated to the cytoplasm through the RRE (Knight et al, 1987; Fornerod et al, 1997a; Stade et al, 1997). In the cytoplasm, translation of structural proteins and assembly of new virions ensue (Figure 1.2B).

THE Tat PROTEIN: A MODULAR STRUCTURE

Tat is encoded in two exons. The first exon is located just 5′ of *env* and encodes 72 amino acids. The second exon is located 3′ of *env*, and is of variable size, from 14 to 29 amino acids, depending on the viral isolate (Arya et al, 1984; Sodroski et al, 1985). Within early transcripts, three species of multiply spliced mRNAs encode Tat. Within

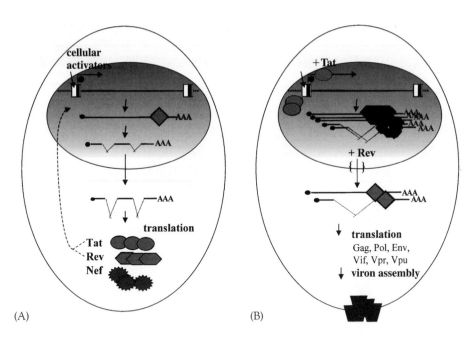

Figure 1.2 *The proviral life cycle*
(A) Early events. Proviral transcription is activated by cellular transcription factors acting on
the 5′ LTR. Low levels of mRNA accumulate in the nucleus (shaded area), from which only the
multiply spliced mRNAs (2 kb) are detected in the cytoplasm. These code for Tat (circles), Rev
(hexagons) and Nef (porcupines). Tat and Rev relocalize into the nucleus.
(B) Late events. In the nucleus, Tat binds to TAR and modifies the transcription complex,
indicated by the drawn-out arrow. This results in increased levels of mRNAs. In addition, Rev
binds to the RRE and allows for the unspliced and singly spliced mRNAs to be exported to the
cytoplasm. After translation, virion assembly and extrusion occur.

late transcripts, Tat is encoded by a singly spliced first exon (Schwartz et al, 1990b). In
addition, Tat is expressed as a hybrid molecule, Tev, the product of the first exon of *tat*
spliced to 114 nucleotides from *env* and the second exon of *rev* (see Figure 1.1) (Benko
et al, 1990).

Transient transfection assays into tissue culture cells, using plasmids containing the
HIV-1 LTR fused to a reporter gene (chloramphenicol acetyltransferase or luciferase)
and expression plasmids containing either wild-type or mutated *tat*, have demonstrated
that the first exon of Tat suffices for high levels of transactivation (Garcia et al, 1988;
Kuppuswamy et al, 1989). The second exon mediates cellular functions, such as lymphocyte activation (Howcroft et al, 1993; Chang et al, 1995; Ott et al, 1997).

At the amino acid level, the first exon of *tat* is highly conserved among different isolates (Myers et al, 1991), as well as among different lentiviruses (Peterlin et al, 1993).
This similarity led to the preliminary delineation of five structural domains, namely the
N-terminal, cysteine-rich, core, basic and C-terminal domains of Tat (Figure 1.3A).
Their roles have been defined by mutagenesis followed by functional analysis. Like
well-known DNA transactivator proteins, Tat has a modular structure, with one domain
that interacts with cellular factors and one domain that binds to its cognate RNA
sequence. However, the RNA-binding domain of Tat cannot be separated from its activation domain for efficient interaction with TAR.

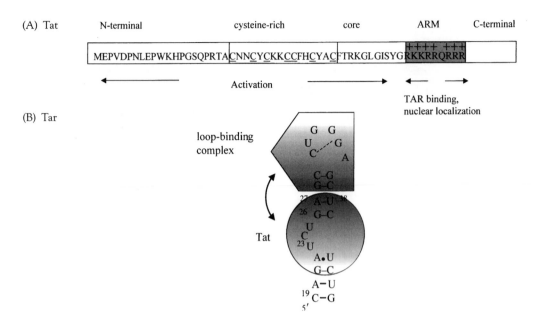

Figure 1.3 *Domain structure of Tat and RNA structure of TAR*
(A) The first exon of Tat. The five structural domains with their relevant amino acid sequences are indicated. The activation domain extends from amino acids 1 to 47. Positive charges in the ARM domain are indicated above the corresponding amino acids. A solid unfilled block represents the C-terminal domain.
(B) The minimal functional TAR sequence, from positions +19 to +43, and its proposed secondary structure are depicted. Within TAR, the lower stem, the bulge, the upper stem and the loop are indicated. The loop is shown with base pairing of C30·G34 and a bulged-out A35 (Jones and Peterlin, 1994). A22·U40 are thought to pair after binding, which is indicated by a dot. Tat binds to the U23 in the major groove, occupying the bulge and 2 base pairs above and below the bulge. A putative loop-binding complex is shown which may interact and stabilize Tat binding to TAR, indicated by the arrow.

The activation domain of Tat

The first three structural domains of Tat are necessary for transcriptional activation. The N-terminal domain corresponds to amino acids 1 to 21. It contains PXXXP repeats, and acidic amino acids at positions 2, 5, 9 which appear to be necessary for transactivation. These amino acids can be replaced, provided the overall structure remains amphipathic (Kuppuswamy et al, 1989; Rappaport et al, 1989; Tiley et al, 1992). The cysteine-rich domain corresponds to amino acids 22 to 37, and contains seven cysteines, of which three cysteine pairs are indispensable (Garcia et al, 1988; Sadaie et al, 1988; Kuppuswamy et al, 1989; Rice and Carlotti, 1990). The core domain corresponds to amino acids 38 to 48. It is essential and has no known similarities to other activation domains (Kuppuswamy et al, 1989; Rice and Carlotti, 1990). The use of chimeric Tat proteins on chimeric HIV-1–LTRs has corroborated that these 47 amino acids correspond to the activation domain. These amino acids were fused to each one of two known RNA-binding proteins: the coat protein of the bacteriophage MS2 and Rev; and TAR was replaced by the MS2 operator RNA target or the stem-loop IIB (SLIIB) of the RRE (Selby and Peterlin, 1990; Southgate et al, 1990). These fusion proteins were able to transactivate the HIV-1 LTR from these hybrid promoters, thus confirming that the activation domain is contained within these 47 N-terminal residues, and that Tat acts by binding to RNA. Further domain swapping with Tat from the equine infectious anemia virus (EIAV) determined that a minimal lentiviral Tat can be constructed with the cysteine-rich and core domains of Tat of EIAV (15 amino acids) and the basic domain of Tat from HIV-1 (Derse et al, 1991).

To prove that Tat can also function at the level of initiation of transcription, Tat was tethered to the promoter as a DNA-binding protein (Berkhout et al, 1990; Kamine et al, 1991; Southgate and Green, 1991). Although these attempts have also been successful, the mechanism of Tat action via DNA and RNA is different. RNA-targeted Tat inhibits the synthesis of short transcripts and enhances that of elongated transcripts, whereas DNA-targeted Tat increases both processes (Pendergrast and Hernandez, 1997). It should be noted that Tat acts via RNA binding in preference to DNA binding.

The basic domain of Tat (arginine-rich motif)

The basic domain of Tat extends from amino acids 49 to 57, and consists of an arginine-rich motif (ARM), RKKRRQRRR. This motif has a dual role: it acts as a nuclear localization signal (Hauber et al, 1989; Ruben et al, 1989; Siomi et al, 1990), and binds to TAR (Dingwall et al, 1989; Roy et al, 1990a; Weeks et al, 1990). In cells, both the ARM and the activation domain of Tat are required to bind TAR efficiently, suggesting that a cellular protein interacts simultaneously with the activation domain of Tat and the loop in TAR (Jones and Peterlin, 1994) (see pp. 8–13).

The C-terminal domain of Tat

The C-terminal domain of Tat, which extends from amino acids 58 to 72, has no known motifs. Since deletions in the region past amino acid 60 cause only a small decrease in the activity of Tat, this domain is thought to act as an auxiliary domain (Muesing et al, 1987; Garcia et al, 1988; Frankel et al, 1989; Kuppuswamy et al, 1989).

Tat structure

Nuclear magnetic resonance (NMR) spectroscopy (Bayer et al, 1995) has solved the structure of Tat from HIV-1Z2, an isolate with 86 amino acids. This analysis has revealed that the N-terminus of Tat is sandwiched between core and C-terminal domains, leaving a flexible cysteine and basic domain accessible for protein–protein and protein–RNA interactions. Bayer et al found no α-helices in Tat, in contrast with the α-helical structure observed with the minimal lentiviral Tat (Mujeeb et al, 1994).

Cellular factors that bind to Tat

Since the activation domain of Tat must interact with some component of the transcription complex, biochemical approaches have been used to identify the factor(s) that interacts with Tat. About 15 proteins, which range from transcription factors to kinases, have been isolated. Among the former are TBP (Kashanchi et al, 1994, 1996); the p62 subunit of TFIIH (Blau et al, 1996); TatSF1 (Zhou and Sharp, 1996); the largest subunit of RNAPII (Mavankal et al, 1996) and Sp1 (Jeang et al, 1993). The latter include CAK, the CDK-activating kinase associated with the TFIIH complex (Cujec et al, 1997; Garcia-Martinez et al, 1997) and CDK9/PITALRE, a kinase associated with the Tat-associated kinase/positive transcription elongation factor b complex (TAK/P-TEFb) (Herrmann and Rice, 1995; Yang et al, 1996; Mancebo et al, 1997; Zhu et al, 1997). The relevance of these kinases in mediating transactivation by Tat is discussed on pp. 11–13.

TRANSACTIVATION RESPONSE ELEMENT, TAR: A DUAL ROLE

The TAR element is the target sequence required for Tat transactivation. This sequence is unique, among all other *cis*-acting elements utilized in transcription, in that it forms an RNA stem loop. Early suggestions that TAR was an RNA element included the following observations: TAR was active only in the sense orientation (Peterlin et al, 1986), computer modeling demonstrated that TAR could form a stable RNA hairpin (Muesing et al, 1987; Feng and Holland, 1988), mutations that affected folding of this putative structure inhibited transactivation, which could be rescued by compensatory mutations (Feng and Holland, 1988; Hauber and Cullen, 1988; Jakobovits et al, 1988; Garcia et al, 1989; Selby et al, 1989; Roy et al, 1990b), Tat could be tethered to the transcription complex via heterologous RNA-binding domains (Selby and Peterlin, 1990; Southgate et al, 1990), and RNA but not DNA TAR decoys could inhibit transactivation (Graham and Maio, 1990; Sullenger et al, 1990, 1991; Lisziewicz et al, 1991).

Computer modeling of TAR exposed two distinctive elements in the TAR sequence (Figure 1.3B): a three-nucleotide bulge between positions +23 and +25 and a six-nucleotide loop between positions +30 and +35. The rest of the sequence provides a stable stem that makes TAR a highly nuclease-resistant RNA structure. Deletions and single point mutations in TAR narrowed the minimal sequences required for transactivation to those located between positions +19 and +43. Within this minimal structure, both the U23 in the bulge and the loop are required for Tat transactivation (Berkhout and Jeang, 1989; Dingwall et al, 1989; Cordingly et al, 1990; Roy et al, 1990a,b; Calnan et al, 1991a,b).

The bulge in TAR

Final proof that Tat binds TAR RNA came with the use of the electrophoretic mobility shift assay (EMSA), which revealed that recombinant Tat protein or Tat peptides bound to ^{32}P-labeled TAR RNA. Binding occurred in a ratio of one TAR to one Tat, was mediated by the ARM, and was specific for U23. In addition, two base pairs in the stem above the bulge, G26·C39 and A27·U38, two base pairs below the bulge, A22·U40 and G21·C41, and two phosphates located between positions 22, 23 and 24 were also involved in binding (Calnan et al, 1991a,b; Weeks and Crothers, 1991; Delling et al, 1992; Tao and Frankel, 1992; Churcher et al, 1993; Hamy et al, 1993; Pritchard et al, 1994).

When in solution, the A22 and U40 in TAR are unpaired making a four-nucleotide bulge, and U23 is stacked between C24 and A22 (Figure 1.3B) (Colvin and Garcia-Blanco, 1992; Critchley et al, 1993). The bulge bends the TAR RNA stem and introduces local distortions that widen the major groove, making it available for Tat binding

(Weeks and Crothers, 1991; Delling et al, 1992). Structural studies using circular dichroism (CD) spectrum analysis suggested that the ARM is unstructured and adopts a fixed conformation only after its binding to TAR (Calnan et al, 1991a). The fact that only one arginine from the ARM is sufficient for binding, albeit at low specificity, led to several studies using NMR spectroscopy, TAR and argininamide, an arginine analog (Calnan et al, 1991b; Puglisi et al, 1992). A simple binding model, the 'arginine fork' was proposed. Upon the binding of arginine, the bases in the bulge become separated and U23 and A27·U38 form a base-triple interaction that stabilizes hydrogen bonding of arginine to G26 and P22·P23 (Puglisi et al, 1993). Recently this model has been challenged by NMR data obtained with TAR and a Tat peptide containing both basic and core domains, which provides maximal specificity of binding (Churcher et al, 1993; Aboul-ela et al, 1995). Contrary to the findings of Frankel and co-workers (Puglisi et al, 1992, 1993), Varani and co-workers (Aboul-ela et al, 1995) found that U23 does not participate in the formation of a base triplet with A27·U38. Instead the conformational change is brought about by interactions of an arginine residue with both U23 and G26, thus repositioning the phosphate groups P22, P23 and P40 on the surface of the molecule.

The loop in TAR

The nucleotides in the loop interact through non-Watson–Crick base pairing, where C30 and G34 base pair and A35 bulges out (Figure 1.3B) (Colvin et al, 1993; Critchley et al, 1993). Notably, although these nucleotides are essential for Tat transactivation, they do not mediate the binding of Tat to TAR (Dingwall et al, 1989, 1990; Cordingly et al, 1990; Roy et al, 1990a; Weeks et al, 1990; Calnan et al, 1991a,b; Weeks and Crothers, 1991; Delling et al, 1992; Churcher et al, 1993). This finding further suggests that cellular factors bind the loop and in conjunction with Tat form an essential part of the modified transcription complex.

Cellular proteins that bind to the loop in TAR have been sought, both biochemically and genetically. Partially purified nuclear extracts from HeLa cells contain two fractions that bind TAR: TRP-185/TRP-1, which binds to the loop, and TRP-2, which binds to the bulge (Sheline et al, 1991; Wu et al, 1991). Their exact roles in transcription mediated by Tat are still under investigation. The fraction TRP-185 binds to TAR through cellular cofactors, of which three – EF-12 – PTB and SRB, have been identified. RNAPII also binds to TAR using these same cofactors (Wu-Baer et al, 1995, 1996).

Genetic evidence indicating that loop-binding proteins are required for Tat transactivation comes from studies in rodent cells which support very low levels of Tat transactivation (Hart et al, 1989; Newstein et al, 1990). Bypassing the requirement of TAR loop-binding proteins, with heterologous tethering of Tat via the coat protein, the EIAV TAR and Gal4 DNA-binding domain, restored Tat transactivation in these cells (Alonso et al, 1992; Madore and Cullen, 1993). Binding studies in vivo suggested that a complex involving Tat, the loop in TAR and a protein encoded in human chromosome 12 are all required for efficient transactivation (Alonso et al, 1994). An 83 kDa protein that binds to the loop and is specific for chromosome 12 has been described (Hart et al, 1995).

Tat AND THE TRANSCRIPTION COMPLEX

Tat acts on non-processive transcription complexes which are set up by the HIV-1 TATA box (Olsen and Rosen, 1992; Lu et al, 1993). In the absence of Tat, these complexes stall and gradually abort transcription 100–200 nucleotides downstream of the initiation site (Kao et al, 1987; Laspia et al, 1989, 1990; Feinberg et al, 1991). Thus, Tat might have effects both on promoter clearance and transcription elongation.

The promoter region

The viral promoter is located in the U3 region of the 5′ LTR (Figure 1.4). It contains upstream regulatory sequences that include binding sites for USF, and T cell-specific activators such as ETS1 and LEF. In addition it contains sites for extracellular signal transducers, such as NF-AT (position −300) and NFkB. All these elements set up the minimal transcription levels that can break cellular latency of the provirus (see The proviral life cycle; Jones and Peterlin, 1994).

The core promoter, which extends from position −80 to the site of initiation of transcription, contains a TATA box, an initiator site (SSR) and three tandem Sp1 sites. The

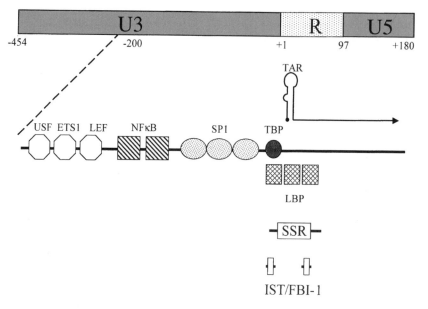

Figure 1.4 *The HIV-1 promoter region*
The 5′ LTR is schematically represented at the top of the Figure. The coordinates given under U3, R and U5 define the limits of these regions with respect to +1, the first transcribed nucleotide. Below, the DNA sequence and the transcription factors that interact with the sequences, from positions −200 to +100, are shown. USF, ETS1 and LEF are upstream transcription factors; sites for NFkB, an inducible activator, are located within positions −100 to −80. The core promoter is composed of a TATA box (positions −30 to −20), and three Sp1 sites (−80 to −45). Additionally it contains three LBP/UBP sites (−30 to +20) which overlap with the initiator site (SSR). The IST (inducer of short transcripts) is located between −5 and +26 and +40 to +59; FBI-1 binds to IST. Transcription is indicated by the arrow, beginning with TAR, which forms at the 5′ end of all RNAs.

Sp1 sites are required to facilitate the assembly of non-processive complexes, but can be functionally replaced by other activator sequences (Southgate and Green, 1991; Berkhout and Jeang, 1992). Regulatory sequences downstream of the promoter include LBP-1/UBP-1 binding sites at position -38 to $+27$, which colocalize with the SSR, and the inducer of short transcripts element (IST). The IST is located between position -5 and $+26$ and position $+40$ to $+59$ and overlaps TAR. The role of the LBP-1/UBP-1 complex is still unclear; however, the IST mediates the synthesis of the short TAR transcripts, which recruit Tat to the promoter (Sheldon et al, 1993; see Tat phenotype) Recently, a factor that binds to the IST, FBI-1, has been described (Pessler et al, 1997).

Transcription

Initiation of transcription occurs through binding of a very stable RNAPII holoenzyme containing RNAPII, the general transcription factors (GTFs) TFIIF, TFIIH and mediator (a multisubunit complex containing suppressors of RNAP B proteins and other activators and repressors) onto a preassembled TFIID and TFIIB at the TATA box (Koleske and Young, 1995; Ossipow et al, 1995; Maldonado et al, 1996). After incorporation of a few nucleotides, the complex will pause until a critical regulatory step, the phosphorylation of the C-terminal domain (CTD) of the largest subunit of RNAPII, is achieved. This step represents the transition from initiation to elongation complexes. The CTD contains 52 repeats of a YSPTSPS domain in which serines, threonines and tyrosines are potential targets of phosphorylation. RNAPII thus exists in a non-phosphorylated form in initiation complexes (RNAPIIA) and in a hyperphosphorylated form in elongation complexes (RNAPIIO). Two steps are controlled by phosphorylation of the CTD, promoter clearance (when GTFs and mediator disengage from RNAPII) and elongation. Kinases that control these events are presently under study, and are targeted by Tat.

Cellular kinases

The original evidence that suggested that Tat affected rates of elongation of RNAPII by recruiting a cellular kinase was based on the observation that the purine nucleoside analog 5,6–dichloro-1B-D-ribofuranosylbenzimidazole (DRB) blocked Tat transactivation. DRB inhibits elongation by inhibiting cellular kinases (Marciniak et al, 1990; Marciniak and Sharp, 1991). The search for the kinase, however, split into two fields. While one set of data implicated TFHII, which plays a pivotal role in promoter clearance and has CAK-associated kinase activity, other studies implicated a new kinase complex, TAK. Tat can interact directly with the p62 subunit of TFIIH, and increases the ability of TFIIH to phosphorylate the CTD (Blau et al, 1996; Parada and Roeder, 1996). On the other hand, Tat interacts, through its activation domain, with TAK. TAK is more DRB sensitive than TFHII and can hyperphosphorylate the CTD (Herrmann and Rice, 1995; Yang et al, 1996).

Two subsequent findings have established that Tat targets both kinases. First, Tat has been shown to bind directly to CAK (Cujec et al, 1997; Garcia-Martinez et al, 1997). Second, the 42 kDa kinase subunit in TAK has been identified as the cyclin-dependent kinase CDK9, and is the human homolog of the kinase subunit of the *Drosophila* elongation complex P-TEFb (Zhu et al, 1997); P-TEFb acts at an early elongation step by preventing RNAPII arrest, is DRB sensitive and can phosphorylate the CTD (Marshall et al, 1996). TAK is therefore the human homolog of P-TEFb.

The role of CDK9 in mediating Tat transactivation has been reinforced by studies in which a significant correlation was found between inhibitors of CDK9 and decreased levels of Tat transactivation (Mancebo et al, 1997).

It has recently been demonstrated that cyclin T, the associated CDK9 cyclin, is a subunit of the TAK/P-TEFb complex. Cyclin T interacts specifically with the activation domain of Tat. This new complex binds to TAR with bulge and loop specificity. Cyclin T has been localized to chromosome 12, and rescues activity in rodent cells (Wei et al, 1998).

Mechanism of Tat transactivation

An attractive two-step regulatory model has been suggested which takes into account all these data (Figure 1.5) (Jones, 1997). First, Tat is recruited to the transcription complex as part of the RNAPII holoenzyme, through its interaction with CAK/TFHII. In this case, Tat increases promoter clearance. In addition, Tat is recruited to the transcription complex via TAR as part of the TAK complex. In this case, either by increasing the local

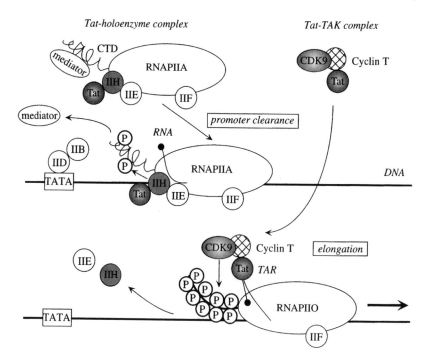

Figure 1.5 *Proposed mechanism of action of Tat*
Tat exists in two complexes in cells, as Tat-RNAPIIA holoenzyme complex (Tat-holoenzyme complex) and Tat-TAK/P-TEFb complex (Tat-TAK complex). The former complex is part of the preinitiation complex on the HIV-1 LTR, where TFIIH and its kinase CDK7 are brought into proximity of the CTD of RNAPII. At this point, RNAPII is hypophosphorylated (RNAPIIA and RNAPIIO) and clears the promoter. Several general transcription factors (GTFs) and mediator leave RNAPII. TAR is transcribed and recruits the Tat-TAK complex, which binds to RNA. CDK9 hyperphosphorylates the CTD of RNAPII and possibly other proteins, thus converting a non-processive to processive RNAPII (RNAPIIO). TAK, which is a multisubunit complex, is shown as consisting minimally of CDK9 and its associated cyclin T.

concentration of TAK or hyperphosphorylating the CTD, Tat enhances the processivity of the early elongation complex. Initially, Tat travels with the early transcription complex bound to TAR; at later stages it disengages from TAR and stays with the modified transcription complexes (Keen et al, 1996, 1997). It remains to be determined whether the complex formed by Tat and TAK binds TAR with specificity for the loop, and, if so, whether a component of TAK is the factor encoded by human chromosome 12.

THE Rev PROTEIN: A MODULAR STRUCTURE

Rev is encoded by two exons, which overlap those encoding Tat, but are read in a different translation frame. Whereas the first exon codes for 25 amino acids, the second exon codes for 91 amino acids. The final protein has a molecular mass of 13 kDa and is modified by phosphorylation (Feinberg et al, 1986; Sodroski et al, 1986; Hauber et al, 1988). Six doubly or triply spliced mRNAs can direct the translation of Rev at early stages of replication (see Figure 1.1) (Schwartz et al, 1990a).

Transient transfection assays, using target plasmids containing an RRE/reporter gene flanked by an HIV-1 splice donor (SD) and splice acceptor (SA) sites, were utilized to map domains for Rev activity. As with Tat, extensive mutageneses studies identified two functional domains within Rev, a multifunctional domain and an effector domain. These domains, however, are located within dispensable regions (Figure 1.6A). The multifunctional domain is found between residues 14 and 56, and the effector domain between amino acids 75 and 84. Amino acids 1 to 8 and 85 to 91 are dispensable for Rev function (Malim et al, 1991).

The multifunctional domain of Rev

The multifunctional domain is responsible for three Rev functions: localization of Rev to the nucleus, binding of Rev to the RRE and the oligomerization of Rev. The nuclear localization domain is contained within amino acids 40 to 45 and can target proteins efficiently to the nucleus (Malim et al, 1989a,b; Perkins et al, 1989). An ARM from positions 35 to 50, which includes the nuclear localization domain, forms the RNA-binding domain. This ARM can functionally replace that of Tat; however, the tryptophan at position 45 is a strict requirement for binding to the RRE (Hope et al, 1990; Malim et al, 1990; Olsen et al, 1990; Subramanian et al, 1990; Zapp et al, 1991).

Unlike Tat, Rev binds to its target sequence, the RRE, as a multimer, in a ratio of up to 8 Revs to 1 RRE. In fact, Rev can even multimerize in the absence of the RRE both in vitro and in vivo. Sequences involved in oligomerization lie on either side of the arginine-rich motif (amino acids 14 to 34 and 51 to 56) (Heaphy et al, 1990, 1991; Olsen et al, 1990; Malim and Cullen, 1991; Zapp et al, 1991). Deletions of the oligomerization domain result in mutant Rev proteins that bind the RRE in a 1 to 1 ratio, and are inactive in vivo. This indicates that oligomerization is essential for Rev activity (see Oligomerization of Rev on the RRE and Mechanism of Rev action) (Malim and Cullen, 1991).

The effector domain of Rev (nuclear export signal)

The effector domain of Rev, i.e. the domain that interacts with the cellular factor(s) that mediate the nuclear export of RRE/Rev-containing mRNAs, comprises amino acids 73

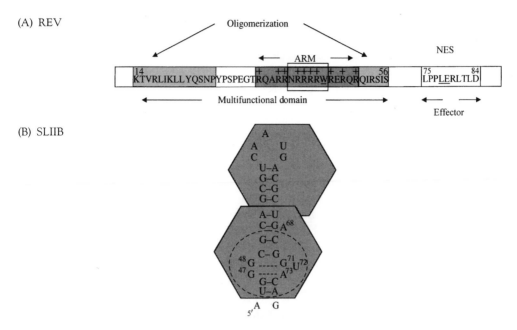

Figure 1.6 *Domain structure of Rev and RNA structure of the SLIIB*
(A) The two functional domains in Rev are indicated below the rectangle that represents Rev.
Inside are the relevant amino acid sequences. The bipartite oligomerization domain spans
amino acids 14 to 27 and 51 to 56; the ARM spans amino acids 35 to 50, the positive charges
are indicated above the respective amino acids, and tryptophan 45 is underlined. Boxed amino
acids within the ARM (40 to 45) comprise the nuclear localization sequence. The NES spans
amino acids 75 to 84; underlined are the amino acids that when replaced by DL eliminate the
function of Rev (RevM10).
(B) The sequence and proposed secondary structure of the SLIIB, from nucleotides 44 to 76 in
the RRE. The purine-rich loop (bubble) is indicated by the dotted circle, non-Watson–Crick
base pairing is indicated by a dotted line. Rev binds to the major groove of the SLIIB and
interacts with bases on either side of the groove. Two Rev molecules are positioned on the
SLIIB to indicate oligomerization occurring after initial binding.

to 84. This is a leucine-rich region (LQLPPLERLTLD). Mutations in these residues
abolished the activity of Rev without compromising its binding to RRE, nuclear localization or oligomerization (Perkins et al, 1989; Hope et al, 1990; Venkatesh and
Chinnadurai, 1990; Malim et al, 1991). This effector domain has recently been shown
to act as a nuclear export signal (NES), and is referred to here as Rev NES (see Nuclear
export signals). Data obtained from domain swapping experiments have strengthened
the notion that Rev NES interacts with cellular factor(s), and suffices for Rev activity.
Rev NES was enough to confer Rev activity when it was assayed as a fusion protein with
the MS2 coat protein and its target RNA operator (McDonald et al, 1992). Likewise,
Rev NES was interchangeable with the equivalent domain from the Rev-like protein Rex
(Hope et al, 1991; Weichselbraun et al, 1991).

Missense mutations in Rev NES, such as those introduced in RevM10 (a change in
positions 78 and 79 from LE to DL), resulted in a *trans*-dominant negative protein
(Malim et al, 1989a; Venkatesh and Chinnadurai, 1990). RevM10 can inhibit HIV-1
replication; however, instead of interacting with a cellular factor and squelching it, the

mutated proteins acted in *trans* by oligomerizing with wild-type Rev, effectively generating non-functional oligomers (Hope et al, 1992).

Cellular factors that bind to Rev

Biochemical approaches aimed at identifying cellular factors that bind to Rev have identified four proteins. One, p32 (a factor that copurifies with the splicing factor ASF/SF2) binds to the ARM. The other three proteins bind to the NES: the translation initiation factor 5A (eIF-5A) (Ruhl et al, 1993); a nucleoporin-homolog protein named Rab/hRIP (Bogerd et al, 1995; Fritz et al, 1995) and Exportin 1 complexed with RanGTP (Fornerod et al, 1997a). Conflicting data as to whether eIF-5A mediates Rev activity exists (Fischer et al, 1994), and since p32 does not bind the NES, only the relevance of Rab/hRIP and genetic and biochemical data supporting Exportin 1/RanGTP as the physiological targets of Rev NES are discussed in the section on Rev and the nuclear transport of mRNAs.

THE Rev RESPONSE ELEMENT: A NUCLEATION SITE

The Rev response element

Deletion of the RRE results in a phenotype identical to that of Rev deletions, i.e. abrogation of HIV-1 replication (Hadzpoulou-Cladaras et al, 1989; Malim et al, 1990). The RRE can be positioned at different places within *env*, preserving its function, but like an RNA element it is orientation-dependent. Computer modeling predicted that the 240 nucleotides could form a very stable series of stem-loop structures (known as SLI through SLV), protruding from a long central stem, stem I (Malim et al, 1989a). However, recent studies have determined that the biologically active RRE is longer and measures 351 nt, which include an extra 58 nt on the 5' end and 59 nt on the 3' end beyond the original model. This new structure effectively fuses SLIII and SLIV, and has a longer SLI (Mann et al, 1994).

The stem loop IIB of RRE

The use of EMSA readily demonstrated that purified Rev bound to the predicted 240 nt RRE (Daly et al, 1989; Zapp and Green, 1989). However, mutational analysis followed by functional assays determined that not all the predicted RRE was necessary for Rev function. While SLIII, IV and V were irrelevant for Rev function, SLI was required for full activity, and only SLII was essential for Rev binding and function (Heaphy et al, 1990; Malim et al, 1990; Bartel et al, 1991). Further binding assays with purified recombinant Rev protein or a peptide containing the ARM demonstrated that in stem loop II there is one high-affinity binding site. This site is called SLIIB, and spans nucleotides 44 to 76 (Figure 1.6B) (Heaphy et al, 1991; Iwai et al, 1992; Kjems et al, 1992; Tiley et al, 1992b).

Nuclease mapping experiments, chemical interference analyses and NMR spectroscopy concurred that SLIIB contained a purine-rich 5 nt loop (also known as the bubble). This loop contains two non-Watson–Crick base pairs, G47·A73 and G48·G71, spaced by a looped-out U72; an unpaired A68 is also looped out (Figure 1.6B).

Although there is conflicting evidence as to whether the bubble is formed in the absence of, or induced by the presence of, the ARM in Rev, as is the case in TAR, this base pairing opens up the major groove and permits recognition and binding by Rev (Heaphy et al, 1991; Iwai et al, 1992; Kjems et al, 1992; Battiste et al, 1996; Peterson and Feigon, 1996). In contrast with interactions mediated by arginine or Tat ARM and TAR, for binding to the RRE to occur, the Rev ARM has to have an α-helical structure, and a threonine, an asparagine and four arginine residues are required (Tan et al, 1993). Two arginines interact with U66, G67 and G70, on one side of the groove, and one asparagine and one arginine interact with U45, G46, G47 and A73 on the opposite side of the groove. A threonine and several arginine residues contact the phosphate backbone (Tan et al, 1993; Battiste et al, 1996). CD spectrum analyses on Rev indicate that it has an α-helical structure and probably contains a helix-loop-helix motif, in sharp contrast to the structure of Tat (Auer et al, 1994). Within wild-type Rev, the 20 N-terminal amino acids might contribute to specific binding by providing an α-helical structure (Daly et al, 1995).

Oligomerization of Rev on the RRE

The application of EMSA to wild-type Rev proteins or those mutated in the oligomerization domain demonstrated that the binding of a Rev monomer to the high-affinity site nucleates the cooperative oligomerization of Rev along flanking RNA sites in SLI (Heaphy et al, 1990, 1991; Malim and Cullen, 1991; Mann et al, 1994). Interestingly, Rev oligomerization is dependent not only on the oligomerization motif of Rev (Olsen et al, 1990; Heaphy et al, 1991; Malim and Cullen, 1991; Zapp et al, 1991), but also on an imperfect duplex structure on SLI, which allows for the flexibility of the RRE (Zemmel et al, 1996).

Rev AND THE NUCLEAR TRANSPORT OF mRNAs

Rev mediates the export of RRE-containing RNAs from the nucleus to the cytoplasm. Rev was originally reported to be primarily localized to the nucleus (Malim et al, 1989b; Perkins et al, 1989). However, in an elegant series of biochemical experiments, it was demonstrated that Rev shuttles in and out of the nucleus in an energy- and Rev NES-dependent manner. After the inhibition of RNA synthesis with actinomycin D, immunofluorescence analysis performed on HeLa cells containing Rev or RevM10 clearly demonstrated a relocalization of Rev but not of RevM10 in the cytoplasm (Kalland et al, 1994; Meyer and Malim, 1994).

The nuclear pore complex

Transport between the nucleus and the cytoplasm occurs through the nuclear pore complex (NPC), an aqueous channel embedded in the nuclear envelope that permits the bidirectional traffic of both proteins and RNAs. A major constituent of the NPC is a family of proteins called nucleoporins, characterized by short, degenerate FXFG repeats. Ions, metabolites and small proteins diffuse passively across the NPC. However, passage of RNAs and larger proteins requires an energy- and signal-dependent saturable process. This process includes pore docking, translocation that utilizes the GTPase Ran, subsequent release of the cargo and recycling of the components (Davis, 1995; Gorlich and Mattaj, 1996; Ullman et al, 1997).

Nuclear export signals

Signal sequences that mediate nuclear export of proteins and ribonucleoprotein complexes have just begun to be defined. In fact, the Rev-effector domain and an NES present in the protein kinase inhibitor (PKI) were the first such sequences described (Fischer et al, 1995; Wen et al, 1995). The experimental approach used in this case involved making a conjugate with Rev NES coupled to bovine serum albumin. When the conjugate was injected into either HeLa or oocyte nuclei, it relocalized to the cytoplasm. This indicated that Rev NES could move proteins, as well as its viral cargo, across the nuclear pore. Furthermore, an excess of this conjugate blocked the export of 5S rRNA and U1 snRNA, suggesting that these RNAs and RRE-containing RNAs use the same transporter protein (Fischer et al, 1995). The Rev NES can be functionally replaced with the PKI NES, implying that the transport of the target RRE-RNA can use the same transporter as a protein (Fridell et al, 1996; Fritz and Green, 1996). This is not extremely surprising since RNA appears to be bound to proteins at all times; if each class of RNA were complexed with specific proteins, then equivalent NES would use the same transporter proteins.

Rab/hRIP and nucleoporins

By using the yeast two-hybrid protein interaction trap (where interactions of two proteins can be detected, in yeast, by the consequent activation of a reporter gene), two different groups identified a new protein named Rab/hRIP (Rev activation-domain binding or Rev interacting protein). The interaction between Rab/hRIP and Rev was mapped to the NES domain. An indirect interaction of these proteins in mammalian cells was shown through a one-hybrid transcription assay. A Tat/Rab fusion, in the presence of Rev, was capable of activating transcription from an SLIIB-dependent HIV-1 LTR, thus making Tat transactivation Rab/Rev-dependent (Bogerd et al, 1995; Fritz et al, 1995). A demonstration of a direct interaction between Rev and purified Rab/hRIP is missing to date. Rab/hRIP is similar to nucleoporins in that it has an abundance of FG motifs. Immunofluorescence studies, however, indicate that Rab/hRIP is a nuclear protein and not localized to the NPC. The identification of a yeast nucleoporin homolog to Rab/hRIP, yRIP, which interacts with Rev (Stutz et al, 1995), and the demonstration that Rev can interact with a number of nucleoporins through their FG repeats (Fritz and Green, 1996; Stutz et al, 1996), suggested that a Rab/hRIP–Rev NES complex mediates the entry of unspliced or singly spliced viral ribonucleoproteins into a pre-existing nuclear export pathway. However, this model has been superseded by the demonstration that the direct partner of Rev NES is Exportin 1 (CRM1/Xpo1) complexed with RanGTP (Fornerod et al, 1997a; Stade et al, 1997).

Exportin 1

The finding of Wolff and colleagues (Wolff et al, 1997) that the antibiotic leptomycin B (LMB) inhibited Rev-mediated transport and actinomycin D-dependent translocation of Rev has proved to be pivotal in uncovering the role of Exportin 1 as a nuclear export factor. Previously, an LMB-resistant mutant in the yeast *Schizosaccharomyces pombe* had been described, and mapped to the *exportin 1* locus. Exportin 1 had been shown to bind the nucleoporin CAN/nup214 and other proteins within the pore (Fornerod et al, 1997b). Thus, the next logical step, to test whether Exportin 1 was transporting Rev through the NPC, was undertaken. In vitro binding assays with purified proteins have demonstrated a direct interaction between Exportin 1 and LMB, as well as a cooperative interaction among Exportin 1, the NES peptide and the GTPase Ran (Fornerod et al, 1997a). Similar findings were also reported in *S. pombe* (Stade et al, 1997).

Mechanism of Rev action

These data can be put together in the following working model. The Rev ARM binds to the high-affinity site (SLIIB) in the RRE, present in the primary and singly spliced HIV-1 mRNA. Rev oligomerizes on the mRNA, forming a ribonucleoprotein complex. Within this complex, the Rev NES binds to Exportin 1/RanGTP, which then interacts with Rab/hRIP or another nucleoporin, which localizes the complex to the nuclear pore. The protein–RNA complex is translocated to the cytoplasm. Once in the cytoplasm, GTP is hydrolyzed to GDP, and RanGDP, Exportin 1 and Rev are released and the HIV-1 mRNAs are available for translation (Figure 1.7).

PROSPECTS AND THERAPEUTIC INTERVENTION

Studies of Tat and Rev have revealed fundamental aspects of eukaryotic biology. Tat represents the best example of a protein that regulates rates of elongation of transcription and leads to catalysis rather than assembly of transcription complexes. By facilitating the phosphorylation of the CTD of RNAPII and possibly other proteins in the transcription complex, Tat allows for promoter clearance and increased processivity of RNAPII (Jones, 1997). Future studies must focus on the identification of other components of TAK/P-TEFb and TAR binding proteins, the reconstitution of Tat transactivation in a minimal in vitro system and structural studies of Tat complexed to its physiological ligand.

Likewise, investigation of Rev has revealed tantalizing aspects of nuclear export of macromolecules. NES was first defined in Rev (Fischer et al, 1995). Roles of Exportin 1 and RanGTP in this process were implicated because only an activated Exportin 1 can interact with this NES, and, upon the hydrolysis of GTP, Rev and RRE-containing transcripts are released in the cytoplasm (Fornerod et al, 1997a; Stade et al, 1997). Again, other components must be identified in this system, e.g. target nucleoporins, the regulation of Ran, and structural aspects of these protein–protein and protein–RNA interactions.

Kinase inhibitors already block Tat transactivation in vitro and in vivo. These range from analogs of DRB to pseudosubstrate peptides from CDK2 (Mancebo et al, 1997). Structural studies on Tat and its physiological partner promise further testing of compounds that might block effects of Tat and viral replication in cells. Likewise, LMB and related compounds block Exportin 1 (Fornerod et al, 1997a; Stade et al, 1997;

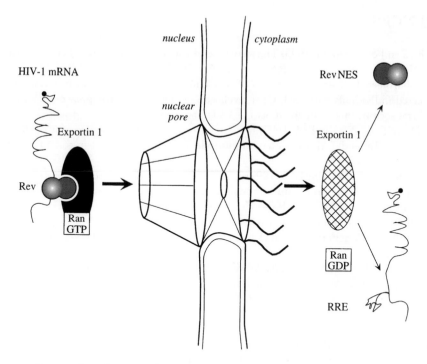

Figure 1.7 *Proposed mechanism of action of Rev*
Rev is represented as two distinct domains: the multifunctional domain is shown in light gray and the effector domain in dark gray. Rev coats the RRE. Only upon the binding of Exportin 1 to RanGTP can Exportin 1 bind to Rev NES. Exportin 1 also binds nucleoporins, which leads to export of HIV-1 transcripts from the nucleus to the cytoplasm. There, GTP is hydrolyzed to GDP, and Exportin 1 and RanGDP separate, releasing Rev in the process. Viral transcripts are now translated by polysomes in the endoplasmic reticulum, which leads to the synthesis of new virions and packaging of genomic mRNA into viral particles.

Wolff et al, 1997). Possibly LMB analogs and/or other compounds that block interactions between Rev, Exportin 1 and RanGTP could be used to block viral replication.

Gene therapies with dominant negative Rev proteins are undergoing clinical trials (Fox et al, 1995; Bonyhadi et al, 1997). It is possible that other dominant negative co-activators and/or TAR and RRE decoys could form the basis of other promising therapeutic interventions (Caputo et al, 1997). Only by studying these fundamental viral and cellular processes further can we hope to arrive at intelligent strategies for defining the Achilles' heel of HIV.

ACKNOWLEDGMENTS

The authors would like to thank Ms Jo Cresswell and Dr Kate Darling for their assistance with the manuscript. B.M.P. would like to thank Anne, Anton Alexander and Sebastian Bogomir, and A.A. would like to thank Peter, for their encouragement and support.

REFERENCES

Aboul-ela F, Karn J & Varani G (1995) The structure of the human immunodeficiency virus type-1 TAR RNA reveals principles of RNA recognition by Tat protein. *J Mol Biol* **253:** 313–332.

Adams M, Sharmeen L, Kimpton J et al (1994) Cellular latency in human immunodeficiency virus-infected individuals with high CD4 levels can be detected by the presence of promoter-proximal transcripts. *Proc Natl Acad Sci USA* **91:** 3862–3866.

Alonso A, Derse D & Peterlin BM (1992) Human chromosome 12 is required for optimal interactions between Tat and TAR of human immunodeficiency virus type 1 in rodent cells. *J Virol* **66:** 4617–4621.

Alonso A, Cujec TP & Peterlin BM (1994) Effects of human chromosome 12 on interactions between Tat and TAR of human immunodeficiency virus type 1. *J Virol* **68:** 6505–6513.

Arya SK, Gallo RC, Hahn BH et al (1984) Homology of genome of AIDS-associated virus with genomes of human T-cell leukemia viruses. *Science* **225:** 927–930.

Auer M, Gremlich HU, Seifert JM et al (1994) Helix-loop-helix motif in HIV-1 Rev. *Biochemistry* **33:** 2988–2996.

Bartel DP, Zapp ML, Green MR & Szostak JW (1991) HIV-1 Rev regulation involves recognition of non-Watson–Crick base pairs in viral RNA. *Cell* **67:** 529–536.

Battiste JL, Mao H, Rao NS et al (1996) α Helix-RNA major groove recognition in an HIV-1 Rev peptide-RRE RNA complex. *Science* **273:** 1547–1551.

Bayer P, Kraft M, Ejchart A, Westendorp M, Frank R & Rosch P (1995) Structural studies of HIV-1 Tat protein. *J Mol Biol* **247:** 529–535.

Benko DM, Schwartz S, Pavlakis GN & Felber BK (1990) A novel human immunodeficiency virus type 1 protein, tev, shares sequences with Tat, env and Rev proteins. *J Virol* **64:** 2505–2518.

Berkhout B & Jeang KT (1989) Trans activation of human immunodeficiency virus type 1 is sequence specific for both the single-stranded bulge and loop of the trans-acting-responsive hairpin: a quantitative analysis. *J Virol* **63:** 5501–5504.

Berkhout B & Jeang KT (1992) Functional roles for the TATA promoter and enhancers in basal and Tat-induced expression of the human immunodeficiency virus type 1 long terminal repeat. *J Virol* **66:** 139–149.

Berkhout B, Silverman RH & Jeang KT (1989) Tat trans-activates the human immunodeficiency virus through a nascent RNA target. *Cell* **59:** 273–282.

Berkhout B, Gatignol A, Rabson AB & Jeang KT (1990) TAR-independent activation of the HIV-1 LTR: evidence that tat requires specific regions of the promoter. *Cell* **62:** 757–767.

Blau J, Xiao H, McCracken S, O'Hare P, Greenblatt J & Bentley D (1996) Three functional classes of transcriptional activation domains. *Mol Cell Biol* **16:** 2044–2055.

Bogerd HP, Fridell RA, Madore S & Cullen BR (1995) Identification of a novel cellular cofactor for the Rev/Rex class of retroviral regulatory proteins. *Cell* **82:** 485–494.

Bonyhadi ML, Moss K, Voytovich A et al (1997) RevM10-expressing T cells derived in vivo from transduced human hematopoietic stem-progenitor cells inhibit human immunodeficiency virus replication. *J Virol* **71:** 4707–4716.

Calnan BJ, Biancalana S, Hudson D & Frankel AD (1991a) Analysis of arginine-rich peptides from the HIV Tat protein reveals unusual features of RNA–protein recognition. *Genes Dev* **5:** 201–210.

Calnan BJ, Tidor B, Biancalana S, Hudson D & Frankel AD (1991b) Arginine-mediated RNA recognition: the arginine fork [published erratum appears in *Science* (1992) **255**(5045): 665]. *Science* **252:** 1167–1171.

Caputo A, Rossi C, Bozzini R et al (1997) Studies on the effect of the combined expression of anti-tat and anti-rev genes on HIV-1 replication. *Gene Ther* **4:** 288–295.

Chang HK, Gallo RC & Ensoli B (1995) Regulation of cellular gene expression and function by the human immunodeficiency virus type 1 Tat protein. *J Biomed Sci* **2:** 189–202.

Churcher MJ, Lamont C, Hamy F et al (1993) High affinity binding of TAR RNA by the human immunodeficiency virus type-1 tat protein requires base-pairs in the RNA stem and amino acid residues flanking the basic region. *J Mol Biol* **230**: 90–110.

Colvin RA & Garcia-Blanco MA (1992) Unusual structure of the human immunodeficiency virus type 1 trans-activation response element. *J Virol* **66**: 930–935.

Colvin RA, White SW, Garcia-Blanco MA & Hoffman DW (1993) Structural features of an RNA containing the CUGGGA loop of the human immunodeficiency virus type 1 *trans*-activation response element. *Biochemistry* **32**: 1105–1112.

Cordingly MG, LaFermina RL, Callahan PL et al (1990) Sequence-specific interaction of tat protein and tat peptides with the transactivation-response sequence element of human immunodeficiency virus type 1 in vitro. *Proc Natl Acad Sci USA* **87**: 8985–8989.

Critchley A, Haneef I, Cousens D & Stockley P (1993) Modeling and solution structure probing of the HIV-1 TAR stem-loop. *J Mol Graph* **11**: 92–97.

Cujec TP, Okamoto H, Fujinaga K et al (1997) The HIV transactivator Tat binds to the CDK-activating kinase and activates the phosphorylation of the C-terminal domain of RNA polymerase II. *Genes Dev* **11**: 2645–2657.

Daly TJ, Cook KS, Gray GS, Maione TE & Rusche JR (1989) Specific binding of HIV-1 recombination Rev protein to the Rev-responsive element in vitro. *Nature* **342**: 816–819.

Daly TJ, Doten RC, Rusche JR & Auer M (1995) The amino terminal domain of HIV-1 rev is required for discrimination of the RRE from nonspecific RNA. *J Mol Biol* **253**: 243–258.

Davis LI (1995) The nuclear pore complex. *Ann Rev Biochem* **64**: 865–896.

Dayton AI, Sodroski JG, Rosen CA, Goh WC & Haseltine WA (1986) The trans-activator gene of the human T cell lymphotropic virus type III is required for replication. *Cell* **44**: 941–947.

Delling U, Reid LS, Barnett RW et al (1992) Conserved nucleotides in the TAR RNA stem of human immunodeficiency virus type 1 are critical for tat binding and trans-activation: model for TAR RNA tertiary structure. *J Virol* **66**: 3018–3025.

Derse D, Carvalho M, Carroll R & Peterlin BM (1991) A minimal lentivirus Tat. *J Virol* **65**: 7012–7015.

Dingwall C, Ernberg I, Gait MJ et al (1989) Human immunodeficiency virus 1 tat protein binds trans-activation-responsive region (TAR) RNA in vitro. *Proc Natl Acad Sci USA* **86**: 6925–6929.

Dingwall C, Ernberg I, Gait MJ et al (1990) HIV-1 tat protein stimulates transcription by binding to a U-rich bulge in the stem of the TAR RNA structure. *EMBO J* **9**: 4145–4153.

Emerman M, Vazeux R & Peden K (1989) The rev gene product of the human immunodeficiency virus affects envelope-specific RNA localization. *Cell* **57**: 1155–1165.

Feinberg MB, Jarrett RF, Aldovini A, Gallo RC & Wong-Staal F (1986) HTLV-III expression and production involve complex regulation at the levels of splicing and translation of viral RNA. *Cell* **46**: 807–817.

Feinberg MB, Baltimore D & Frankel AD (1991) The role of Tat in the human immunodeficiency virus life cycle indicates a primary effect on transcriptional elongation. *Proc Natl Acad Sci USA* **88**: 4045–4049.

Felber BK, Hadzopoulou-Cladaras M, Cladaras C, Copeland T & Pavlakis GN (1989) Rev protein of human immunodeficiency virus type 1 affects the stability and transport of the viral mRNA. *Proc Natl Acad Sci USA* **86**: 1495–1499.

Feng S & Holland EC (1988) HIV-1 tat trans-activation requires the loop sequence within tar. *Nature* **334**: 165–167.

Fischer U, Meyer S, Teufel M, Heckel C, Luhrman R & Rautman G (1994) Evidence that HIV-1 Rev directly promotes the nuclear export of unspliced RNA. *EMBO J* **13**: 4105–4112.

Fischer U, Huber J, Boelens WC, Mattaj IW & Luhrmann R (1995) The HIV-1 REV activation domain is a nuclear export signal that accesses an export pathway used by specific cellular RNAs. *Cell* **82**: 475–483.

Fisher AG, Feinberg MB, Josephs SF et al (1986) The trans-activator gene of HTLV-III is essential for virus replication. *Nature* **320:** 367–371.

Fornerod M, Ohno M, Yoshida M & Mattaj IW (1997a) CRM1 is an export receptor for leucine-rich nuclear export signals. *Cell* **90:** 1051–1060.

Fornerod M, van Deursen J, Reynolds A et al (1997b) The human homolog of yeast CRM1 is in a dynamic subcomplex with CAN/Nup214 and a novel nuclear pore component Nup88. *EMBO J* **16:** 807–816.

Fox BA, Woffendin C, Yang ZY et al (1995) Genetic modification of human peripheral blood lymphocytes with a transdominant negative form of Rev: safety and toxicity. *Hum Gene Ther* **6:** 997–1004.

Frankel AD, Biancalana S & Hudson D (1989) Activity of synthetic peptides from the Tat protein of human immunodeficiency virus type 1. *Proc Natl Acad Sci USA* **86:** 7397–7401.

Fridell RA, Bogerd HP & Cullen BR (1996) Nuclear export of late HIV-1 mRNAs occurs via a cellular protein export pathway. *Proc Natl Acad Sci USA* **93:** 4421–4424.

Fritz CC & Green MR (1996) HIV Rev uses a conserved cellular protein export pathway for the nucleocytoplasmic transport of viral RNAs. *Curr Biol* **6:** 848–854.

Fritz CC, Zapp ML & Green MR (1995) A human nucleoporin-like protein that specifically interacts with HIV REV. *Nature* **376:** 530–533.

Garcia JA, Harrich D, Pearson L, Mitsuyasu R & Gaynor RB (1988) Functional domains required for tat-induced transcriptional activation of the HIV-1 long terminal repeat. *EMBO J* **7:** 3143–3147.

Garcia JA, Harrich D, Soultanakis E, Wu F, Mitsuyasu R & Gaynor RB (1989) Human immunodeficiency virus type 1 LTR TATA and TAR region sequences required for transcriptional regulation. *EMBO J* **8:** 765–778.

Garcia-Martinez LF, Mavankal G, Neveu JM, Lane WS, Ivanov D & Gaynor RB (1997) Purification of a Tat-associated kinase reveals a TFIIH complex that modulates HIV-1 transcription. *EMBO J* **16:** 2836–2850.

Gorlich D & Mattaj IW (1996) Nucleocytoplasmic transport. *Science* **271:** 1513–1518.

Graham GJ & Maio JJ (1990) RNA transcripts of the human immunodeficiency virus transactivation response element can inhibit action of the viral transactivator. *Proc Natl Acad Sci USA* **87:** 5817–5821.

Hadzpoulou-Cladaras M, Felber BK, Cladaras C, Athanassopoulos A, Tse A & Pavlakis GN (1989) The rev (trs/art) protein of human immunodeficiency virus type 1 affects viral mRNA and protein expression via a cis-acting sequence in the env region. *J Virol* **63:** 1265–1274.

Hammarskjold ML, Heimer J, Hammarskjold B, Sangwan I, Albert L & Rekosk D (1989) Regulation of human immunodeficiency virus env expression by the rev gene product. *J Virol* **63:** 1959–1966.

Hamy F, Asseline U, Grasby J et al (1993) Hydrogen-bonding contacts in the major groove are required for human immunodeficiency virus type-1 tat protein recognition of TAR RNA. *J Mol Biol* **230:** 111–123.

Hart CE, Ou CY, Galphin JC et al (1989) Human chromosome 12 is required for elevated HIV-1 expression in human–hamster hybrid cells. *Science* **246:** 488–491.

Hart CE, Saltarelli MJ, Galphin JC & Schochetman G (1995) A human chromosome 12-associated 83-kilodalton cellular protein specifically binds to the loop region of human immunodeficiency virus type 1 *trans*-activation response element RNA. *J Virol* **69:** 6593–6599.

Hauber J & Cullen BR (1988) Mutational analysis of the trans-activation-responsive region of the human immunodeficiency virus type I long terminal repeat. *J Virol* **62:** 673–679.

Hauber J, Bouvier M, Malim M & Cullen B (1988) Phosphorylation of the rev gene product of the human immunodeficiency virus type 1. *J Virol* **62:** 4801–4804.

Hauber J, Malim MH & Cullen BR (1989) Mutational analysis of the conserved basic domain of human immunodeficiency virus tat protein. *J Virol* **63:** 1181–1187.

Heaphy S, Dingwall C, Ernberg I et al (1990) HIV-1 regulator of virion expression (rev) protein binds to an RNA stem-loop structure located within the Rev response element region. *Cell* **60:** 685–693.

Heaphy S, Finch JT, Gait MJ, Karn J & Singh M (1991) Human immunodeficiency virus type 1 regulator of virion expression, rev, forms nucleoprotein filaments after binding to a purine-rich 'bubble' located within the rev-responsive region of viral mRNAs. *Proc Natl Acad Sci USA* **88:** 7366–7370.

Herrmann CH & Rice AP (1995) Lentivirus Tat proteins specifically associate with a cellular protein kinase, TAK, that hyperphosphorylates the carboxyl-terminal domain of the large subunit of RNA polymerase II: candidate for a Tat cofactor. *J Virol* **69:** 1612–1620.

Hope TJ, MacDonald D, Huang X, Low J & Parslow TG (1990) Mutational analysis of the human immunodeficiency virus type 1 Rev transactivator: essential residues near the amino terminus. *J Virol* **64:** 5360–5366.

Hope TJ, Bond BL, McDonald D, Klein NP & Parslow TG (1991) Effector domains of human immunodeficiency type 1 Rev and human T-cell leukemia virus type I Rex are functionally interchangeable and share an essential peptide motif. *J Virol* **65:** 6001–6007.

Hope T, Klein NP, Elder ME & Parslow TG (1992) Trans-dominant inhibition of human immunodeficiency virus type 1 Rev occurs through formation of inactive protein complexes. *J Virol* **66:** 1849–1855.

Howcroft TK, Strebel K, Martin MA & Singer DS (1993) Repression of MHC class I gene promoter activity by two-exon Tat of HIV. *Science* **260:** 1320–1322.

Iwai S, Pritchard C, Mann DA, Karn J & Gait M (1992) Recognition of the high affinity binding site in rev-response element RNA by the human immunodeficiency virus type-1 rev protein. *Nucl Acids Res* **20:** 6465–6472.

Jakobovits A, Smith DH, Jakobovits EB and Capon DJ (1988) A discrete element 3' of human immunodeficiency virus 1 (HIV-1) and HIV-2 mRNA initiation sites mediates transcriptional activation by an HIV trans activator. *Mol Cell Biol* **8:** 2555–2561.

Jeang KT, Chun R, Lin NH, Gatignol A, Glabe CG & Fan H (1993) In vitro and in vivo binding of human immunodeficiency virus type 1 Tat protein and Sp1 transcription factor. *J Virol* **67:** 6224–6233.

Jones KA (1997) Taking a new TAK on Tat transactivation. *Genes Dev* **11:** 2593–2599.

Jones KA & Peterlin BM (1994) Control of RNA initiation and elongation at the HIV-1 promoter. *Annu Rev Biochem* **63:** 717–743.

Kalland KH, Szilvay A, Brokstad M, Saetrevik KA & Haukenes G (1994) The human immunodeficiency virus type 1 Rev protein shuttles between the cytoplasm and nuclear compartments. *Mol Cell Biol* **14:** 7436–7444.

Kamine J, Subramanian T & Chinnandurai G (1991) Sp1-dependent activation of a synthetic promoter by human immunodeficiency virus type 1 tat protein. *Proc Natl Acad Sci USA* **88:** 8510–8514.

Kao SY, Calman AF, Luciw PA & Peterlin BM (1987) Anti-termination of transcription within the long terminal repeat of HIV-1 by tat gene product. *Nature* **330:** 489–493.

Kashanchi F, Piras G, Radonovich MF et al (1994) Direct interaction of human TFIID with the HIV-1 transactivator Tat. *Nature* **367:** 295–299.

Kashanchi F, Khleif SN, Duvall JF et al (1996) Interaction of human immunodeficiency virus type 1 Tat with a unique site of TFIID inhibits negative cofactor Dr1 and stabilizes the TFIID–TFIIA complex. *J Virol* **70:** 5503–5510.

Keen NJ, Gait MJ & Karn J (1996) Human immunodeficiency virus type-1 Tat is an integral component of the activated transcription–elongation complex. *Proc Natl Acad Sci USA* **93:** 2505–2510.

Keen NJ, Churcher MJ & Karn J (1997) Transfer of Tat and release of TAR RNA during the activation of the human immunodeficiency virus type-1 transcription elongation complex. *EMBO J* **16:** 5260–5272.

Kim S, Byrn R, Groopman J & Baltimore D (1989) Temporal aspects of DNA and RNA synthesis during human immunodeficiency virus infection: evidence for differential gene expression. *J Virol* **63**: 3708–3713.

Kjems J & Sharp PA (1993) The basic domain of rev from human immunodeficiency virus type 1 specifically blocks the entry of U4/U6–U5 small nuclear ribonucleoprotein in spliceosome assembly. *J Virol* **67**: 4769–4776.

Kjems J, Calnan BJ, Frankel AD & Sharp PA (1992) Specific binding of a basic peptide from HIV-1 Rev. *EMBO J* **11**: 1119–1129.

Knight DM, Flomerfelt FA & Ghrayeb J (1987) Expression of the art/trs protein of HIV and study of its role in viral envelope synthesis. *Nature* **236**: 837–840.

Koleske AJ & Young RA (1995) The RNA polymerase II holoenzyme and its implications for gene regulation. *Trends Biochem Sci* **20**: 113–116.

Kuppuswamy M, Subramanian T, Srinivasan A & Chinnadurai G (1989) Multiple functional domains of Tat, the trans-activator of HIV-1, defined by mutational analysis. *Nucl Acids Res* **17**: 3551–3561.

Laspia MF, Rice AP & Mathews MB (1989) HIV-1 Tat protein increases transcriptional initiation and stabilizes elongation. *Cell* **59**: 283–292.

Laspia MF, Rice AP & Mathews MB (1990) Synergy between HIV-1 Tat and adenovirus E1A is principally due to stabilization of transcriptional elongation. *Genes Dev* **4**: 2397–2408.

Lisziewicz J, Rappaport J & Dhar R (1991) Tat-regulated production of multimerized TAR RNA inhibits HIV-1 gene expression. *New Biol* **3**: 82–89.

Lu X, Welsh TM & Peterlin BM (1993) The human immunodeficiency virus type 1 long terminal repeat specifies two different transcription complexes, only one of which is regulated by Tat. *J Virol* **67**: 1752–1760.

Madore SJ & Cullen BR (1993) Genetic analysis of the cofactor requirement for human immunodeficiency virus type 1 Tat function. *J Virol* **67**: 3703–3711.

Maldonado E, Shiekhattar R, Sheldon M et al (1996) A human RNA polymerase II complex associated with SRB and DNA-repair protein. *Nature* **381**: 86–89.

Malim MH & Cullen BR (1991) HIV-1 structural gene expression requires the binding of multiple rev monomers to the viral RRE: implications for HIV-1 latency. *Cell* **65**: 241–248.

Malim MH, Bohnlein S, Hauber J & Cullen BR (1989a) Functional dissection of the HIV-1 Rev trans-activator: derivation of a trans-dominant repressor of Rev function. *Cell* **58**: 205–214.

Malim MH, Hauber J, Le SY, Maizel JV & Cullen BR (1989b) The HIV-1 rev trans-activator acts through a structured target sequence to activate nuclear export of unspliced viral mRNA. *Nature* **338**: 254–257.

Malim MH, Tiley LS, McCarn DF, Rusche JR, Hauber J & Cullen BR (1990) HIV-1 structural gene expression requires binding of the Rev trans-activator to its RNA target sequence. *Cell* **60**: 675–683.

Malim MH, McCarn DF, Tiley LS & Cullen BR (1991) Mutational definition of the human immunodeficiency virus type 1 Rev activation domain. *J Virol* **65**: 4248–4254.

Mancebo H, Lee G, Flygare J et al (1997) P-TEFb kinase is required for HIV Tat transcriptional activation in vivo and in vitro. *Genes Dev* **11**: 2633–2644.

Mann DA, Mikaelian I, Zemmel RW et al (1994). A molecular rheostat: co-operative Rev binding to stem I of the Rev-response element modulates human immunodeficiency virus type-1 late gene expression. *J Mol Biol* **241**: 193–207.

Marciniak RA & Sharp PA (1991) HIV-1 Tat protein promotes formation of more-processive elongation complexes. *EMBO J* **10**: 4189–4196.

Marciniak RA, Calnan BJ, Frankel AD & Sharp PA (1990) HIV-1 Tat protein trans-activates transcription in vitro. *Cell* **63**: 791–802.

Marshall NF, Peng J, Xie Z & Price DH (1996) Control of RNA polymerase II elongation by a novel carboxyl-terminal domain kinase. *J Biol Chem* **271**: 27176–27183.

Mavankal G, Ignatious Ou SH, Oliver H, Sigman D & Gaynor RB (1996) Human immuno-deficiency virus type 1 and 2 Tat proteins specifically interact with RNA polymerase II. *Proc Natl Acad Sci USA* **93**: 2089–2094.

McDonald D, Hope TJ & Parslow TG (1992) Posttranscriptional regulation by the human immunodeficiency virus type 1 Rev and human T-cell leukemia virus type 1 Rex proteins through a heterologous RNA binding site. *J Virol* **66**: 7232–7238.

Meyer BE & Malim MH (1994) The HIV-1 Rev trans-activator shuttles between the nucleus and the cytoplasm. *Genes Dev* **8**: 1538–1547.

Muesing MA, Smith DH & Capon DJ (1987) Regulation of mRNA accumulation by a human immunodeficiency virus trans-activator protein. *Cell* **48**: 691–701.

Mujeeb A, Bishop K, Peterlin BM, Turck C, Parslow TG & James TL (1994) NMR structure of a biologically active peptide containing the RNA-binding domain of human immuno-deficiency virus type 1 Tat. *Proc Natl Acad Sci USA* **91**: 8248–8245.

Myers G, Berzofsky JA, Rabson AB, Smith TF & SF, W (1991) *Human Retroviruses and AIDS.* Los Alamos National Laboratories.

Neumann M, Harrison J, Saltarelli M et al (1994) Splicing variability in HIV type 1 revealed by quantitative RNA polymerase chain reaction. *AIDS Res Hum Retroviruses* **10**: 1531–1542.

Newstein M, Stanbridge EJ, Casey G & Shank PR (1990) Human chromosome 12 encodes a species-specific factor which increases human immunodeficiency virus type 1 tat-mediated transactivation in rodent cells. *J Virol* **64**: 4565–4567.

Olsen HS & Rosen CA (1992) Contribution of the TATA motif to Tat-mediated transcriptional activation of human immunodeficiency virus gene expression. *J Virol* **66**: 5594–5947.

Olsen HS, Cochrane AW, Dillon PJ, Nalin CM & Rosen CA (1990) Interaction of the human immunodeficiency virus type 1 Rev protein with a structured region in env mRNA is dependent on multimer formation mediated through a basic stretch of amino acids. *Genes Dev* **4**: 1357–1364.

Ossipow V, Tassan JP, Nigg EA & Schibler U (1995) A mammalian RNA polymerase II holo-enzyme containing all components required for promoter-specific transcription initiation. *Cell* **83**: 137–146.

Ott M, Emiliani S, Van Lint C et al (1997) Immune hyperactivation of HIV-1-infected T cells mediated by Tat and the CD28 pathway. *Science* **275**: 1481–1485.

Parada CA & Roeder RG (1996) Enhanced processivity of RNA polymerase II triggered by Tat-induced phosphorylation of its carboxy-terminal domain. *Nature* **384**: 375–378.

Pendergrast PS & Hernandez N (1997) RNA-targeted activators, but not DNA-targeted activators, repress the synthesis of short transcripts at the human immunodeficiency virus type 1 long terminal repeat. *J Virol* **71**: 910–917.

Perkins A, Cochrane A, Rubens S & Rosen C (1989) Structural and functional characterization of the human immunodeficiency virus rev protein. *J AIDS* **2**: 256–263.

Pessler F, Pendergrast PS & Hernandez N (1997) Purification and characterization of FBI-1, a cellular factor that binds to the human immunodeficiency virus type 1 inducer of short transcripts. *Mol Cell Biol* **17**: 3786–3798.

Peterlin BM, Luciw PA, Barr PJ and Walker MD (1986) Elevated levels of mRNA can account for the transactivation of human immunodeficiency virus (HIV). *Proc Natl Acad Sci USA* **183**: 9734–9738.

Peterlin BM, Adams M, Alonso A et al (1993) Tat trans-activator. In: Cullen, B. ed. *Human Retroviruses*. Oxford: Oxford University Press.

Peterson RD & Feigon J (1996) Structural change in Rev responsive element RNA of HIV-1 on binding Rev peptide. *J Mol Biol* **264**: 863–877.

Pritchard CE, Grasby JA, Hamy F et al (1994) Methyl-phosphonate mapping of phosphate contacts critical for RNA recognition by the human immunodeficiency virus Tat and Rev proteins. *Nucl Acids Res* **22**: 2592–2600.

Puglisi JD, Tan R, Calnan BJ, Frankel AD & Williamson JR (1992) Conformation of the TAR RNA–arginine complex by NMR spectroscopy. *Science* **257**: 76–80.

Puglisi JD, Chen L, Frankel AD & Williamson JR (1993) Role of RNA structure in arginine recognition of TAR RNA. *Proc Natl Acad Sci USA* **90**: 3680–3684.

Rappaport J, Lee SJ, Khalili K & Wong-Staal F (1989) The acidic amino-terminal region of the HIV-1 Tat protein constitutes an essential activating domain. *New Biol* **1**: 101–110.

Rice AP & Carlotti F (1990) Mutational analysis of the conserved cysteine-rich region of the human immunodeficiency virus type 1 Tat protein. *J Virol* **64**: 1864–1868.

Robert-Guroff M, Popovic M, Gartner S, Markham P, Gallo RC & Reitz MS (1990) Structure and expression of tat-, rev-, and nef-specific transcripts of human immunodeficiency virus type 1 in infected lymphocytes and macrophages. *J Virol* **64**: 3391–3398.

Rosen CA, Sodroski JG & Haseltine WA (1985) The location of cis-acting regulatory sequences in the human T cell lymphotropic virus type III (HTLV-III/LAV) long terminal repeat. *Cell* **41**: 813–823.

Rosen CR, Terwilliger E, Dayton AI, Sodroski JG & Haseltine WA (1988) Intragenic cis-acting art gene-responsive sequences of the human immunodeficiency virus. *Proc Natl Acad Sci USA* **85**: 2071–2075.

Roy S, Delling U, Chen CH, Rosen CA & Sonenberg N (1990a) A bulge structure in HIV-1 TAR RNA is required for Tat binding and Tat-mediated trans-activation. *Genes Dev* **4**: 1365–1373.

Roy S, Parkin NT, Rosen C, Itovitch J & Sonenberg N (1990b) Structural requirements for trans activation of human immunodeficiency virus type 1 long terminal repeat-directed gene expression by tat: importance of base pairing, loop sequence, and bulges in the tat-responsive sequence. *J Virol* **64**: 1402–1406.

Ruben S, Perkins A, Purcell R et al (1989) Structural and functional characterization of human immunodeficiency virus tat protein. *J Virol* **63**: 1–8.

Ruhl M, Himmelspach M, Bahr GM et al (1993) Eukaryotic initiation factor 5A is a cellular target of the human immunodeficiency virus type 1 Rev activation domain mediating transactivation. *J Cell Biol* **123**: 1309–1320.

Sadaie MR, Benter T & Wong-Staal F (1988) Site-directed mutagenesis of two trans-regulatory genes (tat-III, trs) of HIV-1. *Science* **239**: 910–913.

Schwartz S, Felber BK, Benko DM, Fenyo E & Pavlakis GN (1990a) Cloning and functional analysis of multiply spliced mRNAs species of human immunodeficiency virus type 1. *J Virol* **64**: 2519–2529.

Schwartz S, Felber BK, Fenyo EM & Pavlakis GN (1990b) Env and Vpu proteins of human immunodeficiency virus type 1 are produced from multiple bicistronic mRNAs. *J Virol* **64**: 5448–5456.

Selby MJ & Peterlin BM (1990) Trans-activation by HIV-1 Tat via a heterologous RNA binding protein. *Cell* **62**: 769–776.

Selby MJ, Bain ES, Luciw PA & Peterlin BM (1989) Structure, sequence, and position of the stem-loop in tar determine transcriptional elongation by tat through the HIV-1 long terminal repeat. *Genes Dev* **3**: 547–558.

Sheldon M, Ratnasabapathy R & Hernandez N (1993) Characterization of the inducer of short transcripts, a human immunodeficiency virus type 1 transcriptional element that activates the synthesis of short RNAs. *Mol Cell Biol* **13**: 1251–1263.

Sheline CT, Milocco LH & Jones KA (1991) Two distinct nuclear transcription factors recognize loop and bulge residues of the HIV-1 TAR RNA hairpin. *Genes Dev* **5**: 2508–2520.

Siomi H, Shida H, Maki M & Hatanaka M (1990) Effects of a highly basic region of human immunodeficiency virus tat protein on nucleolar localization. *J Virol* **64**: 1803–1807.

Sodroski J, Patarca R, Rosen C, Woong-Staal F & Haseltine W (1985) Location of the transactivating region on the genome of human T-cell lymphotropic virus type III. *Science* **229**: 79.

Sodroski J, Goh WC, Rosen C, Dayton A, Terwilliger E & Haseltine W (1986) A second post-transcriptional trans-activator gene is required for HTLV-III replication. *Nature* **321:** 412–417.

Southgate CD & Green MR (1991) The HIV-1 Tat protein activates transcription from an upstream DNA-binding site: implications for Tat function. *Genes Dev* **5:** 2496–2507.

Southgate C, Zapp ML & Green MR (1990) Activation of transcription by HIV-1 Tat protein tethered to nascent RNA through another protein. *Nature* **345:** 640–642.

Stade K, Ford CS, Guthrie C & Weiss KW (1997) Exportin 1 (Crm1p) is an essential nuclear export factor. *Cell* **90:** 1041–1050.

Stutz F, Neville M & Rosbach M (1995) Identification of a novel nuclear-pore-associated protein as a functional target of the HIV-1 Rev protein in yeast. *Cell* **82:** 495–506.

Stutz F, Izaurralde E, Mattaj IW & Rosbach M (1996) A role for nucleoporin FG repeat domains in export of human immunodeficiency virus type 1 Rev protein and RNA from the nucleus. *Mol Cell Biol* **16:** 7144–7150.

Subramanian T, Kuppuswamy M, Venkatesh L, Srinivasan A & Chinnadurai G (1990) Functional substitution of the basic domain of the HIV-1 trans-activator, Tat, with the basic domain of the functionally heterologous Rev. *Virology* **176:** 178–183.

Sullenger BA, Gallardo HF, Ungers GE & Gilboa E (1990) Overexpression of TAR sequences renders cells resistant to human immunodeficiency virus replication. *Cell* **63:** 601–608.

Sullenger BA, Gallardo HF, Ungers GE & Gilboa E (1991) Analysis of trans-acting response decoy RNA-mediated inhibition of human immunodeficiency virus type 1 transactivation. *J Virol* **65:** 6811–6816.

Tan R, Chen L, Buettner JA, Hudson D & Frankel AD (1993) RNA recognition by an isolated α helix. *Cell* **73:** 1031–1040.

Tao J & Frankel AD (1992) Specific binding of arginine to TAR RNA. *Proc Natl Acad Sci USA* **89:** 2723–2726.

Tiley LS, Madore SJ, Malim MH & Cullen BR (1992a) The VP16 transcription activation domain is functional when targeted to a promoter-proximal RNA sequence. *Genes Dev* **6:** 2077–2087.

Tiley LS, Malim MH, Tewary HK, Stockley PG & Cullen BR (1992b) Identification of a high-affinity RNA-binding site for the human immunodeficiency virus type 1 Rev protein. *Proc Natl Acad Sci USA* **89:** 758–762.

Toohey MG & Jones KA (1989) In vitro formation of short RNA polymerase II transcripts that terminate within the HIV-1 and HIV-2 promoter-proximal downstream regions. *Genes Dev* **3:** 265–282.

Ullman KS, Powers MA & Forbes DJ (1997) Nuclear export receptors: from Importin to Exportin. *Cell* **90:** 967–970.

Venkatesh LK & Chinnadurai G (1990) Mutants in a conserved region near the carboxy-terminus of HIV-1 Rev identify functionally important residues and exhibit a dominant negative phenotype. *Virology* **178:** 327–330.

Weeks KM & Crothers DM (1991) RNA recognition by Tat-derived peptides: interaction in the major groove? [published erratum appears in *Cell* (1992) **70(6):** following 1068]. *Cell* **66:** 577–588.

Weeks KM, Ampe C, Schultz SC, Steitz TA & Crothers DM (1990) Fragments of the HIV-1 Tat protein specifically bind TAR RNA. *Science* **249:** 1281–1285.

Wei P et al (1998) A novel CDK9-associated C-type cyclin interacts directly with HIV-1 Tat and mediates its high-affinity, loop-specific binding to TAR RNA. *Cell* **92:** 451–462.

Weichselbraun I, Farrington GK, Rusche JR, Bohnlein E & Hauber J (1991) Definition of the human immunodeficiency virus type 1 Rev and human T-cell leukemia virus type 1 Rex protein activation domains by functional exchange. *J Virol* **66:** 2583–2587.

Wen W, Meinkoth JL, Tsien RY & Taylor SS (1995) Identification of a signal for rapid export of proteins from the nucleus. *Cell* **82:** 463–473.

Wolff B, Sanglier JJ & Wang Y (1997) Leptomycin B is an inhibitor of nuclear export: inhibition of nucleo-cytoplasmic translocation of the human immunodeficiency virus type 1 (HIV-1) Rev protein and Rev-dependent mRNA. *Chem Biol* **4:** 139–147.

Wu F, Garcia J, Sigman D & Gaynor R (1991) Tat regulates binding of the human immunodeficiency virus trans-activating region RNA loop-binding protein TRP-185. *Genes Dev* **5:** 2128–2140.

Wu-Baer F, Sigman D & Gaynor RB (1995) Specific binding of RNA polymerase II to the human immunodeficiency virus trans-activating region RNA is regulated by cellular cofactors and Tat. *Proc Natl Acad Sci USA* **92:** 7153–7157.

Wu-Baer F, Lane SW & Gaynor RB (1996) Identification of a group of cellular cofactors that stimulate the binding of RNA polymerase II and TRP-185 to human immunodeficiency virus TAR RNA. *J Biol Chem* **271:** 4201–4208.

Yang X, Herrmann CH & Rice AP (1996) The human immunodefiency virus Tat proteins specifically associate with TAK in vivo and require the carboxyl-terminal domain of RNA polymerase II for function. *J Virol* **70:** 4576–4584.

Zapp ML & Green MR (1989) Sequence-specific RNA binding by the HIV-1 Rev protein. *Nature* **342:** 714–716.

Zapp ML, Hope TJ, Parslow TG & Green MR (1991) Oligomerization and RNA binding domains of the type 1 human immunodeficiency virus Rev protein: a dual function for an arginine-rich binding motif. *Proc Natl Acad Sci USA* **88:** 7734–7738.

Zemmel RW, Kelley AC, Karn J & Butler PJG (1996) Flexible regions of RNA structure facilitate co-operative Rev assembly on the Rev-responsive element. *J Mol Biol* **258:** 763–777.

Zhou Q & Sharp PA (1996) Tat-SF1: cofactor for stimulation of transcriptional elongation by HIV-1 Tat. *Science* **274:** 605–610.

Zhu Y, Peery T, Peng J et al (1997) Transcription elongation factor P-TEFb is required for HIV-1 tat transactivation in vitro. *Genes Dev* **11:** 2622–2632.

Chapter 2

VIRAL EVOLUTION AND VARIATION IN THE HIV PANDEMIC

Andrew J. Leigh Brown

INTRODUCTION

High levels of genetic variability have been associated with HIV since the virus was first sequenced (Sanchez Pescador et al, 1985; Wain-Hobson et al, 1985). It was recognized at an early stage that this variability could have a major impact on the efficacy of any vaccine and this was the original motivation for much of the work on viral variation. More recently it has been recognized as being integral to analysis of immunological responses to HIV infection, of transmission, of pathogenesis, drug resistance and epidemiology (Leigh Brown and Holmes, 1994; Hu et al, 1996).

GLOBAL VARIATION IN THE HIV-1 PANDEMIC

Early studies of individuals with AIDS-like symptoms in West Africa led to the identification of two distinct human lentiviruses capable of giving rise to a similar immunodeficiency syndrome. Human immunodeficiency virus type 2 (HIV-2) has remained associated with West Africa and people with West African connections. Studies on the immunodeficiency viruses of non-human primates are revealing the processes that are likely to have given rise to their human equivalents.

Early analyses of HIV sequences suggested dates for the separation of HIV-1 and HIV-2 which varied from as little as 40 years (Smith et al, 1988) to as much as 1000 years (Gojobori et al, 1990). The rapidly increasing database has allowed some bounds to be placed on this. We know now that the origin of HIV-1 and HIV-2 was not in a human but in a non-human primate and thus could not be in the recent past (Gao et al, 1992). Within HIV-1, the virus responsible for the HIV pandemic, more extensive sampling in Africa and elsewhere has revealed much greater diversity than previously known. Sequencing of the *gag* and *env* genes has led to the identification of two major groups of HIV-1 viruses, M and O (Myers et al, 1995). The difference between these groups is so large that sequences from certain chimpanzees appear to separate them in phylogenetic analyses. The O group is confined to Cameroon and Gabon and to individuals with connections in those countries (Gurtler et al, 1994; Janssens et al, 1994), but even there it accounts for only about 5% of HIV infections (Zekeng et al, 1994).

Phylogenetic analyses of *gag* and *env* gene sequences have revealed up to nine clusters of the main, M, group of HIV-1 (Louwagie et al, 1993; Myers et al, 1995). The greatest

HIV and the New Viruses Second Edition
ISBN 0-12-200741-7

diversity, in terms of numbers of subgroups present, is found in sub-Saharan Africa, in particular western Central Africa, including Zaïre (Democratic Republic of Congo), Congo, Gabon and neighbouring countries (Hu et al, 1996). In East Africa (Uganda, Kenya) the D and A subgroups are most common, while to the south (Zimbabwe, South Africa) the C subgroup is more abundant. The major epidemics in other continents are mostly also associated with a subset of these variants (Leigh Brown and Holmes, 1994), with the B subtype characterizing the epidemic in North America, Europe and Australia almost exclusively. In fact it was only in 1996 that the first reports of non-B subgroup sequences were confirmed in the USA (Brodine et al, 1995). With the locus of greatest diversity in western Central Africa comes a strong indication that it was in that region that the HIV-1 epidemic originated. In general, classification into subgroups based on *gag* and *env* genes gives similar results, although analyses of a number of exceptions have suggested that recombination between viruses of different subgroups may be more widely distributed in both HIV types 1 and 2 (Robertson et al, 1995).

Although the O group of viruses is less significant in terms of numbers of infections, sequencing studies have revealed an even more startling level of diversity. Whereas most subgroups of the M group represent many thousands or even millions of infected individuals, most of the *individual* sequences from the O group are as different from each other as are the subgroups of group M (Loussert-Ajaka et al, 1995), possibly signifying an older – or even independent – origin for the O group epidemic (Gurtler et al, 1996).

The classification of HIV-1 into subgroups has received much attention because of the possibility that any vaccine raised against one strain (or subgroup) may not protect against others (Moore and Anderson, 1994). There are a number of possible explanations for the clusters: first, they could reflect nothing more than the poor sample of viruses currently available, and as sampling improves so the structure of the phylogenetic trees will change and reduce their distinctiveness; secondly, subtypes could represent biologically differentiated strains; thirdly, it is also possible that the subgroups are real, genetically distinct populations, but their origin and present distribution is caused by the chance invasion of a particular virus strain into a susceptible population – in other words one or a series of 'founder effects'. To some extent this issue has been highlighted by the description of a sequence obtained from the earliest known confirmed HIV-1-positive human sample – obtained in Congo (Zaïre) in 1959 (Nahmias et al, 1986). Phylogenetic analysis has placed this sequence close to the root of the major B and D (and F) subtypes, suggesting that these groups had diverged only a relatively short time earlier (Zhu et al, 1998). This important observation supports the view that analysis of sequence data can help elucidate the origin of the HIV pandemic and extend it to include the timing. This suggests that a major expansion of HIV-1 M group diversity had begun perhaps 10 or 15 years earlier, placing the origin of the pandemic in central West Africa in the late 1940s. The dynamics of the epidemic since then have certainly not been regular and this is also revealed by molecular studies – the B subtype is almost unknown in Africa and yet forms the dominant strain in the USA and Europe (Hu et al, 1996).

GENETIC VARIATION IN LOCAL HIV EPIDEMICS

Thailand

The recent epidemic of HIV-1 in Thailand, about which we know a great deal, well illustrates the possible course of such an epidemic. Thailand experienced an exponential increase in the frequency of HIV infections in high-risk groups in the late 1980s; by mid-1991 approximately 0·5% of the population were estimated to be infected (Ou et al, 1993). Two independent epidemics occurred involving different subgroups of HIV-1. In 1988, an unusual variant of subgroup B began to spread in the central and southern regions (including Bangkok) among injecting drug users, which was not observed among the small numbers of HIV-1 seropositive subjects known before that date (Kalish et al, 1995). In 1989 an epidemic involving mainly heterosexual transmission developed in northern Thailand leading to the spread of subgroup E. Because of the very different global distributions of these subgroups, it is likely that they entered the Thai population independently. The two subgroups are also associated with different risk groups: in 1991 it was found that 85% of the patients infected through sexual transmission possessed viruses of genotype B, while only 24% of individuals infected through injecting drug use had viruses of this genotype (Ou et al, 1993). By 1994, this pattern had changed: subgroup E formed a higher proportion of recent infections, even among drug users in Bangkok, simply because there were many more individuals infected with subgroup E in total (Kalish et al, 1995). While it has been suggested from culture in vitro that the Thai E subgroup shows differences in biological properties (Rubsamen Waigmann et al, 1994), this may not be relevant. It is possible that high rates of sexual transmission of HIV in Thailand may be explained by the high prevalence of other sexually transmitted diseases acting as cofactors (Mastro et al, 1994). One aspect of the Thailand study which also suggests a founder effect is that, despite the highly divergent genotype of the virus found, very little divergence was seen between sequences from different individuals early in the epidemics (Ou et al, 1993).

Injecting drug users in Scotland

A similar level of detail is available for HIV-infected individuals in Scotland as a result of extensive studies of the molecular epidemiology following detailed serosurveillance. The epidemic in this country was heavily driven by major, localized epidemics among injecting drug users (IDUs) (Robertson et al, 1986). The earliest occurred in Edinburgh where retrospective testing revealed almost 90% of a study group of IDUs centred in a western suburb became infected between mid-1983 and the end of 1984. The doubling time of this epidemic was only about 8 weeks at its peak – thus most individuals were becoming infected before the index had seroconverted. This is now known to be a time of high viral load and there is also evidence of high infectivity at this time (Palasanthiran et al, 1993). The first studies of the molecular epidemiology revealed two clusters of sequences in the area – one in a haemophiliac cohort and one in IDUs (Holmes et al, 1995). It appeared that there was no direct connection between the two clusters, although the epidemics coincided in time. More detailed study confirmed this and extended the IDU cluster to include members of this risk group from several other cities in Scotland as well as Dublin in the Republic of Ireland (Leigh Brown et al, 1997).

Epidemics of HIV in IDUs that occurred in Dundee and Dublin somewhat later than in Edinburgh have been described, and sporadic infection of Glasgow-based IDUs has continued, but the demonstration of a direct link between all these infections is surprising. In fact, sequences from IDUs resident in different cities do not form substantial clusters within this group, but are frequently intermingled with those from other cities.

The conclusion from this study is that HIV was transmitted so rapidly among IDUs in these cities that virtually no change in the viral consensus sequence occurred before the epidemics were established. More recently, an outbreak of HIV infection was observed in a Scottish prison, in association with injecting drug use and needle sharing. Molecular epidemiological follow-up of the initial investigation revealed that 13 out of 14 individuals belonged to a single cluster, and that their viral strains were so similar that, in samples taken in the first months of infection, identical sequences of the 360 base pair long p17 coding sequence in *gag* were obtained from 9 of the 14 individuals (Yirrell et al, 1997b). It is likely that the circumstances involved in this epidemic were similar to those that occurred in Edinburgh in late 1983, and the evolutionary processes the virus underwent were also similar.

Uganda

The major HIV epidemic in Uganda has had severe consequences for a country with a very low per capita health budget. As part of the Medical Research Council programme on AIDS a natural history cohort was recruited in a rural area near Masaka in south-western Uganda. Molecular epidemiological investigations have been carried out both in this essentially stable community where the prevalence is about 8% and in a neighbouring trading town on the trans-African highway with a much higher prevalence (about 40%). In Uganda the A and D subtypes are both present in a ratio of about 1 : 2 and have been present at a similar frequency for several years. However, striking differences in the nature of the epidemic in these two centres were revealed by the molecular epidemiological analysis. In the trading town, despite detailed analysis of sexual networks, molecular studies added little to the picture, and frequently suggested infection had come from a very different route from that predicted (Yirrell et al, 1997a). In contrast, in the rural area, where individuals with known dates of seroconversion were studied, newly infected individuals more frequently had virus similar to that of their regular sexual partner. In addition, evidence for spread within this community of a local variant of the A subtype was obtained (A. J. Leigh Brown et al, unpublished observations). Concentration of future sequencing efforts on such well-characterized study groups is likely to be more informative than surveys of seroprevalent cases.

EVOLUTION OF HIV WITHIN AN INDIVIDUAL

At the earliest stages of the infection, the first 7–10 weeks, the virus is able to replicate to high titre levels, reaching 10^6 or 10^7 viral particles per ml of plasma (Piatak et al, 1993a,b). Studies of viral populations at this phase of infection have revealed low levels of sequence variability, especially in parts of the envelope glycoprotein gp120 (Zhang et al, 1993; Zhu et al, 1993), whereas variation is detectable in *gag* (Zhang et al, 1993). Such a rapid increase in viral population size, coupled with the general sequence similarity (in contrast to sequence diversity often observed later in the infection), might be expected to imply strong stabilizing selection for the most rapidly replicating virus.

Although there is a dramatic drop in viral titre soon after seroconversion (Koup et al, 1994; Safrit et al, 1994), the virus continues to replicate in the presence of active cellular and humoral immune responses (Piatak et al, 1993b, Zhang et al, 1991). The availability of quantitative data on plasma viral load in patients entering potent antiviral therapy has revealed that the half-life of plasma virions is only 6 hours and the great majority of the virus population turns over within a few days (Ho et al, 1995; Wei et al, 1995; Perelson et al, 1996). The dynamics of HIV populations are reviewed in Chapter 4 and it is sufficient to note here that continued viral replication at a rate of, on average, 150 replication cycles a year (Coffin, 1995) underlies the rapid evolution of the population in the plasma. Despite this, phylogenetic studies show that lineages of related sequences can be detected over long periods in the circulation (Holmes et al, 1992; Leigh Brown and Cleland, 1996).

The persistence of variation in immunologically significant areas of gp120 and differentiation of subpopulations in solid tissue (Delassus et al, 1992a) have raised a continuing debate over the importance of immune selection, mutation pressure, and chance in its maintenance. These alternatives could be mutually exclusive: Wain-Hobson has argued that the spatial differentiation of the HIV population in solid lymphoid tissue is so extensive that the activation of HIV-infected T cells in response to unrelated antigens could give rise to random temporal fluctuations in the plasma virus population (Wain-Hobson, 1993; Cheynier et al, 1994). On the other hand, Coffin (1995) has argued that the viral populations are so large that small selective differences would inevitably result in the complete replacement of the viral population. Although this issue was raised at the time of the first comparisons of HIV sequence variability (Leigh Brown and Monaghan, 1988; Coffin, 1992), a resolution is still sought.

THE ROLE OF SELECTION IN HIV POPULATIONS – LACK OF PRECISION IN FREQUENCY ESTIMATES

It is reasonable to ask why these issues are still unresolved. Part of the answer relates to inadequacies in population genetics theory for HIV, which are now being addressed and are discussed below. Another part derives from a simple problem inherent to population genetics. It is often essential to be able to describe, and predict, frequency changes in sequence variants. However, the precision of any frequency estimate is limited by the confidence interval of the binomial distribution. To reach any reasonable level of accuracy, a very large number of cases must be counted. This rarely happens, in practice, in sequence-based studies where sample sizes are often 20 or less. The observation of a single copy of a variant in such a sample does not indicate that it is present at a frequency of 5%, but rather anything between 0·15% and 25%.

Two examples illustrate the importance of the imprecision in frequency estimates. Although transmission of zidovudine (azidothymine, AZT) resistance-associated mutations has been observed following the identification of resistant isolates in acutely infected patients (Erice et al, 1993), in one study it was inferred from analysis of four transmissions from patients infected with resistant isolates that these are selected against because a resistant variant was found in only one of the contacts (Wahlberg et al, 1994). Clearly, the claim was not established by this investigation, as 1/4 is not significantly different from 4/4. Another issue concerns the tropism and growth characteristics of the virus from recently infected patients. Viral variants that can grow and form syncytia in T cell lines (Tersmette et al, 1989) are frequently isolated from patients with symptomatic HIV infection, and are present in a proportion of patients before symptoms develop (Schuitemaker et al, 1992; Koot et al, 1993). These variants can be distinguished from sequence data by virtue of a strong correlation with the presence of positively charged amino acids at certain residues in the V3 loop (Fouchier et al, 1992; Milich et al, 1993). Viral genotypes that do not have these features retain the ability to infect macrophages (Chesebro et al, 1992). It is a prominent feature of studies on variation in the *env* gene that levels of variation are much lower very early in infection (Zhang et al, 1993; Zhu et al, 1993); the sequences observed in most of these patients at this time were 'microphage-tropic'.

At the qualitative level, it has been shown that some individuals do become infected with syncytium-inducing (SI) variants (Nielsen et al, 1993; Fiore et al, 1994). Further, the fact that non-syncytium-inducing (NSI) variants are found in all individuals at all stages of infection implies that only a major frequency change could be considered to be evidence for selection. The most convincing result has recently come from a study of an unusual transmission, by intramuscular injection, in which 8 of 10 sequenced clones from plasma virus taken from the index were of the SI genotype. In the first sample from the newly infected contact, taken 21 days following infection before the patient seroconverted, 10 out of 10 were SI. However, in a sample taken 10 weeks later, all 10 viral sequences were NSI (Cornelissen et al, 1995). Only with such dramatic changes in frequency can a convincing case for selection be made on the numbers of sequences that are usually available.

HIV GENETIC DIVERSITY AND DISEASE PROGRESSION

A third area where the difficulty of establishing statistically robust results has been important is in discussions of the relationship between genetic diversity (especially in gp120 sequences) and disease progression. An influential model of the interaction between HIV and the humoral arm of the immune system made specific predictions about the expected pattern of change in diversity of immunologically distinct strains with time (Nowak et al, 1990). An increase in diversity was predicted during the immunologically competent phase, while this would be expected to decrease sharply late in the infection. Despite the clarity of this prediction, there have been difficulties in testing this model, partly because of the difficulty in equating protein sequence diversity with immunological diversity (Nowak et al, 1991).

Studies of the subset of individuals who progress extremely rapidly (symptomatic within 2 years) have suggested that the viral population does not evolve rapidly from the seroconversion sequence, and diversity remains generally low (Wolinsky et al, 1996; Liu et al, 1997). However, when patients with more normal courses of disease progression have been studied, the results are less clear. While some studies have detected no straightforward relationship (McNearney et al, 1992; McDonald et al, 1997), others have claimed a positive relationship (Lukashov et al, 1995). The clearest reasons for differences in rates of progression found to date are generally host factors. Some of these are genetic: particular MHC haplotypes are associated with very rapid progression (Steel et al, 1988; Kaslow et al, 1996) and genotype at coreceptor loci is clearly important (Dean et al, 1996; Smith et al, 1997). Age at infection is also a risk factor for progression (Darby et al, 1990), and a study of two patients infected with virus from the same source found that the virus population of the older, rapid progressor consistently showed about 50% of the genetic diversity of the (younger) slow progressor (Liu et al, 1997) at the same time points. In particular, the viral sequences in these patients accumulated synonymous nucleotide substitutions at similar rates in the gp120 coding region studied, but the very rapid progressor showed about a three-fold lower rate of evolution in amino acid sequence.

The fact that these individuals were infected with virus from a common source controls for any differences in the viral population and strengthens the authors' contention that evolution of amino acid sequence in gp120 is a response to an active immune recognition of HIV. Nevertheless, because advancing HIV disease erodes all immune responses, and such detailed studies can be performed in only a small number of individuals, it remains difficult to establish a specific immune mechanism as being causally associated with the evolution of diversity in general, and the failure to observe consistent results in different patient populations makes a clear conclusion difficult to reach.

DETECTION OF SELECTION FROM THE K_s/K_a RATIO

One approach to the problem of distinguishing selected changes from those due to sampling differences is to make use of comparisons of the ratio of the frequency of synonymous nucleotide substitutions (K_s) to those that result in a change in the amino acid (K_a). This has been used for some time to explore the nature of the selective forces acting on a gene (Kimura, 1981; Li et al, 1985, 1995; Hughes and Nei, 1988, 1989), and previous studies have examined the distribution of this ratio in gp120 coding sequences of HIV and other lentiviruses (Leigh Brown and Monaghan, 1988; Li et al, 1988; Shpaer and Mullins, 1993). The expectation of the ratio of non-synonymous substitutions per non-synonymous site to synonymous substitutions per synonymous site under a strict neutral model is 1. Studies of large numbers of nuclearly encoded proteins revealed a large range in this ratio for different proteins. The variation is due to variation in K_a (up to to 100-fold), as the observed range of synonymous substitution is much lower (Li et al, 1995). The average value for this ratio for nuclearly encoded proteins was about 5, confirming that significant stabilizing selection acts on amino acid sequences. For short regions of certain nuclear proteins, it has been found that this ratio can drop below 1 (Hill and Hastie, 1987; Hughes and Nei, 1988, 1989). In addition, it was observed early on that the ratio obtained for gp120 between patients (but not for other proteins of HIV-1) was close to 1 (Leigh Brown and Monaghan, 1988).

With increasing amounts of data from studies obtaining multiple sequences of HIV from different patients, it has become apparent that there are interesting systematic differences when the ratio is estimated separately for viral sequences found *within* the same hosts from estimates made *between* hosts. In addition, there appear to be differences between different species. In HIV-1 (Simmonds et al, 1991; Bonhoeffer et al, 1995), simian immunodeficiency virus SIV$_{mac}$ (Burns and Desrosiers, 1991), feline immunodeficiency virus (Rigby et al, 1993), SIV$_{agm}$ (Baier et al, 1991) and the yellow baboon (Jin et al, 1994; P. M. Sharp, personal communication) these ratios can be less than 1, i.e. amino acid replacements occur more frequently than silent substitutions within hosts (Table 2.1). However, again the results are not always observed as larger studies have revealed greater diversity in within-sample values (Liu et al, 1997; McDonald et al, 1997; Zhang et al, 1997). In contrast, when the same estimates are made from viruses from *different* hosts (Leigh Brown and Monaghan, 1988; Shpaer and Mullins, 1993; Korber et al, 1994) the ratio rises, indicating that relatively more synonymous substitutions have occurred, and there is no evidence of a further relationship with genetic distance (Korber et al, 1994); however, Shpaer and Mullins (1993) and Sharp et al (1994) have pointed out that different ratios characterize immunodeficiency viruses from different primates.

This approach, comparing the mean values of a large number of estimates, avoids the difficulties of estimating differences in frequency outlined above. The low ratios at lower genetic distances are not consistent with models of intrapatient evolution where genetic drift predominates (Delassus et al, 1992b; Cheynier et al, 1994). Differences in the ratios between immunodeficiency viruses from different species are also not consistent with a strict neutral model. However, the former does not allow us to identify and test the agent responsible, which could be neutralizing antibody, cytotoxic T lymphocytes, or some other mechanism, and the latter, though possibly indicating a relationship with pathogenesis (Shpaer and Mullins, 1993), is an even more difficult effect to explain.

Nevertheless, this approach raises some interesting features and it may be possible to use it to highlight differences in the strength of various selection pressures at different stages in the viral life cycle. Even with the observed fluctuations in within-sample values, several studies have described a systematic difference when samples are compared

Table 2.1 K_s/K_a *ratios estimated within an individual for HIV-1, SIV and FIV in the env gene. The key to the reference numbers is given in the footnote.*

Reference	gp120 K_s/K_a	V3 K_s/K_a
HIV-1		
1, 2, 3, 4		0.67, 0.83, 0.25, 0.62
5		0.6 (T-tropic), 1.4 (M-tropic)
6		0.15–3.3 (3 pts)
7		0.53–2.7 (pt A) 0.7–7.2 (pt B)
8	About 0.5–2.0 (10 pts)	About 0.5–0.77 (10 pts)
SIV		
9	0.24	
10		0.087
11	0.51	
12	2.4	
FIV		
13		0.62

References: 1, Simmonds et al, 1990; 2, Wolfs et al, 1992; 3, Cichutek et al, 1992; 4, Leigh Brown and Cleland, 1996; 5, Bonhoeffer et al, 1995; 6, Zhang et al, 1997; 7, Liu et al, 1997; 8, McDonald et al, 1997; 9, Burns and Desrosiers, 1991; 10, Baier et al, 1991; 11, P. Sharp and B. Hahn, unpublished (vervet); 13, Rigby et al, 1993. Pt, patient.
More data are clearly required to resolve the discrepancy between the result given for the vervet (12) and that published by Baier (10)

with the sequence observed at seroconversion (Liu et al, 1997; Zhang et al, 1997), providing clear evidence for selection affecting amino acid sequences in these regions of gp120.

Despite the rapid and dramatic divergence observed in the V3 region of gp120 within a patient, Kuiken et al (1993) have made the important observation that the V3 region has *not* evolved significantly *between* patients during the 10 years of the Amsterdam HIV-1 epidemic among homosexual men. To accommodate both observations, we do have to infer a selective pressure which constrains the amino acid sequence of the region at transmission – i.e. stabilizing selection. These two inferred selective forces have opposing effects on the rate of amino acid change. If it is assumed that the rate of synonymous substitutions is independent of that of non-synonymous substitutions then we can explain the change in the ratio: synonymous substitutions continue to accumulate with evolutionary distance while amino acid sequence is constrained at transmission, allowing silent sites to 'catch up' with them.

THEORETICAL DEVELOPMENTS IN HIV POPULATION GENETICS

The view of the genetics of HIV populations that has dominated the field has been strongly influenced by the extremely large numbers of particles circulating in the plasma, and the suggestion that very large numbers of cells in solid lymphoid tissue may harbour proviruses (Embretson et al, 1993; Piatak et al, 1993b). The appropriate theory for such a large population is deterministic and this has been developed both in models of the interaction with the immune system (Nowak et al, 1990) and with regard to the important implications for variation at synonymous sites (Coffin, 1995). However, in many respects the evolution of HIV, even in response to a defined selection pressure, is less predictable than expected from a deterministic model (Leigh Brown and Richman, 1997). Applying a specific test of this model, it was shown that a better fit to observations on within-patient sequence variation in *env* is obtained if an effective population size (N_e) of the order of 10^3 is assumed (Leigh Brown, 1997). More recent studies suggest that this may be as low as 10^2 in some patients, especially earlier in infection (S. Frost and A.J. Leigh Brown, unpublished observations).

In the absence of selection, the effective population size is inversely related to the average time to the most recent common ancestor of the population. It can also be envisaged as a quantitation of the average number of ancestors present at a given point in the recent past that contributed to the extant population. The larger this number, the more effective selection will be on the population. Factors that decrease the effective population number below the census size include the clonal expansion of infected T cells, all of which would be activated in response to the same antigen. As there would be no expected correlation between T cell antigen specificity and the genotype of the HIV provirus they might be bearing, this would lead to stochastic temporal fluctuations in frequency of HIV genotypes (Wain-Hobson, 1993; Cheynier et al, 1994). Another major factor reducing effective population number is the fact that most infections appear to be initiated from a single virion, as indicated above (Leigh Brown and Richman, 1997). It takes many replication cycles before the effect on the evolution of the population of this severe bottleneck is eliminated.

The consequence of a low N_e is that the potential for selective evolution of the virus is constrained, and chance may play a role even when selection might have been expected to be the dominant force. However, this may not always reduce the rate of response to selection. When multiple changes are required before a significant advantage is gained, this may occur faster in a subdivided population with a low N_e than in a larger, panmictic equivalent, a principle incorporated by Sewall Wright into his 'shifting balance' theory of evolution (Wright, 1931).

The implications for the field of these observations extend to the identification of appropriate theoretical models and approaches to the analysis of HIV sequence data. These are now being developed and we may expect to see significant developments in this area as further integration of theoretical and experimental results takes place.

ACKNOWLEDGEMENTS

Work in the author's laboratory is supported by the Medical Research Council. This review was completed while at the HIV and Retrovirology Branch, at the Centers for Disease Control and Prevention, Atlanta, Georgia, USA, and supported by the Research Participation Program of the Oak Ridge Institute for Science and Education.

REFERENCES

Baier M, Dittmar MT, Cichutek K & Kurth R (1991) Development in vivo of genetic variability of simian immunodeficiency virus. *Proc Natl Acad Sci USA* **88:** 8126–8130.

Bonhoeffer S, Holmes EC & Nowak M (1995) Causes of HIV diversity. *Nature* **376:** 125.

Brodine SK, Mascola JR, Weiss PJ et al (1995) Detection of diverse HIV-I genetic subtypes in the USA. *Lancet* **346:** 1198-9

Burns DP & Desrosiers RC (1991) Selection of genetic variants of simian immunodeficiency virus in persistently infected rhesus monkeys. *J Virol* **65:** 1843–1854.

Chesebro B, Wehrly K, Nishio J & Perryman S (1992) Macrophage-tropic human immunodeficiency virus isolates from different patients exhibit unusual V3 envelope sequence homogeneity in comparison with T cell-tropic isolates: definition of critical amino acids involved in cell tropism. *J Virol* **66:** 6547–6554.

Cheynier R, Henrichwark S, Hadida F et al (1994) HIV and T cell expansion in splenic white pulps is accompanied by infiltration of HIV-specific cytotoxic T lymphocytes. *Cell* **78:** 373–387.

Cichutek K, Merget H, Norley S et al (1992) Development of a quasispecies of human immunodeficiency virus type 1 in vivo. *Proc Natl Acad Sci USA* **89:** 7365–7369.

Coffin JM (1992) Genetic diversity and evolution of retroviruses. *Curr Top Microbiol Immunol* **176:** 143–164.

Coffin JM (1995) HIV population dynamics in vivo: implications for genetic variation, pathogenesis and therapy. *Science* **267:** 483–489.

Cornelissen M, Mulder-Kampinga G, Veenstra J et al (1995) Syncytium-inducing (SI) phenotype suppression at seroconversion after intramuscular inoculation of a non-syncytium-inducing/SI phenotypically mixed human immunodeficiency virus population. *J Virol* **69:** 1810–1818.

Darby SC, Doll R, Thakrar B, Rizza CR & Cox DR (1990) Time from infection with HIV to onset of AIDS in patients with haemophilia in the UK. *Stat Med* **9:** 681–689.

Dean M, Carrington M, Winkler C et al (1996) Genetic restriction of HIV-1 infection and progression to AIDS by a deletion allele of the CKR5 structural gene. *Science* **273:** 1856–1862.

Delassus S, Cheynier R & Wain-Hobson S (1992a) Nonhomogeneous distribution of human immunodeficiency virus type 1 proviruses in the spleen. *J Virol* **66:** 5642–5645.

Delassus S, Meyerhans A, Cheynier R & Wain-Hobson S (1992b) Absence of selection of HIV-1 variants in vivo based on transcription/transactivation during progression to AIDS. *Virology* **188:** 811–818.

Embretson J, Zupancic M, Ribas JL et al (1993) Massive covert infection of helper T lymphocytes and macrophages by HIV during the incubation period of AIDS. *Nature* **362:** 359–362.

Erice A, Mayers DL, Strike DG et al (1993) Primary infection with zidovudine-resistant human immunodeficiency virus type 1. *N Engl J Med* **328:** 1163–1165.

Fiore JR, Bjorndal A, Peipke KA et al (1994) The biological phenotype of HIV-1 is usually retained during and after sexual transmission. *Virology* **204:** 297–303.

Fouchier RA, Groenink M, Kootstra NA et al (1992) Phenotype-associated sequence variation in the third variable domain of the human immunodeficiency virus type 1 gp120 molecule. *J Virol* **66:** 3183–3187.

Gao F, Yue L, White AT et al (1992) Human infection by genetically diverse SIVsm-related HIV-2 in West Africa. *Nature* **358:** 495–499.

Gojobori T, Moriyama EN, Ina Y et al (1990) Evolutionary origin of human and simian immunodeficiency viruses. *Proc Natl Acad Sci USA* **87:** 4108–4111.

Gurtler LG, Hauser PH, Eberle J et al (1994) A new subtype of human immunodeficiency virus type 1 (MVP-5180) from Cameroon. *J Virol* **68:** 1581–1585.

Gurtler LG, Zekeng L, Tsague JM et al (1996) HIV-1 subtype O: epidemiology, pathogenesis, diagnosis, and perspectives of the evolution of HIV. *Arch Virol* **11** (suppl.): 195–202.

Hill RE & Hastie ND (1987) Accelerated evolution in the reactive centre regions of serine protease inhibitors. *Nature* **326:** 96–99.

Ho DD, Neumann AU, Perelson AS, Chen W, Leonard JM & Markowitz M (1995) Rapid turnover of plasma virions and CD4 lymphocytes in HIV-1 infection. *Nature* **373:** 123–126.

Holmes EC, Zhang LQ, Simmonds P, Ludlam CA & Leigh Brown AJ (1992) Convergent and divergent sequence evolution in the surface envelope glycoprotein of human immunodeficiency virus type 1 within a single infected patient. *Proc Natl Acad Sci USA* **89:** 4835–4839.

Holmes EC, Zhang LQ Robertson P et al (1995) The molecular epidemiology of HIV-1 in Edinburgh, Scotland. *J Infect Dis* **171:** 45–53.

Hu DJ, Dondero TJ, Rayfield MA et al (1996) The emerging genetic diversity of HIV. The importance of global surveillance for diagnostics, research, and prevention. *JAMA* **275:** 210–216.

Hughes AL & Nei M (1988) Pattern of nucleotide substitution at major histocompatibility complex class I loci reveals overdominant selection. *Nature* **335:** 367–370.

Hughes AL & Nei M (1989) Nucleotide substitution at major histocompatibility complex class II loci: evidence for overdominant selection. *Proc Natl Acad Sci USA* **86:** 958–962.

Janssens W, Nkengasong JN, Heyndrickx L et al (1994) Further evidence of the presence of genetically very aberrant HIV-1 strains in Cameroon and Gabon. *AIDS* **8:** 1012–1013.

Jin MJ, Rogers J, Phillips-Conroy JE et al (1994) Infection of a yellow baboon with simian immunodeficiency virus from African Green Monkeys: evidence for cross-species transmission in the wild. *J Virol* **68:** 8454–8460.

Kalish ML, Luo CC, Raktham S et al (1995) The evolving molecular epidemiology of HIV-1 envelope subtypes in injecting drug users in Bangkok, Thailand: implications for HIV vaccine trials. *AIDS* **9:** 851–857.

Kaslow RA, Carrington M, Apple R et al (1996) Influence of combinations of human major histocompatibility complex genes on the course of HIV-1 infection. *Nature Med* **2:** 405–411.

Kimura M (1981) Estimation of evolutionary distances between homologous nucleotide sequences. *Proc Natl Acad Sci USA* **78:** 454–458.

Koot M, Keet IPM, Vos AHV et al (1993) Prognostic value of HIV-1 syncytium-inducing phenotype for rate of CD4+ cell depletion and progression to AIDS. *Ann Intern Med* **118:** 681–688.

Korber BT, MacInnes K, Smith RF & Myers G (1994) Mutational trends in V3 loop protein sequences observed in different genetic lineages of human immunodeficiency virus type 1. *J Virol* **68:** 6730–6744.

Koup R, Safrit J, Cao Y et al (1994) Temporal association of cellular immune responses with the initial control of viremia in primary human immunodeficiency virus type 1 syndrome. *J Virol* **68:** 4650–4655.

Kuiken CL, Zwart G, Baan E, Coutinho RA, van den Hoek JAR & Goudsmit J (1993) Increasing antigenic and genetic diversity of the V3 variable domain of the human immunodeficiency virus envelope protein in the course of the AIDS epidemic. *Proc Natl Acad Sci USA* **90:** 9061–9065.

Leigh Brown AJ (1997) Analysis of HIV-1 *env* gene sequences reveals evidence for a low effective population number in the virus population. *Proc Natl Acad Sci USA* **94:** 1862–1865.

Leigh Brown AJ & Cleland A (1996) Independent evolution of the *env* and *pol* genes of HIV-1 during zidovudine therapy. *AIDS* **10:** 1067–1073.

Leigh Brown AJ & Holmes EC (1994) The evolutionary biology of human immunodeficiency virus. *Annu Rev Ecol Syst* **25:** 127–165.

Leigh Brown A & Monaghan P (1988) Evolution of the structural proteins of human immunodeficiency virus: selective constraints on nucleotide substitution. *AIDS Res Hum Retroviruses* **4:** 399–407.

Leigh Brown AJ & Richman DD (1997) HIV-1: gambling on the evolution of drug resistance? *Nature Med* **3:** 268–271.

Leigh Brown AJ, Lobidel D, Wade CM et al (1997) The molecular epidemiology of human immunodeficiency virus type 1 in six cities in Britain and Ireland. *Virology* **235:** 166–177.

Li WH, Wu CI & Luo CC (1985) A new method for estimating synonymous and nonsynonymous rates of nucleotide substitution considering the relative likelihood of nucleotide and codon changes. *Mol Biol Evol* **2:** 150–174.

Li WH, Tanimura M & Sharp PM (1988) Rates and dates of divergence between AIDS virus nucleotide sequences. *Mol Biol Evol* **5:** 313–330.

Li WH, Luo CC & Wu CI (1995) Evolution of DNA sequences. In: Macintyre RJ, ed. *Molecular Evolutionary Genetics*, pp 1–94. New York: Plenum.

Liu SL, Schacker T, Musey L et al (1997) Divergent patterns of progression to AIDS after infection from the same source: human immunodeficiency virus type 1 evolution and antiviral responses. *J Virol* **71:** 4284–4295.

Loussert-Ajaka I, Chaix ML, Korber B et al (1995) Variability of HIV type 1 group O isolates isolated from Cameroonian patients living in France. *J Virol* **69:** 5640–5649

Louwagie J, McCutchan FE, Peeters M et al (1993) Phylogenetic analysis of *gag* genes from 70 international HIV-1 isolates provides evidence for multiple genotypes. *AIDS* **7:** 769–780.

Lukashov VV, Kuiken CL & Goudsmit J (1995) Intrahost human immunodeficiency virus type 1 evolution is related to length of the immunocompetent period. *J Virol* **69:** 6911–6916.

Mastro TD, Satten GA, Nopkesorn T, Sangkharomya S & Longini IM (1994) Probability of female-to-male transmission of HIV-1 in Thailand. *Lancet* **343:** 204–207.

McDonald RA, Mayers DL, Chung RC et al (1997) Evolution of human immunodeficiency virus type 1 *env* sequence variation in patients with diverse rates of disease progression and T-cell function. *J Virol* **71:** 1871–1879.

McNearney T, Hornickova Z, Markham R et al (1992) Relationship of human immunodeficiency virus type 1 sequence heterogeneity to stage of disease. *Proc Natl Acad Sci USA* **89:** 10247–10251.

Milich L, Margolin B & Swanstrom R (1993) V3 loop of the human immunodeficiency virus type 1 env protein: interpreting sequence variability. *J Virol* **67:** 5623–5634.

Moore J & Anderson R (1994) The WHO and why of HIV vaccine trials. *Nature* **372:** 313–314.

Myers G, Korber B, Hahn BH et al (1995) *Human Retroviruses and AIDS 1995.* Los Alamos National Laboratory.

Nahmias AJ, Weiss J, Yao X et al (1986) Evidence for human infection with an HTLV III/LAV-like virus in Central Africa, 1959. *Lancet* 1279–1280.

Nielsen C, Pedersen C, Lundgren JD & Gerstoft J (1993) Biological properties of HIV isolates in primary HIV infection: consequences for the subsequent course of infection. *AIDS* 7: 1035–1040.

Nowak MA, May RM & Anderson RM (1990) The evolutionary dynamics of HIV-1 quasi-species and the development of immunodeficiency disease. *AIDS* 4: 1095–1103.

Nowak MA, Anderson RM, McLean AR, Wolfs TFW, Goudsmit J & May RM (1991) Antigenic diversity thresholds and the development of AIDS. *Science* 254: 963–969.

Ou CY, Takebe Y, Weniger BG et al (1993) Independent introduction of two major HIV-1 genotypes into distinct high-risk populations in Thailand. *Lancet* 341: 1171–1174.

Palasanthiran P, Ziegler JB, Stewart GJ et al (1993) Breastfeeding during primary maternal human immunodeficiency virus infection and risk of transmission from mother to infant. *J Infect Dis* 167: 441–444.

Perelson AS, Neumann AU, Markowitz M, Leonard JL & Ho DD (1996) HIV-1 dynamics in vivo: virion clearance rate, infected cell life-span and viral generation time. *Science* 271: 1582–1586.

Piatak M, Yang LC, Luk KC et al (1993a) Viral dynamics in primary HIV-1 infection. *Lancet* 341: 1099.

Piatak M, Saag MS, Yang LC et al (1993b) High levels of HIV-1 in plasma during all stages of infection determined by competitive PCR. *Science* 259: 1749–1754.

Rigby M, Holmes EC, Pistello M, Mackay N, Leigh Brown AJ & Neil JC (1993) Evolution of structural proteins of feline immunodeficiency virus; molecular epidemiology and evidence for selection. *J Gen Virol* 74: 425–436.

Robertson JR, Bucknall ABV, Welsby PD et al (1986) Epidemic of AIDS related virus (HTLV-III/LAV) among intravenous drug abusers. *BMJ* 292: 527–529.

Robertson DL, Sharp PM, McCutchan FE & Hahn BH (1995) Recombination in HIV-1. *Nature* 374: 124–126.

Rubsamen Waigmann H, von Briesen H, Holmes H et al (1994) Standard conditions of virus isolation reveal biological variability of HIV type 1 in different regions of the world. *AIDS Res Hum Retroviruses* 11: 1401–1408.

Safrit J, Andrews C, Zhu T, Ho D & Koup R (1994) Characterization of human immunodeficiency virus type 1-specific cytotoxic T lymphocyte clones isolated during acute seroconversion: recognition of autologous virus sequences within a conserved immunodominant epitope. *J Exp Med* 179: 463–472.

Sanchez Pescador R, Power MD, Barr PJ et al (1985) Nucleotide sequence and expression of an AIDS-associated retrovirus (ARV-2). *Science* 227: 484–492.

Schuitemaker H, Koot M, Kootstra N et al (1992) Biological phenotype of human immunodeficiency virus type 1 clones at different stages of infection: progression of disease is associated with a shift from monocytotropic to T-cell tropic virus populations. *J Virol* 66: 1354–1360.

Sharp PM, Robertson DL, Gao F & Hahn BH (1994) Origins and diversity of human immunodeficiency viruses. *AIDS* 8 (suppl. 1): S27–S42.

Shpaer EG & Mullins JI (1993) Rates of amino acid change in the envelope protein correlate with pathogenicity of primate lentiviruses. *J Mol Evol* 37: 57–65.

Simmonds P, Balfe P, Ludlam CA, Bishop JO & Leigh Brown AJ (1990) Analysis of sequence diversity in hypervariable regions of the external glycoprotein of human immunodeficiency virus type 1. *J Virol* 64: 5840–5850.

Simmonds P, Zhang LQ, McOmish F, Balfe P, Ludlam CA & Leigh Brown AJ (1991) Discontinuous sequence change of human immunodeficiency virus (HIV) type 1 *env* sequences in plasma viral and lymphocyte-associated proviral populations in vivo: implications for models of HIV pathogenesis. *J Virol* **65**: 6266–6276.

Smith MW, Dean M, Carrington M et al (1997) Contrasting genetic influence of CCR2 and CCR5 variants on HIV-1 infection and disease progression. *Science* **277**: 959–964.

Smith TF, Srinivasan A, Schochetman G, Marcus M & Myers G (1988) The phylogenetic history of immunodeficiency viruses. *Nature* **333**: 573–575.

Steel CM, Ludlam CA, Beatson D et al (1988) HLA haplotype A1 B8 DR3 as a risk factor for HIV-related disease. *Lancet* **i**: 1185–1188.

Tersmette M, Lange JM, de Goede RE et al (1989) Association between biological properties of human immunodeficiency virus variants and risk for AIDS and AIDS mortality. *Lancet* **i**: 983–985.

Wahlberg J, Fiore J, Angarano G, Uhlen M & Albert J (1994) Apparent selection against transmission of zidovudine-resistant human immunodeficiency virus type 1 variants. *J Infect Dis* **169**: 611–614.

Wain-Hobson S (1993) The fastest genome evolution ever described: HIV variation in situ. *Curr Opin Genet Dev* **3**: 878–883.

Wain-Hobson S, Sonigo P, Danos O, Cole S & Alizon M (1985) Nucleotide sequence of the AIDS virus, LAV. *Cell* **40**: 9–17.

Wei X, Ghosh SK, Taylor ME et al (1995) Viral dynamics in human immunodeficiency virus type 1 infection. *Nature* **373**: 117–122.

Wolfs TFW, Zwart G, Bakker M & Goudsmit J (1992) HIV-1 genomic RNA diversification following sexual and parenteral virus transmission. *Virology* **189**: 103–110.

Wolinsky SM, Korber BTM, Neumann AU et al (1996) Adaptive evolution of human immunodeficiency virus-type 1 during the natural course of infection. *Science* **272**: 537–542.

Wright S (1931) Evolution in Mendelian populations. *Genetics* **16**: 97–159.

Yirrell DL, Pickering H, Palmerini G et al (1997a) Molecular epidemiological analysis of HIV in sexual networks in Uganda. *AIDS* in press

Yirrell DL, Robertson P, Goldberg DJ, McMenamin J, Cameron S & Leigh Brown AJ (1997b) Molecular investigation into outbreak of HIV in a Scottish prison. *BMJ* **314**: 1446–1450.

Zekeng L, Gurtler L, Afane Ze E et al (1994) Prevalence of HIV-1 subtype O infection in Cameroon: preliminary results. *AIDS* **8**: 1626–1628.

Zhang LQ, Simmonds P, Ludlam CA & Leigh Brown AJ (1991) Detection, quantification and sequencing of HIV-1 from the plasma of seropositive individuals and from factor VIII concentrates. *AIDS* **5**: 675–681.

Zhang LQ, MacKenzie P, Cleland A, Holmes EC, Leigh Brown AJ & Simmonds P (1993) Selection for specific sequences in the external envelope protein of human immunodeficiency virus type 1 upon primary infection. *J Virol* **67**: 3345–3356.

Zhang L, Diaz RS, Ho DD, Mosley JW, Busch MP & Mayer A (1997) Host-specific driving force in human immunodeficiency virus type 1 evolution in vivo. *J Virol* **71**: 2555–2561.

Zhu T, Mo H, Wang N et al (1993) Genotypic and phenotypic characterization of HIV-1 in patients with primary infection. *Science* **261**: 1179–1181.

Zhu T, Korber BT, Nahmias AJ, Hooper E, Sharp PM & Ho DD (1998) An African HIV-1 sequence from 1959 and implications for the origin of the epidemic. *Nature* **391**: 594–597.

Chapter *3*

BIOLOGICAL PROPERTIES OF HIV-1 AND THEIR RELEVANCE FOR AIDS PATHOGENESIS

Hanneke Schuitemaker

INTRODUCTION

The human immunodeficiency virus type 1 (HIV-1) has an error-prone reverse transcriptase enzyme, which combined with the absence of proofreading leads to a misincorporation rate of 10^{-4} to 10^{-5} per base or approximately one misincorporation per genome per replication cycle (Preston et al, 1988; Roberts et al, 1988; Pathak and Temin, 1990; Ricchetti and Buc, 1990). With the estimated virion production rate in the range of 10^{10} per day, an HIV-1-infected individual harbours a swarm of closely related viruses. Some mutations can result in phenotypical changes of the virus. Indeed, HIV-1 variants have been shown to differ in biological properties such as replication rate, cell tropism, and syncytium-inducing capacity (Asjo et al, 1986; Cheng-Mayer et al, 1988; Tersmette et al, 1988, 1989; Schuitemaker et al, 1991; Connor et al, 1993; Connor and Ho, 1994; Van 't Wout et al, 1998b). This biological variability is, with immunological, host-genetic, and cofactors, an important contributor to the great variability in the clinical course of infection as observed between individuals.

BIOLOGICAL VARIATION OF HIV-1

Replicative capacity of HIV-1

In 1986, Asjo et al were the first to recognize that HIV isolates from AIDS patients behaved differently from those obtained from asymptomatic individuals (Asjo et al, 1986). Not only did the first day of detection of virus production during primary virus isolation from AIDS patients occur earlier, these viruses also showed increased cytopathicity in culture in vitro and yielded a higher level of virus production. According to their nomenclature used, viruses from AIDS patients were rapid/high, those obtained from asymptomatic individuals were slow/low.

It can be argued that the first day of detection during virus isolation is largely determined by the infectious cellular viral load present in an individual rather than by the replicative capacity of the virus. To elucidate this, viral isolates with equal infectious titres obtained from single individuals at different time points during their course of infection were used to infect peripheral blood mononuclear cells. Despite the same inoculum size, extreme differences in replication kinetics could be observed, both in the first day of detectable virus production and in the total amount of p24 antigen produced (Connor and Ho, 1994; Blaak et al, 1998; Van 't Wout et al, 1998a).

Syncytium-inducing capacity of HIV-1 variants

Virus isolation from patient groups in different phases of their infection showed that not only replicative capacity but also cytopathicity could vary between viruses. Isolates from about half of the AIDS patients induced cell-to-cell fusion in vitro, causing the formation of multinucleated cells or syncytia (Tersmette et al, 1988). These viruses were designated syncytium-inducing (SI). The viruses that lacked this capacity were called non-syncytium-inducing (NSI) (Tersmette et al, 1988). All SI variants were rapid-replicating, as were the NSI variants from AIDS patients. However, NSI viruses obtained from asymptomatic individuals were always slow-replicating (Connor and Ho, 1994; Blaak et al, 1998; Van 't Wout et al, 1997a). In longitudinal studies it appeared that SI variants in general were not present at the moment of seroconversion but emerged during the course of infection (Tersmette et al, 1989; Schuitemaker et al, 1991; Connor et al, 1993).

It is unclear whether SI HIV-1 variants induce syncytia in vivo and whether this mechanism is relevant for CD4+ T cell killing at all.

Cell tropism of HIV-1 variants

The fact that HIV uses the CD4 molecule as its principal receptor makes virtually all CD4+ cells a potential target for infection. At first, however, since CD4 T cells were lost from peripheral blood, it was thought that only, or preferentially, T cells could be infected. However, CD4+ dendritic cells and macrophages also appeared to be potential targets for infection (Levy et al, 1985; Gartner et al, 1986; Ho et al, 1986; Nicholson et al, 1986; Koyanagi et al, 1987; Patterson and Knight, 1987; Popovic and Gartner, 1987; Knight and Macatonia, 1988; Gendelman et al, 1989). In addition, CD4-expressing permanent cell lines, whether of T cell, B cell, or monocyte origin, supported in vitro infection (Montagnier et al, 1984; Levy et al, 1985; Tersmette et al, 1988; Koot et al, 1992; Schuitemaker et al, 1992b). Conclusive evidence for infection of CD34+ stem cells, B cells, CD8+ cells, natural killer (NK) cells, and dendritic cells in vivo has not yet been presented (Molina et al, 1990; Von Laer et al, 1990; Davis et al, 1991). Follicular dendritic cells were shown to trap the virus in the lymph node by carrying the virus on their surface, which would facilitate infection of adhering T cells (Cameron et al, 1992). In the thymus, both lymphocytes and epithelial cells were found to be susceptible to HIV-1 infection (Numazaki et al, 1989). The virus can also infect cells of the central nervous system (CNS), there causing the dementia associated with AIDS. The major target cells for HIV-1 in the CNS are the microglia, although the brain capillary endothelial cells can also be infected (Moses et al, 1996).

Not all virus variants were able to establish infection in all cell types (Asjo et al, 1986; Evans et al, 1987). Non-SI variants, next to their capacity to replicate in primary T cells, in general appeared to be able to replicate in primary macrophages as well

(Schuitemaker et al, 1991). In fact, 80% of NSI variants are macrophage-tropic and 80% of macrophage-tropic viruses are NSI (Fouchier et al, 1994). In general, SI variants lack macrophage tropism but are able to infect both primary T cells as well as permanent T cell lines (Gendelman et al, 1989; Schuitemaker et al, 1991). The basis for this difference in cell tropism was solved with the identification of the coreceptors for HIV-1 (Alkhatib et al, 1996; Choe et al, 1996; Deng et al, 1996; Doranz et al, 1996; Dragic et al, 1996; Feng et al, 1996). The CC or β-chemokine receptor 5 (CCR5) turned out to be the co-receptor for primary HIV-1 variants (Alkhatib et al, 1996; Choe et al, 1996; Deng et al, 1996; Dragic et al, 1996) whereas CXC or α-chemokine receptor 4 (CXCR4) supported infection by T cell line-adapted HIV-1 isolates (Feng et al, 1996). Primary SI variants, which have the intrinsic capacity to replicate in permanent T cell lines, were able to use both CCR5 and CXCR4 as coreceptor (Simmons et al, 1996). Both CCR5 and CCR3 can serve as coreceptor for HIV-1 on microglia cells (He et al, 1997). The lack of T cell line tropism of NSI variants is probably due to the lack of CCR5 expression on these cells. The lack of macrophage tropism by SI variants cannot be explained from the absence of the appropriate coreceptor on macrophages, since both CCR5 and CXCR4 are expressed. It was shown, however, that not coreceptor expression but CD4 expression itself is the limiting factor in macrophage infection by SI variants (Kozak et al, 1997). Apparently, these SI viruses have much lower affinity for CD4 compared with NSI variants, as a consequence of which these variants may require much higher CD4 expression to enter the cells.

When SI variants are present in patient peripheral blood mononuclear cells (PBMCs) they always outgrow the coexisting NSI variants in bulk culture in vitro of the patient PBMCs with phytohaemagglutinin-stimulated target cells. In vivo, however, irrespective of their replicative capacity, NSI and SI HIV-1 clones in general each constitute 50% of the total infectious cellular load (Koot et al, 1996a; Schuitemaker et al, 1992a; Van 't Wout et al, 1998b). The selective growth advantage for SI HIV-1 as observed in vitro is apparently not present in vivo. Since both SI and NSI can use CCR5 as a coreceptor, the equal contributions of SI and NSI may indicate the ability of the latter to compete with the more rapid SI variants for the same CCR5-expressing target cells, mediated for instance by increased affinity of NSI HIV-1 for the CCR5 and/or CD4 receptor.

MOLECULAR BASIS FOR HIV-1 BIOLOGICAL PHENOTYPE VARIABILITY

Mapping of molecular determinants for differences in biological phenotype

Although exchange studies have not been performed to determine the genomic fragment relevant for differences in replicative capacity, accessory genes *tat* (Cheng-Mayer et al, 1991; Dimitrov et al, 1993), *nef* (Miller et al, 1994; Spina et al, 1994), *vpr* (Balliet et al, 1994; Connor et al, 1995), and *vpu* (Balliet et al, 1994) have been demonstrated to influence the replication rates in different cell types. It is unclear whether virus variants with different replicative capacity differ in production or affinity of regulatory proteins or the sequence of the binding site for these proteins.

Differences in cell tropism are predominantly determined at the level of host cell entry (Cann et al, 1990; Groenink et al, 1991; Schuitemaker et al, 1993a). Exchange studies of genomic fragments derived from HIV-1 variants with different biological properties indeed indicated that the *env* gene is predominantly responsible for differences in host range. A 157 bp fragment including the V3 and V4 but not the CD4 binding site was associated with the capacity to infect primary macrophages (O'Brien et al, 1990; Shioda et al, 1991). The HIV-1 tropism for microglial cells is determined by this same region of the envelope glycoprotein (Sharpless et al, 1992). This is in agreement with the observation that most macrophage-tropic HIV-1 isolates replicate equally well in microglia (Strizki et al, 1996). Exchange of a 283 bp domain, again including the V3 loop and only 2 of 11 amino acids of the CD4 binding domain, confers macrophage tropism from the macrophage-tropic variant ADA to the T cell line adapted NL4-3 isolate. Sequence alignment of macrophage-tropic HIV-1 variants identified six amino acid residues as potentially involved in macrophage tropism (Westervelt et al, 1991).

In agreement with the observation that host range and SI capacity generally are coinciding features, approximately the same genomic *env* fragment was identified as important for syncytium induction (Groenink et al, 1993). Exchange of part of the V1 domain together with the V2 and V3 domains confers SI capacity to a NSI variant. Extensive sequence analysis of the different *env* domains involved showed that in the V3 loop two fixed positions (306 and 320) are always negatively charged or uncharged in NSI HIV-1 variants. In SI variants, either one or both of these positions carry a positively charged amino acid (Fouchier et al, 1992). Site-directed mutagenesis indeed confirmed the functional involvement of the amino acids at these positions (De Jong et al, 1992a,b).

Sequence analysis of the V1–V2 domains showed that SI variants and their coexisting NSI variants had elongated V2 domains, coinciding with the relocation or addition of potential N-linked glycosylation site (Groenink et al, 1993). Non-SI isolates from AIDS patients who had never developed SI variants had short V2 domains. Late-stage SI variants, obtained at least 3 months after SI conversion, sometimes had short V2 domains (Fouchier et al, 1995) which shed doubt on the relevance of V2 domain elongation in SI capacity (Wang et al, 1995). However, all SI variants studied at about the moment of their first appearance had elongated V2 domains.

Molecular evolution of SI HIV-1

Elongated V2 domains in recent SI HIV-1 variants but not in later-stage SI viruses suggested that, in the majority of HIV-1 species, elongation of V2 may be required for SI conversion, but that it is not a prerequisite for maintaining the SI phenotype. Extension of the V2 domain may only transiently contribute to the 'fitness' for transition from an NSI to SI envelope configuration (Schuitemaker et al, 1995). Indeed, the pattern with which one observes emergence of SI variants suggested that the evolution towards an SI phenotype is not just a linear accumulation of relevant mutations but more probably the acquisition of mutations in a virus that thus achieves the fitness subsequently to acquire an SI genotype and phenotype. The pattern of evolution of the V3 loop is that of a 'fitness landscape' (Kuiken et al, 1992). Several fitness peaks are formed by viruses that are almost as competent as the parental virus. Viruses that fall in between the fitness peaks generate fewer offspring and therefore exist only at low frequencies. Evolving from one fitness peak to another requires crossing a less competent stage, acquiring mutations that reduce fitness by themselves but that will be advantageous when combined with others. The possibly transient elongation of V2 therefore may create a less fit replicating virus which will be present only as a minor virus population, but which provides the basis for subsequent mutations in V3 that are essential for the development of a full SI phenotype.

In a longitudinal study on the replicative capacity of HIV-1 biological clones in relation to SI phenotype evolution, it appeared that the first SI HIV-1 variants emerged from relatively rapidly replicating NSI HIV-1, in agreement with the possibility of rapidly replicating viruses to accumulate the relevant mutations required for NSI to SI transition.

RELEVANCE OF HIV-1 BIOLOGICAL VARIABILITY IN THE CLINICAL COURSE OF HIV-1 INFECTION

Relevance of macrophage tropism during transmission and the asymptomatic phase of infection

In recently infected individuals the virus population is very homogeneous – at least with respect to V3 sequences (McNearney et al, 1990; Zhang et al, 1993; Zhu et al, 1993; Van 't Wout et al, 1994; Lukashov and Goudsmit, 1997). The basis for this sequence homology is still not understood. It appears, however, that all virus variants in a newly infected individual are preferentially macrophage-tropic (Zhu et al, 1993; Van 't Wout et al, 1994). Hypothetically, selection for macrophage-tropic HIV-1 variants in the donor in specific compartments that are relevant for transmission could have occurred. Differences in the clonal composition of HIV-1 populations in different body compartments have indeed been described (Epstein et al, 1991; Simmonds et al, 1991; Scarlatti et al, 1993; Donaldson et al, 1994). An abundance of macrophage-tropic HIV-1 variants in semen or amniotic fluid would facilitate selective transmission of these variants. Currently, however, there are more data in favour of selection during transmission at the level of the recipient or for a selective elimination of non-macrophage-tropic HIV-1 in the recipient. It has been envisioned that macrophages in the mucosa are the 'port of entry' for HIV-1, thus selecting for macrophage-tropic HIV-1 variants from the inoculum. This was supported by the observation that HIV-1 infection of cervical explants was achieved only when macrophage-tropic HIV-1 strain Ba-L was used. Infection appeared to be restricted to the macrophages present in the explant (Palacio et al, 1994). In favour of this mechanism is that homogeneous virus populations in recently infected individuals can already be observed prior to seroconversion, indeed suggesting that the selection for macrophage-tropic HIV-1 variants may occur before the onset of the humoral immune response. Moreover, the viral population in an individual with acute infection was shown to be homogeneous for a cytotoxic T lymphocyte (CTL) epitope in V3, which was recognized by a CTL clone of that person (Safrit et al, 1994a). This suggested that CTLs can be involved in the suppression of the viraemia during primary infection but are not necessarily involved in the selection of macrophage-tropic HIV-1. Although these observations favour selection at the level of entry in a new individual, and also after parenteral transmission, when the mucosal barrier is bypassed by directly injecting HIV-1 into the circulation, a selective outgrowth of macrophage-tropic HIV-1 has been observed (Van't Wout et al, 1994). Macrophage tropism has therefore been considered to be an escape mechanism enabling the virus to hide at the time of recent infection from the still uncompromised immune system. Interestingly, in one individual infected by parenteral transmission, a heterogeneous population of SI variants was detected in the first sample but a homogeneous NSI HIV-1 population existed later on (Cornelissen et al, 1995). This indicates that macrophage tropism may be advantageous in two ways, at the level of entry and alternatively or additionally at the level of hiding from antiviral immunity.

The mechanism by which macrophage tropism could provide protection from immunity is not clear. Both NSI and SI HIV-1 variants have the same neutralization sensitivity (Groenink et al, 1995). It may be, however, that in the presence of neutralizing antibodies, intercellular transmission is possible only during close cell-to-cell contact, which frequently occurs between macrophages and T cells in the process of antigen presentation. The observation that primary T cells in the early phase of infection carry preferentially macrophage-tropic HIV-1 variants is indeed indicative for recent infection of these T cells by progeny from HIV-1-infected macrophages (Massari et al, 1990). The emergence of preferentially non-macrophage-tropic SI HIV-1 in general occurs only at a stage where CD4+ cell numbers are decreased and T cell function diminished (Koot et al, 1999). This suggests that immune control may be responsible for suppression of highly replicating SI non-macrophage-tropic HIV-1. Even stronger evidence for this possibility came from a study in which it was shown that the amino acid change at position 11 in the V3 loop which was relevant for NSI to SI transition coincided with the loss of a CTL epitope (Safrit et al, 1994b). In this respect, SI HIV-1 may be considered as CTL escape variants.

Although macrophages can be lysed by CTLs, these cells are possibly more resistant to CTL activity than T cells because of their inaccessibility in lymphoid and non-lymphoid tissues. The persistence of HIV-1 variants with CTL epitopes in the presence of recognizing CTLs could then be by virtue of their macrophage tropism. In addition to cytolysis, CD8+ cells can interfere with HIV-1 replication in a MHC-independent manner, via a soluble factor (Brinchmann et al, 1990; Mackewicz and Levy, 1992; Landay et al, 1993). This factor is different from RANTES, MIP-1α and MIP-1β, the natural ligands for CCR5 which can interfere with HIV-1 replication and which are also produced by CD8+ T cells (Cocchi et al, 1995). This CD8+ T lymphocyte-mediated antiviral activity also inhibits HIV-1 replication in primary macrophages (Kootstra et al, 1997). In T cells inhibition of NSI replication was stronger than inhibition of SI replication (Kootstra et al, 1997), indicating that SI variants may again be considered as escape variants.

In agreement with the observation that macrophage-tropic HIV-1 variants use CCR5 as a coreceptor, individuals with a homozygous 32 bp deletion in their CCR5 genes remained without signs of HIV-1 infection despite multiple exposure due to unsafe sexual practice (Dean et al, 1996; Liu et al, 1996; Samson et al, 1996). The white CCR5 genotype distribution with respect to the 32 bp deletion is 80% homozygous wild-type, 19% heterozygous, and 1% homozygous for the deletion. In the Amsterdam cohort, in a group of 38 injecting drug users (IDUs) with high-risk behaviour with respect to HIV-1 transmission, one individual with a homozygous Δ32 CCR5 deletion was identified, whereas none of 365 HIV-1-infected individuals carried the CCR5 Δ32 homozygous genotype (De Roda Husman et al, 1997). The apparent CCR5 restriction during transmission may therefore again not only be determined by the mucosal barrier, since this would be bypassed in IDUs by directly injecting HIV-1 into the circulation. Although some primary SI HIV-1 variants are capable of using both CCR5 and CXCR4, it remains to be established whether CXCR4 coreceptor usage is relevant in vivo (Doranz et al, 1996; Simmons et al, 1996; Zhang et al, 1996). If so, individuals with the homozygous Δ32 CCR5 genotype would still be at risk for infection with SI HIV-1 variants. Indeed, an HIV-1-infected individual with a CCR5 Δ32 homozygous genotype has now been identified (Biti et al, 1997). The absence of homozygous mutants amongst the large group of HIV-1 seropositives (Dean et al, 1996; Huang et al, 1996; Liu et al, 1996; Samson et al, 1996) may then be explained by the fact that SI transmission is a rare event owing to their presence mainly in symptomatic individuals, decreasing their chance of being spread.

Syncytium-inducing HIV-1 and clinical course

In transsectional studies, SI variants are more often isolated from individuals with low CD4 cell counts (Tersmette et al, 1988, 1989). This indicates either that their presence is possible only in severely immunocompromised individuals or that their emergence is directly responsible for the low CD4+ T cell numbers. In a longitudinal study it appeared that SI conversion on rare occasions could occur in individuals with CD4+ T cell counts in the normal range, with an increase in the incidence when CD4+ T cell counts dropped below 400×10^6 l^{-1} (Koot et al, 1999). Conclusive evidence for a role of SI variants in AIDS pathogenesis has come from a prognostic study among 225 HIV-1-infected homosexual men participating in the Amsterdam cohort studies on AIDS, in which the temporal relation between the appearance of SI variants and the kinetics of CD4+ cell depletion was studied (Koot et al, 1993a). Following SI conversion, the rate of CD4+ cell decline was 2·7 times greater than the preswitch decline and the CD4 cell decline in the NSI control group (Koot et al, 1993a). The difference in rate of CD4 cell decline in the period preceding AIDS was reflected in significantly lower CD4 cell numbers at AIDS diagnosis for persons with SI HIV-1, suggesting that AIDS patients with SI isolates are more severely immunocompromised than patients without SI HIV-1 at the onset of disease. The risk for development of AIDS-associated symptoms increases with declining CD4 cell numbers (Koot et al, 1999). Over a 30-month period, the cumulative incidence of AIDS in persons with SI variants at entry was 70·8%, compared with 15·8% of persons with NSI variants at entry.

The relative risk for disease progression of persons with SI HIV-1 adjusted for CD4 cell counts and p24 antigenaemia was 6·7 times higher than the adjusted relative risk of the latter two markers (Koot et al, 1993a).

Mechanism for enhanced CD4 cell decline in patients with SI variants

The mechanism by which SI variants cause accelerated CD4 cell decline has not yet been revealed. Carriers of SI variants in general have a higher viral load, which may be directly responsible for increased CD4 cell turnover. Analysis of changes in infectious cellular load relative to the moment of SI conversion, however, showed that significant increases of viral load could be observed only a year later, whereas the onset of accelerated CD4 cell loss immediately coincided with the emergence of SI variants (Koot et al, 1996). This suggested that not only quantitative differences but also qualitative differences may be responsible for enhanced CD4+ T cell killing. Evidence for this was found upon infection of CD4 T cell clones with either an SI strain or the macrophage-tropic HIV-1 variant Ba-L. Both viruses showed the same kinetics of virus production. Parallel analysis of cell survival in these cultures showed, however, that cells in the SI-infected cultures died, whereas the cell numbers in the NSI Ba-L virus-infected cultures were not affected (Fouchier et al, 1996).

Additional evidence that high cellular infectious load by itself is not sufficient for CD4+ T cell killing came from a subsequent study on individuals who remained healthy for long periods despite HIV-1 infection. These individuals – called long-term survivors (LTSs) – had stable CD4+ T cell counts above 400×10^6 l^{-1} for more than 9 years, no clinical signs of infection and no history of antiviral treatment. In the Amsterdam cohort, all LTSs ($n = 23$) carried solely NSI variants (Hogervorst et al, 1995). Since 50% of AIDS patients also carry only NSI variants, the absence of SI HIV-1 variants cannot explain the benign clinical course in the LTSs. It was hypothesized that LTSs carry only slowly replicating HIV-1 and that individuals with a progressive clinical course carry more rapidly replicating NSI HIV-1. However, from four of eight LTSs, rapidly replicating HIV-1 variants were isolated, and their presence coincided with increased cellular infectious load (Blaak et al, 1998), comparable with levels observed in individuals with a progressive clinical course. However, the presence of rapidly replicating NSI variants and high viral load in these LTSs did not result in CD4+ T cell decline, as was observed in the progressors. Apparently, also within the NSI HIV-1 population, differences in cytopathicity exist, not coinciding with replicative capacity.

Syncytium-inducing phenotype and CCR5 genotype

Individuals heterozygous for the CCR5 Δ32 genotype showed a delayed disease progression compared with carriers of the wild-type CCR5 genotype (Dean et al, 1996). Since primary SI HIV-1 can use both CCR5 and CXCR4 as coreceptor for cell entry, whereas primary NSI variants use only CCR5, it could be envisioned that the beneficial effect of a heterozygous CCR5 Δ32 genotype on the clinical course of infection would be restricted to individuals not carrying SI variants. The delay of disease progression was indeed most evident in – but not restricted to – carriers of only NSI variants (De Roda Husman et al, 1997). This may indicate that SI variants in vivo also preferentially use CCR5 instead of CXCR4 as coreceptor, or that the coexisting NSI variants remaining after the emergence of SI variants (Schuitemaker et al, 1991, 1992a; Koot et al, 1996; Van 't Wout et al, 1998a,b) indeed have a significant role in AIDS pathogenesis (Koot et al, 1996). The large contribution of coexisting NSI variants to the total viral burden supports the latter option.

RELEVANCE FOR VIRUS PHENOTYPE VARIABILITY DURING ANTIVIRAL TREATMENT

In addition to the role of SI HIV-1 in the natural course of infection, the efficacy of zidovudine treatment appeared also to be dependent on the biological phenotype of HIV present in patients receiving treatment (Koot et al, 1993b). Zidovudine (ZDV) treatment significantly delayed clinical progression in individuals who had only NSI at the start of the study and who did not develop SI variants under treatment (Koot et al, 1993b).

Over a 2-year follow-up period, a comparable clinical progression rate was observed in individuals carrying SI variants, irrespective of ZDV treatment. In contrast, in the same period, progression had occurred in none of 14 persons receiving ZDV who did not carry or develop SI variants, compared with a 37% rate of progression in NSI carriers in the untreated control group. A possible mechanism for this viral phenotype-dependent efficacy of ZDV was provided by a study in which the clonal composition of HIV-1 populations was analysed in individuals receiving ZDV treatment (Van 't Wout et al, 1996). Individuals who started ZDV treatment showed a decline in their cellular infectious load which could be attributed mainly to the suppression of NSI HIV-1 variants, whereas the coexisting SI load remained at baseline levels. This difference in response was not due to differential kinetics of development of ZDV resistance, since all viruses still had a wild-type genotype.

In individuals treated with didanosine (ddI) the opposite phenomenon was observed (Van 't Wout et al, 1997b). Individuals who carried SI variants at the start of the study sometimes lost their SI HIV-1; this was not due to a decline in viral load below detection level, since NSI HIV-1 could still be isolated (Delforge et al, 1995; Zheng et al, 1996). Analysis of the clonal composition of HIV-1 populations in individuals that received ddI treatment indeed showed a selective loss of only SI HIV-1 variants, whereas the NSI load did not change. Moreover, in individuals receiving ZDV/ddI combination therapy, both NSI and SI cellular infectious load declined (Van 't Wout et al, 1997b), in agreement with the larger clinical benefit of ZDV/ddI combination therapy (Delta Coordinating Committee, 1996; Hammer et al, 1996; Saravolatz et al, 1996; Schooley et al, 1996). The mechanism for this virus phenotype-dependent efficacy of these nucleoside analogues has not been elucidated yet. The hypothetical model is that both ZDV and ddI require phosphorylation which is dependent on cellular kinases whose expression and activity are cell-type specific. Zidovudine is phosphorylated by kinases that are active only in activated cells, whereas cellular kinases responsible for phosphorylation of ddI are most potent in quiescent cells. The efficacy of ZDV on NSI HIV-1 and the SI-specific effect of ddI would then imply that NSI HIV-1 preferentially replicates in activated cells whereas SI HIV-1 would replicate in quiescent T cells. In agreement with this, CCR5, the coreceptor for NSI HIV-1, is expressed on activated cells, whereas CXCR4, the coreceptor that can be used by SI HIV-1, is also expressed in quiescent cells. In agreement with our model is the finding that ritonavir, a protease inhibitor which does not require intracellular activation, acted equally well on both NSI and SI HIV-1 variants (Van 't Wout et al, 1997).

CONCLUSION

In different phases of HIV-1 infection, different biological properties of HIV-1 variants seem to play a role in AIDS pathogenesis. During transmission, macrophage tropism seems beneficial for the virus. In the early asymptomatic phase, macrophage tropism not only provides the virus with a mechanism to escape immune surveillance but may also be the basis for undermining the immune system since HIV-1-infected macrophages may cause inappropriate antigen presentation and cytokine dysregulation resulting in T cell dysfunction (Meyaard et al, 1993; Schuitemaker et al, 1993b).

The decline of CD4+ T cells seems to be the ultimate event leading to AIDS. From mathematical modelling a huge turnover of both virus and CD4+ T cell population is assumed (Ho et al, 1995; Wei et al, 1995). With a continuous 10-fold increased turnover of CD4+ T cells, it is difficult to understand how SI HIV-1 could accelerate the loss of CD4+ T cells. Biological measurements did not support a large CD4+ T cell turnover (Wolthers et al, 1996), and new data on the number of productively infected cells in vivo indicate that a doubling of CD4+ T cell turnover seems more likely (Haase et al, 1996). Current thinking suggests that CD4+ T cell decline as observed in HIV-1 infection can be explained by a slightly increased loss that cannot be completely compensated for. An increased loss of CD4+ T cells due to increased viral burden as a consequence of the emergence of more rapidly replicating or even SI HIV-1 variants together with a constant and insufficient renewal capacity will then result in an accelerated CD4+ T cell loss. The biological features of HIV-1 therefore seem to play a significant role in the clinical course of HIV-1 infection.

REFERENCES

Alkhatib G, Combadiere C, Broder CC et al (1996) CC CKR5: a RANTES, MIP-1α, MIP-1β receptor as a fusion cofactor for macrophage-tropic HIV-1. *Science* **272:** 1955–1958.

Asjo B, Albert J, Karlsson A et al (1986) Replicative capacity of human immunodeficiency virus from patients with varying severity of HIV infection. *Lancet* **ii:** 660–662.

Balliet JW, Kolson DL, Eiger G et al (1994) Distinct effects in primary macrophages and lymphocytes of the human immunodeficiency virus type 1 accessory genes *vpr*, *vpu*, and *nef*: mutational analysis of a primary HIV-1 isolate. *Virology* **200:** 623–631.

Biti R, French R, Young J, Bennetts B, Stewart G & Liang T (1997) HIV-1 infection in an individual homozygous for the CCR5 deletion allele. *Nature Med* **3:** 252–253.

Blaak H, Brouwer M, Ran LJ, De Wolf F & Schuitemaker H (1998) In vitro replication kinetics of HIV-1 variants in relation to viral load in long-term survivors of HIV-1 infection. *J Infect Dis* **177:** 600–610.

Brinchmann JE, Gaudernack G & Vartdal F (1990) CD8+ T cells inhibit HIV replication in naturally infected CD4+ T cells. Evidence for a soluble inhibitor. *J Immunol* **144:** 2961–2966.

Cameron PU, Freudenthal PS, Barker JM, Gezelter S, Inaba K & Steinman RM (1992) Dendritic cells exposed to human immunodeficiency virus type-1 transmit a vigorous cytopathic infection to CD4+ T cells. *Science* **257:** 383–387.

Cann AJ, Zack JA, Go AS et al (1990) Human immunodeficiency virus type 1 T-cell tropism is determined by events prior to provirus formation. *J Virol* **64:** 4735–4742.

Cheng-Mayer C, Seto D, Tateno M & Levy JA (1988) Biologic features of HIV-1 that correlate with virulence in the host. *Science* **240:** 80–82.

Cheng-Mayer C, Shioda T & Levy JA (1991) Host range, replicative, and cytopathic properties of human immunodeficiency virus type 1 are determined by very few amino acid changes in *tat* and gp120. *J Virol* **65:** 6931–6941.

Choe H, Farzan M, Sun Y et al (1996) The β-chemokine receptors CCR3 and CCR5 facilitate infection by primary HIV-1 isolates. *Cell* **85:** 1135–1148.

Cocchi F, DeVico AL, Garzino-Demo A, Arya SK, Gallo RC & Lusso P (1995) Identification of RANTES, MIP-1α, and MIP-1β as the major HIV-suppressive factors produced by CD8+ T cells. *Science* **270:** 1811–1815.

Connor RI & Ho DD (1994) Human immunodeficiency virus type 1 variants with increased replicative capacity develop during the asymptomatic stage before disease progression. *J Virol* **68:** 4400–4408.

Connor RI, Mohri H, Cao Y & Ho DD (1993) Increased viral burden and cytopathicity correlate temporally with CD4+ T-lymphocyte decline and clinical progression in human immunodeficiency virus type 1 infected individuals. *J Virol* **67:** 1772–1777.

Connor RI, Kuan Chen B, Choe S & Landau NR (1995). Vpr is required for efficient replication of human immunodeficiency virus type-1 in mononuclear phagocytes. *Virology* **206:** 935–944.

Cornelissen M, Mulder-Kampinga GA, Veenstra J et al (1995) Syncytium-inducing (SI) phenotype suppression at seroconversion after intramuscular inoculation of a non-syncytium inducing/SI phenotypically mixed human immunodeficiency virus population. *J Virol* **69:** 1810–1818.

Davis BR, Schwartz DH, Marx JC et al (1991) Absent or rare human immunodeficiency virus infection of bone marrow stem/progenitor cells in vivo. *J Virol* **65:** 1985–1990.

Dean M, Carrington M, Winkler C et al (1996) Genetic restriction of HIV-1 infection and progression to AIDS by a deletion allele of the CKR5 structural gene. *Science* **273:** 1856–1862.

De Jong JJ, De Ronde A, Keulen W, Tersmette M & Goudsmit J (1992a) Minimal requirements for the human immunodeficiency virus type 1 V3 domain to support the syncytium-inducing phenotype: analysis by single amino acid substitution. *J Virol* **66:** 6777–6780.

De Jong JJ, Goudsmit J, Keulen W et al (1992b) Human immunodeficiency viruses type-1 chimeric for the envelope V3 domain are distinct in syncytium formation and replication capacity. *J Virol* **66:** 757–765.

Delforge ML, Liesnard C, Debaisieux L, Tchetcheroff M, Farber CM and Van Vooren JP (1995) In vivo inhibition of syncytium inducing variants of HIV in patients treated with didanosine. *AIDS* **9:** 89–101.

Delta Coordinating Committee (1996) Delta: a randomised double-blind controlled trial comparing combinations of zidovudine plus didanosine or zalcitabine with zidovudine alone in HIV-infected individuals. *Lancet* **348:** 283–291.

Deng HK, Liu R, Ellmeier W et al (1996) Identification of the major co-receptor for primary isolates of HIV-1. *Nature* **381:** 661–666.

De Roda Husman AM, Koot M, Cornelissen M et al (1997) Association between CCRS genotype and the clinical course of HIV-1 infection. *Ann Intern Med* **127:** 882–890.

Dimitrov DS, Willey RL, Sato H, Chang LJ, Blumenthal R & Martin MA (1993) Quantitation of human immunodeficiency virus type 1 infection kinetics. *J Virol* **67:** 2182–2190.

Donaldson YK, Bell JE, Holmes EC, Hughes ES, Brown HK & Simmonds P (1994) In vivo distribution and cytopathology of variants of human immunodeficiency virus type 1 showing restricted sequence variability in the V3 loop. *J Virol* **68:** 5991–6005.

Doranz BJ, Rucker J, Yi Y et al (1996) A dual-tropic primary HIV-1 isolate that uses fusin and the β-chemokine receptors CKR-5, CKR-3 and CKR-2b as fusion cofactors. *Cell* **85:** 1149–1158.

Dragic T, Litwin V, Allaway GP et al (1996) HIV-1 entry into CD4+ cells is mediated by the chemokine receptor CC-CKR-5. *Nature* **381:** 667–673.

Epstein LG, Kuiken C, Blumberg BM et al (1991) HIV-1 V3 domain variation in brain and spleen of children with AIDS: tissue-specific evolution within host-determined quasispecies. *Virology* **180:** 583–590.

Evans LA, McHugh TM, Stites DP & Levy JA (1987) Differential ability of HIV isolates to productively infect human cells. *J Immunol* **138:** 3415–3418.

Feng Y, Broder CC, Kennedy PE & Berger EA (1996) HIV-1 entry cofactor: functional cDNA cloning of a seven-transmembrane, G protein-coupled receptor. *Science* **272:** 872–877.

Fouchier RAM, Groenink M, Kootstra NA et al (1992) Phenotype-associated sequence variation in the third variable domain of the human immunodeficiency virus type 1 gp120 molecule. *J Virol* **66:** 3183–3187.

Fouchier RAM, Brouwer M, Kootstra NA, Huisman JG & Schuitemaker H (1994) Macrophage-tropism of human immunodeficiency virus type 1 is determined by viral and cellular factors at multiple steps of the replication cycle. *J Clin Invest* **94:** 1806–1814.

Fouchier RAM, Broersen SM, Brouwer M et al (1995) Temporal relationship between elongation of the HIV-1 gp120 V2 domain and the conversion towards a syncytium inducing phenotype. *AIDS Res Hum Retroviruses* **11**: 1473–1478.

Fouchier RAM, Meyaard L, Brouwer M, Hovenkamp E & Schuitemaker H (1996) Broader tropism and higher cytopathicity for CD4+ T cells of a syncytium-inducing compared to a non-syncytium inducing HIV-1 isolate as a mechanism for accelerated CD4+ T cell decline in vivo. *Virology* **219**: 87–95.

Gartner S, Markovits P, Markovits DM, Kaplan MH, Gallo RC & Popovic M (1986) The role of mononuclear phagocytes in HTLV-III/LAV infection. *Science* **233**: 215–219.

Gendelman HE, Orenstein JM, Baca LM et al (1989) The macrophage in the persistence and pathogenesis of HIV infection (editorial). *AIDS* **3**: 475–495.

Groenink M, Fouchier RAM, De Goede REY et al (1991) Phenotypic heterogeneity in a panel of infectious molecular HIV-1 clones derived from a single individual. *J Virol* **65**: 1968–1975.

Groenink M, Fouchier RAM, Broersen S et al (1993) Relation of phenotype evolution of HIV-1 to envelope V2 configuration. *Science* **260**: 1513–1516.

Groenink M, Moore JP, Broersen S & Schuitemaker H (1995) Equal levels of gp120 retention and neutralization resistance of phenotypically distinct primary human immunodeficiency virus type-1 variants upon soluble CD4 treatment. *J Virol* **69**: 523–527.

Haase TA, Henry K, Zupanc M et al (1996) Quantitative image analysis of HIV-1 infection in lymphoid tissue. *Science* **274**: 985–989.

Hammer SM, Katzenstein DA, Hughes MD et al (1996) A trial comparing nucleoside monotherapy with combination therapy in HIV-infected adults with CD4 cell counts from 200 to 500 per cubic millimeter. *N Engl J Med* **335**: 1081–1090.

He J, Chen Y, Farzan M et al (1997) CCR3 and CCR5 are co-receptors for HIV-1 infection of microglia. *Nature* **385**: 645–649.

Ho DD, Rota TR & Hirsch MS (1986) Infection of monocyte/macrophages by human T lymphotropic virus type III. *J Clin Invest* **77**: 1712–1715.

Ho DD, Neumann AU, Perelson AS, Chen W, Leonard JM & Markowitz M (1995) Rapid turnover of plasma virions and CD4 lymphocytes in HIV-1 infection. *Nature* **373**: 123–126.

Hogervorst E, Jurriaans S, De Wolf F et al (1995) Predictors for non- and slow progression in human immunodeficiency virus (HIV) type 1 infection: low viral RNA copy numbers in serum and maintenance of high HIV-1 p24-specific but not V3-specific antibody levels. *J Infect Dis* **171**: 811–821.

Huang Y, Paxton WA, Wolinsky SM et al (1996) The role of a mutant CCR5 allele in HIV-1 transmission and disease progression. *Nature Med* **2**: 1240–1243.

Knight SC & Macatonia SE (1988) Dendritic cells and viruses. *Immunol Lett* **19**: 177–182.

Koot M, Vos AHV, Keet RPM et al (1992) HIV-1 biological phenotype in long term infected individuals, evaluated with an MT-2 cocultivation assay. *AIDS* **6**: 49–54.

Koot M, Keet IPM, Vos AHV et al (1993a) Prognostic value of human immunodeficiency virus type 1 biological phenotype for rate of CD4$^+$ cell depletion and progression to AIDS. *Ann Intern Med* **118**: 681–688.

Koot M, Schellekens PThA, Mulder JW et al (1993b) Viral phenotype and T-cell reactivity in human immunodeficiency virus type 1-infected asymptomatic men treated with zidovudine. *J Infect Dis* **168**: 733–736.

Koot M, Van 't Wout AB, Kootstra NA, De Goede REY, Tersmette M & Schuitemaker H (1996) Relation between changes in cellular load, evolution of viral phenotype, and the clonal composition of virus populations in the course of human immunodeficiency virus type 1 infection. *J Infect Dis* **173**: 349–354.

Koot M, Van Leeuwen R, De Goede REY et al (1999) Conversion rate towards a syncytium inducing (SI) phenotype during different stages of HIV infection and prognostic value of SI phenotype for survival after AIDS diagnosis. *J Infect Dis* **179**: 254–258.

Kootstra NA, Miedema F & Schuitemaker H (1997) Analysis of CD8+ T lymphocyte-mediated nonlytic suppression of autologous and heterologous primary human immunodeficiency virus type 1 isolates. *AIDS Res Hum Retroviruses* **13**: 685–693.

Koyanagi Y, Miles S, Mitsuyasu RT, Merrill JE, Vinters HV & Chen ISY (1987) Dual infection of the central nervous system by AIDS viruses with distinct cellular tropisms. *Science* **236**: 819–822.

Kozak SL, Platt EJ, Madani N, Ferro FE, Peden K & Kabat D (1997) CD4, CXCR-4, and CCR5 dependencies for infections by primary patient and laboratory-adapted isolates of human immunodeficiency virus type 1. *J Virol* **71**: 873–882.

Kuiken CL, De Jong JJ, Baan E, Keulen W, Tersmette M & Goudsmit J (1992) Evolution of the V3 envelope domain in proviral sequences and isolates of human immunodeficiency virus type 1 during transition of the viral biological phenotype. *J Virol* **66**: 4622–4627.

Landay AL, Mackewicz CE & Levy JA (1993) An activated CD8+ T cell phenotype correlates with anti-HIV activity and asymptomatic clinical status. *Clin Immunol Immunopathol* **69**: 106–116.

Levy JA, Shimabukuro JM, McHugh T, Casavant C, Stites D & Oshiro L (1985) AIDS-associated retroviruses (ARV) can productively infect other cells besides human T-helper cells. *Virology* **147**: 441–448.

Liu R, Paxton WA, Choe S et al (1996) Homozygous defect in HIV-1 coreceptor accounts for resistance of some multiply-exposed individuals to HIV-1 infection. *Cell* **86**: 1–20.

Lukashov VV & Goudsmit J (1997) Founder virus population related to route of transmission: a determinant of intrahost human immunodeficiency virus type 1 evolution? *J Virol* **71**: 2023–2030.

Mackewicz CE & Levy JA (1992) CD8+ cell anti-HIV activity: nonlytic suppression of virus replication. *AIDS Res Hum Retroviruses* **8**: 1039–1050.

Massari FE, Poli G, Schnittman SM, Psallidopoulos MC, Davey V & Fauci AS (1990) In vivo T lymphocyte origin of macrophage-tropic strains of HIV. Role of monocytes during in vitro isolation and in vivo infection. *J Immunol* **144**: 4628–4632.

McNearney T, Westervelt P, Thielan BJ et al (1990) Limited sequence heterogeneity among biologically distinct human immunodeficiency virus type 1 isolates from individuals involved in a clustered infectious outbreak. *Proc Natl Acad Sci USA* **87**: 1917–1921.

Meyaard L, Schuitemaker H & Miedema F (1993) T-cell dysfunction in HIV infection: anergy due to defective antigen presenting cell function? *Immunol Today* **14**: 161–164.

Miller MD, Warmerdam MT, Gaston I, Greene WC & Feinberg MB (1994) The human immunodeficiency virus-1 *nef* gene product; a positive factor for viral infection and replication in primary lymphocytes and macrophages. *J Exp Med* **179**: 101–113.

Molina J-M, Scadden DT, Sakaguchi M, Fuller B, Woon A & Groopman JE (1990) Lack of evidence for infection of or effect on growth of hematopoietic progenitor cells after in vivo or in vitro exposure to human immunodeficiency virus. *Blood* **76**: 2476–2482.

Montagnier L, Gruest J, Chamaret S et al (1984) Adaption of lymphadenopathy-associated virus (LAV) to replication on EBV-transformed B-lymphoblastoid cell lines. *Science* **225**: 63.

Moses AV, Stenglein SG, Strussenberg JG, Wehrly K, Chesebro B & Nelson JA (1996) Sequences regulating tropism of human immunodeficiency virus type 1 for brain capillary endothelial cells map to a unique region on the viral genome. *J Virol* **70**: 3401–3406.

Nicholson JKA, Cross GD, Callaway CS & McDougal JS (1986) In vitro infection of human monocytes with human T lymphotropic virus type III/lymphadenopathy-associated virus (HTLV-III/LAV). *J Immunol* **137**: 323–329.

Numazaki K, Bai X-Q, Goldman H, Wong I, Spira B & Wainberg MA (1989) Infection of cultured human thymic epithelial cells by human immunodeficiency virus. *Clin Immunol Immunopathol* **51**: 185–195.

O'Brien WA, Koyanagi Y, Namazie A et al (1990) HIV-1 tropism for mononuclear phagocytes can be determined by regions of gp120 outside the CD4-binding domain. *Nature* **348**: 69–73.

Palacio J, Souberbielle BE, Shattock RJ, Robinson G, Manyonda I & Griffin GE (1994) In vitro HIV-1 infection of human cervical tissue. *Res Virol* **145**: 155–161.

Pathak VK & Temin HM (1990) Broad spectrum of in vivo forward mutations, hypermutations, and mutational hotspots in a retroviral shuttle vector after a single replication cycle: deletions and deletions with insertions. *Proc Natl Acad Sci USA* **87**: 6024–6028.

Patterson S & Knight SC (1987) Susceptibility of human peripheral blood dendritic cells to infection by human immunodeficiency virus. *J Gen Virol* **68**: 1177–1181.

Popovic M & Gartner S (1987) Isolation of HIV-1 from monocytes but not T lymphocytes. *Lancet* **ii**: 916.

Preston BD, Poiesz BJ & Loeb LA (1988) Fidelity of HIV-1 reverse transcriptase. *Science* **242**: 1168–1171.

Ricchetti M & Buc H (1990) Reverse transcriptases and genomic variability: the accuracy of DNA replication is enzyme specific and sequence dependent. *EMBO J* **9**: 1583–1593.

Roberts JD, Bebenek K & Kunkel TA (1988) The accuracy of reverse transcriptase from HIV-1. *Science* **242**: 1171–1173.

Safrit JT, Andrews CA, Zhu T, Ho DD & Koup RA (1994a) Characterisation of human immunodeficiency virus type 1-specific cytotoxic T lymphocyte clones isolated during acute seroconversion: recognition of autologous virus sequences within a conserved immunodominant epitope. *J Exp Med* **179**: 463–472.

Safrit JT, Lee AY, Andrews CA & Koup RA (1994b) A region in the third variable loop of HIV-1 gp120 is recognized by HLA-B7 restricted CTLs from two acute seroconversion patients. *J Immunol* **153**: 3822–3830.

Samson M, Libert F, Doranz BJ et al (1996) Resistance to HIV-1 infection in caucasian individuals bearing mutant alleles of the CCR-5 chemokine receptor gene. *Nature* **382**: 722–725.

Saravolatz LD, Winslow DL, Collins G et al (1996) Zidovudine alone or in combination with didanosine or zalcitabine in HIV-infected patients with the acquired immunodeficiency syndrome or fewer than 200 CD4 cells per cubic millimeter. *N Engl J Med* **335**: 1099–1106.

Scarlatti G, Leitner T, Halapi E et al (1993) Comparison of variable region 3 sequences of human immunodeficiency virus type 1 from infected children with the RNA and DNA sequences of the virus populations of their mothers. *Proc Natl Acad Sci USA* **90**: 1721–1725.

Schooley RT, Ramirez-Ronda C, Lange JMA et al (1996) Virologic and immunologic benefits of initial therapy with zidovudine and zalcitabine or didanosine compared with zidovudine monotherapy. *J Infect Dis* **173**: 1354–1366.

Schuitemaker H, Kootstra NA, De Goede REY, De Wolf F, Miedema F & Tersmette M (1991) Monocytotropic human immunodeficiency virus 1 (HIV-1) variants detectable in all stages of HIV infection lack T-cell line tropism and syncytium-inducing ability in primary T-cell culture. *J Virol* **65**: 356–363.

Schuitemaker H, Koot M, Kootstra NA et al (1992a) Biological phenotype of human immunodeficiency virus type 1 clones at different stages of infection: progression of disease is associated with a shift from monocytotropic to T-cell-tropic virus populations. *J Virol* **66**: 1354–1360.

Schuitemaker H, Kootstra NA, Groenink M, De Goede REY, Miedema F & Tersmette M (1992b) Differential tropism of clinical HIV-1 isolates for primary monocytes and promonocytic-cell lines. *AIDS Res Hum Retroviruses* **8**: 1679–1682.

Schuitemaker H, Groenink M, Meyaard L et al (1993a) Early replication steps but not cell-type specific signalling of the viral long terminal repeat determine HIV-1 monocytotropism. *AIDS Res Hum Retroviruses* **9**: 669–675.

Schuitemaker H, Meyaard L, Kootstra NA et al (1993b) Lack of T-cell dysfunction and programmed cell death in human immunodeficiency type-1 infected chimpanzees correlates with absence of monocytotropic variants. *J Infect Dis* **168**: 1140–1147.

Schuitemaker H, Fouchier RAM, Broersen S et al (1995) Envelope V2 configuration and HIV-1 phenotype: clarification. *Science* **268:** 115.

Sharpless NE, O'Brien WA, Verdin E, Kufta CV, Chen ISY & Dubois-Dalq M (1992) Human immunodeficiency virus type 1 tropism for brain microglial cells is determined by a region of the env glycoprotein that also controls macrophage tropism. *J Virol* **66:** 2588–2593.

Shioda T, Levy JA & Cheng-Mayer C (1991) Macrophage and T cell-line tropisms of HIV-1 are determined by specific regions of the envelope gp120 gene. *Nature* **349:** 167–169.

Simmonds P, Zhang LQ, McOmish F, Balfe P, Ludlam CA & Brown AJL (1991) Discontinuous sequence change of human immunodeficiency virus (HIV) type 1 env sequences in plasma viral and lymphocyte-associated proviral populations in vivo: implications for models of HIV pathogenesis. *J Virol* **65:** 6266–6276.

Simmons G, Wilkinson D, Reeves JD et al (1996) Primary, syncytium-inducing human immunodeficiency virus type 1 isolates are dual-tropic and most can use either LESTR or CCR5 as co-receptors for virus entry. *J Virol* **70:** 8355–8360.

Spina CA, Kwoh TJ, Chowers MY, Guaelli JC & Richman DD (1994) The importance of nef in the induction of human immunodeficiency virus type 1 replication from primary quiescent CD4 lymphocytes. *J Exp Med* **179:** 115–123.

Strizki JM, Albright AV, Sheng H, O'Connor M, Perrin L & Gonzalez-Scarano F (1996) Infection of primary human microglia and monocyte-derived macrophages with human immunodeficiency virus type 1 isolates: evidence of differential tropism. *J Virol* **70:** 7654–7662.

Tersmette M, De Goede REY, Al BJM et al (1988) Differential syncytium-inducing capacity of human immunodeficiency virus isolates: frequent detection of syncytium-inducing isolates in patients with acquired immunodeficiency syndrome (AIDS) and AIDS-related complex. *J Virol* **62:** 2026–2032.

Tersmette M, Gruters RA, De Wolf F et al (1989) Evidence for a role of virulent human immunodeficiency virus (HIV) variants in the pathogenesis of acquired immunodeficiency syndrome: studies on sequential HIV isolates. *J Virol* **63:** 2118–2125.

Van 't Wout AB, Kootstra NA, Mulder-Kampinga GA et al (1994) Macrophage-tropic variants initiate human immunodeficiency virus type 1 infection after sexual, parenteral and vertical transmission. *J Clin Invest* **94:** 2060–2067.

Van 't Wout AB, De Jong MD, Kootstra NA et al (1996) Changes in cellular virus load and zidovudine resistance of syncytium-inducing and non-syncytium-inducing human immunodeficiency virus populations under zidovudine pressure: a clonal analysis. *J Infect Dis* **174:** 845–849.

Van 't Wout AB, Blaak H, Ran LJ, Brouwer M, Kuiken C & Schuitemaker H (1998b) Evolution of syncytium inducing and non-syncytium inducing biological virus clones in relation to replication kinetics during the course of HIV-1 infection. *J Virol* **72:** 5099–5107.

Van 't Wout AB, Ran LJ, De Jong MD et al (1997) Selective inhibition of syncytium inducing and non syncytium inducing HIV-1 variants in individuals receiving didanosine or zidovudine respectively. *J Clin Invest* **100:** 2325–2332.

Van 't Wout AB, Ran LJ, Kuiken CL, Kootstra NA, Pals ST & Schuitemaker H (1998a) Analysis of the temporal relationship between human immunodeficiency virus type 1 quasispecies in sequential blood samples and various organs obtained at autopsy. *J Virol* **72:** 488–496.

Von Laer D, Hufert FT, Fenner TE et al (1990) CD34$^+$ hematopoietic progenitor cells are not a major reservoir of the human immunodeficiency virus. *Blood* **76:** 1281–1286.

Wang N, Zhu T & Ho DD (1995) Sequence diversity of V1 and V2 domains of gp120 from human immunodeficiency virus type 1: lack of correlation with viral phenotype. *J Virol* **69:** 2708–2715.

Wei X, Ghosh SK, Taylor ME et al (1995) Viral dynamics in human immunodeficiency virus type 1 infection. *Nature* **373:** 117–122.

Westervelt P, Gendelman HE & Ratner L (1991) Identification of a determinant within the human immunodeficiency virus 1 surface envelope glycoprotein critical for productive infection of primary monocytes. *Proc Natl Acad Sci USA* **88:** 3097–3101.

Wolthers KC, Wisman GBA, Otto SA et al (1996) T-cell telomere length in HIV-1 infection: no evidence for increased CD4+ T cell turnover. *Science* **274:** 1543–1547.

Zhang LQ, MacKenzie P, Cleland A, Holmes EC, Leigh-Brown AJ & Simmonds P (1993) Selection for specific sequences in the external envelope protein of HIV-1 upon primary infection. *J Virol* **67:** 3345–3356.

Zhang L, Huang Y, He T, Cao Y & Ho DD (1996) HIV-1 subtype and second-receptor use. *Nature* **383:** 768.

Zheng NN, McQueen PW, Hurren L et al (1996) Changes in biologic phenotype of human immunodeficiency virus during treatment of patients with didanosine. *J Infect Dis* **173:** 1092–1096.

Zhu T, Mo H, Wang N, Nam DS, Cao Y, Koup RA & Ho DD (1993) Genotypic and phenotypic characterization of HIV-1 in patients with primary infection. *Science* **261:** 1179–1181.

Chapter 4

HIV DYNAMICS:
Lessons from the use of antiretrovirals
A.N. Phillips

INTRODUCTION

Human immunodeficiency virus (HIV) is the cause of AIDS and patients with higher plasma levels of HIV tend to develop AIDS more rapidly than those with lower levels (Mellors et al, 1995, 1996; Phillips et al, 1997a). The process by which HIV causes AIDS is known to be largely one of reducing numbers of CD4 lymphocytes (Lane et al, 1985; Phillips et al, 1991); it is most often assumed that this is due to infection of CD4 lymphocytes by HIV, resulting in cell death, but questions remain about exactly how HIV causes CD4 cell depletion (Weiss, 1993). Treatment of HIV infection with anti-retroviral drugs involves blocking virus replication, so detailed observations in people treated with powerful antiretrovirals can inform us about the role of viral replication in HIV pathogenesis and the dynamics of this process (Coffin, 1995; Ho et al, 1995; Love-day, 1995; Wei et al, 1995; Havlir and Richman, 1996; Perelson et al, 1996, 1997; Autran et al, 1997; Bonhoeffer et al, 1997; Cavert et al, 1997; Finzi et al, 1997; Wong et al, 1997). This chapter outlines the main findings from such studies and discusses some of the implications both for HIV pathogenesis and, more immediately, for understanding how to treat patients more effectively.

EFFECTS OF POTENT ANTIRETROVIRAL THERAPY

In the absence of antiretroviral therapy plasma HIV RNA values vary substantially between individuals, ranging from less than 400 HIV RNA copies per ml to over 1 million copies per ml (Piatak et al, 1993; Phillips et al, 1997a). However, in the absence of any intercurrent illness or event (e.g. influenza vaccination) which may cause immune activation, plasma HIV RNA levels in patients not receiving antiretroviral therapy show little natural short-term variability over time (Hughes et al, 1997; Phillips et al, 1997a). It should be clear that individual virus particles do not persist in the body for weeks at a time, and therefore some continual process of death and renewal is taking place. The fact that overall levels stay constant indicates that the numbers of virus particles being removed in a given period of time must equal the number being produced. Thus it is possible to see that there must be some kind of steady-state or dynamic equilibrium situation. However, simply quantifying the number of virus particles at equilibrium does

HIV and the New Viruses Second Edition
ISBN 0-12-200741-7

not reveal how long each virion lives and how long each cell that produces the virions lives. A crude analogy would be a large shopping centre for which the only available information on customer usage is the total number of customers in the centre at a given point in time, which is found to be constant at, say, 5000. It would be useful to know how many customers visit the centre each day, and for how long they tend to stay. The figure of 5000 people could arise, for example, from 5000 people staying all day or, at the other extreme, several hundred thousand people each visiting for 10 minutes only. One way of finding out about the dynamics underlying the equilibrium would be to shut the doors of the shopping centre to all new visitors at some point during the day, but allow shoppers inside to continue to leave when they wished. Regular monitoring of the number of customers in the centre after this intervention would mean it was possible to track the decrease in total numbers over time. The more rapidly the number of customers declined, the shorter would be the estimated average time that customers spent in the centre, and hence the greater the estimated number of customers visiting per day.

In HIV infection the 'door shutting' intervention is antiretroviral therapy. However, there is an added complication. This relates to the fact that most viral production is believed to take place in lymphoid tissue (Embretson et al, 1993; Pantaleo et al, 1993). Direct sampling of productively infected cells from the peripheral blood is problematic and there is no guarantee that these cells reflect what is happening to cells within the lymphoid tissue – although studies have indicated that similar results would be obtained with peripheral blood mononuclear cells (PBMCs) (Perelson et al, 1997), as explained below. We are left therefore with sampling of cell-free virus in plasma as the measure most likely to reflect events in lymphoid tissue. Even then, changes in cell-free HIV RNA after starting therapy are only an indirect measure of changes in numbers of productively infected cells, which are the subject of most interest. However if, as we believe to be the case, the lifespan of cell-free virus in plasma is small compared with that of the productively infected cells that produce the free virus, then this indirect measure can be considered reliable.

To understand how to interpret changes in plasma HIV RNA and other viral parameters after the institution of antiretroviral therapy it is important to know exactly how the drugs act (Figure 4.1). The two main classes of antiretroviral drugs, reverse transcriptase inhibitors and protease inhibitors, both prevent infection of new cells. For reverse transcriptase inhibitors the viral RNA is unable to reverse transcribe fully into DNA and hence the virus is unable to infect the cell. Protease inhibitors prevent infected cells from producing replication-competent virions and hence new infection of cells is prevented owing to the lack of infectious virus particles (Figure 4.1). Protease inhibitors provide an opportunity directly to study levels of cell-free virions separately from the cells that produce them and hence find out their average lifespan. This follows from the fact that, in the presence of a protease inhibitor, productively infected cells produce defective free virions which still score on the plasma HIV RNA assays but not in infectivity assays, which rely on culturing HIV from plasma. Perelson et al (1996) tracked levels of plasma HIV infectivity in one patient over the hours and days after starting therapy with the protease inhibitor ritonavir (Figure 4.2); the infectious virus titre declined exponentially by 99% in 2 days. 'Exponential' decline means that the *proportionate* decrease is the same each day (or hour or week, etc.). The proportionate loss of plasma virus infectivity was about 90% loss per day, so 90% of the infectious virions

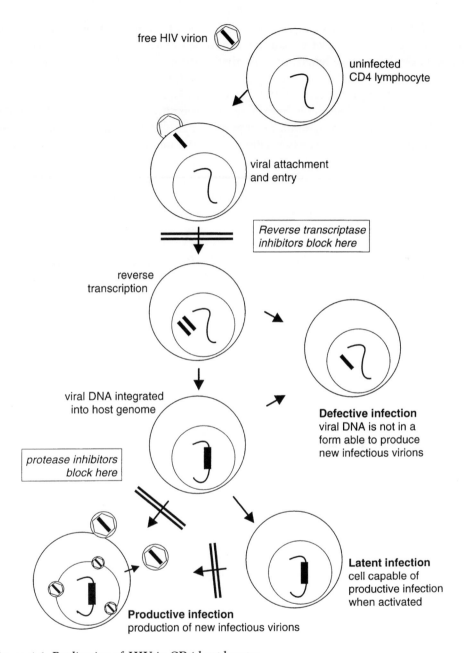

free HIV virion

uninfected
CD4 lymphocyte

viral attachment
and entry

*Reverse transcriptase
inhibitors block here*

reverse
transcription

Defective infection
viral DNA is not in a
form able to produce
new infectious virions

viral DNA integrated
into host genome

*protease inhibitors
block here*

Latent infection
cell capable of
productive infection
when activated

Productive infection
production of new infectious virions

Figure 4.1 *Replication of HIV in CD4 lymphocytes*
Life cycle of HIV replication in CD4 lymphocytes, indicating where different antiretroviral
drugs act.

present at the beginning of the day have gone by the end. The corresponding half-life,
the time taken for each halving of the plasma infectivity, is about 6 hours. This estimate
assumes that the drug is 100% effective at rendering all newly produced virions non-
infectious (Perelson et al, 1996). If this were not the case then the half-life would
be lower than 6 hours. To return to the shopping centre analogy, if some customers

arriving find a way of entering after the doors are supposedly shut then the rate with which the centre would clear of customers would be slower, and this would result in an overestimate of the average length of time that customers spend in the centre. In fact, more recent estimates suggest that the half-life of cell-free virus in plasma could be a matter of only 10 minutes (Ho, 1997), which would mean a proportionate loss of 98·5% per hour. These estimates help with interpretation of findings when tracking plasma HIV RNA levels in patients on therapy. They indicate that the clearance rate observed for plasma HIV RNA essentially reflects the rate of clearance of the productively infected cells which produce the plasma virus.

When a reverse transcriptase inhibitor is given to a patient there is a short period during which intracellular drug concentrations increase to adequate levels. Even then plasma HIV RNA levels do not fall immediately because there is a lag period (corresponding to the time between the points in the viral life cycle blocked by reverse transcriptase inhibitors and protease inhibitors – see Figure 4.1) during which the source of new free virions is not stopped (Herz et al, 1996) – cells that had just completed reverse transcription before the drug came into effect can still carry through the infection process to production of new virions. After this short phase lasting less than a day (Herz et al, 1996), plasma HIV RNA levels tend to fall exponentially for several days (Figure 4.2). For patients treated with protease inhibitors there is also a short delay or 'shoulder' before plasma HIV RNA levels start to drop (in contrast to plasma infectiv-

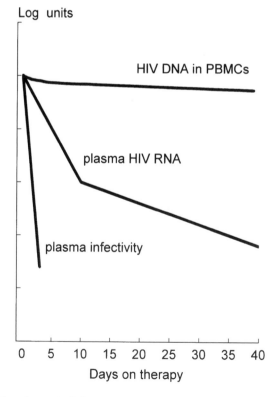

Figure 4.2 *Effects of antiretroviral therapy*
Changes in different viral measures after institution of potent antiretroviral therapy. See text, and Perelson et al (1996, 1997).

ity levels which start to fall almost immediately). This relates to the fact that any virion produced before therapy started is still infectious and hence may find a new cell to infect, which can then produce more free virus, although this latter free virus will be non-infectious owing to the effect of the drug. Thus the length of this shoulder allows another means of estimating the lifespan of cell-free virus. Perelson and colleagues came up with a similar (6 hour) estimate to that from tracking viral infectivity (Perelson et al, 1996). For patients treated with either or both types of drugs there is an exponential phase of decline in plasma HIV RNA after about 1 day. Levels tend to show a proportionate loss of about 35% per day (Perelson et al, 1996). The corresponding half-life is around 1·5 days. If the drug regimen is sufficiently potent, virus levels tend to fall by around 99% within 2 weeks: 35% loss per day for 11 days results in $(1 - 0.65^{11}) \times 100 = 99.1\%$ loss of virus. As pointed out by Ho et al (1995) and Wei et al (1995) in pioneering papers in this area which followed on from earlier work by Loveday (1995), the antiretroviral drugs do not actually kill productively infected cells or stop them from producing virus (albeit defective virus during protease inhibitor therapy); instead they only block infection of new cells, so any decrease in plasma HIV RNA levels must reflect the 'natural' process of removal or killing of productively infected cells by direct viral cytopathic or immune mechanisms. Given that the lifespan of free virus is very short, this rate of decline of plasma HIV RNA (35% per day) indicates the rate with which the productively infected cells which produce most plasma virus are removed or killed. As discussed above, the fact that in the absence of therapy the overall level of plasma HIV RNA is constant at an equilibrium level means that new productively infected cells must be produced at the same rate as they are lost. Thus the productively infected cells responsible for producing half of the plasma virus present at any one time have been infected (or activated from latency) within the past 1·5 days and cells responsible for producing 99% of plasma virus have been infected within the past 11 days (Ho et al, 1995; Wei et al, 1995). A half-life of 1·5 days corresponds to an average cell lifespan of 2·2 days. Perelson et al (1996) have also estimated the average HIV generation time – the time between release of a virion and the time it infects another cell leading to release of a new generation of virions – and found it to be 2·6 days. This indicates that about 140 viral replication cycles occur each year.

Most of this work on viral dynamics has been done by tracking plasma HIV RNA levels and, as described above, making inferences about what was happening to the cells that produced this RNA. More recently Perelson et al (1997) have directly tracked the infectivity of PBMCs in patients starting antiretroviral therapy and confirmed the short half-life (1·1 days) of productively infected cells within the first week of therapy. This indicated that earlier concerns, discussed above, that PBMCs would be unrepresentative of cells producing most plasma virus were probably unjustified.

These estimates for the lifespan of cell-free virus and productively infected cells allow rough estimates of the rate of appearance of productively infected cells and free virus in an infected person. If the half-life of free virus is taken as the maximum estimate of 6 hours then for a person with 100 000 HIV RNA copies per ml in plasma this corresponds to 90 000 HIV RNA copies or 45 000 virions produced (and lost) per day per ml. This means $3 \times 4.5 \times 10^7 = 1.35 \times 10^8$ virions per 3 litres of plasma (for an average adult) per day. In the body as a whole the figure will presumably be several times larger. If the more recent estimate of 10 minutes is taken for the half-life of free virus, this leads to an estimate of 3.5×10^9 new virions arising per day in the plasma alone. If we assume that each productively infected cell produces free virus at a rate of 100 per day, as has been estimated in vivo (Cavert et al, 1997), this latter estimate would mean that $3 \cdot 5 \times 10^7$ new productively infected cells arise per day purely to produce the virus seen in the plasma. These kinds of estimate are approximate and rely on several assumptions, but are useful when considering the likelihood of resistance mutations arising.

These findings have shown that most plasma virus is produced by cells which are short-lived, with an average lifespan of only around 2 days. It seems most likely that the vast majority of these cells are CD4 lymphocytes. Besides the fact that CD4 lymphocytes are known to be infected with HIV, and that in culture the cells die as a result, other evidence consistent with this is the way in which the rate of rise of CD4 lymphocytes after starting therapy mirrors the rate of fall in virus level. This is true at a between-patient level in trials and observational studies, with those patients experiencing more profound declines in plasma HIV RNA tending to be those with the greater CD4 lymphocyte count rises (Hughes et al, 1997). It is also true when considering the average effect of drug combinations; monotherapy with single nucleoside analogue drugs produces a modest and short-term decline in plasma virus level of around $0 \cdot 7$ log ($10^{0 \cdot 7}$ or five-fold) and a rise in CD4 count of only around 30 cells mm^{-3} (Katzenstein et al, 1996; Brun Vezinet et al, 1997). Double nucleoside therapy tends to produce an approximate $1 \cdot 5$ log decrease in plasma virus level which is maintained as a $0 \cdot 5$–1 log decrease for up to 1 year. There is a larger CD4 count rise of around 80 cells mm^{-3} which tends to be held out to 1 year (Eron et al, 1995; Katzenstein et al, 1996). Finally, triple therapy with two nucleosides and a protease inhibitor such as indinavir or ritonavir results in a 2–$2 \cdot 5$ log decrease in virus level which is maintained at 1 year. The CD4 count shows a continual increase which is over 150 mm^{-3} and still rising at 1 year (Hammer et al, 1997).

After around 2 weeks of potent therapy, when the plasma HIV RNA level has fallen by around 2 logs (99%), there appears to be a change in slope of decline to a slower, so-called second phase of decline (Figure 4.2). Using the same logic as above, this slower phase of decline of virus is likely to correspond to the dying out of longer-lived virus-producing cells such as macrophages, to the slow activation of latently infected lymphocytes, or to the release of trapped virions from the processes of follicular dendritic cells. Perelson et al (1997) studied this phase to distinguish whether latently infected lymphocytes or long-lived infected cells were the main producers of plasma virus in this second phase. This was done by tracking PBMC infectivity over time in patients starting therapy in addition to plasma HIV RNA. The PBMC infectivity assay was designed to activate any latently infected cells and hence these would be scored in this limiting dilution assay. The results indicated that latently infected lymphocytes have a half-life of $0 \cdot 5$–2 weeks compared with 1–4 weeks for long-lived infected cells, suggesting that the latter is the major contributor to the second-phase decay. These estimates were used to produce a minimum theoretical estimate of the amount of time that a patient would need to be treated to eliminate HIV from these compartments. This estimate was $2 \cdot 3$–$3 \cdot 1$ years. It was pointed out that to eradicate HIV completely even longer treatment may be needed. This now appears to be the case, as some latently infected cells seem to experience a longer half-life than had previously been noted (Finzi et al, 1997; Wong et al, 1997).

A concern with studies of changes in plasma HIV RNA after therapy has been the degree to which they really do reflect events in lymphoid tissues, the site of most active viral replication. Cavert et al (1997) tracked virus levels in serial tonsil biopsies in patients receiving treatment with reverse transcriptase inhibitors and a protease inhibitor using in situ hybridization and computerized quantitative analysis. They found that the frequency of productively infected mononuclear cells declined with a half-life of about 1 day, and that the amount of HIV RNA trapped on follicular dendritic cells decreased almost as quickly, consistent with findings from studies of plasma HIV RNA. Earlier reports based on the use of one, two or three reverse transcriptase inhibitors had suggested much less effect on virus levels in lymphoid tissue (Haase et al, 1996; Lafeuillade et al, 1996), so the protease inhibitor could exert a qualitatively different effect in lymphoid tissue.

The estimates of numbers of virions produced per day and their lifespan can be usefully compared with similar estimates for hepatitis B virus (HBV) infection. Nowak et al (1996) used a similar approach to that outlined above for the estimates for HIV infection, studying changes in plasma HBV load in patients treated with lamivudine. In contrast to HIV, here the drug acts by inhibiting the production of new virions from infected cells. Measurement of the clearance rate of plasma HBV levels on starting therapy allows estimation of the virion half-life, which was estimated as around 1 per day, greater than that of HIV, and interestingly was found to vary much more between individuals than that of HIV. Using a similar calculation to that described above for HIV, and the fact that the steady-state pretherapy virus level was around 10^{11}, this led to an estimated 10^{11} new virus particles produced each day, higher than corresponding estimates for HIV. The half-life of virus-producing cells was found to vary between 10 days and 100 days, again in marked contrast to the situation for HIV infection. This was thought to reflect the fact that, unlike HIV, HBV is largely non-cytopathic, and hence the wide differences in turnover rates could reflect the different anticellular immune responses in different individuals (Nowak et al, 1996).

As illustrated in Figure 4.2, if another viral parameter, the level of HIV DNA in PBMCs, is tracked in patients starting potent antiretroviral therapy, then a very different picture emerges to those for plasma HIV RNA and plasma infectivity. Instead of a rapid decline in levels over time, there is a much slower rate of decline (Stellbrink et al, 1996; Perelson et al, 1997). The half-life has been estimated to be around 100 days. This means that cells harbouring HIV DNA live, on average, for around 150 days, similar to estimates of the lifespan of uninfected cells. This has been interpreted as suggesting that most HIV DNA in PBMCs is defective and does not affect the lifespan of the cell. Chun et al (1997) showed that the most prevalent form of HIV-1 DNA in CD4 lymphocytes is a full-length, linear, unintegrated form that is not replication competent. The predominance of this form of viral nucleic acid in cells over other forms, such as messenger RNA or integrated HIV-1 DNA, is likely to be a reflection of the fact that cells containing these other forms are more likely to be capable of production of virus with its associated rapid cell death. However, more recent findings have suggested that virus can be cultured from purified resting CD4 cells (Finzi et al, 1997; Wong et al, 1997; Chun et al, 1998) following long-term treatment in patients with undetectable plasma HIV RNA levels.

Besides measures of quantities of virus present in different forms and different compartments, the other main source of information on HIV dynamics has been measures of drug resistance. With each round of HIV replication there is a chance that one or more mutations will arise. In fact, there is an estimated mutation rate of around 10^{-4} to 10^{-5} per base (Mansky and Temin, 1995). Some of these mutations will result in a virus that is better able than its parent virus to infect cells in the presence of drug. This resistant virus will then have a selective advantage and often survive to outgrow the wild-type (McLean and Frost, 1995). This has been clearly seen in the case of the non-nucleoside reverse transcriptase inhibitor nevirapine, which has been found to cause a rapid drop in plasma HIV RNA levels within the first 1–2 weeks as described above, but thereafter levels rebound sharply back close to pretherapy values (Wei et al, 1995). This new virus that arises has been found to be highly resistant to nevirapine. Even single codon changes at certain positions (e.g. position 181) in the reverse transcriptase gene can decrease susceptibility of the virus to nevirapine by over 100-fold. By 4 weeks from start of therapy the plasma HIV RNA consisted of an estimated 100% mutant virus. These data provide independent confirmation of the extremely rapid turnover of plasma virus. When Wei et al (1995) studied PBMC-associated DNA they found that the accumulation of mutant virus was much more prolonged, resulting in an estimate that these PBMCs experienced a half-life of 50–100 days, consistent with the above estimate based on quantification of DNA in PBMCs. Since most virus in PBMCs was wild-type at a time when most plasma virus was mutant, it was concluded that most PBMCs contribute comparatively little to plasma virus load and that other cells in the lympho-reticular system are likely to be the main source. More recent evidence suggests that virus levels in lymph node cells and PBMCs are in fact remarkably similar (Chun et al, 1997). It is probably the case that most cell-free virus, whether in plasma or otherwise, is produced by a minority of HIV DNA-positive cells (most of which are not infected with virus in a form capable of replicating) so the slow accumulation of mutant virus in PBMCs noted by Wei et al would probably pertain also in lymph nodes.

IMPLICATIONS FOR UNDERSTANDING PATHOGENESIS

The findings described above, together with other evidence in vivo of substantial amounts of active HIV replication (Embretson et al, 1993; Pantaleo et al, 1993; Haase et al, 1996), and the strong association between HIV levels and subsequent CD4 cell depletion and risk of AIDS (Mellors et al, 1995, 1996; Hughes et al, 1997; Phillips et al, 1997a), have led to a more widespread belief in a direct pathogenic link between HIV replication and CD4 lymphocyte depletion. However, it is not sufficient merely to invoke HIV replication in CD4 cells, resulting in their death, as an explanation for HIV pathogenesis. Simple mathematical models (McLean et al, 1991; McLean and Nowak, 1992; Perelson et al, 1993; Essunger and Perelson, 1994; Shenzle, 1994; Nowak and Bangham, 1996; Phillips et al, 1997c) have considered numbers of uninfected and (pro-ductively) infected CD4 cells and levels of free virus and assumed that, given a certain number of uninfected cells and free virus, a certain number of productively infected cells will arise. The number of infected cells that will arise depends on a parameter usually called the *infectivity*. Estimates for lifespans of productively infected cells and free virus for use in these models have been derived from the results described above, while lifespans of uninfected CD4 lymphocytes have been estimated by other studies to be of the order of a few hundred days (Michie et al, 1992; McLean and Michie, 1995). These simple models of HIV replication in CD4 lymphocytes do not lead to predictions of the gradual CD4 lymphocyte count decline which is typically seen in HIV infection (Phillips et al, 1997a). Instead, a constant, albeit subnormal, new equilibrium CD4 count is predicted (Essunger and Perelson, 1994; Shenzle, 1994; Phillips et al, 1997c). The lower CD4 count is due to the added source of CD4 cell removal, i.e. HIV infection. Gradual decline in CD4 count would appear to require either a gradual decrease in the rate with which virus-infected cells and/or free virus were removed or an increase in the viral replicative capacity over time, i.e. an increase in the number of infected cells arising from a given number of uninfected cells and free virions.

For many the more likely of these options would be that the rate of removal of virus-infected cells tends to decrease with more advanced infection. If the removal is due to the immune response this makes sense because the immune response (at least that part which depends on CD4 lymphocytes) is weaker in more advanced infection. It was shown above how the decline of virus when starting antiretroviral therapy provides an estimate of the average lifespan of productively infected cells ($2 \cdot 2$ days). Surprisingly, those studies have shown no association between the initial CD4 lymphocyte count and the lifespan of productively infected cells: productively infected cells die at the same rate in those with CD4 counts above 500 $\times$ mm^{-3} as in those with counts below 10 $\times$ mm^{-3} (Ho et al, 1995; Wei et al, 1995; Phillips et al, 1997b; Stellbrink et al, 1996). At first sight this certainly seems at odds with the hypothesis that HIV infection progresses because productively infected cells are removed more and more slowly over time. Klenerman et al (1996) looked closely at this and concluded that it was likely that HIV is cytopathic, but that there were conceivable explanations which would still allow a killing role for the immune system, in particular for cytotoxic T lymphocytes (CTLs). One such explanation related to the fact that estimates of lifetimes of productively infected cells were not based on direct observations of decline in numbers of such cells, but on the free virus they produce. It was suggested that cells that produced most plasma virus might not be subject to CTL killing, but that other productively infected cells are killed by CTLs – which is why they do not produce virus which is found in plasma – and that the proportion killed by CTLs becomes smaller over time. More recent results from Perelson et al (1997) and Cavert et al (1997) in which decay rates of productively infected PBMCs and mononuclear cells in lymphoid tissue were observed directly appear inconsistent with this idea. The other possible explanation was that the rate of decline of plasma virus could reflect the rate with which infected cells became targets for CTLs, which would mean that no correlation between the rate of CTL killing and rate of plasma virus decline would be expected. As pointed out by Feinberg and McLean (1997), if a declining CTL killing rate operated then one might expect to see this in the second phase of plasma HIV RNA decline seen after starting antiretroviral therapy, in the form of a steeper rate of decline in patients with higher CD4 counts. In fact, if anything, the opposite was seen by Perelson et al (1997).

If the gradual decline in CD4 lymphocyte count cannot be attributed to gradual decline in the rate at which productively infected cells are killed or removed, then the spotlight falls on the viral replicative capacity (Phillips et al, 1997c). Does this increase slowly throughout HIV infection? If it does, then models have shown that this could produce a pattern of CD4 cell loss consistent with that observed in practice (Essunger and Perelson, 1994; Shenzle, 1994; Phillips et al, 1997c). There are certainly plausible suggestions why the viral replicative capacity might increase. It has been postulated that this could be due to evolution of viral tropism or virulence (Cheng-Mayer et al, 1988; Tersmette et al, 1989; Miedema et al, 1990; Schuitemaker et al, 1992; Connor et al, 1994, 1997; Asjo et al, 1996; Fouchier et al, 1996), perhaps to use a wider range of coreceptors (Connor et al, 1997); an increase in lymphocyte activation and thus in numbers of cells susceptible to infection (McLean and Nowak, 1992); a decrease in the infectivity of cells due to decline in any CD8 suppressor activity (Walker et al, 1986; Landay et al, 1993; Levy et al, 1996); or some other mechanism. Certainly, virus isolated from patients with low CD4 lymphocyte counts has been found to show a tendency to replicate in vitro more rapidly and to higher titres, and to exhibit a broader cell tropism than virus isolated from those with high counts (Cheng-Mayer et al, 1988; Tersmette et al, 1989; Miedema et al, 1990; Schuitemaker et al, 1992; Connor et al, 1994, 1997; Asjo et al, 1996; Fouchier et al, 1996). Regarding possible changes in the host throughout infection, there is known to be increased immune activation in more advanced infection (Peakman et al, 1995) and the CD8 suppressor factor has been found at lower levels in more advanced infection (Walker et al, 1986; Landay et al, 1993; Levy et al, 1996). Until recently there has been no evidence from studies in vivo showing that these or any other phenomena lead to an increase in viral replicative capacity. Studies of virus levels in patients stopping therapy have been suggested as a possible approach to obtaining such evidence (Phillips et al, 1997c) and recently preliminary evidence from such studies has been reported (Phillips et al, 1999).

Another consequence of the findings described above has been a reconsideration of mechanisms underlying HIV dynamics during primary infection (Phillips, 1996). Several studies have shown that there is a high peak in levels of plasma virus in the first few weeks after infection followed by a marked decline to a stable lower level, which varies between individuals (Clark et al, 1991; Daar et al, 1991). Based on small studies showing the presence of HIV-specific CTL activity at the time of this decline in most patients studied, and more recent reports of escape from CTL epitopes during primary infection, it has usually been assumed that the decline in virus is attributable to this response, thus indicating the strength of the immune response to HIV (Borrow et al, 1994, 1997; Koup et al, 1994; Safrit et al, 1994a,b; Price et al, 1997). The idea is that productively infected cells are killed more rapidly and so have a shorter average lifespan once the HIV-specific CTL activity starts. However, since the lifetime of productively infected cells and free virus seems relatively invariant, as discussed above, it is reasonable to suggest that the same lifetimes *could* operate right from initial infection with HIV and only be slightly, if at all, reduced when the HIV-specific CTL response begins. It was shown by Phillips (1996), using a simple mathematical model similar to that mentioned above, that even if lifetimes of productively infected cells are held constant over time in this way the pattern of viral change that occurs in primary infection, of a rise followed by a sharp decline, is predicted. The decline is due to exhaustion of sufficient numbers of susceptible cells. This study indicated that the decline in virus observed in primary infection cannot, given current knowledge, be assumed to be a true measure of the strength of the HIV-specific immune response.

IMPLICATIONS FOR TREATMENT

The demonstration of the rapid turnover of cell-free virus and most productively infected cells has had major implications for designing therapeutic strategies. As pointed out by Coffin (1995), the large number of new cell infections per day (estimated above to be of the order of 10^7–10^8 rather than the value of 10^9 used by Coffin), combined with the per base mutation rate of around 3×10^{-5} (Mansky and Temin, 1995), means that each and every possible single base mutation (against a replication-competent background) will arise every day, making the appearance of resistant virus likely if only a single base change is required. Taking the lower rate of 10^7 new infections per day, the probability of a single specific mutation, with no other changes (besides synonymous substitutions), arising in 1 day is $^{1-}(1 - [1/3 \times 3 \times 10^{-5} \times 0.82])10^7$ which is more than 0·999. (The probability of no simultaneous non-synonymous substitutions at the 6500 possible bases elsewhere on the viral genome – which would render the virus defective or, at least, at significant selective disadvantage – is $^{1-}[1 - 3 \times 10^{-5}]^{6500} = 0.82$.) However, the probability of a specific two mutations arising against a background of no other non-synonymous changes arising in 1 day is $^{1-}(1 - [1/3 \times 3 \times 10^{-5} \times 1/3 \times 3 \times 10^{-5} \times 0.82])10^7 = 0.0008$. This translates to a probability of 0·006 in 1 week and 0·14 in 6 months. The chance of such a mutation arising would be higher if significant numbers of virus-infected cells contained virus with one or other of the mutations already present (indeed, such a stepped approach towards acquisition of mutations is likely to be the only way in which a virus with a specific three or more mutations can arise). On the other hand, once therapy has started it is unlikely that there will be as many as 10^7 new cells infected each day, even with the weakest antiretroviral drugs.

These estimates are broadly consistent with what has been observed. As noted above, drugs such as nevirapine for which a single base substitution can render the virus highly resistant result in appearance of resistant virus within a few weeks. This is also the case with lamivudine (3TC) monotherapy to which a base change at position 184 on the reverse transcriptase gene renders the virus highly resistant. When resistance to a regimen requires two or more base substitutions, resistant virus tends to arise more slowly (Brun Vezinet et al, 1997). As pointed out by Leigh Brown and Richman (1997), and supported by the above calculations, there is likely to be a strong stochastic element to resistance development and this may explain why there is so much interindividual variability in the time to development of resistance. A further consideration is that virus with resistance mutations could be present before therapy is started. This is dependent on the relative selective disadvantage of the mutants in the absence of drug and the number of infected cells.

These considerations have immediate implications for how to treat patients. If a person's virus level is stable at, say, 400 or even 20 copies per ml, a key clinical question is whether that level is low enough, or if further drugs should be added to therapy (Phillips et al, 1998). Using the information that studies of viral dynamics have given us, it is possible to estimate, for a given plasma HIV RNA level, the probability of any specific mutation arising, including the one(s) that will result in reduced drug susceptibility. Consider, for example, a person with a CD4 count of $100 \times mm^{-3}$ who has been on therapy for several months and has a stable viral load of 50 copies ml^{-1}. Much of the plasma HIV RNA may emanate from long-lived cells and thus does not reflect new rounds of replication. However, if the plasma virus level is stable rather than declining, it seems conservative to assume that at least 10% of the RNA is from short-lived productively infected cells with the typical half-life of 1·5 days. This means that in 0.5 ml plasma there are 2·5 HIV RNA copies from short-lived cells, which corresponds to 2·5 HIV RNA copies per 100 000 CD4 cells in a person with a CD4 count of $100 \times mm^{-3}$ (i.e. 100 000 cells ml^{-1}). Given the estimate of 2×10^{11} CD4 lymphocytes in uninfected individuals (with a typical CD4 count of $1000 \times mm^{-3}$), and hence 2×10^{10} CD4 lymphocytes in those with a blood count of $100 \times mm^{-3}$, this indicates a whole-body load of 5×10^{5} HIV RNA copies which has come from short-lived cells. Assuming 200 copies (100 wrions) (Cavert et al, 1997) produced per productively infected cell per day, this indicates 2500 productively infected short-lived cells. Given the short half-life of these cells, $0 \cdot 35 \times 2500 = 875$ new infected cells must arise per day. With a per base mutation rate of 3×10^{-5}, the probability of one specific mutation occurring in 1 day (in the absence of any other non-synonymous changes) is $1-[1-(1/3 \times 3 \times 10^{-5} \times 0 \cdot 82)]^{875} = 0 \cdot 007$. In 6 months the probability is 0.72. For a plasma HIV RNA level of 1 copy per ml the 6-month probability is $0 \cdot 03$ and for a plasma HIV RNA level of $0 \cdot 1$ copy per ml the corresponding value is $0 \cdot 003$. It should be noted that these figures do not correspond to the probability of a substantial rebound in virus by this time: not every virus with a mutation associated with reduced drug susceptibility survives, and, even if it does, there will usually be more than one mutation required before significant viral rebound can be expected. The figures indicate that if resistance is really to be prevented in practice, over several years the plasma HIV RNA level must probably at least be below 1 copy per ml (Phillips et al, 1998). These are levels far below the current detectability of assays. These kinds of estimate give rough guidance as to the degree of suppression of HIV that will have to be achieved if such suppression is to be durable over several years, or even over the patient's entire lifetime. It indicates that we may need to find still more sensitive assays for detecting ongoing replication and be prepared to use still more drugs in combination in our efforts to keep HIV under control.

CONCLUSION

The advent of reliable assays for quantifying HIV levels and of potent antiretroviral therapies which block different stages of the viral life cycle, together with careful and innovative work carried out between virologists, mathematicians and those from other disciplines, have pushed understanding of mechanisms underlying HIV infection forward dramatically. Although many questions remain unanswered, especially concerning the process underlying CD4 lymphocyte decline, much is now understood about how to treat HIV effectively and there is the real hope that lifelong viral suppression may be attainable in the future. The prospects for the majority of HIV-infected people worldwide who have no access to antiretroviral drugs, however, remain poor. The hope must be that the increased understanding of HIV infection which new therapies have provided will lead to a greater understanding of natural constraints on HIV replication and hence, eventually, to a prophylactic vaccine.

REFERENCES

Asjo B, Albert J & Karlsson L et al (1996) Replicative properties of human immunodeficiency virus from patients with varying severity of HIV infection. *Lancet* **327:** 660–662.

Autran B, Carcelain G, Li TS et al (1997) Positive effects of combined antiretroviral therapy on CD4+ T cell homeostasis and function in advanced disease. *Science* **277:** 112–116.

Bonhoeffer S, May RM, Shaw GM & Nowak MA (1997) Virus dynamics and drug therapy. *Proc Natl Acad Sci USA* **94:** 6971–6976.

Borrow P, Lewicki H, Hahn BH, Shaw GM & Oldstone MBA (1994) Virus specific CD8+ CTL activity associated with control of viremia in primary HIV-1 infection. *J Virol* **68:** 6103.

Borrow P, Lewicki H, Wei X et al (1997) Antiviral pressure exerted by HIV-1–specific cytotoxic T lymphocytes during primary infection demonstrated by rapid selection of CTL escape virus. *Nature Med* **3:** 205–211.

Brun Vezinet F, Boucher C, Loveday C et al (1997) HIV-1 viral load, phenotype, and resistance in a subset of drug-naive participants from the Delta trial. *Lancet* **350:** 983–990.

Cavert W, Notermans DW, Staskus K et al (1997) Kinetics of response in lymphoid tissues to antiretroviral therapy of HIV-1 infection. *Science* **276:** 960–964.

Cheng-Mayer C, Seto D, Tateno M & Levy JA (1988) Biologic features of HIV that correlate with virulence in the host. *Science* **240:** 80–82.

Chun TW, Carruth L, Finzi D et al (1997) Quantification of latent tissue reservoirs and total body virus load in HIV-1 infection. *Nature* **387:** 183–188.

Chun TW, Stuyver L, Mizell SB et al (1998) Presence of an inducible reservoir during highly active antiretroviral therapy. *Proc Natl Acad Sci USA* (in press).

Clark SJ, Saag MS, Decker WD et al (1991) High titres of cytopathic virus in plasma of patients with symptomatic primary HIV-1 infection. *N Engl J Med* **324:** 954.

Coffin JM (1995) HIV population dynamics in vivo: implications for genetic variation, pathogenesis and therapy. *Science* **267:** 483–489.

Connor RI, Ho DD, Cao Y et al (1994) Human immunodeficiency virus type-1 variants with increased replicative capacity develop during the asymptomatic stage before disease progression. *J Virol* **68:** 4400–4408.

Connor RI, Sheridan KE, Ceradini D, Choe S & Landau NR (1997) Change in coreceptor use correlates with disease progression in HIV-1 infected individuals. *J Exp Med* **185:** 621–628.

Daar ES, Mougdil T, Meyer T & Ho DD (1991) Transient high levels of viremia in patients with primary human immunodeficiency virus type 1 infection. *N Engl J Med* **324:** 961.

Embretson J, Zupancic M, Ribas JL et al (1993) Massive covert infection of helper T lymphocytes and macrophages by HIV during the incubation period of AIDS. *Nature* **362:** 359–362.

Eron JJ, Benoit SL, Jemsek J et al (1995) Treatment with lamivudine, zidovudine, or both in HIV-positive patients with 200 to 500 CD4+ cells per cubic millimeter. North American HIV Working Party. *N Engl J Med* **333:** 1662–1669.

Essunger P & Perelson AS (1994) Modelling HIV infection of CD4+ T-cell subpopulations. *J Theor Biol* **170:** 367–391.

Feinberg MB & McLean AR (1997) AIDS: decline and fall of immune surveillance? *Curr Biol* **7:** R136–R140.

Finzi D, Hermankova M, Pierson T et al (1997) Identification of a reservoir for HIV-1 in patients on highly active antiretroviral therapy. *Science* **278:** 1295–1300.

Fouchier RAM, Meyaard L, Brouwer M, Hovenkamp E & Schuitemaker H (1996) Broader tropism and higher cytopathicity for CD4+ T cells of a syncytium-inducing compared to a non-syncytium-inducing HIV-1 isolate as a mechanism for accelerated CD4+ T cell decline in vivo. *Virology* **219:** 87–95.

Haase AT, Henry K, Zupancic M et al (1996) Quantitative image analysis of HIV-1 infection in lymphoid tissue. *Science* **274:** 985–989.

Hammer S, Squires K, Hughes MD et al (1997) A controlled trial of two nucleoside analogues plus indinavir in persons with human immunodeficiency virus infection and CD4+ cell counts of 200 per cubic millimeter or less. *N Engl J Med* **337:** 725–731.

Havlir DV & Richman DD (1996) Viral dynamics of HIV: implications for drug development and therapeutic strategies. *Ann Intern Med* **124:** 984–994.

Herz AVM, Bonhoeffer S, Anderson RM, May RM & Nowak MA (1996) Viral dynamics in vivo: limitations on estimates of intracellular delay and virus decay. *Proc Natl Acad Sci USA* **93:** 7247–7251.

Ho DD (1997) 6th European Conference on Clinical Aspects and Treatment of HIV Infection, Hamburg, Germany, 11–15 Oct 1997.

Ho DD, Neumann AU, Perelson AS et al (1995) Rapid turnover of plasma virions and CD4 lymphocytes in HIV-1 infection. *Nature* **373:** 123–126.

Hughes MD, Johnson VA, Hirsch MS et al (1997) Monitoring plasma HIV-1 RNA levels in addition to CD4+ lymphocyte count improves assessment of antiretroviral therapeutic response. *Ann Intern Med* **126:** 929–938.

Katzenstein DA, Hammer SM, Hughes MD et al (1996) The relation of virologic and immunologic markers to clinical outcomes after nucleoside therapy in HIV infected adults with 200 to 500 CD4 cells per cubic millimeter. *N Engl J Med* **335:** 1091–1098.

Klenerman P, Phillips RE, Rinaldo CR et al (1996) Cytotoxic T lymphocytes and viral turnover in HIV type 1 infection. *Proc Natl Acad Sci USA* **93:** 15323–15328.

Koup RA, Safrit JT, Cao Y et al (1994) Temporal association of cellular immune responses with the initial control of viremia in primary HIV-1 syndrome. *J Virol* **68:** 4650–4655.

Lafeuillade A, Poggi C, Profizi N et al (1996) HIV-1 kinetics in lymph nodes compared with plasma. *J Infect Dis* **174:** 404–410.

Landay AL, Mackewicz C & Levy JA (1993) An activated CD8+ T cell phenotype correlates with antiHIV activity and asymptomatic clinical status. *Clin Immunol Immunopathol* **69:** 106.

Lane HC, Masur H, Gelmann EP et al (1985) Correlation between immunologic function and clinical subpopulations of patients with the acquired immune deficiency syndrome. *Am J Med* **78:** 417–422.

Leigh Brown AJ & Richman DD (1997) HIV-1: gambling on the evolution of drug resistance? *Nature Med* **3:** 268–271.

Levy JA, Mackewicz CE & Barker E (1996) Controlling HIV pathogenesis: the role of the noncytotoxic anti-HIV response of CD8+ T cells. *Immunol Today* **17:** 217.

Loveday C (1995) Quantification of serum HIV-1 RNA load by PCR. Annual Scientific Meeting of the Association of Clinical Pathologists, 20–21 Oct 1994. In: Haeney MR (ed.) *The Association of Clinical Pathologists Yearbook 1995*. London: Association of Clinical Pathologists.

Mansky LM & Temin HM (1995) Lower in vivo mutation rate of HIV-1 than that predicted from the fidelity of purified reverse transcriptase. *J Virol* **69:** 5087–5094.

McLean AR & Frost SDW (1995) Zidovudine and HIV: mathematical models of within-host population dynamics. *Rev Med Virol* **5:**141–147.

McLean AR & Michie CA (1995) In vivo estimates of division and death rates of human T lymphocytes. *Proc Natl Acad Sci USA* **92:** 3707–3711.

McLean AR & Nowak MA (1992) Models of interactions between HIV and other pathogens. *J Theor Biol* **155:** 69–86.

McLean AR, Emery VC, Webster A & Griffiths PD (1991) Population dynamics of HIV within an individual after treatment with zidovudine. *AIDS* **5:** 485–489.

Mellors JW, Kingsley LA, Rinaldo CR et al (1995) Quantitation of HIV-1 RNA in plasma predicts outcome after seroconversion. *Ann Intern Med* **122:** 573–579.

Mellors JW, Rinaldo CR, Gupta P et al (1996) Prognosis in HIV-1 infection predicted by the quantity of virus in plasma. *Science* **272:** 1167–1170.

Michie CA, McLean AR, Alcock C & Beverly PCL (1992) Lifespan of human lymphocyte subsets defined by CD45 isoforms. *Nature* **360:** 264–265.

Miedema F, Tersmette M & van Lier RA (1990) AIDS pathogenesis: a dynamic interaction between HIV and the immune system. *Immunol Today* **11:** 293.

Nowak MA & Bangham CRM (1996) Population dynamics of immune responses to persistent viruses. **272:** 74–79.

Nowak MA, Bonhoeffer S, Hill AM, Boehme R, Thomas H & McDade H (1996) Viral dynamics in hepatitis B virus infection. *Proc Natl Acad Sci USA* **93:** 4398–4402.

Pantaleo G, Graziosi C, Demarest JF et al (1993) HIV infection is active and progressive in lymphoid tissue during the clinically latent stage of disease. *Nature* 362: 355–358.

Peakman M, Mahalingam, Pozwark A et al (1995) Markers of immune cell activation and disease progression. In: Andrieu JM & Lu W (eds) *Cell Activation and Apoptosis in HIV Infection*. New York: Plenum.

Perelson AS, Kirschner DE & De Boer RJ (1993) Dynamics of HIV-1 infection of CD-4 cells. *Math Biosci* 114: 81.

Perelson AS, Neumann A, Markowitz M et al (1996) HIV-1 dynamics in vivo: virion clearance rate, infected cell life-span, and viral generation time. *Science* 271: 1582–1586.

Perelson AS, Essunger P, Cao Y et al (1997) Decay characteristics of HIV-1 infected compartments during combination therapy. *Nature* 387: 188–191.

Phillips AN (1996) Reduction of HIV concentration during acute infection: independence from a specific immune response. *Science* 271: 497–499.

Phillips AN, Lee CA, Elford J et al (1991) Serial CD4 lymphocyte counts and the development of AIDS. *Lancet* 337: 389–392.

Phillips AN, Eron JJ, Bartlett JA et al (1997a) HIV-1 RNA levels and the development of clinical disease. *AIDS* 10: 859–865.

Phillips AN, McLean AR, Loveday C et al (1997b) Lifespan of productively infected cells in early and advanced infection. Abstract 201. *6th European Conference on Clinical Aspects and Treatment of HIV infection*, Hamburg, Germany, 11–15 Oct 1997.

Phillips AN, McLean AR, Loveday C et al (1997c) HIV-1 dynamics after transient antiretroviral therapy: implications for pathogenesis and clinical management. *J Med Virol* 53: 261–265.

Phillips AN, Loveday C, Johnson M (1998) HIV suppression and risk of drug resistance mutations. *AIDS* 12: 1930.

Phillips AN, McLean AR, Loveday C et al (1999) In vivo HIV-1 replicative capacity in early and advanced infection. *AIDS* 13: 67–73.

Piatak M, Saag MS, Yang LC et al (1993) High levels of HIV-1 in plasma during all stages of infection determined by competitive PCR. *Science* 259: 1749–1754.

Price DA, Goulder PJR, Klenerman P et al (1997) Positive selection of HIV-1 cytotoxic T lymphocyte escape variants during primary infection. *Proc Natl Acad Sci USA* 94: 1890–1895.

Safrit JT, Andrews CA, Zhu T, Ho DD & Koup RA (1994a) Characterization of HIV-1 specific cytotoxic and lymphocyte clones isolated during acute seroconversion. Recognition of autologous virus sequences within a conserved immunodominant epitope. *J Exp. Med* 179: 463.

Safrit JT, Lee AY, Andrews CA & Koup RA (1994b) A region of the third variable loop of HIV-1 gp120 is recognized by HLA-BT-restricted CTLs from two acute seroconversion patients. *J Immunol* 153: 3822.

Schuitemaker H, Koot M, Kootstra MW et al (1992) Biological phenotype of HIV-1 clones at different stages of infection: progression of disease is associated with a shift from monocytotropic to T-cell-tropic virus populations. *J Virol* 66: 1354–1360.

Shenzle D (1994) A model for AIDS pathogenesis. *Stat Med* 13: 2067–2079.

Stellbrink HJ, Zoller B, Fenner T et al (1996) Rapid plasma virus and CD4+ T cell turnover in HIV infection: evidence for an only transient interruption by treatment. *AIDS* 10: 849–855.

Tersmette M, Lange JM, de Groede RE et al (1989) Association between biological properties of human immunodeficiency virus variants and risk for AIDS and AIDS mortality. *Lancet* 333: 983–989.

Walker CM, Moody DJ, Stites DP & Levy JA (1986) CD8+ lymphocytes can control HIV infection in vitro by suppressing virus replication. *Science* 234: 1563–1566.

Wei X, Ghosh SK, Taylor ME et al (1995) Viral dynamics in human immunodeficiency virus type-1 infection. *Nature* **373:** 117–122.

Weiss RA (1993) How does HIV cause AIDS? *Science* **260:** 1273–1279.

Wong JK, Hezareh M, Gunthard HF et al (1997) Recovery of replication-competent HIV despite prolonged suppression of plasma viremia. *Science* **278:** 1291–1295.

Chapter 5

THE CYTOTOXIC T LYMPHOCYTE RESPONSE TO THE IMMUNODEFICIENCY VIRUSES

Stephen McAdam and Frances Gotch

INTRODUCTION

Virus-specific cytotoxic T lymphocytes (CTLs) recognize antigen in association with the polymorphic products of the major histocompatibility complex (MHC). The presentation results from the partial degradation of virally encoded proteins within the cytosol, most probably by the proteosome complex. The TAP proteins, TAP1 and TAP2, are encoded for in the MHC as demonstrated by Spies and De Mars (1991), and transport polypeptides into the endoplasmic reticulum where they can bind to and stabilize nascent class I molecules (Nuchtern et al, 1989; Yewdell and Bennink, 1989; Elvin et al, 1991). The mature MHC–peptide complexes transit through the Golgi apparatus where each has three carbohydrate moieties added before they egress to the cell surface, where the complex, which consists of a viral peptide (usually 9 or 10 amino acids in length) bound to a specific class I molecule, may be recognized by the T cell receptor of the CTL. Such viral peptide recognition by CTLs was first demonstrated by Townsend et al (1986) in the influenza system (Figure 5.1). Recognition of an infected cell by a CTL will normally lead to the destruction of the infected cell, via a Fas-mediated pathway or by the release of a pore-forming protein, perforin (Kagi et al, 1996). In addition, CTLs release soluble factors such as interferon gamma (IFN-γ) and tumour necrosis factor alpha (TNF-α) which have been shown to have a powerful antiviral effect.

Cytotoxic T lymphocytes specific for the human immunodeficiency virus (HIV) were first detected directly in 'fresh' assays without any prior culture of the cells ex vivo by Walker et al (1987) and Plata et al (1987). Detection of CTL activity directly without restimulation in vitro is not demonstrable in other viral infections and was an early indication of the unusually high precursor frequencies of CTLs found during HIV infection (discussed below). Soon after these initial observations a restimulation protocol was devised by Nixon et al (1988) which involved the activation of a proportion of the peripheral blood mononuclear cells (PBMCs) from an infected individual, with a mitogen such as phytohaemagglutinin (PHA). The infected, activated CD4+ T cells are thought to express the autologous viral antigens which, when added back to the rest of the PBMC culture, specifically activate the CTL. This method of 'bulk' culturing was used to characterize the first CTL epitope (or specific viral peptide) from HIV which was shown to be presented by HLA-B*27 and to be recognized by CTLs from an HIV-infected

HIV and the New Viruses Second Edition
ISBN 0-12-200741-7

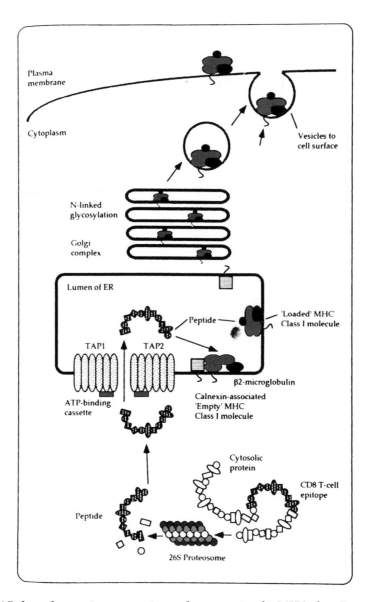

Figure 5.1 TAP-dependent antigen processing and presentation for MHC class I
ER, endoplasmic reticulum; MHC, major histocompatibility complex; TAP, transporter associated with antigen processing.

patient with HLA-B*27. Others, such as McAdam et al (1995) and Klenermann et al (1996) working in Oxford, have successfully produced clonal populations of CTLs from these bulk cultures using peptide pulsed autologous B lymphoblastoid cell lines (BLCLs) and PHA-activated allogeneic PBMCs as feeder cells. As an alternative to PHA blasts expressing autologous antigens, van Baalen et al (1993) and Lubaki et al (1994) have reported the successful use of inactivated, autologous BLCLs infected with recombinant vaccinia virus to restimulate CTLs in vitro. Direct restimulation of PBMCs with high concentrations of peptide has also been shown to be a very sensitive technique for restimulation of HIV-specific CTLs, and this method has been used by Rowland-Jones et al (1995) to detect CTLs from patients thought to have very low precursor frequencies, such as exposed but uninfected commercial sex workers in Africa. The efficiency of this type of restimulation can be further improved by the addition of interleukin 7 (IL-7) into the cultures.

In a different approach CTLs have been isolated by cloning PBMCs at limiting dilution following non-specific activation with anti-CD3 antibodies. One advantage of this method is that the resulting CTL clones are more likely to be free from the specificity bias described by Carmichael et al (1993) following peptide restimulation. Macaque CTLs specific for simian immunodeficiency virus (SIV) antigens have proved much harder to propagate for reasons that remain unclear. Nevertheless, Tsubota et al (1989) and Yamamoto et al (1990) have had considerable success following activation of PBMCs with the mitogen concanavalin A. Alternatively, autologous PHA blasts infected with laboratory strains of SIV have been used by Gotch et al (1991) and later by Gallimore et al (1995) specifically to restimulate CTLs that recognize SIV antigens.

MEASUREMENT OF CTL RESPONSES

The principle behind the measurement of HIV-specific CTL activity is to place CTL effector cells from infected individuals together with appropriate target cells in vitro, and to measure lysis of these target cells by the effector cells. Effector cells may be tested without prior restimulation ('fresh', see above), or may be specifically restimulated using different strategies outlined above. Epstein–Barr virus (EBV)-transformed B cell lines are commonly used as target cells, but any cell that is capable of processing and presenting viral peptides on its surface in conjunction with HLA class I molecules may be used. Such cells include PHA blasts or such cell lines as CIR or CHO which have been transfected with single class I molecules. In order that target cells may be recognized by CTLs it is necessary that they present the relevant viral antigenic peptides on their surfaces. Target cells may be infected with recombinant vectors such as recombinant vaccinia viruses or recombinant adenoviruses expressing different viral proteins, or may be incubated with specific viral peptides. Target cells may be autologous (HLA-matched), matched at single class I alleles or HLA-mismatched. Target cells are often labelled with a radioisotope such as chromium-51 so that lysis may be easily measured.

Most recombinant vectors that are used to infect target cells in experiments where HIV-specific CTL activity is being evaluated contain HIV proteins from well-characterized laboratory-propagated strains of HIV such as MN or IIIB (both of which are B clade viruses). These viruses may differ considerably from autologous wild-type viruses in individual patients and consequently the correct and relevant peptide epitopes may not be presented to the CTL. This fact may be especially important if the patient is infected with a clade of HIV other than B, which is the case in many parts of the world: in Thailand, for example, the predominant infecting HIV is clade E, and in sub-Saharan Africa clades A, C and D are most common.

When CTL activity is being measured, effector cells and target cells are incubated together at different effector to target ratios (commonly, for example, at $100:1$, $50:1$, $25:1$ and $12\cdot5:1$). Samples containing only target cells (to measure spontaneous lysis) and samples containing target cells with detergent (to measure maximal lysis) are included in each series. After a period of incubation (usually 4–6 hours), a small amount of supernatant is removed from the mixtures of effector and target cells, and the amount of radioisotope released is measured. Percentage target lysis can then be calculated from the formula $(E - M)/(D - M) \times 100$; where E is the experimental lysis, M is the spontaneous release in the presence of culture medium with no CTLs present, and D is the maximum release from target cells with detergent.

The CD8+ T cells that recognize specific viral peptides can also be visualized using an enzyme-linked immunospot (ELISPOT) assay which measures the frequencies of cells secreting cytokines (IFN-γ and TNF-α), as originally demonstrated by Czerkinsky et al (1988). As mentioned above, when CTLs recognize specific viral peptides they may release cytokines such as INF-γ and TNF-α. This methodology uses small numbers of cells which can be frozen and thawed before use.

Since the 1980s many viral peptide epitopes from HIV have been identified which are presented by individual class I molecules to HIV-specific CTLs. Peptides (usually 9 or 10 amino acids in length) have been identified from many of the HIV proteins, and different families of peptides with distinct amino acid motifs have been shown to bind to individual class I molecules. Thus at least 10 different HIV peptides from reverse transcriptase (RT), gp120, gp41, gag p17 and nef have been shown to bind to HLA-A2 and to be presented to specific CTLs. All these peptides share a specific motif with particular amino acids found at certain anchor residues which enable them all to bind into the polymorphic pocket of the HLA-A2 molecule. In order to bind to HLA-A2 a peptide will normally have a leucine or isoleucine molecule at the second position and a leucine or valine molecule at the ninth position. Thus the peptide-binding motif for HLA-A2 can be visualized as X L/I X X X X X X L/V. Such conserved anchor residues which are particular for any peptide motif have been shown to be of great importance in binding to the class I molecule, and in fact fit snugly into pockets on the floor of the groove between the α-1 and α-2 helices which comprise the 'top surface' of the class I molecule. Figure 5.2 shows many of the CTL epitopes that have been identified in RT, gp120, gp41, nef, gag p17 and gag p24. All these epitopes bind to one HLA molecule or another and have been shown to be recognized by HIV-specific CTLs.

Limiting dilution analysis is used to derive quantitative estimates of the frequencies of virus-specific CTL precursors (CTLp). This is a method by which the CTL response can be analysed at the level of the individual responding cell as demonstrated by Hoffenbach et al (1989) and by Gotch et al (1990). Using this methodology precursor frequencies of gag-specific CTLp were originally thought to be about 1 in 5×10^3 peripheral blood lymphocytes (PBLs) and precursor frequencies of env-specific CTLp about 1 in 10^4 PBLs. Later work from Carmichael et al (1993) demonstrated frequencies of 1 in 2×10^3 for gag CTL precursors and 1 in 10^4 for pol and env. These precursor frequencies were shown to decline in parallel with the observed decline in CD4 counts as patients infected with HIV progressed towards AIDS. These values for HIV-specific CTL precursor frequencies compare with 1 in 2×10^4 for cytomegalovirus-specific CTLp and 1 in 10^5 for varicella-zoster virus-specific CTLp.

Kalams et al (1994) and Moss et al (1995) have quantitated dominant HIV-specific CTL clones by measuring frequencies of specific T cell receptors using mRNA transcripts. Probes to the CDR3 region of the specific T cell receptor were used and clonal frequencies of 1–5% were demonstrated. This implies that virus-specific CTLs of a single specificity capable of recognizing and lysing virally infected cells may comprise as much as 5% of the PBLs in an HIV-infected person. It was reported by Altman et al

identified CTL epitopes in
RT

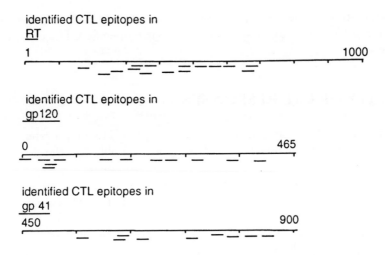

identified CTL epitopes in
nef

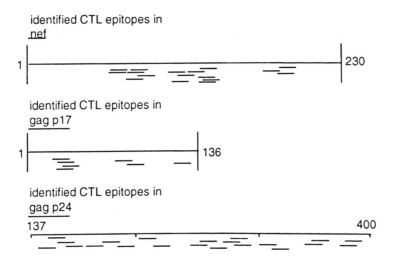

Figure 5.2 *Identified cytotoxic T lymphocyte epitopes*

(1996) that when T cells specific for two epitopes in gag and pol (and restricted through HLA-A2) are directly stained with peptide/HLA complexes using an entirely novel method for direct visualization and phenotypic analysis of virus-specific T lymphocytes, frequencies of between 0·2% and 1·2% of CD8 cells are stained. It was observed that, although the frequency of HIV-specific CTLp measured using this novel method was apparently very high, no 'fresh' responses (without prior culture of the cells ex vivo) could be demonstrated. Such results suggest that memory T cells may be being measured in one assay and naïve T cells in another. Discrepancies between limiting dilution analysis and functional or staining data suggest that many effector cells in blood may be 'terminally differentiated' and unable to divide, perhaps comparable to the expanded population in the acute phase of HIV infection. However, it remains apparent that unusually high precursor frequencies of virus-specific CTLs are found during HIV infection.

SPECIFICITY OF CTL RESPONSES

There is an increasing body of evidence to suggest that CTLs play a crucial role in the control of HIV throughout infection and may even be able to offer some degree of protection.

Following initial infection with HIV there is a burst in viraemia with levels as high as 10^6 RNA copies per ml of serum being detected in some patients using very sensitive polymerase chain reaction (PCR)-based assays. This viraemia peaks 4–8 weeks after infection, and then drops by many logs, sometimes to undetectable levels, as the patient enters the asymptomatic phase of disease. As early as 1992 groups including those of Ariyoshi were suggesting that neutralizing antibodies were not responsible for this drop in viraemia (Ariyoshi et al, 1992). In 1994 Koup and colleagues documented a rise in HIV-specific CTL precursors which coincided with the fall in viraemia, and occurred long before neutralizing antibody could be demonstrated. Five patients were described by Koup et al (1994), four of whom were shown to have CTL precursors specific for Env, Gag or Pol of HIV within 3 weeks of the onset of clinical symptoms of primary infection. Similarly, Borrow et al (1994) demonstrated that four out of five patients studied made responses to Env, Gag or Tat within 2 weeks of the onset of symptoms. Interestingly, both the patients who failed to make a detectable CTL response exhibited a rapid course of infection with high viral loads and a fast decline in CD4+ cell numbers. In contrast, another study from Lamhamedi-Cherradi et al (1995) reported that two out of nine patients failed to make a detectable CTL response to HIV within 4 weeks of seroconversion but had a normal rate of disease progression. It is possible that these latter two patients did in fact make strong CTL responses which were strain-specific for the autologous virus and thus went undetected by the reagents used when evaluating CTL activity. These reagents (as discussed above) are based on well-characterized laboratory strains of HIV which may differ from autologous virus with which individual patients are infected.

Surprisingly, few studies have looked at the kinetics of CTL responses following challenge of non-human primates with SIV. Simian immunodeficiency virus infection of macaques in a laboratory setting produces profound depletion of CD4 cells followed by AIDS-like symptoms, opportunistic infections and death. Gallimore et al (1995) found that CTL activity peaked 14 days after SIV infection while Yasutomi et al (1993) have reported the appearance of Gag-specific CTL precursors within 4 days of infection. Although circumstantial, this evidence strongly supports the idea that CTLs are primarily responsible for control of the initial burst in viraemia following infection.

Numerous studies have evaluated the influence of HLA genotype on HIV susceptibility or disease progression. These studies leave a inconsistent and confused picture. Viral variation between and within individuals will affect which epitopes are presented by particular HLA haplotypes, thereby weakening any underlying associations. A compounding problem stems from the fact that particular population subgroups may have a significantly greater risk of infection, for social, political or financial reasons. If these subpopulations have significantly different HLA gene frequencies from the general population it is likely that infected individuals may differ significantly from the general population, thus making suitable control groups difficult to identify (Hill, 1996). Despite these and other complications there are a few associations between rates of progression and HLA that have been independently confirmed by a number of different groups. The HLA-A*01, B*0801, DR3 haplotype has been linked to rapid progression in several studies, as has HLA-B35. The Twelfth International Histocompatibility Workshop report (Thorsby, 1997) looked at 363 HIV-1-infected individuals and found an association between HLA-B35, Cw4, B39 and A24 and faster rates of progression. In addition this workshop found that HLA-B*27 and HLA-A*32 were associated with slower rates of progression.

If HIV-specific CTLs are important in the control of viral replication, it may be expected that the presence or absence of such activity would have an influence on disease progression. Several groups have quantified either CTLp frequencies or CTL effector (CTLe) lytic capacity to HIV to explore whether these influence either the progression of disease or viral load. One study from Riviere et al (1995) reported that of 38 patients examined almost half had detectable fresh responses specific for targets infected with vaccinia expressing gag. Follow-up of this group showed that patients lacking in this response had a risk to progression to AIDS of 1·89 compared with patients with gag-specific CTLs. However, a study by Rinaldo et al (1995), where 57 HIV-positive men were evaluated for CTLe specific for Gag, Pol, Env and Tat, failed to find CTLe specific for Gag as a marker for disease progression. Indeed, although anti-HIV-1 CTLe responses appeared to be relatively stable in the first 6 years of infection, such responses were not found to correlate with either CD4+ or CD8+ T cell counts, rates of CD4+ T cell loss, HIV-1 infectious viral load, use of antiviral medications, or subsequent progression to AIDS. Three patients from the Multicentre AIDS Cohort Study (the MACS cohort) who have remained asymptomatic for over 8 years and with stable CD4 counts over 1000 cells/mm^3 have been described by Ferbas et al (1995). These patients had low viral loads but no detectable CTLp specific for Gag, Pol, Env or Nef, from which it can be inferred that CTLs are not always necessary for the control of viral replication. In contrast, in studies of HIV-2 infections in West Africa conducted by Ariyoshi et al (1995), a significant negative correlation between proviral DNA load and total CTL activity towards gag, Pol and Nef was demonstrated, this correlation being strongest for gag responses. Coates et al (1992) used cryopreserved PBMCs to compare CTLp frequencies in six rapid progressors and six long-term non-progressors (LTNPs), the latter being patients who remained asymptomatic with stable CD4 counts after more than 8 years of infection. The LTNPs all had persistent Gag-specific CTL responses together with low levels of circulating HIV-1-infected CD4+ T cells, stable and normal CD4+ counts and preserved T cell function (as measured by proliferation following stimulation with anti-CD3 antibody). In contrast, the rapid progressors were shown to have transient CTL responses to gag which were present at the time of an increase in the levels of circulating HIV-infected CD4+ T cells. The decline in gag-specific precursors in the rapid progressors was paralleled by a decline in the T response to anti-CD3 antibody. It seems clear that in these rapid progressors the CTLs failed to control viral replication, and although no explanation is offered as to why this happened, it would be interesting to assess whether viral variation had allowed escape from CTL recognition as has been shown in other patients – originally, for example, by Phillips et al (1991).

When correlations between disease progression or viral load and CTL activity are found, it is often difficult to determine whether this is due to the action of the CTLs or whether the CTLs are merely a marker for more generalized immune function. This issue was avoided in a study by Gallimore et al (1995), in which CTLp frequency specific for Nef was measured in macaques that had been vaccinated with recombinant vaccinia viruses containing the *nef* gene of SIV, so that Nef-specific CTLs were the only measurable SIV-specific immune function generated. The CTLp were measured prior to challenge with SIV. A strong negative correlation between CTLp frequency and the viral load post-challenge was found, and one animal with the highest measurable frequency of CTLp before challenge was protected from infection. These studies must be confirmed in further studies using more animals but they argue that CTLs may be able to control immunodeficiency virus replication directly. Very few studies have investigated the qualitative differences between CTLs that may be important factors in how well the response controls the virus. These differences may include phenotypic and functional differences, T cell receptor repertoire and the selection of epitopes to which the patient responds, all factors that it may be crucial to understand if we are to develop an effective vaccine against HIV.

Since the first reports of cell-mediated responses to HIV in the absence of detectable virus or antibody there has been interest and hope that these responses may be a marker of protection as well as previous exposure to HIV. The initial reports from Clerici et al (1992) were of Env-specific CD4+ T cell responses in seronegatives, which have been estimated to be present in 75% of exposed adults and 35% of uninfected babies born to HIV-infected mothers. It is, however, possible that these responses were the result of exposure to HIV antigens or defective virus rather than to infectious virus. This type of priming is considerably less likely for the CD8+ class I restricted CTL responses detected by Cheynier et al (1992) in three babies born to infected mothers; these children lost maternal antibody and remained PCR-negative but maintained detectable CTL responses up to 3 years of age. Three similar children have since been reported by Rowland-Jones et al (1993) and Aldhous et al (1994) in whom only transient CTL responses were detected lasting a few months after birth.

De Maria et al (1994) reported the occurrence of HIV-specific CTL activity in 6 out of 23 apparently uninfected children born to HIV-infected mothers. High levels of circulating CTLs were found as demonstrated by the fact that responses could be seen directly ex vivo. These responses were seen in the second year of life, possibly indicating that low levels of replicating virus were still present in these children.

There have been several reports of HIV-specific CTL activity in other exposed but uninfected individuals. In one report from Langlade-Demoyen et al (1994), repeated exposure to HIV in the non-infected heterosexual contacts of HIV-infected persons was shown to result in an increase in measurable Nef-specific CTLp. The presence of HIV-specific CTL activity has also been reported in three out of six exposed but uninfected commercial sex workers in The Gambia by Rowland-Jones and her colleagues (Rowland-Jones et al, 1995). This activity was in each case restricted through HLA-B35 and was seen only after in vitro restimulation with relevant viral peptides. No responses were seen in unexposed individuals of the same HLA type from the same ethnic population. Transient Env-specific CTL responses have also been reported by Pinto et al (1995) in health care workers occupationally exposed to body fluids from HIV-infected individuals. No response was seen in healthy controls. There was also a demonstrable increase in CD4+ responders in these exposed individuals when compared with controls.

FAILURE OF CTLs

It is generally held that a progressive decline in HIV-specific cytolytic activity occurs in parallel with the development of HIV-related disease, so that it is rare to be able to demonstrate HIV-specific CTLs in patients with fullblown AIDS. Such decline may be due to a *general* impairment of CD8+ cell cytolytic activity through lack of CD4+ T cell help, although it was noted in 1987 by Lightman and colleagues that mice depleted of CD4 cells to a level where an antibody response could not be mounted were still able to generate a CTL response to influenza (Lightman et al, 1987). Most investigators have not found any abnormality in the general cytolytic capacity of CD8+ cells until very late in the course of disease, as reported by Pantaleo et al (1990a) and Gruters et al (1991). In one small study of influenza-specific CTL responses in HIV-infected people (Shearer et al, 1985) it was observed that there was a decline in influenza-specific CTLs over time in some patients, but strong alloantigen responses were maintained.

Another explanation for the decline in CTL activity in late HIV disease is that there is a loss of HIV-specific CTL activity, with CTLs specific for other pathogens remaining at relatively normal levels. Such specific loss could be due to the loss of precursors in the presence of clonal exhaustion, perhaps due to overstimulation with viral antigen. Tanaka et al (1992) suggested that failure to generate new CD8+ CTLs could also result from infection of CD4+CD8+ CTL precursors in the thymus. Alternatively the CTLs themselves could become infected as has been described for SIV in long-term SIV$_{mac}$-specific CTLs by Tsubota et al (1989), and for HIV in human CD8+ cells. It seems possible that HIV-specific CTLs could become infected at the time they recognize and lyse infected target cells (De Maria et al, 1991).

Evidence for a defect in the HIV-specific CTL population has come from the observation of Pantaleo et al (1990b) that patients with AIDS have an expansion in the CD8+DR+ subset of their T cells which are severely defective in their ability to proliferate to a variety of stimuli, including anti-CD2, anti-CD3 and mitogens such as PHA and PMA (phorbol myristate acetate; pokeweed mitogen). In these experiments cell sorting revealed that HIV-specific cytolytic activity was largely confined to this subset of apparently abnormally activated cells with markedly reduced clonogenic potential. The discrepancy between estimated CTL effector (CTLe) and observed CTL precursor (CTLp) frequencies in healthy seropositive donors, which was discussed above, is consistent with the hypothesis that effectors in the peripheral blood are terminally differentiated cells with low growth potential.

It is also possible that HIV-specific CTL activity in late-stage disease may be being suppressed in some way, and a population of CD8+HNK+CD4−CD16− T cells has been identified by Joly et al (1989) amongst alveolar lymphocytes in AIDS patients. These cells were able to inhibit the activity of HIV-specific CTLs as well as CTLs against HLA alloantigens in a non-MHC-restricted manner.

The generation of variants of HIV which are able to 'escape' from CTL recognition has now been observed in many cases of patients in whom CTL activity can no longer be quantified. Genetic variation is one of the hallmarks of HIV and is seen both in isolates from different individuals and amongst isolates from the same individual followed over time. The reverse transcriptase of HIV is intrinsically prone to error and lacks the capacity for 'proofreading', leading to a base misincorporation rate in vitro of at least 10^{-4}, equivalent to one mismatch per genome per replication cycle, as reviewed by McCune (1991). Because of the high replication rate of the virus (10^9 genomes per day) which has been reported by Ho et al (1995), the initial infecting strain rapidly develops into a population of 'quasispecies' (sets of closely related but distinct genomic sequences). Viruses with variant sequences that have the potential to escape HIV-1-specific CTL recognition have now been demonstrated on numerous occasions, initially by Phillips et al (1991). In each case sequencing of sequential viral isolates has revealed mutations in T cell epitopes which meant they could no longer be recognized by effector cells. The CTLs are generally directed against internal viral proteins, which are usually more conserved than surface glycoproteins recognized by antibodies, but nevertheless mutations frequently occur which render the virus 'invisible' to specific CTLs.

CTLs AND VACCINES

It has been proposed that prophylactic vaccines for the immunodeficiency viruses should be designed to generate high levels of virus-specific CTLs prior to virus exposure. Such vaccines might be expected to reduce virus load after challenge and might even protect against the establishment of infection. Vaccines may have to reflect the clades of virus which are circulating within a given population and the predominant HLA types of the population would also have to be taken into account if a subunit vaccine (a peptide preparation that would interact with single HLA molecules, for example) were to be considered. The ability of antigen delivery systems to produce strong, long-lived virus-specific CTL responses in all vaccinees varies markedly, probably due to selection of different epitopes by the recipient's HLA molecules. Therefore, it cannot be assumed that live recombinant virus vectors presenting particular virus proteins or any other novel vaccine preparations will generate CTL responses of the required specificity and magnitude. Such responses will need to be measured and quantified before a positive or negative effect on protection can be inferred. In any trial undertaken, an assessment will have to be made of the fraction of vaccinees who make the desired CTL response.

The most effective vaccine that has been used in macaques to prevent infection with SIV is still attenuated SIV, as was originally described by Daniel et al (1992), and this vaccine has been shown to induce high levels of SIV-specific CTLs. It is not considered that live attenuated virus will be acceptable as a human vaccine preparation for safety reasons. However, these protected animals offer a real opportunity to define the nature of the protective immune response. Once this is understood, vaccines that induce CTL responses and that mimic this protective capacity can be designed.

REFERENCES

Aldhous MC, Watret KC, Mok JY et al (1994) Cytotoxic T lymphocyte activity and CD8 subpopulations in children at risk of HIV infection. *Clin Exp Immunol* **97(1):** 61–67.
Altman JD, Moss PAH, Goulder PJR et al (1996) Phenotypic analysis of antigen-specific T lymphocytes. *Science* **274(5284):** 94–96.

Ariyoshi K, Harwood E, Chiengsong-Popov R et al (1992) Is clearance of HIV-1 viraemia at seroconversion mediated by neutralising antibodies? *Lancet* **340**: 1257–1258.

Ariyoshi K, Berry N, Jafar S, Sabally S, Corrah T & Whittle H (1995) HIV.2 specific CTL activity is inversely related to proviral load. *AIDS* **9**: 555–559.

Borrow P, Lewicki H, Hahn BH et al (1994) Virus-specific CD8+ cytotoxic T-lymphocyte activity associated with control of viraemia in primary human immunodeficiency virus type 1 infection. *J Virol* **68(9)**: 6103–6110.

Carmichael A, Jin X, Sissons P et al (1993) Quantitative analysis of the human immunodeficiency virus type 1 (HIV-1)-specific cytotoxic T lymphocyte (CTL) response at different stages of HIV-1 infection: differential CTL responses to HIV-1 and Epstein–Barr virus in late disease. *J Exp Med* **177(2)**: 249–256.

Cheynier R, Langlade-Demoyen P, Marescot MR et al (1992) Cytotoxic T lymphocyte responses in the peripheral blood of children born to HIV-1–infected mothers. *Eur J Immunol* **22**: 2211–2217.

Clerici M, Giorgi JV, Chou CC et al (1992) Cell-mediated immune response to human immunodeficiency virus (HIV) type 1 in seronegative homosexual men with recent sexual exposure to HIV-1. *J Infect Dis* **165(6)**: 1012–1019.

Coates RA, Farewell VT, Raboud J, Read SE & Klein M (1992) Using serial observations to identify predictors of progression to AIDS in the Toronto sexual contact study. *J Clin Epidemiol* **45**: 245–253.

Czerkinsky C, Adersson G, Ekre H-P, Nilsson L-A & Ouchterlony O (1988) Reverse ELISPOT assay for clonal analysis of cytokine production (1). Enumeration of gamma interferon secreting cells. *J Immunol Methods* **115**: 31–36.

Daniel MD, Kirchoff F, Czajak SC et al. (1992) Protective effects of a live attenuated SIV vaccine with a deletion in the *nef* gene. *Science* **258(5090)**: 1938–1941.

De Maria A, Pantaleo G, Schnittman SM et al (1991) Infection of CD8 T lymphocytes with HIV. Requirement for interaction with infected CD4 cells and induction of infectious virus from chronically infected CD8 cells. *J Immunol* **146**: 2220–2226.

De Maria A, Cirillo C & Moretta L (1994) Occurrence of HIV-specific CTL activity in apparently uninfected children born to HIV-1–infected mothers. *J Infect Dis* **170**: 1296–1299.

Elvin J, Cerundolo V, Elliott T et al (1991) A quantitative assay for peptide-dependent class I assembly. *Eur J Immunol* **21(9)**: 2025–2031.

Ferbas J, Kaplan AH, Hausner MA et al (1995) Virus burden in long-term survivors of human immunodeficiency virus (HIV) infection is a determinant of anti-HIV CD8 lymphocyte activity. *J Infect Dis* **172(2)**: 329–339.

Gallimore A, Cranage M, Cook N et al (1995) Early suppression of SIV replication by CD8 nef-specific cytotoxic T cells in vaccinated macaques. *Nature Med* **1(11)**: 1167–1173.

Gotch FM, Nixon D, Alp N et al (1990) High frequency of memory and effector gag specific cytotoxic T lymphocytes in HIV seropositive individuals. *Int Immunol* **2**: 707–712.

Gotch FM, Hovell R, Delchambre M et al (1991) Cytotoxic T-cell response to simian immunodeficiency virus by cynomolgus macaque monkeys immunized with recombinant vaccinia virus. *AIDS* **5(3)**: 317–320.

Gruters RA, Terpstra FG, De Goode RE et al (1991) Immunological and virological markers in individuals progressing from seroconversion to AIDS. *AIDS* **5**: 837–844.

Hill AV (1996) HIV and HLA: confusion or complexity? *Nature Med* **2(4)**: 395–396.

Ho DD, Neumann AV, Perelson AS, Chen W, Leonard JN & Makowitz M (1995) Rapid turnover of plasma virions and CD4 lymphocytes in HIV.1 infection. *Nature* **373**: 123–126.

Hoffenbach A, Langlade-Demoyen P, Dadaglio G et al (1989) Unusually high frequencies of HIV-specific cytotoxic T lymphocytes in humans. *J Immunol* **142**: 452–462.

Joly P, Guillon JM, Mayaud C et al (1989) Cell mediated suppression of HIV-specific cytotoxic T lymphocytes. *J Immunol* **143**: 2193–2201.

Kagi D, Ledermann B, Burki K et al (1996) Molecular mechanisms of lymphocyte-mediated cyto-toxicity and their role in immunological protection and pathogenesis in vivo. *Annu Rev Immunol* **14**: 207–232.

Kalams SA, Johnson RP, Trocha AK et al (1994) Longitudinal analysis of TCR gene usage by HIV-1 envelope-specific CTL clones reveals a limited TCR repertoire. *J Exp Med* **179**: 1261–1271.

Klenermann P, Phillips R, Rinaldo C et al (1996) Cytotoxic T lymphocytes and viral turnover in HIV.1 infection. *Proc Natl Acad Sci USA* **15**: 323–328.

Koup RA, Safrit JT, Cao Y et al (1994) Temporal association of cellular immune responses with the initial control of viraemia in primary human immunodeficiency virus type 1 syndrome. *J Virol* **68(7)**: 4650–4655.

Lamhamedi-Cherradi S, Culmann-Penciolelli B, Guy B et al (1995) Different patterns of HIV-1-specific cytotoxic T-lymphocyte activity after primary infection. *AIDS* **9(5)**: 421–426.

Langlade-Demoyen P, Ngo-Giang-Huong N, Ferchal F et al (1994) HIV nef-specific cytotoxic T lymphocytes in noninfected heterosexual contacts of HIV-infected patients. *J Clin Invest* **93**: 1293–1297.

Lightman S, Cobbold S, Waldmann H & Askonas BA (1987) Do L3T4– T cells act as effector cells in protection against influenza virus infection? *Immunology* **62**: 139–144.

Lubaki MN, Egan MA, Siliciano RF et al (1994) A novel method for detection and ex vivo expansion of HIV type 1-specific cytolytic T lymphocytes. *Aids Res Hum Retroviruses* **10(11)**: 1427–1431.

McAdam SN, Klenerman P, Tussey L et al (1995) Immunogenic HIV variants that bind to HLA-B8 but fail to stimulate CTL responses. *J Immunol* **155**: 2729–2736.

McCune JM (1991) HIV.1: the infective process in vivo. *Cell* **64**: 351–363.

Moss PAH, Rowland-Jones SL, Frodsham PM et al (1995) Persistent high frequency of human immunodeficiency virus-specific cytotoxic T cells in peripheral blood of infected donors. *Proc Natl Acad Sci USA* **92(13)**: 5773–5777.

Nixon DF, Townsend AR, Elvin JG & Rizza CR (1988) HIV.1 gag-specific CTL defined with recombinant vaccinia viruses and synthetic peptides. *Nature* **336**: 484–487.

Nuchtern JG, Bonifacino JS, Biddison WE et al (1989) Brefeldin A implicates egress from the endoplasmic reticulum in class I restricted antigen presentation. *Nature* **339**: 223–226.

Pantaleo G, de Maria A, Koenig S et al (1990a) CD8 T lymphocytes of patients with AIDS maintain normal broad cytolytic function despite the loss of immunodeficiency virus-specific cytotoxicity. *Proc Natl Acad Sci USA* **87**: 4818–4822.

Pantaleo G, Koennig S, Baseler M, Lane HC & Fauci AS (1990b) Defective clonogenic potential of CD8 T lymphocytes in patients with AIDS. Expansion in vivo of a nonclonogenic T cell population. *J Immunol* **144**: 1696–1704.

Phillips RE, Rowland-Jones SL, Nixon DF et al (1991) Human immunodeficiency virus genetic variation that can escape cytotoxic T cell recognition. *Nature* **354**: 453–459.

Pinto LA, Sullivan J, Berzofsky JA et al (1995) ENV-specific cytotoxic T lymphocyte responses in HIV seronegative health care workers occupationally exposed to HIV-contaminated body fluids. *J Clin Invest* **96(2)**: 867–876.

Plata F, Autran B, Martins LP et al (1987) AIDS virus specific cytotoxic T lymphocytes in lung disorders. *Nature* **328**: 348–351.

Rinaldo CJ Jr, Beltz LA, Huang XL et al (1995) Anti-HIV type 1 cytotoxic T lymphocyte effector activity and disease progression in the first 8 years of HIV type 1 infection of homosexual men. *Aids Res Hum Retroviruses* **11(4)**: 481–489.

Riviere Y, McChesney MB, Porrot F et al (1995) Gag-specific cytotoxic responses to HIV type 1 are associated with a decreased risk of progression to AIDS-related complex or AIDS. *Aids Res Hum Retroviruses* **11(8)**: 903–907.

Rowland-Jones SL, Nuxon DF, Aldhous MC et al (1993) HIV-specific CTL activity in an HIV-exposed but uninfected infant. *Lancet* **341(8849)**: 860–861.

Rowland-Jones SL, Sutton J, Ariyoshi K et al (1995) HIV-specific cytotoxic T cells in HIV-exposed but uninfected Gambian women. *Nature Med* **1(1):** 59–64.

Shearer G, Salahuddin SZ, Markham PD et al (1985) Prospective study of cytotoxic T lymphocyte responses to influenza and antibodies to HTLVIII in homosexual men. *J Clin Invest* **76:** 1699–1704.

Spies T & DeMars R (1991) Restored expression of major histocompatibility class I molecules by gene transfer of a putative peptide transporter. *Nature* **351:** 323–325.

Tanaka K, Hatch W, Kress Y et al (1992) HIV.1 infection of human thymocytes. *AIDS* **5:** 94–101.

Thorsby E (1997) Invited anniversary review: HLA associated diseases. *Hum Immunol* **53(1):** 1–11.

Townsend AR, Rothbard J, Gotch FM et al (1986) The epitopes of influenza nucleoprotein recognized by cytotoxic T lymphocytes can be defined with short synthetic peptides. *Cell* **44:** 959–968.

Tsubota HC, Lord CI, Watkins D, Morimoto C & Letvin N (1989) A cytotoxic T lymphocyte inhibits acquired immunodeficiency syndrome virus replication in peripheral blood lymphocytes. *J Exp Med* **169:** 1421–1434.

Van Baalen CA, Klein MR, Geretti AM et al (1993) Selective in vitro expansion of HLA class I-restricted HIV-1 Gag-specific CD8+ T cells: cytotoxic T-lymphocyte epitopes and precursor frequencies. *AIDS* **7(6):** 781–786.

Walker BD, Chakrabarti S, Moss B et al (1987) HIV-specific cytotoxic T lymphocytes in seropositive individuals. *Nature* **328(6128):** 345–348.

Yamamoto H, Ringler DJ, Miller MD et al (1990) Studies of cloned simian immunodeficiency virus-specific T lymphocytes. Gag-specific cytotoxic T lymphocytes exhibit a restricted epitope specificity. *J Immunol* **144(9):** 3385–3391.

Yasutomi Y, Reimann KA, Lord CI et al (1993) Simian immunodeficiency virus-specific CD8+ lymphocyte response in acutely infected rhesus monkeys. *J Virol* **67(3):** 1707–1711.

Yewdell YW & Bennink JR (1989) Brefeldin A specifically inhibits presentation of protein antigens to cytotoxic T lymphocytes. *Science* **244:** 1072–1078.

Chapter 6

T HELPER CELLS SPECIFIC FOR RETROVIRAL EPITOPES

Daniela Fenoglio, Giuseppina Li Pira and Fabrizio Manca

INTRODUCTION

Human retroviruses such as human immunodeficiency virus (HIV) and human T cell lymphotropic virus type 1 (HTLV-1) represent a special challenge for the immune system. Their integration in a latent state as provirus in the host cell (Shaw et al, 1988), their ability to hypermutate (Putney and McKeating, 1990) and their capacity to interfere with functions of the immune system (Koenig and Fauci, 1990) allow retroviruses to elude the immune response. A better knowledge of retroviral epitopes may permit the response to be mobilized in a more protective direction and lead to the design and construction of more appropriate immunogens.

Two major protective mechanisms are at work in viral infections: neutralizing antibodies that prevent the spread of infection from the virus-replicating cells to the virgin cells (Goudsmit et al, 1988; Palker et al, 1988; Rusche et al, 1988; Emini et al, 1990), and cytotoxic lymphocytes (CTL) that kill the virus-infected cells and therefore eradicate the infectious source (Sissons and Oldstone, 1980; Koszinowski et al, 1991; Doherty et al, 1992). Both mechanisms depend on the presence of specific CD4+ T helper (Th) lymphocytes that provide the appropriate lymphokines for clonal expansion and differentiation of antibody-producing B cells and CTLs. Characterization of virus-specific CD4 Th cells and of Th epitopes present on the retroviral proteins is therefore important for understanding ongoing responses and for vaccine design.

EPITOPE PREDICTION

It is well established that antigenic proteins can be internalized by antigen-presenting cells (APCs) by phagocytosis or by endocytosis (Schwartz, 1985). Following endosomal proteolysis (Brodsky and Guagliardi, 1991) the derived peptides can associate with MHC class II molecules and can be exported on the surface of the APCs that presents the antigenic peptide displayed by the histocompatibility molecule (Rothbard and Gefter, 1991). If a (naïve or memory) CD4 Th cell in the repertoire can recognize the MHC–peptide complex with its specific T cell receptor (TCR), T cell activation, proliferation and lymphokine production ensue. The exact rules that allow MHC–peptide association are not yet known (Rothbard and Gefter, 1991). Furthermore, the allelic

HIV and the New Viruses Second Edition
ISBN 0-12-200741-7

polymorphism of MHC molecules in the peptide-binding site is an additional variable. Nevertheless, predictive guidelines have been proposed as algorithms that take into account conformation, charge and the hydrophobicity or hydrophilicity of the candidate peptide (Berkower et al, 1986; Rothbard, 1986; De Groot et al, 1997; Lamb et al, 1987). According to these algorithms, predictions have been made for retroviral protein regions that are most likely to carry Th epitopes (De Groot et al, 1997). These studies predict the MHC binding sequences, but cannot anticipate whether a T cell is available in the repertoire with specificity for the relevant epitope. Only the empirical approach can test for the actual availability of such a cell. Epitope screening studies have been performed in experimental animals and in humans.

Epitope screening in animals

Several animal species have been immunized with HIV proteins and studied for fine specificity of proliferating Th cells in vitro. These studies may provide useful models, but are not necessarily predictive of human responses.

Mice immunized with env gp120 recognize overlapping synthetic peptides (Cease et al, 1987) with profiles that differ among H2 haplotypes (Hale et al, 1989), as in the case of mice immunized with reverse transcriptase p66 that respond to different peptides with H2-related patterns (Haas et al, 1991). Antigenic peptides have also been included in immunogens linked to B epitopes of the V3 loop to promote antibody response in mice (Ahlers et al, 1993).

Primates have been extensively used to test candidate vaccines, since in these animals HIV or simian immunodeficiency virus (SIV) can be used for challenge to test protection. Synthetic gp120 peptides have been used in priming experiments to induce neutralizing antibodies (Hart et al, 1990; Hosmalin et al, 1991; Haynes et al, 1993) in the context of different synthetic constructs.

A proliferative response to synthetic peptides has also been demonstrated in chimpanzees immunized with the nef protein, with the demonstration of an immunodominant peptide (Estaquier et al, 1992).

Epitope screening in immune humans

Epitope analysis has been performed initially in seropositive individuals and subsequently in infected or healthy individuals given candidate vaccines.

A poor proliferative response was reported in infected individuals tested in vitro with whole virions (Wahren et al, 1987) or with gp120 (Ahearne et al, 1988). A better response was seen with p24 as compared with gp120, possibly due to gp120 binding to T cell CD4 (Wahren et al, 1987). Stronger responses were seen in asymptomatic individuals with recombinant gp120, whereas normal controls did not respond (Torseth et al, 1988). T cell lines specific for gp120 were also generated from a seropositive individual, suggesting that this protocol can be used to study fine specificity in patients (Walker et al, 1988).

Association of a B and a T immunodominant epitope on the transmembrane moiety of the HIV env glycoprotein has been reported in seropositive subjects (Schrier et al, 1988). This region is immunodominant with respect to antibody production, whereas only a quarter of the patient group exhibited a proliferative response to the peptide (Schrier et al, 1988).

More detailed studies on responses to peptides from different viral proteins have also been produced showing both common and patient-unique responses (Wahren et al, 1989). T cell responses were observed to peptides derived from env, gag and pol in natural infection (Schrier et al, 1989). A parallel study of antibody and T cell response to gag and env peptides in asymptomatic seropositives showed a dissociation between the two sets of epitopes (Mathiesen et al, 1989).

A better analysis of T cell recognition of synthetic peptides by lymphocytes from asymptomatic individuals can be performed by testing interleukin 2 (IL-2) production rather than proliferation (Clerici et al, 1989). This work also showed that a large fraction of individuals respond to two antigenic regions of gp120 located on the basis of amphipathicity (Clerici et al, 1989).

Poor proliferation in response to envelope peptides by seropositive individuals was enhanced by conjugation to liposomes, suggesting that enhanced uptake by APCs can disclose a potential T cell response (Krowka et al, 1990).

The HIV Tat protein also contains epitopes recognized by Th cells of infected individuals (Blazevic et al, 1993). Interestingly, one of these peptides was DR2 restricted, whereas the other two were promiscuously recognized in association with different DR alleles.

Overlapping synthetic peptides of gp120 were used to monitor T cell specificity in seropositive individuals over a period of time. Fluctuations in response intensity and peptide specificity to conserved and hypervariable regions were detected (Geretti et al, 1994), suggesting that epitope specificity is a dynamic process.

Identification of Th epitopes on HIV reverse transcriptase has been undertaken, showing that one peptide that was immunodominant in a mouse strain was also recognized by the majority of seropositive individuals with different MHC class II alleles (De Groot et al, 1991).

An extensive screening of envelope-derived peptides with lymphocytes from seropositive and seronegative donors showed that infection is not an absolute prerequisite for a response to several HIV peptides, suggesting that pre-existing cross-reactive immunity can influence response to HIV irrespective of previous contact with the virus (Mutch et al, 1994). This observation accords with the successful generation of human T cell lines specific for HIV antigens from individuals who had no previous contact with the virus, as described below.

Definition of T cell specificities in humans treated with candidate vaccine preparations has been extensively described in a review which reports in detail the different approaches used so far for HIV vaccination (Cease and Berzofsky, 1994).

Epitope screening in non-immune humans

The possibility of generating human CD4 T cell lines specific for HIV gp120 by repeated in vitro stimulation with APCs and IL-2 from lymphocytes of non-immune donors was reported simultaneously by Lanzavecchia et al (1988) and Siliciano et al (1988). This finding opened up the possibility of working on human T cells specific for HIV antigens obtained from non-immune donors, and enabling investigation of the gp120-specific immune-naïve repertoire in vitro (Manca et al, 1991a). This system is also a useful model of vaccination, when an immune system is confronted with antigen for the first time. Since experimental immunization in vivo cannot be proposed in humans, it is expected that this approach can anticipate responsiveness, immunodominant epitopes and requirements for APCs. A more extensive characterization of immunodominant epitopes in naïve individuals was performed by using T cell lines induced by in vitro stimulation with gp120 and with reverse transcriptase p66 (Manca et al, 1991b, 1993, 1994, 1995a). The general finding was that a large fraction of non-immune individuals can generate T cell lines specific for p66, whereas a smaller fraction of individuals can generate T cell lines specific for gp120.

The wide variety of Th epitopes on p66 recognized by CD4 cells in non-immune individuals and the remarkable clonal heterogeneity of such responses suggest that a previous encounter with reverse transcriptase from other undefined retroviruses has probably occurred; therefore this may not be a bona fide primary immune response.

On the other hand, it is interesting to speculate that Th cells specific for p66 may not only provide help for B lymphocytes specific for the same protein: anti-p66 antibodies may in fact play a negligible role in terms of protection. Interstructural help, as reported in the influenza (Scherle and Gerhard, 1986) and hepatitis (Milich et al, 1988) virus systems, may allow cooperation between T cells specific for internal viral proteins and B cells specific for exposed proteins (Figure 6.1). This may result in the production of potentially neutralizing antibodies. This possibility has been investigated (Manca et al,

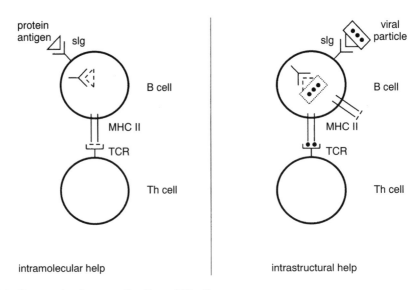

Figure 6.1 Cooperation between B cells and T cells
Left panel: B cell specific for the relevant protein can internalize the antigen with high efficiency thanks to the surface immunoglobulin (sIg). Upon intracellular processing, a T cell epitope is displayed in the context of an MHC class II molecule for recognition by T cell receptor (TCR) on the specific T helper (Th) cell. This depicts *intramolecular help*, since both the B and T epitopes sit on the same molecule. *Right panel*: B cell specific for an exposed protein on a virion (e.g. gp120) takes up the whole viral structure. Intracellular processing results in MHC class II restricted presentation of T epitopes derived from both external (e.g. gp120) and internal (e.g. p66) proteins. This depicts *intrastructural help*, since B and T epitopes sit on different proteins on the same viral structure.

1995b) and is further supported by experiments showing that CD4 cells specific for a virus-like particle (VLP) protein carrying the V3 loop sequence of gp120 can interact with APCs if the recombinant VLP-V3 antigenic particle is complexed with anti-V3 antibodies, thus mimicking an interaction with V3-specific B cells (Sun et al, 1994).

Another aspect of these studies was that different individuals, when immunized in vitro with the viral proteins, consistently recognized the same immunodominant peptides, even though such peptides are different in different individuals. Since these immunodominant peptides can be conserved or variable, T cell responses may focus on consensus sequences or on hypervariable sequences (Manca et al, 1993). If this reflects the situation in vivo it is evident that immunization with viral proteins may have different outcomes in different individuals with respect to recognition of conserved or mutated Th epitopes. Therefore escape may involve epitopes recognized by CD4 cells, in addition to escape mutants for B and for CTL epitopes (Manca, 1994).

In responding individuals the frequency of gp120 and p66 specific precursors is around one CD4 cell out of 10^6 (Li Pira et al, 1998). Furthermore, these precursors are often clonotypically identical in each individual according to Vβ gene usage, to spectratyping and to CDR3 sequence of specific T cell lines generated at intervals of several months (Li Pira et al, 1998).

By using T cell lines obtained by in vitro priming, it was also observed that soluble peptides are effective immunogens, but the T cells that proliferate in response to the peptide do not always respond to the whole protein. This poses some queries about the effectiveness of peptides as immunogens (Manca et al, 1995a,b). This limitation could be overcome by inserting the relevant peptides in the frame of carrier proteins, but also in this case the topology of the Th epitopes relative to the carrier is an important parameter for preservation of antigenicity (Manca et al, 1996).

THERAPEUTIC POTENTIAL

The information gained on HIV-specific CD4 cells recognizing various epitopes on viral proteins is not only useful for the design and construction of candidate vaccines, but it may also help restore the immune competence of the CD4 repertoire which is lost during disease progression. In fact, an appropriate Th response for HIV antigens may contribute to antibody and CTL responses relevant to the control of infection. The possibility of treating CD4-depleted individuals with their own CD4 cells expanded in vitro non-specifically (Levine et al, 1996) or with specific antigens (Manca et al, 1997) is a realistic opportunity to restore a defective repertoire. In this context, the reinfused, antigen-specific CD4 cells can be made resistant to HIV productive infection with a gene therapy approach. This has been tested successfully in vitro with T cells specific for opportunistic pathogens, and preliminary experiments demonstrated also that CD4 cells specific for different retroviral antigens can be transduced with appropriate vectors. Thus these cells can be proposed for immunotherapeutic trials aimed at restoration of HIV-specific Th responses, shown to be associated with slow disease progression (Rosenberg et al, 1997).

CONCLUSION

The definition of Th epitopes on HIV antigens can provide useful information for the design and construction of new-generation vaccines. A number of CD4 Th epitopes on HIV have been accurately listed by Siliciano (1993). One of the goals of these studies is to identify epitopes that are conserved among different strains (Myers et al, 1993), that are recognized as immunodominant by the majority of the population (Panina-Bordignon et al, 1989), and that can prime in vivo for Th cells that also respond to the whole protein, or to the same protein in the context of the viral particle. These requirements may not be so difficult to fulfil, thanks to the large variety of potential epitopes and to the remarkable clonal diversity of the human Th repertoire (Manca et al, 1995b).

ACKNOWLEDGMENTS

The authors' work is supported by the National Health Institute, Rome (AIDS Project), by the Italian Cancer Research Association (AIRC, Milan), and by the EU Concerted Action grant BHH4-CT97-2055 and by the EU grant FAIR-CT97-3046.

REFERENCES

Ahearne PM, Matthews TJ, Lyerly HK, White GC, Bolognesi DP & Weinhold KJ (1988) Cellular immune response to viral peptides in patients exposed to HIV. *AIDS Res Hum Retroviruses* **4:** 259–267.

Ahlers JD, Pendleton CD, Dunlop N, Minassian A, Nara PL & Berzofsky JA (1993) Construction of an HIV-1 peptide vaccine containing a multideterminant helper peptide linked to a V3 loop peptide 18 inducing strong neutralizing antibody responses in mice of multiple MHC haplotypes after two immunizations. *J Immunol* **150:** 5647–5665.

Berkower I, Buckenmeyer GK & Berzofsky JA (1986) Molecular mapping of a histocompatibility-restricted immunodominant T cell epitope with synthetic and natural peptides: implications for T cell antigenic structure. *J Immunol* **136:** 2498–2503.

Blazevic V, Ranki A, Mattinen S et al (1993) Helper T-cell recognition of HIV-1 tat synthetic peptides. *J AIDS* **6:** 881–890.

Brodsky FM & Guagliardi LE (1991) The cell biology of antigen processing and presentation. *Annu Rev Immunol* **9:** 707–744.

Cease KB & Berzofsky JA (1994) Toward a vaccine for AIDS: the emergence of immunobiology-based vaccine development. *Annu Rev Immunol* **12:** 923–989.

Cease KB, Margalit H, Cornette JL et al (1987) Helper T-cell antigenic site identification in the acquired immunodeficiency syndrome virus gp120 envelope protein and induction of immunity in mice to the native protein using a 16-residue synthetic peptide. *Proc Natl Acad Sci USA* **84:** 4249–4253.

Clerici M, Stocks NI, Zajac RA et al (1989) Interleukin-2 production used to detect antigenic peptide recognition by T-helper lymphocytes from asymptomatic HIV-seropositive individuals. *Nature* **339:** 383–385.

De Groot AS, Clerici M, Hosmalin A et al (1991) Human immunodeficiency virus reverse transcriptase T helper epitopes identified in mice and humans: correlation with a cytotoxic T cell epitope. *J Infect Dis* **164:** 1058–1065.

De Groot AS, Jesdale BM, Szu E, Schafer JR, Chicz RM & Deocampo G (1997) An interactive web site providing MHC ligand predictions: application to HIV research. *AIDS Res Hum Retroviruses* **13:** 529–531.

Doherty PC, Allan W & Eichelberger M (1992) Roles of αβ and γδ T cell subsets in viral immunity. *Annu Rev Immunol* **10:** 123–151.

Emini EA, Nara PL, Schleif WA et al. (1990) Antibody-mediated in vitro neutralization of human immunodeficiency virus type 1 abolishes infectivity for chimpanzees. *J Virol* **64:** 3674–3678.

Estaquier J, Boutillon C, Ameisen JC et al (1992) T helper cell epitopes of the human immunodeficiency virus (HIV-1) nef protein in rats and chimpanzees. *Mol Immunol* **29(4):** 489–499.

Geretti AM, Van Baalen CA, Borleffs JCC, Van Els CACM, & Osterhaus ADME (1994) Kinetics and specificities of the T helper-cell response to gp120 in the asymptomatic stage of HIV-1 infection. *Scand J Immunol* **39:** 355–362.

Goudsmit J, Debouck C, Meloen RH et al (1988) Human immunodeficiency virus type 1 neutralization epitope with conserved architecture elicits early type-specific antibodies in experimentally infected chimpanzees. *Proc Natl Acad Sci USA* **85:** 4478–4481.

Haas G, David R, Frank R et al (1991) Identification of a major human immunodeficiency virus-1 reverse transcriptase epitope recognized by mouse CD4+ T lymphocytes. *Eur J Immunol* **21:** 1371–1377.

Hale PM, Cease KB, Houghten RA et al (1989) T cell multideterminant regions in the human immunodeficiency virus envelope: toward overcoming the problem of major histocompatibility complex restriction. *Int Immunol* **4:** 409–415.

Hart MK, Palker TJ, Mattehews TJ et al (1990) Synthetic peptides containing T and B cell epitopes from human immunodeficiency virus envelope gp120 induce anti-HIV proliferative responses and high titers of neutralizing antibodies in rhesus monkeys. *J Immunol* **145:** 2677–2685.

Haynes BF, Torres JV, Langlois AJ et al (1993) Induction of HIV-MN neutralizing antibodies in primates using a prime-boost regimen of hybrid synthetic gp120 envelope peptides. *J Immunol* **151:** 1646–1653.

Hosmalin A, Nara PL, Zweig M et al (1991) Priming with T helper cell epitope peptides enhances the antibody response to the envelope glycoprotein of HIV-1 in primates. *J Immunol* **146:** 1667–1673.

Koenig S & Fauci AS (1990) AIDS immunopathogenesis and immune response to HIV. In: De Vita VT, Hellman S & Rosenberg SA (eds) *AIDS Etiology, Diagnosis, Treatment, and Prevention*, pp 61–77. Philadelphia: JB Lippincott.

Koszinowski UH, Reddehase MJ & Jonijc S (1991) The role of CD4 and CD8 T cells in viral infections. *Curr Opin Immunol* **3:** 471–475.

Krowka J, Stites D, Debs R et al (1990) Lymphocyte proliferative responses to soluble and liposome-conjugated envelope peptides of HIV-1. *J Immunol* **144:** 2535–2540.

Lamb JB, Ivanyi J, Rees ADM et al (1987) Mapping of T cell epitopes using recombinant antigens and synthetic peptides. *EMBO J* **6:** 1245–1249.

Lanzavecchia A, Roosnek E, Gregory T, Berman P & Abrignani S (1988) T cells can present antigens such as HIV gp120 targeted to their own surface molecules. *Nature* **334:** 530–532.

Levine BL, Mosca JD, Riley JL et al (1996) Antiviral effect and ex vivo CD4+ T cell proliferation in HIV-positive patients as a result of CD28 costimulation. *Science* **272:** 1939–1943.

Li Pira G, Oppezzi L, Seri M et al (1998) Repertoire breadth of human CD4 cells specific for primary (HIV gp120 and p66) and for secondary (PPD and tetanus toxoid) antigens. *Hum Immunol* **59:** 137–148.

Manca F (1994) Immune escape mutants of HIV: a hypervariable vaccine for a hypervariable virus. *Vacc Res* **3:** 93–100.

Manca F, Habeshaw J & Dalgleish A (1991a) The naive repertoire of human T helper cells specific for gp120, the envelope glycoprotein of HIV. *J Immunol* **146:** 1964–1971.

Manca F, Habeshaw JA & Dalgleish AG (1991b) The naive repertoire of human T cells specific for HIV envelope glycoprotein gp120. *J Immunol* **146**: 1964–1971.

Manca F, Habeshaw JA, Dalgleish AG, Fenoglio D, Li Pira G & Sercarz E (1993) Role of flanking variable sequences in antigenicity of consensus regions of HIV gp120 for recognition by specific human T helper clones. *Eur J Immunol* **23**: 269–274.

Manca F, Li Pira G, Fenoglio D et al (1994) Dendritic cells are potent antigen presenting cells for the in vitro induction of primary human T cell lines specific for HIV gp120. *J AIDS* **7**: 15–23.

Manca F, Fenoglio D, Valle M et al (1995a) Human CD4+ T cells can discriminate the molecular and structural context of T epitopes of HIV gp120 and HIV p66. *J AIDS* **9**: 227–237.

Manca F, Fenoglio D, Valle M et al (1995b) Human T helper cells specific for HIV reverse transcriptase: possible role in intrastructural help for HIV envelope specific antibodies. *Eur J Immunol* **25**: 1217–1223.

Manca F, De Berardinis PG, Fenoglio D et al (1996) Antigenicity of HIV-derived T helper determinants in the context of carrier recombinant proteins: effects on T helper repertoire selection. *Eur J Immunol* **26**: 2461–2469.

Manca F, Fenoglio D, Franchin E et al (1997) Anti HIV genetic therapy of antigen specific human CD4 lymphocytes for adoptive immunotherapy of AIDS. *Gene Ther* **4**: 1216–1224.

Mathiesen T, Broliden P-A, Rosen J & Wahren B (1989) Mapping of IgG subclass and T-cell epitopes on HIV proteins by synthetic peptides. *Immunology* **67**: 453–459.

Milich DR, Hughers JL, McLachlan A, Thornton GB & Moriarty A (1988) Hepatitis B synthetic immunogen comprised of nucleocapsid T-cell sites and an envelope B-cell epitope. *Proc Natl Acad Sci USA* **85**: 1610–1614.

Mutch D, Underwood J, Geysen M & Rodda S (1994) Comprehensive T-cell epitope mapping of HIV-1 env antigens reveals many areas recognized by HIV-1-seropositive and by low-risk HIV-1-seronegative individuals. *J AIDS* **7**: 879–890.

Myers G, Korber B, Wain-Hobson S, Smith RF & Pavlakis G (1993) *Human Retrovirus and AIDS. A compilation and analysis of nucleic acid and amino acid sequences.* Los Alamos: Los Alamos National Laboratory.

Palker TJ, Clark ME, Langlois AJ et al (1988) Type-specific neutralization of the human immuno-deficiency virus with antibodies to env-encoded synthetic peptides. *Proc Natl Acad Sci USA* **85**: 1932–1935.

Panina-Bordignon P, Tan A, Termijtelen A, Demotz S, Corradin G & Lanzavecchia A (1989) Universally immunogenic T cell epitopes: promiscuous binding to human MCH class II and promiscuous recognition by T cells. *Eur J Immunol* **19**: 2237–2242.

Putney SD & McKeating JA (1990) Antigenic variation in HIV. *AIDS* **4**: 129–136.

Rosenberg ES, Billingsley JM, Caliendo AM et al (1997) Vigorous HIV-1 specific CD4+ T cell responses associated with control of viremia. *Science* **278**: 1447–1450.

Rothbard J (1986) Peptides and the cellular immune response. *Ann Inst Past* **137E**: 518–526.

Rothbard JB & Gefter ML (1991) Interactions between immunogenic peptides and MHC proteins. *Annu Rev Immunol* **9**: 527–565.

Rusche JR, Javaherian K, McDanal C et al (1988) Antibodies that inhibit fusion of human immunodeficiency virus-infected cells bind a 24-amino acid sequence of the viral envelope gp120. *Proc Natl Acad Sci USA* **85**: 3198–3202.

Scherle PA & Gerhard W (1986) Functional analysis of influenza-specific helper T cell clones in vivo. T cells for internal viral proteins provide cognate help for B cell responses to haem-agglutinin. *J Exp Med* **164**: 1114–1127.

Schrier RD, Gnann JW Jr, Langlois AJ, Shriver K, Nelson JA & Oldstone MBA (1988) B- and T-lymphocyte responses to an immunodominant epitope of human immunodeficiency virus. *J Virol* **62**: 2531–2536.

Schrier RD, Gnann JW Jr, Landes R et al (1989) T cell recognition of HIV synthetic peptides in a natural infection. *J Immunol* **142**: 1166–1176.

Schwartz RH (1985) T-lymphocyte recognition of antigen in association with gene products of the major histocompatibility complex. *Annu Rev Immunol* **3**: 237–261.

Shaw GM, Wong-Staal F & Gallo RC (1988) Etiology of AIDS: virology, molecular biology, and evolution of human immunodeficiency viruses. In: DeVita VT, Hellman S & Rosenberg SA (eds) *AIDS Etiology, Diagnosis, Treatment, and Prevention*, pp 11–31. Philadelphia: JB Lippincott.

Siliciano RF (1993) CD4+ T cell epitopes in HIV-1 proteins. *Chem Immunol* **56**: 127–149.

Siliciano RF, Lawton T, Knall C et al (1988) Analysis of host–virus interactions in AIDS with anti-gp120 T cell clones: effect of HIV sequence variation and a mechanism for CD4+ cell depletion. *Cell* **54**: 561–575.

Sissons JGP & Oldstone MBA (1980) Killing of virus-infected cells by cytotoxic lymphocytes. *J Infect Dis* **142**: 114–118.

Sun P, Li Pira G, Fenoglio D et al (1994) Enhanced activation of human T cell clones specific for virus-like particles expressing the HIV V3 loop in the presence of V3 loop-specific polyclonal antibodies. *Clin Exp Immunol* **97**: 361–366.

Torseth JW, Berman PW & Merigan TC (1988) Recombinant HIV structural proteins detect specific cellular immunity in vitro in infected individuals. *AIDS Res Hum Retroviruses* **4**: 23–30.

Wahren B, Morfeldt-Månsson L, Biberfeld G et al (1987) Characteristics of the specific cell-mediated immune response in human immunodeficiency virus infection. *J Virol* **61**: 2017–2023.

Wahren B, Rosen J, Sandström E, Mathiesen T, Modrow S & Wigzell H (1989) HIV-1 peptides induce a proliferative response in lymphocytes from infected persons. *J AIDS* **2**: 448–456.

Walker CM, Steimer KS, Rosenthal KL & Levy JA (1988) Identification of human immunodeficiency virus (HIV) envelope type-specific T helper cells in an HIV-infected individual. *J Clin Invest* **82**: 2172–2175.

Chapter 7

MOLECULAR CONTROL OF PROGRAMMED CELL DEATH IN HIV INFECTION:
Contribution to the dysregulation of T cell homeostasis and to CD4 T cell depletion

Marie-Lise Gougeon and Klaus-Michael Debatin

INTRODUCTION

Human immunodeficiency virus type 1 (HIV-1) infection is a dynamic process involving high virus expression in lymphoid organs such as spleen and lymph nodes, rapid turnover of CD4 T cells and increased cell death in vivo (Pantaleo et al, 1993; Finkel et al, 1995; Ho et al, 1995). A number of pathophysiological mechanisms appear to contribute to the dramatic depletion of CD4 T lymphocytes in AIDS. Although viral load increases with disease progression, direct killing of CD4+ lymphocytes by HIV most probably cannot account for the magnitude of the loss of these cells during the course of HIV infection. Several years ago, it was suggested that CD4 T lymphocyte depletion may involve mechanisms whereby HIV infection primes a series of processes leading to programmed cell death (PCD) (Ameisen and Capron, 1991; Gougeon et al, 1991; Laurent-Crawford et al, 1991; Terai et al, 1991; Gougeon and Montagnier, 1993). Programmed cell death mediated by a process termed 'apoptosis' is the physiological cell death that occurs during embryogenesis, metamorphosis, normal tissue turnover and homeostasis. Apoptosis also plays a crucial role in negative growth control of the immune system. For example, thymocytes that have failed to rearrange their T cell receptor genes, or those that are autoreactive, are eliminated by apoptosis. The homeostatic control of peripheral lymphocytes is maintained by apoptosis which kills potentially autoreactive lymphocytes and limits the clonal expansion of lymphocytes during an immune response (Boise and Thompson, 1996). Exaggerated PCD can induce pathological situations and cause tissue destruction such as in fulminant hepatitis, in other diseases involving cytotoxic T lymphocyte (CTL)-induced tissue destruction, or in AIDS. The apoptotic pathways described below may account for depletion of CD4 T lymphocytes in AIDS patients. These pathways could be mediated either directly by virus replication as a consequence of viral gene expression, or indirectly through priming of uninfected cells to apoptosis when triggered by different agents. In addition to

HIV and the New Viruses Second Edition
ISBN 0-12-200741-7

these pathways, a complementary cytopathic effect is probably provided by the immune system, since infected cells may be killed by HIV-specific CTLs or antibody-dependent cell-mediated cytotoxicity (ADCC). The relevance of apoptosis for AIDS pathogenesis is also discussed.

APOPTOSIS IN PERIPHERAL T CELL DELETION

Regulation of cell survival and death is essential for T cell homeostasis during precursor cell development and termination of an immune response in the periphery. Cell survival may be regulated by default mechanisms in which the expression of anti-apoptotic genes, such as proteins of the Bcl-2 family, is regulated by exogenous survival factors, e.g. cytokines such as interleukin 2 (IL-2) (Broome et al, 1995; Yang and Korsmeyer, 1996; Kroemer, 1997; Reed, 1997). While the expression of survival genes seems to be critical for further development of precursor cells (positive selection) and T cell survival (Veis et al, 1993; Linette et al, 1994), elimination of T cells in the periphery to downregulate the immune response may rather involve active induction of apoptosis through interaction of 'death' receptors with their respective ligands (Nagata, 1997). Triggering of apoptosis through cell surface receptors in peripheral T cells seems critically to involve the CD95 (APO-1/Fas) system (Krammer et al, 1994; Debatin, 1996).

CD95-mediated cell death

The finding that CD95 and CD95 ligand are mutated in mouse strains suffering from severe autoimmune diseases and lymphoproliferation has greatly facilitated the understanding of the physiological role of the CD95 system in T cell homeostasis (Nagata and Golstein, 1995). Thus, mutations of the CD95 molecule in lpr mice and mutations of the CD95 ligand in gld mice constitute the first genetically defined syndromes of defective apoptosis. Recently human counterparts of the *lpr* mutation in mice have been identified (Fisher et al, 1995; Rieux-Laucat et al, 1995; Drappa et al, 1996). The clinical presentation of the patients resembles the phenotype of the lpr syndrome in mice with marked lymphadenopathy, hepatosplenomegaly and autoantibodies. As a consequence of CD95 deficiency, 'deathless' CD4+ and CD8+ single positive T lymphocytes from patients fail to undergo apoptosis following stimulation by CD95 or T cell receptor (TCR) triggering.

The 48 kD CD95 molecule is a cell surface receptor of the tumour necrosis factor receptor (TNFR) superfamily which includes various molecules involved in immune regulation, such as the TNF receptors I and II, CD27, CD30 and CD40 (Trauth et al, 1989; Itoh et al, 1991; Oehm et al, 1992). The CD95 protein structure is characterized by three extracellular cysteine-rich domains (CRDs) found in all family members, a single transmembrane spanning region and an intracellular part that contains a 70 amino acid region highly homologous to the p55 TNFR. This intracellular 'death domain' has been shown to transduce signals for apoptosis in the TNFR and the CD95 molecule (Cleveland and Ihle,1995; Peter et al, 1996). In order to induce apoptosis, CD95 receptors on the cell surface have to oligomerize (Cleveland and Ihle, 1995; Peter et al, 1996). Following multimerization of CD95, the death signal is mediated by activation of IL-1β-converting enzyme (ICE)/Ced-3 proteases or caspases, a family of cysteine proteases with a specificity for aspartic acid residues (Cleveland and Ihle, 1995; Fraser and

Evan, 1996; Peter et al, 1996). Critical elements of the CD95 pathway that link multi-merization of the receptor and downstream activation of caspases in the signal pathway have now been identified (Boldin et al, 1996; Fraser and Evan, 1996; Muzio et al, 1996; Nagata, 1997). Following receptor multimerization, the death domain-containing molecule Fas-associating protein with death domain (FADD) is attracted to the death-inducing signalling complex (DISC) (Kischkel et al, 1995). After the DISC has formed, FLICE (FADD-like ICE) is recruited into the DISC. FLICE or MACH-α is a chimeric molecule that contains an adapter domain able to bind to FADD as well as a proteolytic domain similar to ICE proteases (Boldin et al, 1996; Muzio et al, 1996). Thus, this Janus-like molecule is able to link the death domain associating molecules of the DISC to the proteolytic cascade that exerts the death signal.

The CD95 ligand is a type II transmembrane protein produced by activated T cells and constitutively expressed in a variety of tissues. A soluble form of CD95L is produced by proteolytic cleavage. A fundamental concept for the importance and the role of the CD95 system in growth control of peripheral T cells has been the demonstration of autocrine and paracrine mechanisms of CD95 ligand-mediated death (Alderson et al, 1995; Brunner et al, 1995; Dhein et al, 1995; Ju et al, 1995). Triggering TCRs in activated peripheral T cells may induce apoptosis involving autocrine suicide or paracrine death mediated by CD95 receptor/ligand interaction (Figure 7.1). Although the concept of activation-induced triggering of CD95 ligand/receptor interaction has initially been demonstrated and developed as a key mechanism for elimination of peripheral T cells, triggering of CD95 ligand/receptor interaction may occur in various cell types under experimental and pathological conditions. The recent discovery of induction of CD95 ligand expression and activation of CD95-mediated autocrine and paracrine death by cytotoxic drugs used in chemotherapy of tumours or cellular stress demonstrate that CD95 ligand production may represent a common cellular response to a variety of different stimuli (Friesen et al, 1996; Herr et al, 1997).

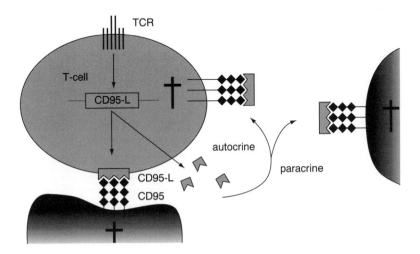

Figure 7.1 *Activation-induced T cell death*
Triggering of the T cell receptor (TCR) in activated peripheral T cells induces expression of CD95L which is found in a membrane-bound form and as a soluble molecule. Membrane-bound CD95L may mediate direct target cell cytotoxicity; the soluble form of the ligand may cause autocrine suicide or paracrine death.

CD95-mediated death depends on an apoptosis-sensitive phenotype. While CD95 is readily expressed in peripheral T cells after activation, sensitivity for CD95-mediated apoptosis requires prolonged activation of T cells (Klas et al, 1993). In addition, the majority of CD95+ human thymocytes are resistant to CD95-mediated apoptosis (Debatin et al, 1994a). Sensitivity or resistance towards CD95-induced apoptosis may depend on the ability to transmit the death signal (DISC formation) or may be modulated by differential expression of antiapoptotic proteins of the Bcl-2 family (Broome et al, 1995; Kroemer, 1997; Peter et al, 1997).

INFLUENCE OF HIV-1 GENES ON THE INDUCTION OF THE APOPTOTIC PROCESS

Role of Tat, Vpr, and gp120 viral proteins

Numerous studies have examined the effects of various HIV-1 gene products on cellular survival. Tat, a viral transcription factor, was shown to affect transcription of genes involved in cell survival. Tat was found to upregulate Bcl-2 expression, protecting cells from apoptosis (Zauli et al, 1995). In contrast, establishment of stable Tat-expressing cell lines or addition of exogenous Tat has been reported to sensitize cells to apoptosis induced by CD95, anti-TCR and anti-CD4 (Li et al, 1995, Westendorp et al, 1995). In these studies, Tat alone was insufficient to induce apoptosis but it appeared to sensitize cells to apoptosis triggered by a second signal, such as CD95 or TCR signalling. Moreover, Tat was found strongly to increase TCR-induced expression of CD95L, thereby accelerating TCR-induced suicide and paracrine death (Westendorp et al, 1995). The *Vpr* gene was also recently found to induce apoptosis (Stewart et al, 1997). Vpr is required for productive infection of non-dividing cells (Hattori et al, 1990) and it was recently shown to induce arrest of cells in the G2 phase of the cell cycle (Bartz et al, 1996). Following the arrest of cells in G2, Vpr induces apoptosis in human T cells, peripheral blood lymphocytes and fibroblasts.

The cytopathic effect of HIV in CD4 T cell cultures, manifested by ballooning of cells and formation of syncytia, was shown to be associated with apoptosis (Laurent-Crawford et al, 1991; Terai et al, 1991). This apoptosis is triggered by the viral envelope glycoprotein gp160 and it is a late process occurring when cells are actively involved in synthesis of viral proteins. Single-cycle infection experiments in the presence of zidovudine (azidothymidine, AZT) have indicated that virus adsorption and entry do not induce apoptosis and that virus replication is required in order to produce viral proteins. Cell surface expression of the viral envelope glycoprotein and accessible CD4 receptors are involved in the induction of apoptosis, and during the fusion process a specific region in the gp120–gp41 complex may become unmasked and thus mediate the onset of apoptosis. Both gp120 and gp41 are required for triggering apoptosis and no other gene besides the envelope is involved (Laurent-Crawford et al, 1993).

Chronically infected cells trigger apoptosis in bystander cells

Chronically HIV-infected cells were shown to serve as effector cells to induce apoptosis in uninfected target CD4 cells. Indeed, coculturing of chronically infected cells with uninfected cells results in the formation of syncytia, apoptosis and also cell-to-cell spread of HIV infection (Laurent-Crawford et al,1993). During this process AZT blocks the spread of HIV infection without any apparent effect on apoptosis. On the other hand, cyclosporin A, a powerful suppressor of the immune system, and cycloheximide, which inhibits protein synthesis, do not affect apoptosis. Therefore, by virtue of expression of the gp120–gp41 complex, HIV-producing cells should be considered as potent effector cells for two independent pathological consequences: the first is the cell-to-cell spread of HIV infection, which is inhibited by AZT; the second is the triggering of apoptosis, which is not affected by AZT.

These interesting observations raise the important question in HIV-1 pathogenesis: is virus killing limited to infected T cells? Non-selective apoptosis in uninfected T cells may significantly contribute to their depletion in AIDS. Indeed, recent studies in vivo and in vitro have shown that an important fraction of peripheral lymphocytes in lymph nodes and blood from HIV-infected persons is primed for apoptosis through the expression of some death factors such as CD95/CD95L, preapoptotic markers such as the enzyme tissue transglutaminase (tTG), or the downregulation of survival factors such as Bcl-2. This preapoptotic status can lead to the destruction of corresponding lymphocytes, unless survival factors prevent this process. As discussed below, an alteration in the expression of some of these factors has been reported in AIDS.

APOPTOSIS OF BYSTANDER LYMPHOCYTES: CORRELATION WITH LYMPHOCYTE ACTIVATION AND WITH DISEASE PROGRESSION

Homeostatic role of apoptosis during acute infections

Homeostasis is maintained by a complex set of regulatory processes which differ markedly in quiescent and activated cells. For example, during primary viral infection by Epstein–Barr and varicella-zoster viruses, T cell lymphocytosis is rapidly detected in the blood of patients, but it is transient as the absolute number of circulating T lymphocytes and the relative proportion of CD4+ and CD8+ cells return to normal upon resolution of the disease. It probably occurs through rapid clearance by apoptosis of the majority of activated T cell blasts in vivo, since the expanded circulating T cells expressing HLA-DR and CD45R0 in these patients die from apoptosis following short-term culture (Akbar et al, 1993). This apoptosis is thought to contribute to the rapid clearance of activated T cell blasts in vivo and to regulate the T cell lymphocytosis associated with viral infections. In order to study the relationship between hyperlymphocytosis during acute HIV infection and apoptosis, we followed in parallel ex vivo apoptosis and the respective representation of blood CD4 and CD8 T cell subsets during HIV primary infection. The increased rate of apoptosis in CD4 and CD8 T cells was associated with the control of T cell lymphocytosis and moreover apoptosis was essentially detected in activated T cells expressing the activation markers CD45R0, CD38 and HLA-DR (Figure 7.2). These observations further support the hypothesis that apoptosis plays a crucial role in homeostatic control of cell numbers following antigenic stimulation, ensuring the clearance of primed lymphocytes that are no longer required

(Akbar and Salmon, 1997). Nevertheless, this normal process of cell elimination may be detrimental for the immune system in the case of a chronic infection such as that induced by HIV.

Indeed, a general state of immune activation is rapidly observed in the asymptomatic phase of HIV infection both in lymphoid tissue and peripheral blood lymphocytes, and persists throughout the entire course of HIV disease. This is reflected by follicular hyperplasia in lymphoid tissue and the expression of activation markers such as HLA-DR, CD45R0 and CD38 in CD4 and CD8 T cells in the lymph nodes (Bofill et al, 1995; Muro-Cacho et al, 1995) and in the peripheral blood (Levacher et al, 1992; Giorgi et al, 1993). Although HIV replication is dramatically downregulated following the appearance of the specific immune response, HIV is never eliminated and its persistence associated with the unceasing expression of HIV antigens is probably the primary mechanism of the chronic stimulation of the immune system. In this regard, it has been shown that the early expansion of activated CD8 T cell subsets displaying the phenotype CD8+CD38+HLA-DR+ correlated with viral load (Bouscarat et al, 1996). This cellular activation is required for productive infection and HIV fully subverts the activation machinery for its own purposes. However, this chronic immune stimulation leads to anergy to TCR-mediated stimulation and priming for apoptosis.

Deleterious role of apoptosis during the chronic phase of HIV infection

It was reported several years ago that peripheral blood T cells from HIV-infected persons were highly prone to apoptosis induced in vitro (Gougeon et al, 1991, 1993; Groux

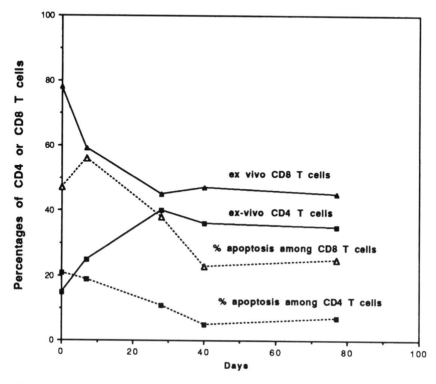

Figure 7.2A

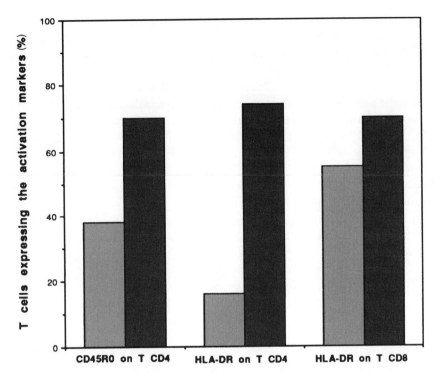

Figure 7.2B

Figure 7.2 *Apoptosis in T cells during acute HIV infection*
(A) Peripheral blood mononuclear cells from an HIV-1-infected donor were characterized for their composition in CD4 and CD8 T cell subsets and their propensity to undergo apoptosis at various times following HIV infection. Time 0 corresponds to the diagnosis of acute primary HIV infection. The transient decrease in CD4 T cells and increase in CD8 T cells was already observed. Apoptosis was measured ex vivo after short-term culture (Gougeon et al, 1996).
(B) The expression of activation markers CD45R0 and HLA-DR was followed on living ■ and apoptotic ■ cells and quantified by FACScan analysis (Gougeon et al, 1996).

et al, 1992; Meyaard et al, 1992). Indeed, while freshly isolated peripheral blood mononuclear cells (PBMCs) from HIV-infected individuals showed a low level of apoptosis (measured by different approaches detecting alteration in membrane permeability, drop in mitochondrial membrane potential, chromatin condensation or DNA fragmentation), comparable with that of control donors (Gougeon et al,1996), their incubation in medium alone induced a rapid spontaneous apoptosis which was detected after a few hours of culture. This premature cell death could concern more than 30% of lymphocytes from an HIV-infected subject, whereas it concerned only 2–5% of lymphocytes from control subjects. Moreover, the rate of apoptosis in blood lymphocytes from HIV-infected persons could be significantly increased following stimulation by various stimuli, including ionomycin, mitogens, superantigens or anti-TCR antibodies, whereas these stimuli had marginal effect on the majority of lymphocytes from control donors (Gougeon et al, 1991, 1993; Groux et al, 1992; Meyaard et al, 1992).

Although it was first reported that this priming for apoptosis in HIV-infected patients exclusively concerned the CD4 subset (Groux et al, 1992), it became rapidly clear that the CD8 subset was similarly primed for apoptosis (Meyaard et al, 1992; Gougeon et al, 1993; Lewis et al, 1994). Moreover, a phenotypic study of apoptotic cells on a large cohort of HIV-positive patients revealed that not only T cells but all mononuclear cells, including B cells, natural killer cells, granulocytes and monocytes, had an increased fragility upon short-term culture (Gougeon et al, 1996). These observations were confirmed by recent analyses of lymph nodes from patients demonstrating that apoptosis occurred in vivo and was detected not only in CD4 but also in CD8, B cells and dendritic cells (Muro-Cacho et al, 1995; Amendola et al, 1996). The relationship between increased lymphocyte apoptosis in AIDS and the activation state of the immune system was strongly suggested by the fact that the majority of apoptotic cells exhibit an activated phenotype (Meyaard et al, 1994; Gougeon et al, 1996), and in vivo the expanded activated subsets expressing CD45R0, HLA-DR or CD38 molecules appeared more prone to apoptosis in infected persons compared with controls. A statistically significant correlation was found for both CD4 and CD8 subsets between the intensity of TCR-triggered apoptosis and their activation state in vivo (Gougeon et al, 1996). Therefore in vivo lymphocyte activation may be the primary mechanism responsible for the premature cell death in patients' lymphocytes.

Although the rate of apoptosis in T cells from HIV-infected patients appears independent of viraemia (Rothen et al, 1997; M. Gougeon and H. Lecoeur, unpublished observations), it is clearly correlated with disease evolution (Gougeon et al, 1996, 1997). In support of this correlation, apoptosis was investigated in two particular groups of patients: the long-term non-progressors (LTNPs) with a CD4 T cell count above 1000 µl and no AIDS-associated symptoms at least 10 years after primary infection with HIV, and the rapid progressors (RPs) with a CD4 T cell count below 200 per µl and appearance of AIDS within 2 years of HIV infection. In both groups, all the mononuclear subsets were involved in the apoptotic process, but the degree of apoptosis was strikingly different: a background level of apoptosis comparable with that of HIV-negative persons was detected in LTNPs, which contrasted with the high degree of apoptosis detected in the RP group (M.L. Gougeon et al, unpublished data). Moreover, several studies in animal models, including macaques or African green monkeys infected with simian immunodeficiency virus (SIV) (Gougeon et al, 1993; Estaquier et al, 1994), chimpanzees infected with HIV or SIV_{cpz} (Gougeon et al, 1993, 1997), cats infected with feline immunodeficiency virus (Bishop et al, 1993), and murine AIDS (Cohen et al, 1993), revealed that increased lymphocyte apoptosis was observed only in pathogenic models of lentiviral infection.

MOLECULAR CONTROL OF APOPTOSIS IN HIV INFECTION

After activation by an antigen, the survival of a lymphocyte is tightly regulated. In the absence of appropriate signals for progression through the cell cycle or differentiation into effectors, antigen receptor engagement may induce apoptosis. Antigen activation induces the expression of cell death effectors such as CD95 and CD95L and leads to sensitivity to CD95-mediated apoptosis. However, since antigen activation does not immediately cause cell death, signal transduction pathways must exist to inhibit apoptosis. The coreceptors such as CD28 on T cells enhance lymphocyte survival inducing high level of the survival factors, Bcl-2 and Bcl-xL (Boise et al, 1995). Some cytokines, such as IL-2, also provide survival signals (Broome et al, 1995). However, whether co-stimulatory receptors will enhance lymphocyte survival or cell death effectors will induce lymphocyte apoptosis depends on the timing and on the level of signal transduction, suggesting a hierarchical control of lymphocyte survival (Boise and Thompson, 1996). As detailed below, in the case of HIV infection the downregulation of Bcl-2 expression associated with an upregulation of CD95 and CD95L expression and alteration in cytokine production will favour the apoptotic pathway rather than lymphocyte survival.

Downregulation of Bcl-2 expression in potential cytotoxic CD8 T cells: functional consequences

In mammals, the proto-oncogene *bcl-2*, involved in human tumours, has been found to function as a repressor of apoptosis in multiple cell types (Vaux, 1993; Yang and Korsmeyer, 1996; Kroemer, 1997; Reed, 1997). It appears to be unique among oncogenes because of the particular localization of its 26 kDa encoded protein to mitochondria, endoplasmic reticulum and perinuclear membranes, and in that it inhibits PCD by enhancing cell survival rather than by accelerating the rate of cellular proliferation (Hockenbery et al, 1990; Jacobson et al, 1993). Both Bcl-2 and its homologous proteins play a key role in the control of cell death of T and B cell lineages during lymphoid development, ensuring their appropriate selection (Linette et al, 1994; Nunez et al, 1994; Reed, 1997). Recent findings also suggest that regulation of Bcl-2 expression in mature T lymphocytes may be crucial for the development and persistence of a memory T cell response following an immune activation (Akbar and Salmon, 1997).

Recent studies suggested that PCD in HIV infection is under the control of the Bcl-2 family. Indeed, consistent decrease of Bcl-2 molecule expression ex vivo in a fraction of freshly isolated CD8+ T cells from HIV-positive persons was reported, and the proportion of low Bcl-2 CD8 T cells was particularly important in HIV-infected subjects at advanced stage of the disease. This downregulation in vivo of Bcl-2 expression was not detected in lymphocytes from control donors. Interestingly, the downregulation of Bcl-2 in CD8+ T lymphocytes was correlated with the propensity of these cells to undergo spontaneous apoptosis in short-term culture, which was enhanced through ligation of the Fas antigen. Ex vivo phenotypic characteristics of these cells suggested that they were cytotoxic since they were in an activated state, expressed the TIA-1 granules and were CD28 negative (Boudet et al, 1996). A parallel study performed by Bofill et al (1995) in lymph nodes of HIV-infected patients confirmed the existence of these cells in lymphoid organs in which a high proportion of LN CD8+ CD45R0+ cells display low levels of Bcl-2 and express TIA-1.

A strong expansion of circulating HIV-specific CTLs in HIV-positive donors associated with expression of activation markers has been described by several groups (Walker and Plata, 1990; Carmichael et al, 1993; Autran et al, 1996). This cytotoxic response was reported to be markedly lost on the onset of symptoms (Pantaleo et al, 1990), and in particular the absence of anti-HIV Gag CTL activity in HIV-positive people correlated with a reduced ability of these cells to expand in vitro (Watret et al, 1993). Such a decline in cytotoxic function in patients with advanced disease may therefore be associated with their priming for apoptosis, including a reduction of Bcl-2 expression. The molecular mechanisms by which these cytotoxic lymphocytes could be deleted remains unknown but the downregulation of Bcl-2 in patients' CD8 T cells was found to correlate with an increased expression of CD95 and a high susceptibility to apoptosis triggered through CD95 ligation (Boudet et al, 1996). It is thus likely that in vivo encounter of such CD95+CD8+ T cells with the CD95L would result in their cytolysis following CD95/CD95L interaction. Apoptosis might occur as a paracrine process or as autocrine suicide through a cell-autonomous Fas/Fas-L interaction (Brunner et al, 1995; Dhein et al, 1995). Studies have reported upregulation of CD95L on CD4 and CD8 cells from patients (Baümler et al, 1996) and soluble CD95L was found in greater amounts in plasma from patients (Sloand et al, 1997). In addition to the influence of CD95/CD95L death factors on the deletion of low Bcl-2 CD8 T cells, the lack of survival factors may also contribute to the apoptosis of this subset. For example, it was reported that IL-2 could upregulate Bcl-2 expression in activated lymphocytes from patients infected with Epstein–Barr virus (Akbar et al, 1993).The defective production of IL-2 linked to the progressive depletion of the CD4 T cell subset in HIV infection (Ledru et al, 1998) would therefore prevent an upregulation of Bcl-2 protein in vivo, so that these cells could not be rescued from apoptosis. Therefore, the cellular dysfunction of the CD8 subset and the further loss of CD8+ cytotoxic activity in the course of HIV infection may be related to an abnormal priming for apoptosis following persistent immune stimulation and the gradual loss of growth factors.

Influence of the CD95 system on the deletion of CD4 T cells

The CD95 system has an important role in the termination of an immune response, deletion of autoreactive T cell clones and elimination of virus-infected cells, and deregulation in the CD95 pathway is associated with pathological conditions in mice (Nagata and Golstein, 1995; Nagata and Suda, 1995) and humans (Fisher et al, 1995; Rieux-Laucat et al, 1995; Debatin, 1996; Drappa et al, 1996). Hyperactivation of the CD95 pathway during viral infection may result in the elimination of infected cells as well as the destruction of non-infected cells, and it was proposed to be one of the mechanisms involved in T lymphocyte depletion in AIDS. The involvement in vivo of the CD95 pathway in T cell apoptosis during HIV infection is supported by reports showing an increase on patients' lymphocytes of CD95 and CD95L molecules. The CD95 molecule (Debatin et al, 1994b; Katsikis et al, 1995), CD95L encoding transcripts (Baümler et al, 1996) and CD95L cell surface molecules (Sloand et al, 1997) were detected in both CD4 and CD8 T cells from HIV-infected patients. This was associated with an increased susceptibility of patients' T cells to apoptosis following ligation of CD95 by agonist antibodies or soluble human CD95L (Katsikis et al, 1995; Baümler et al, 1996; Boudet et al, 1996; Gougeon et al, 1997). Moreover, the proportion of compliant CD95-expressing cells dramatically increased with disease progression (Katsikis et al, 1995; Böhler et al, 1997c; Gougeon et al, 1997). Interestingly, strongly increased CD95 expression and CD95 sensitivity seem to be confined to primed/memory (CD45R0) T cells, suggesting that virus-driven T cell activation is responsible for the increased apoptosis (Böhler et al, 1997a). This is also supported by findings of a strong downregulation of increased CD95L expression in T cells from patients following effective decrease in viral load by antiretroviral combination therapy (Böhler et al, 1997b).

In addition to mechanisms of accelerated suicide or paracrine death, CD95-induced apoptosis during HIV infection may be mediated by cytotoxic effector cells. The magnitude of the CTL response during HIV infection and the high proportion of CD95+ compliant target cells among patients' lymphocytes argue for a deleterious role of CTLs on activated T cells (Garcia et al, 1997). Two pathways for CTL-mediated cytotoxicity were described, the perforin/granzyme pathway being rather involved in antiviral host defence and the CD95L pathway being involved in the elimination of activated cells during an immune response (Kagi et al, 1995; Lynch et al, 1995). However, Rouvier et al (1993) demonstrated the coexistence of these two cytotoxic mechanisms in a single murine clone. In order to evaluate whether specific antiviral professional CTLs are potential effectors of the destruction of CD95+ compliant cells, we tested the ability of an anti-Nef CTL clone (derived from PBMCs from an HIV-infected patient) to kill CD95+ targets while maintaining its HLA-restricted perforin-mediated cytotoxic activity. Indeed both cytotoxic activities were found to be coexpressed in this clone, associated with the dual expression of perforin and CD95L (Garcia et al, 1997). In the context of HIV pathogenesis, these observations suggest that continuously activated antiviral CTLs would constantly express CD95L and therefore be able to kill both virus-infected cells and non-infected, activated compliant CD95+ target cells.

RELEVANCE OF APOPTOSIS FOR AIDS PATHOGENESIS: IS PCD THE CAUSE OF AIDS OR ITS CONSEQUENCE?

One strategy to approach this question is to study chimpanzees which – in contrast to humans – maintain immunological integrity in the face of persistent infection (Heeney, 1995). Studies performed on lymphocytes from HIV-1-infected chimpanzees showed that the intensity of T cell apoptosis was very low (Gougeon et al, 1993; Estaquier et al, 1994). Interestingly, in spite of CD95 expression, lymphocytes from chimpanzees are totally resistant to apoptosis following CD95 ligation. This resistance is correlated with the lack of inappropriate immune activation in vivo and a restricted viral load, less than a tenth of that in humans (Gougeon et al, 1997). Several mechanisms can be evoked arguing for apoptosis as a consequence of the chronicity of a viral infection.

1. Continuous production of viral proteins would induce apoptosis either directly, by inducing a cell death signal, or indirectly, by influencing activation of the immune system.
2. The suggested rapid turnover of CD4 T cells in HIV-infected persons due to an active regenerative process may contribute significantly to the rate of apoptosis in patients (Ho et al, 1995). Owing to an absence of CD4 depletion in chimpanzees and a low viral load (Heeney, 1995), this rapid CD4 cell turnover might not occur in infected chimpanzees.
3. The impaired production of T helper cell type 1 (Th1) cytokine, such as IL-2 and IL-12, would prevent cell rescue from apoptosis (Gougeon et al, 1993; Clerici et al, 1994; Ledru et al, 1998). In chimpanzees, no alteration of the Th1 subset was detected (Gougeon et al, 1997).
4. Inappropriate signalling by MHC class II antigen-presenting cells may contribute to anergy and apoptosis of T cells in infected humans (Meyaard et al, 1994). While the integrity of the Th MHC class II microenvironment is altered in lymphoid tissues of infected humans, it is preserved in infected chimpanzees (Heeney, 1995).

Apoptosis can significantly contribute to AIDS pathogenesis. Since it was shown in vivo to involve mostly non-infected lymphocytes (Finkel et al, 1995), it could be the mechanism responsible for the clearance of activated but healthy T cells and consequently contribute to the impoverishment of the pool of effectors and memory cells. The exacerbation of the CD95 system would drive antiviral CTLs in a deleterious role and particularly in the physiological elimination of CD4 T cells, irrespective of the infected status. This would contribute to the collapse of the immune system.

REFERENCES

Akbar AN & Salmon M (1997) Cellular environments and apoptosis: tissue microenvironments control activated T-cell death. *Immunol Today* **18:** 72.

Akbar AN, Borthwick N, Salmon M et al (1993) The significance of low bcl-2 expression by CD45R0 T cells in normal individuals and patients with acute viral infections. The role of apoptosis in T cell memory. *J Exp Med* **178:** 427.

Alderson MR, Tough TW, Davis-Smith T et al (1995) Fas ligand mediates activation induced cell death in human T lymphocytes. *J Exp Med* **181:** 71.

Ameisen JC & Capron A (1991) Cell dysfunction and depletion in AIDS: the program cell death hypothesis. *Immunol Today* **12:** 102.

Amendola A, Gougeon ML, Poccia F, Bondurand A, Fesus L & Piacentini M (1996) Induction of tissue transglutaminase in HIV pathogenesis. Evidence for a high rate of apoptosis of CD4 T lymphocytes and accessory cells in lymphoid tissues. *Proc Natl Acad Sci USA* **93:** 11057.

Autran B, Hadida F & Haas G (1996) Evolution and plasticity of CTL responses against HIV. *Curr Opin Immunol* **8:** 546.

Bartz SR, Rogel ME & Emerman M (1996) HIV-1 cell cyle control: vpr is cytostatic and mediates G2 accumulation by a mechanism which differs from DNA damage checkpoint control. *J Virol* **70:** 2324.

Bäumler CB, Böhler T, Herr I, Benner A, Krammer PH & Debatin KM (1996) Activation in the CD95 (APO-1/Fas) system in T cells from human immunodeficiency virus type-1-infected children. *Blood* **88:** 1741.

Bishop SA, Gruffydd-Jones TJ, Harbour DA & Stokes CR (1993) Programmed cell death as a mechanism of cell death in PBMC from cats infected with feline immunodeficiency virus (FIV). *Clin Exp Immunol* **93:** 65.

Bofill M, Gombert W, Borthwick NJ et al (1995) Presence of CD3+CD8+Bcl-2^low lymphocytes undergoing apoptosis and activated macrophages in lymph nodes of HIV-1+ patients. *Am J Pathol* **146:** 1542.

Böhler T, Nedel S & Debatin KM (1997a) CD95-induced apoptosis contributes to loss of primed/memory but not resting/naive T cells in HIV-1 infected children. *Pediatr Res* **41:** 878.

Böhler T, Herr I, Geiss M, Haas J & Debatin KM (1997b) Downregulation of increased CD95 (APO-1/Fas) ligand expression in T cells following antiretroviral therapy in HIV-1 infected children. *Blood* **90:** 886.

Böhler T, Bäumler C, Herr I, Groll A, Kurz & Debatin KM (1997c) Activation of the CD95 system increases with disease progression in HIV1-infected children and adolescents. *Pediatr Infect Dis J* **16:** 754.

Boise LH & Thompson CB (1996) Hierarchical control of lymphocyte survival. *Science* **274:** 67.

Boise LH, Minn AJ, Noel PJ, June CH, Accavitti MA, Lindsten T & Thompson CB (1995) CD28 costimulation can promote T cell survival by enhancing the expression of Bcl-XL. *Immunity* **3:** 87.

Boldin MP, Goncharov TM, Goltsev YV & Wallach D (1996) Involvement of MACH, a novel MORT1/FADD-interacting protease, in Fas/APO-1- and TNF receptor-induced cell death. *Cell* **85**: 803.

Boudet F, Lecoeur H & Gougeon ML (1996) Apoptosis associated with ex vivo down-regulation of bcl-2 and up-regulation of Fas in potential cytotoxic CD8+ T lymphocytes during HIV infection. *J Immunol* **156**: 2282.

Bouscarat F, Levacher-Clergeot M, Dazza C et al (1996) Correlation of CD8 lymphocyte activation with cellular viremia and plasma HIV RNA levels in asymptomatic patients infected by HIV-1. *AIDS Res Hum Retroviruses* **12**: 17.

Broome HE, Dargan CM, Krajewski S & Reed JC (1995) Expression of Bcl-2, Bcl-x and Bax after T cell activation and IL-2 withdrawal. *J Immunol* **155(5)**: 2311.

Brunner T, Mogil RJ, LaFace D et al (1995) Cell-autonomous Fas (CD95)/Fas ligand interaction mediates activation induced apoptosis in T cell hybridomas. *Nature* **373**: 441–444.

Carmichael A, Jin X, Sissons P & Borysiewicz L (1993) Quantitative analysis of the human immunodeficiency virus type 1 (HIV-1)-specific cytotoxic T lymphocyte (CTL) response at different stages of HIV-1 infection: differential CTL responses to HIV-1 and Epstein–Barr virus in late disease. *J Exp Med* **177**: 249.

Clerici M, Sarin A, Coffman RL et al (1994) Type 1/type 2 cytokine modulation of T cell programmed cell death as a model for HIV pathogenesis. *Proc Natl Acad Sci USA* **91**: 11811.

Cleveland JL & Ihle JN (1995) Contenders in FasL/TNF death signaling. *Cell* **81**: 479.

Cohen DA, Fitzpatrick EA, Barve SS et al (1993) Activation-dependent apoptosis in CD4+ T cells during murine AIDS. *Cell Immunol* **151**: 392.

Debatin KM (1996) Disturbances of the CD95 (APO-1/Fas) system in disorders of lympho-hematopoetic cells. *Cell Death Diff* **3(2)**: 185.

Debatin KM, Süss D & Krammer PH (1994a) Differential expression of APO-1 on human thymocytes: implications for negative selection. *Eur J Immunol* **24**: 753.

Debatin KM, Fahrig-Faissner A, Enenkel-Stoodt S, Kreuz W, Benner A & Krammer PH (1994b) High expression of APO-1 (CD95) on T lymphocytes from HIV-infected children. *Blood* **83**: 3101.

Dhein J, Walczak H, Baumler C, Debatin KM & Krammer PH (1995) Autocrine T-cell suicide mediated by APO-1/(Fas/CD95). *Nature* **373**: 438.

Drappa J, Vaishnaw AK, Sullivan KE, Chu JL & Elkon KB (1996) Fas gene mutations in the Canale-Smith syndrome, an inherited lymphoproliferative disorder associated with autoimmunity. *N Engl J Med* **335**: 1643–1649.

Estaquier JT, Idziorek F, De Bels F et al (1994) Programmed cell death and AIDS: significance of T cell apoptosis in pathogenic and nonpathogenic primate lentiviral infections. *Proc Natl Acad Sci USA* **91**: 9431.

Fisher GH, Rosenberg FJ, Straus SE et al (1995) Dominant interfering Fas gene mutations impair apoptosis in a human autoimmune lymphoproliferative syndrome. *Cell* **81**: 935.

Fraser A & Evan G (1996) A license to kill. *Cell* **85**: 781.

Friesen C, Herr I, Krammer PH & Debatin KM (1996) Involvement of the CD95 (APO-1/Fas) receptor/ligand system in drug induced apoptosis in leukemia cells. *Nature Med* **2(5)**: 574.

Finkel TH, Tudor-Williams G, Banda NK et al (1995) Apoptosis occurs predominantly in bystander cells and not in productive cells of HIV- and SIV-infected lymphnodes. *Nature Med* **1**: 129.

Garcia S, Fevrier M, Dadaglio G, Lecoeur H, Riviere Y & Gougeon ML (1997) Potential deleterious effect of anti-viral cytotoxic lymphocytes through the CD95 (Fas/APO-1)-mediated pathway during chronic HIV infection. *Immunol. Lett* **57**: 53.

Giorgi JV, Liu Z, Hultin LE, Cumberland WG, Hennessey K & Detels R (1993) Elevated level of CD38+CD8+ cells in HIV infection add to the prognostic value of low CD4+T cell level: results of 6 years follow-up. *J AIDS* **6**: 904.

Gougeon ML (1995) Does apoptosis contribute to CD4 T cell depletion in HIV infection? *Cell Death Diff* **2**: 1.

Gougeon ML & Montagnier L (1993) Apoptosis in AIDS. *Science* **260**: 1269.

Gougeon ML, Olivier R, Garcia S et al (1991) Demonstration of an engagement process towards cell death by apoptosis in lymphocytes of HIV infected patients. *C R Acad Sci* **312**: 529.

Gougeon ML, Garcia S, Heeney J et al (1993) Programmed cell death in AIDS-related HIV and SIV infections. *AIDS Res Hum Retroviruses* **9**: 553.

Gougeon ML, Lecoeur H, Dulioust A et al (1996) Programmed cell death in peripheral lymphocytes from HIV-infected persons: the increased susceptibility to apoptosis of CD4 and CD8 T cells correlates with lymphocyte activation and with disease progression. *J Immunol* **156**: 3509.

Gougeon ML, Lecoeur H, Boudet F et al (1997) Lack of chronic immune activation in HIV-infected chimpanzees correlates with the resistance of T cells to Fas/Apo-1 (CD95)-induced apoptosis and preservation of a Th1 phenotype. *J Immunol* **158**: 2964.

Groux H, Torpier G, Monté D, Mouton Y, Capron A & Ameisen JC (1992) Activation-induced death by apoptosis from human immunodeficiency virus-infected asymptomatic individuals. *J Exp Med* **175**: 331.

Hattori N, Michaels F, Fargnoli K et al (1990) The human immunodeficiency virus type 2 *vpr* gene is essential for productive infection of human macrophages. *Proc Natl Acad Sci USA* **87**: 8080.

Heeney JL (1995) AIDS: a disease of impaired Th-cell renewal? *Immunol Today* **16**: 515.

Herr I, Böhler T, Wilhelm D, Angel P & Debatin KM (1997) Activation of the CD95 (Apo-1/Fas) pathway by ceramide mediates radio- and chemotherapy induced apoptosis. *EMBO J* **16**: 6200.

Ho DD, Neumann AU, Perelson AS, Chen W, Leonard JM & Markowitz M (1995) Rapid turnover of plasma virions and CD4 lymphocytes in HIV-1 infection. *Nature* **373**: 123.

Hockenbery D, Nunez G, Milliman C, Schreiber RD & Korsmeyer SJ (1990) Bcl-2 is an inner mitochondrial membrane protein that blocks programmed cell death. *Nature* **348**: 334.

Itoh N, Yonehara S, Ishii A et al (1991) The polypeptide encoded by the cDNA for human cell surface antigen Fas can mediate apoptosis. *Cell* **66**: 233.

Jacobson MD, Burne JF, King MP, Miyashita T, Reed JC & Raff MC (1993) Bcl-2 blocks apoptosis in cells lacking mitochondrial DNA. *Nature* **361**: 365.

Ju ST, Panka DJ, El-Khatib M, Sheer DH, Stanger BZ & Marshak-Rothstein A (1995) Fas (CD95)/FasL interactions required for programmed cell death after T cell activation. *Nature* **373**: 444.

Kagi D, Ledermann B, Burki K, Zinkernage IRM & Hengartner H (1995) Lymphocyte-mediated cytotoxicity in vitro and in vivo: mechanisms and significance. *Immunol Rev* **146**: 95–115.

Katsikis PD, Wunderlich ES, Smith CA, Herzenberg LA & Herzenberg LA (1995) Fas antigen stimulation induces marked apoptosis of T lymphocytes in human immunodeficiency virus-infected individuals. *J Exp Med* **181**: 2029.

Kischkel FC, Hellbardt S, Behrmann I et al (1995) Cytotoxicity-dependent APO-1(Fas/CD95) associated proteins from a death-inducing signaling complex (DISC) with the receptor. *EMBO J* **14**: 5579.

Klas C, Debatin KM, Jonker RR & Krammer PH (1993) Activation interferes with the APO-1 pathway in mature human T cells. *Int Immunol* **5**: 625.

Krammer PH, Dhein J, Walczak H et al (1994) The role of APO-1 mediated apoptosis in the immune system. *Immunol Rev* **142**: 175.

Kroemer G (1997) The proto-oncogene Bcl-2 and its role in regulating apoptosis. *Nature Med* **3**(6): 614.

Laurent-Crawford AG, Krust B, Muller S et al (1991) The cytopathic effect of HIV is associated with apoptosis. *Virology* **185**: 829.

Laurent-Crawford AG, Krust B, Riviere Y et al (1993) Membrane expression of HIV envelope glycoproteins triggers apoptosis in CD4 cells. *AIDS Res Hum Retroviruses* **9**: 761.

Ledru E, Lecoeur H, Garcia S, Roué R & Gougeon ML (1998) Differential susceptibility to activation-induced apoptosis among peripheral Th1 subsets. Correlation with Bc1-2 expression and consequences for AIDS pathogenesis. *J Immunol* **160**: 3194.

Levacher M, Hulstaert F, Tallet S, Ullery S, Pocidalo JJ & Bach BA (1992) The significance of activation markers on CD8 lymphocytes in human immunodeficiency syndrome: staging and prognostic value. *Clin Exp Immunol* **90**: 376–382.

Lewis DE, Tang DS, Adu-Oppong A, Schober W & Rodgers JR (1994) Anergy and apoptosis in CD8+ T cells from HIV-infected persons. *J Immunol* **153**: 412.

Li CJ, Friedman DJ, Wang C, Metelev V & Pardee AB (1995) Induction of apoptosis in uninfected lymphocytes by HIV-1 Tat protein. *Science* **268**: 429.

Linette GP, Grusby MJ, Hedrick SM, Hansen TH, Glimcher LH & Korsmeyer SJ (1994) Bcl-2 is upregulated at the CD4+CD8+ stage during positive selection and promotes thymocyte differentiation at several control points. *Immunity* **1**: 197.

Lynch DH, Ramsdell F & Alderson MR (1995) Fas and FasL in the homeostatic regulation of immune responses. *Immunol Today* **16**: 569.

Meyaard L, Otto SA, Jonker RR, Mijnster RJ, Keet RP & Miedema F (1992) Programmed death of T cells in HIV-1 infection. *Science* **257**: 217.

Meyaard L, Otto SA, Keet IPM, Roos MTL & Miedema F (1994) Programmed death of T cells in HIV infection: no correlation with progression to disease. *J Clin Invest* **93**: 982.

Motoyama N, Wang F, Roth KA et al (1995) Massive cell death of immature hematopoetic cells and neurons in Bcl-xL-deficient mice. *Science* **267**: 1506.

Muro-Cacho, CA, Pantaleo G & Fauci A (1995) Analysis of apoptosis in lymph nodes of HIV-infected persons. Intensity of apoptosis correlates with the general state of activation of the lymphoid tissue and not with the stage of disease or viral burden. *J Immunol* **154**: 5555.

Muzio M, Chinnaiyan AM, Kischkel FC et al (1996) FLICE, a novel FADD-homologous ICE/Ced-3-like protease, is recruited to the CD95 (Fas/APO-1) death inducing signaling complex. *Cell* **85**: 817.

Nagata S (1997) Apoptosis by death factor. *Cell* **88**: 355.

Nagata S & Golstein P (1995) The Fas death factor. *Science* **267**: 1449.

Nagata S & Suda T (1995) Fas and Fas ligand: *lpr* and *gld* mutations. *Immunol Today* **16**: 39.

Nunez G, Merino R, Grillot D & Gonzalez-Garcia M (1994) Bcl-2 and Bcl-x: regulatory switches for lymphoid death and survival. *Immunol Today* **15**: 582.

Oehm A, Behrmann I, Falk W et al (1992) Purification and molecular cloning of the APO-1 cell surface antigen, a member of the tumor necrosis factor/nerve growth factor receptor superfamily. Sequence indentity with the Fas antigen. *J Biol Chem* **267**: 10709.

Pantaleo G, De Maria A, Koenig S et al (1990) CD8+ T lymphocytes of patients with AIDS maintain normal broad cytolytic function despite the loss of human immunodeficiency virus-specific cytotoxicity. *Proc Natl Acad Sci USA* **87**: 4818.

Pantaleo G, Graziosi C, Demarest JF et al (1993) HIV infection is active and progressive in lymphoid tissue during the clinically latent stage of disease. *Nature* **362**: 355.

Peter ME, Kischkel FC, Hellbardt S, Chinnaiyan AE, Krammer PH & Dixit VM (1996) CD95 (APO-1/Fas) associating proteins and other members of this protein family. *Cell Death Diff* **3**: 161.

Peter ME, Kischkel FC, Scheuerpflug CG, Medema JP, Debatin KM & Krammer PH (1997) Resistance of cultured peripheral T cells towards activation-induced cell death (AICD) involves a lack of recruitment of FLICE to the death-inducing signaling complex (DISC). *Eur J Immunol* **27**: 1207.

Reed JC (1997) Double identity for proteins of the Bcl-2 family. *Nature* **387**: 773.

Rieux-Laucat F, Le Deist F, Hivroz C et al (1995) Mutations in Fas associated with human lymphoproliferative syndrome and autoimmunity. *Science* **268:** 1347.

Rothen M, Gratzl S, Hirsc HH & Moroni C (1997) Apoptosis in HIV-infected individuals is an early marker occurring independently of high viremia. *AIDS Res Hum Retroviruses* **13:** 771.

Rouvier E, Luciani MF & Golstein P (1993) Fas involvement in Ca(2+)-independent T cell-mediated cytotoxicity. *J Exp Med* **177:** 195.

Sloand EM, Young NS, Kumar P, Weichold FF, Sato T & Maciejewski JP (1997) Role of Fas ligand and receptor in the mechanism of T-cell depletion in AIDS: effect on CD4+ lymphocyte depletion and HIV replication. *Blood* **89:** 135.

Stewart SA, Poon B, Jowett JBM & Chen ISY (1997) HIV-1 Vpr induces apoptosis following cell cyle arrest. *J Virol* **71:** 5579.

Terai C, Kornbluth RS, Pauza CD, Richman DD & Carson DA (1991) Apoptosis as a mechanism of cell death in cultured T lymphoblasts acutely infected with HIV-1. *J Clin Invest* **87:** 1710.

Trauth BC, Klas C, Peters AMJ et al (1989) Monoclonal antibody-mediated tumor regression by induction of apoptosis. *Science* **245:** 301.

Vaux DL (1993) Toward an understanding of the molecular mechanisms of physiological cell death. *Proc Natl Acad Sci USA* **90:** 786.

Veis DJ, Sorenson CM, Shutter JR & Korsmeyer SJ (1993) Bcl-2 deficient mice demonstrate fulminant lymphoid apoptosis, polycystic kidneys, and hypopigmented hair. *Cell* **75:** 229.

Walker BD & Plata F (1990) Cytotoxic T lymphocytes against HIV. *AIDS* **4:** 177.

Watret KC, Whitelaw JA, Froebel KS & Bird AG (1993) Phenotypic characterization of CD8+ T cell populations in HIV disease and in anti-HIV immunity. *Clin Exp Immunol* **92:** 93.

Wei X, Ghosh SK, Taylor ME et al (1995) Viral dynamics in human immunodeficiency virus type 1 infection. *Nature* **373:** 117.

Westendorp MO, Frank R, Oschsenbauer C et al (1995) Sensitization of T cells to CD95-mediated apoptosis by HIV-1 Tat and gp120. *Nature* **375:** 497.

Yang E & Korsmeyer SJ (1996) Molecular thanatopsis: a discourse on the Bcl-2 family and cell death. *Blood* **88:** 386.

Zauli G, Gibellini D, Caputo A et al (1995) The HIV-1 tat protein upregulates bcl-2 gene expression in Jurkat T-cell lines and primary peripheral blood mononuclear cells. *Blood* **86:** 3823.

Chapter 8

THE RELATIONSHIPS BETWEEN CYTOKINES, COMPLEMENT AND HIV INFECTION

Mario Clerici, Alberto Clivio and Gene M. Shearer

INTRODUCTION

The immunologic hallmark of successful infection with the human immunodeficiency virus (HIV) is a progressive decline in CD4 T cell count which ultimately results in the appearance of infections characteristic of acquired immune deficiency syndrome (AIDS). However, prior to this decline in CD4+ T cell number, impairment of T helper cell (Th) function is observed. Because a major function of Th is to produce cytokines that activate different effector components, analysis of the changes in cytokine production permits identification of multiple Th defects which result from HIV infection and are predictive of disease progression. In particular, defects in antigen and mitogen-stimulated interleukin (IL) 2 production were shown to be predictive for clinical end-points such as time to AIDS and time to death. Based on these early immunologic findings, we suggested that a decrease in IL-2, IL-12 and interferon gamma (IFN-γ) and an increase in IL-4, IL-5, IL-6 and IL-10 production are critical in HIV-infected individuals progressing to AIDS. We also suggested that IL-2, IL-12 and IFN-γ be grouped together (type 1 cytokines) as those cytokines that mainly stimulate cell-mediated immunity. In contrast, IL-4, IL-5, IL-6 and IL-10 (type 2 cytokines) primarily enhance antibody production and humoral immunity. Thus, a cytokine shift from type 1 to type 2 was postulated to be the immunologic consequence of HIV. Alterations in cytokine production could contribute to HIV disease progression by modulating antigen-induced cell death (AICD), which has recently been reported to occur preferentially in Th1 cells. Additionally, type 1 and type 2 cytokines could influence disease progression secondarily via a differential ability to regulate the production of soluble mediators, such as β chemokines (CC) and the CD8-derived antiviral factor (CAF), which are soluble factors that are capable of downregulating HIV infection in vitro.

DEFECTIVE IL-2 PRODUCTION IN HIV INFECTION

Progression of HIV infection is characterized by complex dysregulation of Th that can involve the T cells themselves and/or affect antigen-presenting cells (APCs) (Lane et al, 1985; Smolen et al, 1985; Giorgi et al, 1987; Miedema et al, 1988; Clerici et al, 1989). A

HIV and the New Viruses Second Edition
ISBN 0-12-200741-7

spectrum of functional defects has been reported in HIV-infected individuals. The defects associated with the interaction between Th and APCs are multiple and sequential (Clerici et al, 1989), and can be summarized as follows: Th defects (analysed as the ability of Th lymphocytes to proliferate and to secrete IL-2) to soluble antigens appear early in the disease and can be detected in approximately 60% of asymptomatic, HIV-seropositive individuals. Soluble antigens are processed and presented to CD4+ Th on the surface of autologous APCs in association with class II molecules (Allen, 1987; Germain and Marguiles, 1993). Thus, an alteration in the ability to recognize soluble antigens is the most sensitive and accurate index of dysfunction of CD4+ T lymphocytes. Because immune response to soluble antigens involves the collaboration between two cell types, APCs and Th (Noelle and Snow, 1991), the primary defect could be due to an impaired ability of the APCs to process or present antigenic peptides or to provide other accessory functions. This hypothesis appeared to be excluded by experiments in which soluble antigen-sensitized APCs from HIV-infected monozygotic twins were shown to be able to stimulate in vitro CD4+ Th function of the HIV-uninfected twins (Clerici et al, 1990; Blauvelt et al, 1995). Nevertheless, APCs exhibit defects in ability to produce IL-12 even in asymptomatic HIV-infected individuals (Chehimi et al, 1994), suggesting that APCs function is not normal even at early stages of disease. Defects in the ability of APCs to present antigen in such experiments are observed in the late phases of HIV infection and are the result of diminished MHC class II expression on the surface of these cells (Clerici et al, 1991). Alterations in the ability of CD4+ T lymphocyte to recognize soluble antigens have also been shown not to be provoked by selective anergy/destruction of antigen-specific memory T lymphocytes and to be independent of reduction in the precursor frequencies of antigen-specific T lymphocytes (Schulick et al, 1993).

As the immune system continues to decline, defects in Th function in response to stimulation with allogeneic lymphocyte in a mixed lymphocyte reaction (MLR) appear next, in patients in whom the ability to respond to soluble antigens has already been lost. Because (a) both CD4 and CD8 T lymphocytes are stimulated in MLR, (b) both autologous and heterologous APCs can stimulate T lymphocytes and (c) CD4 and CD8 T lymphocytes can be stimulated either by allogeneic peptides processed and presented in association with self MHC molecules, or by direct recognition of non-self MHC molecules on the surface of the allogeneic lymphocytes, this Th defect is not selective for CD4+ Th lymphocytes and is independent of alterations in the ability of APCs to process or present antigens (Via et al, 1990). Finally, phytohemagglutinin (PHA)-stimulated proliferation and IL-2 production become defective in HIV-seropositive individuals in whom Th defects to soluble antigens and alloantigens are already present. Because both CD4 and CD8 T cells are stimulated by PHA and because the ability of this T cell mitogen to stimulate lymphocytes is only marginally dependent on APCs, the inability to respond to mitogens is an index of a profound impairment in Th function in HIV-infected patients. The above-mentioned functional defects are (a) observed in the majority of HIV-seropositive individuals even during the phase of clinical latency; (b) sequential (i.e. loss of response to soluble antigens precedes loss of response to alloantigens which precedes the inability to respond to PHA); and (c) independent of changes in CD4 cell counts (Clerici et al, 1989).

Expression of the ζ chain of the T cell receptor (TCR) has been shown to be down-regulated in HIV infection (Trimble and Lieberman, 1997). The expression of the αβ heterodimer of the TCR on the surface of T lymphocytes is dependent on its association with both the CD3 complex and a $\zeta\zeta$ homodimer or a $\zeta\varepsilon$ heterodimer. In addition, the synthesis of the ζ chain and its association to the $\alpha\beta$ heterodimer plus the CD3 complex are the limiting factors in the assembly of a functional and surface-expressed TCR (Davis and Bjorkman, 1988; Fowlkes and Pardoll, 1989). Therefore, downregulation of the synthesis or expression of the TCR ζ chain could reduce the amount of TCR on the surface of T lymphocytes, impede their activation, and be involved in the genesis of the defective T cell function observed in HIV infection. It remains to be determined whether cytokines can also regulate the expression of TCR.

DYSREGULATION OF T HELPER CELLS

Two functionally diverse T helper populations in humans and their dysregulation in HIV infection

A report by Maggi et al (1987) observed that reduced production of IFN-γ and IL-2 as well as enhanced helper activity for IgG synthesis were present in cloned CD4+ T cells of patients with AIDS. At approximately the same time murine cloning studies by Coffman and Mosmann (1987, 1989) showed that CD4+ T helper lymphocytes could be divided in two major families, Th1 and Th2, on the basis of cytokine production. The Th1 responses are characterized by production of IFN-γ and IL-2 and enhance mainly cell-mediated immunity. In contrast, Th2 responses are characterized by secretion of IL-4 and IL-5, with subsequent activation of humoral immunity and generation of antibodies (Mosmann and Coffman, 1987, 1989). Interestingly, Th1 and Th2 are cross-regulatory in that IFN-γ suppresses the activation of Th2 lymphocytes, whereas IL-4 suppresses the production of Th1 cytokines (Mosmann and Coffman, 1987, 1989). The possibility that the defects in IL-2 production reported in HIV infection, and described in detail above, could be accompanied by augmented IL-4 production was investigated. Because of the artifacts induced by the technique of cloning, direct cytokine production by the cells circulating in peripheral blood was analyzed. The cytokines produced by peripheral blood mononuclear cells were redefined as type 1 cytokines, which mainly induced cell-mediated immunity and were more efficient for defence against intracellular microorganisms, and type 2 cytokines, which mainly induced humoral immunity and were more effective for defence against extracellular pathogens (Clerici and Shearer, 1993, 1994). In summary, Th1 and Th2 are clonally defined cytokines produced by clones of CD4+ T cells; type 1 and type 2 are functionally defined cytokines and are independent of the cell type or types that produce them.

Because cytokines were now defined on a functional basis, the group of type 1 cytokines was enlarged to comprise IFN-γ, IL-2, IL-12 and IL-15 (cytokines that enhance cell-mediated immunity but are not mainly or exclusively produced by clones of CD4+ T lymphocytes). Similarly, IL-6, IL-10 and IL-13, which also stimulate B cell activity and antibody generation, were considered to be type 2 cytokines, although they are produced by cells other than CD4+ T lymphocytes. Alterations in cytokine production were reported to be associated with the progression of HIV infection (reviewed by Clerici and Shearer, 1993, 1994). Thus, a progressive decline in type 1 cytokines and a parallel increase in the generation of type 2 cytokines, and a subsequent shift away from cell-mediated immunity and in favor of humoral immunity, were suggested to characterize the course of HIV infection (Clerici and Shearer, 1993, 1994). Although these data were obtained by measuring the relative amounts of cytokines in the supernatant of mitogen-stimulated cell cultures, these findings were confirmed by a study measuring intracellular cytokines on a single-cell level (Klein et al, 1997). Thus, a continuous decrease in the percentage of cells producing IL-2 and IFN-γ was reported to be accompanied by an increase in the frequency of cells producing type 2 cytokines in disease progression. The type 1/type 2 hypothesis of susceptibility to HIV-1 infection and disease progression has been confirmed by different lines of investigation from a number of laboratories. To summarize:

1. Production of IL-2 and IL-12 is reduced in HIV-seropositive patients and both IL-12 and IL-15 restore in vitro defective soluble antigen-stimulated proliferation of Th cells from HIV-infected individuals (reviewed by Clerici and Shearer, 1993, 1994).
2. Increased IL-4, IL-6, IL-10 and/or IL-13 production is present in HIV infection, and anti-IL-4 as well as anti-IL-10 neutralizing antibodies increase antigen-stimulated proliferation of HIV-infected lymphocytes (reviewed by Clerici and Shearer, 1993, 1994).
3. The ability to respond to common antigens in vivo in a delayed hypersensitivity reaction – a classical type 1 cytokine-driven reaction – is lost in, and is predictive for, disease progression (Blatt et al, 1993; Gording et al, 1994).
4. In vivo, the presence of hematological parameters associated with type 2 cytokine production such as IL-4-driven hyper-IgE (Israel-Biet et al, 1992) and IL-5-driven hypereosinophilia (Fleury-Feith et al, 1992; Smith et al, 1994) are unfavourable prognostic factors and are clinically associated with rapid disease progression.

The type 1/type2 cytokine balance is illustrated in Figure 8.1.

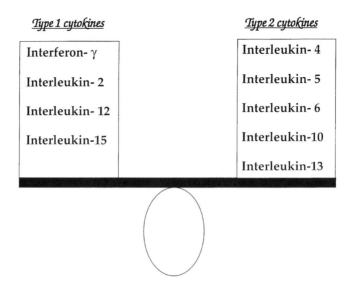

Figure 8.1 *Type 1 and type 2 cytokines that regulate dominant cellular and humoral immune responses*

Type 1 and type 2 patterns and HIV disease progression

The correlation between type 1 and type 2 cytokine production and disease progression was analyzed by studying cohorts of HIV-infected adults and children with different patterns of disease progression. Cytokine production was evaluated in patients with undetectable or delayed disease progression, as well as in patients with more rapid disease progression. The results showed high levels of mitogen-stimulated IL-2 and IFN-γ production in long-term asymptomatic patients but not in HIV-infected individuals with progressing disease (Vigano' et al, 1995; Clerici et al, 1996). These data were confirmed by phenotypic analyses indicating that CD4+CD7− lymphocytes (preferentially producing IL-10) (Autran et al, 1995) are increased in progressing patients, whereas CD4+/Surface Lymphocyte Activation Marker (SLAM)+ T lymphocytes (cells that produce IFN-γ) (Cocks et al, 1995) are augmented in long-term asymptomatic patients. That the presence of strong HIV-specific cell-mediated immunity is correlated with slower disease progression was also supported by the observations that preserved Th function is detected in patients with stable CD4 counts, and that disease-free survival correlates with more potent cytotoxic T lymphocyte activity. Additionally, it was shown that cytokine impairment and defects in Th cell function are not present in HIV-infected chimpanzees, in which (a) seroconversion does not usually result in disease progression (Heeney et al, 1993; Gougeon et al, 1997), (b) cytokine production by antigen- and mitogen-stimulated peripheral blood mononuclear cells is characterized by the generation of type 1 cytokines (Heeney et al, 1993; Gougeon et al, 1997), and (c) strong virus-specific cytotoxic T lymphocyte activity is present (Heeney et al, 1993; Gougeon et al, 1997). Further support for a protective role of cell-mediated immunity in modulating disease progression was provided by data showing that development of the murine retrovirus-induced AIDS-like syndrome and appearance of Th cell dysfunction were delayed subsequent to treatment of mice in vivo with antibodies to IL-4 and IL-10, and accelerated by the use of anti-IFN-γ antibodies (Wang et al, 1994; Doherty et al, 1997). More recent results from experiments in which macaques were infected with either normally pathogenic simian immunodeficiency virus (SIV) strains or with non-pathogenic Nef-deleted strains of SIV showed infection with the latter strains of virus to be associated with the preferential development of a type 1 cytokine pattern, and this pattern was predictive of a more favorable disease outcome (Zou et al, 1997). Finally, infection of macaques with a Nef-deleted, IFN-γ-expressing SIV vector was shown to be followed by lower viral load and reduced disease development (Giavedoni et al, 1997). Therefore evidence seems to suggest that detection of strong cell-mediated immunity may be the main immunologic correlate associated with delayed disease progression.

One important virologic sign of progression of HIV infection is the emergence of HIV variants that induce the generation of syncytia in vitro. Syncytium-inducing (SI) HIV variants are isolated in 50–60% of HIV-seropositive individuals with AIDS, but in less than 10% of HIV-seropositive asymptomatic individuals (Tersmette et al, 1989). Approximately 70% of HIV-seropositive individuals in whom SI HIV variants could be isolated, but only 16% of HIV-positive individuals in whom SI variants could not be isolated, progressed to AIDS in a 30-month period (Koot et al, 1993). Finally, isolation of HIV SI variants is associated with an increased decline in CD4+ T lymphocytes (Connor et al, 1993). Data reinforcing the interdependence between immune dysfunction and disease progression demonstrated that the fast-replicating SI HIV variants are preferentially isolated in HIV-infected individuals exhibiting a significant reduction in type 1 cytokine production, a significant increase in type 2 cytokine production, and the lowest CD4 counts. In the same study, strong type 1 cytokine production and weak type 2 cytokine production as well as higher CD4 counts were associated with failure to isolate virus and/or with the isolation of HIV variants with low replicative ability and non-SI phenotype (Clerici et al, 1996). It remains to be resolved whether the appearance of HIV SI variants induces the defects in antiviral immune response, or if it occurs secondarily to such immune alterations.

CYTOKINES AND HIV-SPECIFIC IMMUNITY IN EXPOSED SERONEGATIVE INDIVIDUALS

Exposure to HIV in the absence of infection has been reported in most if not all groups of seronegative at-risk individuals. These groups include injecting drug users, homosexual people, neonates, prostitutes, and accidentally exposed health workers (reviewed by Rowland-Jones and McMichael, 1995; Shearer and Clerici, 1996). Absence of infection in exposed seronegative individuals was suggested to be correlated with the detection of HIV-specific and type 1 cytokine-secreting Th in the peripheral blood. Subsequently HIV-specific cytotoxic T lymphocytes were also observed in many of these cohorts, reinforcing the hypothesis that the exposure to HIV which selectively activates type 1 cytokine secretion and cell-mediated immunity may not be followed by seroconversion (i.e. activation of B lymphocyte and antibody secretion) and disease (reviewed by Rowland-Jones and McMichael, 1995; Shearer and Clerici, 1996). A study recently confirmed this hypothesis and demonstrated that severe combined immunodeficient (SCID) mice reconstituted with peripheral blood mononuclear cells of HIV-seronegative individuals exposed to HIV were protected against in vivo challenge with multiple strains of HIV (Zhang et al, 1996). Additionally, reconstitution of SCID mice with peripheral blood mononuclear cells of healthy volunteers immunized with a gp160 candidate vaccine, and in whom elicitation of both humoral and cell-mediated systemic HIV immunity was measured, showed protection against in vivo challenge with HIV to be correlated with detection of HIV-specific Th responses but not with the presence of HIV-specific neutralizing antibodies (Mosier et al, 1993). Finally, protection against SIV infection was reported in macaques by exposure to low doses of virus and consequent selective activation of SIV-specific cell-mediated immunity (reviewed by Rowland-Jones and McMichael, 1995; Shearer and Clerici, 1996). However, cell-mediated immunity may not provide the complete immunologic picture for protection because HIV-specific, IgA-mediated mucosal immunity is also locally activated in HIV-seronegative women sexually exposed to HIV (Mazzoli et al, 1997). That the immune response can be tightly regulated and compartmentalized is demonstrated by the observation that systemic HIV-specific cell-mediated immunity and HIV-specific mucosal humoral immunity are simultaneously activated in these at-risk individuals, raising the possibility that HIV infection could be prevented by a synergy between the two different arms of the immune response (Mazzoli et al, 1997). These data are in agreement with other observations in the animal models, in which it was shown that protective mucosal immunity elicited in macaques by lymph node immunization or in chimpanzees by oral immunization is correlated with an increase in the number of IgA-secreting plasma cells

(Lehner et al, 1996; Miller and McGhee, 1996), and also that neutralization of HIV is achieved by secretory (fecal) IgA induced by oral immunization (reviewed by Miller and McGhee, 1996).

CYTOKINES AND ANTIVIRAL FACTORS

Reports that a CD8-generated factor (cell antiviral factor or CAF) inhibited HIV infection of CD4+ T cells in vitro provided the first demonstration that HIV infection may be influenced and controlled by a cellular immune response and its products (Walker et al, 1986). Subsequently, it was reported that the CC chemokines macrophage inflammatory protein (MIP)-1α, MIP-1β, and RANTES (CC) can prevent HIV infection of target lymphocytes in vitro (Cocchi et al, 1995). Cell antiviral factor and the anti-HIV chemokines are distinct entities because:

- CAF is exclusively produced by CD8+ T lymphocytes, whereas CC chemokines are pleiotropic factors produced by different cell types in diverse physiologic and patho-logic conditions;
- CAF blocks the transcription of viral RNA whereas CC chemokines are active at a pretranscriptional level;
- the concentration of CAF and that of chemokines are independent variables in HIV-suppressing culture media;
- chemokine-specific neutralizing antibodies do not block the activity of CAF;
- HIV strains that are susceptible to CAF may or may not be inhibited by chemokines (Cocchi et al, 1995; Levy et al, 1996).

CD8+ T lymphocytes of HIV-seropositive long-term- non-progressors were shown to generate higher amounts of CAF compared with patients who progressed in the disease (reviewed by Levy et al, 1996). Because both a type 1/type 2 shift in cytokine production and a diminished production of CC chemokines and CAF are observed in HIV pro-gression, possible correlations were sought. The findings are summarized as follows:

1. In vitro exposure of CD8 T lymphocyte of long-term non-progressing individuals to type 2 cytokines downmodulates CAF production, whereas type 1 cytokines aug-ment CAF production by CD8 T cells of progressing individuals (reviewed by Levy et al, 1996).
2. Synthesis of CC chemokines by peripheral blood lymphocytes is stimulated by type 1 cytokines and CC generation is detected in Th1 but not in Th2 clones (Schrum et al, 1996).
3. IL-10 suppresses RANTES production by human monocytes (Marfaing-Koka et al, 1996).

Additionally, Loetscher et al (1996) showed that surface expression of CCR5 is posi-tively influenced by IL-2, further underlying the different and multiple levels of com-plexity present in HIV infection.

THE ROLE OF COMPLEMENT AND COMPLEMENT REGULATORY MOLECULES

The human retroviruses, including HIV, are intrinsically resistant to lysis by human complement, even in the presence of a specific antibody response (Banapour et al, 1986), although they are able to activate the early steps of the complement cascade (Ebenbichler et al, 1991; Susal et al, 1994). Activation of complement does not in this case result in virus and infected cell destruction, as would be expected; instead its effect is to drive the virus efficiently towards the immune system. Several complement components are apparently involved in almost all stages of the HIV infection cycle. Lymph node tropism, choice of the target cell, activation of proviral DNA via complement receptor signaling, virus budding, extracellular survival and, ultimately, depression of the cell-mediated response as well as polyclonal B cell activation, may all involve complement components in different ways. These components include fragments derived from complement activation (C3b, iC3b, C3d) as well as the complement regulatory molecules CD55, CD46 and factor H.

Human immunodeficiency virus coated with complement activation fragments can be driven – as was observed in the case of SIV-infected macaques (Reimann et al, 1994) – to the lymph nodes, where it can interact with follicular dendritic cells expressing CR1, CR2 and CR3 (Reynes et al, 1985; Sölder et al, 1989). It has in fact been shown that dendritic cells bind HIV by using receptors for complement fragments (Crowe and Kornbluth, 1994; Weissman et al, 1995), and that these cells can transmit HIV and trigger efficient viral replication when they interact with CD4+ T cells.

Virus coated with complement fragments generated by activation through the alternate pathway can interact efficiently with cells bearing various types of complement receptors, at a time when a specific immune response is not yet evident. This mechanism was shown to be particularly relevant in the case of low virus loads (Sölder et al, 1989; Reisinger et al, 1990). Direct interaction of HIV envelope proteins with C3 receptors, owing to homology with the C3 molecule, and indirect interactions through fixed C3 fragments or through immune complexes on the HIV surface, enable the virus to enhance the efficiency of infection of CD4+ cells (Sölder et al, 1989; Delibrias et al, 1993) and broaden its host range of infectable cells to include those that do not express CD4.

Replication of HIV is strictly dependent on activation and proliferation of infected cells. Virus production requires, in fact, stimulation of infected cells with antigens and cytokines (Folks et al, 1987; Koyanagi et al, 1988; Poli et al, 1990; Kalter et al, 1991). The activating effect of these stimuli is due to the induction of cellular transcription factors which interact with viral long terminal repeats (LTRs) (Vlach and Pitha, 1992; Granelli-Piperno et al, 1995; Briant et al, 1996). Recently, activation of NFkB and Ap1 by the induction of the CR3 and CR4 intracellular signaling pathway was observed (Messika et al, 1995; Noti et al, 1996). It is noteworthy that activation of integrated proviral DNA in macrophages can be triggered by CR3-specific antibodies or iC3b-coated viral particles, which stimulate the CR3 signaling pathway (Thieblemont et al, 1995).

During the process of budding from an infected cell, HIV carries over a coat of plasma membrane, including proteins derived from the host cell. Among other proteins such as MHC products, adhesion molecules and integrins (Hildreth and Orentas, 1989; Arthur et al, 1992; Meerloo et al, 1993), it was shown (Marschang et al, 1995) that HIV particles are selectively enriched in the complement regulatory protein CD55 (decay-accelerating factor, DAF). The presence of CD55 on the virus was shown partially to protect the virus from complement damage (Stoiber et al, 1996). Another relevant mechanism by which HIV avoids lysis is by focusing the soluble regulatory protein factor H on its surface through interaction with specific regions of the two envelope glycoproteins (Pintér et al, 1995a,b; Stoiber et al, 1995a,b). It was shown that CD55 and factor H provide an efficient protection against complement-mediated lysis of both the virus itself and the infected cells (Stoiber et al, 1996). The ability of HIV to interact with soluble complement regulators such as factor H is therefore an important feature which influences survival of retroviruses and infected cells in the bloodstream of infected subjects.

As mentioned above, IL-12 produced by monocytes and macrophages is important for the generation of an efficient cell-mediated response. It was recently shown that cross-linking of the complement regulatory protein CD46 with specific antibodies, with C3b, or with the measles virus, results in a reduced production of IL-12 by macrophages (Karp et al, 1996). Virus particles opsonized with complement fragments may therefore interfere with IL-12 production, causing immunosuppression.

There is no evidence so far of peripheral B cell infection by HIV in vivo. However, significant changes in B cell responses were observed, which result in hypergammaglobulinemia (Boyd and James, 1992). The polyclonal B cell activation induced by HIV glycoproteins is most probably T cell-dependent (Chirmule et al, 1993), and cytokines such as IL-6 (Chirmule et al, 1993) and IL-15 (Kacani et al, 1997) result in an enhanced effect. The increased serum immunoglobulin levels observed upon HIV infection in vivo result in the formation of immune complexes coated with C3b. Further cleavage of C3b on these complexes leads to the formation of iC3b and C3d, the natural ligand of CR2 on B cells (Fearon and Carter, 1995). Aggregated C3d was described as being a potent B cell activator (Melchers et al, 1985; Dempsey et al, 1996). Therefore, interaction of C3d-bearing immune complexes with CR2 expressed at the surface of B cells has a synergistic effect on B cell stimulation.

CYTOKINES, ANTIGEN-INDUCED CELL DEATH AND PREFERENTIAL DESTRUCTION OF Th1 T CELLS

Although the accelerated viral replication characteristic of HIV infection is involved in the decline in CD4+ T cells observed in disease progression (Ho et al, 1995), factors not directly dependent on viral infection of target cells are important in this process. For example:

- the percentage of T lymphocytes in which HIV genetic material is detectable is less than 10% of the circulating pool even in the most advanced phases of the disease (Haas, 1996);
- increases in the number of circulating CD4 T cells in patients undergoing antiretroviral therapy are only partial, even in patients whose HIV viral loads are reduced virtually to zero (Connors et al, 1997);
- the persistent suppression of HIV viral load, often under the limit of detectability, is associated with increases in CD4 T cell counts that are only observed in the first weeks or months of therapy (Connors et al, 1997).

The T lymphocytes of HIV-infected patients undergo apoptosis either when un-stimulated or when stimulated with antigens or mitogens, referred to here as antigen-induced cell death (AICD) (reviewed by Ameisen, 1994; Oyazu and Pahwa, 1995). The involvement of AICD in disease progression is underlined by the following observations:

1. The percentage of lymphocytes spontaneously undergoing apoptosis is directly pro-portional to the clinical severity of the infection (Gougeon et al, 1996).
2. Susceptibility of antigen- and mitogen-stimulated peripheral lymphocytes to apop-tosis increases in disease progression (Gougeon et al, 1996).
3. Serum concentrations of tumour necrosis factor beta (TNF-β) and sAPO-1/Fas (sAPO/Fas was suggested to prevent AICD) (Cheng et al, 1994) are predictors of dis-ease progression independent of HIV viral load (Medrano FJ et al, 1998).
4. Apoptosis is not present in HIV-infected chimpanzees, in whom AIDS development is usually not observed (see above) (Heeney et al, 1993).

Antigen-induced cell death in HIV infection is mediated by at least two different com-ponents of the nerve growth factor/tumour necrosis factor receptor family: TNF-b (Clerici et al, 1996) and APO-1/Fas (Fas) (Katsikis et al, 1995). Indeed, lymphocytes of HIV-seropositive individuals express Fas in large quantities (Debatin et al, 1994; Kat-sikis et al, 1995; McCloskey et al, 1995), and are more prone to undergo programmed cell death upon ligation of this receptor (Debatin et al, 1994; Katsikis et al, 1995; McCloskey et al, 1995), and the magnitude of anti-Fas-induced death is inversely cor-related with absolute CD4 cell counts (Debatin et al, 1994; Katsikis et al, 1995; McCloskey et al, 1995).

The possible correlations between impairment of cytokine production, AICD and disease progression were analyzed. The background for this line of research stems from results showing that cytokines can influence TCR-induced AICD. Thus, IL-2 can block α-CD3-induced AICD of both CD4+ and CD8+ resting thymocytes (Nieto et al, 1990), and can induce Bcl-2-mediated increased survival of memory T cells (Groux et al, 1993). Results showed that in vitro antigen- and mitogen-stimulated AICD of lymphocytes of HIV-infected individuals is influenced by cytokines. Briefly, type 1 cytokines block T lymphocyte AICD, whereas type 2 cytokines have either no effect or enhance AICD in vitro (Clerici et al, 1994; Estaquier et al, 1995; Raddrizzani et al, 1995). Additionally, antigen- and mitogen-stimulated AICD can be inhibited by antibodies against IL-4 and IL-10, and enhanced by anti-IL-12 (Clerici et al, 1994; Estaquier et al, 1995). In particu-lar, IL-12 inhibits apoptosis induced in human Th1 clones and can prevent activation-induced and Fas-mediated apoptosis of CD4 T lymphocytes of HIV-infected patients (Clerici et al, 1994; Estaquier et al, 1995). Finally, type 1 – but not type 2 – cytokine can prevent TNF-β-mediated AICD of antigen-stimulated, HIV-infected CD4+ T lymphocytes (Clerici et al, 1996). Recent results further underline the inter-play between AICD, Th1/Th2, Fas, and progression of HIV infection. Thus, Th1 but not Th2 cells were shown to undergo rapid Fas-mediated AICD upon antigen stimula-tion (Varadhachary et al, 1997; Zhang et al, 1997); Th2 clones were also seen to survive preferentially in cell cultures in vitro (Varadhachary et al, 1997; Zhang et al, 1997). These findings were verified by the observation that, whereas both types of clones express Fas, only Th2 clones express high levels of a Fas-associated phosphate that inhibits Fas signaling (Zhang et al, 1997). Because HIV was shown to replicate preferentially into Th2 lymphocytes, it is possible to envision a scenario in which HIV progression, AICD-mediated preferential destruction of Th1 T lymphocytes, type 1 to type 2 cytokine shift, augmented replication of HIV into Th2 T lymphocytes, increased viral load, and disease progression are linked. The resulting cytokine imbalance would then be due to the opposing but unequal and mechanistically distinct types of Th1 and Th2 death.

CONCLUSION

This brief review summarizes the evidence that cytokines influence the progression of HIV disease, as well as protection against HIV infection. Cytokines also appear to have an effect on AICD, a model for HIV immunopathogenesis, and on the production of cell-generated anti-HIV factors. It remains to be determined whether cytokine-driven immune-based therapies and cytokine-directed prophylactic immunization against HIV and AIDS can be effectively achieved without creating new problems of immune dys-regulation. The immunologic components considered likely to contribute to protection against HIV infection and progression to AIDS are listed in Table 8.1. It remains to be determined whether all of these factors of potential immune protection can be opti-mized simultaneously.

Table 8.1 *Forms of immunity likely to contibrute to protection against HIV infection and AIDS progression*

Systemic Th-1-like help
Dominant type 1 cytokine profile
HIV-specific cytotoxic T lymphocytes
T cell-derived antiviral factors
IgA antibodies at mucosal site of exposure

REFERENCES

Allen PM (1987) Antigen processing at the molecular level. *Immunol Today* **8**: 270–273.

Ameisen JC (1994) Programmed cell death (apoptosis) and cell survival regulation: relevance to AIDS and cancer. *AIDS* **8**: 1197–1213.

Arthur LO, Bess J Jr, Sowder R et al (1992) Cellular proteins bound to immunodeficiency viruses: implications for pathogenesis and vaccines. *Science* **258**: 1935–1938.

Autran B, Legac E, Blanc C & Debré P (1995) A Th0/Th2-like function of CD4+ CD7+ T helper cells from normal donors and HIV infected patients. *J Immunol* **154**: 1408–1417.

Banapour B, Sernatinger J & Levy JA (1986) The AIDS-associated retrovirus is not sensitive to lysis or inactivation by human serum. *Virology* **152**: 268–271.

Blatt SP, Hendrix CW, Butzin CA, Lucey DR & Boswell RN (1993) Delayed type hypersensitiv-ity skin testing predicts progression to AIDS in HIV-infected patients. *Ann Intern Med* **119**: 177–184.

Blauvelt A, Clerici M, Lucey DL, Shearer GM & Katz S (1995) Functional studies of epidermal

Langerhans cells and blood monocytes in human immunodeficiency virus-infected individuals. *J Immunol* **154:** 3506–3515.

Boyd JE & James K (1992) B cell responses to HIV and the development of human monoclonal antibodies. *Clin Exp Immunol* **88:** 189–202.

Briant L, Coudronniere N, Robert-Hebmann V, Benkirane M & Devaux C (1996) Binding of HIV-1 virions or gp120 anti-gp120 immune complexes to HIV-1 infected quiescent peripheral blood mononuclear cells reveals latent infection. *J Immunol* **156:** 3994–4004.

Chehimi J, Trinchieri G, & Frank I (1994) IL-12 deficiency in HIV-infected patients. *J Exp Med* **179:** 1361–1366.

Cheng J, Zhou T, Liu C et al (1994) Protection from Fas-mediated apoptosis by a soluble form of the Fas molecule. *Science* **263:** 1759–1762.

Chirmule N, Kalyanaraman VS, Lederman S et al (1993) HIV-gp160-induced T cell-dependent B cell differentiation. Role of T cell–B cell activation molecule and IL-6. *J Immunol* **150:** 2478–2486.

Clerici M, Sarin A, Coffman RL et al (1994) Type1/type2 cytokine modulation of T cell programmed cell death as a model for HIV pathogenesis. *Proc Natl Acad Sci USA* **91:** 11811–11815.

Clerici M, Balotta C, Meroni L et al (1996) Type 1 cytokine production and low prevalence of viral isolation correlate with long term non progression in HIV infection. *AIDS Res Hum Retroviruses* **12:** 1053–1061.

Cocchi F, De Vico AL, Garzino-Demo A, Arya SK, Gallo RC & Lusso P (1995) Identification of RANTES, MIP-1 alpha, and MIP-1 beta as the major HIV-suppressive factors produced by CD8+ T cells. *Science* **270:** 1811–1815.

Cocks BG, Chang JCC, Carballido JM, Yssel H, de Vries JE & Aversa G (1995) A novel receptor involved in T cell activation. *Nature* **376:** 260–264.

Coffman RI & Mosmann TR (1987) Two types of mouse T helper cell clone: implication for immune regulation. *Immunol Today* **8:** 223–226.

Connor RI, Mohri H & Ho DD (1993) Increased viral burden and cytopathicity correlate temporally with CD4+ T-lymphocyte decline and clinical progression in human immunodeficiency virus type 1-infected individuals. *J Virol* **67:** 772–777.

Connors M, Kovacs JA, Krevat S et al (1997) HIV infection induces changes in CD4+ T cell phenotype and depletions within the CD4+ T cell repertoire that are not immediately restored by antiviral or immune based therapies. *Nature Med* **5:** 533–540.

Crowe SM & Kornbluth RS (1994) Overview of HIV interactions with macrophages and dendritic cells: the other infection in AIDS. *J Leukoc Biol* **56:** 215–217.

Davis MM & Bjorkman PJ (1988) T cell antigen receptor genes and T cell recognition. *Nature* **334:** 395–402.

Debatin K-M, Fahrig-Faissner A, Enenkel-Stoodt S, Kreuz W, Benner M & Krammer PH (1994) High expression of Apo-1 (CD95) on T lymphocytes from HIV-infected children. *Blood* **83:** 3101–3103.

Delibrias CC, Kazatchkine MD & Fischer E (1993) Evidence for the role of CR1 (CD35), in addition to CR2 (CD21), in facilitating infection of human T cells with opsonized HIV. *Scand J Immunol* **38:** 183–189.

Dempsey PW, Allison MED, Akkaraju S et al (1996) C3d of complement as a molecular adjuvant: bridging innate and acquired immunity. *Science* **271:** 348–350.

Doherty TM, Giese N, Morse HC & Coffman RL (1997) Modulation of murine AIDS-related pathology by concurrent antibody treatment and coinfection with *L. major*. *J Virol* **71:** 3702–3709.

Ebenbichler CF, Thielens NM, Vornhagen R, Marschang P, Arlaud GJ & Dierich MP (1991) Human immunodeficiency virus type 1 activates the classical pathway of complement by direct

C1 binding through specific sites in the transmembrane glycoprotein gp41. *J Exp Med* **174:** 1417–1424.

Estaquier J, Idziorek T, Zou W et al (1995) T helper 1/T helper 2 cytokines and T cell death: preventive effect of IL-12 on activation-induced and CD95 (Fas/Apo-1)-mediated apoptosis of CD4+ T cells from human immunodeficiency virus-infected person. *J Exp Med* **182:** 1759–1767.

Fearon DT & Carter RH (1995) The CD19/CR2/TAPA-1 complex of B lymphocytes: linking natural to acquired immunity. *Annu Rev Immunol* **13:** 127–149.

Fleury-Feith J, Van Nheieu JT, Picard C, Escudier E & Bernaudin JF (1992) Bronchoalveolar lavage eosinophilia associated with *Pneumocystis carinii* pneumonia in AIDS patients. Comparative study with non-AIDS patients. *Chest* **95:** 1198–1201.

Folks TM, Justement J, Kinter A, Dinarello CA & Fauci AS (1987) Cytokine-induced expression of HIV-1 in a chronically infected promonocyte cell line. *Science* **238:** 800–802.

Fowlkes BJ & Pardoll DM (1989) Molecular and cellular events of T cell development. *Adv Immunol* **44:** 207–264.

Germain RH & Marguiles DM (1993) The biochemistry and cell biology of antigen processing and presentation. *Annu Rev Immunol* **11:** 403–450.

Giavedoni L, Ahmad S, Jones L & Ylma T (1997) Expression of IFNg by simian immunodeficiency virus increases attenuation and reduces postchallenge virus load in vaccinated rhesus macaques. *J Virol* **71:** 866–872.

Giorgi JV, Fahey JL & Smith DC (1987) Early effects of HIV on CD4 lymphocytes in vivo. *J Immunol* **138:** 3725–3730.

Gording FM, Hartigan P M, Klimas NG, Zolla-Pazner SB, Simberkoff MS & Hamilton DJ (1994) Delayed type hypersensitivity skin tests are an independent predictor of HIV disease progression. *J Infect Dis* **169:** 893–897.

Granelli-Piperno A, Pope M, Inaba K & Steinman RM (1995) Coexpression of NF-kappa B/Rel and Sp1 transcription factors in human immunodeficiency virus 1-induced, dendritic cell–T-cell syncytia. *Proc Natl Acad Sci USA* **92:** 10944–10948.

Gougeon ML, Lecoeur H, Dulioust A et al (1996) Programmed cell death in PBL from HIV-infected persons: the increased susceptibility to apoptosis of CD4 and CD8 T cells correlates with lymphocyte activation and with disease progression. *J Immunol* **156:** 3509–3519.

Gougeon ML, Lecoeur H, Boudet F et al (1997) Lack of chronic immune activation in HIV infected chimpanzees correlates with the resistance of T cells to Fas/Apo-induced apoptosis and preservation of a T helper 1 phenotype. *J Immunol* **158:** 2964–2976.

Groux H, Monte D & Plouvier B (1993) CD3-mediated apoptosis of human medullary thymocytes and activated T cells: respective roles of interleukin-1 interleukin-2, interferon gamma, and accessory cells. *Eur J Immunol* **23:** 1623–1630.

Haas A (1996) In: *Summary report of the public meeting of the NIH panel to define principles of therapy of HIV infection.* Washington, DC, 13–14 November 1996, pp 3–5.

Heeney J, Jonker R, Koornstra W, Garcia S & Gougeon ML (1993) The resistance of HIV-infected chimpanzees to progression to AIDS correlates with absence of HIV-related T cell dysfunction. *J Med Primatol* **22:** 194–200.

Hildreth JE & Orentas RJ (1989) Involvement of a leukocyte adhesion receptor (LFA-1) in HIV-induced syncytium formation. *Science* **244:** 1075–1078.

Ho DD, Neumann AU, Percison AS, Chen W, Leonard JM & Markowitz, M (1995) Rapid turnover of plasma virions and CD4 lymphocytes in HIV-1 infection. *Nature* **373:** 123–126.

Israel-Biet D, Labrousse F, Tourani J-M, Sors H, Andrieu JM & Even P (1992) Elevation of IgE in HIV-infected subjects: a marker of poor prognosis. *J Allergy Clin Immunol* **89:** 68–74.

Kacani L, Stoiber H & Dierich MP (1997) Role of IL-15 in HIV-1-associated hypergammaglobulinaemia. *Clin Exp Immunol* **108:** 14–28.

Kalter DC, Nakamura M, Turpin JA et al (1991) Enhanced HIV replication in macrophage colony-stimulating factor-treated monocytes. *J Immunol* **146**: 298–306.

Karp CL, Wysocka M, Wahl LM et al (1996) Mechanism of suppression of cell-mediated immunity by measles virus. *Science* **273**: 228–231.

Katsikis P D, Wunderlich ES, Smith CA, Herzenberg LA & Herzenberg LA (1995) Fas antigen stimulation induces marked apoptosis of T lymphocytes in human immunodeficiency virus infected individuals. *J Exp Med* **181**: 2029–2036.

Klein SA, Dobmeyer JM, Dobmeyer TS et al (1997) Demonstration of the Th1/Th2 cytokine shift during the course of HIV infection using cytoplasmic cytokine detection on single cell level by flow cytometry. *AIDS* **11**: 758–765.

Koot M, Keet IPM & Vos AHV (1993) Prognostic value of HIV-1 syncytium-inducing phenotype for rate of CD4+ cell depletion and progression to AIDS. *Ann Intern Med* **118**: 681–688.

Koyanagi Y, O'Brien WA, Zhao JQ, Golde DW, Gasson JC & Chen IS (1988) Cytokines alter production of HIV-1 from primary mononuclear phagocytes. *Science* **241**: 1673–1675.

Lane HC, Depper JM, Greene WC & Fauci AS (1985) Qualitative analysis of immune function in patients with the acquired immunodeficiency syndrome: evidence for a selective defect in soluble antigens recognition. *N Engl J Med* **313**: 79–84.

Lehner TA, Wang Y, Cranage M et al (1996) Protective mucosal immunity elicited by targeted iliac lymphnode immunization with a subunit SIV envelope and core vaccine in macaques. *Nature Med* **2**: 767–775.

Levy JA, Mackewicz CE & Barker E (1996) Controlling HIV pathogenesis: the role of the non-cytotoxic anti-HIV response of CD8+ T cells. *Immunol Today* **17**: 217–224.

Loetscher P, Seitz M, Baggiolini M & Moser B (1996) Interleukin-2 regulates CC receptor expression and chemotactic responsiveness in T lymphocytes. *J Exp Med* **184**: 569–577.

Maggi E, Macchia D, Parronchi P et al (1987) Reduced production of interleukin 2 and interferon gamma and enhanced helper activity for IgG synthesis by cloned CD4+ T cells from patients with AIDS. *Eur J Immunol* **17**: 1685–1690.

Marfaing-Koka A, Maravic M, Humbert M, Galanaud P & Emilie D (1996) Contrasting effects of IL-4, IL-10, and RANTES production by human monocytes. *Int Immunol* **10**: 1587–1594.

Marschang P, Sodroski J, Würzner R & Dierich MP (1995) Decay-accelerating factor (CD55) protects human immunodeficiency virus type 1 from inactivation by human complement. *Eur J Immunol* **25**: 285–290.

Mazzoli S, Trabattoni D, Lo Caputo S et al (1997) HIV-specific mucosal and cellular immunity in HIV-seronegative partners of HIV-seropositive individuals. *Nature Med* **3**: 1250–1257.

McCloskey TW, Oyazu N, Kaplan M & Pahwa S (1995) Expression of the Fas antigen on patients infected with HIV. *Cytometry* **22**: 111–114.

Melchers F, Erdei A, Schulz T & Dierich MP (1985) Growth control of activated, synchronized murine B cells by the C3d fragment of human complement. *Nature* **317**: 264–267.

Medrano FJ, Leal M, Arienti D et al (1998) Tumor necrosis factor beta and soluble APO-1/Fas independently predict progression to AIDS in HIV-seropositive patients. *AIDS Res Hum Retroviruses* **14**: 835–843.

Meerloo T, Sheikh MA, Bloem AC et al (1993) Host cell membrane proteins on human immunodeficiency virus type 1 after in vitro infection of H9 cells and blood mononuclear cells. An immuno-electron microscopic study. *J Gen Virol* **74**: 129–135.

Messika EJ, Avni O, Gallily R, Yefenof E & Baniyash M (1995) Identification and characterization of a novel protein associated with macrophage complement receptor 3. *J Immunol* **154**: 6563–6570.

Miedema F, Petit AJ & Terpstra FG (1988) Immunological abnormalities in human immunodeficiency virus (HIV)-infected asymptomatic homosexual men. HIV affects the immune system before CD4+ T helper cell depletion occurs. *J Clin Invest* **82**: 1908–1916.

Miller CJ & McGhee JR (1996) Progress toward a vaccine to prevent sexual transmission of HIV. *Nature Med* **2:** 751–752.

Mosier DE, Gulizia RJ, MacIsaac PD, Corey L & Greenberg PD (1993) Resistance to HIV-I infection of SCID mice reconstituted with peripheral blood leukocytes from donors vaccinated with vaccinia gp160 and recombinant gp160. *Proc Natl Acad Sci USA* **90:** 2443–2447.

Mosmann TR & Coffman RI (1989) TH1 and TH2 cells: different patterns of lymphokine secretion lead to different functional properties. *Annu Rev Immunol* **7:** 145–168.

Nieto MA, Gonzalez A & Lopez-Rivas A (1990) IL-2 protects against anti-CD3-induced cell death in human medullary thymocytes. *J Immunol* **145:** 1364–137.

Noelle RJ & Snow C (1991) T helper cell-dependent B cell activation. *FASEB J* **5:** 2770–2776.

Noti JD, Reinemann BC & Petrus MN (1996) Regulation of the leukocyte integrin gene CD11c is mediated by AP1 and its transcription factors. *Mol Immunol* **33:** 115–127.

Oyazu N & Pahwa S (1995) Role of apoptosis in HIV disease pathogenesis. *J Clin Immunol* **15:** 227–231.

Pintér C, Siccardi AG, Lopalco L, Longhi R & Clivio A (1995a) Direct interaction of complement factor H with the C1 domain of HIV type 1 glycoprotein 120. *AIDS Res Hum Retroviruses* **11:** 577–588.

Pintér C, Siccardi AG, Lopalco L, Longhi R & Clivio A (1995b) HIV glycoprotein 41 and complement factor H interact with each other and share functional as well as antigenic homology. *AIDS Res Hum Retroviruses* **11:** 971–980.

Poli G, Bressler P, Kinter A et al (1990) Interleukin 6 induces human immunodeficiency virus expression in infected monocytic cells alone and in synergy with tumor necrosis factor alpha by transcriptional and post-transcriptional mechanisms. *J Exp Med* **172:** 151–158.

Raddrizzani M, Accornero P, Amidei A et al (1995) IL-12 inhibits apoptosis induced in a human Th1 clone by gp120/CD4 cross-linking and CD3/TcR activation or by IL-2 deprivation. *Cell Immunol* **161:** 14–21.

Reimann KA, Tenner-Racz K, Racz P et al (1994) Immunopathogenic events in acute infection of rhesus monkeys with simian immunodeficiency virus of macaques. *J Virol* **68:** 2362–2370.

Reisinger EC, Vogetseder W, Berzow D et al (1990) Complement-mediated enhancement of HIV-1 infection of the monoblastoid cell line U937. *AIDS* **4:** 961–965.

Reynes M, Aubert JP, Cohen JH et al (1985) Human follicular dendritic cells express CR1, CR2, and CR3 complement receptor antigens. *J Immunol* **135:** 2687–2694.

Rowland-Jones SL & McMichael A (1995) Immune responses in HIV-exposed seronegatives: have they repelled the virus? *Curr Opin Immunol* **7:** 448–455.

Schrum S, Probst P, Fleischer B & Zipfel PF (1996) Synthesis of the CC chemokines MIP-1 alpha, MIP-1 beta, and RANTES is associated with a type 1 immune response. *J Immunol* **157:** 3598–3604.

Schulick RD, Clerici M, Dolan MJ et al (1993) Limiting dilution analysis of interleukin-2-producing T cells responsive to recall and alloantigens in human immunodeficiency virus-infected and uninfected individuals. *Eur J Immunol* **23:** 412–417.

Shearer GM & Clerici M (1996) Protective immunity against HIV infection: has nature done the experiment for us? *Immunol Today* **17:** 21–24.

Smith KJ, Skelton HG, Drabick JJ, McCarthy WF, Ledsky R & Wagner KF (1994) Hypereosinophilia secondary to immunodysregulation in patients with HIV-1 disease. *Arch Dermatol* **130:** 119–125.

Smolen JS, Bettleheim P, Koller U et al (1985) Deficiency of the autologous mixed lymphocyte reaction in patients with classic hemophilia treated with commercial factor VIII concentrate: correlation with T cell subset distribution, antibodies to lymphodenopathy-associated or human T lymphotropic virus, and analysis of the cellular basis of the deficiency. *J Clin Invest* **75:** 1828–1834.

Sölder BM, Reisinger EC, Koefler D, Bitterlich G, Wachter H & Dierich MP (1989) Complement receptors: another port of entry for HIV. *Lancet* **ii:** 271–272.

Stoiber H, Ebenbichler C, Schneider R, Janatova J & Dierich MP (1995a) Interaction of several complement proteins with gp120 and gp41, the two envelope glycoproteins of HIV-1. *AIDS* **9:** 19–26.

Stoiber H, Schneider R, Janatova J & Dierich MP (1995b) Human complement proteins C3b, C4b, factor H and properdin react with specific sites in gp120 and gp41, the envelope proteins of HIV-1. *Immunobiology* **193:** 98–113.

Stoiber H, Pintér C, Siccardi AG, Clivio A & Dierich MP (1996) Efficient destruction of human immunodeficiency virus in human serum by inhibiting the protective action of complement factor H and decay accelerating factor (DAF, CD55). *J Exp Med* **183:** 307–310.

Susal C, Kirschfink M, Kropelin M, Daniel V & Opelz G (1994) Complement activation by recombinant HIV-1 glycoprotein gp120. *J Immunol* **152:** 6028–6034.

Tersmette M, Gruters RA & De Wolf F (1989) Evidence for a role of virulent human immuno-deficiency virus (HIV) variants in the pathogenesis of acquired immunodeficiency syndrome: studies on sequential isolates. *J Virol* **63:** 2118–2225.

Thieblemont N, Haeffner-Cavaillon N, Haeffner A, Cholley B, Weiss L & Kazatchkine MD (1995) Triggering of complement receptors CR1 (CD35) and CR3 (CD11b/CD18) induces nuclear translocation of NF-kappa B (p50/p65) in human monocytes and enhances viral repli-cation in HIV-infected monocytic cells. *J Immunol* **155:** 4861–4867.

Trimble LA & Lieberman J (1997) CD8+ T lymphocytes in the blood of HIV-infected individ-uals may have impaired function because they lack CD3zeta, the signaling chain of the T cell receptor complex. *Fourth Conference on Retroviruses and Opportunistic Infections*, Washington, DC, 1997 (abstr 109).

Varadhachary AS, Perdow SN, Hu C, Ramanarayanan M & Slagame P (1997) Differential ability of T cell subsets to undergo activation-induced cell death. *Proc Natl Acad Sci USA* **94:** 5778–5783.

Via CS, Tsokos GC, Stocks NI, Clerici M & Shearer GM (1990) Human in vitro allogeneic responses: demonstration of three pathways of T helper cell activation. *J Immunol* **144:** 2524–2528.

Vigano' A, Principi N, Villa ML et al (1995) Immunologic characterization of children vertically infected with human immunodeficiency virus, with slow or rapid disease progression. *J Pediatr* **126:** 368–374.

Vlach J & Pitha PM (1992) Activation of human immunodeficiency virus type 1 provirus in T-cells and macrophages is associated with induction of inducer-specific NF-kappa B binding proteins. *Virology* **187:** 63–72.

Walker CM, Moody DJ, Stites DP & Levy JA (1986) CD8+ lymphocytes can control HIV infec-tion in vitro by suppressing virus replication. *Science* **234:** 1563–1566.

Wang Y, Ardestani SK, Liang B, Becham C & Watson RR (1994) Anti IL-4 monoclonal antibody and IFNg administration retards development of immune dysfunction and cytokine dysregu-lation during murine AIDS. *Immunology* **83:** 384–389.

Weissman D, Li Y, Orenstein JM & Fauci AS (1995) Both a precursor and a mature population of dendritic cells can bind HIV. However, only the mature population that expresses CD80 can pass infection to unstimulated CD4+ T cells. *J Immunol* **155:** 4111–4117.

Zhang C, Yan C, Houston S & Chang LJ (1996) Protective immunity to HIV-1 in SCID/beige mice reconstituted with peripheral blood lymphocytes of exposed but uninfected individuals. *Proc Natl Acad Sci USA* **93:** 14720–14725.

Zhang X, Brunner T, Carter L et al (1997) Unequal death in T helper (Th) 1 and Th2 effectors;

Th1, but not Th2 effectors undergo rapid Fas/FasL-mediated apoptosis. *J Exp Med* **185:** 1837–1849.

Zou W, Lackner AA, Simon M et al (1997) Early cytokine gene expression in lymphnodes of macaques infected with SIV is predictive of disease outcome and vaccine efficacy. *J Virol* **71:** 1227–1236.

Chapter 9

CHEMOKINE RECEPTORS AND HIV-1 PATHOGENESIS:
A viral fatal attraction

Michael A. Norcross

INTRODUCTION

When we consider the devastation caused by acquired immune deficiency syndrome (AIDS), it may be difficult to imagine that this infectious disease usually begins with a meeting between a single viral particle and an unsuspecting immune-system cell patroling a mucosal border of the human body. Like other viruses, human immuno-deficiency virus type 1 (HIV-1) must invade and parasitize human cells to ensure its survival. However, before it can enter a target cell, HIV-1 must first evade the immune defenses and attach to the host cell membrane. The attachment of the virus to specific receptors on the cell surface, including the CD4 protein, expressed by certain T cells and macrophages of the human immune system, triggers a complex cascade of molecular interactions between viral proteins and molecules of the host cell. These events allow the virus to fuse with and to pass through the cell membrane and then to release its deadly genetic cargo.

Once inside the cell, the HIV-1 RNA is converted into DNA copies by the viral enzyme reverse transcriptase, a process prone to errors that can eventually help the virus to escape from continuing immune surveillance or therapeutic control. The DNA copies of the viral genome are then integrated into human chromosomes, and thereby come to dominate the mission and fate of the host cell. New viral RNA, produced from the integrated HIV-1 DNA, is coated with core proteins and then assembled together with other essential viral proteins at the periphery of the infected cell; it then buds from the cell membrane, taking with it embedded viral envelope proteins required to infect the next cell as well as some host proteins normally found on the surface of T cells or macrophages. Some of these host proteins also may help the virus adhere to target cells.

In biological fluids, HIV-1 is present either free in solution or associated with inflammatory or lymphoid cells. The vigorous antibody and cellular immune response that occurs in infected individuals probably means that most virus particles are coated with antibodies in addition to other host proteins, all of which may act as camouflage to protect HIV-1 during its journey to a new host cell.

The first cells to become infected, either macrophages or dendritic cells (Spira et al, 1996), migrate from the affected tissues to the regional immune centers, the lymph

HIV and the New Viruses Second Edition
ISBN 0-12-200741-7

nodes. Progeny viruses are then assembled and released, leaving the carcass of the host cells as they repeat this deadly cycle with bystander cells until the infection is spread throughout the immune system.

Usually, a strong immune response involving the production of both antibodies and T helper and CD8+ killer cells achieves partial control of the initial infection. However, the result of this battle is that the virus establishes a chronic, steady-state infection. The extent to which viral production is controlled at this steady state is predictive of the future course of the disease and is determined by both viral and host factors. Even in the asymptomatic phase of the infection, which can last for 10 years, billions of viral particles are produced each day, and this viral replication is accompanied by the continual loss and regeneration of CD4-positive cells.

HIV-1 CELLULAR TROPISM

We now know that not all HIV-1 viruses are alike in their biological properties, a point that is fundamental to understanding how the virus enters cells. Thus, strains of HIV-1 differ markedly in their preferences for target cells, a property referred to as 'cell tropism'. On the basis of cell tropism, HIV-1 viruses can be classified into two main categories: viruses that infect normal activated T cells and transformed (immortal) T cell lines are referred to as T cell-tropic (or T-tropic); viruses that infect macrophages and normal quiescent T cells are termed macrophage-tropic (or M-tropic). The tropism for different cell types is determined predominantly by the amino acid content and charge of the third variable (V3) region of gp120 (De Jong et al, 1992; Hwang et al, 1992), although other envelope regions can affect tropism. This region of gp120 is also an important target for neutralizing antibodies and has been referred to as the principal neutralizing domain of the viral envelope.

The T-tropic viruses were the first to be isolated from AIDS patients and were easily grown in the laboratory in transformed lymphocyte cell lines. These viruses characteristically show aggressive properties in vitro, being highly cytopathic and inducing, through cell fusion, the formation of large multinucleated cells; this has led to the term 'syncytium-inducing viruses'. These viruses tend to be detected later in the course of HIV-1 infection and are a hallmark of disease progression (Schuitemaker et al, 1992), as reflected in the rapid loss of CD4 cells. Positively charged amino acids in the V3 region of gp120 are important in conferring the T-tropic phenotype (De Jong et al, 1992). In the laboratory, the replication of T-tropic viruses is sensitive to antibody neutralization, to a soluble version of CD4, and to negatively charged small compounds such as dextran sulfate. Because of the relative ease with which these viruses can be grown, they have served as the major focus of HIV-1 studies and much has been learned about their structure and biological characteristics. Thus, many of the studies underlying the development of antiviral drugs and vaccines have been based on T-tropic viruses. These viruses are now referred to as X4 viruses because the use of the CXCR4 chemokine receptor for virus entry (Berger et al 1998) (see below).

However, it became clear that T-tropic viruses were not the type of virus found in most individuals shortly after infection or during the asymptomatic period. Instead, these 'primary' viruses were shown to be predominantly M-tropic (Zhu et al, 1993; van 't Wout et al, 1994). Unlike T-tropic viruses, M-tropic viruses are resistant to antibody neutralization and to the inhibitory effects of soluble CD4 and negatively charged compounds. The M-tropic viruses are now classified as R5 viruses based on the preferential use of the CCR5 chemokine coreceptor.

Electron microscopy has revealed that the HIV-1 envelope proteins resemble spikes on the surface of the particle's membrane coat and are the principal viral players in the process of cell binding and entry. The envelope glycoproteins are composed of two subunits: gp41, which has a molecular mass of 41 kDa, extends across the viral membrane, and contains the 'fusion peptide' needed to fuse the viral membrane with the target cell membrane; and gp120, which has a molecular mass of 120 kDa, is non-covalently attached to gp41, and contains sites that mediate binding to CD4 and other receptors on the target cell during virus entry. Analysis of the amino acid sequences of gp120 from viruses derived from infected individuals has revealed that the proteins contain both conserved and variable regions and that they contain multiple carbohydrate attachment sites. This sugar coating makes up 50% of the weight of gp120 and probably protects it from protein-degrading enzymes and immune attack by antibodies. The gp120–gp41 complex exists in groups (oligomers) of three on the surface of the virus particle. As a result of the close packing of these subunits and the attached carbohydrate residues, probably only minute portions of the protein backbone of gp120 are exposed.

CELL SURFACE MOLECULES MEDIATING HIV INFECTION

The CD4 protein was the first cell surface molecule discovered to bind HIV-1 through the gp120 envelope protein (Dalgleish et al, 1984; Klatzmann et al, 1984). Antibodies specific for CD4 were shown to block infection by and replication of viruses in culture, which led to initial excitement that a soluble form of CD4 (sCD4) also might inhibit HIV-1 infection and serve as a basis both for treatment of infected individuals and for preventing new infections. Unfortunately, this excitement was misplaced, mainly because the M-tropic viruses present in infected people differed from the T-tropic viruses studied in the laboratory in their affinity for sCD4 (Moore et al, 1992).

The molecular cloning of CD4 and its engineered synthesis in different cell types revealed that non-human cells expressing this protein could bind HIV-1 but would not permit fusion of the viral membrane with the cell membrane (Maddon et al, 1986). This finding supported the idea that host cell proteins other than CD4 also were needed for virus infection. Subsequent studies have shown that several cellular molecules contribute to the interaction or fusion of HIV-1 with cells. The normal functions of many of these proteins, including leukocyte functional antigen[1] (LFA-1) (Hildreth and Orentas, 1989), intercellular adhesion molecule 1 (ICAM-1) (Butini et al, 1994), and CD44 (Dukes et al, 1995), proteoglycans (Patel et al, 1993; Roderiquez et al, 1995) and glycolipids (Harouse et al, 1991) are related to cell adhesion or attachment.

At the time of these findings, my laboratory was interested in the antiviral properties of small, negatively charged compounds – such as dextran sulfate and the natural anticoagulant heparin – that could inhibit the infection of cultured cells by T-tropic viruses (Callahan et al, 1991). These compounds mimicked the action of a group of cell-surface proteins termed heparan sulfate proteoglycans (HSPGs), which contain sugar and sulfur and are related to heparin. The T-tropic viruses were observed to attach to the cell surface HSPGs through the positively charged V3 region of gp120 at the same time that other regions of the envelope protein bind to CD4 (Patel et al, 1993; Roderiquez et al, 1995). Negatively charged molecules such as dextran sulfate (Callahan et al, 1991), as well as antibodies that bind to the V3 site, prevent the interaction of gp120 with HSPGs and also can block interactions of HIV-1 with specific chemokine coreceptors (see below). The HSPGs facilitate virus binding to CD4 and its coreceptors, but alone do not mediate HIV-1 infection.

An obstacle to studying the biology of M-tropic HIV-1 was the lack of a system to grow the viruses reproducibly in the laboratory. Most early attempts at virus culture relied on peripheral blood cells and macrophages as host cells. Fortunately, Paolo Lusso, working in the laboratory of Robert Gallo at the National Cancer Institute, while trying to clone cells able to grow another virus – human herpesvirus 7 – serendipitously produced a cell clone (PM-1) that would support the replication of both T-tropic and M-tropic HIV-1 (Lusso et al, 1995). Using PM-1 cells as hosts, Bou-Habib et al (1994) were able to define the mechanism by which M-tropic viruses resist antibody neutralization: it appears that V3 and other regions of gp120 targeted by neutralizing antibodies are less exposed in M-tropic viruses than in T-tropic isolates, probably as a result of tighter subunit packing in envelope oligomers.

CHEMOKINES AND HIV SUPPRESSION

It was assumed that PM-1 cells must express a specific receptor in addition to CD4 that mediates the entry of M-tropic viruses. The first clue to the identity of this second receptor came in 1995 from research led by Lusso. It was known that CD8-positive cells from individuals infected with HIV-1 secreted soluble factors that suppressed the replication of M-tropic viruses in CD4+ cells. With the use of the PM-1 cell culture system, Lusso and colleagues showed that these inhibitory factors comprised a mixture of the CC chemokines known by the acronyms RANTES (regulated on activation, normal T expressed and secreted), macrophage inflammatory protein 1α (MIP-1α), and MIP-1β (Cocchi et al, 1995). The purified chemokines inhibited predominantly M-tropic isolates in both the PM-1 cell line and in T lymphocytes. The T-tropic strain IIIB was not inhibited; however, MN, a classic version of a T-tropic virus, was blocked in lymphocyte cultures, suggesting some overlap in phenotype specificity. A related observation by Paxton et al (1996) found that CD4+ cells from exposed uninfected individuals were resistant to M-tropic infection but not to T-tropic viruses and that resistance to infection correlated with the production of chemokines.

Chemokines (chemoattractant cytokines) are secreted from cells in an inflammatory reaction and attract lymphocytes, macrophages, and other leukocytes from the blood and tissues to the site of inflammation – see reviews by Murphy (1996) and Baggiolini et al (1997). They induce a variety of effects in the attracted cells – including rearrangement of the cytoskeleton and changes in the expression of adhesion molecules – that regulate both cell movement and interactions with the extracellular matrix and other cells. Chemokines are classified on the basis of conserved amino acid sequences as well as the number and position of specific cysteine residues. The CXC chemokines, which contain two conserved cysteine residues separated by any amino acid, mostly attract neutrophils, whereas the CC chemokines, which contain two adjacent cysteines, attract both monocytes and lymphocytes. Several receptors for CXC and CC chemokines have been identified and are referred to collectively as CXCR and CCR, respectively. These proteins are members of a large family of receptors – including those for many hormones and neurotransmitters – that contain seven domains which span the cell membrane and signal by coupling to G proteins inside the cell.

CHEMOKINE RECEPTORS AND ENVELOPE FUSION

Early in 1996, Ed Berger and colleagues at the National Institutes of Health (NIH) took another important step toward identifying the HIV-1 coreceptor. These researchers cloned a gene that encodes a chemokine receptor-like protein (Feng et al, 1996). When this protein was expressed in non-human cells together with CD4, the cells would allow not only the binding of T-tropic viruses but also their fusion. The researchers adopted the term 'fusin' for this HIV-1 coreceptor, although it was later renamed CXCR4 when its normal ligand was identified as the CXC chemokine stromal cell-derived factor 1 (SDF-1) (Bleul et al, 1996; Oberlin et al, 1996). This finding suggested to a number of groups that the as yet unidentified receptor for the chemokines RANTES, MIP-1α and MIP-1β was probably the coreceptor for M-tropic viruses. A receptor with exactly this ligand specificity was subsequently cloned by Marc Parmentier's group at the Free University of Brussels in Belgium (Samson et al, 1996a). Several groups then engineered expression of this receptor, which was named CCR5, together with CD4 in cultured cells, and found that such cells could be infected with M-tropic HIV-1 – thus demonstrating that CCR5 is indeed the coreceptor for M-tropic (R5) viruses (Alkhatib et al, 1996; Choe et al, 1996; Deng et al, 1996; Doranz et al, 1996; Dragic et al, 1996). The M-tropic virus infection of cells transfected with CCR5 was inhibitable with the chemokines RANTES, MIP-1α and MIP-1β. Groups contributing to this research included the laboratories of Richard Koup, Nathanial Landau, and John Moore at the Aaron Diamond AIDS Research Center in New York, Dan Littman at the Skirball Institute in New York, Robert Doms at the University of Pennsylvania, Ed Berger at the NIH and Joseph Sodroski of Dana Farber Cancer Institute in Boston.

Doranz et al (1996) reported that the envelope of a strain of virus capable of growing in both cell lines and macrophages, 89.6, was able to fuse with CD4+ cells expressing either CXCR4, CCR5, CCR2b or CCR3. Choe et al (1996) found that envelopes from different M-tropic subtypes used CCR3 and CCR5 and that the envelope V3 region controlled coreceptor envelope fusion specificity. Although other chemokine receptors have subsequently been shown to mediate viral fusion, most primary HIV-1 isolates use CCR5. In most instances, both CD4 and a chemokine receptor are needed for envelope fusion. Primary strains of viruses from various parts of the world have been shown to use CCR5 or CXCR4 as coreceptors (Zhang et al, 1996), and some dual-tropic strains are able to infect cells expressing either coreceptor (Simmons et al, 1996).

Confirming the important role of CCR5 in HIV-1 infection, several individuals who have been exposed to the virus but have remained uninfected have been shown to carry a form of the CCR5 gene that is missing 32 base pairs (Liu et al, 1996; Samson et al, 1996b). This deletion results in the production of a defective, truncated protein which remains in the cytoplasm of the cells. The CD4+ lymphocytes and macrophages isolated from several of these healthy exposed individuals did not express CCR5 on their surface and could not be infected with M-tropic viruses – although the same cells

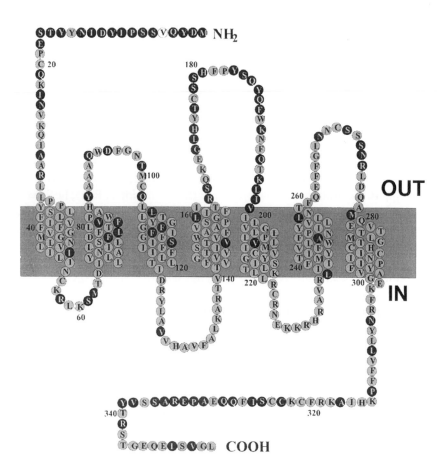

Figure 9.1 *CCR5 chemokine receptors (Courtesy of Dr P Murphy NIH)*

remained susceptible to infection by T-tropic HIV-1 (Connor et al, 1996). These findings confirmed the primary role for CCR5 coreceptors in infection and suggested that other chemokine receptors capable of mediating virus fusion in vitro may not be utilized efficiently for virus infection. The fact that these individuals do not show other immunological abnormalities suggests that the defect is not detrimental to the immune system. Although people in whom both copies of the CCR5 gene are defective are protected from HIV-1 infection, heterozygotes remain susceptible to infection but may show a slower progression of disease (Dean et al, 1996; Michael et al, 1997a). The truncated receptor may complex with full-length CCR5 resulting in reduced surface expression of CCR5 (Benkirane et al, 1997). Individuals with two mutant CCR5 genes should be susceptible to infection by T-tropic viruses, although it is possible that these viruses are not present in their infected partners at the time of exposure or are transmitted less efficiently. Several homozygote mutant CCR5 individuals have been infected with HIV although the phenotypes of the viruses have not been reported. Although 20% of the white population in western Europe are heterozygous for this receptor defect, the defective gene has not been found in African, Venezuelan, or Japanese populations.

Other mutations in the CCR2 and SDF-1 genes have been described with possible relevance to HIV disease progression. The CCR2 variant (V64I) is a conservative mutation in a transmembrane region which does not affect HIV-1 entry (Michael et al, 1997b; Smith et al, 1997). A linkage disequilibrium with a point mutation in the CCR5 regulatory region (Kostrikis et al, 1998) and possibly a lower level of expression of the receptor in these individuals may partially explain the association with delayed disease progression. It is also not clear how the mutation in the 3′ untranslated region of the SDF-1 gene (Winkler et al, 1998) is related to disease progression, although it is possible that this change leads to higher levels of SDF expression and enhanced antiviral effects.

With the discovery of chemokine receptors as coreceptors for HIV, the search for other HIV-compatible chemokine receptors has progressed at a rapid pace (Table 9.1). A number of chemokine receptor-like molecules have been identified which can mediate envelope fusion and in some cases virus infection of CD4+ cells. Five additional molecules with coreceptor activity have been characterized. STRL33 and an identical clone BONZO (Deng et al, 1997; Liao et al, 1997) are expressed in activated lymphocytes and mediate fusion of M-tropic (R5),T-tropic (X4) and dual-tropic (R5X4) HIV-1 and simian immunodeficiency virus (SIV) strains. GPR15 and BOB (Deng et al, 1997; Farzan et al, 1997a) mediate some strains of HIV and SIV. The receptors V28 (Reeves et al, 1997), a receptor for Fractakine, and CCR8 (Rucker et al, 1997), a receptor for I-309, mediate viral entry of all virus phenotypes. The US28 gene from CMV codes for a chemokine receptor capable of mediating fusion of multiple virus phenotypes (Pleskoff et al, 1997). The role of these receptors in mediating HIV or SIV infection in vivo is not clear. The resistance of T cells and macrophages from Δ32 CCR5 individuals to infection by M-tropic viruses, which can fuse with many of the chemokine receptors mentioned above, suggests that these other structures may not participate in virus infection in vivo. The fusion assays in vitro based on high expression of chemokine receptors and envelope proteins may be more sensitive to low-affinity interactions than cells expressing naturally occurring levels of chemokine receptors.

Table 9.1 *HIV and SIV chemokine coreceptors*

Chemokine receptor	Virus type	Chemokine specificity
CXCR4	X4, R5, R5X4	SDF-1
CCR2	R5X4	MCP-1, MCP-2, MCP-3
CCR3	R5X4	Eotaxin, RANTES
CCR5	R5, R5X4, HIV-2, SIV	RANTES, MIP-1α, MIP-1β
CCR8	R5X4, HIV-2, SIV	I-309
STRL33/BONZO	R5X4, SIV	?
GPR-15/BOB	R5X4, SIV	?
GPR-1	SIV	?
V28	R5X4, HIV-2, SIV	Fractalkine
US28	R5X4	RANTES, MIP-1α, MCP-1

X4, T-tropic virus using primarily CXCR4; R5, M-tropic virus using only CCR5; R5X4, dual-tropic virus

Several observations suggest that some viruses can interact with coreceptors independent of cell surface CD4. Type 2 HIV can infect CD4-negative cells through the CXCR4 receptor and can be blocked by antibody to CXCR4 (Endres et al, 1996), while some strains of SIV have been selected to infect through CCR5 without CD4 (Edinger et al, 1997). These results suggest that virus may have initially used chemokine receptors for entry and then evolved to use CD4 as a cofactor.

REGIONS ON CHEMOKINE RECEPTORS INTERACTING WITH HIV

A number of studies have addressed the sites on chemokine receptors which interact with virus envelope and mediate membrane fusion. Most investigators have constructed chimeric chemokine receptors with different members of the CCR or CXCR families. The CCR5 and CXCR4 restricting elements have been mapped by generating chimeric and mutant constructs, including mouse–human CCR5 chimeras (Bieniasz et al, 1997; Picard et al, 1997), hCCR5–CCR2 (Rucker et al, 1996; Wu et al, 1997a), CXCR4– CXCR2 (Lu et al, 1997), human–rat CXCR4 (Brelot et al, 1997), CCR5–CXCR4 (Lu et al, 1997) and chimeras with combinations of CCR and CXCR (Doranz et al, 1997b) The results demonstrate that multiple sites on CCR5 and CXCR4 influence interactions with virus and that the contact sites vary depending on the specific virus isolate tested. The amino terminus of CCR5 including aspartic acids at position 2 and 11 and a glutamic acid at 18 is required for M-tropic HIV and SIV envelope-mediated membrane fusion (Dragic et al, 1998). Tyrosine amino acids in the amino terminus are also critical to the function of the coreceptor (Farzan et al, 1998).

MECHANISM OF CHEMOKINE INHIBITION OF HIV-1 RECEPTOR INTERACTIONS

The mechanism of chemokine antiviral activity in disrupting HIV-1 coreceptor interactions has been investigated by several research groups. Dragic et al (1996) using a membrane fusion assay demonstrated that chemokines blocked envelope-mediated fusion in T cell lines but had only weak inhibitory activity on macrophage fusion. CD4+ target cells required a 6 hour incubation with chemokines to prevent fusion. Oravecz et al (1996) observed that chemokines blocked fusion of M-tropic viruses with target cells, but did not prevent virus binding. The inhibitory activity of chemokines could be reversed after their removal from the culture medium, suggesting that the mechanism of inhibition was competitive. The V3 region of gp120 was found to be critical for both the interaction of virus with the chemokine receptor and chemokine blocking effects. Replacement of the gp120 V3 region of a T-tropic virus with the corresponding region from an M-tropic strain resulted in a virus that was sensitive to inhibition by CC chemokines (Oravecz et al, 1996). Transmission of biochemical signals (G protein and tyrosine kinase) through chemokine receptors was not required for either virus infection or chemokine inhibition of infection. From these results it was suggested that chemokines blocked infection of cells with M-tropic viruses by competing with the V3 region of gp120 for binding to their cognate receptors or by downmodulating receptor expression. Cocchi et al (1996) confirmed the association of the M-tropic V3 region in gp120 with chemokine blocking of infection and that G protein signaling was not required for inhibition. Several groups have found that virus fusion and infection can occur using mutant forms of chemokine receptors devoid of signaling activity (Atchison et al, 1996; Alkhatib et al, 1997; Aramori et al, 1997; Farzan et al, 1997b).

Oravecz et al (1996) proposed that the structure of the V3 region of gp120, either alone or in the context of other regions of the protein, mimics the portion of the chemokine molecule that interacts with the chemokine receptor. It was also suggested that the virus envelope may function through the chemokine receptor to induce chemotaxis of virus-susceptible cells. Weissman et al (1997) showed that gp120 from a M-tropic virus induced calcium signals and chemotaxis in normal cells, possibly through chemokine receptor signaling. In addition to the G protein signaling common to chemokine receptors, the envelope has been found to stimulate tyrosine phosphorylation through the pyk-2 kinase (Davis et al, 1997). These findings support a direct role for envelope in modifying the function of cells through interactions with chemokine receptors along with other associated molecules.

Several studies have demonstrated that free gp120 from an M-tropic isolate, once bound to CD4, interacts with the envelope-specific chemokine receptor and prevents specific chemokine (MIP-1β) binding to CCR5 (Trkola et al, 1996; Wu et al, 1996). The M-tropic gp120 alone had a weak affinity for CCR5, which was enhanced by CD4 binding. An envelope protein missing the amino and carboxy terminal ends, and portions of V1-V2 could still bind to CCR5 with sCD4; however, further deletion of the V3 loop prevented envelope interactions with the coreceptor (Wu et al, 1996). Virus neutralizing antibodies that recognize the V3 or adjacent sites on gp120 that become exposed on binding to CD4 prevent binding of gp120 to the coreceptor. Collectively the data demonstrate that the V3 site together with adjacent gp120 regions along with CD4 form a conserved conformational structure capable of interacting with the coreceptor. CD4 may also directly bind to the coreceptor as demonstrated by an interaction of the D1–D2 domain of CD4 with the CCR5 receptor (Wu et al, 1996).

The interaction of T-tropic gp120 with CXCR4 in principle may be similar to the M-tropic envelope interactions with CCR5. Using an immunoprecipitation method, Lapham et al (1996) showed that gp120 can coassociate with CXCR4, although Sattentau's group has reported difficulty in reproducing this result (Ugolini et al, 1997). Indirect evidence for an envelope-induced association of CD4 and the coreceptor comes from the observation that gp120 from a T-tropic virus will partially block the binding of a monoclonal antibody (12G5) to CXCR4 and that CD4 and CXCR4 colocalize in the presence of gp120 using confocal microscopy (Ugolini et al, 1997).

On the basis of these various observations, a mechanistic model can be presented for the events that lead to the fusion of HIV-1 with target cells (Figure 9.2). The initial interaction occurs between gp120 in the viral oligomeric envelope and CD4; this interaction is stronger for T-tropic viruses than for M-tropic viruses. The binding to CD4 and to heparan sulfate proteoglycan (HSPG) triggers conformational changes in the viral envelope complex that expose regions responsible for the interaction with chemokine receptors. These newly exposed regions of gp120, and possibly CD4 as well, then make contact with the coreceptor. In the case of T-tropic viruses, the V3 region and adjacent sites of gp120 interact with both CXCR4 and HSPGs, whereas comparable sites in gp120 of M-tropic viruses bind to CCR5. The positively charged amino acids in the V3

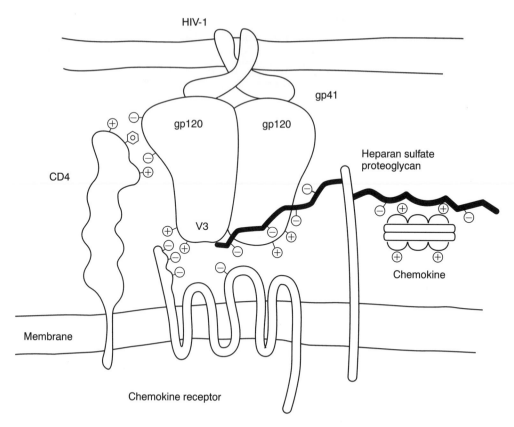

Figure 9.2 Model for HIV-1 and chemokine interactions with molecules on the CD4+ cell surface
Oligomeric envelope gp120 binds to CD4, and heparan sulfate proteoglycans to stabilize viruses on the cell surface. Conformational changes are induced in the envelope, exposing sites that interact with the chemokine coreceptor (CXCR4 or CCR5). The gp41 fusion peptide is uncovered and inserts into the cell membrane to mediate viral-cell membrane fusion. Positive charges in the V3 are shown to interact with negatively charged amino acids in the chemokine receptor and with HSPG. HSPG binds chemokines and facilitates binding to the chemokine receptor.

region of T-tropic viruses may directly bind to negatively charged regions of CXCR4 and with the negatively charged sugar chains on HSPGs. As gp120 docks with the chemokine receptor, the hydrophobic fusion peptide and its supporting structure, previously buried within gp41, are also exposed and insert into the target cell membrane – or, possibly, into the membrane-embedded portion of the coreceptor – thereby initiating the actual fusion of the viral and cell membranes.

CHEMOKINE RECEPTORS AND VIRAL TRANSMISSION

Most HIV-1 strains that are transmitted sexually and from mothers to infants are of the M-tropic type. The correlation of specific coreceptors with viral tropism suggests that such preferential transmission may be due to the selective expression of CCR5 on macrophages and dendritic cells that line the reproductive and intestinal mucosa. Alternatively, mucosal sites may be naturally rich in inhibitors of T-tropic viral fusion, such as the CXCR4 ligand SDF-1 and free HSPGs. The sensitivity of T-tropic viruses to antibody neutralization may also reduce their transmission.

The expression and function of coreceptors on cells of the macrophage and dendritic cell lineages has been studied by several groups. Most published reports find CXCR4 expression on the cell surface of monocytes, although these cells are resistant to T-tropic infection (Di Marzio et al, 1998; Yi et al, 1998). In contrast, CCR5 expression is not found on resting monocytes as measured by specific ligand signaling, monoclonal anti-CCR5 binding, and mRNA analysis (Wu et al, 1997b; Di Marzio et al, 1998; Wang et al, 1998). One group claims to detect CCR5 on monocytes using a rabbit antiserum to a large CCR5 N-terminal peptide, however, the specificity of binding is only 50% and it is possible that the antibodies cross-react with other proteins (Zaitseva et al, 1997).

Differentiation of monocytes to macrophages induces CCR5 expression (Oravecz et al, 1997a) and susceptibility to M-tropic virus infection (Di Marzio et al, 1998; Naif et al, 1998). Dendritic cells express CXCR4 and CCR5 and in some cases are susceptible to virus infection (Garcia-Zepeda et al, 1996; Granelli-Piperno et al, 1996). The finding that monocytes and macrophages are resistant to laboratory-adapted T-tropic virus isolates even though they express the CXCR4 coreceptor has led some investigators to speculate that this phenomenon underlies the preferential selection of M-tropic viruses during viral transmission. The fact that primary T-tropic viruses, unlike laboratory adapted viruses, can infect macrophages through CXCR4 argues that other mechanisms contribute to M-tropic selection (Yi et al, 1998). The fact that macrophages are deficient in HSPG expression may partially explain the resistance to HSPG-dependent laboratory-adapted T-tropic isolate infection.

Another contributing factor in macrophage resistance to T-tropic infection relates to the effect of receptor concentration on virus entry. Primary T-tropic viruses require higher levels of cell surface CD4 for infection than the laboratory-adapted T-tropic or M-tropic viruses (Kabat et al, 1994). Macrophage-tropic viruses despite bearing low-affinity CD4-binding envelopes are able to infect cells with low CD4 levels: this observation suggested that T-tropic infection of macrophages may require higher levels of CD4 than are naturally present on the cell surface.

The T cells are also likely to be initial targets for virus infection and transmission as a consequence of regulated expression of coreceptors and CD4. Chemokine coreceptor expression on lymphocyte subsets has been examined using monoclonal antibodies and mRNA analysis. Expression of CCR5 is low on resting naïve CD45RA, $CD26^{low}$ T cells (Wu et al, 1997b). Memory cells characterized by high expression of CD45R0+ and CD26 are the most positive for membrane expression of CCR5, and cell activation slowly increases CCR5 expression. HIV-1 infectibility by M-tropic isolates correlates with CCR5 levels. In contrast, CXCR4 expression is lower on memory cells but high on naïve resting cells with the $CD26^{low}$ CD45RA+RO− phenotype (Bleul et al, 1997). Cell activation rapidly induces CXCR4 expression on T cells. The reciprocal expression of CXCR4 and CCR5 on T cell subsets suggests that T cells will differ in susceptibility to T cell-tropic (X4) and monocytotropic (R5) viruses and that alterations in the cell environment such as through cytokine and T cell receptor stimulation will regulate virus infection.

CHEMOKINE RECEPTORS, PATHOGENESIS AND DISEASE PROGRESSION

A critical step in disease progression in many HIV-1-infected individuals is the transition from a predominance of M-tropic viruses, which interact primarily with the CCR5 coreceptor, to a preponderance of T-tropic viruses, which bind to CXCR4 or bind to both coreceptors. The factors responsible for this transition are not known but may include (a) reduced production of, or the escape of T-tropic viruses from, neutralizing antibodies or killer T cells; (b) decreased synthesis of chemokines, such as SDF-1, that inhibit infection of cells with T-tropic viruses; or (c) increased viral replication as a result of immune system activation caused by opportunistic infections, virus resistance to chemokine blocking or decreased suppressor factor/chemokine synthesis. As HIV-1 replication increases, so does the frequency of mutant viruses generated randomly by the error-prone reverse transcriptase that copies the viral RNA into DNA. These errors may ultimately lead to the production of viruses resistant to any intervention, whether it be drug therapy or a protective immune response.

Studies characterizing primary viruses have shown wide isolate variability in sensitivity to chemokines (Jansson et al, 1996; Trkola et al, 1998). Resistance to CC chemokine blocking is often seen in isolates from patients with progressive disease and is usually associated with an expansion of viral tropism from CCR5 to the CXCR4 coreceptor (Connor et al, 1997; Scarlatti et al, 1997). However, viral resistance to CC chemokines can be detected in some progressing patients even in the absence of CXCR4 virus types (Scarlatti et al, 1997), suggesting evolution in viral pathogenicity without associated changes in viral coreceptor usage. Chemokine resistance is not characteristic of all late-stage virus isolates, and examples of viruses from AIDS patients have been reported which can be inhibited by low concentrations of chemokines (Jansson et al, 1996).

Some T-tropic viruses isolated from HIV-1-infected individuals retain affinity for CCR5, suggesting that viruses of dual tropism may be intermediate forms during disease progression (Simmons et al, 1996). However, 50% of patients who progress to AIDS do not appear to contain T-tropic strains; rather, the M-tropic strains with which they are infected may show increased virulence, possibly because of increased envelope affinity for either CD4 or CCR5.

REGULATION OF CHEMOKINE ANTIVIRAL ACTIVITY: ROLE OF CELL SURFACE PROTEOGLYCANS

In addition to their role in facilitating virus attachment to cells, HSPGs play an important role in regulating the antiviral activity of chemokines (Oravecz et al, 1997b). The potency of chemokines in blocking viral fusion depends on the type of cell expressing the specific chemokine receptor and CD4 and on the specific chemokine. For example, macrophages are particularly resistant to the antiviral activity of CC chemokines, even though CCR5 functions as the primary coreceptor for infection (Dragic et al, 1996). In addition, the noted hierarchy in the chemokine potency on HIV, with RANTES having the highest activity followed by MIP-1β with intermediate activity and MIP-1α with low activity, is not explained by the direct affinity of these chemokines for CCR5. This suggested that other molecules were involved in the interactions of chemokine with the cell surface. Following reports that chemokines could bind to tissue proteoglycans, the role of heparan sulfate in mediating chemokine antiviral effects was examined. The antiviral activity of RANTES and MIP-1β was found to depend on the expression of heparan sulfate on the cell surface. Enzymatic removal of HSPGs from cells results in marked resistance to the blocking effects of RANTES and MIP-1β. Loss of activity correlated with a reduction in chemokine cell surface binding and thus RANTES must bind to HSPGs for efficient subsequent interactions with CCR5. RANTES does bind to chondroitin sulfate proteoglycans on the cell surface, but in this form is unable to inhibit virus infection. Macrophages are relatively deficient in cell-surface HSPGs, and thus bind fewer RANTES molecules than HSPG-rich CCR5-expressing cells. Adding exogenous HSPGs to macrophage cultures together with RANTES enhanced the antiviral effects of the chemokine (Wagner et al, 1998).

Several possible mechanisms can be considered for the function of HSPGs in enhancing chemokine activity. They may bind chemokines and present a multivalent array of ligands or induce oligomerization, thereby increasing the affinity or avidity of binding to the receptor. Recent structural data support a role for proteoglycan binding in inducing chemokine oligomerization (Hoogewerf et al, 1997). The sensitivity of cells to chemokine antiviral effects therefore depends on the appropriate expression of proteoglycans on the cell surface and suggests that strategies based on proteoglycan-mediated pathways may enhance chemokine effects on virus replication. Chemokines may exist in cytolytic cells in a complex with proteoglycans and other molecules involved in cytolytic effector functions (Wagner et al, 1998) which naturally potentiate chemokine activities.

THE ROLE OF CD26 AND REGULATION OF CHEMOKINE STRUCTURE AND FUNCTION

Considerable controversy has surrounded the role of CD26 in HIV biology. The report from the Hovanessian laboratory that CD26 mediated the infection of T-tropic viruses (Callebaut et al, 1993) was met with great skepticism and was not reproducible. However, CD26 expression does correlate with infection of M-tropic viruses in the PM-1 cell line (Oravecz et al, 1995). Cells expressing high levels of CD26 are easily infectable and support rapid virus replication and cytopathicity. CD26 low-expressing cells are resistant to M-tropic entry and infection, but were easily infectable with T-tropic viruses. The association of CD26 expression primarily with M-tropic infection suggested a relationship to chemokine receptor restriction and chemokine blocking effects. In fact, recent evidence suggests that CD26 expression inversely correlates with the production of the RANTES chemokine. Cells with low levels of CD26 secrete high levels of RANTES compared with cells with high levels of CD26, which secrete low amounts of RANTES. We believe that CD26 regulates chemokine secretion through its enzymatic activity directly on chemokines.

CD26 is a leukocyte activation marker which possesses dipeptidyl-peptidase IV (DPPIV) activity but whose natural substrates and immunologic functions have not been clearly defined. This enzyme is also a leukocyte differentiation antigen, expressed on the cell surface and as a secreted form mostly by activated memory T lymphocytes and macrophages. DPPIV cleaves the first two amino acids from peptides with penultimate proline or alanine residues, although no natural substrate with immune function has been identified. A number of chemokines share a conserved NH$_2$-X-Pro sequence (X is any amino acid) at the NH$_2$-terminus, which conforms to the substrate specificity of DPPIV. Several chemokines, including RANTES, MCP-2, eotaxin, and IP-10 were tested as substrates for recombinant soluble human CD26 (sCD26) (Oravecz et al, 1997a). Following incubation with CD26 all of these chemokines were cleaved after the proline in the second position as demonstrated by mass spectrometric analysis. The cleaved form of RANTES was synthesized and compared with the parental molecule for biological activity. Interestingly, the truncated form of RANTES(3–68) lacked the ability of native RANTES(1–68) to increase the cytosolic calcium concentration in human monocytes, but it still induced this response in macrophages activated with macrophage colony-stimulating factor (M-CSF). Analysis of chemokine receptor messenger RNAs and patterns of desensitization of chemokine responses showed that the differential activity of the truncated molecule results from an altered receptor specificity. RANTES(3–68) showed reduced activity, relative to that of full-length RANTES(1–68), with cells expressing the CCR1 chemokine receptor, but it retained the ability to stimulate CCR5 receptors and to block HIV-1 infection. These results indicate that CD26-mediated processing of RANTES, and possibly other chemokines, may regulate differential cell recruitment into inflammatory sites.

The CD26 cleavage of chemokines may have an impact on HIV infection by several mechanisms. First, CD26-cleaved chemokines may have altered HIV blocking activity because of differences in the affinity of binding to specific coreceptors; for example, a synthetic form of SDF-1 missing two N-terminal amino acids, consistent with the product of CD26-mediated cleavage, was found to lose HIV-1 blocking activity and may function as an antagonist (Crump et al, 1997). Alternatively, chemokines with altered chemokine receptor specificity may be defective in recruiting cells to sites of virus infection. Finally, chemokines that enhance virus replication through cell signaling may be inactivated after CD26 cleavage. It will be important to determine whether the activity of other HIV-1 blocking chemokines such as MIP-1β and eotaxin are altered as a consequence of CD26-mediated cleavage.

CHEMOKINE HIV SUPPRESSOR FACTOR

It has long been known that CD8+ cells from individuals infected with HIV are able to suppress the replication of HIV in CD4+ cells and that part of this activity was due to secreted factors (Mackewicz and Levy, 1992). CD8+ cell-derived factors were able to suppress both T- and M-tropic viruses and differed from the antiviral activity attributed to CC chemokines (Moriuchi et al, 1996). Although part of the suppressive effects could be explained by the presence of a combination of CC chemokines, it became clear that antibodies that neutralize RANTES, MIP-1α and MIP-1β did not inhibit suppression of T-tropic viruses in CD4+ lymphocytes. Recently, a truncated form of macrophage-derived chemokine (MDC) missing two amino acids from the N-terminus was purified from transformed CD8+ cells and found to suppress both viral phenotypes in CD4+ cells (Pal et al, 1997). Several CD8+ clones expressing the message for MDC did not secrete an active factor, suggesting that post-translational modification of the factor was required for either secretion or activation. Based on our results that CD26

DPPIV cleaves chemokines containing an X-proline at the N-terminus, we hypothesized that cleavage by CD26 may activate the suppressive function of MDC. Recently, we have confirmed that soluble CD26 is capable of cleaving two amino acids from the N-terminus of MDC to generate a form of MDC that suppresses HIV. The full-length form of MDC, although capable of activating calcium flux in T cells, did not suppress HIV replication. Our observation indicates that control of HIV by suppressor factors may be intimately associated with the expression of the CD26 molecule. It is interesting to note that progression of HIV correlates with the emergence of activated CD8+ cells bearing DR+ and CD38+, which lack CD28 and CD26 molecules. CD8+ suppressor cell activity is associated with cells expressing CD28. We believe that the expression of CD26 is a key regulator of MDC HIV suppressive activity and that strategies that modify CD26 expression may potentiate viral suppression.

CHEMOKINES FOR POTENTIAL HIV-1 THERAPY

Chemokines may prove effective both in preventing initial HIV-1 infection and in inhibiting ongoing infection. Because individuals with defective CCR5 receptors appear normal and receptor signaling is not required for chemokine antiviral activity, therapeutic strategies that target these receptors may not have detrimental or toxic side-effects. Antagonistic forms of CC chemokines, which do not initiate receptor signaling, are thus promising candidates for anti-HIV-1 drugs. Several forms of antagonist chemokines have be reported, including a truncated RANTES form missing eight amino acids (Arenzana-Seisdedos et al, 1996), and a RANTES with an AOPentane on the N-terminus (Simmons et al, 1997). The AOP-RANTES shows high-affinity binding to cells expressing transfected CCR5 and is able to inhibit macrophage HIV-1 infection. Topical formulations of such agents able to coat a mucosal surface could possibly prevent viral transmission. However, it must be remembered that, although RANTES prevents membrane fusion, it does not block virus binding to cells in culture (Oravecz et al, 1996) and, after its removal, viral replication quickly recovers. Thus, therapy with chemokine analogs alone may require the continuous administration of high doses, which eventually may exert toxic effects.

Other strategies targeting chemokine receptors have been considered. Several peptides have been reported to block the CXCR4 receptor. The peptide T22 (Murakami et al, 1997) and the compound ALX40–4C (Doranz et al, 1997a) both specifically block infection and fusion of CXCR4 viruses and SDF-1 activation through CXCR4. These peptides are both cationic which may explain their specificity for CXCR4, which has more negatively charged amino acids in the N-terminal region than CCR5. Small molecules have also been reported to inhibit CXCR4. The bicyclam, AMD3100, binds to CXCR4 and blocks envelope and SDF-1 interactions with the receptor (Schols et al, 1997; Donzella et al, 1998).

Our new understanding of the interactions of HIV-1 with CD4 and chemokine receptors hopefully will help us to devise strategies of viral receptor blockade that can prevent infection after an individual is exposed to the virus. Furthermore, the uniting of such strategies with the recent breakthroughs in combination drug therapy – in which drugs that inhibit the HIV-1 reverse transcriptase are administered together with those that inhibit the viral protease – may totally deplete residual virus from infected individuals. Perhaps a more formidable task is the development of an effective vaccine that will permanently prevent HIV-1 infection. A clearer understanding of the mechanism of virus entry into human cells and the role of coreceptor interactions with the viral envelope can only help achieve this goal.

ACKNOWLEDGEMENTS

I would like to acknowledge and am grateful for the contributions of my colleagues, Tamas Oravecz, Greg Roderiquez, Jinhai Wang, Mahesh Patel, Marina Pall, Justin Koffi, Masaki Yanagishita, Robert Boykins, Mary Ditto, Larry Callahan, Dumith Chequer Bou-Habib, Ennan Guan, Penny Robbins and Shigeru Suga, to the work from this laboratory. I thank Phil Murphy for generously providing the diagram of CCR5.

REFERENCES

Alkhatib G, Combadiere C, Broder CC et al (1996) CC CKR5: a RANTES, MIP-1alpha, MIP-1beta receptor as a fusion cofactor for macrophage-tropic HIV-1. *Science* **272**(5270): 1955–1958.

Alkhatib G, Ahuja SS, Light D et al (1997) CC chemokine receptor 5-mediated signaling and HIV-1 co-receptor activity share common structural determinants. Critical residues in the third extracellular loop support HIV-1 fusion. *J Biol Chem* **272**(32): 19771–19776.

Aramori I, Zhang J, Ferguson SS et al (1997) Molecular mechanism of desensitization of the chemokine receptor CCR-5: receptor signaling and internalization are dissociable from its role as an HIV-1 co-receptor. *EMBO J* **16**(15): 4606–4616.

Arenzana-Seisdedos F, Virelizier JL, Rousset D et al (1996) HIV blocked by chemokine antagonist (letter). *Nature* **383**(6599): 400.

Atchison RE, Gosling J, Monteclaro FS et al (1996) Multiple extracellular elements of CCR5 and HIV-1 entry: dissociation from response to chemokines. *Science* **274**(5294): 1924–1926.

Baggiolini M, Dewald B, Moser B et al (1997) Human chemokines: an update. *Annu Rev Immunol* **15**: 675–705.

Benkirane M, Jin DY Chun RF et al (1997) Mechanism of transdominant inhibition of CCR5-mediated HIV-1 infection by ccr5delta32. *J Biol Chem* **272**(49): 30603–30606.

Berger EA, Doms RW, Fenyo EM et al (1998) A new classification for HIV-1 (letter). *Nature* **391**(6664): 240.

Bieniasz PD, Fridell RA, Aramori I et al (1997) HIV-1-induced cell fusion is mediated by multiple regions within both the viral envelope and the CCR-5 co-receptor. *EMBO J* **16**(10): 2599–2609.

Bleul CC, Farzan M, Choe H et al (1996) The lymphocyte chemoattractant SDF-1 is a ligand for LESTR/fusin and blocks HIV-1 entry. *Nature* **382**(6594): 829–833.

Bleul CC, Wu L, Hoxie JA et al (1997) The HIV coreceptors CXCR4 and CCR5 are differentially expressed and regulated on human T lymphocytes. *Proc Natl Acad Sci USA* **94**(5): 1925–1930.

Bou-Habib DC, Roderiquez G, Oravecz T et al (1994) Cryptic nature of envelope V3 region epitopes protects primary monocytotropic human immunodeficiency virus type 1 from antibody neutralization. *J Virol* **68**(9): 6006–6013.

Brelot A, Heveker N, Pleskoff O et al (1997) Role of the first and third extracellular domains of CXCR-4 in human immunodeficiency virus coreceptor activity. *J Virol* **71**(6): 4744–4751.

Butini L, De Fougerolles AR, Vaccarezza M et al (1994) Intercellular adhesion molecules (ICAM)-1 ICAM-2 and ICAM-3 function as counter-receptors for lymphocyte function-associated molecule 1 in human immunodeficiency virus-mediated syncytia formation. *Eur J Immunol* **24**(9): 2191–2195.

Callahan LN, Phelan M, Mallinson M et al (1991) Dextran sulfate blocks antibody binding to the principal neutralizing domain of human immunodeficiency virus type 1 without interfering with gp120–CD4 interactions. *J Virol* **65**(3): 1543–1550.

Callebaut C, Krust B, Jacotot E et al (1993) T cell activation antigen, CD26, as a cofactor for entry of HIV in CD4+ cells. *Science* **262**(5142): 2045–2050.

Choe H, Farzan M, Sun Y et al (1996) The beta-chemokine receptors CCR3 and CCR5 facilitate infection by primary HIV-1 isolates. *Cell* **85**(7): 1135–1148.

Cocchi F, DeVico AL, Garzino-Demo A et al (1995) Identification of RANTES, MIP-1 alpha, and MIP-1 beta as the major HIV-suppressive factors produced by CD8+ T cells. *Science* **270**(5243): 1811–1815.

Cocchi F, DeVico AL, Garzino-Demo A et al (1996) The V3 domain of the HIV-1 gp120 envelope glycoprotein is critical for chemokine-mediated blockade of infection. *Nature Med* **2**(11): 1244–1247.

Connor RI, Paxton WA, Sheridan KE et al (1996) Macrophages and CD4+ T lymphocytes from two multiply exposed, uninfected individuals resist infection with primary non-syncytium-inducing isolates of human immunodeficiency virus type 1. *J Virol* **70**(12): 8758–8764.

Connor RI, Sheridan KE, Ceradini et al (1997) Change in coreceptor use correlates with disease progression in HIV-1-infected individuals. *J Exp Med* **185**(4): 621–628.

Crump MP, Gong JH, Loetscher P et al (1997) Solution structure and basis for functional activity of stromal cell-derived factor-1; dissociation of CXCR4 activation from binding and inhibition of HIV-1. *EMBO J* **16**(23): 6996–7007.

Dalgleish AG, Beverly PC, Clapham PR et al (1984) The CD4 (T4) antigen is an essential component of the receptor of the AIDS retrovirus. *Nature* **312**: 20–27.

Davis CB, Dikic I, Unutmaz D et al (1997) Signal transduction due to HIV-1 envelope interactions with chemokine receptors CXCR4 or CCR5. *J Exp Med* **186**(10): 1793–1798.

Dean M, Carrington M, Winkler C et al (1996) Genetic restriction of HIV-1 infection and progression to AIDS by a deletion allele of the CKR5 structural gene. Hemophilia Growth and Development Study, Multicenter AIDS Cohort Study, Multicenter Hemophilia Cohort Study, San Francisco City Cohort, ALIVE Study. Published erratum appears in *Science* (1996) **274**(5290): 1069. *Science* **273**(5283): 1856–1862.

De Jong JJ, De Ronde A, Keulen W et al (1992) Minimal requirements for the human immunodeficiency virus type 1 V3 domain to support the syncytium-inducing phenotype: analysis by single amino acid substitution. *J Virol* **66**(11): 6777–6780.

Deng H, Liu R, Ellmeier W et al (1996) Identification of a major co-receptor for primary isolates of HIV-1. *Nature* **381**(6584): 661–666.

Deng HK, Unutmaz D, Kewal Ramari VN et al (1997) Expression cloning of new receptors used by simian and human immunodeficiency viruses. *Nature* **388**(6639): 296–300.

Di Marzio P, Tse J, Landau NR et al (1998) Chemokine receptor regulation and HIV type 1 tropism in monocyte-macrophages. *AIDS Res Hum Retroviruses* **14**(2): 129–138.

Donzella GA, Schols D, Lin SW et al (1998) AMD3100, a small molecule inhibitor of HIV-1 entry via the CXCR4 co-receptor. *Nature Med* **4**(1): 72–77.

Doranz BJ, Rucker J, Yi Y et al (1996) A dual-tropic primary HIV-1 isolate that uses fusin and the beta-chemokine receptors CKR-5, CKR-3, and CKR-2b as fusion cofactors. *Cell* **85**(7): 1149–1158.

Doranz BJ, Grovit-Ferbas K, Sharron MP et al (1997a) A small-molecule inhibitor directed against the chemokine receptor CXCR4 prevents its use as an HIV-1 coreceptor. *J Exp Med* **186**(8): 1395–1400.

Doranz BJ, Lu ZH, Rucker J et al (1997b) Two distinct CCR5 domains can mediate coreceptor usage by human immunodeficiency virus type 1. *J Virol* **71**(9): 6305–6314.

Dragic T, Litwin V, Allaway GP et al (1996) HIV-1 entry into CD4+ cells is mediated by the chemokine receptor CC-CKR-5. *Nature* **381**(6584): 667–673.

Dragic T, Trkola A, Lin SW et al (1998) Amino-terminal substitutions in the CCR5 coreceptor impair gp120 binding and human immunodeficiency virus type 1 entry. *J Virol* **72**(1): 279–285.

Dukes CS, Yu Y, Rivadeneira ED et al (1995) Cellular CD44S as a determinant of human immunodeficiency virus type 1 infection and cellular tropism. *J Virol* **69**(7): 4000–4005.

Edinger AL, Mankowski JL, Doranz BJ et al (1997) CD4-independent, CCR5-dependent infection of brain capillary endothelial cells by a neurovirulent simian immunodeficiency virus strain. *Proc Natl Acad Sci USA* **94**(26): 14742–14747.

Endres MJ, Clapham PR, Marsh M et al (1996) CD4-independent infection by HIV-2 is mediated by fusin/CXCR4. *Cell* **87**(4): 745–756.

Farzan M, Choe H, Martin K et al (1997a) Two orphan seven-transmembrane segment receptors which are expressed in CD4-positive cells support simian immunodeficiency virus infection. *J Exp Med* **186**(3): 405–411.

Farzan M, Choe H, Martin KA et al (1997b) HIV-1 entry and macrophage inflammatory protein-1beta-mediated signaling are independent functions of the chemokine receptor CCR5. *J Biol Chem* **272**(11): 6854–6857.

Farzan M, Choe H, Vaca L et al (1998) A tyrosine-rich region in the N terminus of CCR5 is important for human immunodeficiency virus type 1 entry and mediates an association between gp120 and CCR5. *J Virol* **72**(2): 1160–1164.

Feng Y, Broder C, Kennedy PE et al (1996) HIV-1 entry cofactor: functional cDNA cloning of a seven-transmembrane, G protein-coupled receptor. *Science* **272**(5263): 872–877.

Garcia-Zepeda EA, Combadiere C, Rothenberg ME et al (1996) Human monocyte chemoattractant protein (MCP)-4 is a novel CC chemokine with activities on monocytes, eosinophils, and basophils induced in allergic and nonallergic inflammation that signals through the CC chemokine receptors (CCR)-2 and -3. *J Immunol* **157**(12): 5613–5626.

Granelli-Piperno A, Moser B, Pope M et al (1996) Efficient interaction of HIV-1 with purified dendritic cells via multiple chemokine coreceptors. *J Exp Med* **184**(6): 2433–2438.

Harouse JM, Bhat S, Spitalnik SL et al (1991) Inhibition of entry of HIV-1 in neural cell lines by antibodies against galactosyl ceramide. *Science* **253**(5017): 320–323.

Hildreth JE & Orentas RJ (1989) Involvement of a leukocyte adhesion receptor (LFA-1) in HIV-induced syncytium formation. *Science* **244**(4908): 1075–1078.

Hoogewerf AJ, Kuschert GS, Proudfoot AE et al (1997) Glycosaminoglycans mediate cell surface oligomerization of chemokines. *Biochemistry* **36**(44): 13570–13578.

Hwang SS, Boyle TJ, Lyerly HK et al (1992) Identification of envelope V3 loop as the major determinant of CD4 neutralization sensitivity of HIV-1. *Science* **257**(5069): 535–537.

Jansson M, Popovic M, Karlsson A et al (1996) Sensitivity to inhibition by beta-chemokines correlates with biological phenotypes of primary HIV-1 isolates. *Proc Natl Acad Sci USA* **93**(26): 15382–15387.

Kabat D, Kozak SL, Wehrly K et al (1994) Differences in CD4 dependence for infectivity of laboratory-adapted and primary patient isolates of human immunodeficiency virus type 1. *J Virol* **68**(4): 2570–2577.

Klatzmann D, Champagne E, Chamaret S et al (1984) T-lymphocyte T4 molecule behaves as the receptor for human retrovirus LAV. *Nature* **312**: 20–27.

Kostrikis LG, Huang Y, Moore JP et al (1998) A chemokine receptor CCR2 allele delays HIV-1 disease progression and is associated with a CCR5 promoter mutation. *Nature Med* **4**(3): 350–353.

Lapham CK, Ouyang J, Chandrasekhar B et al (1996) Evidence for cell-surface association between fusin and the CD4–gp120 complex in human cell lines. *Science* **274**(5287): 602–605.

Liao F, Alkhatib G, Peden KW et al (1997) STRL33, a novel chemokine receptor-like protein, functions as a fusion cofactor for both macrophage-tropic and T cell line-tropic HIV-1. *J Exp Med* **185**(11): 2015–2023.

Liu R, Paxton WA, Choe S et al (1996) Homozygous defect in HIV-1 coreceptor accounts for resistance of some multiply-exposed individuals to HIV-1 infection. *Cell* **86**(3): 367–377.

Lu Z, Berson JF, Chen Y et al (1997) Evolution of HIV-1 coreceptor usage through interactions with distinct CCR5 and CXCR4 domains. *Proc Natl Acad Sci USA* **94**(12): 6426–6431.

Lusso P, Cocchi F, Barlotta C et al (1995) Growth of macrophage-tropic and primary human immunodeficiency virus type 1 (HIV-1) isolates in a unique CD4+ T-cell clone (PM1): failure to downregulate CD4 and to interfere with cell-line-tropic HIV-1. *J Virol* **69**(6): 3712–3720.

Mackewicz C & Levy JA (1992) CD8+ cell anti-HIV activity: nonlytic suppression of virus replication. *AIDS Res Hum Retroviruses* **8**(6): 1039–1050.

Maddon PJ, Dalgleish AG, McDougal JS et al (1986) The T4 gene encodes the AIDS virus receptor and is expressed in the immune system and the brain. *Cell* **47**(3): 333–348.

Michael NL, Chang G, Louie LG et al (1997a) The role of viral phenotype and CCR-5 gene defects in HIV-1 transmission and disease progression. *Nature Med* **3**(3): 338–340.

Michael NL, Louie LG, Rohrbaugh AL et al (1997b) The role of CCR5 and CCR2 polymorphisms in HIV-1 transmission and disease progression. *Nature Med* **3**(10): 1160–1162.

Moore JP, McKeating JA, Huang YX et al (1992) Virions of primary human immunodeficiency virus type 1 isolates resistant to soluble CD4 (sCD4) neutralization differ in sCD4 binding and glycoprotein gp120 retention from sCD4-sensitive isolates. *J Virol* **66**(1): 235–243.

Moriuchi H, Moriuchi M, Combadiere C et al (1996) CD8+ T-cell-derived soluble factor(s), but not beta-chemokines RANTES, MIP-1 alpha, and MIP-1 beta, suppress HIV-1 replication in monocyte/macrophages. *Proc Natl Acad Sci USA* **93**(26): 15341–15345.

Murakami T, Nakajima T, Koyanagi Y et al (1997) A small molecule CXCR4 inhibitor that blocks T cell line-tropic HIV-1 infection. *J Exp Med* **186**(8): 1389–1393.

Murphy PM (1996) Chemokine receptors: structure, function and role in microbial pathogenesis. *Cytokine Growth Factor Rev* **7**(1): 47–64.

Naif HM, Li S, Alali M et al (1998) CCR5 expression correlates with susceptibility of maturing monocytes to human immunodeficiency virus type 1 infection. *J Virol* **72**(1): 830–836.

Oberlin E, Amara A, Bachelerie F et al (1996) The CXC chemokine SDF-1 is the ligand for LESTR/fusin and prevents infection by T-cell-line-adapted HIV-1. *Nature* **382**(6594): 833–835.

Oravecz T, Roderiquez G, Koffi J et al (1995) CD26 expression correlates with entry, replication and cytopathicity of monocytotropic HIV-1 strains in a T-cell line. *Nature Med* **1**(9): 919–926.

Oravecz T, Pall M, Norcross MA et al (1996) Beta-chemokine inhibition of monocytotropic HIV-1 infection. Interference with a postbinding fusion step. *J Immunol* **157**(4): 1329–1332.

Oravecz T, Pall M, Roderiquez G et al (1997a) Regulation of the receptor specificity and function of the chemokine RANTES (regulated on activation, normal T cell expressed and secreted) by dipeptidyl peptidase IV (CD26)-mediated cleavage. *J Exp Med* **186**(11): 1865–1872.

Oravecz T, Pall M, Wang J et al (1997b) Regulation of anti-HIV-1 activity of RANTES by heparan sulfate proteoglycans. *J Immunol* **159**(9): 4587–4592.

Pal R, Garzino-Demo A, Martham PD et al (1997) Inhibition of HIV-1 infection by the beta-chemokine MDC. *Science* **278**(5338): 695–698.

Patel M, Yanagishita M, Roderiquez G et al (1993) Cell-surface heparan sulfate proteoglycan mediates HIV-1 infection of T-cell lines. *AIDS Res Hum Retroviruses* **9**(2): 167–174.

Paxton WA, Martin SR, Tse D et al (1996) Relative resistance to HIV-1 infection of CD4 lymphocytes from persons who remain uninfected despite multiple high-risk sexual exposure. *Nature Med* **2**(4): 412–417.

Picard L, Simmons G, Power CA et al (1997) Multiple extracellular domains of CCR-5 contribute to human immunodeficiency virus type 1 entry and fusion. *J Virol* **71**(7): 5003–5011.

Pleskoff O, Treboute C, Brelot A et al (1997) Identification of a chemokine receptor encoded by human cytomegalovirus as a cofactor for HIV-1 entry. *Science* **276**(5320): 1874–1878.

Reeves JD, McKnight A, Potempa S et al (1997) CD4-independent infection by HIV-2 (ROD/B): use of the 7-transmembrane receptors CXCR-4, CCR-3, and V28 for entry. *Virology* **231**(1): 130–134.

Roderiquez G, Oravecz T, Yanagishita M et al (1995) Mediation of human immunodeficiency virus type 1 binding by interaction of cell surface heparan sulfate proteoglycans with the V3 region of envelope gp120–gp41. *J Virol* **69**(4): 2233–2239.

Rucker J, Samson M, Doranz BJ et al (1996) Regions in beta-chemokine receptors CCR5 and CCR2b that determine HIV-1 cofactor specificity. *Cell* **87**(3): 437–446.

Rucker J, Edinger AL, Sharron M et al (1997) Utilization of chemokine receptors, orphan receptors, and herpesvirus-encoded receptors by diverse human and simian immunodeficiency viruses. *J Virol* **71**(12): 8999–9007.

Samson M, Labbe O, Mollereau C et al (1996a) Molecular cloning and functional expression of a new human CC-chemokine receptor gene. *Biochemistry* **35**(11): 3362–3367.

Samson M, Libert F, Doranz BJ et al (1996b) Resistance to HIV-1 infection in caucasian individuals bearing mutant alleles of the CCR-5 chemokine receptor gene. *Nature* **382**(6593): 722–725.

Scarlatti G, Tresoldi E, Bjorndal et al (1997) In vivo evolution of HIV-1 co-receptor usage and sensitivity to chemokine-mediated suppression. *Nature Med* **3**(11): 1259–1265.

Schols D, Struyf S, van Damme J et al (1997) Inhibition of T-tropic HIV strains by selective antagonization of the chemokine receptor CXCR4. *J Exp Med* **186**(8): 1383–1388.

Schuitemaker H, Koot M, Kootstra NA et al (1992) Biological phenotype of human immunodeficiency virus type 1 clones at different stages of infection: progression of disease is associated with a shift from monocytotropic to T-cell-tropic virus population. *J Virol* **66**(3): 1354–1360.

Simmons G, Wilkinson D, Reeves JD et al (1996) Primary, syncytium-inducing human immunodeficiency virus type 1 isolates are dual-tropic and most can use either Lestr or CCR5 as coreceptors for virus entry. *J Virol* **70**(12): 8355–8360.

Simmons G, Clapham PR et al (1997) Potent inhibition of HIV-1 infectivity in macrophages and lymphocytes by a novel CCR5 antagonist. *Science* **276**(5310): 276–279.

Smith MW, Dean M, Carrington M et al (1997) Contrasting genetic influence of CCR2 and CCR5 variants on HIV-1 infection and disease progression. Hemophilia Growth and Development Study (HGDS), Multicenter AIDS Cohort Study (MACS), Multicenter Hemophilia Cohort Study (MHCS), San Francisco City Cohort (SFCC), ALIVE Study. *Science* **277**(5328): 959–965.

Spira AI, Marx PA, Patterson BK et al (1996) Cellular targets of infection and route of viral dissemination after an intravaginal inoculation of simian immunodeficiency virus into rhesus macaques. *J Exp Med* **183**(1): 215–225.

Trkola A, Dragic T, Arthos J et al (1996) CD4-dependent, antibody-sensitive interactions between HIV-1 and its co-receptor CCR-5. *Nature* **384**(6605): 184–187.

Trkola A, Paxton WA, Monard SP et al (1998) Genetic subtype-independent inhibition of human immunodeficiency virus type 1 replication by CC and CXC chemokines. *J Virol* **72**(1): 396–404.

Ugolini S, Moulard M, Mondor I et al (1997) HIV-1 gp120 induces an association between CD4 and the chemokine receptor CXCR4. *J Immunol* **159**(6): 3000–3008.

van 't Wout AB, Kootstra NA, Mulder-Kampinga GA et al (1994) Macrophage-tropic variants initiate human immunodeficiency virus type 1 infection after sexual, parenteral, and vertical transmission. *J Clin Invest* **94**(5): 2060–2067.

Wagner L, Yang OO, Garcia-Zepeda EA et al (1998) Beta-chemokines are released from HIV-1-specific cytolytic T-cell granules complexed to proteoglycans. *Nature* **391**(6670): 908–911.

Wang J, Roderiquez G, Oravecz T et al (1998) Cytokine regulation of human immunodeficiency virus type 1 entry and replication in human monocyte/macrophages through modulation of CCR5 expression. *J Virol* **72**(9): 7642–7647.

Weissman D, Rabin RL, Arthos J et al (1997) Macrophage-tropic HIV and SIV envelope proteins induce a signal through the CCR5 chemokine receptor. *Nature* **389**(6654): 981–985.

Winkler C, Modi W, Smith MW et al (1998) Genetic restriction of AIDS pathogenesis by an SDF-1 chemokine gene variant. ALIVE Study, Hemophilia Growth and Development Study (HGDS), Multicenter AIDS Cohort Study (MACS), Multicenter Hemophilia Cohort Study (MHCS), San Francisco City Cohort (SFCC). *Science* **279**(5349): 389–393.

Wu L, Gerard NP, Wyatt R et al (1996) CD4-induced interaction of primary HIV-1 gp120 glycoproteins with the chemokine receptor CCR-5. *Nature* **384**(6605): 179–183.

Wu L, LaRosa G, Kassam N et al (1997a) Interaction of chemokine receptor CCR5 with its ligands: multiple domains for HIV-1 gp120 binding and a single domain for chemokine binding. *J Exp Med* **186**(8): 1373–1381.

Wu L, Paxton WA, Kassam N et al (1997b) CCR5 levels and expression pattern correlate with infectability by macrophage-tropic HIV-1, in vitro. *J Exp Med* **185**(9): 1681–1691.

Yi Y, Rana S, Turner JD et al (1998) CXCR-4 is expressed by primary macrophages and supports CCR5-independent infection by dual-tropic but not T-tropic isolates of human immunodeficiency virus type 1. *J Virol* **72**(1): 772–777.

Zaitseva M, Blauvelt A, Lee S et al (1997) Expression and function of CCR5 and CXCR4 on human Langerhans cells and macrophages: implications for HIV primary infection. *Nature Med* **3**(12): 1369–1375.

Zhang L, Huang Y, He T et al (1996) HIV-1 subtype and second-receptor use (letter). *Nature* **383**(6603): 768.

Zhu T, Mo H, Wang N et al (1993) Genotypic and phenotypic characterization of HIV-1 patients with primary infection. *Science* **261**(5125): 1179–1181.

Chapter 10

HLA AND HIV-1 INFECTION

Dean L. Mann

INTRODUCTION

The pathogenesis of an infectious disease involves interactions between the infectious agent and the host's immunologic response to the challenge. In human immunodeficiency virus type 1 (HIV-1) infection, these interactions appear to have fundamental consequences for many different aspects of the disease.

Major advances in our understanding of the complexities of the immune system that have taken place since the 1980s are relevant to HIV-1 infection (Cresswell and Howard, 1997). A significant portion of this advance was embodied in the determination of the structure, function, and interrelated activities of products encoded by genes in the major histocompatibility complex (MHC). During this same period, the world has experienced the HIV-1 epidemic. Studies of the genetic structure of HIV-1 and the function of the viral gene products have been exhaustive. More recently, the rate at which the virus replicates in the infected host has been measured and the relationship of the relative viral load in the host to the rates of disease progression has been described (Ho et al, 1995; Mellors et al, 1995, 1996; Wei et al, 1995). In spite of the accumulation of knowledge about the immune system and about the virus, there remains a poor understanding of the host response to the infection, particularly the inability of the host to control and overcome the infection despite an apparently robust immune response (Paul, 1995; Haynes et al, 1996).

The central role that the MHC gene products play in cellular and humoral immunity suggests the possibility that one or more of the highly polymorphic MHC genes or gene products contribute to variability in pathogen–host interaction.

The discovery of MHC gene products resulted from their properties as determinates of allograft rejection. Since tissue transplantation is not a natural process, investigators examined the possibility that HLA polymorphism might be related to disease processes. Indeed, most human diseases have been tested for their association with HLA types and a number of associations have been observed. Considering our present knowledge of the function of the MHC gene products as regulatory elements of the immune response, it is not surprising that these associations were with diseases commonly designated as having an autoimmune pathogenesis. However, the contribution that specific alleles play in a particular pathologic process remains to be determined for many of these diseases, despite the fact that some of these associations have been observed for years.

In the past, relatively few HLA associations were described with infectious diseases.

HIV and the New Viruses Second Edition
ISBN 0-12-200741-7

This is probably due to a number of factors that potentially confound an analysis of associations or potential associations: these include determination of the proportion of individuals exposed to an agent who develop an infectious process, the clinical and laboratory manifestations of the infection, therapeutic intervention, and outcome. These same factors apply to HLA studies in HIV-1 infection.

Within a relatively short time after the first cases of acquired immunodeficiency were described and before the etiologic agent had been isolated, HLA associations with Kaposi's sarcoma were reported (Pollack et al, 1983a,b; Prince et al, 1984). Subsequent reports of apparent associations of various HLA alleles with different aspects of HIV-1 infection have been published and are summarized in several reviews (Kaslow and Mann, 1994; Mann et al, 1994; Just, 1995). Rather than recatalogue these observations, this chapter discusses the relationship of the function of MHC genes to associations that have been observed. In order to appreciate the potential significance of these observations, the genetic organization of the MHC is reviewed, the extensive polymorphism in seemingly redundant genes is described and the structural/functional properties of the gene products and their interrelated activities are highlighted.

MHC GENETIC ORGANIZATION

The human histocompatibility complex is composed of genes on the short arm of chromosome 6. The relative position and relationship of genes encoding HLA class I and class II molecules, as well as those encoding the antigen-processing and transport genes, are illustrated in Figure 10.1. A region between class I and class II is designated as HLA class III and contains a number of genes whose products are also involved in different aspects of the immune response cascade. These include genes that encode some complement components, heat shock protein 70, tumor necrosis factor alpha (TNF-α) and TNF-β. These genes are not detailed in Figure 10.1 and are considered only briefly in this chapter.

The class I genes, HLA-A, C and B, encode 44 000 M_r heavy chains that combine with $β_2$-microglobulin (light chain) in the endoplasmic reticulum. The molecule is stabilized by incorporation of a peptide and then transported to and expressed on the cell surface. HLA class I molecules are expressed on all nucleated cells. The peptide fragments that are incorporated in the molecule are derived from the degradation of endogenously synthesized cellular proteins or products of intercellular infectious agents (Cresswell

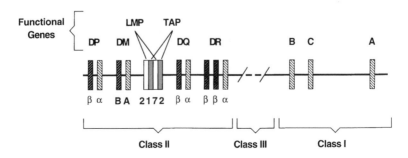

Figure 10.1 *Schematic map of the HLA and antigen-processing genes in the MHC*

and Howard, 1997). According to data from studies in vitro, empty class I molecules are also known to be present for some time at the cell surface; they are, however, relatively unstable and short-lived.

Genes mapping in the class II region encode products that contribute to the generation and transport of the peptides that complex with the HLA class I heavy and light chains (Powis et al, 1992). These peptides are generated by enzymatic processes in the cytosol by proteosomes. The *LMP2* and *LMP7* gene products are subunits of the proteosome (Koopmann et al, 1997). Once the peptides are generated, they are actively transported across the endoplasmic reticulum by the products of the transporter associated with antigen-processing genes, *TAP1* and *TAP2*.

The HLA class II molecules are heterodimers of α and β chains encoded by their respective genes in the class II region. These molecules are constitutively expressed on classic antigen-presenting cells (APCs) such as monocyte/macrophages, dendritic cells and B cells, and are also expressed on T cells and endothelial cells when activated. The class II molecules are assembled with the invariant chain, a portion (CLIP) of which occupies the peptide-binding groove of the molecule (Pieters, 1997). In the acidic endosomal/lysosomal compartments of the cell, class II molecules acquire antigenic peptides which are degraded products of foreign substances that have been ingested by the cell (Wilson, 1996). Prior to incorporation of an antigenic peptide, the CLIP peptide is removed, facilitated by products of two other MHC genes, *DMA* and *DMB*. The products of these genes combine to form bimolecular non-classical MHC class II molecules; they have editorial function in the selection of peptides and catalyze the peptide or epitope integration into the molecule (Pieters, 1997).

HLA POLYMORPHISM, MOLECULAR STRUCTURE AND FUNCTION

HLA gene polymorphism was recognized very early on in studies of this system. Historically, alleles of the various genes were identified by serologic techniques. However, advances in molecular biology now allow allele determination at the molecular level. Using these techniques, the extent of polymorphism recognized in HLA genes has greatly expanded. A comparison of the number of alleles currently detected by serology and molecular techniques is shown in Table 10.1. The majority of the reported HLA–HIV associations have used serologic techniques to identify the different alleles. In some of the more recent studies, alleles of the HLA class II genes were determined by molecular techniques (Kaslow et al, 1996). Genes involved in antigen processing and presentation are also polymorphic, but to a more limited degree. The polymorphism in the HLA molecules dictates their functional diversity.

The sequences in the HLA class I genes that generate the different alleles are clustered in several regions of the encoded protein (Parham et al, 1995). The polymorphism in the αI domain of the molecule at amino acid positions 77–83 can be serologically distinguished as the HLA-Bw4 and HLA-Bw6 epitopes. This portion of the molecule is external to the antigen-binding groove in an α-helix and serves as a differential recognition unit for receptors on natural killer (NK) cells (Moretta et al, 1997). The more extensive polymorphism is in the αII and αIII regions of the molecule. These regions form the floor and sides of the peptide-binding groove of the class I molecule and dictate the

Table 10.1 *Polymorphism of major histocompatibility complex genes*

Locus	Number of alleles detected by: Serology	Molecular methods
	HLA class I	
A	28	83
B	61	187
C	10	42
	HLA class II	
DRB1	21	182
DRB3	1 (DR52)	11
DRB4	1 (DR53)	9
DRB5	1 (DR51)	1
DQA1	0	18
DQB1	9	31
DPA1	0	10
DPB1	6[a]	77
	Antigen processing/presentation	
DMA	0	4
DMB	0	5
TAP1	0	4
TAP2	0	6

[a]Detected by cellular methods.

composition of the peptide(s) that can best be accommodated or bound. Studies of the crystal structure of the class I molecule and sequencing peptides eluted from these molecules provide insight into the relationship of the polymorphism of these molecules with potential functional diversity (Bjorkman et al, 1987; Rammensee et al, 1995).

Peptides found in the HLA class I molecules are generally 8–10 amino acids in length. The various HLA alleles preferentially bind peptides with different amino acids that anchor the peptide to the allele. For most alleles, primary anchor residues are those that occur at position 2 at the carboxy terminus of the peptide and at position 9 at the amino terminus. Other amino acids in the peptide constitute secondary anchors and influence the binding affinity of the peptide to the allele.

In generating an immune response, the class I molecule containing the peptide engages the T cell receptor on CD8+ T cells. Recognition of the antigenic peptide by the specific T cell receptor is dependent on the amino acid composition of the peptide. Interaction of other ligand–receptor combinations on the APC and the T cell signal the latter to expand and proliferate. Once expanded, the CD8+ cell can recognize and kill a cell that presents a specific antigenic peptide that induced expansion in a particular T cell clone.

The structure of the class II molecule is very similar to HLA class I (Brown et al, 1993; Stern et al, 1994). The polymorphisms characterizing the different alleles are also in the regions of the molecule that interface with the antigenic peptides occupying the molecular groove. The optimal peptide length is of the order of 13–24 amino acids and, like the HLA class I molecules, particular amino acid residues are preferred by the different alleles with selected residues serving as anchors. Compared with the class I molecule, HLA class II molecules appear to accommodate a greater degree of redundancy at the anchor positions (Rammensee et al, 1995).

The HLA class II–peptide complex on the APC interacts with receptors on CD4+ lymphocytes. Other receptor–ligand interactions between the APC and T cell trigger expansion of the T cell clone and the generation and release of cytokines that enhance and augment an immune response. A subset of CD4+ T cells has the potential to mediate cytotoxic activity against target cells expressing the class II molecule with the cognate peptide.

As can be seen from Table 10.1, the genes that encode products involved in antigen processing are also polymorphic, but to a much lesser degree. No functional diversity can yet be ascribed to the molecular polymorphism. However, these genes and their products have only recently been described and future studies may show that molecular polymorphisms translate into functional differences. This possibility is discussed below in context with the observations that combinations of alleles of different MHC genes with complementary biologic activity show associations with HIV-1-related events that were not found when the alleles were present independently.

The extensive polymorphism in the HLA genes far exceeds that of any other known genetic system in the human species. Based on our knowledge of their function, current theory favors the concept that this polymorphism protects the species from potential annihilation by an infectious agent. It is conceivable that historic epidemics caused by bacterial or viral organisms were controlled in part (or in whole) by the diversity of the immune response to the infecting agent made possible by the heterogeneity of HLA types present in the exposed population. The HIV-1 epidemic is substantially different from other known epidemics in that the virus encountered can mutate very rapidly in a short period and still maintain the capacity to cause disease in the host and infect other individuals. This makes it difficult for the infected host to develop a sustained effective immune response. When a response is developed in the host, its effectiveness may be short-lived if the rate of virus mutation exceeds that of the ability of the MHC alleles to accommodate the changing antigens (Phillips et al, 1991; Goulder et al, 1997). Thus, it is in the context of an ever-changing virus that a polymorphic immune system must react in order for this infection to be controlled by an individual or by the population at large (Wolinsky et al, 1996).

MHC AND HIV-1

The immune regulatory capacity of MHC genes and gene products may influence HIV infection in three general areas: susceptibility to infection, the natural history of the disease process, and the outcome or AIDS-defining illness. HLA associations have been described in all three areas but should be considered independently since there are a multiplicity of factors that are unique to each circumstance. The associations observed should be viewed in context with these factors as well as the biologic properties of the MHC alleles and how they may be involved in each of the processes.

Susceptibility to infection

Relationship between MHC alleles and susceptibility to infection is the most problematic of any of the three areas. This is primarily due to the problem of evaluating the level of virus exposure (or inoculum) that leads to infection, given the different routes whereby the virus is acquired. Moreover, the route of exposure may determine the relative amount of virus that is likely to be cell-free or cell-associated. This in turn may dictate the most likely target cell which has the appropriate cell surface structures to permit infection. These structures include CD4, as well as several chemokine receptors (Choe et al, 1996; Deng et al, 1996; Dragic et al, 1996; Murphy, 1996). At least one of these receptors (CCR5) is known to be polymorphic with a 35 base pair deletion in the structural gene. Cells homozygous for this deletion cannot be infected with macrophage-tropic virus strains; however, they can be infected with the T cell-tropic strain which uses another receptor, CXCR4 (Alkhatib et al, 1996; Bleul et al, 1996; Liu et al, 1996; Paxton et al, 1996; Samson et al, 1996). With these confounding factors, it is easy to see why there have been no consistent findings describing resistant or permissive HLA phenotypes. The lack of consistency should not be considered as a negative result: sufficient numbers of individuals stratified on the various parameters that might control infection independently of HLA would be required for meaningful data to be generated in the context of an extensively polymorphic HLA genetic system.

Mother-to-infant HIV transmission is a circumstance in which virus exposure may be relatively constant over a long period and where only about 25–30% of the offspring become infected. Even in this situation, HLA typing of limited numbers of infected and non-infected infants has not yielded striking associations (reviewed by Just, 1995). This is not surprising considering the rate of infection in the infants born to infected mothers and the extent of HLA polymorphism in the general population. However, the immunologic function of the MHC has been shown to be operative in some studies of virus-exposed individuals (including infants) where evidence for persistent infection is lacking. The pioneering studies of Shearer and Clerici demonstrated clearly that some exposed non-infected individuals had an antigenic recall response to viral proteins that could be measured in vitro (reviewed by Shearer and Clerici, 1996). Virus-specific cytotoxic T lymphocyte (CTL) responses were reported in infants born of infected mothers (Cheynier et al, 1992; Rowland-Jones et al, 1993; Aldhous et al, 1994). These infants had lost maternal HIV antibodies, and when studied were virus-negative by culture and/or polymerase chain reaction (PCR). No particular HLA alleles appeared to be preferentially associated with the response, but the number of individuals tested was such that meaningful statistical data would not be expected. There are, however, a number of laboratories that have demonstrated HLA allele-specific CD8+ CTL activity in exposed seronegative individuals (reviewed by Rowland-Jones and McMichael, 1995). HLA-A2 restricted CTL to the HIV-1 Nef protein was found in several seronegative individuals whose sexual partners were HIV-1 positive (Langlade-Demoyen et al, 1994). In a study of Gambian uninfected sex workers repeatedly exposed to HIV-1 and HIV-2, Rowland-Jones et al (1995) found vigorous CTL reactivity that was restricted to peptide sequences considered common to the two viruses that could be presented by the HLA-B35, a common class I allele in Gambia. These data indicate that a CTL response probably accounted for protection from infection although there is no evidence that HLA-B35+ individuals are at an advantage in this regard. While it was not directly demonstrated, these studies imply exposure to viral antigens in such a manner that an immunologic response can be generated. Whether or not these individuals had transient infections was not determined.

In other studies transient viral infection was documented in individuals who did not seroconvert, e.g. in homosexual men (Imagawa et al, 1989). MHC alleles were determined in this cohort and the *TAP1.4* or *TAP2.3* genotypes were found to be significantly ($P = 0.006$) different in these men compared with the frequency of these genes in a comparable seroconverting population (Detels et al, 1996). Cytotoxic T lymphocyte activity was not measured in these individuals; however, the association observed implicates HLA class I function, given the role of the *TAP* gene products in this regard.

Studies demonstrating a potential immunologic mechanism whereby HIV-1 is cleared are extremely important in formulating vaccine strategies. Further insight into the issue is provided in some of the studies of HLA associations with disease progression.

Disease progression

The majority of studies that examined HLA in HIV infection have determined the HLA profile in individuals who were seroprevalent and had progressed to a disease end-point over a relatively short period (reviewed by Kaslow and Mann, 1994; Mann et al, 1994; Just, 1995). In this context, a number of different HLA alleles, both class I and class II, were reported to be associated with longer or shorter periods to different end-points. These end-points varied depending on the study and included death, AIDS-defining illness and/or decreased CD4+ T cell count to levels below a certain threshold. With some exceptions, the findings of different investigators have lacked consistency. This lack of consistency can be accounted for by a number of factors that include methodological differences in HLA typing, the small numbers of individuals studied, differences between ethnic groups, and the natural history of HIV infection.

The HLA typing in these studies was mostly done by serologic assays which have been found to be somewhat unreliable in comparison with molecular techniques that define the alleles at the DNA level. This is particularly true in HLA class II serologic typing where the error rate in assigning alleles may be as high as 25%. Typing for class I by serology appears to be more reliable; however, the technology for typing HLA class I at the DNA level is only now emerging. Studies of HIV-1 where HLA class I alleles are identified by molecular techniques are forthcoming. Given the recognized polymorphism in the MHC, evaluation of an observed difference in allele frequency at the level usually observed is difficult and associations may be spurious. Moveover, allele frequencies differ among ethnic and racial groups and must be considered when comparing study results.

In the majority of the studies, the follow-up period to an end-point was relatively short, based on our current knowledge of the natural history of the disease progression after infection. This potentially influences analysis in the following way. If it is hypothesized that the MHC (HLA) contributes to the rate of disease progression and an infected population is selected for study, the hypothesis may be rejected out of hand if the duration of infection in the individuals studied is not known. Moreover, if a cohort is not studied in its entirety from the point of acquisition of infection, HLA alleles can be "lost" from the population owing to their association with or influence on rapid disease progression. It is conceivable that a study of a seroprevalent population may reflect

HLA genes associated with relative survival, depending on when the infected population was sampled relative to the subjects' duration of infection.

These factors were difficult to appreciate in the early stages of the epidemic, when it appeared that the majority of infected individuals would develop AIDS rapidly (over 2–5 years). Current data (in the absence of antiviral therapy) suggest that AIDS will develop in about 50% of the infected population after 9–11 years. This implies that the most informative data on the potential effect of MHC alleles or combinations of alleles on disease progression will come from studies of populations where the dates of seroconversion can be established with some degree of certainty and the patients closely followed to a disease end-point.

One of the best ways to show genetic associations with disease is in families. In HIV infection this is difficult except in the cases of hemophilic sibships where more than one member of a family was infected with therapeutic blood products prior to the availability of viral testing. Moreover, closely approximated dates of seroconversion can be established in this population based on epidemiologic modeling (Kroner et al, 1994). An HLA influence on disease progression was established in a study of 95 infected sib pairs, which examined disease progression to AIDS status or to CD4 counts that fell below 20% (Kroner et al, 1995). Sibs sharing one or two HLA haplotypes were concordant in both parameters. By the nature of the diversity of HLA in these sibships, specific alleles and/or haplotypes that might be associated with the relative different rates of disease progression were not defined.

As already indicated, the extensive polymorphism in the MHC genes and the different allele frequencies in the population create difficulties in identifying alleles or combination of alleles that may be associated with accelerated rates of disease progression or relatively long-term non-progression. In light of this diversity, the most convincing evidence for HLA association is confirmation of observations made in different cohorts with similar histories of acquiring infection and where both cohorts have relatively well-documented dates of seroconversion and are closely followed. It is also important that the cohort be captured in its entirety, because loss to follow-up (which could be death due to the viral disease) might distort gene frequencies in the different populations and thus affect the analysis. In a study constructed with these considerations in mind, two HIV-infected cohorts were typed for HLA class I by serology and HLA class II alleles by molecular techniques (Kaslow et al, 1996). In addition, the sites of variation in the *TAP* genes were determined. The rationale for investigating *TAP* was that the function of these gene products contributed to the availability of the antigenic peptides that are incorporated in the class I molecule (Powis et al, 1992; reviewed by Koopmann et al, 1997). It was reasoned that combinations of specific *TAP* genotypes with class I alleles might show more informative associations when considered together rather than as individual determinants (Barron et al, 1995). In addition, certain of the *TAP* alleles had been observed to be in linkage disequilibrium with some HLA class II alleles, and an association with an HLA class II allele might be a surrogate marker for a more meaningful association with a linked gene (Carrington et al, 1993, 1994). Both cohorts studied had closely estimated dates of seroconversion and had been followed for at least 9 years. The measure of disease progression or the end-point of the analysis was the time from infection to an AIDS diagnosis. Analysis of the association utilized a Cox proportional hazard model where a relative hazard (RH) value for association with different rates of disease progression was determined for individual HLA class I alleles, class II alleles and haplotypes, and the sites of variation in the *TAP* genes (*TAP* genotypes) that configure the various alleles. Using the HLA typing and clinical data from one of the cohorts (Multicenter AIDS Cohort Study), a relative hazard for each of the alleles and haplotypes was established and compared with the same data generated from a different cohort (DCG). With few exceptions, the MHC alleles associated with relatively rapid progression in one cohort were also found to be associated with the different rates of disease progression in the other cohort. The next step was to determine if combinations of alleles confer differential risks that are not apparent from the RH of individual alleles or haplotypes. This approach offered an explanation for the inconsistencies for some HLA allele associations with disease progression. Some investigators have reported that HLA-B35 is associated with rapid disease progression, while others have not found this association (Scorza et al, 1988; Itescu et al, 1991). Similar disparate observations have been made with another allele, HLA-B8 or its commonly associated haplotype (A1,B8,DR3) (Steel et al, 1988). On analysis of the data, in the two cohorts neither of these individual alleles (or the A1,B8,DR3 haplotype) appeared to be associated with a substantial risk for an increased rate of disease progression. The progression rate of HLA-B35+ individuals was slightly but not significantly more rapid than in individuals without this allele (Figure 10.2, top panel). However, the rate of progression of individuals with this allele and the *TAP2.1* genotype was markedly and significantly faster than that of controls (Figure 10.2, bottom panel). In analysis of the HLA-B8 association with disease progression, HLA-B8 had an RH of 1·35 – a modest value for an association with more rapid rate of disease progression. When this allele was present with *TAP2.1* the RH increased to 2·4, indicating an association with an accelerated disease course. In addition to offering an explanation for the discrepancy in the previous studies, the data suggest that evaluating combinations of unlinked genes with interrelated biologic function may reveal associations not heretofore recognized.

In evaluating associations of combinations of MHC alleles with different rates of disease progression, we considered the possibility that several genes may appear to be associated with the same event owing to linkage disequilibrium. This was factored into the analysis so that the combination of alleles were analyzed as independent events in the absence of other alleles with known or observed associations that could be explained on this basis.

Table 10.2 summarizes the relative hazards of the combination of alleles that were associated with either very rapid disease progression or a relatively prolonged disease-free interval. Not surprisingly, the majority of alleles and haplotypes (not listed here) had RH values that were intermediate to those listed in this table. Using these data, further analysis was carried out and results compared between the two cohorts. Individual alleles or the combinations of alleles associated with either extremes of disease progression (as listed in Table 10.2) were assigned at integer value of −1 if the association was with rapid progression, and +1 if associated with long-term non-progression. Other alleles received a value of zero. Based on the HLA phenotype, the integers were summed in each individual in both cohorts. The result showed a marked consistency between the groups, indicating that the alleles and combinations of alleles identified in one group were also associated with the comparable rates of disease progression in the other cohort. These values were quite predictive of the frequency of AIDS in this population. The percentage of the population with AIDS with the different scores is

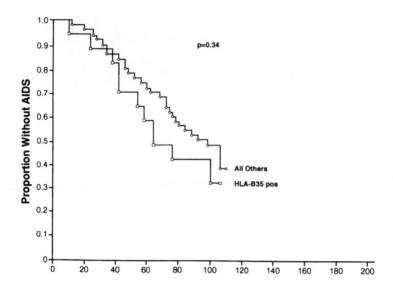

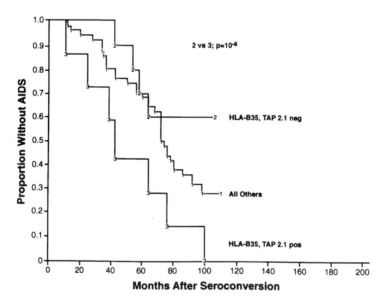

Figure 10.2 *HLA allele association with disease progression*
Disease progression in individuals with and without HLA-B35 (top panel) and in individuals with HLA-B35 and the *TAP2.1* genotype (bottom panel).

shown in Figure 10.3. This figure shows that the incidence of AIDS with scores of −2 or less was almost 100% and decreased in frequency as the scores increased.

One of the observations made in this study was that certain HLA class I alleles (with and without *TAP2.3*) were associated with long-term non-progression (LTNP). An indication as to the alleles that might be associated with LTNP was given by the finding that individuals with the HLA-Bw4 epitope had a retarded CD4+ T cell decline early in the disease course (Figure 10.4, top panel) and a slower progression to AIDS (Figure 10.4,

Table 10.2 *Relative hazard (RH) of HLA alleles and combinations for different rates of disease progression*

HLA	RH
Prolonged disease-free interval	
B27	0.23
B51	0.41
B57	0.30
A25 and TAP2.3	0.30
A26 and TAP2.3	0.36
A32 and TAP2.3	Indeterminate[a]
B18 and TAP2.3	Indeterminate
Rapid disease progression	
B37	1.9
B49	1.9
A28 and TAP2.3	2.9
A29 and TAP2.1	2.1
B8 and TAP2.1	2.4
A23 and not TAP2.3	3.1
A24 and (TAP2.1 or TAP2.3)	2.3
B60 and (TAP2.1 or TAP2.3)	2.0
DRB1*0401–DQA1*0300–DQB1*0301 and TAP1.2	3.2
DRB1*1200–DQA1*0501–DBQ1*0301 and TAP1.2	3.1
DRB1*1300–DQA1*0102–DBQ1*0604 and TAP1.2	2.2
DRB1*1400–DQA1*0101–DBQ1*0503 and TAP1.2	4.8

[a]No individual with this phenotype developed AIDS during the follow-up period.

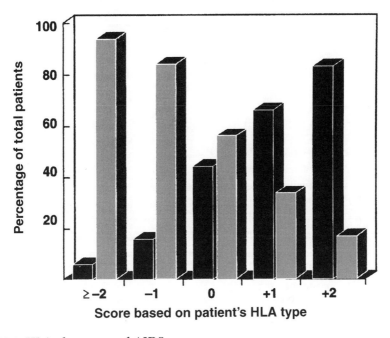

Figure 10.3 HLA phenotype and AIDS
Percentage of individuals with AIDS (light shading) or without AIDS (dark shading), stratified by HLA phenotype score (see text).

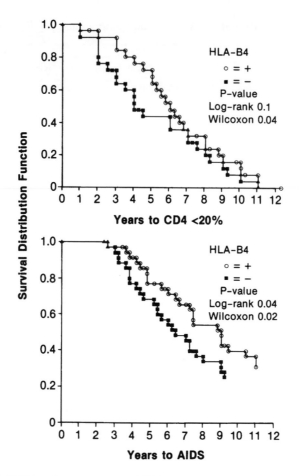

Figure 10.4 *HLA-BW4 and progression to AIDS*
Comparison of the rates of decline of CD4+ T cells (top panel) and time to AIDS (bottom panel) in individuals with and without HLA-Bw4.

bottom panel). This epitope is shared by the HLA-B27, B51, and B57 alleles and is the ligand for one of the receptors (NKB1) on NK cells. Given the important functions of these cells in the control of other viral disease, it seems likely that the association observed predicts a functional correlate. In addition to the Bw4–NKB1 interaction, studies have demonstrated other class I-restricted recognition by other NKB1 receptors (Moretta et al, 1997). Other findings relevant to these interactions are that the peptide composition of the groove of the class I molecule can change an inhibitory signal to cytotoxic activation (Malnati et al, 1995; Peruzzi et al, 1996). Together these data suggest a re-evaluation of the role of NK cells in HIV-1 infection.

A number of investigators have shown that a vigorous CTL response is associated with relative LTNP (Carmichael et al, 1993; Borrow et al, 1994; Cao et al, 1995; Harrer et al, 1996a,b; Paxton et al, 1996). In some instances, the alleles found to be associated with LTNP have also been shown to present HIV-derived peptides to CTLs. These studies have been summarized by McMichael and Walker (1994). One of these alleles, HLA-B27, has been examined extensively as a target for CTL activity (Gotch et al, 1990; Goulder et al, 1997). In a study of HLA-B27+ individuals who were long-term non-progressors, the role of HLA class I function in controlling infection is most informative. In this study, serial sampling of blood from the patients with this allele allowed viral quantitation and the determination of the sequence of the predominant viral epitope that was presented by the B27 molecule. The sequence of the peptide was found to be stable over the period of years during which these individuals were essentially disease-free. However, two individuals progressed to AIDS after 9–12 years, and the sequence of the virus isolated from these individuals showed an amino acid change at a position in the epitope sequence that anchors the peptide to the HLA-B27 molecule that was accompanied by loss of CTL activity. This study demonstrates that changes in the virus due to mutation can alter the immune response by abrogating the capacity of MHC class I molecules in the host to present a viral antigenic peptide effectively.

These data, together with reports of previous studies by these investigators and others in which HLA-restricted CTL activity in HIV-infected individuals was demonstrated but subsequently lost owing to changes in the virus, establish the paradigm for HLA function in the host response to HIV-1 infection. One may hypothesize that HLA alleles associated with LTNP are those that present peptides from regions in the virus that tend to be conserved during virus replicative cycles in the host. Since the TAP proteins are involved in peptide transport, the different genotypes may act as preliminary determinants of the array of peptides that are available for class I presentation (Hammond et al, 1995). Thus it is not surprising that a TAP genotype may be associated with very different rates of disease progression depending on the class I allele that occurs in the same individual.

In order to advance our understanding of the role of HLA alleles and CTLs in regulating relative rates of disease progression, future studies must also evaluate situations where an apparently protective allele is found in individuals with a rapid disease course. As an example, it would be interesting to test HLA-B27+ individuals with accelerated disease progression to determine whether the viral epitope presented was the same as that found in the Goulder study or in regions of the virus that tend to change more rapidly.

Studies such as these are relevant not only to our understanding of the immunologic response by the host to the viral infection and the ability to control the virus, but also in vaccine development. It is easy to see how a single-component vaccine with a fixed amino acid sequence would not be adequate to generate a protective cytotoxic T cell response in populations with highly polymorphic MHC molecules who might be challenged with a polymorphic virus.

AIDS diagnosis

Associations of HLA with a specific disease outcome in HIV-1 infection (AIDS diagnosis) have been reported in a number of studies (reviewed by Just, 1995). A major difficulty in evaluating these data is the control population which constituted the comparison group. In many instances the associations were established with control populations who were not infected with HIV; in other instances a control population was used that had been infected with HIV, but for an unknown duration. This again raises the issue of comparing potential differences in the gene pool if HLA confers any selective pressure on susceptibility or resistance to infection or on the rate of disease progression in the infected individuals. The results of the studies reporting HLA associations with an outcome must be evaluated accordingly. As an example, some of the early papers reported HLA-DR5 associations with Kaposi's sarcoma (Pollack et al, 1983a,b; Prince et al, 1984). The controls for most of these studies were populations whose infectious status was unknown. Aside from the fact that HLA-DR5 was notoriously difficult to determine using serologic reagents, the high frequency of this allele in these cases was subsequently accounted for by the relative higher frequency of this allele in the ethnic population being evaluated (Pollack et al, 1983b). Another explanation that may account for the observed HLA-DR5 association is the possibility that this allele is a risk factor for more rapid disease progression. One of the molecular equivalents of the serologic determined HLA-DR5 is HLA-DRB1*1200 which we found associated with rapid progression in our study of HIV-1 seroconverters. In addition, some of the class I alleles (A23, B49, B37) that were found to be associated with different outcomes in studies of seroprevalent subjects were found to be associated with a more rapid disease progression in seroconverters (Just, 1995; Kaslow et al, 1996). Thus, HLA associations that have been described with specific outcomes may reflect the rapidity of disease progression rather than a contributing factor to the disease itself. Given the multifactorial etiology of disease processes and outcomes in immunosuppressed individuals, it is unlikely that an HLA association can be established with one or more of the AIDS-related disease entities unless large numbers of individuals are studied.

CONCLUSION

The ultimate goal of establishing HLA associations with any disease and particularly with HIV infection is to provide direction and focus for studies to discern their role in the disease process. The observations that have been made regarding MHC genes in HIV infection begin to provide that direction and in addition have added substantive information regarding the potential importance of including MHC gene products with interactive function. Future studies will need to focus on expanding the numbers of individuals studied and determining associations by evaluating the effects of combinations of MHC genes on the natural history and outcome of HIV-1 infection. Using this information, hypotheses concerning the role of the MHC in HIV-1 infection can be generated and tested. Expanding our knowledge in this area will undoubtedly provide fundamental insights into the immune response to HIV-1 and in turn will contribute to a rational construction of effective vaccines.

ACKNOWLEDGMENT

The author wishes to thank Ms Millie Michalisko for excellent editorial assistance and Dr Petra Lenz for reading the manuscript critically.

REFERENCES

Aldhous MC, Watret KC, Mok JY, Bird AG & Froebel KS (1994) Cytotoxic T lymphocyte activity and CD8 subpopulations in children at risk of HIV infection. *Clin Exp Immunol* **97:** 61–67.

Alkhatib G, Combadiere C, Broder CC et al (1996) CCCKR5: a RANTES, MIP-1a, MIP-1B receptor as a fusion cofactor for macrophage-tropic HIV-1. *Science* **272**: 1955–1958.

Barron KS, Reveille JD, Carrington M, Mann DL & Robinson MA (1995) Susceptibility to Reiter's syndrome is associated with alleles of *TAP* genes. *J Arthritis Rheum* **38**: 684–689.

Bjorkman PJ, Saper MA, Samraoui B, Bennett WS, Strominger JL & Wiley DC (1987) The foreign antigen binding site and T cell recognition regions of class I histocompatibility antigens. *Nature* **329**: 512–518.

Bleul CC, Farzan M, Choe H et al (1996) The lymphocyte chemoattractant SDF-1 is a ligand for LESTR/fusin and blocks HIV-1 entry. *Nature* **382**: 829–833.

Borrow P, Lewicki H, Hahn BH, Shaw GM & Oldstone MBA (1994) Virus-specific CD8+ cytotoxic T-lymphocyte activity associated with control of viremia in primary human immunodeficiency virus type 1 infection. *J Virol* **68**: 6103–6110.

Brown JH, Jardetzky TS, Gorga JC et al (1993) Three-dimensional structure of the human class II histocompatibility antigen HLA-DR1. *Nature* **364**: 33–39.

Cao Y, Qin L, Zhang L, Safrit J & Ho DD (1995) Virological and immunological characterization of longterm survivors of human immunodeficiency virus type 1 infection. *N Engl J Med* **332**: 201–208.

Carmichael A, Jin X, Sissons P & Borysiewicz L (1993) Quantitative analysis of the human immunodeficiency virus type 1 (HIV-1)-specific cytotoxic T lymphocyte (CTL) response at different stages of HIV-1 infection: differential CTL responses to HIV-1 and Epstein–Barr virus in late disease. *J Exp Med* **177**: 249–256.

Carrington M, Colonna M, Spies T, Stephens JC & Mann DL (1993) Haplotypic variation of the transporter associated with antigen processing (TAP) genes and their extension of HLA class II region haplotypes. *Immunogenetics* **37**: 266–273.

Carrington M, Stephens JC, Klitz W, Begovich AB, Erlich HA & Mann D (1994) MHC class II haplotypes and linkage disequilibrium values observed in the CEPH families. *Hum Immunol* **41**: 234–240.

Cheynier R, Langlade-Demoyen P, Marescot MR et al (1992) Cytotoxic T lymphocyte responses in the peripheral blood of children born to HIV-1-infected mothers. *Eur J Immunol* **22**: 2211–2217.

Choe H, Farzan M, Sun Y et al (1996) The β-chemokine receptors CCR3 and CCR5 facilitate infection by primary HIV-1 isolates. *Cell* **85**: 1135–1148.

Cresswell P, Howard JC (1997) Antigen recognition Currzzopin. *J Immunol* **9(1)**: 71–74

Deng H, Liu R, Ellmeier W et al (1996) Identification of a major co-receptor for primary isolates of HIV-1. *Nature* **381**: 661–666.

Detels R, Mann D, Carrington M et al. (1996) Persistently seronegative men from whom HIV-1 has been isolated are genetically and immunologically distinct. *Immunol Lett* **51**: 29–33.

Dragic T, Litwin V, Allaway GP et al (1996) HIV-1 entry into CD4+ cells is mediated by the chemokine receptor C-C CKR-5. *Nature* **381**: 667–673.

Gotch FM, Nixon DF, Alp N, McMichael AJ & Bprysiewicz LK (1990) High frequency of memory and effector gag specific cytotoxic T lymphocytes in HIV seropositive individuals. *Int Immunol* **2**: 707–712.

Goulder PJR, Phillips RE, Colbert RA et al (1997) Late escape from an immunodominant cytotoxic T-lymphocyte responses associated with progression to AIDS. *Nature Med* **3**: 212–217.

Hammond SA, Johnson RP, Kalams SA et al (1995) An epitope-selective, transporter associated with antigen presentation (TAP)-1/2-independent pathway and a more general TAP-1/2-dependent antigen-processing pathway allow recognition of the HIV-1 envelope glycoprotein by CD8+ CTL. *J Immunol* **154**: 6140–6156.

Harrer T, Harrer E, Kalams SA et al (1996a) Cytotoxic T lymphocytes in asymptomatic long-term nonprogressing HIV-1 infection. *J Immunol* **156**: 2616–2623.

Harrer T, Harrer E, Kalams SA et al (1996b) Strong cytotoxic T cell and weak neutralizing antibody responses in a subset of persons with stable nonprogressing HIV type 1 infection. *AIDS Res Hum Retroviruses* **12**: 585–592.

Haynes BF, Pantaleo G & Fauci AS (1996) Toward an understanding of the correlates of protective immunity to HIV infection. *Science* **271**: 324–327.

Ho DD, Neumann AU, Perelson AS, Chen W, Leonard JM & Markowitz M (1995) Rapid turnover of plasma virions and CD4 lymphocytes in HIV-1 infection. *Nature* **373**: 123–126.

Imagawa DT, Lee MH, Wolinsky SM et al (1989) HIV-1 infection in homosexual men who remain seronegative for prolonged periods. *N Engl J Med* **320**: 1458–1462.

Itescu S, Mathur-Wagh U, Skovron ML et al (1991) HLA-B35 is associated with accelerated progression to AIDS. *J AIDS* **5**: 37–45.

Just J (1995) Genetic predisposition to HIV-1 infection and acquired immune deficiency syndrome: review of the literature examining associations with HLA. *Hum Immunol* **156**: 156–169.

Kaslow R & Mann DL (1994) The role of histocompatibility complex I HIV infection – ever more complex? *J Infect Dis* **169**: 1332–1333.

Kaslow RA, Carrington M, Apple R et al (1996) Influence of combination of genes in the major histocompatibility complex and the course of HIV-1 infection. *Nature Med* **2**: 405–511.

Koopmann JO, Hämmerling GJ & Momburg F (1997) Generation, intracellular transport and loading of peptides associated with MHC class I molecule. *Curr Opin Immunol* **9**: 80–88.

Kroner BL, Rosenberg PS, Aledort LM et al (1994) HIV-1 infection incidence among persons with hemophilia in the United States and western Europe, 1978–1990. *J AIDS* **7**: 279–286.

Kroner BL, Goedert JJ, Blattner WA, Wilson SE, Carrington MN & Mann DL (1995) Concordance of HLA haplotype sharing, CD4 decline and AIDS status in hemophilic siblings. *AIDS* **9**: 275–280.

Langlade-Demoyen P, Ngo-Giang-Huong N, Ferchal F & Oksenhendler E (1994) Human immunodeficiency virus (HIV) nef-specific cytotoxic T lymphocytes in noninfected heterosexual contact of HIV-infected patients. *J Clin Invest* **93**: 1293–1297.

Liu R, Paxton WA, Choe S et al (1996) Homozygous defect in HIV-1 coreceptor accounts for resistance of some multiply-exposed individuals to HIV-1 infection. *Cell* **86**: 367–377.

Malnati MS, Peruzzi M, Parker KC et al (1995) Peptide specificity and recognition of MHC class I by natural killer cell clones. *Science* **264**: 1016–1018.

Mann DL, Carrington M & Kroner BL (1994) The major human histocompatibility complex and HIV-1 pathogenesis. *AIDS* **8**: 53–60.

McMichael AJ & Walker BD (1994) Cytotoxic T lymphocyte epitopes: implications for HIV vaccines. *AIDS* **8**: S155–S173.

Mellors JW, Kingsley LA, Rinaldo CR Jr et al (1995) Quantitation of HIV-1 RNA in plasma predicts outcome after seroconversion. *Ann Intern Med* **122**: 573–579.

Mellors JW, Rinaldo CR, Gupta P, White RM, Todd JA & Kingsley LA (1996) Prognosis in HIV-1 infection predicted by the quantity of virus in plasma. *Science* **272**: 1167–1170.

Moretta A, Biassoni R, Bottino C et al (1997) Major histocompatibility complex class I-specific receptors on human natural killer and T lymphocytes. *Immunol Rev* **155**: 105–117.

Murphy PM (1996) Chemokine receptors: structure, function and role in microbial pathogenesis. *Cytokine Growth Fact Rev* **7**: 147–164.

Parham P, Adams EJ & Arnett KL (1995) The origins of HLA-A,B,C polymorphism. *Immunol Rev* **143**: 141–180.

Paul WE (1995) Can the immune response control HIV infection? *Cell* **82**: 177–182.

Paxton WA, Martin SR, Tse D et al (1996) Relative resistance to HIV-1 infection of CD4 lymphocytes from persons who remain uninfected despite multiple high-risk sexual exposures. *Nature Med* **2(4)**: 412–417.

Peruzzi M, Parker KC, Long EO & Malnati MS (1996) Peptide sequence requirements for recognition of HLA-B 2705 by specific natural killer cells. *J Immunol* **157**: 3350–3356.

Phillips RE, Rowland-Jones S, Nixon DF et al (1991) Human immunodeficiency virus genetic variation that can escape cytotoxic T cell recognition. *Nature* **354**: 453–459.

Pieters J (1997) MHC class II restricted antigen presentation. *Curr Opin Immunol* **9**: 89–96.

Pollack MS, Safai B & Dupont B (1983a) HLA-DR5 and DR2 are susceptibility factors for acquired immunodeficiency syndrome with Kaposi's sarcoma in different ethnic subpopulations. *Dis Markers* **1**: 135–139.

Pollack MS, Safai B, Myskowski PL, Gold JWM, Pandey J & Dupont B (1983b) Frequencies of HLA and Gm immunogenetic markers in Kaposi's sarcoma. *Tissue Antigens* **21**: 1–8.

Powis SJ, Deverson EV, Coadwell WJ et al (1992) Effect of polymorphism of an MHC-linked transporter on the peptides assembled in a class I molecule. *Nature* **357**: 211–215.

Prince HE, Schroff RW, Ayoub G, Han S, Gottlieb MS & Fahey JL (1984) HLA studies in acquired immune deficiency syndrome patients with Kaposi's sarcoma. *J Clin Immunol* **4**: 242–245.

Rammensee HG, Friede T & Stevanovic S (1995) MHC ligands and peptide motifs: first listing. *Immunogenetics* **41**: 178–228.

Rowland-Jones SL & McMichael A (1995) Immune responses in HIV-exposed seronegatives: have they repelled the virus? *Curr Opin Immunol* **7**: 448–455.

Rowland-Jones SL, Nixon DF, Aldhous MC et al (1993) HIV-specific CTL activity in an HIV-exposed but uninfected infant. *Lancet* **341**: 860–861.

Rowland-Jones S, Sutton J, Ariyoski K et al (1995) HIV-specific cytotoxic T-cells in HIV-exposed but uninfected Gambian women. *Nature Med* **1**: 59–64.

Samson M, Libert F, Doranz BJ et al (1996) Resistance to HIV-1 infection in caucasian individuals bearing the mutant alleles of the CCR-5 chemokine receptor gene. *Nature* **382**: 722–725.

Scorza ?, Smeraldi R, Fabio G, Lazzarin A et al (1988) HLA-associated susceptibility to AIDS: HLA-B35 is a major risk factor for Italian HIV-infected intravenous drug addicts. *Hum Immunol* **22**: 73–79.

Shearer GM & Clerici M (1996) Protective immunity against HIV infection: has nature done the experiment for us? *Immunol Today* **17**: 21–24.

Steel CM, Ludlam CA, Beatson D et al (1988) HLA haplotype A1B8DR3 as a risk factor for HIV-related disease. *Lancet* **i**: 1185–1188.

Stern LJ, Brown JH, Jardetzky TS et al (1994) Crystal structure of the human class II MHC protein HLA-DR1 complexed with an influenza virus peptide. *Nature* **368**: 215–224.

Wei X, Ghosh SK, Taylor ME et al (1995) Viral dynamics in human immunodeficiency virus type 1 infection. *Nature* **373**: 117–122.

Wilson IA (1996) Another twist to MHC-peptide recognition. *Science* **272**: 973–974.

Wolinsky SM, Korber BTM, Neumann AU et al (1996) Adaptive evolution of human immunodeficiency virus-type 1 during the natural course of infection. *Science* **272**: 537–542.

Chapter 11

THE POTENTIAL ROLE OF HLA MIMICRY IN THE PATHOGENESIS OF AIDS

A.G. Dalgleish, E.F. Hounsell and H. Baum

INTRODUCTION

Even though the human immunodeficiency virus (HIV) is understood at the genomic level and clearly causes acquired immune deficiency syndrome (AIDS) in the majority of humans, exactly how it does so is still far from clear. Initially it seemed obvious that as HIV could enter through the CD4 receptor and kill CD4-positive cells in vitro, this was the explanation for the slow decline of CD4 cells measured in the peripheral blood in vivo. Experiments whereby CD4 was transfected into a variety of cells clearly showed that a second cofactor was involved in the infection of human cells and this has subsequently been identified as the family of chemokine receptors covered elsewhere in this volume (Maddon et al, 1986; Barcellini et al, 1996). Initially it appeared that mutations in these receptors accounted for the different susceptibility of individuals to HIV infection. Although there are mutations that appear to confer a protective effect, this accounts for only a small percentage of those infected individuals who progress very slowly.

A number of other cofactors have been suggested, from other infections through to different immunogenetic genes such as the human leukocyte antigen (HLA) (Dalgleish, 1996). The latter is particularly interesting as a number of studies have shown certain types or certain HLA groupings to be associated with rapid progression or slow progression of disease, while some even appear to confer a resistance to progression of disease (long-term non-progression, LTNP) (Westby et al, 1996). In spite of the fact that hundreds of different HLA alleles exist in the human population, it is of interest that HLA-B8 stands out as being associated with fast progression and HLA-B27 with non-progression or at least slow progression.

MHC MIMICRY AND AIDS PATHOGENESIS

Differences between rapid progressors and long-term non-progressors

The single most important differentiating feature that predicts rapid progression is probably determined shortly after infection when a high virus load can be detected, a

HIV and the New Viruses Second Edition
ISBN 0-12-200741-7

low CD4 count and high levels of activation markers. Following the initial infection, the virus load is rapidly contained and the CD4 count rises to near its preinfection level. Although it has been shown that the virus load reached at this stage is a good indicator of progression, the activation markers such as soluble tumour necrosis factors (TNF) and its receptors are equally good and indicate widespread activation of the immune system (Barcellini et al, 1996). It is particularly noteworthy that patients with HLA-B8 who progress more rapidly than many other haplotypes have a higher baseline activation of the immune system than slow progressors such as those with HLA-B27. The association with progression and basic activation of the immune system is reinforced by studies in chimpanzees, who become infected but rarely progress to disease (Gougeon et al, 1997). There is a complete absence of activation markers and with it another indirect marker of activation apoptosis. The relationship with activation and viruses such as HIV is interesting in that retroviruses are not very efficient replicators within host cells, and it has been known for many years that HIV replicates much better in activated cells than in non-activated cells. The virus's ability to establish a foothold in the host would be enhanced if the virus itself was able to activate resting cells. This appears to be the case for many viruses, including retroviruses, which may employ a range of different mechanisms to do so. The mouse mammary tumour virus (MMTV) encodes a superantigen in the open reading frame next to the envelope which acts like a superantigen in that it is capable of activating all cells that bear a specific T cell receptor variable region. This has the effect of causing a panactivation of the T cell response, thus giving the virus many more activator cells to infect and establish a virus load. This case represents clearly the importance of the immunogenetics of the host with regard to the pathogenicity of the virus. Mice that do not have the variable region of the T cell receptor that is recognized by MMTV do not develop disease, because the immune system is not activated. As the name of this virus suggests, it causes mammary tumours, and it can do so only if the virus load is built up by activating a large number of cells in the immune system, thus allowing the virus to replicate and infect other tissues including breast (Acha-Orbea and Palmer, 1991; Huber et al, 1994).

The ability to measure the T cell receptor variable genes in the family led initially to suggestions that HIV is able to act as a superantigen and activate T cell receptors through specific variable genes. However, the earlier studies that strongly suggested that this was the case were limited by technical problems; many other groups have investigated this aspect of pathogenesis and, with a few exceptions, the majority concluded that there is a general perturbation of the T cell receptor without a specific variable region being either stimulated or deleted (Dalgleish et al, 1992; Westby et al, 1998). Moreover, this perturbation has recently been shown to be correlated with the CD4 count and strongly suggests that the variability seen is due to a random deletion of CD4 cells; hence it is unlikely that they are being activated and deleted by a superantigen encoded by HIV (Westby et al, 1998).

Another way of activating the immune system in an antigen non-specific way is seen following allogeneic transplantation of bone marrow or an organ. The immune system is able to recognize and reject an organ with alarming ferocity. In addition, in the presence of immunosuppression, lymphocytes that have been taken in with the transplant can activate the host's immune system in a non-specific manner, causing marked generalized activation associated with increased cytokine profiles and autoimmune features, a process known as graft-versus-host disease (GVHD). There are two major manifestation of GVHD: the first is an acute reaction which is associated with a strong cell-mediated T helper cell 1 (Th1) activation; the second is a more insidious and chronic version which, paradoxically, is associated with the attenuation of cell-mediated responses and a strong humorally mediated response (Th2) with high TNF levels. The features of chronic GVHD are remarkably similar to HIV-induced AIDS in that the disease is characterized by wasting, opportunistic infections and B cell lymphomas, along with a variety of other manifestations often seen in HIV infections such as hepatosplenomegaly and marrow disorders (Habeshaw et al, 1992).

The importance of the clinical and immunological similarities between HIV and chronic GVHD is made more interesting by the fact that HIV encodes several regions of HLA-like sequences. This raises the possibility that HIV encodes HLA-like sequences for the purpose of activating the immune system. This would occur only where the sequences were seen as foreign or allogeneic, and this would be likely to occur therefore in the majority of infected individuals. As if to demonstrate a proof of concept for this suggestion, Ter-Grigorov and colleagues recently published a murine model of AIDS which required allogeneic stimulation in order to activate enough cells for an otherwise very weak virus to replicate to a high enough load for the infection to be self-sustaining and disease to be induced (Ter-Grigorov et al, 1997). In the discussion of this paper, they referred to suggestions made by other workers that HIV encodes its own allogeneic sequences for this purpose. It is important to note that the list of viruses that have 'hijacked' immune response genes in order to evade or mimic the host's immune system continues to grow. Other viruses have HLA-like sequences, most notably cytomegalovirus which uses this sequence to defend itself from natural killer cell attack (Farrell et al, 1997). The herpesviruses encode both cytokines (interleukin 10 in the case of Epstein–Barr virus) and chemokines; an example is the relatively recently identified Kaposi's sarcoma-associated herpesvirus, HHV8 (Moore et al, 1996). It should therefore not be too surprising if these HLA sequences play a role in the pathogenesis of HIV. Although to date there is no evidence that these sequences definitively play a role, an expanding body of evidence is consistent with this hypothesis.

The HLA regions of HIV

The first region to be identified was a region on gp41 which had homologies with MHC class II β1 domain which was recognized by cross-reacted monoclonal antibodies. However, there are several more homologies present on the gp120 envelope (Figure 11.1). Indeed, all the conserved regions of the gp120 envelope have some HLA homology, but two – mainly the C2 and C5 – have the largest and most consistent similarities. Peptides from both these regions have been reported to perturb immune recognition (Clerici et al, 1993).

Molecular modelling of the carboxy terminus of gp120 by Hounsell and colleagues revealed the C5 region to have a large α-helix which has similarities with both HLA class I and class II (Figures 11.2 and 11.3). Baretta's group identified an antibody to this part of the carboxy terminus which cross-reacted with HLA class I (HLA C) and which could inhibit HIV-induced fusion (Lopalco et al, 1993). The carboxy terminus has been recognized as a highly antigenic site, but it has been ignored because it does not induce

Amino acid no.	HIV conserved regions with HLA class I or class II homology	
	R　V	
247	OCTHGIK<u>PIVSTQLLNGS</u>LAE	
432	YAPPI	(conserved minor helix)
	LI	
444	<u>SNITGILTRDGG</u>	DR-β1
	D　R	
468	<u>GGG</u>NMKDNW	HLA-A$_2$
	V	
479	EL<u>YKYK</u>VI	
	V　　R　　　O	
488	IEPLGIAP<u>TKAK<u>RR</u>VV<u>E</u>RE<u>KR</u></u>A	
837	<u>EGTDRVI</u>	

Figure 11.1 *The HLA homologous regions of HIV*
Data from Golding et al (1988), Grassi et al (1991), Hounsell et al (1991) and Beretta (unpublished observations). Underlining indicates absolute homology with HLA.

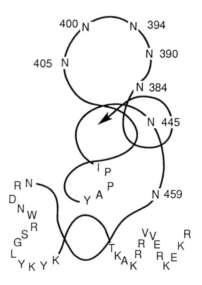

Figure 11.2 *Molecular modelling for the gp120 carboxy terminus*
N denotes the N glycosylation sites; YAPPI is part of the CD4 binding loop; NRDNWRSGL is part of the minor α helix; YKYK is recognized by an anti-HLA antibody M38; TKAKRRVVEREKEKR is the sequence leading to the gp120/gp41 cleavage site which forms an α-helical structure.

The amino acid sequence of gp120 from the so-called V3 loop to the carboxy terminus gp120–gp40 cleavage site also contains a domain homologous with immunoglobulin light-chain folds. This sequence containing amino acids 416–442 was modelled by homology to immunoglobulin RE1, the crystallographic coordinates for which were obtained from the Brookhaven database. The secondary structure for amino acids 443–508 was from prediction methods and molecular dynamics. The amino acids 375–415 describe a highly glycosylated loop. Three high mannose (445, 390 and 384) and four complex (459, 405, 400 and 394) N-linked chains are shown attached in Figure 11.3 (Renouf and Hounsell, 1995).

Figure 11.3
A snapshot of the molecular modelling giving the view explained in figure 11.2. The darkest area is the exposed protein, the amino acids of which are spelt out in figure 11.2 and the first paragraph of its legend. The lighter areas show different motifs of the oligosaccharides linked at the asparagine sites (N) indicated in figure 11.2. The figure illustrates the accessibility of protein epitopes in the carboxy terminus quarter of the molecule (amino acids 374 to 508 with attached glycosylation) including along the bottom of the molecule (as shown) an area of comparable molecular size and conformation to MHC minor and major alpha-helices.

neutralizing antibodies. However, the recognition of this site by antibodies appears to be more dominant in patients who have been exposed to HIV but have not seroconverted and in those about to seroconvert (L. Brown, unpublished observations).

Since this modelling was performed, a number of other studies have noted the potential for structural homology. However, it was believed that crystallization would reveal the true structure of gp120. Several years of trying were met with failure until the component that binds to CD4 was successfully crystallized by Sodrowski and Henderson's groups (Lopalco et al, 1993). Unfortunately, the HLA-like regions did not remain attached and therefore the true structure of these HLA-like sequences remains unclear. In an attempt to see whether this possible structure could have any functional significance, Brian Austen and colleagues at St George's Hospital Medical School predicted a peptide-binding site and were able to show that gp120 could bind peptides as shown by ultraviolet cross-linking in a similar manner to HLA molecules, and that this binding was relatively specific in that it could be competed out with non-labelled peptides (J. Sheikh et al, 1995). Subsequent studies have shown that the gp120-bound peptide complex is recognized by the T cell receptor in a similar way to HLA-DR1 binding the same peptide (Sheikh et al, unpublished observation). This strongly suggests that there are enough HLA-like regions within HIV gp120 to confer some functional similarity. However, what this actually means in vivo is unclear. It may be that the ability to bind peptides leads to enhanced stimulation as the virus buds from a cell surface or, paradoxically, it may mean that the peptides are bound to gp120 thus preventing the normal loading of peptides on to the normal host's self HLA molecule.

The association between HIV and HLA is further complicated by the fact that HIV incorporates HLA molecules into its envelope following budding through the cell membrane (Cantin et al, 1997). Indeed, of all the ligands on the cell surface, the HIV variant has a predilection for HLA molecules. It has been reported that this has functional significance in that virions which have budded through class II-producing cells are able to activate and infect other cells more efficiently than viruses that have budded through class II-negative cells (Cantin et al, 1997). Bearing in mind that gp120 and MHC cluster on the surface of antigen-presenting cells, it is possible that it is the constellation of gp120 and HLA that can activate cells. The role of HLA is further highlighted by the fact that antibodies to HLA can protect macaques from simian immunodeficiency virus (SIV) infection if the challenged virus has been grown through the same cell used to vaccinate the monkey (Chan et al, 1997). This finding, which is readily reproducible and elicits some of the strongest protection against SIV, has been largely ignored because of the association with the HLA molecules budding from the vaccinated cell. Nevertheless, the fact that other ligands do not confer any protection and the ability of the envelope and HLA to coassociate may be behind this phenomenon.

A complex of the envelope and MHC may be a minimum requirement for activation. It is likely that other factors, particularly Nef, also play a role in the activation process as *nef*-deleting mutants do not activate and are poorly infectious. The *nef* gene is in the same region as superantigens are in other retroviruses. Although it does not appear to encode a specific superantigen it – like the envelope – does encode HLA-like regions. In the mouse mammary tumour virus the superantigen effect is seen only when it is co-presented with the envelope and this may well be the case with the *nef* gene of HIV.

The aforementioned scenario would depend on the complex resembling allogeneic MHC/HLA to activate an infecting cell and may be necessary to initiate infection. However, the activation pattern seen in HIV infection is much more consistent with chronic stimulation, which would in turn be more compatible with an indirect mechanism of allorecognition, i.e. one where HLA-like regions are presented on self MHC and the whole appears as an 'allo' complex. Evidence for such a possibility for the peptide regions from C2 and C5 has been reported (Atassi and Atassi, 1992; Clerici et al, 1993). It may well be the constant presentation of these epitopes by antigen-presenting cells in the lymph node that fans the embers of chronic activation, as it were. In this regard, it is of interest that the C5 region is very similar to the self peptides presented by HLA-B27, which leads to the prediction that patients with HLA-B27 would be slow progressors as their systems would not activate against something that resembled their own HLA (Habeshaw et al, 1994). Furthermore, the prediction was applied to explain why chimpanzees do not develop AIDS, in that they have HLA sequences resembling those associated with HIV. Surprisingly, this is the case with all the cohorts containing HLA-B27 patients, showing that they are statistically very slow or non-progressors, and that most of the chimpanzee sequences to date in the Primate Research Centre in the Netherlands are HLA-B27-like sequences (R. Bontrop, unpublished observations). This is particularly interesting considering that chimpanzees have normal CD4 molecules and gp120-binding properties as well as wild-type chemokine receptors, and hence this cannot be the explanation of why they do not become ill.

Implications for treatment and vaccines

If AIDS is fuelled by a chronic activation of the immune system, then it would make sense to immunosuppress these patients. It is superficially oxymoronic to immunosuppress people with an acquired immune deficiency syndrome. However, out of desperation, several physicians have added significant immunosuppression, superficially treating the side-effects of HIV infection. The use of high-dose steroids was reported in a patient who could no longer take retroviral treatment and who was suffering from wasting, hepatosplenomegaly, marrow aplasia and opportunistic infections. The patient had a marked response to steroids, which was titratable with the effect starting to disappear when the dose of prednisone was reduced below 40 mg (Aitken et al, 1992). Since this observation, a further study has looked at the use of prednisone in AIDS patients and documented a reduction in virus load with prednisone alone (Kilby et al, 1997). At earlier stages of disease, a beneficial effect of prednisone on CD4 counts has been reported at 1 year and a shorter-term benefit from cyclosporin by a French group (Andrieu et al, 1995). Thalidomide treatment was reported to inhibit TNF production in patients with asymptomatic HIV with marked improvement in the patients' chronic diarrhoea, resulting in the disappearance of the causative organism and return to normal histology of the gut mucosa (Marriott et al, 1997). Other groups have used thalidomide for cachexia in later-stage disease with considerable benefit. The fact that the most dramatic clinical effects are on the oral mucosa and bowel suggests that thalidomide is much more effective at lowering mucosal TNF than systemic TNF, and biopsies of the mucosa taken before and after thalidomide treatment strongly support this, with marked reductions in the face of only minor systemic reductions of TNF production after activation.

The overall benefit from adding immunosuppression to the treatment of HIV patients strongly suggests that this should be investigated further. Patients who are on highly active antiretroviral therapy often relapse in a very few weeks after it is discontinued, and their CD4 count and viral loads return to pretreatment levels. It is interesting to note, where the data are available, that these patients continue to have high activation markers even though their viral loads have been reduced to very low levels on treatment. It may be necessary to add immunosuppression prior to reducing the antiviral treatment and to maintain it for some months after the antiviral treatment has been withdrawn.

The finding that patients who appear to be protected have antibodies to the carboxy terminus or other HLA-like epitopes strongly suggests that, if nothing else, this is a surrogate marker for a protective immune response to HIV, which also includes a strong cell-mediated response and high immunoglobulin A response in the mucosa (Mazzoli et al, 1997). It would therefore appear reasonable to try to mimic this protective response by using a strong Th1 adjuvant and the epitopes from the HLA-like regions of HIV which, although they do not neutralize, may well be involved in T cell priming. An excellent place to try such studies would be in the very patients referred to above, i.e. those who have to abandon antiretroviral therapy for reasons of toxicity.

AUTOIMMUNITY AND AIDS PATHOGENESIS

A number of authors have written about the potential for the cascade or idiotype network involving CD4 and MHC class II to invoke a response that would kill CD4 cells and other activated cells, resulting in general autoimmunity not too dissimilar from that described in GVHD. It is not clear whether these explanations are just the other side of the coin, as it were, but there is still a need to explain the other autoimmune process seen in HIV infection. It is interesting to note that a number of autoimmune conditions involving the skin, joints, marrow and nervous system have been described in HIV patients, but it rare for these phenomena to affect more than 10% of any HIV-infected cohort. It may be that these phenomena are a result of an extra cross-reactivity, and we now describe another mechanism whereby mimicry of MHC peptides can lead to a response directed at a particular organ, for instance the skin, the joints or the marrow.

MHC mimicry and autoimmune disease

Molecular mimicry, at the level of cross-reacting antibodies or cross-reacting T cells, is popularly invoked as a mechanism for autoimmune disease, and both CD4+ and CD8+ T cells have been isolated in a number of such diseases that recognize epitopes of both pathogens and target autoantigens, with varying degrees of sequence similarity. In turn it has been suggested that the association of autoimmune diseases with particular MHC haplotypes relates to restricted presentation by the susceptibility allele, of class I or class II, of the respective mimicking peptides.

A further level of mimicry has recently been suggested, which offers an additional explanation for MHC disease association (Baum et al, 1996). This involves mimicry with peptides derived from MHC-coded molecules themselves. In the 'three-way mimicry' hypothesis, it is proposed that the CD4+ T cell repertoire is biased to recognize MHC-derived peptides (class I or class II, invariant or allele-specific) in class II MHC association, and that this results in the dominant T cell response to foreign antigens being towards epitopes that mimic such self-complexes. If a tissue-specific antigen in turn mimics the 'common epitope' (class II MHC:MHC peptide and class II MHC:mimicking foreign peptide) then an autoimmune cascade is initiated.

The rationale for this concept is the fact that dominant amongst self-peptides extracted from class II MHC are those deriving from MHC molecules themselves, both class I and class II (including the invariant, class II γ chain) (Rammensee et al, 1995). In thymic ontogeny, therefore, the resultant epitopes, presented on thymic antigen-presenting cells, would play a dominant role in CD4+ T cell development, cells with moderate to low affinity being positively selected for survival into the periphery. (Class I MHC also bind MHC-derived peptides, but with lower frequency, and the hypothesis has not yet been developed in terms of CD8+ T cells.)

The role of MHC haplotype in disease susceptibility is thus seen not only in terms of presentation of mimicking peptides, but also in terms of the origin of MHC-derived peptide epitopes themselves, if they are allele-specific. Also, the requirement for 'three-way mimicry' increases the chances that one of the mimicking peptides (derived respectively from MHC molecule, foreign antigen and target autoantigen) may, by virtue of slight structural differences, partition between available class II heterodimers differently from the other two. This (and other permutations of such a scenario) would account for the effect of protective alleles.

The model also allows for 'two-way mimicry'. The tissue target itself might be an MHC-derived peptide mimicking a pathogen, and presented by class II MHC heterodimers whose expression is locally upregulated, e.g. by interferon gamma (IFN-γ) following viral infection. Alternatively, under specific pathogen-free conditions, T cells selected to recognize a dominant MHC-derived peptide in class II association might directly attack a tissue presenting a mimicking self-peptide.

This last concept is well illustrated by the non-obese diabetic (NOD) mouse, an animal model of human autoimmune diabetes (Zekzer et al, 1998). The NOD mouse has a single, unique class II MHC gene product I-A(g)^7. The spontaneous onset of autoimmune diabetes seen in virtually all female NOD mice under specific pathogen-free conditions is prevented by site-directed mutagenesis of a single, characteristic amino acid of that class II molecule. The first indication of the autoimmune disease is the appearance of CD4+ T cells recognizing an I-A(g)^7 presented peptide derived from the islet autoantigen glutamic acid decarboxylase-65 (GAD-65). The first 10 amino acids of that epitope share seven identities and two 'conservative substitutions' with a characteristic peptide of I-A(g)^7, and injection of that MHC-derived peptide into neonatal NOD mice prevents the onset of diabetes.

Similar evidence can be given in a few cases (rheumatoid arthritis, primary biliary cirrhosis, autoimmune uveitis) for the operation of 'three-way mimicry'. However, the hypothesis, though attractive, still rests primarily on circumstantial evidence, expressed in terms of the statistics of sequence similarity (Baum, 1997).

MOLECULAR MIMICRY AND IDIOTYPE NETWORKS IN HIV INFECTION

The specific infection of CD4+ T cells by HIV rests upon the binding of gp120 to CD4. In that respect gp120 mimics the CD4 receptor of class II MHC. There is also some evidence that the variable region domain of Fab may also mimic the CD4-binding region of HIV gp120, and hence also mimics MHC. In consequence of these similarities it would follow, for example, that anti-Fab antibody should bind to gp120, that anti-gp120 antibody should bind to MHC class II, that anti-Fab antibody should bind to anti-anti-gp120 antibody, and so forth, in an idiotype network that could disturb normal self–non-self recognition and account in part for the complex time course of events in HIV pathogenesis. There is some evidence that such a humoral network is in operation; for example, during the evolution of AIDS there is an inverse association between anti-Fab antibody and anti-anti-gp120 antibody, the onset of AIDS itself being marked by a major increase in the former (Susal et al, 1995).

However, such complex equilibria at the humoral level, while important in understanding disease progression, do not relate to MHC mimicry at the level of T cell epitopes, and only derive from one particular conformational similarity between gp120 and the CD4 receptor of class II MHC molecules.

Primary sequence similarities between HIV and MHC peptides

Peptides presented in the groove of class II MHC heterodimers retain very little secondary structure, and their primary sequence is a major determinant of epitope recognition by T cell receptors. However, the result of binding to anchor regions, and other side-chain interactions with the presenting molecule, is such that it is difficult – without very specific molecular modelling – accurately to predict functional mimicry between epitopes. This is an important caveat in the discussion that follows.

Table 11.1 *Primary sequence similarities between HIV and MHC peptides*

Source	Sequence
Env	85 EDVWHLFE
HLA-DP(w4)a	71 ETVWHLEE
Env	258 PVVSTQLLLNG
HLA-DPβ	168 GVVSTNLIRNG
Env	270 GFNGTRAE
HLA-DQ(w3)β	99 VLEGTRAE
Env	FKGKWKDAM
Calnexin	379 YKGKWKPPM
Env	507 EKRAVG
HLA-DP(w4)β	97 EKRAVP
Env	512 GVFVLGFL
HLA-DRα	5 GVPVLGFL
Env	709 QGYSPL
MAD-3	249 QGYSPY
Env	734 ETEEDGGS
Calnexin	555 DAEEDGGT
Env	757 EDLRSL
HLA-A2α	152 EDLRSW
Gag	1 GARASV
HLA-DQ(w1.1)	102 GARASV
Gag	65 PSLQTG
HLA-DQ3β	210 PSLQNP
Gag	144 QAISPRT
HLA-B35α	63 DAASPRT
Nef	181 VSQEAED
Calnexin	563 VSQEEED
Nef	218 LAYDYKAFIL
HLA-B49α	140 LAYDGKDYIA
Pol	5 DLAFLQ
MAD-3	100 DLAFLN
Pol	100 PKMIGGIGGFI
HLA-DQ1β	229 SKMLSGIGGFV
Pol	352 LEPFR
Calnexin	411 LEPFR
Pol	689 SASOPT
HLA-E1α	297 PASQPT
Pol	1026 LEGAR
HLA-DQ(w1.1)β	100 LEGAR
Rev	4 RADEEGLQRKLRL
HLA-C4α	260 HMQHEGLQEPLTL
Rev	62 TYLGRSAE
HLA-DXβ	83 TELGRSIE
Vif	185 LALTALIA
HLA-DP(w4)β	13 VALTALLM
Vif	222 GSHTMN
HLA-Hα	115 GSHTMQ

Nevertheless, strong similarity between primary sequences constitutes prima facie evidence of the possibility of such mimicry, and from that viewpoint the extent of mimicry between HIV and MHC is remarkable. Some examples are listed in Table 11.1. Some of these similarities may be stochastic, but others may well be examples of true mimicry, in the evolutionary sense, selected for the benefit of the survival of the pathogen.

There is some controversy over the statistical significance of sequence similarities. However, stated in simplistic terms (assuming equal frequency of the different amino acids), the probability of a perfect match with a specified five amino acid motif is $(0·05)^5$, or about 3×10^{-7}. A major protein database, incorporating virtually all proteins that have ever been sequenced, will contain up to approximately $1·5 \times 10^7$ amino acids, so only about five proteins in the entire set will bear that particular sequence, unless it is an evolutionarily selected functional motif. The odds against a perfect match decrease by a factor of n if one is looking for a match with *any* 5-mer in a protein n amino acids long; they increase 20-fold for each additional perfect match, and around 4-fold for each additional 'conservative substitution'. The odds against a match also increase considerably with the addition of an extra constraint, for example that the matching motif must be known to bind to class II MHC, which is the case for several sequences in Table 11.1 (for example, the HLA-DRα motif was the first self-peptide ever to be identified on elution from class II MHC).

Pathogenic implications of MHC–HIV mimicry

Any cell infected with HIV will present HIV peptides in class I association. If there is any bias in the cytotoxic T lymphocyte (CTL) repertoire towards recognizing class I-presented, MHC-derived peptides – by analogy with what is proposed in the 'three-way mimicry' hypothesis for CD4+ T cell epitopes – then initial CTL responses might conceivably be affected by mimicry between those MHC peptides and HIV peptides. If the MHC peptide that determined the CTL repertoire in such a scenario were allele-specific, then this might contribute to an HLA-haplotype-linked bias in initial responses to HIV infection.

Even if CTL selection is not affected in such a way, it is known that MHC peptides, class I and class II, may occasionally be presented in class I association. Hence any cell that did present such peptides would, in principle, be subject to 'bystander' attack by CTLs initially stimulated by infected cells presenting, in class I MHC, mimicking HIV peptides. Such an autoimmune response could be on any tissue. However, MHC class II peptides (the major 'mimics') are more likely to be available for class I presentation when class II expression is upregulated, as it would be in the antigen-presenting cell (APC) and T cell population in HIV infection. Again, the extent of any such autoimmune reactivity might be determined by HLA haplotype.

The 'three-way mimicry' hypothesis is much more firmly rooted in regard to class II MHC-presented peptides, and the activation of CD4+ T cells. The model would lead to the prediction that the dominant CD4+ T cell epitopes against HIV-derived peptides presented on APCs would correspond to mimicking MHC peptides (class I and class II), and that this might subsequently bias B cell responses, which would, in part, thus reflect HLA haplotypes.

Of much greater significance in terms of pathogenicity might be circumstances where class II-presented, MHC-derived peptides were recognized by CD4+ T cells, already activated by a mimicking HIV epitope (whether or not such cells had been particularly selected in thymic ontogeny). Such presentation would presumably occur on APCs themselves, and it is difficult to predict what the outcome of such further stimulation might be, particularly in view of the likely differing affinity for the respective mimicking epitopes. Might this modulate Th1–Th2 switching? More complex still would be the scenario where, in the cytokine environment of a developing HIV infection, bystander T cells, both CD4+ and CD8+, expressed class II MHC, presenting mimicking, MHC-derived peptides. Would CD4+ T cells, activated by viral peptide epitopes and then recognizing the mimicking epitopes on other T cells, be themselves subject to apoptosis through the 'veto' effect? Would the T cell targets for such recognition be stimulated or suffer autoimmune killing? It would certainly seem that some kind of immune chaos would occur, and with some features of GVHD, involving a population of cells not directly infected with the virus. If so, then perhaps immunosuppression might be paradoxically therapeutic in modulating the onset of full-blown AIDS, as discussed earlier.

The various potential cellular interactions described above are summarized in Table 11.2, but it should be stressed that there is little direct evidence that any of these actually contributes to AIDS pathogenesis.

Possible implications of mimicry for antibody profile

As implied above, a specific stimulation of CD4+ T cells recognizing the viral epitopes that mimicked class II-presented MHC peptides may subsequently stimulate antibody production against the parent viral protein (or possibly against the mimicking MHC antigen itself). In principle, of course, a single T cell clone, recognizing a single peptidyl epitope, could drive the production of antibody recognizing any exposed conformational epitope of a parent molecule or macromolecular complex. However, it is frequently the case that B cell epitopes overlap corresponding T cell epitopes. This may be because immunoglobulin receptors in B cells remain bound to their ligands on internalization and 'chaperone' them for preferential presentation on class II MHC.

Such considerations lead to the testable prediction that HIV infection might lead to the preferential production of antibody-recognizing epitopes containing the kinds of mimicking sequences shown in Table 11.1. The pattern of such antibody production should also reflect HLA haplotype, and could even be a marker of disease prognosis.

CONCLUSION

A number of models have now been demonstrated for the induction of autoimmune disease in mice whereby MHC (HLA in humans) mimicry by an exogenous agent and self peptide results in disease. HIV encodes enough HLA-like sequences to interfere with a normal self immune recognition as well as to panactivate the immune system in an HLA (allogeneic) like manner. It remains to be demonstrated whether and how this leads to pathogenesis in vivo.

Table 11.2 *Possible T cell epitope mimicries in HIV infection*

Effector cell	Activated/selected by	Presented on	Target cell	Presenting	On
CTL	MHC-derived peptides	Class I MHC in thymus	HIV-infected	HIV peptides	Class I MHC
CTL	HIV peptides	Class I MHC on infected cells	Any	MHC peptides (I or II)	Class I MHC
CD4+ T cells	MHC-derived peptides	Class II MHC in thymus	HIV-infected	HIV peptides	Class II MHC
CD4+ T cells	HIV peptides	Class II MHC on infected cells	APC	MHC peptides (I or II)	Class II MHC
CD4+ T cells	HIV peptides	Class II MHC on infected cells	T cells	MHC peptides (I or II)	Class II MHC
	(+ MHC peptides?)	(+ class II MHC on APCs?)	(CD4+/CD8+?)		

APC, antigen-presenting cell; CTL, cytotoxic T lymphocyte; HIV, human immunodeficiency virus; MHC, major histocompatibility complex.

REFERENCES

Acha-Orbea H & Palmer E (1991) MIS – a retrovirus exploits the immune system. *Immunol Today* **12**(10): 356–361.

Aitken C, Dalgleish AG, Salama A & Forster G (1992) Pancytopenia and hepatosplenomegaly in an AIDS patient responding to high dose steroids. *Int J STD AIDS* **3**(6): 450–451.

Andrieu JM, Lu W & Levy R (1995) Sustained increase in CD4 cell counts in symptomatic human immunodeficiency virus type-1 seropositive patients treated with prednisolone for 1 year. *J Infect Dis* **171**(3): 523–530.

Atassi H & Atassi MZ (1992) HIV envelope protein is recognised as an alloantigen by human DR-specific alloreactive T cells. *Hum Immunol* **34**(1): 31–38.

Barcellini W, Rizzardi GP, Poli G et al (1996) Cytokines and soluble receptor changes in the transition from primary to early chronic HIV type 1 infection. *Aids Res Hum Retroviruses* **12**(4): 325.

Baum H (1997) Molecular mimicry with MHC-derived peptides: does this underlie MHC association with autoimmune disease? *Biochem Soc Trans* **25**(2): 636–642.

Baum H, Davies H & Peakman M (1996) Molecular mimicry in the MHC: hidden clues to autoimmunity? *Immunol Today* **17**(2): 64–70.

Cantin R, Fortin JF, Lamontagne G & Tremblay M (1997) The acquisition of host-derived major histocompatibility complex class II glycoproteins by human immunodeficiency virus type 1 accelerates the process of virus entry and infection in human T-lymphoid cells. *Blood* **90**(3): 1109–1100.

Chan WL, Rodgers A, Grief C et al (1997) Immunisation with class 1 human histocompatibility leukocyte antigen can protect macaques against challenge infection with SIV. *Aids Res Hum Retroviruses* **13**(11): 923–931.

Clerici M, Shearer GM, Hounsell EF, Jameson B, Habeshaw J & Dalgleish AG (1993) Alloactivated cytotoxic T cells recognise the carboxy terminal domain of HIV-1 gp120 envelope glycoprotein. *Eur J Immunol* **23**: 2022–2025.

Dalgleish AG (1996) Pathogenesis of AIDS and the importance of HLA. *Semin Clin Immunol* **12**: 31–39.

Dalgleish AG, Wilson S, Gompels M et al (1992) T cell receptor variable gene products and early HIV infection. *Lancet* **339**: 824–828.

Farrell HE, Vally H, Lynch DM et al (1997) Inhibition of natural killer cells by a cytomegalovirus MHC class 1 homologue in vivo. *Nature* **386**(6624): 446–447.

Golding H, Robey FA, Gales FT et al (1998) Identification of homologous regions in HIV gp41d MHL class II beta I domain I. *J Exp Med* **167**(3): 914–923.

Gougeon ML, Lecoeur H, Boudet F et al (1997) Lack of chronic immune activation in HIV-infected chimpanzees correlates with the resistance of T cells to Fas/Apo-1 (CD95)-induced apoptosis and preservation of a T helper 1 phenotype. *J Immunol* **158**(6): 2964–2976.

Grassi F, Meneveri R, Fullberg M et al (1991) HIV gp120 mimics a hidden monomorphic epilope borneby class I MHC heavy chains. *J Exp Med* **174**(1): 53–62.

Habeshaw J, Hounsell EF & Dalgleish AG (1992) Does the HIV envelope induce a chronic graft-versus-host like disease. *Immunol Today* **13**(6): 356–358.

Habeshaw J (1994) HLA mimiary by HIV gp120 and the pathogenesis of AIDS. *Immunol Today* **15**(1): 39–40.

Hounsell EF, Renouf DV, Liney D, Dalgleish AG & Habeshaw JA (1991) A proposed molecular model for the carboxy terminus of HIV-1 gp120 showing structural features consistent with the presence of a T-cell alloepitope. *Mol Aspect Med* **12**: 283–296.

Huber BT, Beutner U & Subramanyam M (1994) The role of superantigens in the immunology of retroviruses. *CIBA Found Symp* **187**: 132–140.

Kilby JM, Tabereaux PB, Mulanovich V, Shaw GM, Bucy RP & Saag MS (1997) Effects of tapering doses of oral prednisone on viral loading among HIV-infected patients with unexplained weight loss. *Aids Res Hum Retroviruses* **13**(17): 1533–1537.

Kurong PD, Wyatt R, Tobinson J, Sweet RW, Sodroski J & Hendrickson WA (1998) Structure of an HIV gp120 envelope glycoprotein in complex with the CD4 receptor and a neutralising antibody. *Nature* **393**: 648–659.

Lopalco L, Longhi R, Ciccomascolo F et al (1993) Identification of human immunodeficiency virus type 1 glycoprotein gp120/gp41interacting sites by the idiotypic mimicry of two monoclonal antibodies. *Aids Res Hum Retroviruses* **9**(1): 33–39.

Maddon PJ, Dalgleish AG, McDougal JS, Chapman P, Weiss RA & Axel R (1986) The T4 gene encodes the AIDS virus receptor and is expressed in the immune system and the brain. *Cell* **47**: 333–348.

Marriott JB, Cookson S, Carlin E et al (1997) A double-blind placebo-controlled phase II trial of thalidomide in asymptomatic HIV-positive: clinical tolerance and effect on activation markers and cytokines. *Aids Res Hum Retroviruses* **13**(18): 1625.

Mazzoli S, Trabattoni D, Lo Caputo S et al (1997) HIV-specific mucosal and cellular immunity in HIV-seronegative partners of HIV-seropositive individuals. *Nature Med* **3**(11): 1250–1257.

Moore PS, Boshoff C, Weiss RA & Chang Y (1996) Molecular mimicry of human cytokine and cytokine response pathway genes by KSHV. *Science* **274**(5292): 1739–1744.

Rammensee HG, Friede T & Stevanoviic S (1995) MHC ligands and peptide motifs: first listing. *Immunogenetics* **41**(4): 78–228.

Renouf DJ & Hounsell EF (1995) Molecular modelling of glycoproteins by homology with non-glycosylated protein domains, computer simulated glycosylation and molecular dynamics. In: *Glycoimmunology*, ch. 4, London: Plenum Press eds: JS Axford and A Alavi.

Sheikh J, Ongradi J, Austen B & Dalgleish AG (1995) The potential importance of MHC mimicry by HIV in the pathogenesis of AIDS. *Biochem Soc Trans* **23**: 471S.

Susal C, Daniel V & Opelz G (1995) Does AIDS emerge from a disequilibrium between two complementary groups of molecules that mimic MHC? *Immunol Today* **17**(3): 114–119.

Ter-Grigorov VS, Krifuks O, Liubashevsky E, Nyska A, Trainin Z & Toder V (1997) A new transmissible AIDS like disease in mice induced by alloimmune stimuli. *Nature Med* **3**(1): 37–41.

Westby M, Manca F & Dalgleish AG (1996) The role of the host's immune response in determining the outcome of HIV infection. *Immunol Today* **17**(3): 120–126.

Westby M, Vaughan AN, Balotta C, Massimo G, Clerici M & Dalgleish AG (1998) Low CD4 counts rather than superantigenic like effects account for the differences in expressed T cell receptor (TCR) repertoires between HIV-1 seropositive long-term non-progressors and individuals with progressive disease. *Br J Haematol* **102**(5): 1187–1196.

Zekzer D, Wong FS, Ayalon O et al (1998) GAD-reactive CD4+ Th1 cells induce diabetes in NOD/SCID mice. *J Clin Invest* **101**(1): 68–73.

Chapter 12

THE IMPACT OF ANTIVIRAL THERAPY ON HIV DISEASE

Ian V. D. Weller

INTRODUCTION

The first randomized placebo-controlled trial of an antiviral agent in symptomatic human immunodeficiency virus (HIV) disease was published in 1987 (Fischl et al, 1987). In this trial 282 patients with acquired immune deficiency syndrome (AIDS) or AIDS-related complex (ARC) were randomized to receive zidovudine or placebo. The trial was stopped, with an average follow-up of about 4 months, because significant survival benefit had been demonstrated with one death in the treatment group and 19 in the placebo group. There was also a reduction in the probability of developing an opportunistic infection. These results led to the licensing of zidovudine for symptomatic disease in 1987. The next questions were whether the use of this drug earlier in infection would be beneficial and how long any benefit would last.

In 1989 two studies showed that zidovudine delayed the progression to severe ARC and AIDS in patients with early symptomatic disease and in those who were symptomless with CD4 counts below 500 μl^{-1} (Fischl et al, 1990; Volberding et al, 1990). The average follow-up of both trials was about 1 year. The trial in symptom-free patients (AIDS Clinical Trials Group protocol 019, ACTG 019) was interrupted in those with CD4 counts below 500 μl^{-1} because of a significant delay in disease progression. There were a total of 74 AIDS and ARC events amongst 1338 patients randomized to placebo, low-dose zidovudine (500 mg daily) or high-dose zidovudine (1500 mg daily). There were 38 events, 17 events and 19 events in the placebo, low-dose and high-dose arms respectively. Because of a high incidence of haematological side-effects in the high-dose arm, 500 mg of zidovudine was recommended for the treatment of symptom-free patients with CD4 counts below 500 μl^{-1}.

In 1992 a study was completed in 1013 patients who had been taking zidovudine for an average of 2 years and who were randomized either to continue zidovudine or to switch to didanosine (ddI) (Kahn et al, 1992). There were two didanosine arms, one low-dose (500 mg daily) and one high-dose (750 mg daily). There was a significant delay in the progression time to a new AIDS diagnosis in the low-dose group compared with the zidovudine group, but no detectable effect on mortality. The relative risk of progression to a primary end-point in the zidovudine group compared with the low-dose ddI group was 1·39 (95% confidence interval 1·06 to 1·82, P = 0·015).

These studies fuelled the hope that zidovudine in the mid- to long-term would continue

HIV and the New Viruses Second Edition
ISBN 0-12-200741-7

to benefit those with symptomless infection. In addition, it appeared that, at some stage after 6 months of therapy, switching to didanosine in zidovudine-exposed patients was better than continuing zidovudine in terms of delaying disease progression. The results also encouraged the belief that transient and modest rises in CD4 counts and decrease in virus levels, as measured then by p24 antigenaemia, would be shown to translate into important mid- to long-term clinical benefits. Indeed, it was hoped that at some stage controlled trials in HIV infection might be shortened by using biological marker changes rather than clinical end-points.

However, in 1992 a study comparing immediate versus deferred zidovudine therapy in patients with early symptomatic disease and CD4 counts of 200–500 μl^{-1} showed that there was no significant difference between the groups in survival, progression to AIDS or death after a mean follow-up of more than 2 years (Hamilton et al, 1992). Progression to AIDS was reported to have been delayed in the immediate group when all deaths before AIDS were censored, an analysis that assumes that such deaths are unrelated to HIV or drug.

In 1993 the first results of the Concorde study were announced (Albouker and Swart, 1993). The main report was published in 1994 (Concorde Coordinating Committee, 1994). This study in symptomless patients followed for just over 3 years detected no important clinical benefit when immediate and deferred treatment policies were compared, i.e. starting zidovudine immediately after randomization, or delaying it until the onset of symptomatic disease or a CD4 count at which an onset of symptomatic disease was imminent.

Following the Concorde study, three European and Australian trials were published comparing zidovudine with placebo in symptomless infection. Two studies had too few end-points to draw any firm conclusions (Mannucci et al, 1994; Mulder et al, 1994). The other European/Australian study (protocol 020) (Cooper et al, 1993) had only 16 severe ARC or AIDS events, partly because patients had to have a CD4 count above 400 μl^{-1} at entry and also the study was not analysed on a conventional 'intention to treat' basis. An end-point of 'clinical HIV disease' was used, which included oral candidiasis, leucoplakia and a CD4 end-point of time to a CD4 count below 350 μl^{-1}. Zidovudine was shown to delay progression time to this end-point.

In the Concorde study the transient, modest (approximately 30 cells) rise in CD4 counts and a persistent difference of around 30 cells between the immediate and deferred treatment groups over 3 years did not translate into an important clinical benefit. However, if a CD4 end-point was used to assess efficacy, such as time to a CD4 count below 200 μl^{-1}, below 350 μl^{-1} or less than 50% of baseline values, then a highly significant difference was seen between the two policies in favour of immediate therapy. The number of clinical end-points (AIDS and death) in Concorde (347) outnumbered the total of those in all other published trials in symptom-free and early symptomatic infection. The results questioned the uncritical use of such modest changes in markers with monotherapy as a surrogate end-point for mid- to long-term clinical efficacy. However, they strongly supported their value as a prognostic marker for disease progression.

One month before the full Concorde report appeared, a retrospective quality of life analysis of the ACTG 019 trial in symptomless patients was published using 'Twist' methodology (time without symptoms of disease or toxicity) (Lenderking et al, 1994). It showed that the increase in quality of life associated with a small delay in progression of disease at approximately 18 months was offset by the decrease in quality of life associated with the earlier onset of severe side-effects in the treatment group. Furthermore, in August of the same year further follow-up of the group of 1565 patients in ACTG 019 who entered with a CD4 count below 500 μl^{-1} followed for an average of 2·6 years showed that the delay in progression of disease diminished over time. However, 24·5% of patients were lost to follow-up before they reached a primary end-point (Volberding et al, 1994).

There was much confusion following the Concorde results. Those who were certain that starting monotherapy with zidovudine early was the best strategy became less certain, or spent time trying to find fault with the study to explain a result that was in someways difficult to digest. The results also raised questions about the longer-term benefits of monotherapy even in symptomatic disease, even though the trial did not address that question. The patients in Concorde and the three European/Australian trials (OPAL) have been subsequently followed by the Medical Research Council (MRC) HIV Clinical Trials Centre in London and the coordinating centre at INSERM SC10 in Paris.

Taken separately, the Concorde and OPAL follow-up results to 31 March 1995 indicated a significant increase in mortality in one trial and a non-significant decrease in mortality in the other in the groups who received zidovudine immediately. The combined analysis, which provides a more reliable estimate of the effect, shows a small and non-significant relative increase at 5 years of 8% (95% CI $^{-6}$ to 24) and, although not all centres continued to follow participants after March 1995, these findings have been confirmed in the 2 years subsequent to 1995 (Joint Concorde and OPAL Coordinating Committee, 1998).

We can learn some lessons from this history of monotherapy. Firstly, small early differences which emerge between groups in trials can be misleading. Changes, albeit modest, in biological markers that have a good prognostic value may not translate into important clinical benefits in the long term (Walker AS et al, 1998). Important information can be obtained from the long-term follow-up of trial participants. In clinical trials efficacy data are reported first and safety issues often appear later, i.e. the risk–benefit balance often changes with time. Clinical practice is driven by hope, belief and extrapolation from the data at hand, as well as by sound evidence.

COMBINATION CLINICAL END-POINT TRIALS

Two nucleoside analogues versus monotherapy

The depression amongst the therapeutic community did not last long. With the developments in nucleic acid amplification technologies for measuring plasma HIV RNA, promising effects of double nucleoside combinations on viral load, superior in the short term to those seen with monotherapy, were emerging in zidovudine-naïve and experienced patients (Katlama et al, 1996; Staszewski et al, 1996) and a combination therapy wish list was generated. It was hoped that using more than one drug would produce addictive or synergistic effects; have activity in different cell populations; prevent or delay the development of resistance, even reverse resistance; compromise the virus (i.e. decrease its fitness); decrease toxicity by the use of lower doses of drugs with non-overlapping toxicity profiles; and perhaps advantageously exploit pharmacokinetic interactions.

In August and September 1995 the results of the Delta trials and ACTG 175 were announced (Delta Coordinating Committee, 1996; Hammer et al, 1996) (Table 12.1). Delta-1 was a study in zidovudine-naïve patients, and in Delta-2 patients had received at least 3 months' treatment. In both trials patients were randomized to zidovudine monotherapy 600 mg daily *or* zidovudine plus didanosine 400 mg daily *or* zidovudine plus zalcitabine 2·25 mg daily. The mean CD4 count at entry was 205 μl^{-1} and the median follow-up was 30 months. A total of 3207 patients were included in the main analysis. Combination therapy improved survival over monotherapy. Of 1055 patients in the zidovudine monotherapy arm, 275 (26%) died compared with 196 of 1080 (18%) and 228 of 1072 (21%) in the didanosine and zalcitabine combination groups respectively (global log rank P = 0·00006). Combination therapy also significantly delayed progression to AIDS or death. Of 917 patients without AIDS at entry, 348 (38%) in the zidovudine monotherapy arm developed AIDS or died compared with 278 of 933 (30%) and 312 of 915 (34%) in the didanosine and zalcitabine combination groups respectively (global log rank P < 0·002). There was a suggestion that the didanosine combination group fared better than the zalcitabine group. In Delta-2 there were no significant differences detected between the monotherapy and combination groups in terms of survival or disease progression. However, a benefit from combination therapy in this group was not excluded. The combined results of both Delta-1 and Delta-2 revealed a relative reduction in mortality rate of 33% (P < 0·001) for zidovudine plus didanosine and 21% (P = 0·008) for zidovudine plus zalcitabine.

The AIDS Clinical Trials Group protocol 175 differed from Delta in that the trial included both zidovudine-naïve and exposed patients and a second monotherapy arm, namely didanosine alone. The mean baseline CD4 count of the 2493 participants was higher than in Delta (mean 352 μl^{-1}) so there were fewer clinical events in the longer median follow-up time of 36 months and the primary end-point included time to a 50% decline in CD4 count as well as to new or recurrent AIDS events and death. Progression to the primary end-point (including the CD4 end-point) was more frequent with zidovudine alone (32%) than with the combinations with didanosine (18%) or zalcitabine (20%) or didanosine alone (22%). Considering progression to AIDS or death without the CD4 end-point, zidovudine with didanosine and didanosine alone were superior to zidovudine monotherapy.

In a third trial comparing the same combinations as ACTG 175 and Delta, 1102 patients with AIDS or CD4 counts less than 200 μl^{-1} were randomized to receive zidovudine alone or zidovudine combined with either didanosine or zalcitabine (Saravolatz et al, 1996) (Table 12.1). After a median follow-up of 35 months no difference was detected in progression to AIDS or death in the three arms. Sixty-two per cent of the 363 patients randomized to receive zidovudine and didanosine, 63% of the 367 who received zidovudine and zalcitabine, and 66% of the 372 monotherapy patients progressed to AIDS or death.

A systematic overview of trials with zidovudine (ZDV) alone and in combination with didanosine (ddI) and zalcitabine (ddC) in HIV infection has been carried out by the HIV Triallists Collaborative Group and coordinated by the MRC HIV Clinical Trials Centre cooperating with the MRC clinical trial service unit in Oxford and the Division of AIDS, National Institute of Allergy and Infectious Diseases. In the overview of immediate versus deferred zidovudine monotherapy, immediate treatment with zidovudine was not associated with a reduction in mortality or disease progression over the period

Table 12.1 Summary of combination trials with clinical end-points

Trial	Reference	Drugs	N	Duration of previous treatment (months)	Mean CD4 count (μl^{-1})	AIDS or death hazard ratio (95% CI)	Death hazard ratio (95% CI)	Follow-up (months)
Delta-1, 2	Delta Coordinating Committee (1996)	ZDV/ddI	1080	17 (Delta-2)	205	0·76 (0·65–0·89)	0·67 (0·56–0·80)	30
		ZDV/ddC	1072				0·88 (0·75–1·02)	0·79 (0·66–0·94)	
		ZDV	1055						
ACTG 175	Hammer et al (1996)	ZDV/ddI	613	3	352	0·64 (0·46–0·87)	0·55 (0·36–0·86)	36
		ZDV/ddC	615				0·77 (0·57–1·04)	0·71 (0·47–1·07)	
		ddI	620				0·69 (0·51–0·94)	0·51 (0·32–0·8)	
		ZDV	619						
Nucombo	Saravolatz et al (1996)	ZDV/ddI	366	12 (23% naïve)	119	0·86 (0·71–1·03)	0·88 (0·71–1·08)	35
		ZDV/ddC	372				0·92 (0·76–1·10)	0·96 (0·78–1·17)	
		ZDV	375						
CAESAR	CAESAR Coordinating Committee (1997)	3TC ± loviride	1895	28	126	0·42 (0·32–0·57)	0·40 (0·23–0·69)	12
	Cameron et al (1998)	Ritonavir	543	24–36	20	0·53 (0·42–0·66)	0·6 (0·40–0·89)	7
		Placebo	547						
Pisces (SV14604)	Stellbrink et al (1997)	Saquinavir/ZDV/ddC	955	<16	170	0·50* (0·38–0·66)	(not presented)	18
		ddC/ZDV	942						
		SQV/ZDV	935						
		ZDV	653						
ACTG 320	Hammer et al (1997)	ZDV/d4T + indinavir	577	21	87	0·50 (0·33–0·76)	0·43 (0·19–0·99)	9·5
		ZDV/d4T + 3TC	579						

*Saquinavir/ZDV/ddC compared with ZDV/ddC. The difference in mortality alone did not reach a conventional level of significance.
CI, confidence interval; ddC, zalcitabine; ddI, didanosine; d4T, stavudine; 3TC, lamivudine; SQV, saquinavir; ZDV, zidovudine.

of follow-up but did increase the probability of AIDS-free survival in the first year. In the overview of nucleoside combinations versus zidovudine alone, combinations of ZDV and ddI and of ZDV and ddC were both associated with significant reductions in disease progression and death compared with ZDV alone (A. Babiker, 1998, personal communication; HIV Trialists Collaborative Group, 1998).

In the CAESAR study (Table 12.1), 1895 patients receiving zidovudine monotherapy or zidovudine plus zalcitabine or didanosine combination therapy with CD4 counts between 25 and 250 μl^{-1} were randomized to receive placebo, lamivudine (3TC) or loviride, a non-nucleoside reverse transcriptase inhibitor (CAESAR Coordinating Committee, 1997). The study was stopped after an interim analysis and in the 1840 patients in the analysis, progression to AIDS or death occurred in 95 (20%) of the 471 placebo-treated patients, 86 (9%) of 907 lamivudine-treated patients and 42 (9%) of 462 who received both lamivudine and loviride. The study did not have sufficient power to detect a clinical benefit from the addition of loviride.

While these trials were in progress other nucleoside analogues, non-nucleoside reverse transcriptase inhibitors and protease inhibitors were in clinical development (Table 12.2). Furthermore, there were rapid developments in the understanding of the kinetics of HIV replication. Phase II studies were beginning to show superior effects of triple combination regimens including protease inhibitors or non-nucleoside reverse transcriptase inhibitors over double and monotherapy arms in terms of changes of CD4 counts and viral load over 24–52 weeks (Collier et al, 1996; D'Aquila et al, 1996; Gulick et al, 1997).

Protease inhibitors

The licensed protease inhibitors are structurally related molecules and most contain a synthetic analogue of the cleavage site for HIV protease, i.e. they are peptide substrate mimetics and prevent cleavage of Gag and Pol protein precursors, arrest maturation and reduce the infectivity of virions (Flexner 1998). They are all metabolized by cytochrome P-450 enzymes, mainly the 3A4 isoform. There are limited data on central nervous system penetration. The new formulation of saquinavir (soft gel capsules) has increased its bioavailability.

All the protease inhibitors inhibit cytochrome P-450 enzymes – the most potent being ritonavir – and therefore increase the plasma levels of drugs also metabolized by the same pathway. This can lead to potentially toxic effects such as uveitis with rifabutin (Sun et al, 1996); potential toxicity may also result from the interaction with antihistamines, benzodiazepines and other drugs. Nelfinavir and ritonavir can also reduce the level of drugs by hepatic enzyme induction, e.g. oral contraceptives. The cytochrome P-450 interaction can be used to enhance antiviral activity. Ritonavir and saquinavir given together enhance the bioavailability of saquinavir by increasing steady-state plasma levels by 20–30-fold (Cohen et al, 1996). Indinavir has a similar but less potent effect.

A number of amino acid mutations in the protease protein are associated with the emergence of resistant clinical isolates (Schinazi et al, 1996). Single or primary mutations do not lead, in the main, to high levels of resistance, but subsequent secondary ones do. These secondary or compensatory mutations may arise in order to restore viral replicative fitness (Roberts et al, 1998). Furthermore, resistance may arise not only to the protease inhibitors being given but also to other protease inhibitors (Condra et al, 1995).

Table 12.2 Antiretroviral agents

Nucleoside analogues		Protease inhibitors		Non-nucleoside RT inhibitors	
Drug	*Manufacturer*	*Drug*	*Manufacturer*	*Drug*	*Manufacturer*
Zidovudine (ZDV)	Glaxo-Wellcome	Saquinavir	Roche	Nevirapine	Boehringer
Lamivudine (3TC)	Glaxo-Wellcome	Ritonavir	Abbott	Delavirdine	Pharmacia-Upjohn
Abacavir (1592U89)	Glaxo-Wellcome	Indinavir	Merck	Efavirenz (DMP266)	Dupont Pharma/Merck
Zalcitabine (ddC)	Roche	Nelfinavir	Roche/Agouron		
Didanosine (ddI)	Bristol Myers Squibb	Amprenavir	Glaxo-Wellcome		
Stavudine (d4T)	Bristol Myers Squibb	ABT-378	Abbott		

Protease inhibitors are a potent addition to the antiviral therapy of HIV infection. In a study of the effect of ritonavir, 1090 antiretroviral-experienced patients with CD4 counts less than 101 μl^{-1} were randomized to receive ritonavir or placebo (Cameron et al, 1998) (see Table 12.1). The blinded phase of the study was stopped after a median of 7 months, because the numbers of patients in the ritonavir group who had progressed to AIDS or death – 119 (21·9%) – or death alone – 26 (4·8%) – were significantly lower than those in the placebo group: 205 (37·5%) for AIDS or death and 35 (6·4%) for death. This study provided the first evidence for clinical benefit of protease inhibitors, even when given as a single agent in advanced stage, heavily pretreated patients.

In the Pisces study, 3485 patients with a mean CD4 count of 200 μl^{-1} who had received no more than 16 weeks of zidovudine therapy were randomized to receive a combination of saquinavir, zalcitabine and zidovudine; saquinavir and zidovudine alone; zalcitabine and zidovudine; or zidovudine alone (Stellbrink et al, 1997) (see Table 12.1). Patients in the zidovudine monotherapy arm were switched to the triple combination regimen following the publication of the Delta and ACTG 175 results. Over 14 months 76 (8%) of the patients randomized to receive the triple combination progressed to a new AIDS-defining event or death, compared with 142 (15%) in the zalcitabine and zidovudine arm, 116 (12%) in the saquinavir and zidovudine arm, and 116 (8%) in those originally randomized to zidovudine alone. The difference between the triple therapy arm and the zalcitabine/zidovudine arm was significant (P = 0·0001).

In a randomized, double-blind trial in 1156 patients not previously treated with lamivudine or protease inhibitors and with CD4 counts below 200 μl^{-1} followed for a mean of 38 weeks, a triple combination of either zidovudine or stavudine (d4T) with lamivudine and indinavir, compared with two nucleoside analogues, reduced the progression to AIDS or death and to death alone over 38 weeks (Hammer et al, 1997) (see Table 12.1). In the double nucleoside arm 63 patients (11%) progressed to AIDS or death, compared with 33 patients (6%) in the triple therapy arm.

It is clear from all these clinical end-point trials that long-term clinical benefit has not been clearly demonstrated for triple therapy compared with double therapy in naïve, symptomless patients.

Toxicity

Up to 50% of patients receiving protease inhibitors have been reported, on clinical grounds, to have developed peripheral fat wasting (lipodystrophy). The incidence varies in different reports because of the lack of a precise case definition. This fat loss occurs from the face, limbs and upper trunk. There is central adiposity ('Crix belly') with accumulation of intra-abdominal fat. In women there is also fat accumulation in the breasts, and in both sexes over the cervical spine ('buffalo hump'). Hyperlipidaemia, insulin resistance and in some cases type 2 diabetes mellitus occur (Carr et al, 1998a).

The mechanism for these effects is still unexplained. A hypothesis has been put forward that, because of about 60% homology between the amino sequence of the catalytic site of HIV-1 protease and two lipid regulatory proteins, the protease inhibitors may inhibit these proteins and cytochrome P-450 3A, leading to the disorder of lipid metabolisms seen (Carr et al, 1998b). The protease inhibitors may also vary in the extent to which they are associated with this syndrome.

Other adverse events seen with the protease inhibitors include gastrointestinal effects such as nausea and diarrhoea, perioral paraesthesia with ritonavir, nephrolithiasis with indinavir, and some cases of hepatitis (Flexner, 1998).

IMMUNE RESTORATION

For some time the only evidence of immune restoration was the effect on CD4+ cell numbers in blood and disease progression. It now appears that much of the early increase in CD4 cell counts is probably due to a redistribution of CD4+ memory cells from their trapped position in lymphoid tissues into the circulation (Fleury et al, 1998; Pakker et al, 1998). This is accompanied early on with a decrease in activated CD8+ cells. This is followed by slow repopulation with naïve T cells and improved CD4 responses to recall antigens (Li et al, 1998). To what extent the T cell receptor repertoire depletion is corrected is unclear. Some studies suggest that there may be a partial repletion (Gorochov et al, 1998); other early findings suggest that the CD4+ T cell pool in adults is maintained mostly by the division of mature T cells rather than by differentiation of prethymic stem cells. Thus, after elements of the T cell repertoire are lost through HIV infection they may be difficult to replace (Walker RE et al, 1998). A body of evidence is growing that potent combination therapy can hold or reverse certain opportunistic conditions such as Kaposi's sarcoma, cryptosporidiosis and progressive multifocal leucoencephalopathy. In addition, phenomena are now being seen that may also be a reflection of immune restoration when potent combination therapy is started – namely, worsening of the signs and symptoms of tuberculosis and *Pneumocystis carinii* pneumonia, granulomatous inflammatory reactions in patients infected with atypical mycobacteria, macular oedema and vitritis in patients with cytomegalovirus (CMV) retinopathy, and a worsening of hepatitis in patients with chronic hepatitis B virus infection (Carr and Cooper, 1997).

SURROGACY

The initiation of trials with clinical end-points early in infection has been hampered by the move towards earlier use of more potent combination antiviral therapy and consequently a lower incidence of disease progression; the rising number of combinations requiring evaluation; and the need to assess the activity of new antiviral agents quickly. As with other diseases with a long clinical latent period, both CD4 cell count and plasma HIV RNA are being used as more proximal surrogate end-points by patients, clinicians, triallists and licensing authorities.

The ideal surrogate marker should be related to the pathogenesis of the disease, be detectable in the majority of patients at all stages of the disease, change quickly in response to treatment, respond in the same way to different drugs and be a simple and reliable measurement. Both the CD4+ cell count and plasma HIV RNA levels therefore are potentially ideal marker candidates. However, changes in the marker should capture as fully as possible the effect of treatment on the true end-point, i.e. disease progression and death. A measurement can be regarded as a valid surrogate for treatment if:

1. The existence of a treatment effect on the surrogate implies the existence of a treatment effect on the true end-point.
2. The lack of a treatment effect on the surrogate implies the lack of a treatment effect on the true end-point.
3. There is no residual treatment effect on the true end-point after accounting for the effect of the treatment on the surrogate (Prentice, 1989).

Serum HIV RNA level was shown to be a robust prognostic marker in the Multicenter AIDS Cohort Study (MACS) (Mellors et al, 1997), although the confidence limits around the estimate of risk in some subgroups in this study were quite wide (Table 12.3).

The early suppression of HIV RNA levels below the level of detection of the most sensitive assays now available has become a therapeutic goal. The initial duration of viral suppression appears to be correlated with the baseline viral load, the slope of its descent and the nadir reached in the early phase of therapy (Kemf et al, 1998). However, good prognostic markers are not always good surrogates, and in a number of diseases, although treatments have produced promising changes in the surrogate in the right direction, these changes have been accompanied by unexpected and sometimes harmful clinical effects (Fleming and De Mets, 1996).

Essential requirements in trials to evaluate a marker's ability to capture fully a treatment benefit are significant treatment differences, many clinical end-points, and marker measurements in as many participants as possible. The proportion of the treatment effect (PTE) explained by the marker changes is expressed as PTE = $(U - A)/U$, where U is the log hazard ratio for the net effect of the treatment compared with its comparator, e.g. combination versus monotherapy, and A is the log hazard ratio adjusted for the marker. If all the treatment effect is lost with adjustment for the marker, $A = 0$ and 100% of the PTE is accounted for by the marker changes.

Changes in CD4 count have been shown to be incomplete surrogates in early clinical end-point trials before plasma HIV RNA measurement was available (Choi et al, 1993; Lin et al, 1993; Concorde Coordinating Committee, 1994).

A number of studies have shown that absolute values of HIV RNA at baseline and at various intervals in the early weeks of treatment are independent and highly significant predictors of disease progression or death (Katzenstein et al, 1996; Brun-Vézinet et al, 1997; Montaner et al, 1998). However, studies on the earlier clinical end-point trials have shown variable results in the proportion of treatment effect explained by changes in HIV RNA (O'Brien et al, 1996; Babiker et al, 1997; Montaner et al, 1998).

An analysis of data from a trial of immediate versus deferred zidovudine therapy in patients with symptomatic infection revealed that a decrease of at least 75% in the plasma level of HIV RNA over the first 6 months of zidovudine therapy accounted for 59% of the benefit of treatment (O'Brien et al, 1996). However, the treatment effect was small, the number of clinical events was low and the estimate of the PTE had wide confidence limits (13% to 112%).

An analysis of the CAESAR trial suggested that HIV RNA and CD4 were good surrogate markers, but follow-up in this trial was short with clinical end-points only reported up to 52 weeks (Montaner et al, 1998). Again, the confidence limits around the estimate were wide. A surrogate marker that captures short-term clinical effects is not much use as a surrogate marker, since its use would only shorten trials marginally with

Table 12.3 **Baseline HIV RNA and CD4+ T cell count and the probability of AIDS**

	HIV RNA 10 001–30 000 copies ml⁻¹				HIV RNA > 30 000 copies ml⁻¹				
	CD4 count (μl^{-1})				CD4 count (μl^{-1})				
	> 750	351–750	201–350	< 200	> 750	501–750	351–500	201–350	< 200
N	64	259	53	20	45	96	121	104	70
AIDS	42	194	45	18	32	73	111	92	67
Percentage with AIDS (95% CI)									
3 years	9·5	16·1	36·4	50·0	20·8	37·9	47·9	64·4	85·5
	(4–20)	(12–21)	(25–51)	(31–73)	(11–36)	(29–48)	(39–57)	(55–74)	(76–93)
6 years	36·7	54·9	72·2	75·0	50·4	74·2	77·7	89·3	97·9
	(26–50)	(49–61)	(60–84)	(55–91)	(36–66)	(65–83)	(70–85)	(82–95)	(91–100)
9 years	62·4	76·3	84·5	90·0	70·3	79·2	94·4	92·9	100·0
	(50–75)	(71–82)	(73–93)	(73–98)	(56–84)	(70–87)	(89–98)	(86–97)	(91–100)

Modified from Mellors et al (1997) and Report of the National Institutes of Health (NIH) panel to define principles of therapy of HIV infection and guidelines for the use of antiretroviral agents in HIV-infected adults and adolescents. *Ann Intern Med* 1998; **128**(12) (Part 2): 1057–1100.
A branched DNA (bDNA) assay was used on the stored serum samples. Initially it was thought that there might be two-fold loss of RNA and a correction factor of × 2 was used. However, this was later considered to be unnecessary so the values in the table represent the original bDNA levels × 2 (RT-PCR values are approximately double those obtained by bDNA assay).

uncertainty remaining about whether it captured treatment effects fully in the longer term.

Probably the most robust analysis presented to date, in terms of degree of treatment effect, number of end-points and viral load measurements, and length of follow-up, is that of the Delta trial (Babiker et al, 1997) in which the clinical benefit from combinations of zidovudine and didanosine or zidovudine and zalcitabine was underestimated by CD4 counts and overestimated by RNA levels and by the two markers combined. It would seem from the data available that CD4 counts and HIV RNA levels are incomplete surrogates for clinical outcome. There may be a number of reasons for this. To expect changes in these markers over (say) the first year of therapy to capture clinical outcome fully at 2–3 years and beyond is a tall order, taking into account the possible development of resistance, the high pill burden and poor compliance with therapy, changes in therapy and above all the possible effects of antivirals (other than those mediated through the markers) on the virus or on immune response and toxicity. Furthermore, it may be that with more potent triple combinations, CD4 count and HIV RNA alone or combined are more complete surrogates.

Nevertheless, licensing authorities have moved towards earlier licensing of drugs on the basis of antiviral activity. In Europe an ad hoc group met in September 1997 to advise the Committee for Proprietary Medicinal Products (CPMP). The group brought together industry, academia and a group of experts on antiretroviral medicinal products. They prepared a revision of the previous 'points to consider' document, which was subsequently adopted by the CPMP. For approval under exceptional circumstances if there are no comprehensive data on clinical efficacy and safety, the following licensing criteria now apply to anti-HIV medicinal products:

- short-term (e.g. 24 weeks) data on antiretroviral activity and safety should be provided;
- antiretroviral activity should be substantiated by relevant effects on biological markers, particularly viral load, but also a consistency of CD4 cell response;
- there should be an acceptable tolerability profile;
- added values should be identified: these may be clinically demonstrated activity in patients failing therapy, a unique safety profile, or applicability to areas of unmet medical need, for example patients with central nervous system disease, children or pregnant women.

For ordinary approval, licensing will require the demonstration of durability of virological response, consistency of CD4 cell response and comprehensive information on safety. The duration of clinical trials should be defined by the presumed event rates in the control group and may thus differ considerably between trials and groups of patients. The duration of therapy should be at least 48 weeks for the last recruited patient, but in cases of non-inferiority studies, conducted in patients with favourable prognostic factors at baseline, may be considerably longer. It is recommended that a broad spectrum of HIV-infected patients be included in the drug development programme and the recruitment of naïve, therapy experienced and advanced patients is encouraged. Whenever appropriate (e.g. in patients with progressive or advanced disease), clinical events such as opportunistic infections should be documented systematically. This may be relevant for the efficacy assessment (CPMP, 1997). In essence, drugs will be licensed both in exceptional circumstances and as part of normal approval for their antiviral and immunological activity rather than for clinical effectiveness, provided that the risk–benefit assessment is appropriate. A similar position has been adapted by the Food and Drug Administration in the USA.

FROM CLINICAL TRIALS TO CLINICAL PRACTICE

Epidemiological and observational cohort studies indicate that the clinical benefits observed in trials are being seen in populations treated in clinical practice. Although there are inherent methodological difficulties with these sorts of studies, which may produce bias, it is clear from the volume of reports and clinical experience that combination therapy has had a dramatic impact on HIV disease. Several reports have described reductions in mortality and rate of hospitalization of HIV-infected patients (Brodt et al, 1997; Egger et al, 1997; Mouton et al, 1997; Update, 1997). In a cohort of 1255 patients with CD4 counts below 100 μl^{-1} mortality declined from 29·4 per 100 person years in 1995 to 8·8 per 100 person years in the second quarter of 1997, with a reduction in the incidence of major opportunistic infections (*Pneumocystis carinii* pneumonia, *Mycobacterium avium* complex disease and cytomegalovirus retinitis) from 21·9 per 100 person years in 1994 to 3·7 per 100 person years by mid-1997. These benefits were shown to be associated with the intensity of antiviral therapy (Palella et al, 1998).

WHEN SHOULD ANTIVIRAL THERAPY BEGIN?

If a decision is to be evidence-based, it seems that the available evidence comes from three sources. The first is controlled clinical trials; the second is extrapolations from what we have learnt about the pathogenesis of HIV; and the third is 'expert opinion'. A number of bodies have drawn up guidelines for the antiretroviral therapy of HIV infection which include advice on when to begin therapy; these recommendations need continual updating (Table 12.4). A very early start to therapy would be in primary infection (Kinloch-de Loës et al, 1995), or in patients with a low viral load or high CD4 count. A late start may be regarded as the treatment of patients with minor symptoms and signs. Between these two positions is a grey area in the natural history of HIV infection in which arguments can be put on both fronts for an early or a deferred start (Table 12.5).

Because of the marker effects and the clinical data from triple therapy studies, most physicians now begin therapy with a triple combination consisting of two nucleosides and either a protease inhibitor or a non-nucleoside reverse transcriptase inhibitor. Preliminary data have been presented on a direct comparison between these two options (Staszewski et al, 1998). Although patients with more advanced infection are likely to be started on a protease inhibitor, important studies to determine the advantages of protease-sparing initial regimens in the longer term on virological, immunological and clinical end-points are planned.

Physicians who opt for early therapy do not wish to withhold what they believe in the long term will be shown to be the correct treatment for their patients early in their infection. The 'late starters' are concerned that if they expose their patients early on to all drug classes they may run out of options before illness appears. A randomized trial

Table 12.4 *Initiation of therapy guidelines in symptomless patients*

	US DHHS (1997)	IAS–USA (*Carpenter et al, 1997*)	BHIVA (1997)
CD4 count (μl^{-1})	< 500	< 500	< 300*
Viral load[†]	> 20 000	5000–10 000[‡]	> 10 000–50 000[§]

*Modified to > 350 μl^{-1} in 1998 revision (Gazzard and Moyle, 1998).
[†]RT-PCR (copies/ml).
[‡]Regardless of CD4 count, and 'therapy should be considered for all subjects with detectable plasma HIV RNA who request it'.
[§]Or in range detectable to 10 000 copies ml^{-1} with falling CD4 count (modified to 'viral load value associated with risk of disease progression' in 1998 revision).
BHIVA, British HIV Association; US DHSS, US Department of Health and Human Sciences; IAS–USA, International Aids Society, USA panel.

Table 12.5 *Should antiretroviral therapy begin early?*

'I think so'	'I don't know'
Viral turnover is so rapid that replication should be stopped before the infecting viral population is too diverse to control. Early therapy also slows the relentless destruction of lymphoid tissue	Biologically this makes sense, but do we have enough of the right drugs to begin treatment early and continue for life? Will all drugs penetrate potential 'sanctuary sites'?
Viral load is a good marker of prognosis	That appears to be true, but its surrogacy – i.e. the extent to which it captures treatment effect – remains incompletely defined
Current regimens are potent and suppress viral load to unprecedented levels, so resistant virus is less likely to emerge	Results so far must be characterized as short-term. Current regimens do not always suppress replication completely, so resistant virus will emerge eventually
Immune reconstitution becomes more and more difficult as HIV disease progresses	Late intervention can still be very effective and immune reconstitution with antiretrovirals alone may be incomplete in the short- to mid-term, even early in the course of disease
In the short- to mid-term the risk–benefit assessment of intervention with current regimens is acceptable	It is too early to conclude that the risk–benefit assessment is acceptable because the long-term side-effects of drugs such as the protease inhibitors are just emerging, and those of others may appear

comparing these policies will be difficult to initiate until there is enough uncertainty about this important question and sufficient resources to conduct it.

REFERENCES

Aboulker JP & Swart AM (1993) Preliminary analysis of the Concorde trial. *Lancet* **341**: 889–890.

Babiker AG on behalf of the Delta Coordinating Committee (1997). Can HIV RNA viral load be used as a surrogate for clinical endpoints in HIV disease? Abstract 103. *Programs and Abstracts of the VI European Conference on Clinical Aspects and Treatment of HIV Infection*, Hamburg, October 1997.

British HIV Association Guidelines Coordinating Committee (1997) British HIV Association guidelines for antiretroviral treatment of HIV seropositive individuals. *Lancet* **349**: 1086–1092.

Brodt HR, Kamps BS, Gute P, Knupp B, Staszewski S & Helm EB (1997) Changing incidence of AIDS-defining illnesses in the era of antiretroviral combination therapy. *AIDS* **11**: 1731–1738.

Brun-Vézinet F, Boucher C, Loveday C et al (1997) HIV-1 viral load, phenotype, and resistance in a subset of drug-naive participants from the Delta trial. *Lancet* **350**: 983–990.

CAESAR Coordinating Committee (1997) Randomized trial of addition of lamivudine or lamivudine plus loviride to zidovudine-containing regimens for patients with HIV-1 infection: the CAESAR trial. *Lancet* **349**: 1413–1421.

Cameron DW, Heath Chiozzi M & Danner S (1998) Randomized placebo-controlled trial of ritonavir in advanced HIV-1 disease. *Lancet* **351**: 543–549.

Carpenter CCJ, Fischl MA, Hammer SM et al (1997) Antiretroviral therapy for HIV in 1997: updated recommendations of the International AIDS Society – USA Panel. *JAMA* **277**: 1962–1969.

Carr A & Cooper DA (1997) Restoration of immunity to chronic hepatitis B infection in HIV-infected patient on protease inhibitor. *Lancet* **349**: 995–996.

Carr A, Samaras K, Burton S et al (1998a) A syndrome of peripheral lipodystrophy hyperlipidaemia and insulin resistance in patients receiving HIV protease inhibitors. *AIDS* **12**: F51–58.

Carr A, Samaras K, Chisholm DJ & Cooper DA (1998b) Pathogenesis of HIV-1 protease inhibitor associated peripheral lipodystrophy, hyperlipidaemia and insulin resistance. *Lancet* **351**: 1881–1883.

Choi S, Lagakos SW, Schooley RTA & Volberding PA (1993) CD4 lymphocytes are an incomplete surrogate marker for clinical progression in persons with asymptomatic HIV infection taking zidovudine. *Ann Intern Med* **118**: 674–680.

Cohen C, Sun E, Cameron W et al (1996) Ritonavir–saquinavir combination treatment in HIV-infected patients. Abstract 8. *Addendum to Program and Abstracts of 36th Interscience Conference on Antimicrobial Agents and Chemotherapy*, New Orleans, 15–18 September 1996.

Collier AC, Coombs RW, Schoenfeld DA et al (1996) Treatment of human immunodeficiency virus infection with saquinavir, zidovudine, and zalcitabine. *N Engl J Med* **334**: 1011–107.

Committee for Proprietary Medicinal Products (1997) *Points to Consider in the Assessment of Anti-HIV Medicinal Products*, November 1997. London: European Agency for the Evaluation of Medicinal Products.

Concorde Coordinating Committee (1994) Concorde: MRC/ANRS randomised double-blind controlled trial of immediate and deferred zidovudine in symptom-free HIV infection. *Lancet* **343**: 871–882.

Condra JH, Schleif WA, Blahy OM et al (1995) In vivo emergence of HIV-1 variants resistant to multiple protease inhibitors. *Nature* **374**: 569–571.

Cooper DA, Gatell JM, Kroon S et al (1993) Zidovudine in persons with asymptomatic HIV infection and CD4+ cell counts greater than 400 per cubic millimetre. *N Engl J Med* **329**: 297–303.

D'Aquila RT, Hughes MD, Johnson VA et al (1996) Nevirapine, zidovudine, and didanosine compared with zidovudine and didanosine in patients with HIV-1 infection. *Ann Intern Med* **124:** 1019–1030.

Delta Coordinating Committee (1996) Delta: a randomised double-blind controlled trial comprising combinations of zidovudine plus didanosine or zalcitabine with zidovudine alone in HIV infected individuals. *Lancet* **348:** 283–291.

Department of Health and Human Sciences (1997) *Guidelines for the use of antiretroviral agents in HIV-infected adults and adolescents.* 5 November 1997. CDC National Clearing House (1-800-458-5231). Website http://www/hivatis/org.

Egger M, Hirschel B, Francioli P et al (1997) Impact of new antiretroviral combination therapies in HIV infected patients in Switzerland: prospective multicentre study. *BMJ* **315:** 1194–1199.

Fischl MA, Richman DD, Grieco MH et al (1987) The efficacy of azidothymidine (AZT) in the treatment of participants with AIDS and AIDS-related complex: a double-blind, placebo-controlled trial. *N Engl J Med* **317:** 185–191.

Fischl MA, Richman DD, Hansen N et al (1990) The safety and efficacy of zidovudine (AZT) in the treatment of subjects with mildly symptomatic human immunodeficiency virus type 1 (HIV) infection. *Ann Intern Med* **112:** 727–737.

Fleming TR & De Mets DL (1996) Surrogate endpoints in clinical trials – are we being misled? *Ann Intern Med* **125:** 605–613.

Fleury S, De Boer RJ, Rizzardi GP et al (1998) Limited CD4+ T-cell renewal in early HIV-1 infection: effect of highly active antiretroviral therapy. *Nature Med* **4:** 794–801.

Flexner C (1998) HIV protease inhibitors. *N Engl J Med* **338:** 1281–1292.

Gazzard B & Moyle G on behalf of the BHIVA Guidelines Writing Committee (1998). 1998 revision to the British HIV Association guidelines for antiretroviral treatment of HIV seropositive individuals. *Lancet* **352:** 314–316.

Gorochov G, Neumann AU, Kereveur A et al (1998) Perturbation of CD4 and CD8 T-cell repertoire during progression to AIDS and stabilisation of the CD4 repertoire during antiviral therapy. *Nature Med* **4:** 215–221.

Gulick RM, Mellors JW, Havlir D et al (1997) Treatment with indinavir, zidovudine and lamivudine in adults with human immunodeficiency virus infection and prior antiretroviral therapy. *N Engl J Med* **337:** 734–739.

Hamilton JD, Hartigan PM, Simberkoff MS et al (1992) A controlled trial of early versus late treatment with zidovudine in symptomatic human immunodeficiency virus infection. *N Engl J Med* **326:** 437–443.

Hammer S, Katzenstein D, Hughes M et al (1996) A trial comparing nucleoside monotherapy with combination therapy in HIV infected adults with CD4 cell counts from 200–500 per cubic millimeter. *N Engl J Med* **335:** 1081–1090.

Hammer SM, Squires KE & Hughes MD (1997) A controlled trial of two nucleoside analogues plus indinavir in persons with human immunodeficiency virus infection and CD4 cell counts of 200 per cubic millimeter or less. *N Engl J Med* **337:** 725–733.

HIV Trialists Collaborative Group (1998) Systematic overview of trials of zidovudine alone, and of combinations with didanosine and zalcitabine in HIV infection (submitted for publication).

Joint Concorde and OPAL Coordinating Committee (1998) Long-term follow-up of randomised trials of immediate versus deferred zidovudine in symptom-free HIV infection. *AIDS* **12:** 1259–1265.

Kahn JO, Lagakos SW, Richman DD et al (1992) A controlled trial comparing continued zidovudine with didanosine in human immunodeficiency virus infection. *N Engl J Med* **327:** 581–587.

Katlama C, Ingrand D, Loveday C et al (1996) Safety and efficacy of lamivudine–zidovudine combination therapy in antiretroviral-naive patients. *JAMA* **276:** 118–125.

Katzenstein DA, Hammer SM, Hughes MD et al (1996) The reaction of virologic and immunologic markers to clinical outcomes after nucleoside therapy in HIV-infected adults with 200 to 500 CD4 cells per cubic millimeter. *N Engl J Med* **335:** 1091–1098.

Kemf D, Rode R, Xu Y et al (1998) The duration of viral suppression during protease inhibitor therapy from HIV-1 infection is predicted by plasma HIV-1 RNA at the nadir. *AIDS* **12:** F9–F14.

Kinloch-de Loës S, Hirschel BJ, Hoen B et al (1995) A controlled trial of zidovudine in primary human immunodeficiency virus infection. *N Engl J Med* **333:** 408–413.

Lenderking WR, Gelber RD, Cotton DJ et al (1994) Evaluation of the quality of life associated with zidovudine treatment in asymptomatic human immunodeficiency virus infection. *N Engl J Med* **330:** 738–743.

Li TS, Tabiana R, Katlama C et al (1998) Long lasting recovery in CD4 T-cell function and viral-load reduction after highly active antiretroviral therapy in advanced HIV-1 disease. *Lancet* **351:** 1682–1686.

Lin DY, Fischl MA & Schoenfeld DA (1993) Evaluating the role of CD4 lymphocyte counts as surrogate endpoints in HIV clinical trials. *Stat Med* **12:** 835–842.

Mannucci PM, Gringeri A, Savidge G et al (1994) Randomised double-blind, placebo-controlled trial of twice-daily zidovudine in symptomatic haemophiliacs infected with the human immunodeficiency virus type 1. *Br J Haematol* **86:** 174–179.

Mellors JW, Munoz A, Giorgi J et al (1997) Plasma viral load and CD4+ lymphocytes as prognostic markers of HIV-1 infection. *Ann Intern Med* **126:** 946–954.

Montaner J, De Masi R & Hill A (1998) The effects of lamivudine treatment on HIV-1 disease progression are highly correlated with plasma HIV-1 RNA and CD4 cell count. *AIDS* **12:** F23–28.

Mouton Y, Cartier F, Dellamanica P et al (1997) Dramatic cut in AIDS defining events and hospitalisation for patients under protease inhibitors (PI) and tritherapies (TTT) in 9 AIDS reference centres (ARC) and 7,931 patients. Abstract 208. *Program and Abstracts of the Fourth Conference on Retroviruses and Opportunistic Infections*, Washington, DC, 22–26 January 1997.

Mulder JW, Cooper DA, Mathiesen L et al (1994) Zidovudine twice daily in asymptomatic subjects with HIV infection and a high risk of progression to AIDS: a randomised, double-blind placebo-controlled study. *AIDS* **8:** 313–321.

O'Brien WA, Hartigan PM, Martin D et al (1996) Changes in plasma HIV-1 RNA and CD4+ lymphocyte counts and the risk of progression to AIDS. *N Engl J Med* **334:** 426–431.

Pakker NG, Notermans DW, DeBoer RJ et al (1998) Biphasic kinetics of peripheral blood T cells after triple combination therapy in HIV-1 infection: a composite of redistribution and proliferation. *Nature Med* **4:** 208–214.

Palella FJ, Delaney KM, Moorman AC et al (1998) Declining morbidity and mortality among patients with advanced immunodeficiency virus infection. *N Engl J Med* **338:** 853–860.

Prentice RL (1989) Surrogate endpoints in clinical trials: definition and operation criteria. *Stat Med* **8:** 431–440.

Roberts NA, Craigh JC & Sheldon J (1998) Resistance and cross-resistance with saquinavir and other protease inhibitors: theory and practice. *AIDS* **12:** 453–460.

Saravolatz LD, Winslow DC, Collins G et al (1996) Zidovudine alone or in combination with didanosine or zalcitabine in HIV infected patients with the acquired immunodeficiency syndrome or fewer than 200 CD4 cells per cubic millimeter. *N Engl J Med* **335:** 1099–1106.

Schinazi RF, Larder BA & Mellors JW (1996) Mutations in retroviral genes associated with drug resistance. *Int Antiviral News* **4:** 95–107.

Staszewski S, Loveday C, Picazo JJ et al (1996) Safety and efficacy of lamivudine–zidovudine combination therapy in zidovudine-experienced patients. *JAMA* **276:** 111–117.

Staszewski S, Marales-Ramirez J, Tashima K et al (1998) A phase II multicentre randomised, open label study to compare the antiretroviral activity and tolerability of efavirenz (EFV) +

indinavir (IDV) versus EFV + zidovudine (ZDV) + lamivudine (3TC) versus IDV + ZDV + 3TC at 24 weeks [study DMP 266-006]. Abstract 22336. *Conference Record, 12th World AIDS Conference*, Geneva, June 28–July 3, 1998.

Stellbrink HT on behalf of the Invirase International Phase III Trial (SV-14604) Group (1997). Clinical and survival benefit of saquinavir (SQV) in combination with zalcitabine (ddC) and zidovudine (ZDV) in untreated/minimally treated HIV-infected patients. Abstract 212. *Sixth European Conference on Clinical Aspects and Treatment of HIV Infection*, Hamburg, 1997.

Sun E, Heath-Chiozzi M, Cameron DW et al (1996) Concurrent ritonavir and rifabutin increases risk of rifabutin-associated adverse events. Abstract 18. Vol. 1. *Program and Abstracts of the 11th International Conference on AIDS*, Vancouver, 7–12 July 1996.

Update (1997) Trends in AIDS incidence, deaths and prevalence – United States 1966. *MMWR Morb Mortal Wkly Rep* **46:** 165–173.

Volberding PA, Lagakos SW, Koch MA et al (1990) Zidovudine in asymptomatic human immunodeficiency virus infection. *N Engl J Med* **322:** 941–949.

Volberding PA, Lagakos SW, Grimes JM et al (1994) The duration of zidovudine benefit in persons with asymptomatic HIV infection. Prolonged evaluation of protocol 019 of the AIDS Clinical Trials Group. *JAMA* **272:** 437–442.

Walker AS, Peto TEA, Babiker AG & Darbyshire JH on behalf of the Concorde Co-ordinating Committee (1998). Markers of HIV infection in the Concorde trial. *Q J Med* **91:** 423–438.

Walker RE, Carter CS, Muul L et al (1998) Peripheral expansion of pre-existing mature T-cells is an important means of CD4+ T-cell regeneration in HIV-infected adults. *Nature Med* **4:** 852–856.

Chapter 13

HIV-1 AND HIV-2 MOLECULAR DIAGNOSIS

Neil Berry and Richard Tedder

INTRODUCTION

Human immunodeficiency viruses type 1 (HIV-1) and type 2 (HIV-2) are human lentiviruses belonging to the *Retroviridae* family. Infection with HIV-1, HIV-2 or in some cases coinfection with both viruses is now widely accepted to be causally associated with the subsequent progression to severe immunodeficiency and acquired immune deficiency syndrome (AIDS). Type 1 HIV is recognized as the agent of the global AIDS pandemic, with HIV-2 mostly confined to the countries of West Africa (De Cock et al, 1993). The accurate and reliable diagnosis of lentivirus infections, including the differentiation between HIV-1, HIV-2 and dual infections, is an important issue, having not only implications for studies into the natural history, pathogenesis and treatment of HIV infections, but also considerable consequences for the individuals concerned.

The diagnosis of HIV infections may be considered, as for most virus infections, under a number of headings including the direct demonstration of virus (electron microscopy), virus isolation by culture techniques, demonstration of a virus-specific antibody response, detection of viral antigens or the presence of the viral genome. Although routine diagnosis of HIV remains serology-based (virus antibody/antigens), molecular techniques are gaining in use owing to advances in molecular biology. Traditional molecular approaches (Southern or dot-blot hybridization and autoradiography) have been largely superseded by amplification techniques often using non-radioactive protocols. This chapter highlights the principal applications and advantages of viral genomic detection procedures in current use for HIV diagnosis where a molecular approach may be desirable. These are viewed in the context of the global HIV pandemic, highlighting both the advantages and potential problems associated with a molecular approach to HIV diagnosis.

BIOLOGICAL AND EPIDEMIOLOGICAL ASPECTS

HIV-1 infections

Diagnosis of HIV using molecular methods exploits the two biological forms of HIV in the body, detection of either of which may result in a confirmation of infection. The

HIV and the New Viruses Second Edition
ISBN 0-12-200741-7

virus exists in the virion form as two copies of single-stranded, positive-sense RNA surrounded by a protein coat. This comprises the inner core and proteins encapsulated by an external envelope containing glycoproteins. The virion is detectable in the cell-free component of peripheral blood. Like all other retroviruses, HIV also exists and persists as proviral DNA (usually integrated, but in some cases in unintegrated forms) representing the cellular part of the HIV life cycle. Detection of both viral RNA and cellular proviral DNA using molecular methods is usually taken to be definitive evidence for infection with HIV. Diagnostic techniques for HIV may be applied at different stages of infection (Figure 13.1); however, a number of factors should be considered before the application of such methodologies, particularly relating to the sensitivity of detection and the possibility of false-negative results.

HIV-1 sequence diversity

Infections with HIV-1 are characterized in the majority of individuals by a high level of replication in vivo (Ho et al, 1995; Wei et al 1995), which combined with the error-prone nature of reverse trancriptase results in a myriad of virus strains, and hence viral sequences, circulating both in any given community and in a given host. Genetically, HIV-1 infections have been divided into two major groups: group M (major) and group O (outlier, other) viruses. At least nine members (subtypes A–H, J) and U (unassigned) have been identified to date; these cluster phylogenetically within group M but are

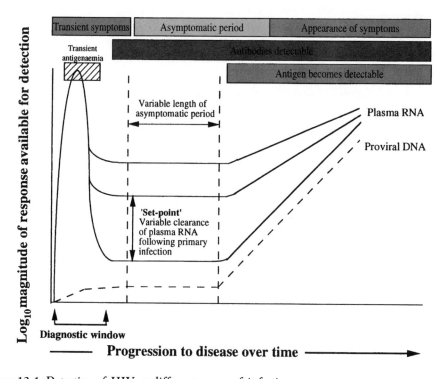

Figure 13.1 *Detection of HIV at different stages of infection*
Time course and magnitude of HIV infection indicating the relative level of viral genome abundance in the peripheral circulation available for detection at different stages of infection.

genetically distinct from each other. Certain subtypes, often referred to as clades, have been associated with particular countries or regions of the world: subtypes A, C and D with Africa (although all HIV-1 subtypes identified to date are present on the African continent), subtype B with North America and western Europe, subtype E with Thailand, subtype O with Cameroon, and so forth. However, with global demographic movements this nationality-focused interpretation should not be relied upon to indicate freedom from divergent genovars in a patient population. Most non-clade-B HIV-1 infections (A–D, G, H) have now been identified in Europe (Alaeus et al, 1997a), where previously almost exclusively all HIV-1 infections were clade B. Thus the design of strategies for the detection of HIV genome on a population basis should aim to include as large a repertoire of known HIV sequences as possible and should be applied irrespective of the demography of a study population.

HIV-2 infections

In both epidemiological and diagnostic terms HIV-2 may be viewed as a separate virus group. Although HIV-2 is clearly capable of causing AIDS (Clavel et al, 1986; Romieu et al, 1990; Naucler et al, 1992), clinical and epidemiological studies of HIV-1 and HIV-2 have demonstrated that HIV-2 infections generally take longer to result in symptomatic disease (Marlink et al, 1994; Ricard et al, 1994; Whittle et al, 1994) and exhibit reduced mortality ratios (Poulsen et al, 1997) compared with HIV-1. By analogy with HIV-1, genetic analysis of individuals infected with HIV-2 in different countries in West Africa has demonstrated at least six different HIV-2 subtypes (subtypes A–F). Subtype A infections have been most frequently identified, representing the dominant HIV-2 subtype, with relatively fewer subtype B isolates. Subtypes C–F were only identified by molecular techniques (Gao et al, 1992, 1994; Chen et al, 1996). Where HIV-2 infections have been identified outside West Africa – in India, South America and Europe – these have mostly been subtype A. Though HIV-2 has been generally regarded as having reduced pathogenicity in vivo compared with HIV-1, pathogenic HIV-2 undoubtedly exists, and its prevalence seems likely to increase outside the traditional endemic regions. Thus, the need for a molecular diagnosis of HIV-2 is also likely to increase.

MOLECULAR METHODS USED TO DIAGNOSE HIV INFECTIONS

Such a broad genetic diversity of HIV poses significant problems for molecular diagnostic techniques, where the plasticity of the genome can easily circumvent attempts to design diagnostically secure molecular probes. This may be overcome by using one or more genomic targets which are least likely to be susceptible to sequence differences and to design oligonucleotide primers and probes which are complementary to the diversity of sequences found in the target population. Detection of nucleic acid comprises three essential components and it is prudent to think of these separately, as these are in essence interchangeable within detection strategies which may be termed 'front end', 'middle' and 'back end' (Figure 13.2).

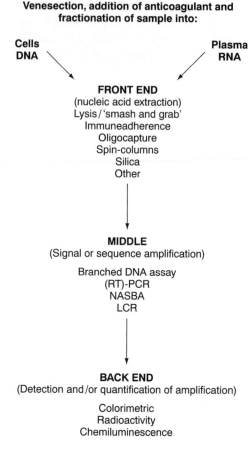

Figure 13.2 *Algorithm for virus quantification*
The process is separated into three distinct components of nucleic acid extraction, signal or sequence amplifications and the detection or quantification of the resulting product. LCR, ligase chain reaction; NASBA, nucleic acid-sequence based amplification; RT-PCR, reverse transcription and polymerase chain reaction.

Nucleic acid extraction ('front end')

Some of the techniques used for extraction of RNA or DNA are listed in Figure 13.2; there is some overlap in any given protocol.

DNA

DNA for analysis can be prepared from whole blood lysates, buffy coats and peripheral blood mononuclear cells (PBMCs). The PBMCs are usually prepared from anticoagulated whole blood following venesection by Ficoll–Hypaque or similar methodologies, and preserved at −80°C in the presence of an antifreezing medium. Mixtures of glycerol and gelatin, 'Glycigel', have been successfully used as an alternative to preserve PBMC material in HIV diagnosis (Kaye et al, 1991) to prevent cell lysis on storage. Varying degrees of purity of DNA can be attained by methods ranging from classical sodium dodecyl sulfate (SDS)/proteinase K and phenol–chloroform/ethanol precipitation to more rapid extraction protocols (Higuichi, 1989) and spun-column silica preparations (Qiagen, Hilden, Germany). The speed of sample preparation and quality of sample must be balanced. Estimation of the total cellular DNA concentration by fluorometry or spectrophotometry may be desirable.

RNA

RNA may be prepared from separated serum or plasma by the use of:

- chaotropic agents ('smash and grab')
- immune adherence
- pelleting
- oligocapture (Figure 13.2)

Heparin is a widely used anticoagulant, which can sometimes be copurified with RNA and be inhibitory to the polymerase chain reaction (PCR) process, and therefore ethylenediaminetetraacetic acid (EDTA) is now the preferred anticoagulant; the calcium chelation achieved by EDTA appears to protect virions. Separated plasma is usually stored in aliquots at −80°C or below, although viral RNA appears stable at −20°C or even 4°C. Chaotropes such as guanidinium isothiocyanate (Chomczynski and Sacchi, 1987; Boom et al, 1990) are used to lyse serum or plasma samples and release (viral) RNA for purification, and also serve to inhibit ribonucleases; inclusion of additional ribonuclease inhibitors at later stages is favoured by many investigators. Most RNA isolation protocols are laborious and automation of this procedure would be an obvious advantage; commercial systems are now being developed.

Amplification of genomic target ('middle')

Conceptually, the amplification of signals recognized by molecular probes (branched DNA signal assay) or sequence amplification by a derivative process (PCR, NASBA, LCR – see below) are employed as outlined in Figure 13.3.

Branched DNA assay

The branched DNA (bDNA) signal amplification assay (Pachl et al, 1995) employs multiple DNA probes linked to an enzymatic system for visualization of captured target molecules by sequential oligonucleotide hybridization, and is an alternative to sequence amplification. The bDNA assay (Quantiplex; Chiron Corporation, Emeryville, USA) has potential advantages over other molecular approaches to HIV genome detection, by virtue of its ability to employ a range of DNA probes which recognize a wide spectrum of HIV sequence variants. The enhanced sensitivity (ES) modification of the prototype assay (Kern et al, 1996) is capable of detecting as few as 400 copies per millilitre (copies per ml) of HIV-1 RNA in serum or plasma by using preamplification molecules and a cruciform configuration of target probes. Derivatization of probe sequences to reduce background binding in the most recent version of the assay is likely to increase sensitivity to around 50 genome equivalents per ml of plasma.

Polymerase chain reaction

The PCR is the most widely used sequence amplification technique and represents perhaps the single most important advance in the application of molecular techniques

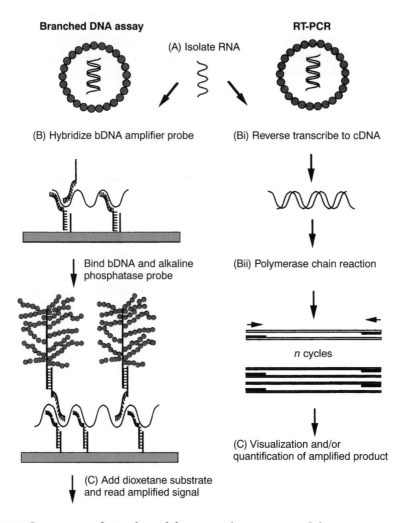

Figure 13.3 *Comparison of signal amplification and sequence amplification*
Comparison of the conceptual differences of signal amplification (branched DNA assay) and sequence amplification (reverse transcription, polymerase chain reaction) methodologies starting from virion HIV RNA.

to the diagnosis of viral infections, including HIV, in recent years. The basic principles and applications of the PCR process have been extensively reviewed elsewhere (Ehrlich, 1989; Kellog and Kwok, 1990; Clewley, 1995). The amplification of HIV sequences from clinical samples using the PCR process can be performed using one of two approaches depending on whether proviral DNA or viral RNA is to be detected. A plethora of DNA polymerases in addition to *Taq* are now readily available (e.g. *Pfu* (Stratagene, Cambridge, UK), *Pwo* (Roche Diagnostics, Lewes, UK)), although empirical optimization of each primer set and reaction conditions for the purpose of HIV detection (or any other sequence) should be employed. The versatility of PCR gives it 'added value' by generating usable PCR amplicons for subsequent DNA characterization, for example in DNA sequencing reactions.

For the detection of cell-free virus, virion RNA must first be reverse transcribed into complementary DNA (cDNA) followed by PCR (RT-PCR). Reverse transcription and PCR can be performed in one or two steps. Optimized reverse transcription reactions typically using reverse transcriptase derived from Moloney murine leukaemia virus (MMLV) or avian myeloblastosis virus (AMV) followed by separate PCR amplification of cDNA constitute a typical two-step RT-PCR. Dual-function enzymes able to function as both reverse transcriptases and DNA polymerases such as *Tth* (Perkin-Elmer) and the Titan dual enzyme (AMV and *Taq/Pwo*) system (Boehringer Mannheim) offer a more simplified approach to RT-PCR. Detection and quantification of HIV-1 RNA by competitive RT-PCR (Mulder et al, 1994) is also commercially available (Amplicor Monitor; Roche Diagnoctics, Lewes, UK).

The increased sensitivity of PCR, however, presents potential problems of contamination by previously amplified sequences feeding back into the system. This may be minimized by the strict enforcement of stringent laboratory practices and guidelines (Higuichi and Kwok, 1989). The risk may be further reduced by the substitution of dUTP into reaction mixes, allowing degradation of uracil by the enzyme uracil N-glycosylase (UNG), preventing carry-over of previously amplified product (Longo et al, 1990), as in the 'Amplicor' HIV DNA detection assay (Roche Molecular Systems).

Nucleic acid-based sequence amplification

Nucleic acid-based sequence amplification (NASBA) is an isothermal amplification process (Kievits et al, 1991) involving alternate steps of DNA synthesis from an RNA template and RNA synthesis from a DNA template adapted for HIV-1 RNA quantification (van Gemen et al, 1993, 1994). The cyclical use of a reverse transcriptase (AMV) in conjunction with RNAase H and T7 RNA polymerase to derive alternate templates from a known sequence template with synthetic oligonucleotide primers provides the basis for the assay. The process is internally controlled for quantification purposes with three distinguishable RNA templates (Q_A, Q_B and Q_C) utilizing HIV-1 *gag*-specific sequences (Figure 13.4).

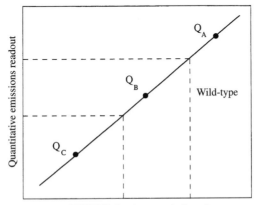

HIV RNA (copies per ml plasma)

Figure 13.4 *Nucleic acid-based signal amplification*
Quantification of HIV-1 RNA by the NASBA process, exemplifying an internally calibrated assay relating the proportion of signal in the wild-type to a previously determined number of input molecules (Q_A, Q_B, Q_C) which are simultaneously amplified.

Ligase chain reaction

The ligase chain reaction (LCR) represents an alternative amplification method to PCR and NASBA. Exponential amplification is achieved by repetitive cycles of ligation with two adjacent pairs of oligonucleotides to form longer ligated products using a thermo-stable ligase in a template-dependent manner. Although LCR has not gained widespread use in HIV diagnostic settings, it has the added application of being able to confirm the identity of a sequence amplicon and is suited to the detection of short sequences, requiring as it does adjacent annealing of probes.

Sequence target selection

Genomic hybridization reactions are necessary for all genomic detection approaches. Prior knowledge of target sequences is essential; in the case of HIV, the wide sequence base – reflected in the complexity of viral subtypes and recombination events in infected individuals – can cause severe problems in the design of reliable and accurate oligonucleotide primers and probes. Sequence amplification techniques such as PCR and NASBA offer increased sensitivity and specificity over conventional diagnostic procedures, molecular or otherwise, although one caveat must be that they remain prey to the inherent genetic diversity of HIV infections. In general, when considering the choice of target region of the HIV genome, *env* sequences are most variable and are not usually used for diagnostic purposes, with *gag* and *pol* being constitutively more conserved. Target sequences and amplicon probe sites for PCR have been sited in *gag* ('Amplicor' utilizes SK431/SK462 and an internal probe SK102), with similar *gag* sequences used in the NASBA technique. Examples of PCR 'misses' with *gag* primers have been reported where non-clade-B HIV-1 infections have been present, exemplified by identification of a clade G virus (Arnold et al, 1995), leading manufacturers to modify their assays. In contrast, multiple HIV *pol* sequences are employed in the bDNA assay and this target is inherently less prone to sequence divergence in the study population. Sequences within the HIV long terminal repeat (LTR) are also functionally highly conserved, and have been applied in HIV diagnostic settings. Detection of more than one gene amplification product by PCR has been a criterion recommended by some investigators to confirm the presence of genome, although this approach is not entirely satisfactory and can lead to significant false-negative results. Alternatively, pan-lentivirus-specific primer sets which detect all putative lentivirus strains (Gelman et al, 1992) have been developed, but have not been widely applied to HIV diagnosis. Application of selected *gag*, *pol* or LTR sequences has demonstrated LTR to be the most secure diagnostically.

Detection of amplification signal ('back end')

The detection of amplified sequences represents the 'back end' of the diagnostic algorithm and is often linked by a necessity to provide additional specificity and/or sensitivity. Conventional Southern and dot blots represent one additional hybridization step. For PCR reactions, a nested approach with additional internal sequences usually represents two more hybridizations and is generally considered to be more sensitive than a single PCR for either HIV-1 (Simmonds et al, 1990) or HIV-2 (Grankvist et al, 1992; Berry et al, 1994). Products of PCR are generally viewed by ethidium bromide staining on agarose gels where a combination of the expected molecular weight (size of amplified product) and unambiguous amplification provides a diagnosis. Single target molecule amplification should be the aim of the investigator. Probe binding may also be visualized, particularly for quantitative measurement, either by colorimetric means using enzymatically labelled probes (alkaline phosphatase or horseradish peroxidase) or by detection of emissions in radiometric, magnetizable or (electro)-chemiluminescence reactions (Figure 13.5). All have been successfully applied to HIV diagnostic assays, although the methods in the latter group have the additional advantage of an expanded dose response.

DETECTION OF VIRUS IN THE PERIPHERAL BLOOD

The two components of peripheral blood (cellular or cell-free) may be used to detect the two distinct phases of the HIV life cycle. Depending on the stage of infection, the approach to HIV diagnosis may differ, although in general the detection of proviral DNA in cellular material has been used to diagnose HIV infections by molecular methods in cases of conflicting serology, particularly during the 'diagnostic window' (see Figure 13.1). For most clinical purposes the detection of HIV proviral DNA is sufficient to provide a diagnosis, although there are circumstances in which HIV RNA detection may be desirable, if for example cellular components are not available for testing. The general constraints of HIV sequence variation apply equally to both proviral and virion detection.

Detection of proviral DNA

Detection of HIV proviral sequences in PBMCs by PCR remains the principal method of diagnosing HIV infection by molecular techniques. In general, the more protein and other PCR-inhibitory substances (e.g. haemoglobin) that are removed (such as contaminating red blood cells), the less likelihood there is that false-negative reactions will occur. This is particularly relevant to HIV provirus detection where a potentially low peripheral load exists, consisting of rare sequences in a complex mixture of human genomic sequences, particularly during asymptomatic infection where a molecular diagnosis is perhaps of most use. Modifications to standard PCR protocols to reduce non-virus-specific amplifications, thus improving the efficiency of detection, may be achieved by adopting 'hot start' protocols which separate constituent components at key parts of the amplification process, or the substitution of *Taq* DNA polymerase with '*Taq* Gold' to achieve the same result. Inclusion of 'housekeeping' genes which are ubiquitous in human genomic DNA, such as β-globin, and the addition of internal spiking or calibrators may serve to assess the quality of the DNA preparation and its suitability for amplification; these are particularly important to validate negative results.

Detection of viral RNA

In the majority of individuals infected with HIV-1 not receiving antiretroviral treatment, HIV virion RNA is detectable in the cell-free fraction of peripheral blood (Piatek et al, 1993; Mellors et al, 1996; Bruisten et al, 1997). The presence of HIV-1 RNA in serum or plasma is readily detectable with sensitive assays (i.e. those that measure down

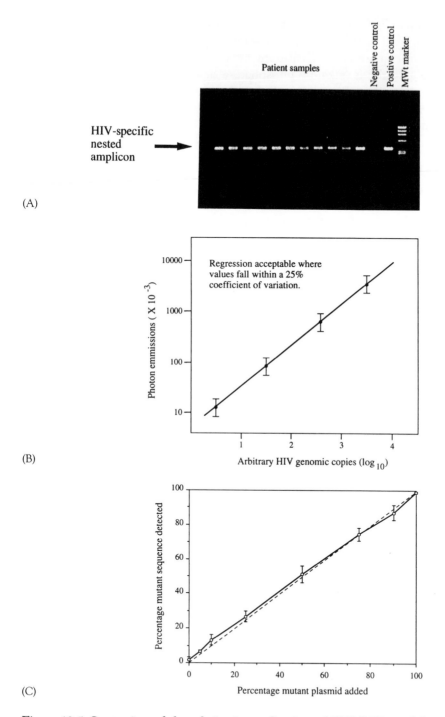

Figure 13.5 *Comparison of three derivative applications of HIV PCR amplification*
(A) Visualization of nested PCR amplicons by ethidium bromide staining on agarose gels for detection purposes. (B) Quantification of PCR amplicons by measuring photon emissions of decreasing concentrations of PCR-amplified HIV DNA in a quantitative chemiluminescence assay. (C) Application of PCR amplicons to quantify the relative proportions of single point mutations associated with HIV drug resistance.

to at least 500 copies of HIV RNA, or their equivalent, per ml of serum or plasma) in the majority of individuals, and thus would provide a diagnosis in the majority of cases. However, there is an absence of readily available, internationally recognized HIV RNA quantification standards, and values obtained from different sera with different assays (RT-PCR, NASBA and bDNA) may vary (Schuurman et al, 1996). Furthermore, performances of sequence amplification assays (NASBA and RT-PCR) also differ in an idiosyncratic fashion (Alaeus et al, 1997b) with divergent genovars. Despite attempts to circumvent sequence bias, the bDNA assay suffers from other potential problems such as inhibitory substances in the sample. The choice of assay thus remains somewhat arbitrary, and the detection of HIV RNA should be reserved in most instances for the investigation of patients known to be infected.

Samples from long-term survivors with a very low peripheral HIV load (Schwartz et al, 1994; Cao et al, 1995) are likely to fall below the threshold of detectability in such assays, and detection of the proviral component to provide a diagnosis becomes essential. Patients on effective antiretroviral treatment also pose similar problems with respect to sensitivity and raise the question of what goes on 'below threshold'. The need for ultrasensitive RNA assays is now recognized and a switch to proviral DNA detection in order to monitor the infection in these patients becomes inevitable.

The lower levels of replication of HIV-2 in the asymptomatic period, reflected in low peripheral HIV-2 RNA (Berry et al, 1998), effectively makes RNA detection alone unsuitable for the diagnosis of HIV-2 infections. Since this appears to be the more usual situation in HIV-2 infection, particularly in those showing no signs of symptomatic disease, provirus detection is therefore the more appropriate measurement (Berry et al, 1994, 1998; Ariyoshi et al, 1996). However, HIV-2 viraemia does become readily detectable in the later stages of infection where disease is at a more advanced state, thus a positive RNA result provides important diagnostic information in the management of HIV-2-infected individuals in such circumstances. The failure to detect HIV-2 RNA as a sole diagnostic genomic target per se must be treated with extreme caution.

DETECTION OF VIRUS AT OTHER SITES

Human immunodeficiency virus replicates in lymph nodes and microglial cells, and the detection of HIV provirus in tissues at central sites of the body may further confirm the presence of HIV (Embretson et al, 1993; Pantaleo et al, 1993). While these are not front-line diagnostic procedures, the possible clearance of detectable virus from peripheral sites in HIV-infected individuals receiving potent combination antiretroviral therapy indicates that the requirement to detect virus at central sites may be likely to increase. However, the potential presence of inhibitors of PCR in these tissue samples emphasizes the need for careful validation of the molecular procedures used to amplify target sequences from such specimens.

APPLICATION TO OTHER SITUATIONS

In some cases the detection of virus-specific antibodies is either inappropriate or extremely difficult to interpret, thus molecular techniques become an alternative option.

Primary infection and the seroconversion period

Detection of HIV RNA may be particularly appropriate when applied in cases of acute seroconversion, perhaps if HIV infection is suspected on other grounds, and may provide a diagnosis before any other HIV marker becomes detectable. This relies on the observation that HIV viral load is highest immediately following exposure (see Figure 13.1), before the initial clearance of viral infection by host immunity has occurred. Therefore, application of molecular techniques to detect viral sequences directly either as viral RNA or proviral DNA in the peripheral circulation in such circumstances would be appropriate, and perhaps the only means of diagnosing acute infections.

Perinatal and postnatal transmission

The application of PCR to diagnose vertical transmission events in the perinatal and postnatal periods has several advantages over detection of virus-specific antibodies which may take up to 18 months to yield an unequivocal diagnosis, owing to the presence of passively acquired maternal antibodies. Transmission of HIV during pregnancy may be diagnosed by HIV provirus detection by PCR in infants as early as 6 weeks, although confirmation with later samples is usually required to exclude or confirm transmission. Vertical transmission of HIV-2 appears to occur only rarely (frequency 0–4%) (Adjorolo-Johnson et al, 1994; O'Donovan et al, 1996), although a much longer period may be required before the absence of detectable HIV-2 genome and hence infection may be excluded. Genomic tests should be performed in conjunction with antibody assays. The timing of sampling is open to debate, although a simple rate of three – 3 days, 3 weeks and 3 months – will usually enable identification of a child infected in utero or in the immediate postnatal period. Late infection is rare but can be identified by seroconversion and genome detection.

Discrimination between HIV-1 and HIV-2 infections

Discrimination between HIV-1 and HIV-2 infections may be required in certain circumstances, particularly where coinfection with both viruses is suspected. These have been described principally in west Africa, notably the Ivory Coast (Peeters et al, 1992, 1994), as well as in India (Grez et al, 1994) and Brazil (Pieniazek et al, 1991), where both virus types have been identified. The extensive cross-reactions on heterologous antigens makes diagnosis and speciation of infection by serology alone difficult (Tedder et al, 1988; De Cock et al, 1991). Diagnosis of dual infections by PCR requires selection of primer sequences that detect predominantly non-clade-B HIV-1 infections and HIV-2 with comparable efficiency, thus providing a differential diagnosis with virus type-specific primer sets for HIV-1 and HIV-2.

Genomic characterization

In certain circumstances some characterization of the genome may be required, for example where knowledge of the sequence of the infecting virus is desirable, such as in transmission cases or where the patient has been receiving antiretroviral therapy.

Genetic identification of the infecting strain

Where more detailed information about the infecting HIV strain is required, DNA sequencing has been used to map phylogenetically its relatedness to other sequences to deduce or infer an epidemiologically relevant event, such as in a transmission case (Arnold et al, 1995). This approach is often facilitated by automation. The heteroduplex mobility assay (HMA) has also been used to type HIV DNA in clinical specimens based upon the relative migration of PCR-derived *env* fragments on gels following hybridization with clade-specific DNA probes (Delwart et al, 1993). This assay is usually used in combination with DNA sequencing, its principal application with regard to HIV diagnosis being the identification of a non-clade-B virus infection (Novitsky et al, 1996).

Drug resistance

Different approaches have been used to measure genotypic drug resistance in individuals receiving antiretroviral chemotherapy. DNA sequencing of PCR-amplified fragments of the *pol* gene; the quantitative point mutation assay (PMA) which determines the percentage of mutation for a given resistance codon (Kaye et al, 1992); and differential PCR (Larder et al, 1996) or line-probe assay (InnoLipa; Murex Diagnostics, Dartford, UK) which provides qualitative data (Stuyver et al, 1997) have all been applied to this problem (see Figure 13.5). Where the biological nature of virus infection at the genomic level is required, this may be investigated using the recombinant virus assay (RVA) which combines genotypic and phenotypic approaches and obviates the need to derive clinical isolates (Kellam and Larder, 1994).

FUTURE SCENARIOS

The changing nature of HIV infections in different communities and regions of the world exerts a strong pressure on the performance and accuracy of diagnostic assays. With the application of combinations of potent antiretroviral chemotherapy, the possibility of clearing virus from the individual has been suggested (Perelson et al, 1996), and viral reservoirs in such patients have been identified (Chun et al, 1997; Finzi et al, 1997; Wong et al, 1997). Clarification of the nature of HIV infection on an individual basis presents a considerable challenge to diagnostic methodologies and approaches. Improved diagnostic techniques therefore assume a heightened significance in terms of specificity, sensitivity and performance as greater pressure is placed upon the method of choice. Moreover, taking a global perspective, the equivalent detection of all HIV genomes irrespective of their geographical location or the nationality of individuals infected should remain a priority.

REFERENCES

Adjorolo-Johnson G, De Cock KM, Ekpini E et al (1994) Prospective comparison of mother-to-child transmission of HIV-1 and HIV-2 in Abidjan, Ivory Coast. *JAMA* **272**: 462–466.

Alaeus A, Leitner T, Lidman K & Albert J (1997a) Most HIV-1 genetic subtypes have entered Sweden. *AIDS* **11**: 199–202.

Alaeus A, Lidman K, Sonnerborg A & Albert J (1997b) Subtype-specific problems with quantification of plasma HIV-1 RNA. *AIDS* **11**: 859–865.

Ariyoshi K, Berry NJ, Wilkins A et al (1996) A community-based study of HIV-2 proviral load in a rural village in West Africa. *J Infect Dis* **173**: 245–248.

Arnold C, Barlow KL, Kaye S, Loveday C, Balfe P & Clewley JP (1995) HIV type 1 sequence subtype G transmission from mother to infant: failure of variant sequence species to amplify in the Roche amplicor test. *AIDS Res Hum Retroviruses* **11**: 999–1001.

Berry NJ, Ariyoshi K, Jobe O et al (1994) HIV type 2 proviral load measured by quantitative polymerase chain reaction correlates with CD4+ lymphopenia in HIV type 2-infected individuals. *AIDS Res Hum Retroviruses* **10**: 1031–1037.

Berry NJ, Ariyoshi K, Jaffar S et al (1998) Low peripheral blood viral HIV-2 RNA in individuals with high CD4% differentiates HIV-2 from HIV-1 infections. *J Hum Vir* **1**(7): 457–468.

Boom R, Sol CJ, Salimans MM, Jansen CL, Wertheim-van Dillen PM & Van der Noordaa J (1990) Rapid and simple method for purification of nucleic acids. *J Clin Microbiol* **28**: 495–503.

Bruisten SM, Frissen PH, Van Swieten P et al (1997) Prospective longitudinal analysis of viral load and surrogate markers in relation to clinical progression in HIV type 1-infected persons. *AIDS Res Hum Retroviruses* **13**: 327–335.

Cao Y, Qin L, Zhang L, Safrit J & Ho DD (1995) Virologic and immunologic characterisation of long-term survivors of human immunodeficiency virus type 1 infection. *N Engl J Med* **332**: 201–208.

Chen Z, Tefler P, Gettie A et al (1996) Genetic characterisation of new West African simian immunodeficiency virus SIVsm: geographic clustering of household-derived SIV strains with human immunodeficiency virus type 2 subtypes and genetically diverse viruses from a single feral sooty mangabey troop. *J Virol* **70**: 3617–3627.

Chomczynski P & Sacchi N (1987) Single-step method of RNA isolation by guanidinium thiocyanate–phenol–chloroform extraction. *Anal Biochem* **162**: 156–159.

Chun TW, Carruth L, Finzi D et al (1997) Quantification of latent tissue reservoirs and total body viral load in HIV-1 infection. *Nature* **387**: 183–188.

Clavel F, Guetard D, Brun-Vezinet F et al (1986) Isolation of a new human retrovirus from West African patients with AIDS. *Science* **223**: 343–346.

Clewley JP, ed. (1995) *The Polymerase Chain Reaction (PCR) for Human Viral Diagnosis*. London: CRC Press.

De Cock K, Porter A, Kouadio J et al (1991) Cross-reactivity on Western blot in HIV-1 and HIV-2 infections. *AIDS* **5**: 859–863.

De Cock KM, Adjorlolo G, Ekpini E et al (1993) Epidemiology and transmission of HIV-2: why there is no HIV-2 pandemic. *JAMA* **270**: 2083–2086.

Delwart E, Shpaer EG, Louwagie J et al (1993) Genetic relatedness determined by a DNA heteroduplex mobility assay: analysis of HIV-1 env genes. *Science* **262**: 1257–1261.

Ehrlich HA (ed.) (1989) *PCR Technology: Principles and Applications for DNA Amplification*. New York: Stockton Press.

Embretson J, Zupancic M, Ribas J et al (1993) Massive covert infection of helper T lymphocytes and macrophages by HIV during the incubation period of AIDS. *Nature* **362**: 359–362.

Finzi D, Hermankova M, Pierson T et al (1997) Identification of a reservoir for HIV-1 in patients on highly active antiretroviral therapy. *Science* **278**: 1295–1300.

Gao F, Yue L, White A et al (1992) Human infection by genetically diverse SIVsm-related HIV-2 in West Africa. *Nature* **358**: 495–499.

Gao F, Yue L, Robertson DL et al (1994) Genetic diversity of human immunodeficiency virus type 2: evidence for distinct sequence subtypes with differences in virus biology. *J Virol* **68**: 7433–7447.

Gelman IH, Zhang J, Hailman E, Hanafusa H & Morse SS (1992) Identification and evaluation of new primer sets for the detection of lentivirus proviral DNA. *AIDS Res Hum Retroviruses* **8**: 1981–1989.

Grankvist O, Bredberg-Raden U, Gustafsson A et al (1992) Improved detection of HIV-2 DNA in clinical samples using a nested primer-based polymerase chain reaction. *J AIDS* **5**: 286–293.

Grez M, Dietrich U, Balfe P et al (1994) Genetic analysis of HIV-1/HIV-2 mixed infections in India reveals a recent spread of HIV-1 and HIV-2 from a single ancestor for each virus. *J Virol* **68**: 2161–2168.

Higuichi R (1989) Simple and rapid preparation of samples for PCR. In: Ehrlich H (ed.) *PCR Technology: Principles and Applications for DNA Amplification*, pp 31–38. New York: Stockton.

Higuichi R & Kwok S (1989) Avoiding false-positives with PCR. *Nature* **339**: 237.

Ho DD, Neumann AU, Perelson AS, Chen W, Leonard JM & Markovitz M (1995) Rapid turnover of plasma virions and CD4 lymphocytes in HIV-1 infection. *Nature* **373**: 123–126.

Kaye S, Loveday C & Tedder R (1991) Storage and preservation of whole blood samples for use in detection of human immunodeficiency virus type-1 by the polymerase chain reaction. *J Virol Methods* **35**: 217–226.

Kaye S, Loveday C & Tedder R (1992) A microtitre format point mutation assay: application to the detection of drug resistance in human immunodeficiency virus type-1 infected patients treated with zidovudine. *J Med Virol* **37**: 241–246.

Kellam P & Larder B (1994) Recombinant virus assay: a rapid, phenotypic assay for assessment of drug susceptibility of human immunodeficiency virus type 1 isolates. *Antimicrob Agents Chemother* **38**: 23–30.

Kellog DE & Kwok S (1990) In Innis MA, Gelfand DH, & Sninsky JJ (eds) *PCR Protocols: a guide to methods and applications*. London: Academic Press.

Kern D, Collins M, Fultz T et al (1996) An enhanced-sensitivity branched-DNA assay for quantification of human immunodeficiency virus type 1 RNA in plasma. *J Clin Microbiol* **34**: 3196–3202.

Kievits T, van Gemen B, van Strijp D et al (1991) NASBA isothermal enzymatic in vitro nucleic acid amplification optimised for the diagnosis of HIV-1 infection. *J Virol Methods* **35**: 273–286.

Larder BA, Kohli BA, Bloor S et al (1996) Human immunodeficiency virus type 1 drug susceptibility during zidovudine (AZT) monotherapy compared with AZT plus 2′,3′-dideoxycytidine combination therapy. *J Virol* **70**: 5922–5929.

Longo MC, Berninger MS & Hartley JL (1990) Use of uracil DNA glycosylase to control carryover contamination in polymerase chain reactions. *Gene* **93**: 125–128.

Marlink R, Kanki P, Thior I et al (1994) Reduced rate of disease development after HIV-2 infection as compared to HIV-1. *Science* **265**: 1587–1590.

Mellors JW, Rinaldo CR, Gupta P, White R, Todd JA & Kingsley LA (1996) Prognosis of HIV-1 infection predicted by the quantity of virus in plasma. *Science* **272**: 1167–1170.

Mulder J, McKinney N, Christopherson C, Sninsky J, Greenfield L & Kwok S (1994) Rapid and simple PCR assay for quantitation of human immunodeficiency virus type 1 RNA in plasma: application to acute retroviral infection. *J Clin Microbiol* **32**: 292–300.

Naucler A, Albino P, Andersson S et al (1992) Clinical and immunological follow-up of previously hospitalised HIV-2 seropositive patients in Bissau, Guinea-Bissau. *Scand J Infect Dis* **24**: 725–731.

Novitsky V, Arnold C & Clewley JP (1996) Heteroduplex mobility assay for subtyping HIV-1: improved methodology and comparison of phylogenetic analysis of sequence data. *J Virol Methods* **59**: 61–72.

O'Donovan D, Ariyoshi K, Yamuah L et al (1996) Perinatal transmission of HIV-2 in The Gambia, West Africa. *Eleventh International Conference on Aids*, Vancouver (abstr. Tu C.2577).

Pachl C, Todd JA, Kern DG et al (1995) Rapid and precise quantification of HIV-1 RNA in plasma using a branched DNA signal amplification assay. *J AIDS Hum Retroviruses* **8**: 446–454.

Pantaleo G, Graziosi C, Demarest J et al (1993) HIV infection is active and progressive in lymphoid tissue during the clinically latent stage of disease. *Nature* **362**: 355–358.

Peeters M, Gershy-Damet G-M, Fransen K et al (1992) Virological and polymerase chain reaction studies of HIV-1/HIV-2 dual infection in Cote d'Ivoire. *Lancet* **340:** 339–340.

Peeters M, Katrien F, Gershy-Damet G-M et al (1994) Effect of methodology on detection of HIV-1/HIV-2 dual infections in Cote d'Ivoire. *J Virol Methods* **48:** 23–30.

Perelson AS, Neumann AU, Markovitz M, Leonard JM & Ho DD (1996) HIV-1 dynamics *in vivo*: virion clearance rate, infected cell life-span and viral generation time. *Science* **271:** 1582–1585.

Piatek MJ, Saag M, Yang L et al (1993) High levels of HIV-1 in plasma during all stages of infection determined by competitive PCR. *Science* **259:** 1749–1754.

Pieniazek D, Peralta J, Ferreira J et al (1991) Identification of mixed HIV-1 and HIV-2 infections in Brazil by polymerase chain rection. *AIDS* **5:** 1293–1299.

Poulsen A-G, Aaby P, Larsen O et al (1997) 9-year HIV-2-associated mortality in an urban community in Bissau, West Africa. *Lancet* **349:** 911–914.

Ricard D, Wilkins A, Ngom PT et al (1994) The effects of HIV-2 infection in a rural area of Guinea Bissau. *AIDS* **8:** 977–982.

Romieu I, Marlink R, Kanki P, M'Boup S & Essex M (1990) HIV-2 link to AIDS in West Africa. *J AIDS* **3:** 220–230.

Schuurman R, Descamps D, Weverling GJ et al (1996) Multicenter comparison of three commercial methods for quantification of human immunodeficiency virus type 1 RNA in plasma. *J Clin Microbiol* **34:** 3016–3022.

Schwartz D, Sharma U, Busch M et al (1994) Absence of recoverable infectious virus and unique immune response in an asymptomatic HIV+ long-term survivor. *AIDS Res Hum Retroviruses* **10:** 1703–1711.

Simmonds P, Balfe P, Peutherer J, Ludlam C, Bishop J & Leigh-Brown A (1990) Human immunodeficiency virus-infected individuals contain provirus in small numbers of peripheral mononuclear cells and at low copy numbers. *J Virol* **64:** 864–872.

Stuyver L, Wyseur A, Rombout A et al (1997) Line probe assay (LiPA) for the detection of drug-selected variants in the HIV-1 reverse transcriptase gene. *Int Antiviral News* **5:** 38–40.

Tedder R, O'Connor T, Hughes A, N'Jie H, Corrah T & Whittle H (1988) Envelope cross-reactivity in Western blot for HIV-1 and HIV-2 may not indicate dual infection. *Lancet* **ii:** 927–930.

Van Gemen B, Kievits T, Nara P et al (1993) Qualitative and quantitative detection of HIV-1 RNA by nucleic acid sequence-based amplification. *AIDS* **7:** S107–S110.

Van Gemen B, van Beuningen R, Nabbe A et al (1994) A one-tube quantitative HIV-1 RNA NASBA nucleic acid amplification assay using electrochemiluminescent (ECL) probes. *J Virol Methods* **49:** 157–168.

Wei X, Ghosh SK, Taylor ME et al (1995) Viral dynamics in human immunodeficiency virus type 1 infection. *Nature* **373:** 117–122.

Whittle H, Morris J, Todd J et al (1994) HIV-2-infected patients survive longer than HIV-1-infected patients. *AIDS* **8:** 1617–1620.

Wong JW, Hezarah M, Gunthard HF et al (1997) Recovery of replication-competent HIV despite prolonged suppression of plasma viraemia. *Science* **278:** 1291–1295.

Chapter 14

HIV RESISTANCE TO ANTAGONISTS OF VIRAL REVERSE TRANSCRIPTASE

Mark A. Wainberg

INTRODUCTION

Human immunodeficiency virus type 1 (HIV-1) resistance against antagonists of the viral reverse transcriptase (RT) and protease enzymes is due to a series of mutations in the genes that encode these respective enzymes. Mutations occur owing to the error-prone nature of the HIV RT itself, which is responsible for copying the HIV genome. The error rate of HIV RT is estimated at approximately 10^{-4}. This means that one mutation in RT-directed synthesis of viral DNA will occur on average once during every virus replication cycle, given a genomic length of 9·2 kb. While many such mutations may be inconsequential or silent, many may be lethal. In either situation it may be impossible to determine whether a mutation has arisen, since in the first case there may be no substitution in amino acid sequence, while in the second case viral replication may not take place.

The most effective way to recognize mutagenesis is under conditions of selective pressure, as is exerted during the course of HIV disease by both antiviral drugs and the immune system. In the first case, spontaneously occurring mutations in either the HIV RT or protease genes may be amplified under conditions of drug pressure to yield drug-resistant forms. In the case of immunological pressure, HIV variants may emerge that are no longer reactive against neutralizing antibodies and/or cytotoxic T lymphocytes, owing to mutations in epitopes that might have otherwise been recognized by these immune effector mechanisms. In this context, the principles that govern the emergence of drug-resistant viruses are similar to those that define bacterial resistance to antibiotics and the resistance of cancer cells to oncochemotherapeutic regimens.

Of course, HIV mutagenesis is an ongoing process. This means that mutations that encode drug-resistant variants can take place both prior to the administration of antiviral drugs as well as during antiviral therapy, as long as the drugs utilized are unable to shut down virus replication completely. Mutated viruses that are not amplified by conditions of drug pressure may never be recognized, since they may be present only as a tiny fraction of the much larger quasispecies. In addition, mutated viruses in untreated individuals are probably unable to replicate faster than wild-type viruses, since they would otherwise become the dominant population in the absence of antiviral therapy.

HIV and the New Viruses Second Edition
ISBN 0-12-200741-7

HIV REVERSE TRANSCRIPTASE

The enzyme HIV reverse transcriptase acts to convert viral genomic RNA into double-stranded proviral DNA. This essential role has made this enzyme a primary target of anti-retroviral chemotherapy. Reverse transcriptase inhibitors can be classified into two major groups based on structural considerations: nucleoside analogs and non-nucleoside RT inhibitors (NNRTIs); these act in different ways to block HIV-1 RT polymerase activity. Nucleoside analogs are 2′,3′-dideoxy derivatives of the natural substrates, i.e. 2′-deoxy-nucleosides, of DNA polymerases. All 2′,3′-dideoxynucleoside analogs are thought to act in similar fashion to inhibit RT activity (Furman et al, 1986; Mitsuya and Broder, 1986; St Clair et al, 1987; Yarchoan et al, 1989; Hart et al, 1992). Following intracellular phosphorylation to 5′-triphosphorylated forms (ddNTP), they competitively interfere with the binding of natural substrates to RT and are incorporated into elongating DNA strands. Chain termination then results from the lack of a hydroxyl motif at the 3′ carbon of the pentose ring of the nucleoside antagonist, without which a 3′–5′ phosphodiester bond cannot be formed with an incoming substrate.

Non-nucleoside RT inhibitors are a group of highly specific inhibitors of HIV-1 RT which bind to a hydrophobic pocket near the polymerase catalytic site of RT (Wu et al, 1991; Kohlstaedt et al, 1992; Ding et al, 1995). Although the precise mechanisms of action of NNRTIs are unclear, recent studies showed that the binding of NNRTIs to HIV-1 lowered the catalytic rate of polymerization without affecting nucleotide binding or nucleotide-induced conformational change (Spence et al, 1995). Binding of NNRTIs to RT had little effect on interactions with template/primer or deoxynucleotide sub-strates/dideoxynucleotide inhibitors (Rittinger et al, 1995; Gu et al, 1995c). Kinetic studies have revealed that NNRTIs are non-competitive inhibitors of RT.

Reverse transcriptase inhibitors can decrease the viral load of HIV-1-infected individuals (Yarchoan et al, 1986, 1988; Fischl et al, 1987, 1989). However, patients commonly suffer clinical deterioration, once drug-resistant viruses appear during prolonged therapy (Fischl et al, 1987; Larder et al, 1989; Rooke et al, 1989; Moore et al, 1991).

In the case of zidovudine (azidothymidine, AZT) monotherapy studies, most evidence indicates that HIV drug resistance is probably causally associated with treatment failure and death. However, such findings may be much harder to demonstrate in the case of patients receiving three or more antiviral drugs, owing to higher numbers of confounders associated with more complicated regimens.

HIV RESISTANCE TO NUCLEOSIDE ANALOGS

Resistance of HIV to nucleoside analogs follows prolonged therapy of infected individuals with these drugs. Drug-resistant viruses were first isolated from individuals treated with 3′-azido-3′-deoxythymidine (azidothymidine or zidovudine) (Larder et al, 1989; Rooke et al, 1989; Moore et al, 1991). Although resistance to NNRTIs can develop quickly, resistance to nucleoside analogs usually takes about 6 months to develop in vivo, with the exception of the (−)-enantiomer of 2′,3′-dideoxy-3′-thiacytidine, lamivudine or 3TC (Wainberg et al, 1995). The extent of drug resistance is usually expressed phenotypically by comparing the 50% inhibitory concentration (IC_{50}) values of pretreatment isolates with those of viruses isolated after initiation of therapy.

To date, resistance-conferring mutations in HIV RT have been reported for all nucleoside and NNRTI compounds used to treat patients infected with the virus. Different patterns of drug sensitivity and resistance have emerged, however, even among members of the same drug class. In some cases HIV may remain relatively sensitive even after mutations develop. For example, a 100-fold drop in sensitivity to AZT may develop after 6 months of monotherapy, while only several weeks of monotherapy with lamivudine may result in high-level resistance to the latter agent. In regard to the three other commonly employed nucleoside analogs, didanosine (ddI), zalcitabine (ddC) and stavudine (d4T), HIV may remain fairly sensitive even after the appearance of specific resistance-conferring mutations.

The IC_{50} is the concentration of drug that can inhibit viral replication in culture by 50%. Thus, low IC_{50} values denote susceptibility while high values denote loss of susceptibility or resistance. An increase in IC_{50} is usually demonstrated over time, concomitant with an increase in the relative proportion of drug-resistant viruses, as analyzed by selective polymerase chain reaction (PCR) for detection of mutated sequences in the RT gene. Sometimes the development in vivo of resistance to nucleoside analogs can occur independently of drug dose. In the case of AZT, less than a threefold increase in IC_{50} is considered to be within the normal range of variation, while a 10–50-fold increase in IC_{50} denotes partial resistance. Increases in IC_{50} over 50-fold represent high-level resistance.

Viral drug resistance can also be selected in cell culture by gradually increasing the concentrations of any of nucleoside analogs, NNRTIs or protease inhibitors (Larder et al, 1991; Richman et al, 1991b; Gao et al, 1992; Mellors et al, 1992; El-Farrash et al, 1994). Drug resistance commonly appears faster in cell culture under drug pressure than in the clinic. This may be due, in part, to the fact that HIV can replicate faster in transformed cell lines than in healthy T lymphocytes. However, the same resistance-conferring mutations that arise in cell culture are also seen clinically (Larder et al, 1991; Richman et al, 1991b; Gao et al, 1992). Thus, tissue culture selection provides a fast and efficient way to study HIV-1 drug resistance and identify resistance-conferring mutations, even before the initiation of clinical trials with the same drugs.

As stated above, RTs are low-fidelity DNA polymerases in comparison with other prokaryotic and eukaryotic DNA polymerases. This error-prone nature is due, in part, to a lack of a 3′-5′ exonuclease proofreading activity (Roberts et al, 1988), which is present in the DNA polymerases of higher organisms. Based on crystal structure analysis, HIV RT has a wider catalytic cleft than that of the *Escherichia coli* Klenow fragment (KF) of DNA polymerase I (Kohlstaedt et al, 1992). This suggests that RT may have less accuracy for binding of substrates, but abnormal strand transfers during reverse transcription have also been proposed as being responsible for the error-prone nature of this enzyme.

As with other DNA polymerases, the HIV RT is processive, and may make a higher frequency of mutations at pause sites (Ricchetti and Buc, 1990; Bebenek et al, 1993). In this context, low processivity of HIV RT might help to explain the low fidelity of this enzyme. However, the mutation rate of HIV in vivo may be much lower than that predicted from assays in vitro with purified RT (Mansky and Temin, 1995).

Variants of HIV are quasispecies, owing to the high turnover of HIV-1 replication in vivo (Ho et al, 1995; Wei et al, 1995), and include a high frequency of defective particles (Boulerice et al, 1990). High variability of HIV replication spontaneously generates drug resistance-related mutations, and drug-resistant variants are selected rather than induced under drug pressure (Coffin, 1995). Thus, singly mutated drug-resistant variants arise in culture and in vivo both prior to therapy and before identification of multiply mutated drug-resistant variants. The latter must be less frequent than singly mutated isolates in the HIV pool. However, patients with advanced disease generally have high viral load and a broader range of quasispecies than patients with early-stage illness. The former patients generally yield drug-resistant variants faster than asymptomatic individuals (Richman et al, 1990). However, low levels of viral burden had little effect on the rate of emergence of resistance to at least one NNRTI, nevirapine, which was detected as early as 1 week after initiation of treatment (Richman et al, 1994; Havlir et al, 1995).

Different phenotypes of HIV, e.g. syncytium-inducing (SI) versus non-syncytium-inducing (NSI) strains, may also display different kinetics of appearance of drug resistance, possibly due to differential viral turnover. However, pretreatment HIV isolates commonly display a narrow range of IC_{50} values, regardless of disease stage. This is because drug resistance is due to selective pressure exerted by the compounds and because resistant variants initially exist as a minority within the quasispecies. Hence resistant viruses generally cannot be detected in drug susceptibility assays in vivo in the case of treatment-naïve patients. In some cases, however, substitutions in RT may reversibly convert to wild-type after long-term replication in the absence of AZT, both in vivo and in vitro, and quickly revert to mutated forms under drug pressure (St Clair et al, 1991).

It is not surprising, therefore, that many virologists and clinicians have recommended that therapy be initiated early, so that viral replication can be curbed when the quasispecies are less diverse (Ho, 1995). Nucleoside monotherapy with RT inhibitors commonly reduces viral replication to only about 10% of usual levels, making it likely that resistant isolates will emerge.

Factors other than selective pressure can also determine how quickly resistance emerges. These include faster viral replication rates that lead to larger viral populations. Since the entire quasispecies will grow faster in patients with advanced disease and weakened immune systems, such individuals may be especially prone to develop drug resistance.

MOLECULAR BASIS OF HIV RESISTANCE TO ANTIVIRAL NUCLEOSIDES AND NNRTIs

General considerations

Resistance of HIV to nucleoside analogs and NNRTIs is associated with the appearance of certain mutations in RT. Many mutations in RT, once identified by sequencing of the RT of drug-resistant variants, have been shown by site-directed mutagenesis to encode HIV resistance to both nucleoside analogs and NNRTIs (Larder and Kemp, 1989; Nunberg et al, 1991; Richman et al, 1991b, 1994; St Clair et al, 1991; Fitzgibbon et al, 1992; Gu et al, 1992, 1994b; De Clercq, 1994; Shirasaka et al, 1995). Mutations that confer resistance to NNRTIs are found in residues that make contact with these compounds within their binding pocket, as shown by crystallographic and chemical studies (Kohlstaedt et al, 1992; Nanni et al, 1993; De Clercq, 1994; Ding et al, 1995; Ren et al, 1995; Arts and Wainberg, 1996).

Mutated RTs can be expressed in *E. coli* systems and purified. Mutated RTs that contain relevant mutations have often been shown to have reduced sensitivity to both nucleoside analog triphosphates, i.e. the intracellular active form of nucleoside drugs, and to NNRTIs in cell-free enzyme assays (Loya, et al, 1994). Mutations that encode resistance to nucleoside analogs are found in different regions of RT. This is probably because nucleoside incorporation is a complex process which involves recognition/binding, formation of a phosphodiester bond and translocation. Also, the incorporation of a modified nucleotide may affect the dissociation of RT and elongated template/primer. While the mechanism of action through which each individual mutation works is not well understood, mutations affecting any of the above steps may lead to decreased sensitivity of RT to nucleoside analogs. Table 14.1 provides a summary of prominent mutations that encode resistance to nucleoside analogs. Crystallographic analysis suggests that HIV-1 resistance to nucleoside analogs is mostly conferred by residues in the p66 subunit (Tantillo et al, 1994). Cross-resistance among nucleoside analogs and NNRTIs has not been widely reported, suggesting that these mutations act in specific ways. A fuller listing of RT mutations that encode resistance to nucleoside analogs and NNRTIs, as well as resistance-conferring substitutions in the protease gene, can be found elsewhere (Schinazi et al, 1996).

Multimutational basis of resistance to AZT

Patterns of resistance to AZT are distinct from those of other nucleoside agents since decreased sensitivity to ddI, ddC, 3TC and d4T are commonly encoded by single mutations. Decreased susceptibility to AZT is associated with a variety of mutations in RT, including those first described at codons 41, 67, 70, 215 and 219. Ability to detect resistance to AZT in clinical samples normally requires the accumulation of at least two mutations.

These differences in resistance patterns may be attributable to the fact that AZT, unlike 2′,3′-dideoxycytidine (zalcitabine, ddC), 2′,3′-dideoxyinosine (didanosine, ddI) and 2′,3′-didehydro-2′,3′-dideoxythymidine (stavudine, d4T), is a deoxynucleoside derivative in which a 3′-hydroxyl moiety has been replaced by an azido. This fact probably gives AZT different characteristics of inhibition and resistance from other analogs. A series of five substitutions in RT, i.e. methionine→leucine (M41L), aspartic acid→asparagine at codon 67 (D67N), lysine→arginine at codon 70 (K70R), threonine→tyrosine/phenylalanine at residue 215 (T215Y/F), and arginine→glutamine at amino acid 219 (K219Q), were shown to occur individually or in combination in HIV isolated from patients treated with AZT for over 1 year. Site-directed mutagenesis, in which the wild-type amino acids were replaced by those present in resistant isolates, confirmed that these mutations were responsible for AZT resistance in tissue culture (Larder and Kemp, 1989). These five mutations may act additively or synergistically to confer varying levels of AZT resistance (Larder and Kemp, 1989; Kellam et al, 1992).

These mutations seem to appear in ordered fashion, with increasing levels of resistance related to the manner in which the mutations accumulate (Larder et al, 1991; Boucher et al, 1992b). This is consistent with observations that increased resistance to zidovudine occurs gradually, with the K70R substitution often appearing first followed by T215Y or F, or D67N, M41L and K219Q (Boucher et al, 1992b; Kellam et al, 1992). The most commonly observed mutations in regard to resistance to AZT are T215Y and

Table 14.1 Prominent mutations that encode HIV resistance to nucleoside analogs

Position	Amino acid		Location of mutation in RT crystal structure	Drugs against which resistance has been demonstrated using:	
	Wild-type	Mutant		Viral replication assay	Cell-free RT assay
41	Met (M)	Leu (L)	αA	AZT	AZTTP
62	Ala (A)	Val (V)	β3	AZT, ddI, ddC d4T, ddG	ddCTP, 3TCTP, ddATP, PMEApp, AZTTP
65	Lys (K)	Arg (R)	β3–β4 loop	ddC, 3TC, ddI, PMEA	ddCTP, 3TCTP, ddATP, ddITP, PMEApp, AZTTP
67	Asp (D)	Asn (N)	β3–β4 loop	AZT	
69	Thr (T)	Asp (D)	β3–β4 loop	ddC	
70	Lys (K)	Arg (R)	β3–β4 loop	AZT	
74	Leu (L)	Val (V)	β4	ddI, ddC	ddATP, ddCTP, AZTTP
75	Val (V)	Thr (T)	β4	d4T	
75	Val (V)	Ile (I)	β4	AZT, ddI, ddC, d4T, ddG	AZTTP
77	Phe (F)	Leu (L)	β4	AZT, ddI, ddC, d4T, ddG	AZTTP
89	Glu (E)	Gly (G)	β5a	NOa	ddGTP
116	Phe (F)	Tyr (Y)	αC	AZT, ddI, ddC, d4T, ddG	AZTTP
151	Gln (Q)	Met (M)	β8–αE loop	AZT, ddI, ddC, d4T, ddG	AZTTP
184	Met (M)	Val/Ile (V/I)	β9–β10 turn	ddI, ddC, 3TC	3TCTP
215	Thr (T)	Tyr/Phe (Y/F)	β11a	AZT	
219	Lys (K)	Gln (Q)	β11b	AZT	

AZT, azidothymidine (zidovudine); AZTTP, AZT 5′-triphosphate; ddATP, 2′,3′-dideoxyadenosine triphosphate, 2′,3′-dideoxyxytidine (zalcitabine); ddCTP, ddC triphosphate; ddG, 2′,3′-dideoxyguanosine; ddGTP, ddG triphosphate; ddI, 2′,3′ dideoxyinosine (didanosine); d4T, 2′,3′-didehydro-2′,3′-dideoxythymidine (stavudine); PMEApp, 9(2-phosphonylmethoxyethyl)adenine diphosphate; 3TC, 2′,3′-dideoxy-3′-thiacytidine (lamivudine); 3TCPP, 3TC triphosphate.

K70R (Kellam et al, 1992). It is interesting that these mutations only show cross-resistance to other 2',3'-dideoxynucleoside analogs that contain a 3'-azido motif, e.g. 3'-azido-2',3'-dideoxyguanosine (AZG), but not to other 2',3'-dideoxynucleoside analogs, e.g. ddC, ddI (Larder et al, 1990).

Resistance to AZT commonly begins within 2 months of initiation of therapy and may persist for longer than 1 year after the termination of treatment. As mentioned, the 70 mutation usually appears first but is commonly superseded by the 215 mutation. The 41 mutation then appears, and with continued treatment the 70 mutation reappears, followed by mutations at codons 67 and 219. Initially, the site 70 mutation was perceived to be transient and only to facilitate the appearance of other substitutions, but it now appears that this change alone may suffice to account for reduced sensitivity to AZT, as measured by viral burden levels in plasma (D'Aquila et al, 1995). Disease stage can strongly predict the rate at which resistance to AZT will emerge. After 1 year of treatment, AZT-resistant isolates were found in 89% of subjects with less than 100 CD4 cells per μl, but in only 41% and 27% of those with counts of 100–400 cells per μl, and above 400 cells per μl, respectively. Eighty-nine per cent of patients with symptomatic disease had resistant isolates after 1 year of AZT therapy compared with 31% of asymptomatic patients (D'Aquila et al, 1995).

Several studies strongly support the correlation of AZT resistance with disease progression (D'Aquila et al, 1995). First, a Canadian study that followed 72 initially asymptomatic patients on AZT monotherapy showed that about 60% of these individuals developed resistance to the drug over 4 years (Montaner et al, 1993). One study, ACTG 116B/117, studied the effects of switching to ddI among patients with advanced disease who had received AZT for at least 16 weeks (Kahn et al, 1992; D'Aquila et al, 1995). It was found that disease progression or death was delayed among patients who switched to ddI from AZT. Similar results were reported in a Canadian study performed with relatively asymptomatic patients followed over 2 years (Montaner et al, 1995). In both studies, patients with HIV that was resistant to AZT were more than twice as likely to progress to a new AIDS-defining event or death than patients with wild-type isolates. Finally, disease progression was observed most commonly in individuals who harbored drug-resistant viruses at the same time. In one study, high-level resistance to AZT was found in only 7% of patients with high CD4 counts but in 28% of patients with CD4 counts below 50 cells per μl (D'Aquila et al, 1995).

These studies also showed that clinical benefit in the ddI group was not restricted to those with AZT-resistant virus, but that high-level AZT resistance was predictive for accelerated disease progression or death, independent of other factors. In addition, switching from AZT to ddI was beneficial, even in subjects who did not have high-level resistance to the former drug. Patients benefited by switching to ddI regardless whether they had only one or both of the codon 41 and 215 mutations (D'Aquila et al, 1995). The Canadian study also showed that CD4 count increased in those who switched to ddI and that patients continuing AZT therapy were more likely to have diminished CD4 counts (Montaner et al, 1995). Resistance at baseline to AZT was associated with a subsequent drop in CD4 count. Only low-level resistance to ddI was seen in patients who switched to that drug, while resistance to AZT increased progressively in patients who remained on AZT monotherapy. In contrast, resistance values for AZT remained stable or diminished after switching to ddI.

The mechanism of resistance to AZT is unclear. Amino acids 41, 67 and 70 are located in the fingers subdomain of RT, and residue 41 is located on strand αA. Amino acids D67 and K70 are found on the β3–β4 connecting loop. Although the RT crystal structure shows that the latter region is not involved in binding of dNTP substrates or ddNTP inhibitors (Nanni et al, 1993), enzyme kinetic work showed that this region might play a role in interactions between RT and dNTP/ddNTPs. Amino acid 219 is located on strand β11b of the palm region of the p66-subunit of RT. The K219Q mutation may directly affect the binding of nucleoside substrates/inhibitors and indirectly affect interactions of RT with template/primer (Tantillo et al, 1994; Arts and Wainberg, 1996). However, T215, located on strand β11a, does not have a direct effect on binding of dNTP/ddNTP.

Enzyme kinetic analysis, using homopolymeric template/primer, revealed that neither E. coli-derived recombinant HIV-1 RT containing these mutations nor mutated RT purified from AZT-resistant virus had significantly different catalytic efficiency for natural substrate dNTPs, or altered recognition efficiency for either AZT 5′-triphosphate (AZTTP) or other ddNTPs compared with wild-type enzymes (Larder and Kemp, 1989; Wainberg et al, 1990; Lacey et al, 1992). Nor were differences in chain termination efficiency observed between wild-type and mutated RTs in chain elongation/dNTP incorporation assays using synthetic heteropolymeric template/primer (Lacey et al, 1992). In contrast, increased K_m values for template/primer were observed with AZT-resistant RT (Pokholok et al, 1993). In addition, HIV-1 containing these mutations had increased replication potential or fitness in unstimulated peripheral blood mononuclear cells (PBMCs) than wild-type virus, and recombinant mutated RTs displayed higher processivity than wild-type RT (Caliendo et al, 1995). These considerations suggest that mutated RTs may have had altered interaction with template/primer.

Resistance to lamivudine (3TC)

A methionine→valine substitution at RT codon 184 (M184V) was initially identified in culture-selected ddI-resistant isolates (Gu et al, 1992) and later shown to be present as well in viruses resistant to ddC and 3TC (Boucher et al, 1993; Gao et al, 1993a, 1994; Schinazi et al, 1993; Tisdale et al, 1993). This substitution differs from others that encode resistance to nucleoside analogs, being characterized by:

1. High-level resistance to 3TC (about 1000-fold increased IC_{50}) but low-level resistance to both ddI and ddC – this is the only mutation known to encode as much as 1000-fold resistance to a single drug;
2. Rapid appearance in both cell culture and in patients receiving therapy with 3TC but not ddI or ddC (Boucher et al, 1993; Tisdale et al, 1993; Kavlick et al, 1995; Schuurman et al, 1995);
3. Location in or close to the polymerase active site (Wakefield et al, 1992);
4. About 50-fold diminished sensitivity to 3TC 5′-triphosphate (3TCTP) on the part of recombinant M184V-containing RT, but only insignificant differences when compared with wild-type RT in regard to K_m and V for ddCTP (Faraj et al, 1994; Quan et al, 1996).

Patients on prolonged therapy with 3TC had low virus burden in plasma in spite of the appearance of phenotypically resistant virus containing the M184V mutation (Schuurman et al, 1995). A second mutation, i.e. methionine to isoleucine at the same residue

(M184I), also confers about 1000-fold resistance to 3TC (Boucher et al, 1993). The latter substitution has been observed in 3TC treatment prior to M184V but is overwhelmed in patient samples by the latter after several weeks. Viruses containing M184V can grow faster, i.e. have greater fitness, than viruses containing the 184I substitution (Larder et al, 1995; Back et al, 1996).

Amino acid M184 is located in the center of a highly conserved YMDD motif (residues 183–186) of RT, in which two DD residues are thought to be involved in polymerase activity (Gu et al, 1992; Gao et al, 1993a; Schinazi et al, 1993); The M184V substitution seems to cause diminished RT processivity and also results in decreased viral fitness in comparison with wild-type virus (Faraj et al, 1994; Larder et al, 1995; Back et al, 1996; Quan et al, 1996). Both ddC and 3TC are 2',3'-dideoxy derivatives of deoxycytidine that differ in that the 3'-carbon of the pentose ring is replaced by a sulfur in the case of 3TC. The M184V mutation can result in more than 1000-fold decreased sensitivity to 3TC but only a five-fold resistance to ddC. The sugar moiety must therefore play an important role in interactions between nucleoside analogs and RT.

The molecular basis of resistance encoded by M184V is a subject of debate. Although M184V-mutated RT had around 50-fold diminished sensitivity to 3TCTP, no alterations in recognition of dNTP substrates were observed (Faraj et al, 1994; Quan et al, 1996). Crystal structure observations have reported that residue M184 is involved in contact with the template/primer but not in the binding of dNTPs. The M184V substitution caused neither changed recognition of template/primer nor altered polymerase function (Quan et al, 1996). However, M184V-mutated RT may have three-fold to five-fold resistance to ddC 5'-triphosphate and more than 200-fold resistance to 3TCTP in endogenous RT assays, especially with longer DNA products (Quan et al, 1996). Thus, M184V results in altered RT sensitivity to ddNTPs. While synthesis of short DNA products did not reveal significant differences between M184V and wild-type RTs, the use of longer templates did. In this context, the utility of HIV-1 genomic RNA as a template in endogenous RT assays may be essential to show that M184V RT has decreased processivity.

The RNA copy number in the plasma of HIV-1-infected individuals treated with 3TC remained low despite the appearance of M184V-mutated virus (Schuurman et al, 1995). It may be difficult to select for resistance to AZT when studying M184V cloned virus (Larder et al, 1995; Wainberg et al, 1996). The M184V substitution may cause conformational change of the polymerase catalytic site, causing increased fidelity of the enzyme, which may limit genotypic variation in some cases (Preston et al, 1988; Temin, 1993; Pandey et al, 1996; Wainberg et al, 1996). This may be important in patients with early-stage disease in whom host immune responses can act to keep viral load in check.

The rapid emergence of high-level resistance to 3TC is correlated with a substantial rebound in initially suppressed viral load (Schuurman et al, 1995). However, combinations of 3TC and AZT consistently produced higher and more sustained increases in CD4 counts and greater reductions in viral load than either agent alone (Eron et al, 1995; Katlama et al, 1996; Staszewski et al, 1996). Treatment with 3TC plus AZT also produced higher increases in CD4 count than did ddC plus AZT, but viral load reductions were equivalent with the two regimens.

Viral isolates in these studies remained susceptible to AZT but not to 3TC after 24 weeks of therapy. The 184V mutation was apparently able to resensitize virus containing the 215 substitution to AZT in these clinical studies, consistent with results obtained earlier with AZT and ddI. The sustained antiviral effect of AZT plus 3TC in clinical trials may be due in part to interactions between the 184 and 215 mutations that render AZT-resistant viruses sensitive to AZT. However, another explanation is that viruses containing the M184V mutation replicate more slowly than wild-types (Larder et al, 1995; Back et al, 1996). In addition, replication of the 184 mutant virus may still be partially suppressed by 3TC. Certain drug combinations may be effective because they pressure the virus into mutating into multiresistant variants that are less replication-competent or fit than wild-type viruses.

If the poor replication capacity of 184 mutant virus is responsible for much of the antiretroviral effect of 3TC plus AZT, a similar benefit might be conferred by combining 3TC with other drugs. Clinical trials in progress are evaluating 3TC plus ddI or d4T, both in the presence and absence of protease inhibitors. Viruses resistant to both 3TC and AZT have also emerged in both clinical trials and tissue culture. Such doubly resistant viruses have also been generated by recombination (Gu et al, 1995b).

MUTATIONS ENCODING RESISTANCE TO OTHER NUCLEOSIDE ANALOGS

Resistance to ddC

Mutations in RT that confer resistance to other nucleosides, including ddI, ddC and d4T, are commonly single amino acid alterations. Many of these mutations are located on the β3–β4 connecting loop and β4 strand of the p66 fingers subdomain (Nanni et al, 1993; Tantillo et al, 1994). None of them encodes cross-resistance to AZT but they do against other 2′,3′-dideoxynucleoside analogs in tissue culture assays. In addition, these mutations encode only moderate levels of drug resistance, and cause decreased sensitivity of recombinant RT enzymes to nucleoside analog triphosphates in cell-free RT assays. They also have a low rate of appearance in viral samples obtained from patients on prolonged drug therapy.

A codon 65 mutation, i.e. lysine→arginine (K65R), in RT was identified in culture-selected ddC-resistant HIV variants. This mutation also confers cross-resistance among ddI, 3TC and 9(2-phosphonylmethoxyethyl)adenine (PMEA) (Gu et al, 1994b, 1995d; Zhang et al, 1994). The K65R substitution has been observed in viral samples from patients on prolonged treatment with either ddC or ddI (Gu et al, 1994b; Zhang et al, 1994). It is located in the center of a conserved IKKK (residues 63–66) motif within the β3–β4 connecting loop of the fingers subdomain of RT.

A substitution (threonine→aspartic acid) at codon 69 (T69D) was shown to encode resistance to ddC in viruses isolated from a patient after prolonged treatment with this drug. It is not clear whether this mutation encodes cross-resistance to other compounds.

The mechanism of drug resistance in regard to the K65R mutation has been extensively studied. Enzyme kinetic assays demonstrated that recombinant mutated RT, containing the K65R substitution, increased K_m values for dCTP and dATP but not dTTP and dGTP, in comparison with wild-type RT. ('d' designates 'deoxy', which is an abbreviation for 'deoxyribose', i.e. the normal sugar motif found in nucleoside bases that are present in DNA. In contrast, 'dd' designates 'dideoxy', the altered structure commonly found in nucleoside drugs used to combat viral replication.) This research was performed using either synthetic homopolymeric template/primers or a heteropolymeric template derived from HIV-1 genomic RNA and containing primer binding sequences, repeated (R) and 5' unique (U5) regions. The V/K_m ratio for dctp and datp, a measure of catalytic efficiency, was increased about two-fold to three-fold. In addition, the K_i and K_i/K_m ratio, used to normalize changes in substrate recognition of the K65R mutant enzyme, were increased 10- to 20-fold for each of ddCTP, 3TCTP and ddA 5'-triphosphate (ddATP, the intracellular active form of ddI) in comparison to wild-type enzyme (Gu et al, 1994a). Hence, K65R has a selectively altered recognition of dNTPs/ddNTPs corresponding to HIV resistance to these antiviral nucleosides. The K65R mutated RT also showed an increased K_i for AZTTP and may be indirectly or directly involved in interactions with dNTPs/ddNTPs during polymerization. This may help to explain why this mutation shows cross-resistance to different nucleoside analogs.

Reverse transcriptase polymerase activity is competitively inhibited by dideoxynucleoside analog triphosphates, in a manner restricted to their homologous dNTPs and complementary template (Gu et al, 1994a). Therefore, chain termination must be the major means whereby nucleoside analogs inhibit RT polymerase activity. Chain elongation/dNTP incorporation analysis also showed that K65R-mutated RT had decreased sensitivity to a number of nucleoside analog triphosphates, including ddCTP, 3TCTP, ddATP, ddI 5'-triphosphate (ddITP), PMEA diphosphate (PMEApp) as well as AZTTP, in reactions that employed the HIV-1 genome-derived template and either deoxyoligonucleotide or tRNA$^{Lys.3}$ (Gu et al, 1995a,d) as a cognate primer to initiate RT reactions.

The K65R-mutated RT also had altered chain termination frequencies at different complementary bases on the RNA template compared with wild-type RT (Gu et al, 1995a). This is possibly related to the secondary structure of the template used as well as to pausing. In addition, although K65R mutant RT had similar processivity to wild-type, pausing frequencies were diminished at certain sites with increased concentrations of ddNTPs in comparison with wild-type enzyme. This could also be related to drug resistance encoded by K65R. Regions of decreased pausing and diminished frequency of chain termination may also play a role in K65R-mediated resistance to nucleoside analogs, i.e. K65R may directly or indirectly cause changes in interactions between template and RT.

Resistance to ddI

Resistance to ddI is usually moderate and of the order of a five- to ten-fold increase in IC_{50} value, i.e. the concentration of drugs needed to inhibit viral replication by 50%. Furthermore, the presence of the L74V mutation, believed to be the principal resistance-conferring mutation for ddI, is not always correlated with resistance to this drug. It is still not known to what extent resistance to ddI is of clinical significance (Abrams et al, 1994; Antonelli et al, 1994).

Other mutations associated with decreased sensitivity to ddI are at codons 184, 75, 69 and 65. As stated, the 184 mutation also confers rapid high-level resistance to 3TC and is associated with a loss of susceptibility to ddC. The 65 and 69 mutations are also shared with ddC while the 75 mutation is shared with d4T.

When resistance to ddI was evaluated in patients who were first treated with AZT before switching to ddI, a mutation at codon 74 was identified that apparently conferred a six- to 26-fold decrease in ddI susceptibility after 6–12 months of treatment. In addition, sensitivity to AZT was restored during such therapy (St Clair et al, 1991).

Among other mutations that confer resistance to one drug while restoring sensitivity to AZT are those at codons 65, 74, 181 (which confers resistance to NNRTIs) and 184.

As stated, resistance to ddI emerges slowly during therapy and is modest in extent. However, the rate at which ddI resistance emerges is probably dependent on disease stage. In AZT-experienced patients who switched to ddI, the 74 mutation was detected as early as 8 weeks after commencement of ddI in patients with CD4 counts in the 100–150 cells per μl range who had a relatively high viral burden (Reichman et al, 1992, 1993). Virus replication is more extensive in patients with high viral burden, accounting for the faster generation of relevant mutations. In contrast, resistance to ddI was infrequent in patients with less advanced disease. Only one case of high-level ddI resistance was found in one study among 50 patients with a median CD4 count of 320 cells per μl who were switched to ddI from AZT monotherapy (Montaner et al, 1995).

A mutation at residue 74 (L74V) encodes resistance to ddI and cross-resistance to ddC and has been observed in tissue culture as well as in patients who received ddI therapy (St Clair et al, 1991). Before ddI, the patients had received AZT. Resistance to AZT was decreased with the concomitant appearance of resistance to ddI. Sequencing showed that these viruses contained both the L74V and T215Y substitutions, associated with AZT resistance. Combinations of L74V and T215Y within a recombinant HIV-1 clone resulted in reversal of resistance to AZT but not to ddI or ddC (St Clair et al, 1991), indicating the complexity of interactions among RT subdomains. Amino acids L74 and T215 are located in the β4 strand of the fingers subdomain and the β11a of the palm subdomain respectively, and are far from each other in the RT crystal structure (Tantillo et al, 1994). How L74V could reduce AZT resistance induced by the T215Y mutation is unknown.

Cell-free reverse transcriptase assays

Mutant RT containing the L74V substitution also has selectively altered recognition of dNTP/ddNTP. Altered recognition of natural substrate deoxynucleotides and dideoxynucleotide inhibitors, mediated by these mutations, helps to explain the molecular basis of resistance to these drugs.

In addition, L74V-mutated RT was found to behave differently depending on the length of templates used. Wild-type RT also showed resistance to nucleoside analog triphosphates if template extension was three nucleotides or shorter. However, when template extension was longer, sensitivity was observed instead. The L74V-mutated RT conferred resistance to both short and long template extensions (Boyer et al, 1994b), indicating that altered interactions between RT and template may partially confer HIV-1 resistance to nucleoside analogs in this case.

While both the K65R and L74V mutant RTs showed increased K_i values for AZTTP in cell-free RT assays, recombinant viruses containing these mutations did not display resistance to AZT in tissue culture assays (Martin et al, 1993; Gu et al, 1994a). This seeming contradiction suggests that the mutated RT may have an altered conformation in cells than in solution. Of course, other factors may also be involved in inhibition and resistance of nucleoside analogs, especially AZT. As an example, HIV-1 nucleocapsid proteins can stimulate RT activity and template switching in a cell-free RT system (Barat et al, 1989; Li et al, 1996).

Mutations encoding AZT resistance have been observed only in cell culture assays, since cell-free RT reactions performed with AZTTP have failed to show resistance (Larder et al, 1989; Lacey et al, 1992; Martin et al, 1993). The 3'-azido moiety of AZT may play an essential role in interactions with the enzyme. At least five other mutations that encode resistance to all known nucleoside analogs, including AZT, have been shown to have diminished sensitivity to AZTTP in cell-free assays performed with viral RT obtained from lysed virus particles (Shirasaka et al, 1995).

Resistance to d4T

A mutation at amino acid 75 (valine→threonine, V75T) has been observed in d4T-resistant viruses selected in cell culture and shown to confer about seven-fold resistance to this compound by molecular mutagenesis (Lacey and Larder, 1994). Although d4T is derived from thymidine, V75T did not display cross-resistance to AZT, but did to other 2',3'-dideoxynucleoside analogs. No enzymatic data are available in regard to mechanisms of resistance encoded by this mutation.

In spite of the existence of the V75T mutation, studies of d4T have failed to link decreased susceptibility to a specific mutation site. Moreover, decreases in susceptibility to d4T have been modest in all reported cases. Studies of patients on d4T have shown a low rate of resistance in viral isolates sampled, e.g. three- to five-fold. In one such study, the site 75 mutation was identified in only one of five isolates that showed decreased susceptibility to d4T after treatment with this compound (Lin et al, 1994). This relatively large study found that long-term monotherapy with d4T resulted in very little loss of sensitivity. The V75T mutation was identified in isolates from only 2 of 32 patients treated with d4T for periods between 6 months and almost 2·5 years. Both patients whose viruses were resistant had only a two-fold decrease in susceptibility to d4T. In contrast, the greatest degree of resistance to d4T was a 12-fold change seen in a patient who did not have the V75T mutation after 1·5 years of monotherapy (Lin et al, 1994).

Interestingly, this same study showed that isolates from 5 of 11 patients showed decreased susceptibility to AZT after monotherapy with d4T, raising the possibility of cross-resistance between the two agents. However, viruses that carry mutations conferring resistance to AZT (codons 67, 70, 215 and 219) remained sensitive to both d4T and ddI.

Conceivably, viruses that are resistant to d4T may uniquely display a broader cross-resistance to other drugs.

Other mutations

In vitro procedures to screen bacterially expressed RTs, which had been mutagenized in *E. coli*, identified a glutamic acid→glycine mutation at codon 89 (E89G) that encoded diminished RT sensitivity to ddGTP, other nucleoside analog triphosphates, and phosphonoformic acid (foscarnet, PFA), a pyrophosphate analog (Prasad et al, 1991). However, in tissue culture, E89G conferred resistance only to PFA but not to ddG or other nucleoside analogs (Kew et al, 1996). Although E89G has also been observed in 3TC-resistant HIV variants selected in cell culture, it did not confer resistance to 3TC in culture and did not alter resistance patterns when combined with M184V within a single construct. Interestingly, a degree of cross-resistance was observed in regard to NNRTIs (Kew et al, 1996). The E89 residue is located on strand β5a in the palm subdomain and is two or three nucleotides downstream of the polymerase active site, at which the template strand makes contact with the enzyme 'template grip' (Tantillo et al, 1994). Mutated RT containing E89G displayed both increased deoxynucleotide triphosphate utilization and reaction velocity and altered recognition of dNTP/ddNTPs (Prasad et al, 1991). Although the effect on utilization of substrates was due to the presence of E89G on the p66 and not the p51 subunit, increased velocity of the mutant RT was conferred by the presence of the mutation in both the p66 and p51 subunits (Kew et al, 1994). As in the case of L74V, E89G-mutated RT also had altered dependence on the length of template used, in comparison with wild-type enzyme. The E89G mutation also led to a loss of divalent cation preference of HIV-1 RT (Prasad et al, 1991).

MULTIDRUG RESISTANCE AND NUCLEOSIDE ANALOGS

Five other mutations in RT, including alanine→valine at codon 62 (A62V), valine→ isoleucine at residue 75 (V75I), phenylalanine→leucine at codon 77 (F77L), phenylalanine→tyrosine at amino acid 116 (F116Y), and glutamine→methionine at codon 151 (Q151M), can also confer resistance to a variety of 2',3'-dideoxynucleoside analogs, including AZT, ddI, ddC, d4T, 3TC and ddG, but not to NNRTIs or PFA (Shirasaka et al, 1995). These substitutions were seen in HIV-1 strains from patients who received combinations of AZT plus ddI or ddC for longer than 12 months. Other mutations associated with resistance to each of AZT, ddI and ddC in monotherapy were not identified in these patients.

Non-appearance of these mutations could be due to limitations of RT structure and function, genotypic features of the viruses induced, or both (Shafer et al, 1994). These mutations appeared in sequential fashion. The substitution Q151M appeared first, in spite of the fact that a double nucleotide substitution was necessary for its generation, followed by F116Y and F77L in patients receiving combination AZT/ddC or AZT/ddI (De Clercq, 1994; Shirasaka et al, 1995). The Q151M single substitution caused low-level resistance to all of the above-mentioned drugs. Combinations of mutations caused cumulative enhanced resistance to all nucleoside analogs that were evaluated.

Amino acids Q151 and F116 are located respectively in the highly conserved A and B motifs of the enzyme. Residue F116 is believed to be located close to the dNTP-binding site. Based on crystal structure, V75, F77 and Q151 are thought to form part of the 'template grip' (Tantillo et al, 1994; Shirasaka et al, 1995) as they are located near the first template base. These residues are close to amino acids E89, L74 and K65 in the RT crystal structure and may simulate the same pattern of drug resistance as the latter mutations. Virus-associated mutated RT, containing these mutations, had increased K_i values for AZTTP in comparison with RT of wild-type virus, but both enzymes behaved with similar catalytic efficiency (Shirasaka et al, 1995).

RESISTANCE TO NNRTIs

Resistance to NNRTIs appears quickly both in culture selection protocols and in patients (Nunberg et al, 1991; Richman et al, 1991b, 1994; Chong et al, 1994; Vandamme et al, 1994); NNRTIs share a common binding site, and mutations that encode NNRTI resistance are located within the binding pocket that makes drug contact (Wu et al, 1991; Kohlstaedt et al, 1992; De Clercq, 1994; Ding et al, 1995; Esnouf et al, 1995; Ren et al, 1995). This explains the finding that extensive cross-resistance is observed among all NNRTIs (De Clercq, 1994), although synergy among NNRTIs has occasionally been reported (Fletcher et al, 1995).

A substitution at codon 181 (tyrosine→cysteine) (Y181C) is a common mutation that encodes cross-resistance among many NNRTIs (Richman et al, 1991b, 1994; Mellors et al, 1992; Balzarini et al, 1993a,c; Byrnes et al, 1993). Replacement of Y181 by a serine or histidine also conferred HIV resistance to NNRTIs (Sardana et al, 1992). This residue is located on the floor of the binding pocket directly involved in interactions with NNRTIs. In contrast, replacement of Y181 by tryptophan or phenylalanine had little effect on resistance to NNRTIs. Aromatic stacking of Y181 may be important for the inhibitory activity of these compounds.

A mutation at amino acid 236 (proline→leucine) (P236L), conferring resistance to a class of NNRTIs, i.e. BHAP compounds, e.g. delavirdine, can diminish resistance to each of TIBO, nevirapine and pyridinones that is encoded by Y181C, if both mutations are present in the same virus (Dueweke et al, 1993). The P236 residue lies on the β3–β4 reverse turn that forms the roof of the NNRTI binding pocket, and is found opposite to Y181 (Smerdon et al, 1994). The enlarged net pocket volume, caused by Y181C, may be decreased by introduction of P236L (Smerdon et al, 1994), causing increased sensitivity to other NNRTIs. Altered chemical interactions between RT and NNRTIs, induced by Y181C, may also cause reversion of viral phenotype.

A substitution at residue 188 (tyrosine→histidine) (Y188H) confers resistance to TIBO, pyridinone and HEPT (Sardana et al, 1992; Balzarini et al, 1993a), while Y188 replaced by cysteine (Y188C) encodes resistance to TIBO, pyridinone and nevirapine (Sardana et al, 1992; Balzarini et al, 1993a; Byrnes et al, 1993). Thus, nevirapine and HEPT apparently interact with RT through different mechanisms.

A mutation at codon 103 (lysine→asparagine) (K103N) is another common substitution encoding resistance to all classes of NNRTIs, except HEPT (Nunberg et al, 1991; Richman et al, 1991b; Balzarini et al, 1993c; Byrnes et al, 1993). The amino acid K103 is located on the β5–β6 connecting loop, distant from the polymerase active site of the enzyme (Kohlstaedt et al, 1992; Jacobo-Molina et al, 1993). The RT crystal structure positions the side-chain of K103 as pointing out at the entrance of the NNRTI binding pocket, thus enabling drug contact to occur. Substitution of K103N results in alteration of interactions between NNRTIs and RT. The K103N mutation shows synergy with Y181C in regard to resistance to NNRTIs (Nunberg et al, 1991), unlike antagonistic interactions involving Y181C and P236L.

A mutation at codon 138 (glutamic acid→lysine) (E138K) confers resistance only to TSAO and some derivatives of TIBO, but not to other NNRTIs (Balzarini et al, 1993a,b,c; Jonckheere et al, 1994). The only substitution that encodes resistance to NNRTIs, contributed by the p51 rather than the p66 subunit, is E138K. This is because E138 of the p51 subunit is located on the connecting loop of β7–β8, adjacent to the

NNRTI binding pocket, and forms part of this site (Nanni et al, 1993). The E138 residue is positioned on the underside of the pocket, and may thus make contact with only some NNRTIs, i.e. TSAO.

A substitution at codon 179 (valine→aspartic acid) (V179D) encodes resistance to HEPT, TIBO and pyridinone (Byrnes et al, 1993; Vandamme et al, 1994) and may also affect binding of NNRTIs by forming a salt bridge with the K103 residue.

Studies have indicated through use of cell-free assays that NNRTIs may act additively or synergistically with nucleoside compounds to block HIV RT activity (Gu et al, 1995c). This confirms earlier results on the combination of NNRTIs with AZT obtained in tissue culture (Richman et al, 1991a; Koup et al, 1993). Inhibition of viral replication by combinations of NNRTIs and either protease inhibitors or cytokines has also been reported (Rusconi et al, 1994; Pagano and Chong, 1995).

It is interesting that both V179D and E138K can cause resistance to two distinct TIBO compounds, R82913 and R86183. The opposite finding was obtained when E138K was present in the p51 subunit (Boyer et al, 1994a). With the exception of the fact that R82913 is a 9-chloro derivative, while R86183 is a 8-chloro derivative, these two compounds are identical. In the crystal structure, V179 of p66 is located close to E138 of p51 (Nanni et al, 1993). Hence, the chloro group may be involved in interaction of drugs with E138 and/or V179.

Resistance to NNRTIs is also observed in cell-free enzyme assays (Sardana et al, 1992; Boyer et al, 1993; Byrnes et al, 1993; Loya et al, 1994). Both Y181I and Y188L mediated decreased sensitivity to NNRTIs without affecting either substrate recognition or catalytic efficiency, supporting the idea that resistance to NNRTIs is attributable to diminished ability of these drugs to be bound by RT. A mutation at L100I caused a 15-fold decrease in sensitivity to pyridinone derivatives, while a two-fold increase in IC_{95} was observed in tissue culture studies. However, L101E had only a minimal effect on RT in cell-free assays in regard to sensitivity to pyridinones but a 10-fold increase in IC_{95} in tissue culture (Byrnes et al, 1993). The L100 residue is located on the same β5–β6 connecting loop as K103 (Nanni et al, 1993). Therefore, a variety of factors contribute directly or indirectly to interactions between NNRTIs and RT.

MULTIDRUG RESISTANCE AND COMBINATION THERAPY

When nucleoside analogs and/or NNRTIs, or drugs directed at other targets, are used in combination, synergistic inhibition of HIV replication both in cell culture and in vivo is commonly the result. Synergy between different drugs in cell culture and in vivo may be due, in part, to the distinctive metabolic pathways involved, at least in the case of dNTPs, since NNRTIs are not metabolized. However, only additive effects re inhibition of HIV-1 RT polymerase activity were seen in cell-free RT assays when nucleoside analogs and NNRTIs were combined (White et al, 1993; Gu et al, 1995c). This suggests that different drugs may act in different ways to inhibit HIV replication in infected cells as distinct from antagonism of RT polymerase activity in cell-free reactions.

Additivism and synergy between AZT and other RT inhibitors could also involve inhibition of RT RNAase H and integrase (IN) activities mediated by metabolites of ddNTPs (Mazumder et al, 1994). The efficiency of inhibition may partially be dependent on the state of HIV-1 infection and the state of activation of the cells themselves.

This is because phosphorylation of nucleoside analogs to their active triphosphate forms may occur faster in some cell types than others and in activated versus latently infected cells. The simultaneous use of different nucleoside analogs, e.g. AZT and ddI, had a more positive effect on patients than did alternating therapy, consistent with the fact that the drugs used were metabolized differently and did not interfere with each other's phosphorylation pathways.

While combination therapy may be a means of preventing or delaying drug resistance, the acquisition of multidrug resistance following combination treatment has been demonstrated both in vivo and in vitro in each of alternating and simultaneous therapy protocols (Eron et al, 1993; Larder et al, 1993; Gao et al, 1994; Shirasaka et al, 1995). Multiple resistance may even be generated more easily following alternating rather than simultaneous use of different drugs (Gao et al, 1994). Synergy among different drugs, leading to more effective inhibition of replication, could account for this observation. Multiply resistant viruses have combinations of mutations that encode resistance to each single drug or may possess different mutation patterns (Eron et al, 1993; Larder et al, 1993; Gao et al, 1994; Shirasaka et al, 1995). Whether generation of different mutations for multidrug resistance may be related to different combinations of drug and regimens, e.g. simultaneous or alternative, or to genotypic features of the viruses involved, is unknown. Multidrug resistance may also be acquired through fusion of host cells which are infected with different drug-resistant viruses, to generate virus particles containing chimeric genomic strands of RNA (Gu et al, 1995b). This could lead to genomic recombination through strand transfer during reverse transcription. Viruses that are doubly resistant to AZT and 3TC and to dNTPs and protease inhibitors have been generated through recombination protocols.

The differential presence of drug resistance-conferring mutations in RT may yield distinct phenotypes. For instance, HIV that contained both L74V and mutations conferring AZT resistance had increased sensitivity to AZT but not to ddI. Combinations of M184V with T215Y plus M41L also had decreased resistance to AZT but not to ddI or 3TC (Larder et al, 1993; Tisdale et al, 1993). Viruses that contained the M41L and T215Y and/or L74V mutations with a NNRTI resistance mutation (V106A) showed no change in resistance to AZT, ddI or nevirapine. In contrast, resistance to AZT was eliminated if the viruses also contained M184V (Larder et al, 1993). Finally, the combination of M41L, L74V and T215Y together with Y181C, encoding resistance to most NNRTIs, resulted in near-elimination of resistance to AZT without affecting resistance to nevirapine and ddI.

Recombinant viruses containing both the M184V and K65R or V75T mutations did not show significantly different drug resistance in comparison with viruses containing only single substitutions (Gu et al, 1994b; Lacey and Larder, 1994). Reactions performed with recombinant RT, containing both the M184V and K65R mutations, displayed similar chain termination characteristics to K65R RT alone (Gu et al, 1995a).

While the coexistence of mutations conferring AZT resistance and other substitutions may eliminate AZT resistance, combinations of AZT and ddI or NNRTIs delayed appearance of ddI- and NNRTI-resistant viruses in patient samples, without affecting the development of AZT resistance in either alternating or simultaneous treatment protocols (Kojima et al, 1995; Staszewski et al, 1995). The fact that AZT phosphorylation is dependent on cell activation, while that of ddI is not, may play a role in this observation

(Gao et al, 1993b). The HIV-1 population in plasma is probably mostly produced by activated CD4+ T lymphocytes. The combination of either L74V or mutations associated with NNRTI resistance, with mutations responsible for resistance to AZT, led to increased AZT sensitivity. This suggests that activated CD4+ cells that are infected with viruses containing L74V or substitutions conferring NNRTI resistance may be inhibited by AZT.

Human immunodeficiency virus may not be able to generate resistance to AZT if the latter is combined with 3TC (Larder et al, 1995). The substitution M184V has also been shown to increase enzyme fidelity (Pandey et al, 1996; Wainberg et al, 1996), a result attributable to its location close to the polymerase catalytic site of RT (Jacobo-Molina et al, 1993; Nanni et al, 1993). The M184V substitution may change enzyme conformation, leading to increased accuracy of binding to and/or incorporation of dNTPs. The M184V substitution also causes decreased RT processivity and viral replication. This helps to explain the clinical success of combination AZT–3TC therapy, even after the appearance of the 184V mutation and high-level resistance to 3TC.

CLINICAL SIGNIFICANCE OF HIV DRUG RESISTANCE

Treatment with nucleoside analogs has proved beneficial in decreasing viral burden in plasma, increasing CD4+ T lymphocyte counts, and reducing incidence of opportunistic disease. Resistance to both nucleoside analogs and NNRTIs has been correlated with clinical progression.

Viral drug resistance is associated with increased HIV load in plasma in many studies (Boucher et al, 1992a; Kozal et al, 1993; Montaner et al, 1993; Shirasaka et al, 1995). While some individuals had mostly drug-resistant viruses in their plasma after treatment, viral titer was not always increased. Sometimes, high levels of viremia correlated with the appearance of drug-resistant variants (Mohri et al, 1993). With regard to multiple mutation-encoded drug resistance, IC_{50} values in culture did not always correlate with viral RNA measurements in plasma. The presence of a first mutation, e.g. K70R or Q151M, which usually confers low-level resistance to nucleoside analogs, was accompanied by an increase in viral burden over pretreatment levels (Shirasaka et al, 1995). In contrast, the generation of viruses containing multiple resistance-associated mutations, conferring synergistic or additive effects, did not lead to a further increase in viral RNA copy number (Kellam et al, 1992; Shirasaka et al, 1995). These results may reflect differences in disease stage, phenotypic and genotypic features of individual viruses, and the state of immune responsiveness in the patients who were followed. Combinations of M41L and T215Y in AZT-resistant viruses may carry a significantly greater risk for progression and death than wild-type and T215Y singly mutated viral variants (Japour et al, 1995). The presence of the SI phenotype may also correlate with a significantly greater decline in CD4+ count during treatment with AZT than is seen in patients with NSI variants (Kozal et al, 1994).

In children with AZT-resistant viruses, disease progression was apparently more rapid in those with AZT-sensitive virus, as demonstrated by decline in CD4+ cell counts, opportunistic infection and failure to thrive (Tudor-Williams et al, 1992; Ogino et al, 1993; Nielsen et al, 1995). Children with low CD4+ counts were also more likely to harbor AZT-resistant isolates, probably reflecting high viral load and disease progression in individuals with low CD4+ counts. Children on AZT monotherapy also had more evidence of disease progression than those who received combination of drugs.

The appearance of M184V-substituted viruses in both adults and children who received 3TC was not always accompanied by increased viral load and decline in CD4+ T cells. Nor was cross-resistance to 3TC and other nucleoside analogs and/or NNRTIs observed in culture, when 3TC was combined with other drugs. As stated above, HIV containing M184V grows less rapidly than wild-type viruses (Larder et al, 1995; Back et al, 1996).

OTHER CONSIDERATIONS

Other considerations in regard to resistance include the concept of the 'genetic barrier'. One reason that high-level resistance to AZT takes long periods (6 months to 2 years) to develop may be the number of mutations that must be present and selected for under drug therapy. The same is true in regard to certain protease inhibitors, e.g. indinavir. In contrast, resistance might be expected to develop much more rapidly in the case of resistance conferred by single substitutions, e.g. M184V and 3TC. In some cases, plasma drug levels are important and may be higher than the susceptibility levels of any resistant viruses that are present.

An additional consideration is that drug-resistant viruses may be responsible for some cases of primary HIV infection. Presumably, both sexual transmission of drug-resistant variants and transmission of such viruses by users of intravenous drugs may occur. The extent to which such viruses may be involved in cases of primary infection may be expected to increase, in view of the high percentage of viral isolates that display resistance to various antiviral drugs. It is apparent that issues of compliance in regard to therapeutic regimens are rapidly becoming important, since incomplete suppression of viral replication, i.e. suboptimal antiviral pressure, may be expected to lead to drug resistance in virtually all cases. This subject is of obvious importance in regard to public health.

To overcome or prevent resistance, it will probably be necessary to suppress viral replication. As an example, resistance mutations were less likely to be present in patients who took AZT plus ddI, and who had sustained reductions in viral load, than in individuals on the same regimens but who had only transient reductions in viral load (Shafer et al, 1995). Resistant viruses cannot outgrow the wild-type population, which is also the most fit, in the absence of drug pressure. Antiretroviral therapy will select for outgrowth of mutated viruses and increased selective pressure may, therefore, increase the likelihood that resistance will occur. However, if viral replication is shut down through use of effective combination therapeutic regimens, the possibility of resistance will be reduced, since non-replicating viruses cannot mutate into resistant forms in the first place.

More efficient suppression of virus replication is now being achieved through use of combinations of drugs, including nucleosides, NNRTIs and protease inhibitors.

REFERENCES

Abrams DI, Goldman AI, Launer C et al (1994) A comparative trial of didanosine or zalcitabine after treatment with zidovudine in patients with human immunodeficiency virus infection. N Engl J Med **330:** 657–662.

Antonelli G, Dianzani F, Bellarosa D et al (1994) Drug combination of AZT and ddI synergism of action and prevention of appearance of AZT-resistance. Antiviral Chem Chemother **5:** 51–55.

Arts EJ & Wainberg MA (1996) Mechanisms of nucleoside analog antiviral activity and resistance during human immunodeficiency virus reverse transcription. Antimicrob Agents Chemother **40:** 527–540.

Back NK, Nijhuis M, Keulen W et al (1996) Reduced replication of 3TC-resistant HIV-1 variants in primary cells due to a processivity defect of the reverse transcriptase enzyme. EMBO J **15:** 4040–4049.

Balzarini J, Karlsson A, Perez-Perez MJ et al (1993a) HIV-1-specific reverse transcriptase inhibitors show differential activity against HIV-1 mutant strains containing different amino acid substitutions in the reverse transcriptase. Virology **192:** 246–253.

Balzarini J, Karlsson A, Perez-Perez MJ et al (1993b) Knocking-out concentrations of HIV-1-specific inhibitors completely suppress HIV-1 infection and prevent the emergence of drug-resistant virus. Virology **196:** 576–585.

Balzarini J, Karlsson A, Perez-Perez MJ et al (1993c) Treatment of human immunodeficiency virus type 1 (HIV-1)-infected cells with combinations of HIV-1-specific inhibitors results in different resistance pattern than does treatment with single-drug therapy. J Virol **67:** 5353–5359.

Barat C, Lullien V, Schatz O et al (1989) HIV-1 reverse transcriptase specifically interacts with the anticodon domain of its cognate primer tRNA. EMBO J **8:** 3279–3285.

Bebenek K, Abbotts J, Wilson SH et al (1993) Error-prone polymerization by HIV-1 reverse transcriptase: Contribution of template-primer misalignment, miscoding, and termination probability to mutational spots. J Biol Chem **268:** 10324–10334.

Boucher CAB, Lange JMA, Miedema FF et al (1992a) HIV-1 biological phenotype and the development of zidovudine resistance in relation to disease progression in asymptomatic individuals during treatment. AIDS **6:** 1259–1264.

Boucher CAB, O'Sullivan E, Mulder JW et al (1992b) Ordered appearance of zidovudine (AZT) resistance mutations during treatment. J Infect Dis **165:** 105–110.

Boucher CAB, Cammack N, Schipper P et al (1993) High-level resistance to (−)enantiomeric 2'-deoxy-3'-thiacytidine in vitro is due to one amino acid substitution in the catalytic site of human immunodeficiency virus type 1 reverse transcriptase. Antimicrob Agents Chemother **37:** 2231–2234.

Boulerice F, Bour S, Geleziunas R et al (1990) High frequency of isolation of defective human immunodeficiency virus type 1 and heterogeneity of viral gene expression in clones of infected U-937 cells. J Virol **64:** 1745–1755.

Boyer PL, Currens MJ, McMahon JB et al (1993) Analysis of non-nucleoside drug-resistance variants of human immunodeficiency virus type 1 reverse transcriptase. J Virol **67:** 2412–2420.

Boyer PL, Ding J, Arnold E et al (1994a) Subunit specificity of mutations that confer resistance to non-nucleoside inhibitors in human immunodeficiency virus type 1 reverse transcriptase. Antimicrob Agents Chemother **38:** 1909–1914.

Boyer PL, Tantillo C, Jaccob-Molina A et al (1994b) The sensitivity of wildtype human immunodeficiency virus type 1 reverse transcriptase to dideoxynucleotides depends on template length; the sensitivity of drug-resistant mutations does not. Proc Natl Acad Sci USA **91:** 4882–4886.

Byrnes VW, Sardana VV, Schleif WA et al (1993) Comprehensive mutant enzyme and viral variant assessment of human immunodeficiency virus type 1 reverse transcriptase resistance to non-nucleoside inhibitors. Antimicrob Agents Chemother **37:** 1576–1579.

Caliendo A, Savara A, An D et al (1995) ZDV resistance mutations increase replication in drug-free PBMC stimulated after infection. *Fourth International Workshop on HIV Drug Resistance, Sardinia, Italy* (abstr. 4:4).

Chong KT, Pagano PJ & Hinshaw RR (1994) Bisheteroarylpiperazine reverse transcriptase inhibitor in combination with 3'-azido-3'-deoxythymidine or 2',3'-dideoxycytidine synergistically inhibits human immunodeficiency virus type 1 replication in vitro. *Antimicrob Agents Chemother* **38**: 288–293.

Coffin JM (1995) HIV population dynamics in vivo: implications for genetic variation, pathogenesis, and therapy. *Science* **267**: 483–489.

D'Aquila RT, Johnson VA, Welles SL et al (1995) Zidovudine resistance and HIV-1 disease progression during antiretroviral therapy. *Ann Intern Med* **122**: 401–408.

De Clercq E (1994) HIV resistance to reverse transcriptase inhibitors. *Biochem Pharmacol* **47**: 155–169.

Ding J, Das K, Moereels H et al (1995) Structure of HIV-1 RT/TIBO R 86183 complex reveals similarity in the binding of diverse non-nucleoside inhibitors. *Nature Struct Biology* **2**: 407–415.

Dueweke TJ, Pushkarskaya T, Poppe SM et al (1993) A mutation in reverse transcriptase of bis(hetroary)piperazine-resistant human immunodeficiency virus type 1 that confers increased sensitivity to other non-nucleoside inhibitors. *Proc Natl Acad Sci USA* **90**: 4713–4717.

El-Farrash MA, Kuroda MJ, Kitazaki T et al (1994) Generation and characterization of a human immunodeficiency virus type 1 (HIV-1) mutant resistant to an HIV-1 protease inhibitor. *J Virol* **68**: 233–239.

Eron JJ, Chow YK, Caliendo AM et al (1993) *pol* Mutations conferring zidovudine and didanosine resistance with different effects in vitro yield multiply resistant human immunodeficiency virus type 1 isolates in vivo. *Antimicrob Agents Chemother* **37**: 1480–1487.

Eron JJ, Benoit SL, Jemsek J et al (1995) Treatment with lamivudine, zidovudine, or both in HIV-positive patients with 200 to 500 CD4+ cells per cubic millimeter. *N Engl J Med* **333**: 1662–1669.

Esnouf R, Ren J, Ross C et al (1995) Mechanism of inhibition of HIV-1 reverse transcriptase by non-nucleoside inhibitors. *Nature Struct Biol* **2**: 303–308.

Faraj A, Agrofoglio LA, Wakefield JK et al (1994) Inhibition of human immunodeficiency virus type 1 reverse transcriptase by the 5'-triphosphate β enantiomers of cytidine analogs. *Antimicrob Agents Chemother* **38**: 2300–2305.

Fischl MA, Richman DD, Grieco MH et al (1987) The efficacy of azidothymidine (AZT) in the treatment of patients with AIDS and AIDS-related complex: a double-blind placebo-controlled trial. *N Engl J Med* **317**: 185–191.

Fischl MA, Richman DD, Causey DM et al (1989) Prolonged zidovudine therapy in patients with AIDS and advanced AIDS-related complex. *JAMA* **262**: 2405–2410.

Fitzgibbon JE, Howell RM, Haberzettl CA et al (1992) Human immunodeficiency virus type 1 *pol* gene mutations which cause decreased susceptibility to 2',3'-dideoxycytidine. *Antimicrob Agents Chemother* **36**: 153–157.

Fletcher RS, Arion D, Borkow G et al (1995) Synergistic inhibition of HIV-1 reverse transcriptase DNA polymerase activity and virus replication in vitro by combinations of carboxanilide non-nucleoside compounds. *Biochemistry* **34**: 10106–10112.

Furman PA, Fyfe JA, St Clair MH et al (1986) Phosphorylation of 3'-azido-3'deoxythymidine and selective interactions of the 5'-triphosphate with human immunodeficiency virus reverse transcriptase. *Proc Natl Acad Sci USA* **83**: 8333–8337.

Gao Q, Gu Z, Parniak MA et al (1992) In vitro selection of variants of human immunodeficiency virus type 1 resistant to 3'-azido-3'-deoxythymidine and 2',3'-dideoxyinosine. *J Virol* **66**: 12–19.

Gao Q, Gu Z, Parniak MA et al (1993a) The same mutation that encodes low-level human immunodeficiency virus type 1 resistance to 2',3'-dideoxyinosine and 2',3'-dideoxycytidine confers high-level resistance to the (−)enantiomer of 2′,3′-dideoxy-3′-thiacytidine. *Antimicrob Agents Chemother* **37**: 1390–1392.

Gao WY, Shirasaka T, Johns DG et al (1993b) Differential phosphorylation of azidothymidine, dideoxycytidine and dideoxyinosine in resting and activated peripheral blood mononuclear cells. *J Clin Invest* **91**: 2326–2333.

Gao Q, Gu Z, Salomon H et al (1994) Generation of multiple drug resistance by sequential in vitro passage of the human immunodeficiency virus type 1. *Arch Virol* **136**: 111–122.

Gu Z, Gao Q, Li X et al (1992) Novel mutation in the human immunodeficiency virus type 1 reverse transcriptase gene that encodes crossresistance to 2′,3′-dideoxyinosine and 2′,3′-dideoxycytidine. *J Virol* **66**: 7128–7135.

Gu Z, Fletcher RS, Arts EJ et al (1994a) The K65R mutant reverse transcriptase of HIV-1 cross-resistance to 2′,3′-dideoxycytidine, 2′,3′-dideoxy-3′-thiacytidine, and 2′,3′-dideoxyinosine shows reduced sensitivity to specific dideoxynucleoside triphosphate inhibitors in vitro. *J Biol Chem* **269**: 28118–28122.

Gu Z, Gao Q, Fang H et al (1994b) Identification of a mutation at codon 65 in the IKKK motif of reverse transcriptase that encodes human immunodeficiency virus resistance to 2′,3′-dideoxycytidine and 2′,3′-dideoxy-3′-thiacytidine. *Antimicrob Agents Chemother* **38**: 275–281.

Gu Z, Arts EJ, Parniak MA et al (1995a) Mutated K65R recombinant HIV-1 reverse transcriptase shows diminished chain termination in the presence of 2′,3′-dideoxycytidine-5′-triphosphate and other drugs. *Proc Natl Acad Sci USA* **92**: 2760–2764.

Gu Z, Gao Q, Faust EA et al (1995b) Possible involvement of cell fusion and viral recombination in generation of human immunodeficiency virus variants that display dual drug resistance. *J Gen Virol* **76**: 2601–2605.

Gu Z, Quan Y, Li Z et al (1995c) Effects of non-nucleoside inhibitors of human immunodeficiency virus type 1 in cell-free recombinant reverse transcriptase assay. *J Biol Chem* **270**: 31046–31051.

Gu Z, Salomon H, Cherrington JM et al (1995d) The K65R mutation of human immunodeficiency virus type 1 reverse transcriptase encodes resistance to 9-(2-phosphonylmethoxyethyl)adenine. *Antimicrob Agents Chemother* **39**: 1888–1891.

Hart GJ, Orr DC, Penn CR et al (1992) Effects of (−)2′-deoxy-3′-thiacytidine (3TC) 5′-triphosphate on human immunodeficiency virus reverse transcriptase and mammalian DNA polymerases alpha, beta and gamma. *Antimicrob Agents Chemother* **37**: 918–920.

Havlir D, McLaughlin MM & Richman DD (1995) A pilot study to evaluate the development of resistance to nevirapine in asymptomatic human immunodeficiency virus-infected patients with CD4 cell counts of >500/mm^3: AIDS clinical trials group protocol 208. *J Infect Dis* **172**: 1379–1383.

Ho DD (1995) Time to hit HIV, early and hard. *N Engl J Med* **333**: 450–451.

Ho DD, Neumann AU, Perelson AS et al (1995) Rapid turnover of plasma virions and CD4 lymphocytes in HIV-1 infection. *Science* **273**: 123–126.

Jacobo-Molina A, Ding J, Nanni RG et al (1993) Crystal structure of human immunodeficiency virus type 1 reverse transcriptase complexed with double-stranded DNA at 3.0 Å resolution shows bent DNA. *Proc Natl Acad Sci USA* **90**: 6320–6324.

Japour AJ, Welles S, D'Aquila RT et al (1995) Prevalence and clinical significance of zidovudine resistance mutations in human immunodeficiency virus isolated from patients after long-term zidovudine treatment. *J Infect Dis* **171**: 1172–1179.

Jonckheere H, Taymans JM, Balzarini J et al (1994) Resistance of HIV-1 reverse transcriptase against [2′,5′-bis-O-(tert-butyldimethylsily)-3′-spiro5″-(4″-amino-1″2″-oxathiole-2″,2″diox-ide)] (TIBO) derivatives is determined by the mutation Glu138–Lys on the p51 subunit. *J Biol Chem* **269**: 25255–25258.

Kahn JO, Lagakos SW, Richman DD et al (1992) A controlled trial comparing continued zidovudine with didanosine in human immunodeficiency virus infection: the NIAID AIDS Clinical Trials Group. *N Engl J Med* **327**: 581–587.

Katlama C, Ingrand D, Loveday C et al (1996) Safety and efficacy of lamivudine–zidovudine combination therapy in antiretroviral-naïve patients: a randomized controlled comparison with zidovudine monotherapy. *JAMA* **276**: 118–125.

Kavlick MF, Shirasaka T, Kojima E et al (1995) Genotypic and phenotypic characterization of HIV-1 isolated from patients receiving (−)-2′,3′-dideoxy-3′-thiacytidine. *Antiviral Res* **28**: 133–146.

Kellam P, Boucher CAB & Larder BA (1992) Fifth mutation in human immunodeficiency virus type 1 reverse transcriptase contributes to the development of high-level resistance to zidovudine. *Proc Natl Acad Sci USA* **89**: 1934–1938.

Kew Y, Song Q & Prasad VR (1994) Subunit-selective mutagenesis of Glu-89 residue in human immunodeficiency virus reverse transcriptase. *J Biol Chem* **269**: 15331–15336.

Kew Y, Salomon H, Olsen LR et al (1996) The nucleoside analog-resistant E89G mutant of human immunodeficiency virus type 1 reverse transcriptase displays a broader cross-resistance that extends to nonnucleoside inhibitors. *Antimicrob Agents Chemother* **40**: 1711–1714.

Kohlstaedt LA, Wang J, Friedman JM et al (1992) Crystal structure at 3.5 Å resolution of HIV-1 reverse transcriptase complexed with an inhibitor. *Science* 256: 1783–1790.

Kojima E, Shirasaka T, Anderson BD et al (1995) Human immunodeficiency virus type 1 (HIV-1) viremia changes and development of drug-related mutations in patients with symptomatic HIV-1 infection receiving alternating or simultaneous zidovudine and didanosine therapy. *J Infect Dis* **171**: 1152–1158.

Koup RA, Brewster F, Grob P et al (1993) Nevirapine synergistically inhibits HIV-1 replication in combination with zidovudine, interferon or CD4 immunoadhesin. *AIDS* **7**: 1181–1184.

Kozal MJ, Shafer RW, Winters MA et al (1993) A mutation in human immunodeficiency virus reverse transcriptase and decline in CD4 lymphocyte numbers in long-term zidovudine recipients. *J Infect Dis* **167**: 526–532.

Kozal MJ, Shafer RW, Winters MA et al (1994) HIV-1 syncytium-inducing phenotype, virus burden, codon 215 reverse transcriptase mutation, and CD4 cell decline in zidovudine-treated patients. *J AIDS* **7**: 832–838.

Lacey S & Larder BA (1994) Novel mutation (V75T) in human immunodeficiency virus type 1 reverse transcriptase confers resistance to 2′,3′-didehydro-2′,3′-dideoxythymidine in cell culture. *Antimicrob Agents Chemother* **38**: 1428–1432.

Lacey SF, Reardon JE, Furfine ES et al (1992) Biochemical studies on the reverse transcriptase and RNase H activities from human immunodeficiency virus strains resistant to 3′-azido-3′-deoxythymidine. *J Biol Chem* **267**: 15789–15794.

Larder BA & Kemp SD (1989) Multiple mutations in HIV-1 reverse transcriptase confer high-level resistance to zidovudine (AZT). *Science* **246**: 1155–1158.

Larder BA, Darby G & Richman DD (1989) HIV with reduced sensitivity to zidovudine (AZT) isolated during prolonged therapy. *Science* **243**: 1731–1734.

Larder BA, Chesebro B & Richman DD (1990) Susceptibility of zidovudine-susceptible and -resistant human immunodeficiency virus isolates to antiviral agents determined by using a quantitative plaque reduction assay. *Antimicrob Agents Chemother* **34**: 436–441.

Larder BA, Coates KE & Kemp SD (1991) Zidovudine-resistant human immunodeficiency virus selected by passage in cell culture. *J Virol* **65**: 5232–5236.

Larder BA, Kellam P & Kemp SD (1993) Convergent combination therapy can select viable multidrug-resistance HIV-1 in vitro. *Nature* **365**: 451–453.

Larder BA, Kemp SD & Harrigan PR (1995) Potential mechanism for sustained antiretroviral efficacy of AZT–3TC combination therapy. *Science* **269**: 696–699.

Li X, Quan Y, Arts EJ et al (1996) HIV-1 nucleocapsid protein (NCp7) directs specific initiation of minus strand DNA synthesis primed by human tRNA$^{Lys.3}$ in vitro: studies of viral RNA molecules mutated in regions that flank the primer binding site. *J Virol* **70**: 4996–5004.

Lin P-F, Samanta H, Rose RE et al (1994) Genotypic and phenotypic analysis of human immuno-deficiency virus type 1 isolates from patients on prolonged stavudine therapy. *J Infect Dis* **170:** 1157–1164.

Loya S, Bakhanashvili M, Tal R et al (1994) Enzymatic properties of two mutants of reverse tran-scriptase of human immunodeficiency virus type 1 (tyrosine 181– and tyrosine 188–Leucine), resistant to non-nucleoside inhibitors. *AIDS Res Hum Retroviruses* **10:** 939–946.

Mansky LM & Temin HM (1995) Lower in vitro mutation rate of human immunodeficiency virus type 1 that predicted from the fidelity of purified reverse transcriptase. *J Virol* **69:** 5087–5094.

Martin JL, Wilson JE, Haynes RL et al (1993) Mechanism of resistance of human immuno-deficiency virus type 1 to 2′,3′-dideoxyinosine. *Proc Natl Acad Sci USA* **90:** 6135–6139.

Mazumder A, Cooney D, Agbaria R et al (1994) Inhibition of human immunodeficiency virus type 1 integrase by 3′-azido-3′-deoxy-thymidine. *Proc Natl Acad Sci USA* **91:** 5771–5775.

Mellors JW, Dutschman GE, Im GJ et al (1992) In vitro selection and molecular characterization of human immunodeficiency virus-1 resistance to non-nucleoside inhibitor of reverse tran-scriptase. *Mol Pharmacol* **41:** 446–451.

Mitsuya H & Broder S (1986) Inhibition of the in vitro infectivity and cytopathic effect of human T-lymphotropic virus type III/lymphadenopathy-associated virus (HTLV-III/LAV) by 2′,3′-dideoxynucleosides. *Proc Natl Acad Sci USA* **83:** 1911–1915.

Mohri H, Singh MK, Ching WT et al (1993) Quantitation of zidovudine-resistant human immunodeficiency virus type 1 in the blood of treated and untreated patients. *Proc Natl Acad Sci USA* **90:** 25–29.

Montaner JSG, Singer J, Schechter MT et al (1993) Clinical correlates of 'in vitro' HIV-1 resistance to zidovudine. Results of the multicentre Canadian AZT trial (MCAT). *AIDS* **7:** 189–196.

Montaner JSG, Schechter MT, Rachlis A et al (1995) Didanosine compared with continued zidovudine therapy for HIV-infected patients with 200 to 500 CD4 cells/mm^3: a double-blind, randomized, controlled trial. *Ann Intern Med* **123:** 561–571.

Moore RD, Hildago J, Sugland BW et al (1991) Zidovudine and the natural history of the acquired immunodeficiency syndrome. *N Engl J Med* **324:** 1412–1416.

Nanni RG, Ding J, Jacobo-Molina A et al (1993) Review of HIV-1 reverse transcriptase three-dimensional structure: implication for drug design. *Perspect Drug Design* **1:** 129–150.

Nielsen K, Wei LS, Sim MS et al (1995) Correlation of clinical progression in human immuno-deficiency virus-infected children with in vitro zidovudine resistance measured by a direct quantitative peripheral blood lymphocyte assay. *J Infect Dis* **172:** 359–364.

Nunberg JH, Schleif WA, Boots EJ et al (1991) Viral resistance to human immunodeficiency virus type 1-specific pyridinone reverse transcriptase inhibitors. *J Virol* **65:** 4887–4892.

Ogino MT, Dankner WM, Spector SA et al (1993) Development of zidovudine resistance in children infected with human immunodeficiency virus. *J Pediatr* **123:** 1–8.

Pagano PJ & Chong KT (1995) In vitro inhibition of human immunodeficiency virus type 1 by a combination of delavirdine (U-90152) with protease inhibitor U-75875 or interferon-α. *J Infect Dis* **171:** 61–67.

Pandey VN, Kaushik N, Rege N et al (1996) Role of methionine 184 of human immunodeficiency virus type-1 reverse transcriptase in the polymerase function and fidelity of DNA synthesis. *Biochemistry* **35:** 2168–2179.

Pokholok DK, Gudima SO, Yesipov DS et al (1993) Interactions of the HIV-1 reverse transcrip-tase AZT-resistant mutant with substrates and AZTTP. *FEBS Lett* **325:** 237–241.

Prasad VR, Lowy I, Santos TDL et al (1991) Isolation and characterization of a dideoxyguano-sine triphosphate-resistant mutant of human immunodeficiency virus reverse transcriptase. *Proc Natl Acad Sci USA* **88:** 11363–11367.

Preston BD, Poiesz BJ & Loeb LA (1988) Fidelity of HIV-1 reverse transcriptase. *Science* **242:** 1168–1171.

Quan Y, Gu Z, Li X et al (1996) Mutated HIV-1 M184V reverse transcriptase displays resistance to the triphosphate of (−)2′,3′-dideoxy-3′-thiacytidine (3TC) in both endogenous and cell-free enzyme assays. *J Virol* **70**: 5642–5645.

Reichman R, Tejani N, Strussenberg J et al (1992) Antiviral susceptibilities of HIV isolates obtained from long-term recipients of dideoxyinosine (ddI). Abstract PoB 3328, *XIII International Conference on AIDS*, Amsterdam, July 19–24.

Reichman RC, Tejani N, Lambert JL et al (1993) Didanosine (ddI) and zidovudine (ZDV) susceptibilities of human immunodeficiency virus (HIV) isolates from long-term recipients of ddI. *Antiviral Res* **20**: 267–277.

Ren J, Esnouf R, Garman E et al (1995) High resolution structures of HIV-1 RT from four RT–inhibitor complexes. *Nature Struct Biol* **2**: 293–302.

Ricchetti M & Buc H (1990) Reverse transcriptase and genomic variability: the accuracy of DNA replication is enzyme specific and sequence dependent. *EMBO J* **9**: 1583–1593.

Richman DD, Grimes JM & Lagakos SW (1990) Effect of stage of disease and drug dose on zidovudine susceptibilities of isolates of human immunodeficiency virus. *J AIDS* **3**: 743–746.

Richman D, Rosenthal AS, Skoog M et al (1991a) BI-587 is active against zidovudine-resistant human immunodeficiency virus type 1 and synergistic with zidovudine. *Antimicrob Agents Chemother* **35**: 305–308.

Richman D, Shih CK, Lowy I et al (1991b) Human immunodeficiency virus type 1 mutants resistant to non-nucleoside inhibitors of reverse transcriptase arise in cell culture. *Proc Natl Acad Sci USA* **88**: 11241–11245.

Richman DD, Havlir D, Corbeil J et al (1994) Nevirapine resistance mutations of human immunodeficiency virus type 1 selected during therapy. *J Virol* **68**: 1660–1666.

Rittinger K, Divita G & Goody RS (1995) Human immunodeficiency virus reverse transcriptase substrate-induced conformational changes and the mechanism of inhibition by non-nucleoside inhibitors. *Proc Natl Acad Sci USA* **92**: 8046–8049.

Roberts JD, Bebenek K & Kunkel TA (1988) The accuracy of reverse transcriptase from HIV-1. *Science* **242**: 1171–1173.

Rooke R, Tremblay M, Soudeyns H et al (1989) Isolation of drug-resistant variants of HIV-1 from patients on long-term zidovudine (AZT) therapy. *AIDS* **3**: 411–415.

Rusconi S, Merill DP & Hirsch MS (1994) Inhibition of human immunodeficiency virus type 1 replication in cytokine-stimulated monocytes/macrophages by combination therapy. *J Infect Dis* **170**: 1361–1366.

Sardana VV, Emini EA, Gotlib L et al (1992) Functional analysis of HIV-1 reverse transcriptase amino acids involved in resistance to multiple non-nucleoside inhibitors. *J Biol Chem* **267**: 17526–17530.

Schinazi RF, Lloyd RM Jr, Nguyen MH et al (1993) Characterization of human immunodeficiency viruses resistant to oxathiolane-cytosine nucleosides. *Antimicrob Agents Chemother* **37**: 875–881.

Schinazi RF, Larder BA & Mellors JW (1996) Mutations in retroviral genes associated with drug resistance. *Int Ant News* **4**: 95–107.

Schuurman R, Nijhuis M, Leeuwen RV et al (1995) Rapid changes in human immunodeficiency virus type 1 RNA load and appearance of drug-resistant virus populations in persons treated with lamivudine (3TC). *J Infect Dis* **171**: 1411–1419.

Shafer RW, Kozal MJ, Winters MA et al (1994) Combination therapy with zidovudine and didanosine selects for drug-resistant human immunodeficiency virus type 1 strains with unique patterns of *pol* gene mutations. *J Infect Dis* **169**: 722–729.

Shafer RW, Iversen AK, Winters MA et al (1995) Drug resistance and heterogeneous long-term virologic responses of human immunodeficiency virus type 1-infected subjects to zidovudine and didanosine combination therapy. *J Infect Dis* **172**: 70–78.

Shirasaka T, Kavlick MF, Ueno T et al (1995) Emergence of human immunodeficiency virus type 1 variants with resistance to multiple dideoxynucleosides in patients receiving therapy with dideoxy-nucleosides. *Proc Natl Acad Sci USA* **92**: 2398–2402.

Smerdon SJ, Jager J, Wang J et al (1994) Structure of the binding site for non-nucleoside inhibitors of the reverse transcriptase of human immunodeficiency virus type 1. *Proc Natl Acad Sci USA* **91**: 3911–3915.

Spence RA, Kati WM, Anderson KS et al (1995) Mechanism of inhibition of HIV-1 reverse transcriptase by non-nucleoside inhibitors. *Science* **267**: 988–992.

St Clair MH, Richards CA, Spector T et al (1987) 3'-Azido-3'-deoxythymidine triphosphate as an inhibitor and substrate of purified human immunodeficiency virus reverse transcriptase. *Antimicrob Agents Chemother* **3**: 1972–1977.

St Clair MH, Martin JL, Tudor-Williams G et al (1991) Resistance to ddI and sensitivity to AZT induced by a mutation in HIV-1 reverse transcriptase. *Science* **253**: 1557–1559.

Staszewski S, Massari FE, Kober A et al (1995) Combination therapy with zidovudine prevents selection of human immunodeficiency virus type 1 variants expressing high-level resistance to L-6097,661, a non-nucleoside reverse transcriptase inhibitor. *J Infect Dis* **171**: 1159–1165.

Staszewski S, Loveday C, Picazo JJ et al (1996) Safety and efficacy of lamivudine–zidovudine combination therapy in zidovudine-experienced patients: a randomized controlled comparison with zidovudine monotherapy. *JAMA* **276**: 111–117.

Tantillo C, Ding J, Jacobo-Molina A et al (1994) Location of anti-AIDS drug binding sites and resistance mutations in the three-dimensional structure of HIV-1 reverse transcriptase. *J Mol Biol* **243**: 369–387.

Temin HM (1993) Retrovirus variation and reverse transcription: abnormal strand transfer results in retrovirus genetic variation. *Proc Natl Acad Sci USA* **90**: 6900–6903.

Tisdale M, Kemp SD, Parry NR et al (1993) Rapid in vitro selection of human immunodeficiency virus type 1 resistance to 3'-thiacytidine inhibitors due to a mutation in the YMDD region of reverse transcriptase. *Proc Natl Acad Sci USA* **90**: 5653–5656.

Tudor-Williams G, St Clair MH, McKinney RE et al (1992) HIV-1 sensitivity to zidovudine and clinical outcome in children. *Lancet* **339**: 15–19.

Vandamme AM, Debyser Z, Pauwels R et al (1994) Characterization of HIV-1 strains isolated from patients treated with TIBO R82913. *AIDS Res Hum Retroviruses* **10**: 39–46.

Wainberg MA, Tremblay M, Rooke R et al (1990) Characterization of reverse transcriptase activity and susceptibility to other nucleosides of AZT-resistant variants of HIV-1. In: St Georgiev V & McGowan JJ (eds) *AIDS: Anti-HIV Agents, Therapies, and Vaccines*, pp 346–355. New York: Ann NY Acad.Sci.

Wainberg MA, Salomon H, Gu Z et al (1995) Development of HIV-1 resistance to (−)2'-deoxy-3'-thiacytidine in patients with AIDS or advanced AIDS-related complex. *AIDS* **9**: 351–357.

Wainberg MA, Drosopoulos WC, Salomon H et al (1996) Enhanced fidelity of 3TC-selected mutant HIV-1 reverse transcriptase. *Science* **271**: 1282–1285.

Wakefield JK, Jablonski SA & Morrow CD (1992) In vitro enzymatic activity of human immunodeficiency virus type 1 reverse transcriptase mutants in the highly conserved YMDD amino acid motif correlates with the infectious potential of the proviral genome. *J Virol* **66**: 6806–6812.

Wei X, Ghosh SK, Taylor ME et al (1995) Viral dynamics in human immunodeficiency virus type 1 infection. *Science* **273**: 117–122.

White EL, Parker WB, Ross SJ et al (1993) Lack of synergy in the inhibition of HIV-1 reverse transcriptase by combinations of the 5'-triphosphates of various anti-HIV nucleoside analogs. *Antiviral Res* **22**: 295–308.

Wu JC, Warren TC, Adams J et al (1991) A novel dipyridodiazepinone inhibitor of HIV-1 reverse transcriptase acts through a nonsubstrate binding site. *Biochemistry* **30**: 2022–2026.

Yarchoan R, Klecker RW, Weinhold KJ et al (1986) Administration of 3'-azido-3'-deoxythymidine, an inhibitor of HTLV-III/LAV replication, to patients with AIDS or AIDS-related complex. *Lancet* **i:** 575–580.

Yarchoan R, Perno CF, Thomas RV et al (1988) Phase I studies of 2',3'-dideoxycytidine in severe human immunodeficiency virus infection as a single agent and alternating with zidovudine (AZT). *Lancet* **i:** 76–81.

Yarchoan R, Mitsuya H, Myers CE et al (1989) Clinical pharmacology of 3'-azido-2',3'-dideoxythymidine (zidovudine) and related dideoxynucleosides. *N Engl J Med* **321:** 726–738.

Zhang D, Caliendo AM, Eron EJ et al (1994) Resistance to 2',3'-dideoxycytidine conferred by a mutation in codon 65 of the human immunodeficiency virus type 1 reverse transcriptase. *Antimicrob Agents Chemother* **38:** 282–287.

Chapter 15

NON-LENTIVIRAL PRIMATE RETROVIRUSES

Lisa Rosenblum and Myra McClure

HISTORICAL BACKGROUND AND CLASSIFICATION

The classification of viruses in general and retroviruses in particular has been problematic. Retroviruses were originally divided into three subfamilies known as the *Spumavirinae*, *Lentivirinae* and *Oncovirinae* (Lowry, 1985). This classification was based on a number of criteria including morphological data, cytopathic effects on cultured cells, and the pathological conditions these viruses induce in their host species. For instance, viruses in the *Spumavirinae* (or foamy viruses) were named for the foamy appearance of cytopathic effects on cultured cells. Although foamy viruses induce persistent infection and frequently occur in neural tissue, they do not cause specific clinical disease (Hooks and Gibbs, 1975). On the other hand, lentiviruses ('lenti-' meaning slow) are generally associated with slow neurological diseases, and were so named because of the extremely long latency period following infection. Finally, retroviruses in the *Oncovirinae* were supposedly capable of inducing neoplasia. However, not all retroviruses could be neatly categorized within the subfamily systematic scheme. Those retroviruses that did not (a) have a foamy type response on cultured cells (spumaviruses), (b) cause a slow disease (lentiviruses), or (c) cause tumours (oncoviruses) were lumped together in the oncovirus subfamily even though they did not cause malignancies. Consolidation of 'orphan' viruses under the *Oncovirinae* was confusing from a systematic point of view.

The International Committee on the Taxonomy of Viruses has now redefined virus systematics to reflect the nucleotide sequence relationships and genome structures of the *Retroviridae* (Coffin, 1992; Murphy et al, 1995). Recognized by the current classification, the *Retroviridae* family comprises seven distinct genera, some of them not specifically named: *Spumavirus*; *Lentivirus*; unnamed mammalian type B group; unnamed mammalian type C group; unnamed avian type C group; unnamed type D group; and unnamed bovine leukaemia virus – human T cell leukaemia virus (BLV-HTLV) group. Viruses are further described by:

- the four basic morphological types known as A type, B type, C type and D type particles which have been identified by electron microscopic examination;
- whether they are transmitted either vertically as a stable heritable trait (endogenous) or horizontally as non-integrated copies of viral DNA until after infection (exogenous);
- the particular malignancy with which they are associated (e.g. carcinoma, leukaemia, sarcoma).

HIV and the New Viruses Second Edition
ISBN 0-12-200741-7

Table 15.1 Summary of non-human primate virus isolates from C type, D type and HTLV-BLV genera. Host species' names, mode of transmission and available GenBank accession numbers are given.

Virus (isolate)	Host species (common name)	Genus	Transmission	Nucleotide Sequence Accession numbers
SSAV	Lagothrix spp. (woolly monkey)	C	Exogenous	X15311, J02396, J02397
GALV	Hylobates spp. (gibbon)	C	Exogenous	U20589
BaEV	Papio cynocephalus (yellow baboon)	C	Endogenous	D10032, D00088, N00088
MAC-1	Macaca arctoides (stumptail macaque)	C	Endogenous	J02244
MMC-1	Macaca mulatta (rhesus macaque)	C	Endogenous	J02253, J02243, V01176
CPC-1	Colobus polykomos (black/white colobus)	C	Endogenous	J02063
	Aotus trivigatus (owl monkey)	C	Endogenous	
	Pan troglodytes (chimpanzee)	C	Endogenous	J02055, J02056
SRV-1	Macaca spp.	D	Exogenous	M11841
SRV-1	Macaca spp.	D	Exogenous	M16605
SRV-3	Macaca spp.	D	Exogenous	M12349
Po-1-Lu	Trachypithecus obscurus (spectacled langur)	D	Endogenous	
SMRV	Saimiri (squirrel monkey)	D	Endogenous	K01706, M23385
	Papio cynocephalus (yellow baboon)	D	Endogenous	U85505, U85506
STLV-I	Pan troglodytes (chimpanzee)	HTLV-BLV	Exogenous	Z46895–Z46900, U86376, U03124, L75784–L75793
STLV-I	Pan troglodytes (chimpanzee)	HTLV-BLV	Endogenous	M33064
STLV-I (PHSul)	Pan paniscus (bonobo)	HTLV-BLV	Exogenous	X83120
STLV-I	Gorilla gorilla (gorilla)	HTLV-BLV	Exogenouse	

Strain	Species	Virus group	Type	Accession numbers
STLV-I	*Cercopithecus aethiops* (green monkey)	HTLV-BLV	Exogenous	U03126–U03132, U11555, U12101–U12103, U03122, AF012730, L00700–L00703, M92845, M92646
STLV-I (sm)	*Cercopithecus atys* (sooty mangabey)	HTLV-BLV	Exogenous	U94516
STLV-I	*Cercopithecus sabaeus* (tantalus monkey)	HTLV-BLV	Exogenous	L00700–L00703, M92845, M92846, AF012728
STLV-I	*Papio anubis* (olive baboon)	HTLV-BLV	Exogenous	Y07616, X88852, Z29673,
STLV-I (PH969)	*Papio hamadrayas* (Hamadryas baboon)	HTLV-BLV	Exogenous	X8886, X88878
STLV-I (IF)	*Papio hamadrayas* (Hamadryas baboon)	HTLV-BLV	Exogenous	S70512, S70595
STLV-I	*Papio hamadrayas* (Hamadryas baboon)	HTLV-BLV	Exogenous	L47161, L42252, L47128,
STLV-I	*Papio cynocephalus* (yellow baboon)	HTLV-BLV	Exogenous	L57519, L60024–L60027, L60528, L58023
STLV-I	*Macaca fascicularis* (long-tailed macaque)	HTLV-BLV	Exogenous	M14875
STLV-I	*M. nemestrina* (pig-tailed macaque)	HTLV-BLV	Exogenous	U76624–U76626
STLV-I (marc1)	*M. arctoides* (stump-tailed macaque)	HTLV-BLV	Exogenous	
STLV-I	*M. mulatta* (rhesus macaque)	HTLV-BLV	Exogenous	
STLV-I	*M. radiata* (bonnet macaque)	HTLV-BLV	Exogenous	
STLV-I	*M. silenus* (lion-tailed macaque)	HTLV-BLV	Exogenous	
STLV-I	*M. sinica* (toque macaque)	HTLV-BLV	Exogenous	
STLV-I	*M. nigrescens* (Dumoga-Bone macaque)	HTLV-BLV	Exogenous	
STLV-I	*M. nigra* (crested black macaque)	HTLV-BLV	Exogenous	
STLV-I	*M. tonkeana* (tonkean macaque)	HTLV-BLV	Exogenous	
STLV-I	*M. maura* (Celebes Moor macaque)	HTLV-BLV	Exogenous	Z46895–Z46900
STLV-I	*M. brunnescens* (booted macaque)	HTLV-BLV	Exogenous	U76624, U76625
STLV-I	*Pongo* spp. (orang-utan)	HTLV-BLV	Exogenous	U53562

Finally, as retroviruses have been identified as pathogens in almost every vertebrate species, they have been characterized on the basis of the animal from which they were isolated (e.g. gibbon ape leukaemia virus, squirrel monkey virus, baboon C type virus).

This section of the chapter describes the genome structure, epidemiology and evolutionary history of non-human primate retroviruses in three genera including the BLV-HTLV group and the C type and D type retroviruses. Many of the numerous retroviruses discovered in monkeys and apes are members of one of the three genera described above. Table 15.1 summarizes the large number of viruses belonging to C type, D type, and BLV-HTLV genera, their origin, and the species of non-human primate from which they were isolated.

Type C viruses

Both exogenous and endogenous C type viruses have been isolated from various non-human primate species. Two isolated exogenous C type viruses, simian sarcoma virus (SSV) and gibbon ape leukaemia virus (GALV), have been clearly shown to be oncogenic. The SSV was found in a South American woolly monkey with fibrosarcomas (Thielen et al, 1971). While experimental infection of other New World monkey species resulted in fibrosarcomas from which the virus could be recovered, this virus has not been found in other woolly monkey individuals (Rabin, 1971; Wolfe et al, 1971; Deinhardt et al, 1972). On the other hand, GALV has been isolated from a number of gibbon individuals with leukaemia or lymphosarcoma from two different colonies (Kawakami et al, 1973; Kawakami and Buckley, 1974). Simian sarcoma virus and GALV are biologically very similar and it may not be a coincidence that the woolly monkey infected with SSV was a pet housed together with other animals including a gibbon.

Endogenous type C viruses have been isolated and characterized from Old and New World monkey species including two isolates from two species of macaques. These isolates are known as MAC-1 from the stump-tailed macaque (*Macaca arctoides*), and MMC-1 from the rhesus monkey (*M. mulatta*) (Rabin et al, 1979; Todaro *et al*, 1978b). However, it has been suggested that these may be two isolates of the same virus (van der Kuyl et al, 1995). The best-characterized type C endogenous virus has been isolated from a baboon and is known as the baboon endogenous virus (BaEV) (Benveniste et al, 1974). This particular virus seems to be chimeric as it contains the *env* gene from a D type virus (Kato et al, 1987). The fourth known endogenous C type virus was isolated from a colobus monkey and is identified as CPC-1 (Sherwin and Todaro, 1979). The only endogenous C type virus that has been isolated from a New World monkey species was found in an established cell culture of owl monkey (*Aotus trivirgatus*) kidney cells (OMC-1) (Todaro et al, 1978c).

D viruses

Type D exogenous retroviruses are generally associated with an acquired immunodeficiency syndrome in macaque monkeys which can cause significant morbidity and mortality in both wild and captive populations (Gravell et al, 1984; Lerche et al, 1984, 1987; Bryant et al, 1986). The first type D exogenous retrovirus was isolated in 1969 from a rhesus macaque mammary tumour at the Mason Institute in Worcester, Massachusetts, and subsequently named Mason–Pfizer monkey virus (MPMV) (Chopra and Mason, 1970; Jenson et al, 1970; Mason et al, 1972). Because MPMV was isolated from a malignancy, it was originally systematically placed, somewhat prematurely, in the *Oncovirinae* subfamily under the old subfamily scheme. In subsequent transmission experiments of MPMV, juvenile rhesus macaques failed to develop tumours, leading researchers to conclude that MPMV does not cause tumours in macaques (Fine and Schochetman, 1978).

To date, five unique serogroups of exogenous D type retroviruses have been isolated from various species of macaques, all of which are genetically closely related to the prototype virus, MPMV (Chopra and Mason, 1970; Daniel et al, 1984; Marx et al 1984, 1986a; Stromberg et al, 1984; Barker et al, 1985; Henderson et al, 1985). While these isolates are generally known as simian retrovirus types 1–5 (SRV-1, SRV-2, SRV-3, SRV-4 and SRV-5), they are properly identified by the different species at the various regional primate centres from which they were isolated. For instance, the original serogroup 2 virus was isolated from a Celebes macaque (M. *nigra*) at the Oregon Regional Primate Research Center (ORPRC) and is thus identified as D2/Cel/OR (Marx et al, 1985). Other isolates of serotype 2 are identified as D2/RHE/OR, from a rhesus macaque (M. *mulatta*) at the ORPRC, and as D2/PTM/WA, from a pig-tailed macaque (M. *nemestrina*) at the Washington Regional Primate Research Center (Stromberg et al, 1984). The prevalence of D type infection in breeding colonies is high. Screening surveys by serology in the late 1980s suggested that up to 20% of captive macaques at USA primate centres were infected with a D type virus (Lairmore et al, 1990).

In addition to various macaque species from which D type viruses have been isolated, reports suggest that some captive baboon populations from Ethiopia (*Papio cynocephalus*) are infected with a D type virus whose serotype is similar to SRV-2 (Benveniste et al, 1993a; Grant et al, 1995). It is not known, however, whether these animals became infected with type D virus during captivity or whether they carry the infection in the wild.

Of the five serotypes, only SRV-1, SRV-2 and MPMV (also known as SRV-3) have been molecularly cloned and completely sequenced (Power et al, 1986; Sonigo et al, 1986; Thayer et al, 1987). Nucleotide sequence data from the two remaining serotypes, SRV-4 and SRV-5, isolated from macaques at the University of California at Berkeley, USA, and at the primate centre in Beijing, China, respectively, have yet to be published (Axthelm et al, 1987; Gardner et al, 1988).

Type D endogenous viruses are not associated with disease and thus far have been isolated from three species of monkeys (Table 15.1). Endogenous type D viruses have been isolated from lung tissue from the spectacled leaf monkey (*Trachypithecus obscurus*) (Todaro et al, 1978a), from peripheral blood mononuclear cells (PBMCs) from yellow baboons (*Papio cynocephalus*) (van der Kuyl et al, 1997), and from a variety of tissues from a squirrel monkey (*Saimiri sciureus*), the only New World monkey species in which D type virus has been found (Heberling et al, 1978).

Simian T cell leukaemia viruses

The first human T cell leukaemia viruses (HTLV) were isolated and characterized in the early 1980s (Poiesz et al, 1980, 1981). Shortly after the isolation and characterization of HTLV-1, similar T cell leukaemia viruses were discovered in non-human primate species, many of which were isolated from the genus *Macaca*. Serological analysis was employed to screen numerous non-human primate species in an attempt to isolate a virus similar to HTLV-1 (Miyoshi et al, 1982). The first T cell leukaemia virus in monkeys was found by demonstrating the presence of anti-HTLV-1 antibodies in Japanese macaques (M. *fuscata*) and it was subsequently called simian T cell leukaemia virus type 1 (STLV-1) (Miyoshi et al, 1982). While serological screening has detected STLV-1 antibodies in many Old World monkey species and apes, STLV-1 does not seem to cause an adult T cell leukaemia (ATL) resembling the human disease in most of the species examined (Homma et al, 1984).

The only New World monkey species known to be infected with an STLV variant were zoo-captive spider monkeys (*Ateles fusciceps*), from whom STLV-2 was isolated (Chen et al, 1994). Two new retroviruses have been isolated from an Ethiopian *Papio hamadrayas* (PTLV-L) (Van Brussel et al, 1996) and from captive colonies of *Pan paniscus* (STLV-PP) (Vandamme et al, 1996). However, not much is known about STLV-2, PTLV-L or STLV-PP infection and its association with disease in these animals. Neither related endogenous viruses nor oncogene-containing members of this genus are known.

Foamy viruses

Foamy viruses are enveloped viruses which are morphologically, serologically and biologically different from other retroviruses (Achong et al, 1971; Parks and Todaro, 1972; Epstein et al, 1974, Johnston, 1974; Hooks and Gibbs, 1975; Hruska and Takemoto, 1975; Liu et al, 1977; Loh et al, 1977, Chiswell and Pringle, 1979; Cavalieri et al, 1981; Hooks and Detrick-Hooks, 1981; Flügel et al, 1987; Maurer et al, 1988; Lewe and Flügel, 1990). The mature particle measures approximately 106–116 nm in diameter and consists of an internal ring-shaped core surrounded by an outer envelope embedded with evenly spaced, characteristically needle-like spikes about 13 nm long (Clark and Attridge, 1968; Malmquist et al, 1969, Clarke and McFerran, 1970; Epstein et al, 1974; Hooks and Gibbs, 1975). These glycoprotein spikes represent one of the morphological features distinguishing the foamy virus from the other subfamilies of retroviruses (Benzair et al, 1985; Hotta and Loh, 1986). A summary of the unique characteristics of the foamy viruses, which distinguishes them from the other retroviruses, is given in Table 15.2.

Table 15.2 *Characteristics of foamy viruses*

	Characteristic
Isolation	From *in vitro* cultivation of cells from original tissues
Morphology	Enveloped virion, with unusually long (13 nm) and evenly spaced spike glycoproteins
Nucleic acid	Virion contains both RNA and DNA; genome longer than that of most other retroviruses ($\sim$ 13 000 kb)
Reverse transcriptase	Optimal RT activity requires Mn^{2+}, in contrast to Mg^{2+} for other retroviruses
Serological relationship	Immunologically distinct from other retrovirus subfamilies
Viral replication	Intracytoplasmic pre-formed nucleoid does not undergo further maturation post-budding
Clinical disease	No association with clinical disease in naturally or accidentally (human) infected host animals

Foamy viruses were first identified by Enders and Peebles (1954) as cytopathogenic agents in cultures of primary monkey kidney cells. Degenerating cultures were characterized by the formation of highly vacuolated syncytia, described as 'foamy' in appearance, hence the nomenclature. First isolated by Gajducek in 1967 from chimpanzees, these viruses are now known to be widely distributed in monkeys, cats, cows and many other animal species (Hooks and Gibbs, 1975; Hooks and Detrick-Hooks, 1981).

Among primate foamy viruses, the simian foamy viruses (SFV) have been serologically classified into SFV types 1–11. Three of these, SFV-1 (from an infected macaque), SFV-3 (from an African green monkey) and SFV_{cpz} from a chimpanzee, have been cloned and completely sequenced, while partial sequences are available for two other chimpanzee isolates, SFV-6 and SFV-7. In 1971 the first human foamy virus (HFV) was reportedly isolated from a Kenyan patient with nasopharyngeal carcinoma. For this isolate, too, the complete sequence is available and infectious molecular clones have been constructed (for references see pp 259–260).

GENETIC STRUCTURE

Much of the genome organization and virion structure is common to all retroviruses and is described only briefly here. For a more comprehensive treatment, there are several excellent books and reviews worth consulting (e.g. Coffin, 1996; Weiss et al, 1984). Members of the retrovirus family have genomes that are 7–11 kilobases (kb) in size, encompassing three genes known as *gag*, *pol* and *env*, which encode all the necessary structural proteins and enzymes for replication. These genes, located on the genome in the above order, are flanked on either end by long terminal repeats (LTRs) several hundred base pairs in length. All genes are transcribed from one promoter located at the 5' end LTR. The *gag* gene encoding the group-specific antigens is translated from the complete RNA sequence to produce a polyprotein. The polyprotein is then cleaved into three to five non-glycosylated structural proteins that comprise the matrix, capsid and nucleocapsid proteins. The protease protein is encoded by the *pro* region whose function is to cleave the *gag* and *pol* polyproteins. In various retroviruses the protease domain is encoded at the end of the *gag* gene or at the beginning of the *pol* gene. In other retroviruses, *pro* designates a separate gene located between *gag* and *pol*. The *pol* gene encodes two proteins, reverse transcriptase (RT) and integrase (IN), which are essential to the virus in the early stages of infection. The reverse transcriptase serves to convert the genetic information from single-stranded RNA to double-stranded DNA. The integrase is responsible for integrating the double-stranded DNA intermediate into the chromosomes of the host cell. In most retroviruses, the *pol* and *env* genes overlap slightly so that there is no untranslated sequence between the two genes. There are two envelope glycoproteins encoded by the *env* gene which are cleaved from a larger precursor. The larger protein, the surface protein (SU), mediates virus entry by interacting with the host cell receptors, and the second protein, the transmembrane protein (TM), is believed to mediate fusion of the viral and host membranes.

While it is clear that there are major structural similarities among retroviruses, there are also a number of appreciable structural differences among the genera worth describing (Figure 15.1).

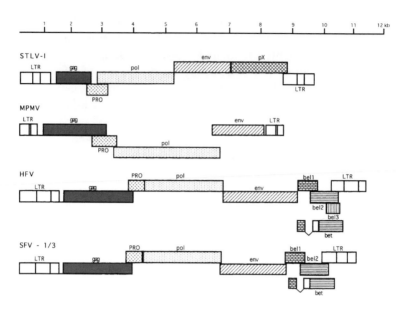

Figure 15.1 *General genome organization of lymphotropic, D type and foamy viruses*

Type D viruses

Because simian retrovirus serotypes 1, 2 and 3 have been completely sequenced, their genomic structure and organization are well characterized. Not much is known about the genomic organization of SRV serotypes 4 and 5. The D type retroviruses are uncomplicated and comprise the *gag, pro, pol* and *env* genes, flanked on either side by an LTR region (Power et al, 1986; Sonigo et al, 1986; Thayer et al, 1987). The SRV-3 *gag* gene encodes a precursor polyprotein which is cleaved during particle maturation by the viral protease to yield the matrix, capsid and nucleocapsid proteins (Bradac and Hunter, 1986a,b). Type D retroviruses, like members of the HTLV group, code the protease gene in a distinct reading frame from the *pol* gene. Additionally, SRV-1 and SRV-3 share 97% amino acid sequence similarity in the *pro* gene, while the SRV-2 shares around 83% sequence similarity in the *pro* gene with SRV-3 (Thayer et al, 1987).

The *env* gene of SRV-3 encodes a glycosylated protein processed into gp70, the surface protein, and gp20, the transmembrane protein (Bradac and Hunter, 1986b). There are large differences in the gp70 domain between SRV-2 and the other SRV serotypes (58% amino acid similarity) and smaller but appreciable differences in the gp70 region between SRV-1 and SRV-3 (83% amino acid similarity) (Thayer et al, 1987). The significance of the tremendous variation in the *env* gene in SRV serotypes has not been established. However, as the portion of the glycoprotein is assumed to be the target of protective immunity, it is suggested that various immune responses may provide selective pressure, thereby increasing variation found in this region (Thayer et al, 1987).

Recent attention has focused on a constitutive transport element (CTE) that has been mapped to a 154 nucleotide region located between the *env* region and the 3' end of the SRV-3 genome (Bray et al, 1994). This is a *cis*-acting element that serves to facilitate nucleocytoplasmic export of intron-containing RNA. Mutational analysis of this region suggests that primary sequences in the secondary structure loop section may contain binding sites for cellular proteins involved in RNA export (Ernst et al, 1997). Other groups suggest that CTE is critical for the regulation of the MPMV gene expression and functions in the viral life cycle by interacting with constitutively present cellular factors analogous to the Rev response element (RRE) regulatory system of the human immunodeficiency virus (Rizvi et al, 1996). The CTE region is highly conserved among SRV-1, SRV-2 and SRV-3 (Tabernero et al, 1996).

STLV

The entire STLV genome is around 9000 bp in length. Its overall genetic organization resembles the genetic structure of HTLV-1 and HTLV-2 isolates without major insertions in the LTR, *gag*, *pol*, *env* and pX regions. The pX portion of the genome, common to all members of the HTLV genus, is a long region overlapping *env*, but in an alternate reading frame. It was originally termed X or lor (for long open reading frame). The additional genes located in this region encode two regulatory proteins which modify the rate and pattern of gene expression of the virus. These proteins are known as Tax and Rex: Tax can transactivate both viral and cellular genes, and the protein resulting from it is essential for any LTR-driven expression of the virus (for review, see Poiesz, 1995); the *rex* gene encodes for a protein which is necessary for the expression of full-length and *env* mRNAs. The Tax protein and the neutralizing domain present in the *env* gene were found to be fully conserved among many STLV and HTLV isolates (Lal et al, 1991; Palker et al, 1992). The entire length of an STLV-1 isolate was sequenced in order to compare its genetic structure with that of HTLV and other partially sequenced isolates of STLV (Ibrahim et al, 1995). Tax was found to be particularly conserved in the first 90 amino acids, which contain active sites for functional activity, including the zinc finger region (Ibrahim et al, 1995). In contrast with Tax, Rex was found to be highly divergent among different isolates of STLV and HTLV (van Brussel et al, 1996). Preliminary data seem to suggest that the pX region encodes for some new proteins in HTLV-2, but it is not clear whether they are present in STLV isolates (Franchini, 1995).

Foamy viruses

While the genome sequence of four isolates has been determined (Flügel et al, 1987; Maurer et al, 1988; Kupiec et al, 1991; Renne et al, 1993; Herchenröder et al, 1994) and infectious molecular clones derived from human and chimpanzee isolates have been constructed (Rethwilm et al, 1990; Löchelt et al, 1991; Herchenröder et al, 1994), most of the molecular data on the genomic structure of this genus of viruses have been derived from analysis of HFV.

Like other complex retroviruses (Cullen, 1991) HFV encodes (in addition to the three structural genes *gag*, *pol* and *env*) a transcriptional transactivator (Bel-1) which is essential for LTR-driven gene expression and for virus replication (Keller et al, 1991; Rethwilm, et al, 1991; Baunach et al, 1993). In contrast, the two other open reading frames at the 3' end of the genome, *bel-2* and *bel-3*, are dispensable for HFV replication in vitro (Baunach et al, 1993; Yu and Linial, 1993). The simian foam viruses SFV-1 and SFV-3 differ from HFV and the chimpanzee isolates in that they have only two accessory open reading frames, designated ORF1 (equivalent to *bel-1* of HFV) and ORF2. (Figure 15.1).

A further accessory protein synthesized in the infected cell is generated by a splicing event from the *bel-1* ORF into the *bel-2* ORF. The resulting protein, called Bet (**Bel** plus **two**), is expressed at higher levels than any other viral protein and conserved in all foamy virus isolates, suggesting functional importance. Since mutations in the *bet* gene decrease cell-free viral transmission in vitro approximately 10-fold (Yu and Linial, 1993), Bet may play a role in efficient cell-free viral transmission, similar to that of the Vif protein in HIV.

Primate foamy viruses carry a second internal promoter (IP) which is dependent on Bel-1 (Löchelt et al, 1994; Campbell et al, 1994) and directs the expression of the Bel-1 transactivator and Bel protein early after infection (Löchelt et al, 1994). Foamy viruses are the only retrovirus for which an IP dependent on the viral transactivator has been defined.

EPIDEMIOLOGICAL CHARACTERISTICS OF INFECTION

Host range

Type C viruses
The species host range of exogenous and endogenous C type viruses is varied. Simian sarcoma virus, isolated from a captive woolly monkey (Rabin, 1971), and gibbon ape leukaemia virus, found in several captive gibbon individuals, have not been detected in other species (Kawakami et al, 1973; Kawakami and Buckley, 1974). However, the host cell range of the SSV/GALV isolates is extensive and there are no cell lines that can demarcate the different isolates (Teich et al, 1975; Weiss and Wong, 1977).

Complete endogenous type C viruses have been described in the stump-tailed macaque (*Macaca arctoides*), the rhesus monkey (*M. mulatta*), a leaf monkey (*Colobus polykomos*), and an owl monkey (*Aotus trivirgatus*) (Rabin et al, 1979; Todaro et al, 1978b; Sherwin and Todaro, 1979; Todaro et al, 1978c, respectively). Additionally, a chimeric type C/type D baboon endogenous virus has been described in various African monkey species (Benveniste et al, 1974; van der Kuyl et al, 1995). The cell host range of BaEV includes other primate (not baboon), dog and bat cells, while MAC-1 isolate replicates well in cat and human cells (Todaro et al, 1978b). The host range of the owl monkey virus (OMC-1) is unusually limited to the cat, bat and owl monkey cells, which makes it the only primate endogenous C or D type virus that can infect cells of its own host species (Todaro et al, 1978c).

Type D viruses
Exogenous D type viruses are widely distributed in a variety of tissues and body fluids including lymph nodes, spleen, salivary glands, bone marrow, mammary gland, brain, peripheral blood, saliva, lacrimal secretions, urine, breast milk, and cerebrospinal fluid (Gravell et al, 1984; Bryant et al, 1986; Lackner et al, 1988). Thus far, infection appears to be limited to species in the genus *Macaca*, and various species of captive baboons. This is particularly significant because macaques represent a large portion of primates housed in breeding colonies used for medical research. Continued use of animals unknowingly infected with D type virus could seriously confound results from investigations related to acquired immunodeficiency syndromes. Type D viruses appeared as a serious pathogen in captive macaques, especially after these viruses were recognized as

the cause of immunodeficiency disease at several primate research centres in the USA (Daniel et al, 1984; Heidecker et al, 1987; Schultz et al, 1989).

Two endogenous D type retroviruses have been isolated. The first was isolated from a leaf monkey (*Trachypithecus obscurus*) and identified as PO-1-Lu (Todaro et al, 1978a). It grows only in bat and human cell lines. The second endogenous D type, known as SMRV, was isolated from a squirrel monkey (Heberling et al, 1978). It has a much broader host range and will grow in cell lines derived from dog, mink, bat, human, chimpanzee and rhesus monkey, but not in a variety of other non-human primate-derived cell lines, including those from New World monkey species (Heberling et al, 1978). As expected, the host ranges of the endogenous D type are distinct from exogenous D type which is known to grow in human and rhesus monkey cells.

STLV

While C type and D type viruses infect a number of different non-human primate species, STLV-1 is the most universally distributed virus among them. Serological screening has detected STLV-1 antibodies in many Old World monkey species and apes including chimpanzees (*Pan troglodytes*), gorillas (*Gorilla gorilla*), grivet monkeys (*Cercopithecus aethiops aethiops*), long-tailed macaques (*Macaca fascicularis*), pig-tailed macaques (*M. nemestrina*), bonnet macaques (*M. radiata*), lion-tailed macaques (*M. silenus*), toque macaques (*M. sinica*) and other species of macaques found exclusively on the island of Sulawesi (*M. nigrescens*, M. *nigra*, M. *hecki*, M. *tonkeana*, M. *maura*, M. *ochreata* and M. *brunnescens*) (Ishida et al, 1983; Miyoshi et al, 1983; Hayami et al, 1984; Ishikawa et al, 1987; Ibrahim et al, 1995). Type 2 STLV has been found in captive spider monkeys (*Ateles fusciceps*) (Chen et al, 1994). Both STLV-1 and STLV-2 are known to replicate in human cells (Chen et al, 1994; Van Brussel et al, 1996).

Foamy viruses

Antibodies to foamy virus are readily detected within diverse primate populations and are relatively species-specific in terms of serotype. Eleven primate foamy virus serotypes are currently documented: simian foamy viruses SFV-1, SFV-2, SFV-3 and SFV-10 are found in Old World monkeys (macaques, cercopithecus monkeys and baboons), SFV-4, SFV-8 and SFV-9 in New World monkeys (squirrel monkeys, spider monkeys and capuchins, and SFV-5 in prosimians. By convention, the foamy viruses found in apes have also been given the SFV nomenclature, although they are not strictly simian: SFV-6, SFV-7 and SFV$_{cpz}$ were isolated from chimpanzees and SFV-11 from orang-utans.

Viruses are readily detected in throat and pharyngeal swabs and easily isolated from most tissues. As for all retroviruses, transmission is both vertical from mother to child and horizontal via scratches and bites. Uninfected animals introduced into an infected primate colony develop foamy virus antibodies within weeks of entry (Ruckle, 1958).

When uninfected natural hosts for foamy virus infection are inoculated with a species-associated virus they seroconvert but remain clinically well (Ruckle, 1958; Malmquist et al, 1969; McKissick and Lamont, 1970). Inoculation of simian viruses into laboratory animals such as rabbits, newborn and adult mice, hamsters, guinea pigs, chicks and embryo hens' eggs, has never resulted in disease (Rustigian et al, 1955; Johnston, 1961, 1971; Stiles et al, 1964; Riggs et al, 1969; McKissick and Lamont, 1970; Hooks et al, 1972, 1973). Interestingly, the infection in the natural host resembles that in the inoculated rabbit, characterized by ease of viral isolation from peripheral blood, mononuclear cells, neutralizing antibody, and persistence of virus in tissues such as kidney, liver, spleen, lung and salivary gland.

Foamy viruses in vitro infect an extensive range of cell lines. Indeed, for the chimpanzee isolates and HFV, we have not been able to identify a cell line refractory to virus infection, including avian cells. Foamy virus replication is dependent on host cell division. Optimal production of plaques and virus titre are known to be obtained in actively dividing cell cultures from a variety of host cell species (Parks and Todaro, 1972; Loh et al, 1977; Hooks and Detrick-Hooks, 1981; Loh and Ang, 1981). Bieniasz et al (1995) demonstrated formally that, in common with the oncoviruses but unlike HIV, foamy viruses fail to establish productive infection in G_1–S or G_2 growth-arrested cells.

Transmission

In general terms, there are two recognized modes of viral transmission from one host animal to another. Retroviruses can be transmitted either horizontally by contact between two animals, or vertically from parent to offspring. There are two distinct ways that vertical transmission can occur:

1. The virus can be transmitted as an infectious agent with an RNA genome via the placenta or milk of an animal.
2. The virus, known as an endogenous virus, can be genetically transmitted as a DNA provirus, as a heritable trait, from one generation to the next.

Very few studies have directly examined transmission of these virus groups in non-human primates. In captive populations especially, type D retrovirus infection is mainly dependent on the history of colony exposure to the virus and on management practices (Lerche, 1992). Most transmission occurs horizontally through direct physical contact between infected animals. Behaviour plays a significant role in the transmission of many of these viruses, since monkeys frequently groom, bite and scratch each other. High titres of SRV type D virus have been measured in saliva, leading to the suggestion that inoculation by biting is the most likely mode of D type infection (Lerche et al, 1986). Additionally, vertical transmission through the placenta and postnatally has been well documented for D type infections (Tsai et al, 1986; Lerche et al, 1987). It is also believed that GALV (Kawakami et al, 1978), STLV-1 and STLV-PP isolates are efficiently transmitted through breast-feeding and sexual contact (Giri et al, 1994).

Clinical consequences of infection

Type C viruses
Type C exogenous retroviruses are associated with a large range of diseases including malignancies, immunosuppression and neurological disorders. Simian sarcoma virus was first isolated from a fibrosarcoma in a pet monkey who had multiple small tumours dispersed throughout the peritoneum (Thielen et al, 1971). Gibbon ape leukaemia virus was found in various captive gibbons that were reported to have malignant lymphomas, lymphosarcomas and granulocytic leukaemias (De Paoli et al, 1973; Kawakami et al, 1972). Experimental inoculation with GALV of two captive gibbons produced chronic

granulocytic leukaemia after 5 months, indicating that GALV isolates were responsible for producing granulocytic leukaemias (Kawakami et al, 1980).

Type D viruses
Like the C type viruses discussed above, an extensive range of clinical outcomes of experimental infection with all SRV serotypes has been reported (Maul et al, 1986). At primate breeding colonies housing macaque species for research, SRV-2 is the predominant serotype and has been associated with severe immunodeficiency and retroperitoneal fibromatosis in Celebes (*Macaca nigra*), pig-tailed (*M. nemestrina*) and long-tailed (*M. fascicularis*) macaques. In fact, SRV-2 is the only serotype that has been associated with a malignant tumour of any kind (Stromberg et al, 1984; Marx et al, 1985). Simian retrovirus type 2 causes severe immunodeficiency and anaemia in Japanese macaques (*M. fuscata*) and low to severe levels of immunodeficiency in rhesus macaques (*M. mulatta*) (Giddens et al, 1985; Axthelm et al, 1987; Gardner and Luciw, 1989). Type D retroviruses are also the cause of a great many fatalities in breeding colonies as a direct consequence of severe immunodeficiency. In the period between 1976 and 1984 infection with D type virus resulted in a tremendous 60% mortality rate in macaques (most of whom were under 2 years old) at the California Regional Primate Research Center (Henrickson et al, 1983). At the New England Regional Primate Research Center in the early 1980s, SRV-1 caused up to 33% mortality in Taiwanese rock macaques (*M. cyclopis*) (Letvin et al, 1983). Another serious consequence of infection is the ability of some infected monkeys to remain apparently healthy where antibody response is not detected by serological screening, while shedding large amounts of virus (Lowenstine and Lerche, 1988).

STLV
A large proportion of wild non-human primate species in Africa are infected with STLV-I but remain in good health for much of their lives (Hunsmann et al, 1983; Ishikawa et al, 1987). However, in at least a few species, including the African green monkeys, baboons and macaques, it has been established that STLV-1 causes a disease resembling adult T cell leukaemia (ATL) (Sakakibara et al, 1986; Yoshimura et al, 1990; Schatzl et al, 1992). To determine whether STLV-1 is associated with neoplasia in monkeys, sera from 13 leukaemic macaque monkeys from three species, *M. mulatta*, *M. fascicularis* and *M. cyclopis*, were examined for antibody reactivity to HTLV-1. Eleven out of 13 monkeys were seropositive, compared with 7 out of 95 normal macaque species (Homma et al, 1984), suggesting that STLV-1 or a related virus may be responsible for leukaemia and lymphoma in non-human primates. In non-human primates STLV-1 is not known to cause neurological diseases resembling those caused by HTLV-1 in humans (Gessain and de The, 1996).

Foamy viruses
Foamy viruses infect diverse mammalian species, e.g. cats, dogs, rabbits, sea lions, cattle, rodents and primates. In these animals they induce persistent infections and high-titre neutralizing antibodies, but no apparent disease (Hooks and Gibbs, 1975; Hooks and Detrick-Hooks, 1981; Weiss, 1988), suggesting strict immunological control in the infected host. Although these viruses have never been reproducibly associated with any disease, transient immunodeficiency in rabbits has been reported following

experimental infection with SFV-7 (Hooks and Detrick-Hooks, 1979). More recently a foamy virus serotype was isolated from an orang-utan with encephalopathy and myopathy (McClure et al, 1994). This was of some interest in the light of a previous report that transgenic mice expressing all or part of the HFV genome developed a severe neurodegenerative disorder (Bothe et al, 1991). However, it should be emphasized that foamy viruses have long been isolated from both healthy and clinically ill animals (Malmquist et al, 1969; McKissick and Lamont, 1970; Achong et al, 1971) and the encephalopathic orang-utan remains an isolated case report.

In considering the pathogenic potential of foamy viruses, it is worth noting that animal keepers who seroconverted to SFV as a result of being bitten or scratched by a range of Old World primates, and a laboratory worker accidentally infected with HFV, have remained completely healthy, some for well over a decade (Heneine et al, 1998).

Vaccines for SRV type D infection

Because of the high degree of morbidity and mortality caused by SRV type D infections in captive breeding colonies, control of type D infection in captive populations of macaques is especially important. To this end, several vaccine preparations have been developed that successfully protect against experimental infection with various SRV serogroups. Even though the first vaccine against D type infection was developed in the 1980s, these vaccines are still not widely used to control type D infection in captive breeding colonies, mainly owing to their expense and lack of commercial availability.

Protection of macaques against infection by SRV-1 and SRV-3 serogroups, as measured by neutralizing antibody response, was first achieved with a formalin-inactivated whole virus vaccine preparation (Marx et al, 1986). It was unknown which antigens were responsible for eliciting protective immunity until Hu and co-workers (Hu et al, 1989) constructed a recombinant vaccinia virus which expressed the envelope glycoproteins of SRV-2. Their results indicated that the recombinant vaccinia virus successfully protected the experimentally infected monkeys from disease, leading them to conclude that the *env* glycoproteins are accountable for the immune response (Hu et al, 1989). However, it was also reported that immunization with SRV-1 envelope antigens expressed in yeast did not elicit neutralizing antibodies in experimentally infected monkeys (Kwang et al, 1988). To explain the apparent discrepancy between the results from the two studies, it was suggested that the yeast-produced SRV proteins have a different structure from the native viral envelope glycoproteins. Protection against SRV-1 and SRV-3 has also been achieved with recombinant live vaccinia SRV-1 and SRV-3 *env* (gp70 and gp22) vaccines (Brody et al, 1992). Finally, it has been shown that long-term protection against SRV-2 can be accomplished (Benveniste et al, 1993b).

EVOLUTIONARY HISTORY OF TYPE D AND LYMPHOTROPIC RETROVIRUSES

Phylogenetic methods are employed to assess hypotheses of ancestor–descendant relationships among many different organisms, including viruses. However, evolutionary analysis of viruses in the past, unlike many other organisms, was limited by the lack of characteristics accessible for study. Most notably, researchers used host range and neutralization by antisera to distinguish, for instance, strains of influenza (Kilbourne, 1978). Antigenic relatedness was widely used as an evolutionary marker for many viruses. With the advent of powerful molecular evolutionary techniques, such as restriction site polymorphism and nucleic acid sequencing, researchers could examine variation on the level of the genome to make inferences regarding viral evolution.

In general, phylogenetic inferences can be made from nucleotide sequence data because it is assumed that most mutations involved in nucleotide substitution are effectively neutral and not the product of Darwinian selection (Sueoka, 1962; Kimura, 1968, 1983; King and Jukes, 1969). Under the traditional neutralist view, allele frequencies in animal populations increase or decrease randomly until alleles are ultimately fixed or lost by chance. Therefore, at any time, loci possess alleles at frequencies that are neither 0% nor 100% and thus most polymorphisms in populations are seen as transient, ongoing dynamic processes in nature maintained by random genetic drift. While selection can operate on population polymorphism, its force is considered too weak to offset the influence of chance effects. On the genome level, the neutral theory predicts that synonymous nucleotide substitutions, changes that do not affect amino acid structure of proteins, are not under the stringent selective constraints experienced by non-synonymous substitutions. It is well documented that the rate of synonymous substitutions is significantly higher than that of non-synonymous substitutions in various RNA, DNA, mitochondria and chloroplast genomes. Given the high ratio of synonymous to non-synonymous changes in various genetic material, the neutral theory of evolution presumably can be widely applied to numerous organisms.

The neutral theory was groundbreaking in traditional evolutionary biology and critical to the development of molecular evolutionary theory. Temin's and Kimura's groups were the first few investigators to adopt the neutral theory to characterize the evolution of viral genes (Temin, 1988; Gojobori et al, 1990). Consistent with the neutral theory, Kimura's group found that the rate of synonymous nucleotide substitutions was higher than that of non-synonymous substitutions in all examined loci of various lentivirus isolates, even though the substitution rate varied considerably among viruses (Gojobori et al, 1990). Additionally, some virus genes have molecular clock tendencies which would allow predictions of divergence time between variants or between higher taxonomic groups. The rate of genetic variation in influenza viruses has been shown to behave in a clock-like manner resulting in a linear relationship between the mutation rate per replication cycle and the number of replication cycles per unit time (Hayashida et al, 1985; Saitou and Nei, 1986). The fact that the neutral theory is supported by observations of the evolutionary rates and patterns of virus genomes allows one to study their phylogenetics in order to determine which hosts or virus characteristics are associated with high diversity in some virus groups or lack of diversity in others.

Despite the fact that a number of studies have used phylogenetic techniques to infer virus evolutionary history, there are a few problems uniquely associated with the application of phylogenetic techniques to virus evolution that should be kept in mind. It is likely that evolutionary rates of viruses are coupled to the rates of host evolution. Thus, it is not surprising that exogenous viruses have been observed to have a much greater number of replication events per year than endogenous viruses, reported (for example) by Gojobori and Yokoyama (1985). Because the replication rate of the endogenous virus is closely bound to host DNA replication, this discrepancy between the two rates can be attributed to an apparent limitation on endogenous virus replication by the replication rate of the host genome. Additionally, the virus does not evolve uniformly so that genes of viruses may have evolved under different constraints, manifested as different rates of mutation. Mutation rates may be very slow for reverse transcriptase, an enzyme which is both ancient and well conserved, whereas *env* and *gag* gene products can change very rapidly in comparison. Undoubtedly, the host immune system may be another mechanism by which hosts can exert influences on the evolution of viruses. While the evolution of viruses is unlikely to be host-independent or free from other problems, phylogenetic analysis can still be a valuable tool for reconstructing the history of a virus, and analysing genetic variants and disease pathogenicity differential.

Type D viruses

Only a handful of studies have investigated any aspect of evolution of SRV type D viruses, their relationship to each other and to other viruses. Originally, D type viruses were considered to be closely related to HIV simply because the immunosuppressive properties of SRV shared similar features with AIDS. However, it was quickly established by comparisons of the complete sequences of SRV-1 (Power et al, 1986) and SRV-3 (MPMV) (Sonigo et al, 1986) with different isolates of HIV represented by lymphadenopathy-associated virus (Wain-Hobson et al, 1985), human T cell lymphotropic virus type III (Muesing et al, 1985; Ratner et al, 1985) and AIDS-associated retrovirus (ARV-2) (Sanchez-Pescador et al, 1985) that the D type viruses were not closely related to HIV. Similarly, a close relationship of D type viruses to simian T cell lymphotropic virus type III (STLV-III), another virus known to cause immunodeficiency in macaques, was excluded by serological studies (Kanki et al, 1985). Higher systematic studies of retroviruses revealed D type viruses to be a statistically well supported monophyletic group distantly related to both the lentiviruses and to the lymphotropic viruses (Xiong and Eickbush, 1990).

While the genome organization of SRV-1, SRV-2 and SRV-3 is highly conserved, there are genetic differences among the serotypes as well as measurable differences among isolates of the same serotype. When sequences of SRV-1, SRV-2 and SRV-3 were compared, SRV-1 and SRV-3 were shown to be very similar throughout their genome while SRV-2 was genetically distant from the two (Thayer et al, 1987; van der Kuyl et al, 1997). Differences have been found among isolates of the same serotype. One study found microchanges in a fragment of the *env* region by sequencing a portion of *env* (gp70 and gp20) from two isolates of SRV-2 from different species of macaques (rhesus and Celebes) (Marracci et al, 1995). From these data, they suggested that the structural differences between the two isolates in the envelope glycoprotein may result in divergent cell or tissue tropisms that ultimately affect the pathogenic ability of the different isolates of SRV-2.

There is some evidence to suggest that D type viruses may be derived by recombination from other retrovirus groups. The core (*gag*) proteins of D type retroviruses are closely related to the B type retroviruses of mouse mammary tumour virus (MMTV), while the envelope gene is closely related to primate and feline C type retroviruses (Gardner et al, 1994). Another study showed that there is about 50% amino acid similarity between the *pol* regions of MMTV and SRV D types (Thayer et al, 1987). Additionally, hybridization results demonstrated that baboon endogenous virus, a type C retrovirus, shares envelope gp70 antigenic determinants with MPMV (Cohen et al, 1982).

A search for endogenous type D viruses among species of primates revealed that partial sequences of endogenous type D viruses are present in many species of *Papio* and *Cercopithecus*, suggesting that these viruses are very old and universal among species of Old World monkeys (van der Kuyl et al, 1997). In this study, the provirus was not detected in humans and apes. From phylogenetic analysis of the endogenous viruses and from what is known about the evolution of primates, the authors were able to infer the history of the type D endogenous virus (van der Kuyl et al, 1997). If monkeys and hominids diverged from each other about 36 million years ago (MYA) (Szalay and Delson, 1979), then integration of the virus into the cercopithecines did not take place until well after the divergent event. Phylogenetic analysis also indicates that the squirrel monkey endogenous virus (SMRV) is genetically distant, most probably the result of an early separation from its counterpart in the cercopithecines. Because New World and Old World monkey lineages diverged from each other more than 55 MYA (Szalay and Delson, 1979), van der Kuyl et al (1997) suspect that type D retroviruses have a long association with monkeys and may have entered the germline more than once.

Other than the most recent study cited here, very little phylogenetic analysis has been done with the other well-characterized endogenous D types. The genome of the endogenous langur virus (PO-1-Lu) is 30% similar to the MPMV genome, while very little similarity exists between it and the SMRV genome (Todaro et al, 1978a). Likewise, SMRV is more distantly related to MPMV (Heberling et al, 1978). It has been suggested that MPMV emerged from an endogenous langur virus whose hosts infected rhesus macaques, and that infection continues among populations of rhesus monkeys by horizontal transmission. However, no complete sequences have been determined from these endogenous viruses and thus no further phylogenetic analyses are available.

STLV

While much more information exists on simian T lymphotropic virus evolution than on D type retrovirus evolution, there is still controversy about the origin of this virus. The first sequence of STLV was analysed from a pig-tailed macaque and showed around 90% similarity with an HTLV-1 strain (Watanabe et al, 1985, 1986). In other studies, STLV-1 and HTLV-1 could not be separated into distinct phylogenetic clades according to their species of origin, regardless of which gene was used in phylogenetic analysis. Sequence and subsequent phylogenetic analysis of portions of the *env* gene of STLV-1 isolates from chimpanzees and African green monkeys suggested that these were more closely related to the cosmopolitan strain of HTLV-1 than to an STLV-1 strain of a pig-tailed macaque from Indonesia (Watanabe et al, 1985). It has been suggested that either one may have evolved from the other or that they may have evolved from a common ancestor (Watanabe et al, 1985). Because STLV and HTLV are not monophyletic with respect to each other in many phylogenetic analyses, some researchers suggested that STLV and HTLV should be collapsed into one group called 'primate T lymphotropic viruses' (PTLV) (Van Brussel et al, 1996).

Some evidence, such as the existence of extremely divergent HTLV-1 and STLV-1 strains in Asia, implies that both strains evolved in Asia and that STLV-1 existed in Asia long before the macaque speciation and subsequent radiation (Song et al, 1994). However, other evidence based on the LTR region supports an African origin of HTLV-1 and STLV-1 strains (Vandamme et al, 1994). This controversy has led to the speculation that

STLV-1 evolved independently in Asia and Africa (Song et al, 1994). Suggestions for an independent evolution of STLV in Asia are difficult to assess. Despite the broad geographical distribution of STLV-infected monkeys in Asia, there are only three sequenced STLV-1 fragments from monkey species from southeast Asia – *Macaca nemestrina* (Watanabe et al, 1985), *M. fascicularis* (Koralnick et al, 1994) and *M. mulatta* (Koralnick et al, 1994) – unfortunately, their exact geographical locations have not been identified.

Whether studies support an Asian or African origin for STLV, much of the evidence is consistent with the hypothesis that several interspecies transmissions have occurred (Song et al, 1994; Vandamme et al, 1994; Ibrahim et al, 1995). In one study, an STLV isolate from a red monkey in Senegal was genetically closer to isolates found in a West African green monkey than to another red monkey individual from Central Africa, indicating multiple occurrences of cross-species transmission (Saksena et al, 1994). Other data indicate viral transmission between humans and monkeys carrying STLV-1 (Gessain and de The, 1996). Phylogenetic analysis of an STLV isolate from a Sulawesi macaque (*M. tonkeana*) based on LTR and *env* regions implies a closer genetic relationship with a Melanesian HTLV-1 isolate than with any other Asian STLV-1 (Ibrahim et al, 1995). The authors raise the possibility that ancestors of the extant Australo-Melanesians migrated across what was then the southeast Asian land mass to the Australian continent. During this migration, multiple transmissions of STLV-1 from monkeys to humans may have occurred, and thus multiple variants of HTLV-1 persist in the present Melanesian population (Ibrahim et al, 1995).

THE SEARCH FOR HUMAN COUNTERPARTS

Human type D retrovirus

While the exogenous forms of SRV type D are ubiquitous among macaques, and endogenous forms have been found in at least two other monkey species, it is difficult to believe that D type retroviruses have yet to be credibly associated with humans. Even though a host of studies reported serological reactivity of human sera with D type antigens (Thiry et al, 1978a,b; Hunsmann et al, 1990; Morozov et al, 1991), only one truly convincing report exists of a D type retrovirus isolated from humans (Bohannon et al, 1991). Bohannon and co-workers reportedly isolated a D type retrovirus from lymphoma tissue and bone marrow from an AIDS patient seropositive for HIV-1. The patient's serum had antibody reactivity against *gag* and *env* viral gene products by Western blot and radioimmunoprecipitation, and DNA extracts of these tissues following polymerase chain reaction (PCR) were positive for integrated proviral DNA for two different regions. Subsequent sequencing of the amplified fragments revealed a close relationship to MPMV and SRV-1 (Ford et al, 1992). The authors could not explain how the patient obtained a D type infection, as there were no apparent risk factors such as contact with monkeys. They suggested that the depressed state of the immune system would have made the patient more susceptible to infections that a person with a healthy immune system could have resisted.

Since this discovery, large-scale searches were launched by various researchers to find conclusive evidence of SRV-like D type infection among human populations. Although most searches included groups of individuals who were at increased risk for other retroviral infections, none of these studies reported successful isolation of D type virus among these groups. One study tested for D type infection in three groups of subjects, none of whom was conclusively shown to be infected with D type virus: 375 patients with lymphoproliferative diseases (75 of whom were infected with HIV-1); 7 people with unexplained low CD4 lymphocyte counts; and 45 blood donors, some of whom were HTLV-1/2 seroindeterminate (Heneine et al, 1993). In a larger study whose cohort of 897 people included SIV research workers, homosexual men from America, and injecting drug users from Thailand, not one sample reacted seropositively against *gag* and *env* viral gene products by Western blot analysis (Lerche et al, 1995). A more recent study examined a small cohort of healthy people from Guinea for evidence of D type infection and found 6 out of 16 samples to be positive by PCR analysis for both MPMV serotype *gag* and *env* sequences (Morozov et al, 1996). In comparison with previous studies which examined huge cohorts of people at risk for other retroviral infections, the fact that this study found almost 38% of the cohort positive for D type virus is surprising. The authors suggested that transmission might have occurred through direct contact with West African dwarf guenon monkeys, which were found to have antibodies to type D viruses (Ilyinskii et al, 1991). However, it is uncertain whether any person from the cohort had had direct contact with those monkeys or with any other monkey. Until larger cohorts of people from populations who live in close contact with monkeys are examined, evidence of human infection with D type viruses is tenuous at best. The only conclusion that can be drawn from these studies is that D type infections in human populations are rare and have occurred in a few isolated cases. Further investigation may reveal that there are tiny segments of human populations, specifically those who come into close and frequent contact with monkeys, who are indeed likely to be infected with D type viruses.

Human foamy virus

The first human foamy virus to be described was isolated from nasopharyngeal carcinoma tissue from a Kenyan patient (Achong et al, 1971; Epstein et al, 1974). Several other foamy virus isolates from human tissues have been reported. A virus related to SFV-1 was cultured from peripheral blood cells taken from a patient with leukaemia (Young et al, 1973), while isolates from patients with de Quervain subacute granulomatous thyroiditis are well documented (Stancek and Gressnerova; 1974; Stancek et al, 1975). The last recorded identification of an HFV was from the brain of a British patient with encephalopathy (Cameron et al, 1978).

Early seroepidemiology carried out on the prototype virus by immunofluorescence using human sera on HFV-infected cells concluded that there was a significant infection rate among Kenyans with tumours, particularly nasopharyngal carcinomas (Achong and Epstein, 1978). Nemo et al (1978) demonstrated by immunofluorescence microscopy (IFM) and neutralization that HFV was closely related to SFV-6 and that HFV, SFV-6 and SFV-7 were indistinguishable by complement fixation tests. Moreover, since the human isolate was not neutralized by human sera taken from patients with nasopharyngeal carcinoma or Burkitt's lymphoma, and since it shared antigenic determinants with the primate virus, it represented a variant of SFV-6. Thus, as early as 1978 it was being suggested that the human isolate represented either chance human infection with a simian foamy virus, or laboratory contamination (Nemo et al, 1978). This premise was later called into question following a more extensive serosurvey of over 600 samples assayed by IFM carried out in East Africa, which indicated that 3·4% of normal individuals had HFV antibodies (Muller et al, 1980). A further study established that 6·9% of serum samples ($n = 1717$) taken from inhabitants of nine Pacific islands were reactive to HFV by both IFM and confirmatory neutralization assays (Loh et al, 1980). Since chimpanzees are not found in this area, these data argued in favour of a natural HFV infection in humans. A more recent serosurvey on samples collected from patients with a wide variety of diseases from African and other countries using a HFV env-specific enzyme-linked immunosorbent assay (ELISA) suggested that, although the prevalence of anti-HFV antibodies worldwide is low (3·2%), the incidence in African patients is significantly higher (6·3%) (Mahnke et al, 1992). In contrast, Schweizer et al (1995) found no evidence for foamy virus infection in 2688 individuals from several African countries using more stringent serological tests and PCR.

The most recent study (Ali et al, 1996), designed to estimate the prevalence of natural HFV infection, examined an extensive bank of serum samples collected from geographically diverse areas of the world. Many of the sera were antibody-positive for HTLV-1 or HIV. This was considered to be an advantage in detecting HFV infection, since patients immunosuppressed by one retroviral infection might be more likely to harbour novel viral infections, as for example was the case when HHV-6 was discovered (Salahuddin et al, 1986), and with the reported isolation of a D type retrovirus from an AIDS patient with a B cell lymphoma (Bohannan et al, 1991).

While none of the serum samples from Europe, the USA or South America tested antibody-positive, a strong ELISA reactivity was found in a significant proportion of the population of the Pacific islands, in central Africa (Malawi) and in Uganda, areas previously cited as having a high level of HFV infection (Achong et al, 1971; Loh et al, 1980; Muller et al, 1980; Mahnke et al, 1992). However, none of the serum samples was confirmed as positive by Western blot, neutralization and PCR, assays which were either unavailable for earlier studies (PCR), carried out on very few samples – four sera tested by IFM by Loh et al (1980) were further assayed for neutralizing antibody – or unconvincing, e.g. Western analysis by Mahnke et al (1992). This study, on a much larger sample size than had been previously considered, suggested that sera in certain regions of the world, e.g. the Solomon Islands and parts of Africa, produce non-specific serological reactions, necessitating a variety of confirmatory assays for accurate and reliable screening of some populations.

It is worth noting that recent sequence analysis of the higher primate foamy virus (Bieniasz et al, 1995) endorses the phylogenetic and antigenic closeness of SFV-6 and SFV-7 to HFV (Brown et al, 1978; Herchenröder et al, 1994). This, taken together with the fact that recent studies have not been able to confirm the early seroepidemiological evidence of HFV infection, would strongly suggest that HFV is not a virus naturally found in the human population. Since only one HFV isolate is currently available to the scientific community and since the human population does not appear to be a natural host for HFV infection, it follows that reports associating HFV with a variety of disease states should be viewed with caution.

REFERENCES

Achong BG & Epstein MA (1978) Preliminary seroepidemiological studies on the human syncytial virus. *J Gen Virol* **40**: 175–181.

Achong BG, Mansell PW, Epstein MA & Clifford P (1971) An unusual virus in cultures from a human nasopharyngeal carcinoma. *J Natl Cancer Inst* **46**: 299.

Ali M, Taylor GP, Pitman RJ et al (1996) No evidence of antibody to human foamy virus in widespread human populations. *AIDS Res Hum Retroviruses* **12**: 1473–1483.

Axthelm MK, Hallick LM, Mcnulty WP, Shiigi SM, Malley A & Marx PA (1987) Characterization of type D simian retroviruses. *Program and Abstracts of the 5th Annual Symposium on Nonhuman Primate Models for AIDS*, p. 46.

Barker CS, Willis JW, Bradac JA & Hunter E (1985) Molecular cloning of the Mason Pfizer monkey virus genome: characterization and cloning of subgenomic fragments. *Virology* **142**: 223–240.

Baunach G, Maurer B, Hahn H, Kranz M & Rethwilm A (1993) Functional analysis of human foamy virus accessory reading frames. *J Virol* **67**: 5411–5418.

Benveniste RE, Lieber MM, Livingston DM, Sherr CJ & Todaro GJ (1974) Infectious C-type virus isolated from a baboon placenta. *Nature (Lond)* **248**: 17–20.

Benveniste RE, Hill RW, Knott WB, Tsai C-C, Kuller L & Morton WR (1993a) Detection of serum antibodies in Ethiopian baboons that cross-react with SIV, HTLV-I, and type D retroviral antigens. *J Med Primatol* **22**: 124–128.

Benveniste RE, Kuller L, Roodman ST, Hu SL & Morton WR (1993b) Long-term protection of macaques against high-dose type D retrovirus challenge after immunization with recombinant vaccinia virus expressing envelope glycoproteins. *J Med Primatol* **22**: 74–79.

Benzair AB, Rhodes-Feuillette A, Lasnevet J, Emanoil-Ravier R & Peries J (1985) Purification and characterization of the major envelope glycoprotein of simian foamy virus type 1. *J Gen Virol* **66**: 1449.

Bieniasz PD, Rethwilm A, Pitman R, Daniel MD, Chrystie I & McClure MO (1995) A comparative study of higher primate foamy viruses, including a new virus from a gorilla. *Virology* **207**: 217–228.

Bohannan RC, Donehower LA & Ford RJ (1991) Isolation of a type D retrovirus from B-cell lymphomas of a patient with AIDS. *J Virol* **65**: 5663–5672.

Bothe K, Aguzzi A, Lassman H, Rethwilm A & Horak I (1991) Progressive encephalopathy and myopathy in transgenic mice expressing human foamy virus genes. *Science* **253**: 555–557.

Bradac J & Hunter E (1986a) Polypeptides of Mason–Pfizer monkey virus. I. Synthesis and processing of the *gag*-gene products. *Virology* **138**: 260–275

Bradac J & Hunter E (1984b) Polypeptides of Mason–Pfizer monkey virus. III. Translational order of proteins on the *gag* and *env* gene specified precursor polypeptides. *Virology* **150**: 503–508.

Bray M, Prasad S, Dubay JW et al (1994) A small element from the Mason–Pfizer monkey virus genome makes human immunodeficiency virus type I expression and replication Rev-independent. *Proc Natl Acad Sci USA* **91**: 1256–1260.

Brody BA, Hunter E, Kluge JD, Lasarow R, Gardner M & Marx PA (1992) Protection of macaques against infection with simian type D retrovirus (SRV-1) by immunization with recombinant vaccinia virus expressing the envelope glycoproteins of either SRV-1 or Mason–Pfizer monkey virus (SRV-3). *J Virol* **66**: 3950–3954.

Brown PW, Nemo GJ & Gajdusek DC (1978) Human foamy virus: further characterisation, seroepidemiology and relationship to chimpanzee foamy viruses. *J Infect Dis* **137**: 421–427.

Bryant ML, Gardner MB, Marx PA et al (1986) Immunodeficiency in rhesus monkeys associated with the original Mason–Pfizer monkey virus. *J Natl Cancer Inst* **77**: 957–965.

Cameron KR, Birchall SM & Moses MA et al (1978) Isolation of foamy virus from patient with dialysis encephalopathy. *Lancet* **2:** 796.

Campbell M, Renshaw-Gegg L, Renne R & Luciw PA (1994) Characteristics of the internal promoter of simian foamy viruses.

Cavalieri F, Rhodes-Feuillett A, Benzair AB, Emanoil-Ravicovitch R & Peries J (1981) Biochemical characterisation of simian foamy virus type 1. *Arch Virol* **68:** 197.

Chen Y-MA, Jang Y-J, Kanki PJ et al (1994) Isolation and characterization of simian T-cell leukemia virus type II from New World monkeys. *J Virol* **68:** 1149–1157.

Chiswell DJ & Pringle CR (1979) Feline syncytium-forming virus: DNA provirus size and structure. *J Gen Virol* **44:** 145.

Chopra HC & Mason MM (1970) A new virus in a spontaneous tumor of a rhesus monkey. *Cancer Res* **30:** 2081–2086.

Clarke JK & Attridge JT (1968) The morphology of simian foamy agents. *J Gen Virol* **3:** 185.

Clarke JK & McFerran JB (1970) The morphology of bovine syncytial virus. *J Gen Virol* **9:** 155.

Coffin JM (1992) Structure and classification of retroviruses. In: Levy JA (ed.) *The Retroviridae,* New York: Plenum. pp 19–50.

Coffin JM (1996) Retroviridae: the viruses and their replication. In: Fields BN, Knipe DM & Howley PM (eds) *Fields' Virology,* 3rd edn, pp 1767–1847. Philadelphia: Lippincott–Raven.

Cohen M, Rice N, Stephans R & O'Connell C (1982) DNA sequence relationship of the baboon endogenous retrovirus genome to the genomes of the other type C and type D retroviruses. *J Virol* **41:** 801–812.

Cullen BR (1991) Human immunodeficiency virus as a prototypic complex retrovirus. *J Virol* **65:** 1053–1056.

Daniel MD, King NW, Letvin NL, Hunt RD, Sehgal PK & Desrosiers RC (1984) A new type of D retrovirus isolated from macaques with an immunodeficiency syndrome. *Science* **223:** 602–605.

Deinhardt F, Wolfe L, Northrop R et al (1972) Induction of neoplasms by viruses in marmoset monkeys. *J Med Primatol* **1:** 29–50.

De Paoli A, Johnsen DO & Noll WW (1973) Granulocytic leukemia in whitehanded gibbons. *J Am Vet Med Assoc* **163:** 624–628.

Enders J & Peebles T (1954) Propagation in tissue cultures of cytopathic agents from patients with measles. *Proc Soc Biol Med* **86:** 277–287.

Epstein MA, Achong BG & Ball G (1974) Further observations on a human syncytial virus from a nasopharyngeal carcinoma. *J Nat Cancer Inst* **53:** 681.

Ernst RK, Bray M, Rekosh D & Hammarskjold ML (1997) Secondary structure and mutational analysis of the Mason–Pfizer monkey virus RNA constitutive transport element. *RNA* **3:** 210–222.

Fine D & Schochetman G (1978) Type D primate retroviruses: a review. *Cancer Res* **38:** 3123–3139.

Flügel RM, Rethwilm A, Maurer B & Darai G (1987) Nucleotide sequence analysis of the *env* gene and its flanking regions of the human spuma-retrovirus reveals two novel genes. *EMBO J* **6:** 2077.

Ford RJ, Donehower LA & Bohannon RC (1992) Studies of a type D retrovirus isolated from an AIDS patient. *AIDS Res Hum Retroviruses* **8:** 742–751.

Franchini G (1995) Molecular mechanisms of human T-cell leukemia/lymphotropic virus type I infection. *Blood* **86:** 3619–3639.

Gardner MB & Luciw PA (1989) Animal models of AIDS. *FASEB J* **3:** 2593–2606.

Gardner MB, Luciw P, Lerche N & Marx P (1988) Nonhuman primate retrovirus isolates and AIDS. *Adv Vet Sci Comp Med* **32:** 171–226.

Gardner MB, Endres M & Barry P (1994) The simian retroviruses: SIV and SRV. In: Levy JA (ed.) *The Retroviridae,* vol. 3, pp 133–276. New York: Plenum.

Gessain A & de The G (1996) Geographic and molecular epidemiology of primate T lymphotropic retroviruses: HTLV-I, HTLV-II, STLVI, STLV-PP, and PTLV-L. *Adv Virus Res* **47:** 377–426.

Giddens WE, Tsai C-C, Morton WR, Ochs HD, Knitter GH & Blakley GA (1985) Retroperitoneal fibromatosis and acquired immunodeficiency syndrome in macaques. *Am J Pathol* **119:** 253–263.

Giri A, Markharn P, Digilio L, Hurteau G, Gallo RC & Franchini G (1994) Isolation of a novel simian T-cell lymphotropic virus from *Pan paniscus* that is distantly related to the human T-cell leukemia/lymphotropic virus types I and II. *J Virol* **68:** 8392–8395.

Gojobori T & Yokoyama S (1985) Rates of evolution of the retroviral oncogene of Maloney murine sarcoma virus and of its cellular homologues. *Proc Natl Acad Sci USA* **82:** 4198–4201.

Gojobori T, Moriyama EN & Kimura M (1990) Molecular clock of viral evolution, and the neutral theory. *Proc Natl Acad Sci USA* **87:** 10015–10018.

Grant RF, Windsor SK, Malinak CJ, Bartz CR, Sabo A, Benveniste RE & Tsai C-C (1995) Characterization of infectious type D retrovirus from baboons. *Virology* **207:** 292–296.

Gravell M, London WT, Hamilton RS et al (1984) Transmission of simian AIDS with type D retrovirus isolate. *Lancet* **i:** 334–335.

Hayami M, Komuro A, Nozawa K et al (1984) Prevalence of antibody to adult T-cell leukaemia virus-associated antigens (ATLA) in Japanese monkeys and other non-human primates. *Int J Cancer* **33:** 179–183.

Hayashida H, Toh H, Kikuno R & Miyata T (1985) Evolution of influenza virus genes. *Mol Biol Evol* **2:** 289–303.

Heberling RL, Barker ST, Kalter SS, Smith GC & Helmke RJ (1978) Oncornavirus: isolation from a squirrel monkey (*Saimiri sciureus*) lung culture. *Science* **195:** 289–292.

Heidecker G, Lerche NW, Lowenstine LJ et al (1987) Induction of simian acquired immunodeficiency syndrome with a molecular clone of a type D retrovirus. *J Virol* **61:** 3066–3071.

Henderson LE, Sawder R, Symthers G, Benveniste RE & Oroszlan S (1985) Purification and N-terminal amino acid sequence comparisons of structural proteins from retrovirus-D/Washington and Mason Pfizer monkey virus. *J Virol* **55:** 778–787.

Heneine W, Lerche NW, Woods T et al (1993) The search for human infection with simian type D retroviruses. *J AIDS* **6:** 1062–1066.

Heneine W, Switzer WM, Sandstrom P et al (1998) Identification of a human population infected with simian foamy viruses. *Nature Med* **4:** 403–407.

Henrickson RV, Osborn KG, Madden DL et al (1983) Epidemic of acquired immunodeficiency in rhesus monkeys. *Lancet* **i:** 388–390.

Herchenröder O, Renne R, Locar D et al (1994) Isolation, cloning and sequencing of simian foamy viruses from chimpanzees (SFVcpz): high homology to human foamy virus (HFV). *Virology* **201:** 187–199.

Homma T, Kanki PJ, King NW et al (1984) Lymphoma in macaques: association with virus of human T-lymphotropic family. *Science* **225:** 716–718.

Hooks JJ & Detrick-Hooks B (1979) Simian foamy virus-induced immunosuppression in rabbits. *J Gen Virol* **44:** 383.

Hooks JJ & Detrick-Hooks B (1981) Spumavirinae. Foamy virus group infections: comparative aspects and diagnosis. In: Kurstak K & Kurstak C (eds) *Comparative Diagnosis of Viral Diseases,* vol. 4, pp 599–618. New York: Academic Press.

Hooks JJ & Gibbs CJ (1975) The foamy viruses. *Bacteriol Rev* **39:** 169–185.

Hooks JJ, Gibbs CJ, Cutchins EC, Rogers NG, Lampert P & Gajdusek DC (1972) Characterization and distribution of two new foamy viruses isolated from chimpanzees. *Arch Gesamte Virusforsch* **38:** 38–55.

Hooks J, Gibbs CJ, Chou S, Howk R, Lewis M & Gajdusek DC (1973) Isolation of a new simian foamy virus from a spider monkey brain culture. *Infect Immun* **8:** 804–813.

Hotta J & Loh PC (1986) The transforming potential of the human syncytium-forming virus (Spumavirus). *Am Soc Microbiol Abstr* 325.

Hruska JF & Takemoto KK (1975) Biochemical properties of a hamster syncytium-forming ('foamy') virus. *J Natl Cancer Inst* **54**: 601.

Hu S-L, Zarling JM, Chinn J et al (1989) Protection of macaques against simian AIDS by immunization with a recombinant vaccinia virus expressing the envelope glycoproteins of simian type D retrovirus. *Proc Natl Acad Sci USA* **86**: 7213–7217.

Hunsmann G, Schneider J, Schmitt J & Yamamoto N (1983) Detection of serum antibodies to adult T-cell leukemia virus in nonhuman primates and people from Africa. *Int J Cancer* **32**: 329–332.

Hunsmann G, Flugel RM & Walder R (1990) Retroviral antibodies in Indians. *Nature* **345**: 120.

Ibrahim F, de The G & Gessain A (1995) Isolation and characterization of a new simian T cell leukemia virus type I from naturally infected Celebes macaques (*Macaca tonkeana*): complete nucleotide sequence and phylogenetic relationship with the Australo-Melanesian HTLV-I. *J Virol* **69**: 6980–6993.

Ilyinskii P, Daniel M, Lerche N & Desrosiers R (1991) Antibodies to type D retroviruses in talapoin monkeys. *J Gen Virol* **72**: 453–456.

Ishida T, Yamamoto K, Kaneko R, Tokita E & Hinuma Y (1983) Seroepidemiological study of antibodies to adult T-cell leukaemia virus-associated antigen (ATLA) in free-ranging Japanese monkeys (*Macaca fuscata*). *Microbiol Immunol* **27**(3): 297–301.

Ishikawa K, Fukasawa M, Tsujimoto H et al (1987) Serological survey and virus isolation of simian T-cell leukemia/T-lymphotropic virus type I (STLV-I) in nonhuman primates in their native countries. *Int J Cancer* **40**: 233–239.

Jenson EM, Zelljadr I, Chopra HC & Mason MM (1970) Isolation and propagation of a virus from a spontaneous mammary carcinoma of a rhesus monkey. *Cancer Res* **32**: 2388–2393.

Johnston PB (1974) Studies on simian foamy viruses and syncytium-forming viruses of lower animals. *Lab Anim Sci* **24**: 159.

Kanki PJ, McLane MF, King NW et al (1985) Serologic identification and characterization of a macaque T-lymphotropic retrovirus closely related to HTLV-III. *Science* **228**: 1199–1201.

Kato S, Matsuo K, Nishimura N, Takahashi N & Takano T (1987) The entire nucleotide sequence of baboon endogenous virus DNA: a chimeric genome structure of murine type C and simian type D retroviruses. *Jpn J Genet* **62**: 127–137.

Kawakami TG & Buckley PM (1974) Antigenic studies on gibbon ape type C viruses. *Transplant Proc* **6**: 193–196.

Kawakami TG, Huff SD, Buckley PM, Dungworth DL, Snyder SP & Gilden RV (1972) C-type virus associated with gibbon lymphosarcoma. *Nature New Biol* **235**: 170–171.

Kawakami TG, Buckley PM, McDowell TS & Depaoli A (1973) Antibodies to simian C-type virus antigen in sera of gibbons (*Hylobates* sp.). *Nature New Biol* **246**: 105–107.

Kawakami TG, Sun L & McDowell TS (1978) Natural transmission of gibbon leukemia virus. *J Natl Cancer Inst* **61**: 1113–1115.

Kawakami TG, Kollias GV & Holmberg C (1980) Oncogenicity of gibbon type-C myelogenous leukemia virus. *Int J Cancer* **25**: 641–646.

Keller A, Partin KM, Löchelt M, Bannert H, Flügel R & Cullen BR (1991) Characterisation of the transcriptional *trans* activator of human foamy retrovirus. *J Virol* **65**: 2589–2594.

Kilbourne ED (1978) The molecular epidemiology of influenza. *J Infect Dis* **127**: 478–487.

Kimura M (1968) Evolutionary rate at the molecular level. *Nature* **217**: 624–626.

Kimura M (1983) *The Neutral Theory of Molecular Evolution*. Cambridge: Cambridge University Press.

King JL & Jukes TH (1969) Non-Darwinian evolution. *Science* **164**: 788–798.

Koralnick IJ, Boeri E, Saxinger WC et al (1994) Phylogenetic associations of human and simian T-cell leukemia/lymphotropic virus type I strains: evidence for interspecies transmission. *J Virol* **68**: 2693–2707.

Kupiec JJ, Kay A, Haya A, Ravier R, Peries J & Galibert F (1991) Sequence analysis of the simian foamy virus type-1 genome. *Gene* **101**: 185–194.

Kwang H-S, Barr PJ, Sabin EA et al (1988) Simian retrovirus-D serotype 1 (SRV-1) envelope glyco-proteins gp70 and gp20: expression in yeast cells and identification of specific antibodies in sera from monkeys that recovered from SRV-1 infection. *J Virol* **62:** 1774–1780.

Lackner AA, Rodriguez MH, Bush CE et al (1988) Distribution of macaque immunosuppressive type D retrovirus in neural, lymphoid, and salivary tissues. *J Virol* **62:** 2134–2142.

Lairmore MD, Lerche NW, Schultz KT et al (1990) SIV, STLV-I and type D retrovirus antibodies in captive macaques and immunoblot reactivity to SIV p27 in human and rhesus monkey sera. *AIDS Res Hum Retroviruses* **6:** 1233–1238.

Lal RB, Rudolf DL, Griffis KP et al (1991) Characterization of immunodominant epitopes of *gag* and *pol* gene-encoded proteins of human T-cell lymphotropic virus type I. *J Virol* **65:** 1870–1876.

Lerche NW (1992) Epidemiology and control of type D retrovirus infection in captive macaques. In: Matano S, Tuttle RH, Ishida H & Goodman M (eds) *Topics in Primatology, vol 3: Evolutionary Biology, Reproductive Endocrinology and Virology*, pp 439–448. Tokyo: University of Tokyo Press.

Lerche NW, Hendrickson RV & Maul DH (1984) Epidemiologic aspects of an outbreak of acquired immunodeficiency in rhesus monkeys (*Macaca mulatta*). *Lab Anim Sci* **34:** 146–150.

Lerche NW, Osborn KG, Marx PA et al (1986) Inapparent carriers of simian acquired immune deficiency syndrome type D retrovirus and disease transmission with saliva. *J Natl Cancer Inst* **77:** 489–496.

Lerche NW, Marx PA, Osborn KG et al (1987) Natural history of endemic type D retrovirus infection and acquired immunodeficiency syndrome in group-housed rhesus monkeys. *J Natl Cancer Inst* **79:** 847–885.

Lerche NW, Heneine W, Kaplan JE, Spira T, Yee JL & Khabbaz RF (1995) An expanded search for human infection with simian type D retrovirus. *AIDS Res Hum Retroviruses* **11:** 527–529.

Letvin NL, Eaton KA, Aldrich WT et al (1983) Acquired immunodeficiency syndrome in a colony of macaque monkeys. *Proc Natl Acad Sci USA* **80:** 2718–2722.

Lewe G & Flügel RM (1990) Comparative analyses of the retroviral *pol* and *env* protein sequences reveal different evolutionary trees. *Virus Genes* **3:** 195.

Liu NT, Naturi T, Chang KSS & Wu AMM (1977) Reverse transcriptase of foamy virus: purification of the enzyme and immunological identification. *Arch Virol* **55:** 1987.

Löchelt M, Zentgraf H & Flugel RM (1991) Construction of an infectious DNA clone of the full-length human spuma retrovirus genome and mutagenesis of the bel-1 gene. *Virology* **184:** 43–54.

Löchelt M, Flügel RM & Aboud M (1994) The human foamy virus internal promoter directs the expression of the functional Bel-1 transactivator and Bel protein early after infection. *J Virol* **68:** 638–645.

Loh PC & Ang KS (1981) Reflection of human syncytium-forming virus in human cells: effects of certain biological factors and selective chemicals. *J Med Virol* **7:** 67–72.

Loh PC, Achong BC & Epstein MA (1977) Further biological properties of the human syncytial virus. *Intervirology* **8:** 204.

Loh PC, Mutsuuru F & Mizumoto C (1980) Seroepidemiology of human syncytial virus: antibody prevalence in the Pacific. *Intervirology* **13:** 87–90.

Lowenstine LJ & Lerche NW (1988) Retrovirus infections of nonhuman primates: a review. *J Zoo Anim Med* **19:** 168–187.

Lowry DR (1985) Transformation and oncogenesis: retroviruses. In: Fields BN (ed.) *Fields Virology*, pp 235–623. New York: Raven.

Mahnke C, Kashaiya P, Rössler J et al (1992) Human spumavirus antibodies in sera from African patients. *Arch Virol* **123:** 243–253.

Malmquist WA, Van der Maaten MJ & Boothe AD (1969) Isolation, immunodiffusion, immunofluorescence and electron microscopy of a syncytial virus of lymphosarcomatous and apparently normal cattle. *Cancer Res* **29:** 188.

Marracci GH, Kelley RD, Pilcher KY et al (1995) Simian AIDS type D serogroup 2 retrovirus: isolation of an infectious molecular clone and sequence analyses of its envelope glycoprotein gene and 3' long terminal repeat. *J Virol* **69**: 2621–2628.

Marx PA, Maul DH, Osborn KG et al (1984) Simian AIDS: Isolation of a type D retrovirus and disease transmission. *Science* **223**: 1083–1086.

Marx PA, Bryant ML, Osborn KG et al (1985) Isolation of a new serotype of simian acquired immune deficiency syndrome type D retrovirus from Celebes black macaques (*Macaca nigra*) with immune deficiency and retroperitoneal fibromatosis. *J Virol* **56**: 571–578.

Marx PA, Pederson NC, Lerche NW et al (1986) Prevention of simian acquired immune deficiency syndrome with a formalin-inactivated type D retrovirus vaccine. *J Virol* **60**: 431–435.

Mason MM, Bogden AE, Ilievski V, Esber HJ, Baker JR & Chopra HC (1972) History of a rhesus monkey adenocarcinoma containing virus particles resembling oncogenic RNA viruses. *J Natl Cancer Inst* **48**: 1323–1331.

Maul DH, Lerche NW & Osborn KG (1986) Pathogenesis of simian AIDS in rhesus macaques inoculated with SRV-1 strain of type D retrovirus. *Am J Vet Res* **47**: 863–868.

Maurer B, Bannert H, Darai G & Flügel RM (1988) Analysis of the primary structure of the long terminal repeat and the *gag* and *pol* genes of the human spumaretrovirus. *J Virol* **62**: 1590–1597.

McClure MO, Bieniasz PD, Schulz TF et al (1994) Isolation of a new foamy retrovirus from orang-utans: association with encephalopathy. *J Virol* **68**: 7124–7130.

McKissick GE & Lamont RH (1970) Characteristics of a virus isolated from a feline fibro-sarcoma. *J Virol* **5**: 247.

Miyoshi I, Yoshimoto S, Fujisjita M et al (1982) Detection of type C virus particles in Japanese monkeys seropositive to adult T-cell leukaemia-associated antigens. *Gann* **73**: 848–849.

Miyoshi I, Fujisjita M, Taguchi H, Matsubayashi K, Miwa N & Tanioka Y (1983) Natural infection in non-human primates with adult T cell leukemia virus or a closely related agent. *Int J Cancer* **33**: 333–336.

Morozov VA, Saal F, Gessain A, Terrinha A & Peries J (1991) Antibodies to *gag* gene-coded polypeptides of Mason–Pfizer monkey virus in healthy people from Guinea Bissau. *Intervirology* **32**: 253–257.

Morozov VA, Lagaye S & ter Meulen J (1996) Type D retrovirus markers in healthy Africans from Guinea. *Res Virol* **147**: 341–351.

Muesing MA, Smith DH, Cabradella CD, Benton CV, Laskey LA & Capon DJ (1985) Nucleic acid structure and expression of the human AIDS/lymphadenopathy retrovirus. *Nature (Lond)* **313**: 250–258.

Muller HK, Ball G, Epstein MA, Achong BG, Lenoir G & Levin A (1980) The prevalence of naturally occurring antibodies to human syncytial virus in East African populations. *J Gen Virol* **47**: 399–406.

Murphy FA, Fauquet CM, Bishop DHL et al (eds) (1995) *Virus Taxonomy Classification and Nomenclature of Viruses.* Sixth Report of the International Committee on Taxonomy of Viruses, pp 193–204. Vienna: Springer.

Nemo GJ, Brown PW, Gibbs CJ & Gajdusek DC (1978) Antigenic relationship of human foamy virus to the simian foamy viruses. *Infect Immun* **20**: 69–72.

Palker TJ, Riggs ER, Spragion DE et al (1992) Mapping of homologous, amino-terminal neutralizing regions of human T-cell lymphotropic virus type I and II gp46 envelope glycoproteins. *J Virol* **66**: 5879–5889.

Parks WP & Todaro GJ (1972) Biological properties of syncytium-forming ('foamy') viruses. *Virology* **47**: 673.

Poiesz BJ (1995) Etiology of acute leukaemia: molecular genetics and viral oncology. In: Wiernik PH, Canellos GP, Dutcher JP & Kyle RA (eds) *Neoplastic Diseases of the Blood*, 3rd edn, pp 159–175. New York: Churchill Livingstone.

Poiesz BJ, Ruscetti FW, Gazdar AF, Bunn PA, Minna JD & Gallo RC (1980) Detection and isolation of type C retrovirus particles from fresh and cultured lymphocytes of a patient with cutaneous T-cell lymphoma. *Proc Natl Acad Sci USA* **77**: 7415–7419.

Poiesz BJ, Ruscetti FW, Reitz MS, Kalyanaraman VS & Gallo RC (1981) Isolation of a new type C retrovirus (HTLV) in primary uncultured cells of a patient with Sezary T-cell leukaemia. *Nature* **294**: 268–271.

Power MD, Marx PA, Bryant ML, Gardner MB, Barr PJ & Luciw PA (1986) Nucleotide sequence of SRV-1, a type D simian acquired immune deficiency syndrome retrovirus. *Science* **231**: 1567–1572.

Rabin RH (1971) Assay and pathogenesis of oncogenic viruses in nonhuman primates. *Lab Anim Sci* **21**: 1032–1049.

Rabin RH, Benton CV, Tainsky MA, Rice NR & Gilden RV (1979) Isolation and characterization of an endogenous type C virus of rhesus monkeys. *Science* **204**: 841–842.

Ratner L, Haseltine W, Patarca R et al (1985) Complete nucleotide sequence of the AIDS virus, HTLV-III. *Nature (Lond)* **313**: 277–284.

Renne R, Mergia A, Renshaw-Gegg LW, Newmann-Haefelin D & Luciw PA (1993) Regulatory elements in the long terminal repeat (LTR) of simian foamy virus type-3 (SFV-3). *Virology* **192**: 365–369.

Rethwilm A, Baunach GFO, Mauser B, Bovisch B & Meulen V (1990) Infectious DNA of the human spumavirus. *Nucleic Acids Res* **18**: 733–738.

Rethwilm A, Erlwein O, Baunach G, Maurer B & ter Meulen V (1991) The transcriptional transactivator of human foamy virus maps to the bel 1 genomic region. *Proc Natl Acad Sci USA* **88**: 941–945.

Riggs JL, Oshire LS, Taylor DON & Lennett EH (1969) Syncytium-forming agents isolated from domestic cats. *Nature* **222**: 1190.

Rizvi TA, Lew KA, Murphy EC & Schmidt RD (1996) Role of Mason–Pfizer monkey virus (MPMV) constitutive transport element (CTE) in the propagation of MPMV vectors by genetic complementation using homologous/heterologous *env* genes. *Virology* **224**: 517–532.

Ruckle G (1958) Studies with the monkey-intra-nuclear-agent (MINIA) and foamy-agent derived from spontaneously degenerating monkey kidney cultures I. Isolation and tissue culture behaviour of the agents and identification of MINIA as closely related to measles virus. *Arch Gesamte Virusforsch* **8**: 167–139.

Rustigian R, Johnston P & Rejhart H (1955) Infection of monkey kidney tissue culture with virus-like agents. *Proc Soc Exp Biol Med* **88**: 8–16.

Saitou N & Nei M (1986) Polymorphism and evolution of influenza A virus genes. *Mol Biol Evol* **3**: 57–74.

Sakakibara I, Sugimoto Y, Sasagawa A et al (1986) Spontaneous malignant lymphoma in an African green monkey naturally infected with simian lymphotropic virus (STLV). *J Med Primatol* **15**: 311–318.

Saksena NK, Herve V, Durano JP et al (1994) Seroepidemiologic, molecular, and phylogenetic analyses of simian T-cell leukemia viruses (STLV-I) from various naturally infected monkey species from Central and Western Africa. *Virology* **198**: 297–310.

Salahuddin SZ, Dharam VA, Markham PD et al (1986) Isolation of a new virus, HBLV, in patients with lymphoproliferative disorders. *Science* **234**: 596–601.

Sanchez-Pescador R, Power MD, Barr PJ et al (1985) Nucleotide sequence and expression of an AIDS-associated retrovirus (ARV). *Science* **227**: 484–492.

Schatzl H, Yakovleva L, Lapin B et al (1992) Detection and characterization of T-cell leukemia virus-like proviral sequences in PBL and tissues of baboons by PCR. *Leukemia* **6**: 158–160.

Schultz KT, Benveniste RE, Bridson WE, Houser WD, Uno H & Warner TFCS (1989) Pathologic and virologic description of three cases of type D retrovirus infection in rhesus monkeys and a brief review of nonhuman primate retroviruses. *Zoo Biol* (suppl. 1): 77–87.

Schweizer M, Turek R, Hahn H et al (1995) Markers of foamy virus (FV) infections in monkeys, apes and accidentally infected humans: appropriate testing fails to confirm suspected FV prevalence in man. *AIDS Res Hum Retroviruses* **11**: 161–170.

Sherwin SA & Todaro GJ (1979) A new primate type C virus isolated from the Old World monkey *Colobus polykomos*. *Proc Natl Acad Sci USA* **76**: 5041–5045.

Song K-J, Nerurkar VR, Saituo N et al (1994) Genetic analysis and molecular phylogeny of simian T-cell lymphotropic virus type I: evidence for independent virus evolution in Asia and Africa. *Virology* **199**: 56–66.

Sonigo P, Barker C, Hunter E & Wain-Hobson S (1986) Nucleotide sequence of Mason–Pfizer monkey virus: an immunosuppressive D-type retrovirus. *Cell* **45**: 375–385.

Stancek D & Gressnerova M (1974) A viral agent from a patient with sub-acute de Quervain type thyroiditis. *Acta Virol* **18**: 365.

Stancek D, Stancekova M, Janotka M, Hnilica P & Oravec D (1975) Isolation and some serological and epidemiological data on the viruses recovered from patients with subacute thyroiditis de Quervain. *Med Microbiol Immunol* **161**: 133–144.

Stiles GE, Bittle JL & Cabasso UJ (1964) Comparison of simian foamy virus strains including a new serological type. *Nature* **201**: 1350.

Stromberg K, Benveniste RE, Arthur LO et al (1984) Characterization of exogenous type D retrovirus from a fibroma of a macaque with simian AIDS and fibromatosis. *Science* **224**: 289–292.

Sueoka N (1962) On the genetic basis of variation and heterogeneity of DNA base composition. *Proc Natl Acad Sci USA* **48**: 582–592.

Szalay FS & Delson E (1979) *Evolutionary History of the Primates*. San Diego: Academic Press.

Tabernero C, Zolotukhin AS, Valentin A, Pavlakis GN & Felber BK (1996) The posttranscriptional control element of the simian retrovirus type I forms an extensive RNA secondary structure necessary for its function. *J Virol* **70**: 5998–6011.

Teich NM, Weiss RA, Salahuddin SZ, Gallagher RE, Gillespie DH & Gallo RC (1975) Infective transmission and characterisation of a C-type virus released by cultured human myeloid leukaemia cells. *Nature* **256**: 551–555.

Temin HM (1988) Evolution of cancer genes as a mutation-driven process. *Cancer Res* **48**: 1697–1701.

Thayer RM, Power MD, Bryant ML, Gardner MB, Barr PJ & Luciw PA (1987) Sequence relationships of type D retroviruses which cause simian acquired immunodeficiency syndrome. *Virology* **157**: 317–329.

Thielen GH, Gould D, Fowler M & Dungworth DL (1971) C-type virus in tumor tissue of a woolly monkey (*Lagothrix* spp.) with fibrosarcoma. *J Natl Cancer Inst* **47**: 881–889.

Thiry L, Sprecher-Goldberger S, Bossens M, Cogniaux-LeClerc J & Vereer-Straeten P (1978a) Neutralization of Mason–Pfizer virus by sera from patients treated for renal disease. *J Gen Virol* **41**: 587–597.

Thiry L, Sprecher-Goldberger S, Bossens M & Neuray F (1978b) Cell-mediated immune response to simian oncornavirus antigens in pregnant women. *J Cancer Inst* **60**: 527–532.

Todaro GJ, Benveniste RE, Sherr CJ, Schlom J, Schidlovsky G & Stephenson JR (1978a) Isolation and characterization of a new type of C retrovirus from the Asian primate, *Prebytis obscurus* (spectacled langur). *Virology* **84**: 189–194.

Todaro GJ, Benveniste RE, Sherwin SA & Sherr CJ (1978b) MAC-1, a new genetically transmitted type C virus of primates: 'low frequency' activation from stumptail monkey cell cultures. *Cell* **13**: 775–782.

Todaro GJ, Sherr CJ, Sen A, King N, Daniel MD & Fleckenstein B (1978c) Endogenous New World primate type C viruses isolated from owl monkey (*Aotus trivirgatus*) kidney cell lines. *Proc Natl Acad Sci* **75**: 1004–1008.

Tsai CC, Follis KE & Warner TFCS (1986) Vertical transmission of simian retroviruses (abstr.). *Lab Invest* **56**: 81A.

Van Brussel M, Goubau P, Rousseau R, Desmyter J & Vandamme AM (1996) The genomic structure of a new simian T-lymphotropic virus STLV-PH969, differs from that of human T-lymphotropic virus types I and II. *J Gen Virol* **77**: 347–358.

Van der Kuyl AC, Dekker JT & Goudsmit J (1995) Distribution of baboon endogenous virus among species of African monkeys suggests multiple ancient cross-species transmission in shared habitats. *J Virol* **69**: 7877–7887.

Van der Kuyl AC, Mang R, Dekker JT & Goudsmit J (1997) Complete nucleotide sequences of simian endogenous type D retrovirus with intact genome organization: evidence for ancestry to simian retrovirus and baboon endogenous virus. *J Virol* **71**: 3666–3676.

Vandamme AM, Liu HF, Goubau P & Desmyter J (1994) Primate T-lymphotropic virus type I LTR sequence variation and its phylogenetic analysis: compatibility with an African origin of PTLV-I. *Virology* **202**: 212–223.

Vandamme AM, Liu HF, van Brussel M, de Meurichy W, Desmyter J & Goubau P (1996) The presence of a divergent T-lymphotropic virus in a wild-caught pygmy chimpanzee (*Pan paniscus*) supports an African origin for the human T-lymphotropic/simian T-lymphotropic group of viruses. *J Gen Virol* **77**: 1089–1099.

Wain-Hobson S, Sonigo P, Danus O, Cole S & Alizon M (1985) Nucleotide sequence of the AIDS virus, LAV. *Cell* **40**: 9–17.

Watanabe T, Seiki M, Tsujimoto H, Miyoshi I, Hayami M & Yoshida M (1985) Sequence homology of the simian retrovirus (STLV) genome with human T-cell leukemia virus type I (HTLV-I). *Virology* **144**: 59–65.

Watanabe T, Seiki M, Hirayama Y & Yoshida M (1986) Human T-cell leukemia virus type I is a member of the African subtype of simian viruses (STLV). *Virology* **148**: 385–388.

Weiss RA (1988) A virus in search of a disease. *Nature* **333**: 497–498.

Weiss RA & Wong AL (1977) Phenotypic mixing between avian and mammalian RNA tumor viruses. I. Envelope pseudotypes of Rous sarcoma virus. *Virology* **52**: 535–552.

Weiss RA, Teich N, Varmus H & Coffin J (1984) *RNA Tumor Viruses*. New York: Cold Spring Harbor.

Wolfe LG, Deinhardt F, Theilen GH, Rabin H, Kawakami T & Bustad LK (1971) Induction of tumors in marmoset monkeys by simian sarcoma virus, type I (*Lagothrix*): a preliminary report. *J Natl Cancer Inst* **47**: 1115–1120.

Xiong Y & Eickbush TH (1990) Origin and evolution of retroelements based upon their reverse transcriptase sequences. *EMBO J* **9**: 3353–3362.

Yoshimura N, Nakamura H, Ishikawa K, Noda Y, Honjo S & Hayauri M (1990) Simian T cell leukemia virus type-I specific killer T cells in naturally infected African green monkey carriers. *J Immunol* **144**: 2173–2178.

Young D, Samuels J & Clarke JK (1973) A foamy virus of possible human origin isolated in BHK-21 cells. *Arch Gesamte Virusforsch* **42**: 228–234.

Yu SF & Linial ML (1993) Analysis of the *bel* and *bet* open reading frames of human foamy virus by using a new quantitative assay. *J Virol* **67**: 6618–6634.

Chapter 16

GENETIC DIVERSITY AND MOLECULAR EPIDEMIOLOGY OF PRIMATE T CELL LYMPHOTROPIC VIRUSES: HUMAN T CELL LEUKAEMIA/LYMPHOMA VIRUSES TYPES 1 AND 2 AND RELATED SIMIAN RETROVIRUSES (STLV-1, STLV-2 PAN-P AND PTLV-L)

Antoine Gessain and Renaud Mahieux

HISTORICAL BACKGROUND AND EPIDEMIOLOGICAL FEATURES

The primate T cell lymphotropic viruses (PTLVs), which include the human T cell leukaemia/lymphoma-lymphotropic viruses type 1 and type 2 and the related simian retroviruses (STLV-1 and STLV-2 pan-p), as well as the recently described PTLV-L, exhibit common features including a characteristic and peculiar microepidemiology with a puzzling distribution throughout the world, similar modes of transmission mainly through breast-feeding and sexual contact, and a high genetic stability over time, which can be used as a molecular guide to the migrations of infected populations in the recent or distant past.

HTLV-1

The first human oncoretrovirus was isolated in 1980 in Dr Gallo's laboratory from cell cultures from a African American patient suffering from a lymphoproliferative disease originally considered to be a cutaneous T cell lymphoma but later characterized as an adult T cell leukaemia. This virus was named human T cell leukaemia/lymphoma virus (HTLV) (Poiesz et al, 1980, 1981). Quite independently, in Japan, Takatsuki and colleagues had described in 1977 a T cell lymphoproliferation which they labelled adult T cell leukaemia/lymphoma (ATL) (Takatsuki et al, 1977). The epidemiological features suggested a strong environmental factor which prompted researchers to characterize the tumour cells and to search for an oncogenic virus, which was isolated in 1981 and named adult T cell leukaemia/lymphoma virus (ATLV) (Hinuma et al, 1981; Miyoshi et

HIV and the New Viruses Second Edition
ISBN 0-12-200741-7

al, 1981; Yoshida et al, 1982). Japanese and American scientists rapidly showed that both isolates were the same virus, and agreed to name it HTLV-1 (Popovic et al, 1982). In parallel, the causal association between ATL and HTLV-1 was established. In 1983, we initiated studies in the French West Indies to investigate the epidemiological and clinical impact of HTLV-1 in this area. This led us to discover the association between this virus and a neuromyelopathy which is endemic in the Caribbean, originally named tropical spastic paraparesis (TSP) (Gessain and Gout, 1992; Gessain et al, 1985). A similar entity was then recognized in Japan and labelled HTLV-1-associated myelopathy (HAN) (Osame et al, 1986, 1987). These two conditions were considered to be the same clinicovirological entity and this myelopathy is now referred to as TSP/HAM. HTLV-1 has also been associated with other clinical conditions including uveitis (mainly in Japan), infective dermatitis (mainly in Jamaica), and some cases of arthritis and polymyositis. The question then arose as to whether the same virus could induce, through different pathways, two different diseases, or, as in the case of murine leukaemia viruses, whether specific mutations in certain structural viral genes could control tissue tropism and direct pathogenesis (see below). Since HTLV-1 – in sharp contrast to HIV-1 and HIV-2 (Goodenow et al, 1989; Boeri et al, 1992; Wain-Hobson, 1992) – exhibits a high genomic stability, it was possible to try to answer that question and also to determine whether or not the HTLV-1 isolates were identical in different regions of the world.

HTLV Type 1, which is not a ubiquitous virus, is present throughout the world with clusters of high endemicity located often near to areas where the virus is nearly absent (Blattner, 1990; Mueller, 1991; Kaplan and Khabbaz, 1993; Blattner and Gallo, 1994; Gessain, 1996, 1996b). These highly endemic areas are the southwestern part of the Japan archipelago (mainly the islands of Okinawa, Kiyushu and Shikoku), the Caribbean area and its surrounding regions, foci in South America including Colombia, French Guyana and parts of Brazil, equatorial Africa (Gabon, Congo), in the Middle East (the Mashhad region in Iran) and isolated clusters in Melanesia. The origin of this puzzling geographical or rather ethnic distribution is not well understood but is probably linked to a founder effect in certain ethnic groups, followed by the persistence of a high viral transmission rate due to favourable environmental and cultural local situations. As an example, the apparent progressive decrease of the HTLV-1 seroprevalence in southern Japan seems to be due to a slow decrease in viral transmission in the second half of the twentieth century. This could be related to important modifications in the health care system, or to nutritional and socioeconomic factors, including the diminution of duration of breast-feeding (Oguma et al, 1992). In all highly endemic areas, and despite different socioeconomic and cultural environments, the HTLV-1 seroprevalence increases gradually with age, especially among women aged 20–40 years. This is possibly due either to an accumulation of sexual exposures with age, or to a cohort effect. The global infected population is estimated to be around 15–20 million people, of whom 2–10% will develop an HTLV-1-associated disease (ATL, TSP/HAM, uveitis) during their lifetime (De Thé and Bomford, 1993). Three modes of transmission have been demonstrated for HTLV-1:

1. Mother-to-child transmission is mainly linked to the persistence of breast-feeding after 4–6 months of age: 15–20% of breast-fed children born to HTLV-1-seropositive mothers will become infected and become HTLV-1-seropositive carriers (Tajima and Ito, 1990; Takahashi et al, 1991; Katamine et al, 1994).

2. Sexual transmission, which occurs mainly from male to female, is thought to be responsible for the increased seroprevalence with age in women (Takezaki et al, 1995).
3. Transmission by contaminated blood products is responsible for an acquired HTLV-1 infection in 15–60% of blood recipients (Sandler et al, 1991; Kaplan and Khabbaz, 1993).

HTLV-2

In 1982, a second human lymphotropic retrovirus, labelled HTLV-2, was also isolated in Dr Gallo's laboratory from a cell line derived from the splenic cells of a patient suffering from a lymphoproliferative disease, originally described as a 'T variant of hairy cell leukemia' (Kalyanaraman et al, 1982). This virus is known to be highly endemic among some disparate New World populations including the Navajo and Pueblo in New Mexico (Hjelle et al, 1993), the Seminole in Florida (Levine et al, 1993), the Guyami in Panama (Lairmore et al, 1990; Pardi et al, 1993a,b), the Cayapo and Kraho in Brazil (Maloney et al, 1992; Black et al, 1994; Ishak et al, 1995), the Wayu (Ijichi et al, 1993; Duenas-Barajas et al, 1993) and Orinoco (Fujiyama et al, 1993) in Colombia, and the Tobas and Matacos in north Argentina (Biglione et al, 1993, 1999; Ferrer et al, 1993, 1996). In these ethnic groups, HTLV-2 seroprevalence varies greatly, but can reach up to 20% of the general adult population and up to 50% in women aged 50 years and above, as described in some Kraho groups (Maloney et al, 1992; Ishak et al, 1995). In the western world, HTLV-2 takes an endemo-epidemic course among injecting drug abusers in the USA (Lee et al, 1989, 1993), in some South and Central American countries, and to a lesser extent in Europe (Italy and Spain) (De Rossi et al, 1991; Zella et al, 1990, 1993). While breast-feeding appears to be the predominant mode of transmission in the developing world (Heneine et al, 1992; Tuppin et al, 1996), infection occurs mainly through sharing contaminated needles and sexual transmission in industrialized countries (Lee et al, 1989; Hall et al, 1992, 1996; Murphy, 1996). Furthermore, some studies are consistent with the hypothesis that heterosexual transmission may play a significant role in the spread of HTLV-2 in both environments (Tuppin et al, 1996). The pathogenic role of the virus remains unclear, despite its presence in some rare cases of CD8 lymphoproliferative diseases and neuromyelopathies (Fouchard et al, 1995; Hall et al, 1996; Murphy, 1996).

STLV-1 and STLV-2 pan-p

Isolated in 1982 by Dr Miyoshi and colleagues, STLV-1 is highly prevalent in a large variety of Old World monkey species (Miyoshi et al, 1982; Komuro et al, 1984). Thus, for example, STLV-1 or antibodies against STLV-1 have been detected in chimpanzees (*Pan troglodytes*), gorillas (*Gorilla gorilla*), grivet monkeys (*Cercopithecus aethiops aethiops*), several baboon subspecies (*Papio anubis, P. cynocephalus, P. hamadryas, P. ursinus*), cynomolgus or crab-eating macaques (*Macaca fascicularis*), pig-tailed macaques (*M. nemestrina*), stump-tailed macaques (*M. arctoides*), rhesus macaques (*M. mulatta*), bonnet macaques (*M. radiata*), lion-tailed macaques (*M. silenus*), toque monkeys (*M. sinica*) and Celebes macaques (*M. nigrescens, M. nigra, M. hecki, M. tonkeana, M. maura, M. ochreata* and *M. brunnescens*) (Miyoshi et al, 1982; Ishida et al, 1983; Hayami et al, 1984; Becker et al,

1985; Ishikawa et al, 1987; Ibrahim et al, 1995). In contrast, New World monkey species (except for a non-confirmed observation; Chen et al, 1994) or prosimians lack infection by STLV-1 (Kaplan et al, 1993; Ibrahim et al, 1995; A. Gessain, unpublished observations). In most of the monkey colonies studied, STLV-1 seroprevalence increases with age, is more elevated in females and, as observed in human species, STLV-1 transmission seems to occur mainly from mother to offspring through breast-feeding and from male to female by sexual contact (Ibrahim et al, 1995). These observations strengthen the view of the great similarities, if not identities, between simian and human T lymphotropic type 1 retrovirus characteristics. However, cases of ATL-like disease, with clonal integration of STLV-1 provirus in the tumoral cells, have been described in only few species such as African green monkeys, macaques and gorillas, while no neurological disease similar to TSP/HAM has been so far reported in monkeys infected with STLV-1.

Two different but closely related strains of a new HTLV-2-related simian retrovirus have been independently isolated and characterized from two different captive colonies of *Pan paniscus* (bonobo chimpanzee), one by a group in Belgium (Liu et al, 1994b), the other in the USA (Giri et al, 1994). The Belgian isolate was named STLV-pp1664, while the American one was named STLV-2 pan-p. A familial study conducted in the US colony suggested, as is the case for PTLV-1, a transmission from mother to offspring and from male to female (Giri et al, 1994).

PTLV-L

A simian retrovirus different from both HTLV-1 and HTLV-2 has been isolated from a single baboon (*Papio hamadryas*) originating from Ethiopia but kept in captivity in Leuven (Goubau et al, 1994). No data are currently available on the transmission of this virus (PTLV-L) or on its prevalence in the wild.

GENETIC STRUCTURE OF HTLV-1 AND HTLV-2

The viruses HTLV types 1 and 2, together with STLV-1/2 and PTLV-L, form the group of primate T lymphotropic exogenous (onco)retroviruses, sharing the same overall genetic organization (Cann and Chen, 1990). In addition to the *gag*, *pol* and *env* coding regions, and to the long terminal repeats (LTRs) which contain major regulatory sequences, these viruses possess, inserted between the *env* gene and the 3' LTR (Figure 16.1), the pX region. This region encodes different regulatory proteins, named Tax (respectively p40/I, p37/II, a viral transcriptional transactivator), Rex (respectively p27/I, p26/24II, a post-transcriptional regulator of viral expression) and a third protein (p21/I, P20–22/II) of yet unknown function (Cann and Chen, 1990; Koralnik et al, 1992; Ciminale et al, 1996; Koralnik, 1996).

The existence of four new alternatively spliced mRNAs, encoded by the open reading frames ORF1 and ORF2 of the 3' pX region of HTLV-1, has now been discovered (Berneman et al, 1992; Ciminale et al, 1992; Koralnik et al, 1992, 1993; Koralnik, 1996) (Figure 16.1). Their translation lead to the p12, p13, p30 proteins which are yet ill-defined, with unknown functions except for the p12 (Franchini et al, 1993; Franchini, 1995; Cereseto et al, 1996). Studies are in progress to search for similar proteins in the other members of the PTLV group. Preliminary data indicate that the HTLV-2 pX

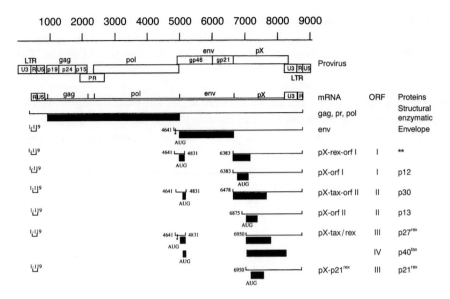

Figure 16.1 *Genomic structure of HTLV-1 provirus and mRNAs with corresponding encoded proteins*

region encodes also for some similar proteins (Ciminale et al, 1992, 1996), named p10, p11 and p28. STLV-1 shares with HTLV-1 the same genetic organization with, however, the absence of ATG for the p12 protein in some Asian STLV-1 strains (Ibrahim et al, 1995; Mahieux et al, 1997b). Regarding the new STLV-2 pan-p and PTLV-L variant strains, additional messages were found in STLV-2 pan-p px region and in the case of PTLV-L that one additional non-homologous ORF exists in the proximal pX region and is accessible for translation through alternative splicing (Van Brussel et al, 1997).

The first HTLV-1 complete sequence was published by Yoshida's group in 1983 from a Japanese ATL patient isolate (labelled ATK) and since considered as the HTLV-1 sequence prototype (Seiki et al, 1983). However, one year earlier the same group had sequenced the LTR of another Japanese ATL isolate named ATM (Seiki et al, 1982). This first report was followed by the sequencing of the LTR of American ATL isolates (Josephs et al, 1984). Sequences from TSP/HAM isolates were then published, starting as early as 1988 (Imamura et al, 1988; Tsujimoto et al, 1988). These data showed the great sequence homologies between those isolates and the original ATK prototype, with minor nucleotide divergence, ranging from 0% to 3% according to the genomic region considered and to the origin of the isolate. Ratner et al published the first sequence of an HTLV-1 molecular variant obtained from a Zaïrean patient suffering from an ATL; this virus exhibited an overall 3·3% divergence when compared with the ATK prototype (Ratner et al, 1985, 1991).

So far, only eight complete HTLV-1 nucleotide sequences have been published, four of them originating from ATL patients: ATK (Seiki et al, 1983), HS35 (Malik et al, 1988), EL (Ratner et al, 1991) and YS (Chou et al, 1995); two of them originating from TSP/HAM patients: TSP1 (Evangelista et al, 1990) and BOI (Bazarbachi et al, 1995); and the final two from healthy seropositive individuals: CH (Ratner et al, 1991) and MEL5 (Gessain et al, 1993a).

Concerning HTLV-2, the first sequence (MO), referred to as the prototype, was published in 1985 (Shimotohno et al, 1985). Since then, only four other complete sequences have been reported: G12 from a healthy Guyami Amerindian from Panama (Pardi et al, 1993b); NRA from an American patient with a concomitant CD8 oligoclonal lymphoproliferation and a B hairy cell leukaemia (Lee et al, 1993); GU from an Italian injecting drug user (Salemi et al, 1996); HTLV-2-GAB, from a healthy individual from Gabon and HTLV-2 Efe originating from an Efe pygmy from D.R. of Congo presenting the two complete HTLV-2 sequences from Africa (Letourneur et al, 1998, Vandamme et al, 1998).

Despite the fact that partial sequences of STLV-1 genomes (Ptm3 from Indonesia) have been available since 1984 (Watanabe et al, 1985, 1986), only one complete STLV-1 nucleotide sequence has been reported (Ibrahim et al, 1995). The complete sequence of the PTLV-L strain (Van Brussel et al, 1997) and of the two original STLV-2 pan-p have been obtained only recently, and demonstrated a similar genomic organization to HTLV-1 and HTLV-2 isolates (Digilio et al, 1997; Vandamme et al, 1996, 1998a). However, only two (instead of three) 21 bp direct repeats with similarity to Tax responsive elements were present in the LTR of both PTLV-L and STLV-2 pan p, which suggests differences in the Tax-mediated transactivation of the viral LTR (Digilio et al, 1997; Van Brussel et al, 1997).

METHODS USED IN PTLV MOLECULAR EPIDEMIOLOGY

Most of the work performed in the PTLV molecular epidemiology field focused on the analysis of fragments of either the *env* gene, mainly in the region coding for the transmembrane (TM) gp21 protein (Gessain et al, 1991, 1992a,b, 1994a, 1996a; Ehrlich et al, 1992; Koralnik et al, 1994; Mahieux et al, 1994, 1997a,b), or the non-coding LTR region (Bangham et al, 1988; Daenke et al, 1990; Kinoshita et al, 1991; Komurian et al, 1991; Komurian-Pradel et al, 1992; Saksena et al, 1992; Miura et al, 1994; Ureta-Vidal et al, 1994a,b; Vandamme et al, 1994; Switzer et al, 1995b; Heneine, 1996; Mahieux et al, 1997a; Voevodin et al, 1997a,b; Murphy et al, 1998), or a small fragment of the *pol* gene (Dube et al, 1993, 1994, 1995; Poiesz et al, 1993). Fewer data are available for the *gag* gene, for most of the *pol* gene or pX region (Gray et al, 1987, 1990; Kwok et al, 1988; Komurian et al, 1991; Nerurkar et al, 1993a,b, 1994a,b; Song et al, 1994, 1995; Vandamme et al, 1994, 1996; Mahieux et al, 1995, 1997b). The HTLV-1, HTLV-2 and STLV-1 sequences published before 1988 were obtained using the classical molecular cloning method in phages followed by subcloning in pBR322 vectors (Seiki et al, 1983; Malik et al, 1988). Since 1988, the use of polymerase chain reaction (PCR) led to a rapid progress in molecular epidemiology, owing to the simplicity, efficiency and rapidity of this technique. Furthermore, PCR represents a powerful tool for studying directly the *ex vivo* genomic variability without cell culture. This makes possible the detection of rare copies of the virus in the genomic DNA of an infected individual, eliminating the possibility of *in vitro* viral selection of a minor variant, as demonstrated in the HIV/ SIV system (Meyerhans et al, 1989). The PCR also enables the identification of different viral strains within an individual at a given time (intrastrain variability or quasispecies), thus allowing the sequencing of multiple clones from a single amplified product (Daenke et al, 1990; Ehrlich et al, 1992; Gessain et al, 1992b; Nerurkar et al, 1993a; Kazanji et al, 1997). In the case of the HIV system, the existence of quasispecies has been established using this technique (Goodenow et al, 1989; Meyerhans et al, 1989).

Direct sequencing of the amplified DNA without cloning enabled the dominant HTLV-1 species to be rapidly sequenced, and represents a procedure which is adapted to the molecular epidemiology. This is especially true when searching for geographical variations of HTLV-1 sequences (Komurian et al, 1991; Major et al, 1993).

The restriction fragment length polymorphism (RFLP) method is another technique that is well adapted to molecular epidemiology, owing to its simplicity and rapidity. This technique has been applied to the study of HTLV-1 LTRs, and has allowed the demonstration of the existence of at least three major geographical HTLV-1 subtypes (Komurian-Pradel et al, 1992; Ureta-Vidal et al, 1994a,b). Similarly, the existence of two major HTLV-2 subtypes was uncovered by RFLP studies of the *env* gene and/or of the LTR (Hall et al, 1992; Eiraku et al, 1995; Switzer et al, 1995b; Heneine, 1996; Murphy et al, 1998). However, owing to the increased availability and power of rapid automatic sequencing processes, RFLP analysis is now only useful for a large screening of samples (Murphy et al, 1998).

Several different techniques could be applied to comparative sequence analysis, including phylogenetic studies (Felsenstein, 1993). The most frequently used are neighbour-joining (NJ) DNA maximum parsimony, the maximum likelihood technique and the UPGA methods (Song et al, 1994; Vandamme et al, 1994, 1996; Giri et al, 1997; Mahieux et al, 1997a,b; Pecon-Slattery et al, 1998; Salemi et al, 1998). The new modified NJ method (Nei et al, 1995; Saitou and Nei, 1987) and the weighted parsimony technique are also incisive methods designed to analyze closely related sequences (around 2–15% nucleotide sequence differences) as encountered with the PTLVs. The use of bootstrap analysis is also important to test the robustness of the obtained trees (Vandamme et al, 1994, 1996).

EXISTENCE OF HTLV-1 TISSUE-SPECIFIC GENOMIC MUTATIONS

The question as to whether specific diseases were linked to specific HTLV-1 sequences arose when it was recognized that HTLV-1 was associated with at least two major diseases of totally different characteristics, namely ATL (Takatsuki et al, 1977) and TSP/HAM (Gessain et al, 1985; Osame et al, 1986, 1987; Gout et al, 1990; Gessain and Gout, 1992). In the experimental murine leukaemia viruses system (Moloney, Friend and BRE strains), haematological or neurological degenerative disorders are specifically associated with viral mutations in the *env* and/or in the LTR sequences (Lenz et al, 1984; Rassart et al, 1986; Li et al, 1987; Szurek et al, 1988). Early British (Daenke et al, 1990) and Japanese (Kinoshita et al, 1991) studies indicated that there were no obvious differences in the HTLV-1 *env* and LTR sequences when comparing isolates obtained from the peripheral blood mononuclear cells (PBMCs) of either ATL or TSP/HAM patients. Further to these early studies which involved only patients from a single country (Jamaica or Japan), a comparison was made of proviral DNA from four leukaemic and five TSP/HAM patients from Japan, the Ivory Coast and the French West Indies (Komurian et al, 1991). The sequences of 1918 bp covering the U3/R region of the LTR, most of gp46 *env*, and pX regions of these nine DNA specimens obtained from PBMCs showed 1·7% variation in *env*, 2·7% in *tax*, and 6·2% in the LTR. While no mutation could be specifically linked to either haematological or neurological diseases, it was suggested that geographical HTLV-1 subtypes might exist (Komurian et al, 1991). Several other studies similarly failed to detect any specific mutations that could be related to the outcome of a specific disease (Gessain et al, 1991, 1992a; Paine et al, 1991; Ratner et al, 1991; Schulz et al, 1991; Major et al, 1993; Ureta-Vidal et al, 1994a,b). The recent claim that a specific *tax* mutation was associated with TSP/HAM (Renjifo et al, 1995) remains to be confirmed, and this finding may be due to a bias in the sampling of the studied specimens (most of them came from the same area in Colombia where only one HTLV-1 molecular subtype is predominant). In fact, this mutation seems to be present only in the HTLV-1 strains that belong to the Cosmopolitan geographical subtype but is not related to any specific clinical outcome: in our studies, all the ATL and healthy carriers infected with the Cosmopolitan HTLV-1 strain also possessed this 'tax-specific TSP/HAM mutation' (Mahieux et al, 1995).

Such negative results concerning tissue-tropic sequences of HTLV-1 do not preclude the possibility of yet undetected differences in HTLV-1 proviruses, targeting either haematological or neurological tissues. Thus, comparison of HTLV-1 strains present in cerebrospinal fluid (CSF) lymphocytes, or possibly in HTLV-1-infected cells from the central nervous system, would be instrumental for the search of a possible neurological specific variant.

The group of Bangham (Niewiesk et al, 1994, 1995) has shown that the *tax* gene is more variable within and between healthy carriers than in patients with TSP/HAM. This suggests that the *tax* sequence heterogeneity, rather than the presence of a particular sequence, distinguishes healthy seropositive carriers individuals from TSP/HAM patients. In contrast, Saito et al (1995) reported the existence of frequent mutations in the pX region of HTLV-1 among isolates obtained from TSP/HAM patients compared with ATL or healthy carriers. However, they did not find any differences between sequences obtained from PBMCs or CSF in a single patient. This also contrasts with the findings of Kira et al (1994) who described sequence heterogeneity of HTLV-1 proviral DNA in the central nervous system of TSP/HAM patients. Studies using transgenic mice, carriers of either the LTR sequence or the *tax* gene derived from either the ATL or TSP/HAM patients, are necessary to investigate the possible functional relevance of the subtle nucleotide differences observed in certain viral strains (Gonzalez-Dunia et al, 1992; Ozden et al, 1996). However, available results do not suggest different functional properties which could be related to the clinical status of the patients (Ozden et al, 1996). Finally, Xu et al (1995) suggested that TSP/HAM patients isolates bear mutations in an eight-base segment of a part of the LTR R region (nucleotides 210–227). These mutations could prevent the binding of cAMP response element binding protein (CREB) and ATF-2 factors to their target sequences. However, a careful analysis of all the 143 Genbank LTR sequences did not reveal any significant difference in the occurrence of such mutations in viruses originating either from healthy seropositive carriers or from TSP/HAM or ATL patients (R. Mahieux et al, unpublished observations). Finally, the study of a complete HTLV-1 sequence obtained from a patient only few weeks after he developed a post-transfusional TSP/HAM did not reveal any specific mutations that could reflect a neurotropic variant (Bazarbachi et al, 1995).

EXISTENCE OF FOUR MAJOR GEOGRAPHICAL MOLECULAR HTLV-1 SUBTYPES

Malik et al (1988), by comparing the HS35 ATL isolate (from Jamaica) with all the available sequences, first suggested the possible existence of closer genetic homologies between HTLV-1 isolates originating from the same geographical area rather than between ATL or TSP/HAM isolates obtained from different locations. This notion was not shared by Gray et al (1990), based on data from the *env* gene of only three isolates. Because of these discordances, and because there were only a few HTLV-1 sequences available from Africa, the West Indies and South America, larger-scale studies of specimens from various geographical areas began during the late 1980s (Fukasawa et al, 1987; Kwok et al, 1988; Shirabe et al, 1990; De et al, 1991; Dekaban et al, 1992; Gessain et al, 1991, 1992a,b, 1993a; Komurian et al, 1991; Paine et al, 1991; Ratner et al, 1991; Schulz et al, 1991; Komurian-Pradel et al, 1992; Sherman et al, 1992, 1993; Major et al, 1993; Mahieux et al, 1994; Miura et al, 1994; Ureta-Vidal et al, 1994a,b; Vandamme et al, 1994; Yanagihara, 1994). This allowed HTLV-1 molecular clusters to be defined from the accumulated data, and demonstrated without ambiguity that the nucleotide changes observed in some fragments of the HTLV-1 genome were specific to the geographical origin of the patients rather than to the type of associated pathology.

In 1998, based on the study of more than 300 HTLV-1 strains originating from nearly all the known HTLV-1 endemic areas, and using either the gp21 *env* gene and/or the LTR genomic fragments for phylogenetic analyses, four major molecular subtypes were identified (Figures 16.2, 16.3).

Subtype A (Cosmopolitan subtype)

The first subtype to be discovered is the most widespread and dispersed one. It has been found in many different geographical areas and among several human populations of very different ethnic backgrounds. As an example, this HTLV-1 subtype is encountered in Japan, Taiwan, China and India (Hashimoto et al, 1993); in the Middle East including Iran (Nerurkar et al, 1995; Voevodin et al, 1995; Voevodin and Gessain 1997), Iraq and Kuwait; in Western and South Africa; in the Americas, including Alaska, Central and South America and the Caribbean area; in Romania, and in most of the infected European immigrants, the great majority of whom originated from the above-cited HTLV-1 endemic regions.

Subtype A is also – surprisingly – the least divergent subtype, representing a group of highly related viruses. The average nucleotide divergence, within the different HTLV-1 strains of this subtype, is less than 2% in the 522 bp gp21 *env* region.

Subtype A comprises several molecular groups which can be characterized either by specific mutations or by phylogenetically well-supported clades. Using the gp21 *env* gene, only two main groups can be defined in the subtype A; however, with low bootstrap values, the use of the complete LTR sequence (more variable and discriminant) has allowed the clear distinction of four molecular groups (Miura et al, 1994; Ureta-Vidal et al, 1994a,b; Vandamme et al, 1994; Yamashita et al, 1996): the North African (Gasmi et al, 1994), the transcontinental (Miura et al, 1994), the Japanese (Miura et al, 1994; Ureta-Vidal et al, 1994a,b) and the West African (Miura et al, 1994; Ureta-Vidal et al, 1994b) (Figure 16.3).

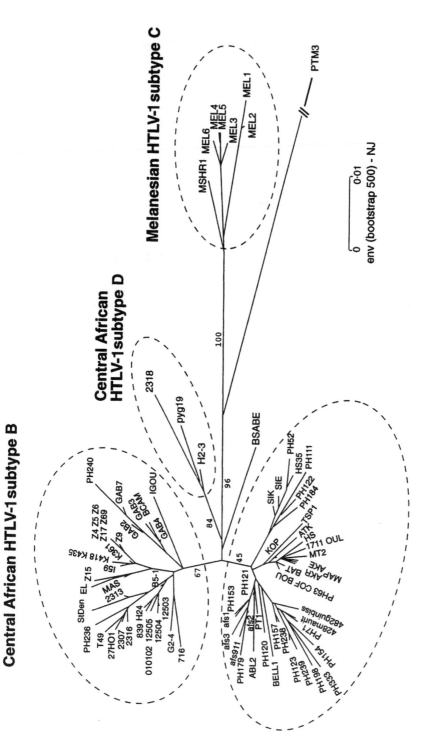

Figure 16.2 *Phylogenetic tree for HTLV-1*

Phylogenetic tree constructed with the neighbour-joining (NJ) method using the Neighbour program with a fragment of 522 bp encompassing most of the gp21 and the end of the carboxy terminus of gp46 on 93 HTLV-1 isolates, including all the 75 African ones. Ptm3 (STLV-1 isolate) was used to root the tree. The numbers indicated at some nodes (bootstrap values) represent their frequencies of occurrence out of 500 trees (adapted from Mahieux et al, 1997a).

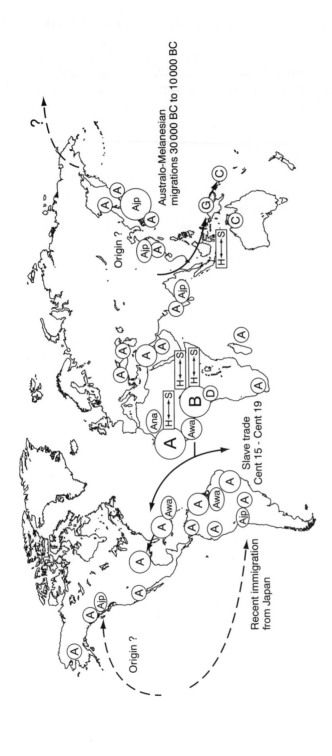

Figure 16.3 *Distribution of HTLV-1 subtypes*
World distribution of the four major HTLV-1 molecular subtypes (genotypes) A–D with their main possible modes of dissemination through migrations of infected populations. The boxed arrows represent probable interspecies transmission episodes of STLV-1 from monkeys (S) to humans (H). A, Transcontinental group (Ajp, Japanese group; Ana, North African group; Awa, West African group); B, Central African subtype; C, Melanesian subtype; D, Central African subtype.

Existence of two distinct HTLV-1 ancestral lineages in Japan with a particular geographical distribution

For as yet unknown reasons, southeast Japan is one of the highest HTLV-1 endemic areas of the world (Mueller, 1991; Blattner and Gallo; 1994; Tajima et al, 1994). Based on historical data, several hypotheses have been proposed, invoking population migrations and the possible introduction of the virus through commercial activities in previous centuries. The possible high prevalence of HTLV-1 in the Ainu population, considered to be one of the two oldest native Japanese populations, argues against the latter hypothesis and suggests that HTLV-1 was present in aboriginal Japanese in prehistoric times, some 2300 years ago. By studying the LTRs of 82 HTLV-1 samples from Japanese residents of different geographical areas of Japan (Hokkaido, Honshu, Kyushu or the Ryukyu Islands) (74 cases) or Americans of Japanese ancestry living in Hawaii (8 cases), using either an RFLP technique and/or sequence analyses, it was demonstrated that all of them belonged to the Cosmopolitan subtype, with, however, two different molecular clades (Ureta-Vidal et al, 1994a). The first one, which corresponded to the Japanese molecular group, was more frequent (67 out of 86, 78%) than the Transcontinental one (19 out of 86, 22%). Furthermore, the geographical distribution of these two molecular groups varied among the 80 samples where the place of residence in Japan was known. It was shown that, while the Japanese molecular group was present in all parts of Japan, the Transcontinental one clustered in the southern islands of the archipelago (i.e. Kyushu and the Ryukyu Islands) as well as in immigrants from those areas who had been living in Hawaii for decades. Historically, at least three distinct ethnic groups constituted the roots of the present Japanese population: the Wajin, the Ainu and the Ryukyuan. The Wajin, considered to be descendants of postneolithic migrants (300 BC to AD 600) from mainland China and South Korea, form the largest part of the population. The two other groups, considered to be descendants of the aboriginal native populations (mongoloid), live principally in the north (Hokkaido) for the Ainu, and in the south (Okinawa) for the Ryukyuan. These two latter groups are considered as having been present in Japan since the prehistoric Jomon period some 2300 years ago. During the seventeenth to nineteenth centuries, Kyushu island remained open to international trade, while the main central and northern Japanese islands were tightly closed to any contact with foreigners. The north–south gradient of the Transcontinental HTLV-1 molecular group may be related to a migration of the Transcontinental infected population group from the south towards the north. It is interesting that the only known HTLV-1 isolate from an Ainu clustered within the Transcontinental group on a phylogenetic tree (Miura et al, 1994). However, these data need to be confirmed on other Ainu samples from Hokkaido and Sachalin, where HTLV-1-infected individuals have recently been found (Gurtsevitch et al, 1995; Yamashita et al, 1995a; Gessain et al, 1996b). Since the Okinawan population, or at least a part of it, is considered to be the oldest Japanese ethnic group, why should the Transcontinental group be the most prevalent there? It was difficult to determine whether the Japanese or the Transcontinental group was the most ancient by using phylogenetic analyses, because the nucleotide divergence between the two groups is very low in comparison with all other HTLV-1 isolates, including variants from Africa and Melanesia. Since the Transcontinental molecular group is found throughout the world and has a very high genomic stability, it is thought to have been dispersed in the relatively recent past (i.e. the last few centuries), from a single origin, through the migration of infected individuals. It might thus have arrived in Okinawa during the latter period by a route that remains to be determined. There are also alternative possibilities. Both molecular groups could have been present in Japan since ancient times, introduced by the original settlers in the preagricultural Jomon period by two waves of populations, each infected by one or the other of these two subtypes. In such a case, the uneven distribution of the two HTLV-1 groups now found in Japan could reflect either a difference in the mode or efficiency of transmission, or a particular migration pattern within Japan, other than that discussed above. Furthermore, the pressure of various environmental and/or genetic cofactors could result in an increase in the prevalence rate of one of the two viral molecular groups in a given population over a long period. These questions may be answered by further molecular studies of HTLV in Japan and other Asian countries, and by linking genetic data on human ethnic populations to viral molecular groups (Tajima et al, 1994; Ureta-Vidal et al, 1994a; Yamashita et al, 1996; Chen et al, Personal Communication).

The origins of HTLV-1 in the American continent

A study of 522 base pairs within the gp21 *env* gene demonstrated that all the 23 available HTLV-1 strains from the western part of Africa (including 18 new strains) belong to the Cosmopolitan genotype (subtype A), with some of them constituting a small separate cluster within this subtype (West African group) (Mahieux et al, 1997a). These 23 viruses originated from seven different countries (Senegal, Mauritania, Cape Verde, Guinea-Bissau, Mali, Burkina Faso, Ivory Coast), most of which were major centers for the slave trade during the sixteenth to nineteenth centuries. These African HTLV-1 are identical or very closely related to the HTLV-1 strains present in the Americas (North, Central, Caribbean and South) which all belong to the Cosmopolitan genotype, as demonstrated by several groups (Malik et al, 1988; Evangelista et al, 1990; Komurian et al, 1991; Paine et al, 1991; Schulz et al, 1991; Gessain et al, 1992a, 1994a; Miura et al, 1994; Bazarbachi et al, 1995; Franchini, 1995; Song et al, 1995; Van Doonen et al, 1998). Therefore, our view, based mostly on these findings, but also on historical background, on the clustering in the Americas of HTLV-1 mainly in populations of African ancestry (Blattner et al, 1990; Mueller, 1991), and on the rarity of HTLV-1 in pure, isolated Amerindian populations without contact with African Americans, is that the great majority of the HTLV-1 strains found in the Americas are of recent African origin (fifteenth to nineteenth centuries) (Figures 16.3, 16.4). Exceptions to this include recent Japanese immigrants to South America and possibly some British Coloumbian Indians (Picard et al, 1995). Such views, already suggested by us and others (Komurian et al, 1991; Paine et al, 1991; Schulz et al, 1991; Gessain et al, 1992b, 1994a; Mukhopadhyaya and Sadie, 1993; Vandamme et al, 1994; Franchini, 1995; Song et al, 1995; Dekaban et al, 1996; Liu et al, 1996; Van Doonen et al, 1998) on smaller series of HTLV-1 African strains, contrast strongly with the hypothesis of another group which still claims that the majority of American HTLV-1 strains originated from ancestral mongoloids who migrated from Asia to the American continents (Miura et al, 1994; Yamashita et al, 1995b, 1996). This hypothesis may be linked to the fact that this group based its theory on results obtained using only a fragment of the LTR, a viral region for which very few African sequences are available, while there are numerous published Japanese and South American HTLV-1 LTR sequences. Recent findings based on the LTR analysis indicate that several HTLV-1 strains of the West African molecular group were also detected in

Figure 16.4 *Najor African subtypes of HTLV-1*
Map of Africa with the geographical distribution of the najor molecular subtypes of HTLV-1.
'A' corresponds to the widespread Cosmopolitan subtype, with 'Ana' being the molecular
group present in North Africa and 'Awa' the molecular group present in West Africa. 'B'
corresponds to the Central African subtype which is dispersed over this large HTLV-1 endemic
region and represents most of the strains present in Central African infected inhabitants. 'D'
corresponds to the subtype which is restricted to some areas of Central Africa, mainly in
Pygmy populations from Cameroon and the Central African Republic (CAR). 'E' and 'F'
compound to new subtypes recently described. The sizes of the circles are roughly
representative of the number of specimens studied for a given subtype.

the Americas and that several HTLV-1 strains from West Africa belong to the Transcontinental group, also frequently encountered in the Americas (Hayami et al, 1997; Picchio
et al, 1997; Van Doonen et al, 1998; Chen et al, in preparation).

Finally, an interesting report also describes in detail the different possible origins of
the HTLV-1 strains present in some Amerindian tribes from British Columbia, and
comments on the difficulties of reaching a definitive answer on this topic, despite a well-
performed phylogenetic analysis (Picard et al, 1995; Dekaban et al, 1996).

Subtype B (Central African subtype)

Subtype B was initially isolated from a Zaïrean patient suffering from an ATL (Ratner et al, 1985, 1991). Later on, we and others reported the presence of this molecular geno-type in other individuals living in different regions of Zaïre (now Democratic Republic Congo) who had ATL or TSP/HAM, or were healthy seropositive carriers (Gessain et al, 1992a; Boeri et al, 1993; Liu et al, 1994a, 1996; Mahieux et al, 1997a). In a large study of 36 new Central African HTLV-1 isolates (including eight from Pygmies), Mahieux et al (1997a) demonstrated a greater HTLV-1 genomic diversity within this Central African subtype than originally thought. Furthermore, within this large Central African clade, subgroups clearly existed, specific to the geographical origin of the strain (Zaïre, Gabon, Cameroon) (Figure 16.4). These groups, which differed by 1·2–3·5% in their nucleotide sequence (*env* gene), are not always highly supported phylogenetically but are charac-terized by specific mutations. Two non-exclusive hypotheses can be proposed to explain such geographically restricted genetic diversity:

1. An ancient episode of STLV-1 transmission to a human group was followed by a long period of independent evolution occurring simultaneously in several separated and isolated human populations, with possibly different rates of genetic variability.
2. Several episodes of interspecies viral transmission from simian reservoirs occurred, leading in each human recipient group to a different molecular group.

We favour the latter hypothesis, because we believe that such discrete episodes of inter-species transmission are still occurring. More recent studies seem, furthermore, to indi-cate the presence of other HTLV-1 molecular divergent subtypes called E and F in some Central African inhabitants (Figure 16.4), (Salemi et al 1998b; Vandamme et al, 1998).

Subtype C (Australo-Melanesian subtype)

Investigating six HTLV-1 DNA specimens obtained by Dr Gajdusek's laboratory (Yanagihara et al, 1991a,b) from Papua New Guinea (PNG) and Solomon Islands inhab-itants, Gessain et al (1991) focused their study on the gp21 *env* region. Surprisingly, both the isolates from healthy Melanesian carriers and from a Solomon Islander suffering from TSP/HAM diverged by 7–8% from the ATK HTLV-1. Furthermore, the HTLV-1 variants from PNG differed by nearly 3–4% from the Solomon Islands strain, indicat-ing the existence of a high HTLV-1 intervariability within this geographical region. In contrast, HTLV-1 strains from two residents of Bellona Island, a Polynesian outlier within the Solomon Islands, were closely related to the Cosmopolitan group (> 97% homology in the 522 gp21 region), suggesting a more recent introduction of HTLV-1 in this area. The complete nucleotide sequence of one of these HTLV-1 variants, namely Mel-5, showed and overall 8·5% nucleotide divergence from the prototype ATK, with 9·5% divergence in the LTR. The U5 and U3 were more variable than the R region (Gessain et al, 1993a). The degree of variability at the amino acid level of the structural genes ranged from 3% (p24) to 11% (p19), being higher (8·5% to 29%) for both the reg-ulatory (Tax and Rex) and the new pX-encoded proteins. Moreover, the conservation of the *env* neutralizing epitope (amino acids 88–98) within the external envelope glyco-protein suggested that a vaccine prepared against the Cosmopolitan prototype of HTLV-1 might also protect against these HTLV-1 variants (subtype C) (Gessain et al,

1993a; Benson et al, 1994). In parallel, Sherman et al (1992) and Saksena et al (1992) have sequenced parts of LTR, *pol* and pX genes of one of these HTLV-1 variants isolated from a Papua New Guinean (mel1/PNG1). Similarly, they showed an overall divergence of 7–8% from the ATK prototype. Nerurkar et al (1993b) have also studied the inter-familial and intrafamilial genomic diversity of such variant strains.

Furthermore, HTLV-1 strains were also isolated from Australian aboriginals (Bastian et al, 1993). Sequence analysis of the *env* gene and segments of the *pol* and pX regions of an ATL isolate confirmed the presence of an HTLV-1 variant belonging to the Melanesian subtype C. This virus exhibited a sequence divergence of roughly 5–7% compared with the ATK prototype, depending on the genes studied. Specific oligonu-cleotide primers have also been generated, allowing rapid genotyping of such Australo-Melanesian HTLV-1 strains (Nerurkar et al, 1994a).

It is important to note here that all these Australo-Melanesian HTLV-1 molecular vari-ants were isolated from individuals whose sera had complete Western blot HTLV-1 seroreactivities, not from subjects whose sera exhibited only *gag*-encoded proteins. The question of the existence of more distantly related HTLVs in the latter individuals, who are also frequently detected in Central Africa, remains open. All attempts at isolation after culture or detection by PCR retroviral markers in these individuals have been neg-ative (Lal et al, 1992; Nerurkar et et al, 1992; Yanagihara, 1994; Gessain et al, 1995a; Dube, 1997; Vandamme et al, 1997b; Mauclère et al, 1997a; Mahieux et al, in preparation).

Subtype D (Central Africa – Pygmies)

The subtype D molecular genotype has been shown to be present in some individuals from Central African countries including Cameroon, the Central African Republic and Gabon (Mahieux et al, 1997a) (Figures 16.2–16.5). Recent unpublished data from our laboratory seem to indicate that this new genotype may represent only a small percent-age of all the HTLV-1 strains present in Central Africa (compared with the common and widely distributed HTLV-1 subtype B strains), with, however, some possible clusters, for example in southern Gabon and the western part of Cameroon. The HTLV-1 subtype D strains exhibit a genetic divergence of 2·2% in the gp21 *env* gene between themselves, while they diverge from subtype A and subtype B strains by 3·5%. The HTLV-1 subtype D clade is phylogenetically well supported both in *env* and LTR analyses, with high bootstrap values (Figures 16.3, 16.5). (Liu et al, 1997; Mahieux et al, 1997a, 1998b; Van-damme et al, 1998b; Pecon-Slattery et al, 1999). This new phylogenetic cluster, charac-terized by specific mutations, occupies in all analyses a unique position between the large Central African genotype (subtype B) and the Australo-Melanesian one (subtype C). Whether some of the recently reported HTLV-1 variants strains from Gabon (sequences available only for the gp46 *env* gene) (Moynet et al, 1995) are related to the molecular subtype remains to be elucidated by further sequencing studies. Further-more, preliminary analysis based on the only available partial LTR sequence seems to indicate that another recently described isolate from a Pygmy living in south Cameroon (Chen et al, 1995) may be related to the HTLV-1 subtype D.

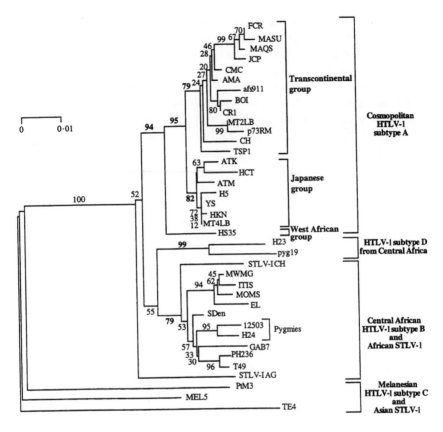

Figure 16.5 *Phylogenetic tree for HTLV-1 and STLV-1*
Phylogenetic tree constructed by the neighbour-joining method using the MEGA program with the complete LTR (755 bp) of available HTLV-1 and STLV-1. The numbers indicated at some nodes (bootstrap values) represent their frequencies of occurrence out of 500 trees (adapted from Mahieux et al, 1997a).

VARIABILITY OF HTLV-1 WITHIN INDIVIDUALS AND FAMILIES

While geographical subtypes differ between themselves, from 2% to 8% depending on the gene of interest, the HTLV-1 intrastrain variability appears to be much lower (< 0·5%). This is in sharp contrast with the human and simian immunodeficiency viruses data (Meyerhans et al, 1989; Boeri et al, 1992; Wain-Hobson, 1992) where the diversity *in vivo* of the genomic strain present in one individual (quasispecies) represents one of the main problem in developing an anti-HIV vaccine. In the case of HTLV-1, a few studies have tried to approach this question (Daenke et al, 1990; Ehrlich et al, 1992; Gessain et al, 1992a; Kira et al, 1994; Saito et al, 1995, 1996; Wattel et al, 1995; Kazanji et al, 1997). By studying multiple clones obtained from one PCR product, Daenke et al (1990) first demonstrated the existence of intrastrain variability by studying the *env* gene of 12 HTLV-1 strains obtained from TSP/HAM patients or healthy carriers originating from the West Indies and observing in five of these individuals some HTLV-1 sequence variants: in four patients, two variants were identified, and in the remaining patient, three variants. However, in all cases a dominant variant emerged. In a second study, Ehrlich et al (1992) studied multiple clones of a sequence of 235 bp of the gp21 *env* gene from either fresh or cultured cells obtained from 19 HTLV-1-infected individuals, mostly originating from the USA. By studying at least two clones for each individual, they demonstrated the presence of multiple HTLV-1 genomic variants. In a third report, Gessain et al (1992a) investigated 71 clones of 522 bp (gp21 *env* gene) obtained from 28 *ex vivo* DNA samples from ATL,TSP/HAM patients and healthy carriers from various geographical areas including the West Indies, West and Central Africa, and South America. In most cases it was possible to identify a prevalent molecular clone within a viral quasispecies population. The nucleotide changes observed in the microvariants within an individual host were exclusively single base substitutions, half of which led to amino acid changes. In the same report, the genetic variability of HTLV-1 was determined, *in vivo* and over time, by studying a well-documented HTLV-1 transmission case involving three individuals: a blood donor who transmitted HTLV-1-infected cells to a cardiac transplant patient, who a few months later developed a TSP/HAM and transmitted the virus to his wife by sexual contact (Gout et al, 1990; Gessain et al, 1992a). The proviral sequences revealed a total identity in the 522 bp (gp21 *env*); this included specific mutations present only within these three individuals. This study demonstrated the lack of any rapid genetic drift after the *in vivo* passage of the virus in three different hosts over a period of 5 years.

As noted above, Niewiesk et al (1994) claimed that the *tax* sequence is more variable within and between healthy carriers than in patients with TSP/HAM. This was suggested by sequencing 20 clones of the full-length proviral *tax* gene obtained from four TSP/HAM patients and four healthy HTLV-1-seropositive individuals, mostly of Jamaican origin.

There are only a few studies that have attempted to define the HTLV-1 intrafamilial genetic variability. The interfamilial sequence variation observed between HTLV-1 strains from the Solomon Islands and those from Papua New Guinea was 3·4–4·2% (in the *env* gene), whereas it was 0·2–0·9% between virus strains from three Solomon Islands families (Nerurkar et al, 1993b; Yanagihara, 1994). In a study performed in an area of Zaïre (D.R. Congo) with a high prevalence of TSP/HAM, Liu et al (1994a) showed that identical sequences were present the two studied families with the exception of a woman infected with two variants, one being the familial strain, the other a mutated one with a single substitution in the 755 bp LTR region. Furthermore, in a Jamaican family living in England (the father having an ATL, the mother a TSP/HAM and three out the five children being healthy HTLV-1 seropositive), Major et al (1993) found that all the family members had identical *tax* nucleotide sequences as determined by direct sequencing of PCR products. However, by studying multiple clones of this same gene, they found more sequence variation (point mutations) in the TSP/HAM patient than in the ATL patient, and demonstrated that there was no conservation of mutations between the two individuals. All these results (as well as some of those described in Chapter 4) showing a greater genetic variability among healthy carriers than in TSP/HAM and in ATL may reflect the importance of the clonal population, frequently seen in HTLV-1-infected individuals, even in absence of malignant ATL (Wattel et al, 1995; Cavrois et al, 1996). Studies on the HTLV-1 sequences genetic diversity within a given clone are in progress (E. Wattel et al, personal communication).

CLOSE PHYLOGENETIC RELATIONSHIP OF STLV-1 WITH HTLV-1, AND ORIGIN OF HTLV-1

The first STLV-1 sequence was obtained from an Asian pig-tailed monkey (Ptm3), and demonstrated an overall nucleotide similarity of 90% with the HTLV-1 ATK prototype in the *env*, pX and LTR regions (Watanabe et al, 1985). Furthermore, Watanabe et al (1986), by sequencing the LTR of two STLV-1 (from an African green monkey, STLV-AG, and from a chimpanzee, STLV-CH), showed that the nucleotide sequence of HTLV-1 (ATK) presented more homologies (95%) with the African STLV-1 strains than with the Asian STLV-1 strain (90%). They proposed the existence of a distinct group of primate retroviruses that may be collectively named primate T cell leukaemia virus type 1 (PTLV-1). Since then, numerous STLV-1 strains have been isolated and their sequences partially characterized (Watanabe et al, 1986; Saksena et al, 1993, 1994; Schätzl et al, 1993; Koralnik et al, 1994; Song et al, 1994; Vandamme et al, 1994; Ibrahim et al, 1995; Engelbrecht et al, 1996; Liu et al, 1996, 1997; Voevodin et al, 1996a, 1997a,b; Mahieux et al, 1997b, 1998a,b; Yamashita et al, 1997).

Regarding the African continent, phylogenetic analyses of fragments of the *env* and/or *pol* genes and/or LTR of both HTLV-1 and STLV-1 from mainly Central and West Africa have provided evidence that multiple discrete episodes of interspecies transmission of STLV-1 have occurred between primates – including humans – in the distant and recent past in these areas. This led to recognizable phylogenetic clades that persist in modern species (Figures 16.5, 16.6) (Koralnik et al, 1994; Mahieux et al, 1998b). Thus, regarding transmission of STLV-1 between different species of monkeys, an initial analysis of all available STLV-1 strains revealed that no STLV-1 clusters recapitulated host species specificity; rather, multiple clades from the same species were closer to clades from other species (*Cercopithecus* and *Papio*, and *Cercopithecus* and *Pan troglodytes*) than to each other. Furthermore, geographical concordance of divergent host species that harbour related viruses reinforces the physical feasibility for such hypothesized interespecies transmission (Koralnik et al, 1994). However, most of the available STLV-1 strains originated from animals kept in captivity, where interspecies transmission could not be excluded, raising the possibility that the studied viruses may not represent the strains present in the same animal species living in the wild. A demonstration of such transmission episodes within *Papio* and *Macaca species*, living in the Sukhumi colony has been reported by Voevodin et al (1996b).

Recent data from our laboratory may provide good evidence of interspecies transmission episodes of STLV-1, occurring in the wild between different monkey species. Thus three STLV-1 strains from wild-caught *Papio ursinus*, from South Africa, form a cluster with high bootstrap support that is affiliated with a monophyletic lineage from two wild-caught South African vervets (*Cercopithecus aethiops pygerythrus*) (Mahieux et al, 1998a).

The hypothesis of the transmission of STLV-1 to humans was based mainly on the original findings of a very close nucleotide relationship (around 97–98% of nucleotide similarity) between the gp21 *env* gene of three STLV-1 isolates from chimpanzees and several HTLV-1 isolates (subtype B) originating from inhabitants of Zaïre (Koralnik et al, 1994). More recently, we reported that some Pygmies and other Central African inhabitants (villagers living in south Cameroon) harbour HTLV-1 isolates which share even closer relationships with the chimpanzee STLV-1 (> 98% nucleotide similarity)

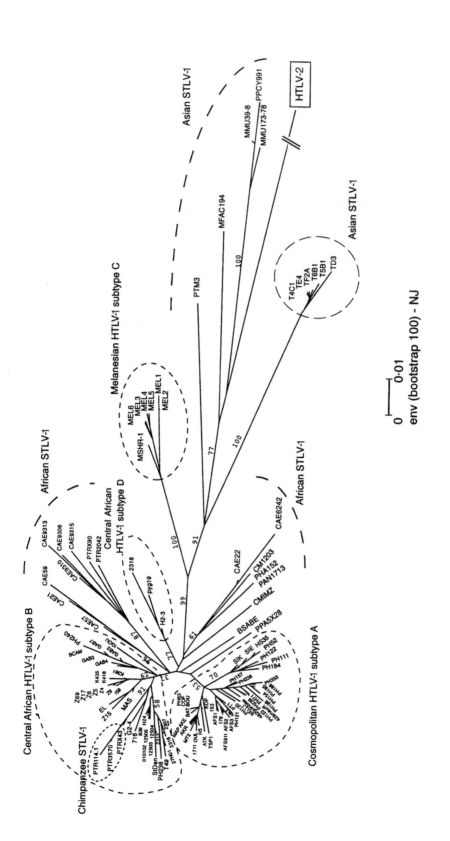

Figure 16.6 *Phylogenetic tree*

Phylogenetic tree constructed using the neighbour-joining method with a fragment of 522 bp of the *env* gene on 124 isolates including 36 available STLV-1. The HTLV-2 MO isolate was used as an outgroup to root the tree. The numbers indicated at some nodes (bootstrap values) represent their frequencies of occurrence out of 100 trees (adapted from Mahieux et al, 1997a).

(Mahieux et al, 1997a) (Table 16.1; see also Figures 16.3, 16.5, 16.6). Another STLV-1 strain obtained from a captive chimpanzee was found to be also very similar to some Central African (subtype B) HTLV-1 strains (Voevodin et al, 1997b). Such data reinforce the possibility of relatively recent interspecies transmission from monkeys to humans in the Central African forest area (Crandal 1996; Pecon Slattery et al, 1998).

Concerning the possible origin of the Cosmopolitan HTLV-1 strains (subtype A), only one STLV-1 (PPA5X28) obtained from an individual *Papio papio* from West Africa was found to be related to such HTLV-1 strains. The nucleotide similarity with the Cosmopolitan isolates was only 96–97%, and in phylogenetic analyses this clade was aligned with the PPA5X28 isolate with, however, a low bootstrap value (34%) (Koralnik et al, 1994).

Concerning the HTLV-1 subtype D, which was found in Central Africa and especially in Pygmies, ongoing studies from our laboratory demonstrate that some STLV-1 strains are highly related to HTLV-1 subtype D strains with a 97–100% nucleotide similarity and a high bootstrap value on phylogenetic analysis (Mahieux et al, 1998b). In Asia, the Australo-Melanesian HTLV-1 strain (subtype C) differs from the other subtypes by roughly 7% of nucleotide substitutions and occupies a unique phylogenetic position between Asian STLV-1 strains and all the other remaining divergent HTLV-1 subtypes (Figure 16.6). It can therefore be hypothesized that, as in Africa, interspecies virus transmission episodes may have taken place. Thus, STLV-1 strains genetically related to HTLV-1 Australo-Melanesian strains may exist in populations of non-human primates inhabiting islands that served as the migratory pathways for the early settlers of Papua New Guinea and Australia 30 000–40 000 years ago (Yanagihara, 1994).

We have isolated an STLV-1 originating from a troop of Celebes macaques (*Macaca tonkeana*), and performed the first complete nucleotide sequence of an STLV-1 from one of these isolates (STLV-1 TE4) (Ibrahim et al, 1995). *Macaca tonkeana* live in the Sulawesi island which is one of the nearest natural monkey habitats to Papua New Guinea. This area is devoid of monkeys, but is also found to be on one of the two migratory pathways

Table 16.1 *Human T cell leukaemia/lymphoma viruses types 1 and 2, and their simian counterparts*

Human	gp21 nucleotide homology (%)	Monkey
HTLV-1		**STLV-1**
Cosmopolitan A	97%	Baboon (*Papio papio*)
Central African B	98–99%	Chimpanzees (Pan troglodytes)
Melanesian C		?
Pygmy, Central Africa D	98–99%	Mandrills (Mandrillus/sphinx)
?		Macaques (*Macaca arctoides*)
HTLV-2		?
2 A		?
2 B		?
2 C		?
2 Pygmy D		?
?		STLV-2 pan p
?		PTLV-L

Based on data available in 1998.

believed to have been taken by the early Australoid migrants. Phylogenetic trees performed on the LTR and the *env* (gp46, gp21) gene demonstrated that this new STLV-1 occupies a unique position, being by most analyses more related to the Australo-Melanesian HTLV-1 topotype than to any other Asian STLV-1 (Figure 16.6). These data suggest new hypotheses on the possible interspecies viral transmission episodes between monkeys carrying STLV-1 and early Australoid settlers, ancestors of the present-day Australo-Melanesian inhabitants, during their migration routes from the southeast Asian land mass to the greater Australian continent. Despite these close phylogenetic relationships between the STLV-1 from M. *tonkeana* and the Melanesian HTLV-1 strains, these two groups of viruses exhibited roughly 10% nucleotide divergence and, to date, there are no STLV-1 found to be as closely related to the HTLV-1 Melanesian strains as are some STLV-1 chimpanzee strains to HTLV-1 subtype B or some STLV-1 mandrill strains to HTLV-1 subtype D from Central Africa. This may be due in part to the long period of independent viral evolution in remote Melanesian populations which took place after the occurrence of the interspecies transmission episodes in Asia, while in Africa some of these interspecies transmissions may have occurred more recently.

Infection of monkeys by all the STLV-1 strains described above elicits, in the natural host, a Western blot serological pattern which is generally indistinguishable from (or very similar to) an HTLV-1 Western blot pattern, i.e. clear reactivities to *gag*-encoded antigens (p19, p24 and pr53, usually with p19 > p24) associated with reactivities to *env*-encoded antigens (gp21-rgp21 and gp46-MTA-1 peptide). A highly divergent STLV-1 has been characterized from three captive monkeys from a colony of *Macaca arctoides* (Mahieux et al, 1997b). Over a 4-year interval, several animals continued to exhibit a peculiar Western blot resembling to an HTLV-2 pattern (p24 *gag* reactivity of equal or greater intensity than that of p19 *gag*, associated with a rgp21 strong reactivity) but exhibiting also, in five out of six cases, a reactivity against MTA-1, an HTLV-1 gp46 peptide. This 'HTLV-2-like' Western blot pattern resembled some of the reactivities obtained with sera of monkeys infected by the recently isolated STLV-2 pan-p and PTLV-L (see below). In a first set of experiments, PCR on DNA extracted from PBMCs using specific HTLV-1 or HTLV-2 LTR, *gag*, *pol*, *env* and *tax* primers yielded negative results. However, highly conserved primers successfully amplified three different gene segments of *env*, *tax* and *env/tax*. Comparative sequence analyses demonstrated that STLV-1*marc1* was not closely related to any known STLV-1 and was the most divergent strain of the PTLV-1 group. Phylogenetic analyses incorporating representative strains of all known HTLV/STLV clades consistently depicted STLV-1*marc1* within the HTLV/STLV type 1 lineage, but probably as an early divergence, being clearly apart from all known viral strains of this group, with a strong support from bootstrap resampling analysis of 100% (Figure 16.7). Genetic distance estimates between STLV-1*marc1* and all other type 1 viruses were of the same level as STLV-2 pan-p relative to all other type 2 viruses. In light of the demonstration of interspecies transmission of some STLV-1 strains described above, these results suggested the existence in Asia of HTLV-1 strains related to this new divergent STLV-1*marc1* which may be derived from a common ancestor early within the evolution of the type 1 viruses, and this could therefore be considered as a prototype of a new HTLV/STLV clade.

Recently Salemi et al, (1998b) requested 2 other HTLV-1 subtypes from individuals, originating from Central Africa whose sera shared indeterminate Western blot pattern (Salemi et al, 1998b).

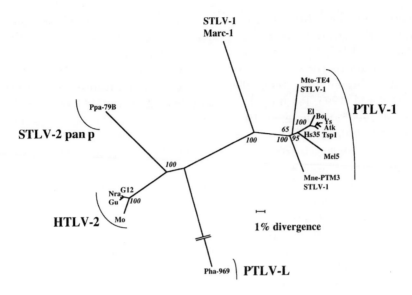

Figure 16.7 *Phylogenetic tree*
Phylogenetic tree constructed by the neighbour-joining method with a fragment of 1612 bp (6149 to 7760 according to HTLV-1 ATK sequence) encompassing most of the gp21 *env* gene and a large portion of the pX region, on all available sequences: 16 isolates including three STLV-1 and the recently described STLV-1*marc1* strain. This NJ tree is based on the Jukes–Cantor model of substitution. Numbers on branches represent bootstrap proportions in support of adjacent node based on 100 resampling iterations. Values greater than 70% are considered strong support for the adjacent node (adapted from Mahieux et al, 1997b).

EXISTENCE OF TWO MAJOR MOLECULAR SUBTYPES OF HTLV-2

The complete sequence of HTLV-2 (MO) was published in 1985 by Shimotohno et al (1985). The comparison of its sequence with that of HTLV-1 ATK demonstrated an overall homology of 60% at the nucleotide level in the coding regions. However, some regions exhibited greater homology: 85% for the p24 *gag* and 78% for the *tax* region, whereas it was less for other regions such as the p19 *gag* (59%), the gp46 *env* (63%) or the *rex* (61%).

In 1990 Lairmore et al published small fragments of the *gag* (107 bp) and *pol* (112 bp) genes of two different HTLV-2 isolates from Guyami Indians living in Panama (Lairmore et al, 1990). The comparison of these small sequences exhibiting a high percentage of homology with HTLV-2 MO (94% and 98% respectively) suggested that, as with HTLV-1, a low genetic variability exists between different HTLV-2 isolates. However, in 1992, Hall et al, by comparing the restriction map analysis of several isolates from US injecting drug users, demonstrated the existence of two HTLV-2 molecular subtypes, tentatively designated HTLV-2A and HTLV-2B. Study of the gp21 *env* gene nucleotide sequence confirmed the existence of these two subtypes, which diverge by roughly 4% in this region (Hall et al, 1992). Furthermore, this study demonstrated that there was a marked sequence conservation within the two subtypes, with a maximum of 0·4% nucleotide divergence within a subtype. This notion was confirmed by the same group after analysing the entire *env* gene, portion of the *pol* and the LTR sequences. The LTRs exhibited the greatest divergence (5·8%) within subtypes, as is the case with HTLV-1 (Hall et al, 1992, 1996; Takahashi et al, 1993). Since then, several other groups have confirmed the existence of these two subtypes (Calabro et al, 1993; Dube et al, 1993, 1994, 1995; Ferrer et al, 1993, 1996; Hjelle et al, 1993; Igarashi et al, 1993; Ijichi et al, 1993; Lee et al, 1993; Pardi et al, 1993a,b; Zella et al, 1993; Gessain et al, 1994b, 1995b; Lal et al, 1994b; Eiraku et al, 1995, 1996; Ishak et al, 1995, Mauclère et al, 1995; Salemi et al, 1995, 1996, 1998; Switzer et al, 1995a,b, 1996; Biggar et al, 1996; Heneine, 1996; Tuppin et al, 1996; Murphy et al, 1998; M. Biglione et al, 1999) (Figures 16.7, 16.8). After the first studies it was clear that the subtype A (prototype MO) was mainly present in US injecting drug abusers and only rarely encountered in Amerindians, whereas subtype B (prototypes NRA and G12) was mainly found in most of the Amerindian groups tested, scattered throughout North, Central and South America, including some remote groups in Brazil, Argentina and Colombia. Subtype B was thus referred to as the Paleo-Amerindian strain. The situation is currently changing since a divergent variant of HTLV-2 subtype A (also labelled by one group as HTLV-2 subtype C) has been found to be the genuine virus present in some Amerindian tribes including the Kayapo of Brazil but also among injecting drug users from several urban areas of Brazil (Ishak et al, 1995; Biggar et al, 1996; Eiraku et al, 1996; Hall et al, 1996; Heneine, 1996; Switzer et al, 1996).

Two different but closely related classifications of HTLV-2 molecular subtypes using RFLP analyses of two large fragments of the LTR have been reported (Eiraku et al, 1995; Heneine, 1996; Switzer et al, 1996). Although based on the analysis of more than 300 HTLV-2 isolates from several geographical origins and risk behaviours, at least 15 different RFLP patterns have been obtained; careful phylogenetic analyses of the LTR region supported at least two subgroups within the HTLV-2A and at least three within the HTLV-2B subtype (Salemi et al, 1998) (Figure 16.8).

A recent study performed on 160 new HTLV-2 strains from either blood donors or injecting drug users from various cities of the USA confirms that the majority of the HTLV-2 strains in the USA belong to the type A0 but also reports the existence of some novel and unusual HTLV-2 RFLP types. Furthermore, epidemiological analysis suggests the existence of an age cohort effect for HTLV-2 RFLP type A0 among older white and black injecting drug users and blood donors (Murphy et al, 1998).

In Europe, although only subtype A has been found in injecting drug users from Norway (Switzer et al, 1995b), both subtypes have been found to be present in injecting drug users from southern Europe (Italy and Spain) (Salemi et al, 1995, 1996, 1998a; Vallejo et al, 1996) with, however, a predominance of HTLV-2 subtype B. A phylogenetic study including nine new European isolates (seven from Italy and two from the UK) strongly suggests that the epidemic HTLV-2 infection among drug users from southern Europe arose from a limited number of infections due to connections with the USA. This study also suggests that, although a certain differentiation by restriction analysis in different subgroups is possible, carefully interpreted phylogenetic analyses remain necessary (Salemi et al, 1998). Furthermore, using the historically documented date for the introduction of the injecting drugs to Europe, the first possible estimation of the evolutionary rate of HTLV-2 among injecting drug users was proposed (Salemi et al, 1998a).

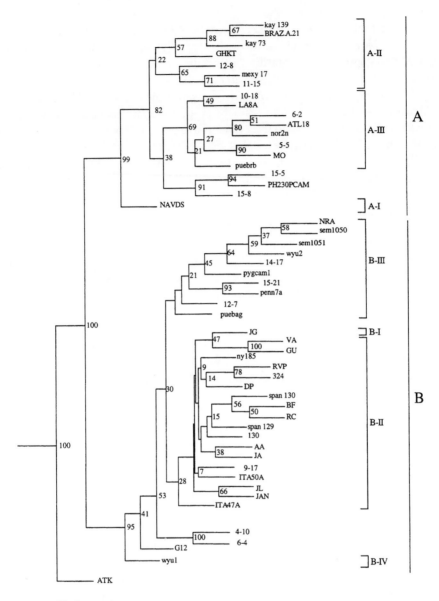

Figure 16.8 *Phylogenetic tree*
Phylogenetic analysis performed on a 625 bp region of the HTLV-2 LTR of 52 available sequences. The method used was the DNAPARS program, from the PHYLIP package, with bootstrap values from a separate 200 iteration bootstrap model inscribed at the branch points. Phylogenetic subgroups on the right of the figure are based upon the classification of Switzer et al (1996) (adapted from Murphy et al, 1998).

EVIDENCE OF AN ANCIENT PRESENCE OF HTLV-2 IN AFRICA?

Epidemiological and phylogenetic analyses indicate that HTLV-1 and STLV-1 have been present in the Old World (Africa and Asia) at least for tens of millennia, whereas

HTLV-2, which is highly endemic in certain native Amerindian tribes, has been considered to be a New World virus, brought from Asia into the Americas some 10 000–40 000 years ago by the migration of populations infected with HTLV-2 over the Bering land bridge (Maloney et al, 1992; Hall et al, 1994; Neel et al, 1994). This view had to be revised in the light of evidence of sporadic serological cases of HTLV-2 infection in West Africa (Gessain et al, 1993c; Igarashi et al, 1993; Bonis et al, 1994; Lal et al, 1994a; Vallejo et al, 1995), Central Africa (Delaporte et al, 1991a,b; Goubau et al, 1992, 1993; Froment et al, 1993; Mauclère et al, 1993, 1995; Dube et al, 1994, Gessain et al, 1994b, 1995b, 1996b; Vallejo et al, 1994, 1995; Tuppin et al, 1996) and East Africa (Buckner et al, 1992), raising the possibility that HTLV-2 or a related retrovirus might have also been endemic in Africa for a long period. This hypothesis was based primarily on the detection of HTLV-2 antibodies in the sera of individuals from two Pygmy tribes, living in remote areas of Zaïre (D.R Congo) (Goubau et al, 1992, 1993) and Cameroon (Froment et al, 1993; Goubau et al, 1993; Gessain et al, 1995b). These populations are considered to be the oldest inhabitants of Central Africa. However, at that time, few data were available on the molecular structure of the African HTLV-2 isolates. An HTLV-2 subtype A virus was first isolated from a female sex worker in Ghana, West Africa, suggesting the possibility of an imported infection (Igarashi et al, 1993). An HTLV-2 subtype B virus, closely related to the Paleo-Amerindian HTLV-2B strains, was also found in a Zaïrean patient but no sociodemographic data were available (Dube et al, 1994). Thus, the presence of indigenous HTLV-2 infection in Africa remained an open question. The present state of HTLV-2 infection in the African continent is discussed below from an epidemiological and a molecular point of view, together with data to support the hypothesis that HTLV-2 is an ancient retrovirus of Central Africa.

Seroepidemiological situation of HTLV-2 in Africa

The first serological evidence of an HTLV-2 infection in Africa was obtained in the early 1990s by Delaporte et al (1991a), who reported antibodies to HTLV-2 in a healthy pregnant woman living in Gabon, West Central Africa. Another case of HTLV-2 infection occurring in an asymptomatic 44-year-old man was further reported by the same group, in Gabon (Delaporte et al, 1991b). In 1993, some prostitutes (mostly also infected with HIV) from Zaïre (Goubau et al, 1993), Ghana (Igarashi et al, 1993) and Cameroon (Mauclère et al, 1993, 1995) were also found to be HTLV-2 infected. Furthermore, HTLV-2 infections were also reported among blood donors from Guinea (Gessain et al, 1993b) in West Africa. Since then, other sporadic cases of HTLV-2 infection have been reported in West, Central and East African countries including Senegal and Ivory Coast (Bonis et al, 1994), Equatorial Guinea (Vallejo et al, 1994), Guinea-Bissau (Vallejo et al, 1995), Somalia and Ethiopia (Buckner et al, 1992). To date, at least 40 sporadic cases have been reported (Figure 16.9) but the major question still remains unanswered: were these infections imported, or do they reflect the presence of a very ancient HTLV-2 infection in Africa? Two recent findings argue for the latter.

First, evidence was reported of an intrafamilial clustering of a new molecular variant of HTLV-2B in southeastern Gabon, with both mother-to-child and sexual transmission (Gessain et al, 1994b; Tuppin et al, 1996). Among 41 family members, seven exhibited specific HTLV-2 antibodies in their sera as demonstrated by high immunofluorescence titres on C19 cells and/or specific Western blot pattern. Four further individuals from this family were detected as being HTLV-1 seropositive. The second husband (58 years old) of the index case (60 years old), two of his sisters (aged 56 and 48 years) and three of their children (aged 21, 34 and 42 years) were HTLV-2 infected, suggesting the presence of this variant virus in this family over more than two generations. Around 1930, the region of Franceville was a remote forest area, rarely visited by foreigners. Moreover, injecting drug abuse could be considered as non-existent at that time, and even now is extremely rare in the area. Blood transfusions have been practised only very recently. Consequently, transmission from mother to child (in the Franceville area, infants are breast-fed for at least 18–24 months) and by sexual intercourse can be considered as the major routes of transmission of these new HTLV-2 isolates.

The second and even more important indication of an ancient presence of HTLV-2 in Africa was the unexpected discovery of endemic HTLV-2 infection in some isolated Pygmy populations. Goubau et al first reported in 1992 the presence of HTLV-2 antibodies in 4 out of 12 Efe Pygmies, a group of the Bambuti tribe. This Pygmy population living in a remote area of the Ituri forest of northeastern Zaïre (D.R. Congo) is considered to be the least admixed Pygmy population (Figure 16.9) (Goubau et al, 1992). In a second study, this group confirmed their first observation in the same population with evidence of HTLV-2 infection in 14 out of 102 sera collected in 1970 for anthropological studies (Goubau et al, 1993). These results were extended to another group of Pygmies living in Cameroon, where 5 out of 214 samples were HTLV-2 positive; however, in that case, the exact origin of the samples collected between 1967 and 1970 was unknown. The presence of HTLV-2 infection in a few Bakola Pygmies – 4 of 89 (Froment et al, 1993), and 1 of 41 (Gessain et al, 1995b) – living in a rainforest and game reserve in the southeastern part of Cameroon near the border with Equatorial Guinea confirms that HTLV-2 is endemic in some Pygmy groups. The notion of an ancient HTLV-2 infection in Bakola Pygmies was reinforced by the fact that such infection was absent in other tribes of different origin, living nearby in contact with the Bakola group (Froment et al, 1993). Furthermore, HTLV-2 appears to be virtually absent in other ethnic groups also living in Cameroon: no antibodies to HTLV-2 were detected in more than 7000 samples of sera from adult individuals of different ethnic groups and from various areas of the country, although HTLV-1 was present in 0·5–2% of the population (Mauclère et al, 1997a). We failed to detect any HTLV-2-infected individuals in representative samples of Biaka Pygmies from the Lobaye region of the Central African Republic and of Twa pygmies from the Lake Tomba area of Zaïre or in the Bantu population of the surrounding areas (Gessain et al, 1993b). Similar observations were reported by others (Goubau et al, 1993). Even though the studied samples were small, the Bambuti and the Bakola Pygmies appear to be not only the solely HTLV-2-infected Pygmy groups but also the only populations yet described in the whole of Africa in whom HTLV-2 is really endemic. The level of endemicity is estimated to be around 2–3% in the adult Bakola Pygmy population (Mauclère et al, 1997b; unpublished observations). It is important to note that the Bambuti and Bakola Pygmies are located at the western and eastern ends, respectively, of the present area of Pygmy habitation in Central Africa, and that they are of different paleoanthropological origins, with a estimated divergence, based on linguistic and cultural features, of around 10 000 years (Bahuchet, 1993; Cavalli-Sforza et al, 1994). Furthermore, genetic studies suggest that about 20 000 years of isolation would have been required for Pygmies to become genetically distinct (Bahuchet, 1993; Cavalli-Sforza et al, 1994). Taken together, these results strongly suggest that isolated

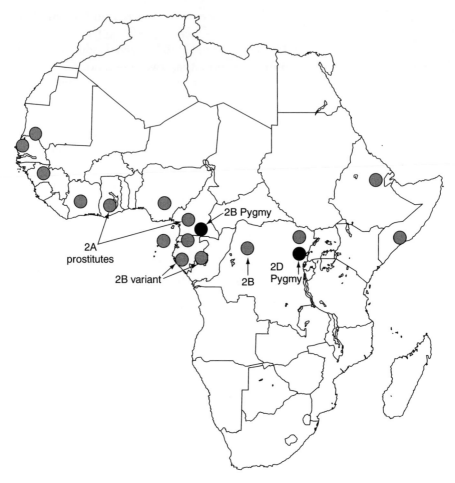

Figure 16.9 *HTLV-2 in Africa in 1998*
Geographical distribution of the different HTLV-2 strains obtained from African inhabitants.
Solid circles represent the aboriginal (Pygmy) HTLV-2 endemic population; shaded circles
represent sporadic cases of HTLV-2 infection.

Pygmy populations have been an African reservoir for HTLV-2 infection for a very long
time.

Genetic diversity and molecular epidemiology of HTLV-2 in Africa

Only six studies have reported molecular characteristics of African HTLV-2 isolates
(Table 16.2). Igarashi et al (1993) first reported the presence of an HTLV-2 subtype A
isolate in a prostitute from Ghana. More recently, a similar viral subtype was observed
in a prostitute from Cameroon (Mauclère et al, 1995). The fact that subtype A is the
most common subtype in injecting drug users in Western countries, and that both these
prostitutes were working in large towns with many outside contacts, initially suggested
that the HTLV-2 subtype A was imported through commercial sexual activities rather
than being a genuine African virus. However, more detailed sequence analysis of these

Table 16.2 *Epidemiological and molecular features of the HTLV-2 strains from Africa*

Case	Age/sex (yr)	Geographical origin	Ethnic/risk group	HTLV serology	Sequence data	Subtype	Reference
GhKT	25/F	Ghana	Prostitute HIV+	HTLV-2 (ELISA)	Fragment LTR 455 bp R/U5	2A	Igarashi et al (1993)
1 patient	–	D.R. Congo (Congo)	Hospitalized patients	HTLV-2 (WB)	*pol* – 140 bp	2B	Dube et al (1994)
HTLV-II GAB JPS	58/M	Gabon	Familial transmission	HTLV-2 (IF-WB)	*env*-pX – 1161 bp	2B variant	Gessain et al (1994)
PYGCAM-1	59/F	Cameroon	Pygmy Bakola	HTLV-2 (IF-WB)	LTR – 715 bp; *env*-pX – 1508 bp	2B	Gessain et al (1995b)
PH230PCAM	32/F	Cameroon	Prostitute HIV+	HTLV-2 (IF-WB)	LTR – 715 bp; *env*-gp21 – 589 bp	2A	Mauclère et al (1995)
HTLV-II GAB JPS	58/M	Gabon	Familial transmission	HTLV-2 (IF-WB)	*env*-pX – 1508 bp; *env*-pX – 519 bp	2B variant	Tuppin et al (1996)
HTLV-IIGAB	44/M	Gabon	–	HTLV-2 (WB)	Complete sequence; pX – 159 bp	2B	Letourneur et al (1998); Goubau et al (1996)
HTLV-II Pygmy2	–	D.R. Congo (congo)	Pygmy Efe	HTLV-2 (WB)	Complete sequence	2D	Vandamme et al (1998)

Adapted from Gessain et al (1996b).
ELISA, enzyme-linked immunosorbent assay; IF, immunofluorescence; WB, Western blot.

viruses revealed that these two HTLV-2 subtype A strains were closely related, and were slightly different (especially the Cameroon isolate) from HTLV-2 A isolates from inject-ing drug users and Amerindians. Thus, the virus detected in these prostitutes may represent a genuine African molecular strain of HTLV-2 subtype A (Mauclère et al, 1995; Switzer et al, 1996; Murphy et al, 1998).

The Pygmy populations studied were thought to have a completely divergent virus, reflecting a very long period of evolution in vivo similar to the HTLV-1 strains present in remote populations of Papua New Guinea and Australia (Gessain et al, 1993b). How-ever, the results of the molecular analyses were unexpected: in 1994, we isolated an HTLV-2 from a healthy 59-year-old Pygmy woman of the Bakola population living in a remote forest area of southeastern Cameroon (Gessain et al, 1995b). Following PBMC culture, a continuous CD8+ cell line containing HTLV-2 proviruses was established. These cells produced HTLV-2 antigens and retroviral particles. Molecular analysis of this HTLV-2 isolate, PYGCAM-1, surprisingly demonstrated very close sequence homologies with the HTLV-2 subtype B prototypes (G12 and NRA) originally isolated respectively from a Guaymi Amerindian in Panama and from a US patient with a 'T hairy cell leukaemia'. Indeed, we were surprised to find an apparently genuine African HTLV-2 so closely related to the Paleo-Amerindian subtype B.

However, our hypothesis that HTLV-2 subtype B is a genuine African virus is still supported by the following arguments. Firstly, two independent groups have reported the presence of HTLV-2 subtype B isolates in Central Africa. Dube et al (1994) found traces of a such a virus by RT/PCR (with only 140 base pairs being sequenced in the *pol* gene) in a Zaïrean patient's plasma collected in 1969 in a hospital in the Equateur region of northwest Zaïre (D.R. Congo). Two other patients from the same series were also found to be HTLV-2 seropositive, suggesting that HTLV-2 has been present in this remote area for at least several decades. Another group reported the complete nucleotide sequence of an HTLV-2 B isolate present in a 44-year-old Gabonese healthy carrier without evidence of any New World connection (Letourneur et al, 1998). This virus was also closely related to the paleoindian strains and to the PYGCAM-1 strain.

Moreover, for as yet unexplained reasons, the genetic variability within both HTLV-2 subtypes seems to be extremely low: a nucleotide divergence of only 0–0·4% (over 1000 base pairs of the *env* gene) was observed between HTLV-2 of different Indian groups in North, Central and South America who have had probably no contact for several thou-sand years (Hall et al, 1994, 1996). Furthermore, preliminary data from Mongolia (not yet confirmed) have shown the presence of a typical HTLV-2 subtype A, similar to the one present in the Americas (Hall et al, 1994). This again supports the concept that the genetic drift of HTLV-2 (assuming a similar evolutionary rate in the different geograph-ical locations) is lower than the roughly estimated 0·2–1% nucleotide divergence in the HTLV-1 *env* and *pol* genes during a 1000-year period of evolution (Poiesz et al, 1993; Fer-rer et al, 1996; Gessain et al, 1996a). Data indicating that viral amplification occurs via clonal cellular expansion, rather than by reverse transcription, would explain – as was well shown for HTLV-1 (Wattel et al, 1995) – this remarkable genetic stability (Cimarelli et al, 1996).

Besides these findings, recent data indicate that some HTLV-2 strains exist only in Central Africa, for example the HTLV-2 subtype B variant in the Gabonese family men-tioned above (Tuppin et al, 1996). Molecular analysis of the total *env* gene (1462 bp) and fragments of the *pol* and pX regions confirmed that this African HTLV-2 isolate was the most divergent HTLV-2 subtype B yet described. It exhibited 2·3% of nucleotide substitutions in the *env* gene (33 bases) compared with the two HTLV-2B prototypes (Tuppin et al, 1996). This sequence divergence was considered unusual as this gene is remarkably conserved within a subtype (Hall et al, 1994, 1996). Furthermore, Goubau et al reported, to the VIIIth International Conference on Human Retrovirology: HTLV, the preliminary molecular characterization of a very divergent HTLV-2 isolate, present in an Efe Pygmy from northeastern Zaïre (Vandamme et al, 1997b). This new HTLV-2 strain, which differs from HTLV-2A and HTLV-2B in the LTR by roughly 7% of nucleotide substitutions, seems slightly related to STLV-2 pan-p and is considered to be a new subtype, provisionally labelled HTLV-2 subtype D (Vandamme et al, 1998).

HTLV-2-RELATED SIMIAN RETROVIRUSES IN *PAN PANISCUS*

Two slightly related isolates of a new simian retrovirus, related to HTLV-2, have been obtained independently from captive colonies of pygmy chimpanzees (*Pan paniscus*) originating from Zaïre (D.R. Congo) but living in primate centres either in the USA (Giri et al, 1994) or in Belgium (Liu et al, 1994b). The Belgian virus was designated STLV-PP1664, and phylogenetic analyses performed on fragments of the *tax* region indicated that, although this virus is more related to HTLV-2 than to the other PTLV types, it is clearly separated from HTLV-2, indicating a long independent evolution (Liu et al, 1994b; Vandamme et al, 1996). Preliminary sequence data from the American isolate obtained after amplification of the proviral DNA with the SK43–SK44 primers (*tax* gene) indicated that this was slightly more related to HTLV-2 (87%) than to HTLV-1 ATK (78%). Furthermore, this isolate appeared to be equally distant from STLV-Ptm3 (STLV-1 prototype) and from the new PTLV-L PH969. The complete sequence of this American isolate has recently been obtained and phylogenetic analyses indicate that this virus is an early divergence within the type 2 lineage, referred to by Digilio et al, (1997) as STLV-2 pan-p (see Figure 16.7 and by Van Brussel 1998 as STLV-pp1664). Furthermore, this new retrovirus displays a host range similar to that demonstrated for other HTLV and STLV strains (Digilio et al, 1997).

A THIRD TYPE OF PRIMATE T LYMPHOTROPIC VIRUS

Another new PTLV designated PTLV-L, with STLV-PH969 as the prototype strain, was isolated from a wild-born baboon (*Papio hamadryas*) from Eritrea (Goubau et al, 1994). A 1802 bp long fragment was first identified from a cDNA library. This sequence extends from the *env* region including the complete transmembrane protein gene (gp21) to part of the *tax/rex* gene. Homologies of STLV-PH969 with HTLV-1 and HTLV-2 were 62% and 64% respectively overall, 65% and 70% in the *env* region, and 80% and 80% in the partial *tax/rex* sequence, at the nucleotide sequence level. A phylogenetic analysis based on the gp21 sequence indicates that PTLV-L represents a PTLV type with a longer independent evolution than any strain within the PTLV-1 or PTLV-2 groups (see Figure 16.7). The entire sequence of this new virus has been published, and all the major genes that are encountered in the PTLVs and their corresponding mRNAs, including appropriate splicing, were identified (Van Brussel et al, 1997). In all coding regions, the similarities tend to be lower between STLV-PH969 and HTLV-1 than with HTLV-2. However, within the LTR, the lowest similarities were found between STLV-PH969 and HTLV-2. Furthermore, while the presence of three 21 bp repeats is conserved within the U3 region of HTLV-1, HTLV-2 and BLV, only two direct repeats were described in the STLV-PH969 sequence (Van Brussel et al, 1997).

It is interesting to note that the animals infected by these new PTLVs (PTLV-L and STLV-2 pan-p/STLV-PP1664) exhibit an HTLV-2-like serology on Western blot with at least a p24 >> p19 and a rgp21 with or without the reactivity against the specific gp46 peptide (K55) (Hadlock et al, 1992) of HTLV-2 but associated in some cases with a reactivity against the specific gp46 peptide (MTA-1) of HTLV-1 (Giri et al, 1994; Goubau et al, 1994; Vandamme et al, 1996; Digilio et al, 1997).

CONCLUSION

Analyses of viral strains of HTLV-1 from throughout the world yield a general pattern of dissemination that includes:

1. Transmission of STLV-1 to humans, as exemplified by the high percentage of sequence homologies between STLV-1 from chimpanzees and mandrills and some HTLV-1 from inhabitants of Central Africa (Table 16.1).
2. The probable transmission of STLV-1 between different simian species as suggested in the wild between *Cercopithecus* and *Papio* in East and South Africa, between *Cercopithecus* and *Pan troglodytes* in West Africa, and as demonstrated in captive colony between *Papio* and *Macaca*.
3. The persistence of the virus over long periods of time in remote areas as represented by HTLV-1 in Melanesia.
4. Global distribution of the virus by large-scale human migration of populations infected with HTLV-1 such as the slave trade from Africa to the New World.

The peculiar geographical distribution of HTLV-2 throughout the world probably results from similar events. The introduction of HTLV-2 to the New World during the original settlement of the American continent by the ancestors of the present-day Amerindians, 10 000–40 000 years ago, was followed by a more recent dissemination by injecting drug users, mostly from an original North American cluster (Hall et al, 1996). However, some important pieces of the puzzle are lacking regarding the origin of HTLV-2:

1. There are no confirmed data regarding the presence of an HTLV-2 endemic population in Asia (Neel et al, 1994; Gessain et al, 1996b).
2. The simian reservoirs of each of the HTLV-2 subtypes have not yet been found (see Table 16.1). However, based on the recent demonstration of the presence of HTLV-2 in some remote Pygmy populations in Central Africa and the discovery of STLV-2 pan-p and PTLV-L, we suggest that the number of PTLV types or variants should be considered open, and that the variety of indigenous viruses in the PTLV group is larger in the African continent (Vandamme et al, 1996; Gessain et al, 1996b). Preliminary data from our laboratory indicate the possible presence of an PTLV related retrovirus in another monkey species living in the western part of Central

Africa, an area also inhabited by the Bakola Pygmy population, recently shown to be endemically infected by HTLV-2 (Mauclère et al, 1997b; A. Gessain and P. Mauclère, unpublished observations).
3. In the case of HTLV-2, the exact origin and significance of the high level of viral endemicity among injecting drug users are still poorly understood (Murphy et al, 1998; Salemi et al, 1998).

It is therefore difficult to draw a clear picture concerning the origin and modes of dissemination of HTLV types 1 and 2 from all the accumulated data (Gessain et al, 1996a; Vandamme et al, 1998b; Pecon-Slattery et al, 1998). Difficulties in the reconstruction of the natural history of these primate retroviruses are due to several factors, including:

1. Discrete viral transmission episodes – not only between monkeys and humans but also between different monkey species – both in Africa and Asia probably occurred at many different times during primate species evolution.
2. Possible differences in evolutionary rates of nucleotide substitutions according to the infected host (humans or monkeys) or the ethnic group, or to different modes of transmission (as suggested for HTLV-2 in injecting drug User).
3. The absence of a good calibration of the rate of molecular divergence, giving a precise time scale for phylogenetic reconstructions (Dekaban et al, 1996).

Some authors have tried to estimate such a time scale. Thus, based on the comparison of available historical, paleoanthropological and sequence data from HTLV-1 of the different Australo-Melanesian strains, Gessain et al (1996a) roughly estimated the evolution rate in vivo of HTLV-1 to be around 0·02–0·1% per century (1% per 1000–5000 years) in the gp21 *env* gene. Another group estimated that 1% of divergence within the SK110/111 *pol* sequence of the PTLVs represents around 500–1000 years of separation of the host population (Poiesz et al, 1993; Ferrer et al, 1996). Another study, based on the known divergence between Japanese and rhesus macaques, estimated the substitution rate for STLV-1 to be about 1% per 20 000–122 000 years (Song et al, 1994). Using samples from injecting drug abusers, Salemi et al estimated the evolutionary rate of HTLV-2 to be around 10^{-4}–10^{-5} nucleotide substitutions per site per year, corresponding to 0·1–1% per 1000 years (Salemi et al, 1998). These authors suggested also that in the case of HTLV-2 the fixation rate might be much lower in populations with predominantly vertical transmission, such as Amerindians, than in populations with viral transmission by needle sharing (Salemi et al, 1998a).

Studies aiming to isolate and characterize new human and simian retroviruses, especially in remote human populations and in wild-caught monkeys, will provide new insights in the origin, genetic diversity, evolution and modes of dissemination of such viruses. This will open new avenues of research on the evolutionary history both of primate retroviruses and of early human groups.

ACKNOWLEDGMENTS

The authors thank Professor Guy de Thé for his continuing support, Fredj Tekaia, Colombe Chappey and Jill Pecon-Slattery for the phylogenetic analyses, and Monique Van Beveren for the preparation of some figures. Financial support from l'Agence

Nationale de Recherches sur le SIDA (ANRS) and the Virus Cancer Prevention Association is acknowledged. Renaud Mahieux was a CANAM fellow.

REFERENCES

Bahuchet S (1993) History of the inhabitants of the central African rain forest: perspectives from comparative linguistics. In: Hladik CM, Hladik A, Linares O, Pagezy H, Semple A, Hadley M (eds) *Tropical Forests: People and Food. Man and the Biosphere*, pp 37–54. Paris: Parthenon–UNESCO.

Bangham CR, Daenke S, Phillips RE, Cruickshank JK & Bell JI (1988) Enzymatic amplification of exogenous and endogenous retroviral sequences from DNA of patients with tropical spastic paraparesis. *EMBO J* **7**: 4179–4184.

Bastian I, Gardner J, Webb D & Gardner I (1993) Isolation of a human T-lymphotropic virus type I strain from Australian aboriginals. *J Virol* **67**: 843–851.

Bazarbachi A, Huang M, Gessain A et al (1995) Human T-cell-leukemia virus type I in posttransfusional spastic paraparesis: complete proviral sequence from uncultured blood cells. *Int J Cancer* **63**: 494–499.

Becker WB, Becker ML, Homma T, Brede HD & Kurth R (1985) Serum antibodies to human T-cell leukaemia virus type I in different ethnic groups and in non-human primates in South Africa. *S Afr Med J* **67**: 445–449.

Benson J, Tschachler E, Gessain A, Yanagihara R, Gallo RC & Franchini G (1994) Cross-neutralizing antibodies against Cosmopolitan and Melanesian strains of human T cell leukemia/lymphotropic virus type I in sera from inhabitants of Africa and the Solomon Islands. *AIDS Res Hum Retroviruses* **10**: 91–96.

Berneman ZN, Gartenhaus RB, Reitz MS et al (1992) Expression of alternatively spliced human T-lymphotropic virus type I pX mRNA in infected cell lines and in primary uncultured cells from patients with adult T-cell leukemia/lymphoma and healthy carriers. *Proc Natl Acad Sci USA* **89**: 3005–3009.

Biggar RJ, Taylor ME, Neel JV et al (1996) Genetic variants of human T-lymphotropic virus type II American Indian groups. *Virology* **216**: 165–173.

Biglione M, Gessain A, Quiruelas S et al (1993) Endemic HTLV-II infection among Tobas and Matacos Amerindians from north Argentina. *J AIDS* **6**: 631–633.

Biglione M, Vidan O, Mahieux R et al (1999) Seroepidemiological and molecular studies of Human T Cell lymphotrophic Virus Type 2, Subtype B, in isolated groups of Mataco and Toba Indians of Northern Argentina. *Aids and HR* (in press).

Black FL, Biggar RJ, Neel JV, Maloney EM & Waters DJ (1994) Endemic transmission of HTLV type II among Kayapo Indians of Brazil. *AIDS Res Hum Retroviruses* **10**: 1165–1171.

Blattner WA (1990) Epidemiology of HTLV-I and associated diseases. *Hum Retrovirol HTLV* 251–265.

Blattner WA, Saxinger C, Riedel D et al (1990) A study of HTLV-I and its associated risk factors in Trinidad and Tobago. *J AIDS* **3**: 1102–1108.

Blattner WA & Gallo RC (1994) Epidemiology of HTLV-I and HTLV-II infection. In: Takatsuki K (ed.) *Adult T-cell Leukaemia*, pp 45–90. New York: Oxford University Press.

Boeri E, Giri A, Lillo F et al (1992) In vivo genetic variability of the human immunodeficiency virus type 2 V3 region. *J Virol* **66**: 4546–4550.

Boeri E, Gessain A, Garin B, Kazadi K, de Thé G & Franchini G (1993) Qualitative changes in the human T-cell leukemia/lymphotropic virus type I *env* gene sequence in the spastic versus nonspastic tropical paraparesis are not correlated with disease specificity. *AIDS Res Hum Retroviruses* **9**: 1–5.

Bonis J, Verdier M, Dumas M & Denis F (1994) Low human T cell leukemia virus type II sero-prevalence in Africa. *J Infect Dis* **169:** 225–227.

Buckner C, Roberts CR, Foung SK et al (1992) Immune responsiveness to the immunodominant recombinant envelope epitopes of human T lymphotropic virus types I and II in diverse geographic populations. *J Infect Dis* **166:** 1160–1163.

Calabro ML, Luparello M, Grottola A et al (1993) Detection of human T lymphotropic virus type II/b in human immunodeficiency virus type 1-coinfected persons in southeastern Italy. *J Infect Dis* **168:** 1273–1277.

Cann AJ & Chen ISY (1990) Human T-cell leukemia virus types I and II. In: Fields BN, Knipe DR et al (eds) *Virology*, pp 1501–1527. New York: Raven.

Cavalli-Sforza LL, Menozzi P & Piazza A (1994) *The History and Geography of Human Genes.* Princeton: Princeton University Press.

Cavrois M, Gessain A, Wain-Hobson S & Wattel E (1996) Proliferation of HTLV-1 infected circulating cells in vivo in all asymptomatic carriers and patients with TSP/HAM. *Oncogene* **12:** 2419–2423.

Cereseto A, Mulloy JC & Franchini G (1996) Insights on the pathogenicity of human T-lymphotropic/leukemia virus types I and II. *J AIDS Hum Retrovirol* **13:** S69–S75.

Chen YM, Jang YJ, Kanki PJ et al (1994) Isolation and characterization of simian T-cell leukemia virus type II from New World monkeys. *J Virol* **68:** 1149–1157.

Chen J, Zekeng L, Yamashita M et al (1995) HTLV type I isolated from a Pygmy in Cameroon is related to but distinct from the known central African type. *AIDS Res Hum Retroviruses* **11:** 1529–1531.

Chou KS, Okayama A, Tachibana N, Lee TH & Essex M (1995) Nucleotide sequence analysis of a full-length human T-cell leukemia virus type I from adult T-cell leukemia cells: a prematurely terminated PX open reading frame II. *Int J Cancer* **60:** 701–706.

Cimarelli A, Angelin Duclos C, Gessain A, Casoli C & Bertazzoni U (1996) Clonal expansion of human T-cell leukemia virus type II in patients with high proviral load. *Virology* **223:** 362–364.

Ciminale V, D'Agostino DM, Zotti L & Chieco-Bianchi L (1996) Coding Potential of the X Region of Human T-Cell Leukaemia/Lymphotrophic Virus Type II. *J AIDS Hum Retrovirol* **13:** S220–S227.

Ciminale V, Pavlakis GN, Derse D, Cunningham CP & Felber BK (1992) Complex splicing in the human T-cell leukaemia virus (HTLV) family of retroviruses: novel mRNAs and proteins produced by HTLV type I. *J Virol* **66:** 1737–1745.

Crandall KA (1996) Multiple interspecies transmissions of Human and Simian T Cell Leukaemia Virus Type I sequences. *Mol Biol Evolution* **13:** 115–131.

Daenke S, Nightingale S, Cruickshank JK & Bangham CR (1990) Sequence variants of human T-cell lymphotropic virus type I from patients with tropical spastic paraparesis and adult T-cell leukaemia do not distinguish neurological from leukemic isolates. *J Virol* **64:** 1278–1282.

De BK, Lairmore MD, Griffis K et al (1991) Comparative analysis of nucleotide sequences of the partial envelope gene (5′ domain) among human T lymphotropic virus type I (HTLV-I) isolates. *Virology* **182:** 413–419.

De Rossi A, Mammano F, Del Mistro A & Chieco-Bianchi L (1991) Serological and molecular evidence of infection by human T-cell lymphotropic virus type II in Italian drug addicts by use of synthetic peptides and polymerase chain reaction. *Eur J Cancer* **27:** 835–838.

de Thé G & Bomford R (1993) An HTLV-I vaccine: why, how, for whom? *AIDS Res Hum Retroviruses* **9:** 381–386.

Dekaban G, Coulthart M & Franchini G (1996) Natural History of HTLVs/STLVs. In: Höllsberg P and Hafler DA (eds) *Human T-cell Lymphotropic Virus Type I*, pp 11–32, John Wiley & Sons Ltd, England.

Dekaban GA, King EE, Waters D & Rice GP (1992) Nucleotide sequence analysis of an HTLV-I isolate from a Chilean patient with HAM/TSP. *AIDS Res Hum Retroviruses* **8:** 1201–1207.

Delaporte E, Louwagie J, Peeters M et al (1991a) Evidence for HTLV-II infection in Central Africa. *AIDS* **5**: 771–772.

Delaporte E, Monplaisir N, Louwagie J et al (1991b) Prevalence of HTLV-I and HTLV-II Infection in Gabon, Africa – Comparison of the Serological and PCR Results. *Int J Cancer* **49**: 373–376.

Digilio L, Givi A, Cho N et al (1997) The Simian T-Lymphotrophic/Leukaemia Virus from *Pan paniscus* belongs to the Type 2 Family and Infects Asian Macaques. *J Virol* **71(5)**: 3684–3692.

Dube DK, Dube S, Erensoy S et al (1994) Serological and nucleic acid analyses for HIV and HTLV infection on archival human plasma smples from Zaïre. *Virology* **202**: 379–389.

Dube DK, Sherman MP, Saksena NK et al (1993) Genetic heterogeneity in human T-cell leukemia/lymphoma virus type II. *J Virol* **67**: 1175–1184.

Dube S, Spicer T, Bryz-Gornia V et al (1995) A rapid and sensitive method of identification of HTLV-II subtypes. *J Med Virol* **45**: 1–9.

Dube S, Bachman S, Spicer T et al (1997) Degenerate and specific PCR assays for the detection of bovine leukaemia virus and primate T cell leukaemia/lymphoma virus *pol* DNA and RNA: phylogenetic comparisons of amplified sequences from cattle and primates from around the world. *J Gen Virol* **78**: 1389–1398.

Duenas-Barajas E, Bernal JE, Vaught DR et al (1993) Human retroviruses in Amerindians of Colombia: high prevalence of human T cell lymphotropic virus type II infection among the Tunebo Indians. *Am J Trop Med Hyg* **49**: 657–663.

Ehrlich GD, Andrews J, Sherman MP, Greenberg SJ & Poiesz BJ (1992) DNA sequence analysis of the gene encoding the HTLV-I p21e transmembrane protein reveals inter- and intraisolate genetic heterogeneity. *Virology* **186**: 619–627.

Eiraku N, Monken C, Kubo T et al (1995) Nucleotide sequence and restriction fragment length polymorphism analysis of the long terminal repeat of human T cell leukemia virus type II. *AIDS Res Hum Retroviruses* **11**: 625–636.

Eiraku N, Novoa P, da Costa Ferreira M et al (1996) Identification and characterization of a new and distinct molecular subtype of human T-cell lymphotropic virus type 2. *J Virol* **70**: 1481–1492.

Engelbrecht S, van Rensburg EJ & Robson BA (1996) Sequence variation and subtyping of human and simian T-cell lymphotropic virus type I strains from South Africa. *J AIDS Hum Retrovirol* **12**: 298–302.

Evangelista A, Maroushek S, Minnigan H et al (1990) Nucleotide sequence analysis of a provirus derived from an individual with tropical spastic paraparesis. *Microb Pathog* **8**: 259–278.

Felsenstein J (1993) PHYLIP: Phylogenetic Interference Package. Version 3.5. Seattle: University of Washington.

Ferrer JF, del Pino N, Esteban E et al (1993) High rate of infection with the human T-cell leukemia retrovirus type II in four Indian populations of Argentina. *Virology* **197**: 576–584.

Ferrer JF, Esteban E, Dube S et al (1996) Endemic infection with human T cell leukemia/lymphoma virus type IIB in Argentinean and Paraguayan Indians: epidemiology and molecular characterization. *J Infect Dis* **174**: 944–953.

Fouchard N, Flageul B, Bagot M et al (1995) Lack of evidence of HTLV-I/II infection in T CD8 malignant or reactive lymphoproliferative disorders in France: a serological and/or molecular study of 169 cases. *Leukemia* **9**: 2087–2092.

Franchini G (1995) Molecular mechanisms of human T-cell leukemia/lymphotropic virus type I infection. *Blood* **86**: 3619–3639.

Franchini G, Mulloy JC, Koralnik IJ et al (1993) The human T-cell leukemia/lymphotropic virus type I p121 protein cooperates with the E5 oncoprotein of bovine papillomavirus in cell transformation and binds the 16-kilodalton subunit of the vacuolar H+ ATPase. *J Virol* **67**: 7701–7704.

Froment A, Delaporte E, Dazza MC & Larouze B (1993) HTLV-II among pygmies from Cameroon. *AIDS Res Hum Retroviruses* **9**: 707.

Fujiyama C, Fujiyoshi T, Miura T et al (1993) A new endemic focus of human T lympho-tropic virus type II carriers among Orinoco natives in Colombia (letter). *J Infect Dis* **168:** 1075–1077.

Fukasawa M, Tsujimoto H, Ishikawa K et al (1987) Human T-cell leukemia virus type I isolates from Gabon and Ghana: comparative analysis of proviral genomes. *Virology* **161:** 315–320.

Gasmi M, Farouqi B, D'Incan M & Desgranges C (1994) Long terminal repeat sequence analysis of HTLV type I molecular variants identified in four north African patients. *AIDS Res Hum Retroviruses* **10:** 1313–1315.

Gessain A (1996) Epidemiology of HTLV-I and associated diseases. In: Höllsberg P & Hafler DA (eds) *Human T-cell Lymphotropic Virus Type I*, pp 33–64. Chichester: John Wiley.

Gessain A & de Thé G (1996a) Geographic and molecular epidemiology of primate lym-photropic retroviruses: HTLV-I, HTLV-II, STLV-I, STLV-PP, and PTLV-L. *Adv Virus Res* **47:** 377–426.

Gessain A & de Thé G (1996b) What is the situation of human T cell lymphotropic virus type II (HTLV-II) in Africa? Origin and dissemination of genomic subtypes. *J AIDS Hum Retrovirol* **13 (suppl. 1):** S228–S235.

Gessain A & Gout O (1992) Chronic myelopathy associated with human T-lymphotropic virus type-I (HTLV-I). *Ann Intern Med* **117:** 933–946.

Gessain A, Barin F, Vernant JC et al (1985) Antibodies to human T-lymphotropic virus type-I in patients with tropical spastic paraparesis. *Lancet* **ii:** 407–410.

Gessain A, Yanagihara R, Franchini G et al (1991) Highly divergent molecular variants of human T-lymphotropic virus type I from isolated populations in Papua New Guinea and the Solomon Islands. *Proc Natl Acad Sci USA* **88:** 7694–7698.

Gessain A, Boeri E, Kazadi K et al (1992a) Variant rétroviral HTLV-I au Zaire chez un patient ayant une neuromyélopathie chronique, séquence nucléotidique du gène d'enveloppe. *C R Acad Sci* **314:** 159–164.

Gessain A, Gallo RC & Franchini G (1992b) Low degree of human T-cell leukemia/lymphoma virus type I genetic drift in vivo as a means of monitoring viral transmission and movement of ancient human populations. *J Virol* **66:** 2288–2295.

Gessain A, Boeri E, Yanagihara R, Gallo RC & Franchini G (1993a) Complete nucleotide sequence of a highly divergent human T-cell leukemia (lymphotropic) virus type I (HTLV-I) variant from Melanesia: genetic and phylogenetic relationship to HTLV-I strains from other geographical regions. *J Virol* **67:** 1015–1023.

Gessain A, Hervé V, Jeannel D, Garin B, Mathiot C & de Thé G (1993b) HTLV-1 but not HTLV-2 found in pygmies from Central African Republic (letter, comment). *J AIDS* **6:** 1373–1374.

Gessain A, Fretz C, Koulibaly M et al (1993c) Evidence of HTLV-II infection in Guinea, West Africa. *J AIDS* **6:** 324–325.

Gessain A, Koralnik IJ, Fullen J et al (1994a) Phylogenetic study of ten new HTLV-I strains from the Americas. *AIDS Res Hum Retroviruses* **10:** 103–106.

Gessain A, Tuppin P, Kazanji M et al (1994b) A distinct molecular variant of HTLV-IIB in Gabon, Central Africa. *AIDS Res Hum Retroviruses* **10:** 753–755.

Gessain A, Mahieux R & de Thé G (1995a) HTLV-I 'indeterminate' western blot patterns observed in sera from tropical regions: the situation revisited. *J AIDS Hum Retrovirol* **9:** 316–319.

Gessain A, Mauclère P, Froment A et al (1995b) Isolation and molecular characterization of a human T-cell lymphotropic virus type II (HTLV-II), subtype B, from a healthy Pygmy living in a remote area of Cameroon: an ancient origin for HTLV-II in Africa. *Proc Natl Acad Sci USA* **92:** 4041–4045.

Gessain A, Mahieux R & de Thé G (1996a) Genetic variability and molecular epidemiology of human and simian T cell leukemia/lymphoma virus type I. *J AIDS Hum Retrovirol* **13 (suppl. 1):** S132–S145.

Gessain A, Malet C, Robert-Lamblin J et al (1996b) Serological evidence of HTLV-I but not HTLV-II infection in ethnic groups of northern and eastern Siberia. *J AIDS Hum Retrovirol* **11:** 413–414.

Giri A, Markham P, Digilio L, Hurteau G, Gallo RC & Franchini G (1994) Isolation of a novel simian T-cell lymphotropic virus from *Pan paniscus* that is distantly related to the human T-cell leukemia/lymphotropic virus types I and II. *J Virol* **68:** 8392–8395.

Giri A, Pecon-Slattery J, Heinene W et al (1997) *Tax* gene sequences form two deeply divergent monophyletic lineages corresponding to types I and II of simian and human T cell leukemia lymphotropic viruses. *Virology* **231:** 96–104.

Gonzalez-Dunia D, Grimber G, Briand P, Brahic M & Ozden S (1992) Tissue expression pattern directed in transgenic mice by the LTR of an HTLV-I provirus isolated from a case of tropical spastic paraparesis. *Virology* **187:** 705–710.

Goodenow M, Huet T, Saurin W, Kwok S, Sninsky J & Wain-Hobson S (1989) HIV isolates are rapidly evolving quasispecies: evidence for viral mixtures and preferred nucleotide substitution. *J AIDS* **2:** 344–352.

Goubau P, Desmyter J, Ghesquiere J & Kasereka B (1992) HTLV-II among Pygmies. *Nature* **359:** 201.

Goubau P, Liu HF, De Lange GG, Vandamme AM & Desmyter J (1993) HTLV-II seroprevalence in pygmies across Africa since 1970. *AIDS Res Hum Retroviruses* **9:** 709–713.

Goubau P, Van Brussel M, Vandamme AM, Liu HF & Desmyter J (1994) A primate T-lymphotropic virus, PTLV-L, different from human T-lymphotropic viruses types I and II, in a wild-caught baboon (*Papio hamadryas*). *Proc Natl Acad Sci USA* **91:** 2848–2852.

Goubau P, Vandamme AM & Desmyter J (1996) Questions on the evolution of primate T-lymphotropic viruses raised by molecular and epidemiological studies of divergent strains. *J AIDS Hum Retrovirol* **13:** S242–S247.

Gout O, Baulac M, Gessain A et al (1990) Rapid development of myelopathy after HTLV-I infection acquired by transfusion during cardiac transplantation. *N Engl J Med* **322:** 383–388.

Gray GS, Bartman T & White M (1987) Nucleotide sequence of the core (*gag*) gene from HTLV-I isolate MT2. *Nucleic Acids Res* **17:** 7998.

Gray GS, White M, Bartman T & Mann D (1990) Envelope gene sequence of HTLV-1 isolate MT-2 and its comparison with other HTLV-1 isolates. *Virology* **177:** 391–395.

Gurtsevitch V, Senyuta N, Shih J et al (1995) HTLV-I infection among Nivkhi people in Sakhalin. *Int J Cancer* **60:** 432–433.

Hadlock KG, Lipka JJ Chow TP, Foung SK & Reyes GR (1992) Cloning and analysis of a recombinant antigen containing an epitope specific for human T-cell lymphotropic virus type II. *Blood* **79:** 2789–2796.

Hall WW, Takahashi H, Liu C et al (1992) Multiple isolates and characteristics of human T-cell leukemia virus type-II. *J Virol* **66:** 2456–2463.

Hall WW, Zhu SW, Horal P, Furuta Y, Zagaany G & Vahlne A (1994) HTLV-II infection in Mongolia (abstr. 2). Program and Abstracts from the 6th International Conference on Human Retrovirology, Absecon, New Jersey. *AIDS Res Hum Retroviruses* **10:** 443.

Hall WW, Ishak R, Zhu SW et al (1996) HTLV-II epidemiology, molecular properties and clinical features of infection. *Nucleic Acids Res* **13:** S204–S214.

Hashimoto K, Lalkaka J, Fujisawa JM et al (1993) Limited sequence divergence of HTLV-I of Indian HAM/TSP patients from a prototype Japanese isolate. *AIDS Res Hum Retroviruses* **9:** 495–498.

Hayami M, Komuro A, Nozawa K et al (1984) Prevalence of antibody to adult T-cell leukemia virus-associated antigens (ATLA) in Japanese monkeys and other non-human primates. *Int J Cancer* **33:** 179–183.

Hayami M, Mboudjeka I, Ohkura S et al (1997) Phylogenetic analysis of 13 HTLV-I isolates in Ghana and Cameroon including those from four Pygmies. Abstract ME03. *VIIIth International Conference on Human Retrovirology HTLV*, 9–13 June 1997, Rio, Brazil.

Heneine W (1996) The phylogeny and molecular epidemiology of human T-cell lymphotropic virus type II. *J AIDS Hum Retrovirol* **13**: S236–S241.

Heneine W, Woods T, Green D et al (1992) Detection of HTLV-II in breastmilk of HTLV-II infected mothers. *Lancet* **340**: 1157–1158.

Hinuma Y, Nagata K, Hanaoka M et al (1981) Adult T-cell leukemia: antigen in an ATL cell line and detection of antibodies to the antigen in human sera. *Proc Natl Acad Sci USA* **78**: 6476–6480.

Hjelle B, Zhu SW, Takahashi H, Ijichi S & Hall WW (1993) Endemic human T cell leukemia virus type II infection in southwestern US Indians involves two prototype variants of virus. *J Infect Dis* **168**: 737–740.

Ibrahim F, de Thé G & Gessain A (1995) Isolation and characterization of a new simian T-cell leukemia virus type 1 from naturally infected Celebes macaques (*Macaca tonkeana*): complete nucleotide sequence and phylogenetic relationship with the Australo-Melanesian human T-cell leukemia virus type 1. *J Virol* **69**: 6980–6993.

Igarashi T, Yamashita M, Miura T et al (1993) Isolation and genomic analysis of human T lymphotropic virus type II from Ghana. *AIDS Res Hum Retroviruses* **9**: 1039–1042.

Ijichi S, Tajima K, Zaninovic V et al (1993) Identification of human T cell leukemia virus type IIb infection in the Wayu, an aboriginal population of Colombia. *Jpn J Cancer Res* **84**: 1215–1218.

Imamura J, Tsujimoto A, Ohta Y et al (1988) DNA blotting analysis of human retroviruses in cerebrospinal fluid of spastic paraparesis patients: the viruses are identical to human T-cell leukemia virus type-1 (HTLV-1). *Int J Cancer* **42**: 221–224.

Ishak R, Harrington WJ, Azevedo VN et al (1995) Identification of human T cell lymphotropic virus type IIa infection in the Kayapo, an indigenous population of Brazil. *AIDS Res Hum Retroviruses* **11**: 813–821.

Ishida T, Yamamoto K, Kaneko R, Tokita E & Hinuma Y (1983) Seroepidemiological study of antibodies to adult T-cell leukemia virus-associated antigen (ATLA) in free-ranging Japanese monkeys (*Macaca fuscata*). *Microbiol Immunol* **27**: 297–301.

Ishikawa K, Fukasawa M, Tsujimoto H et al (1987) Serological survey and virus isolation of simian T-cell leukemia/T-lymphotropic virus type I (STLV-I) in non-human primates in their native countries. *Int J Cancer* **40**: 233–239.

Josephs SF, Wong-Staal F, Manzari V et al (1984) Long terminal repeat structure of an American isolate of type I human T-cell leukemia virus. *Virology* **139**: 340–345.

Kalyanaraman VS, Sarngadharan MG, Robert-Guroff M, Miyoshi I, Golde D & Gallo RC (1982) A new subtype of human T-cell leukemia virus (HTLV-II) associated with a T-cell variant of hairy cell leukemia. *Science* **218**: 571–573.

Kaplan JE & Khabbaz RF (1993) The epidemiology of human T-lymphotropic virus types I and II. *Rev Med Virol* **3**: 137–148.

Kaplan JE, Holland MU, Green DB, Gracia F & Reeves WC (1993) Failure to detect human T-lymphotropic virus antibody in wild-caught New World primates. *Am J Trop Med Hyg* **49**: 236–238.

Kazanji M, Moreau JP, Mahieux R et al (1997) HTLV-I infection in squirrel monkey (*Saïmiri sciureus*) using autologous, homologous or heterologous HTLV-I-transformed cell lines. *Virology* **231**: 258–266.

Katamine S, Moriuchi R, Yamamoto T et al (1994) HTLV-I proviral DNA in umbilical cord blood of babies born to carrier mothers. *Lancet* **343**: 1326–1327.

Kinoshita T, Tsujimoto A & Shimotohno K (1991) Sequence variations in LTR and env regions of HTLV-I do not discriminate between the virus from patients with HTLV-I-associated myelopathy and adult T-cell leukemia. *Int J Cancer* **47**: 491–495.

Kira J, Koyanagi Y, Yamada T et al (1994) Sequence heterogeneity of HTLV-I proviral DNA in the central nervous system of patients with HTLV-I-associated myelopathy. *Ann Neurol* **36**: 149–156.

Komurian F, Pelloquin F & de Thé G (1991) In vivo genomic variability of human T-cell leukemia virus type I depends more upon geography than upon pathologies. *J Virol* **65**: 3770–3778.

Komurian-Pradel F, Pelloquin F, Sonoda S, Osame M & de Thé G (1992) Geographical subtypes demonstrated by RFLP following PCR in the LTR region of HTLV-I. *AIDS Res Hum Retroviruses* 8: 429–434.

Komuro A, Watanabe T, Miyoshi I et al (1984) Detection and characterization of simian retroviruses homologous to human T-cell leukemia virus type I. *Virology* 138: 373–378.

Koralnik I (1996) Genomic structure of HTLV-I. In: Höllsberg P (ed.) *Human T cell Lymphotropic Virus Type 1*. Chichester: John Wiley.

Koralnik IJ, Gessain A, Klotman ME, Lo Monico A & Berneman ZN (1992) Protein isoforms encoded by the pX region of human T-cell leukemia/lymphotropic virus type I. *Proc Natl Acad Sci USA* 89: 8813–8817.

Koralnik IJ, Fullen J & Franchini G (1993) The p12I, p13II, and p30II proteins encoded by human T-cell leukemia/lymphotropic virus type I open reading frames I and II are localized in three different cellular compartments. *J Virol* 67: 2360–2366.

Koralnik IJ, Boeri E, Saxinger WC et al (1994) Phylogenetic associations of human and simian T-cell leukemia/lymphotropic virus type I strains: evidence for interspecies transmission. *J Virol* 68: 2693–2707.

Kwok S, Kellogg D, Ehrlich G, Poiesz B, Bhagavati S & Sninsky JJ (1988) Characterization of a sequence of human T cell leukemia virus type I from a patient with chronic progressive myelopathy. *J Infect Dis* 158: 1193–1197.

Lairmore MD, Jacobson S, Gracia F et al (1990) Isolation of human T-cell lymphotropic virus type-2 from Guaymi Indians in Panama. *Proc Natl Acad Sci USA* 87: 8840–8844.

Lal RB, Rudolph DL, Nerurkar VR & Yanagihara R (1992) Humoral responses to the immunodominant *gag* and *env* epitopes of human T-lymphotropic virus type I among Melanesians. *Viral Immunol* 5: 265–272.

Lal RB, Gongora-Biachi RA, Pardi D, Switzer WM, Goldman I & Lal AF (1993) Evidence for mother-to-child transmission of human T lymphotropic virus type II. *J Infect Dis* 168: 586–591.

Lal RB, Owen SM, Mingle J, Levine PH & Manns A (1994a) Presence of human T lymphotropic virus types I and II in Ghana, West Africa. *AIDS Res Hum Retroviruses* 10: 1747–1750.

Lal RB, Owen SM, Rudolph D & Levine PH (1994b) Sequence variation within the immunodominant epitope-coding region from the external glycoprotein of human T lymphotropic virus type II in isolates from Seminole Indians. *J Infect Dis* 169: 407–411.

Lee H, Swanson P, Shorty VS, Zack JA, Rosenblatt JD & Chen IS (1989) High rate of HTLV-II infection in seropositive i.v. drug abusers in New Orleans. *Science* 244: 471–475.

Lee H, Idler KB, Swanson P et al (1993) Complete nucleotide sequence of HTLV-II isolate RNA: comparison of envelope sequence variation of HTLV-II isolates from US blood donors and US and Italian i.v. drug users. *Virology* 196: 57–69.

Lenz J, Celander D, Crowther RL, Patarca R, Perkins DW & Haseltine WA (1984) Determination of the leukaemogenicity of the murine retrovirus by sequences within the long terminal repeat. *Nature* 308: 467–470.

Letourneur F, d'Auriol L, Dazza MA et al (1998) Complete nucleotide sequence of an African human T-lymphotropic virus type II subtype b isolate (HTLV-II-Gab): molecular and phylogenetic analysis. *J Gen Virol* 79: 269–277.

Levine PH, Jacobson S, Elliott R et al (1993) HTLV-II infection in Florida Indians. *AIDS Res Hum Retroviruses* 9: 123–127.

Li Y, Golemis E, Hartley JW & Hopkins N (1987) Disease specificity of nondefective Friend and Moloney murine leukemia viruses is controlled by a small number of nucleotides. *J Virol* 61: 693–700.

Liu HF, Vandamme AM, Kazadi K, Carton H, Desmyter J & Goubau P (1994a) Familial transmission and minimal sequence variability of human T-lymphotropic virus type I (HTLV-I) in Zaire. *AIDS Res Hum Retroviruses* 10: 1135–1142.

Liu HF, Vandamme AM, Van Brussel M, Desmyter J & Goubau P (1994b) New retroviruses in human and simian T-lymphotropic viruses. *Lancet* **344**: 265–266.

Liu HF, Goubau P, Van Brussel M et al (1996) The three human T-lymphotropic virus type I subtypes arose from three geographically distinct simian reservoirs. *J Gen Virol* **77**: 359–368.

Liu HF, Goubau P, Van Brussel M, Desmyter J & Vandamme AM (1997) Phylogenetic analysis of a STLV-1 from a Hamadryas baboon. *AIDS Res Hum Retroviruses* **13**: 1545–1548.

Mahieux R, Gessain A, Truffert A et al (1994) Seroepidemiology, viral isolation, and molecular characterization of human T cell leukemia/lymphoma virus type I from La Reunion Island, Indian Ocean. *AIDS Res Hum Retroviruses* **10**: 745–752.

Mahieux R, de Thé G & Gessain A (1995) The *tax* mutation at nucleotide 7959 of human T-cell leukemia virus type 1 (HTLV-1) is not associated with tropical spastic paraparesis/HTLV-1-associated myelopathy but is linked to the Cosmopolitan molecular genotype *J Virol* **69**: 5925–5927.

Mahieux R, Ibrahim F, Mauclère P et al (1997a) Molecular epidemiology of 58 new African human T-cell leukemia virus type 1 (HTLV-1) strains: identification of a new and distinct HTLV-1 molecular subtype in Central Africa and in Pygmies. *J Virol* **71**: 1317–1333.

Mahieux R, Pecon-Slattery J & Gessain A (1997b) Molecular characterization and phylogenetic analyses of a new, highly divergent simian T-cell lymphotropic virus type 1 (STLV-1marc1) in *Macaca arctoides*. *J Virol* **71**: 6253–6258.

Mahieux R, Pecon-Slattery J, Chen GM & Gessain A (1998a) Evolutionary Inferences of Novel Simian T Lymphotropic Virus Type I from Wild-caught Chacma (*Papio ursinus*) and Olive Baboons (*Papio anubis*) *J Virol* **251**: 71–84.

Mahieux R et al (1998b) Simian T-cell lymphotropic virus type 1 from *Mandrillus sphinx* as a simian counterpart of human T-cell lymphotropic virus type 1 subtype D. *J Virol* **(12)**: 10316–10322.

Major ME, Nightingale S & Desselberger U (1993) Complete sequence conservation of the human T cell leukaemia virus type 1 *tax* gene within a family cluster showing different pathologies. *J Gen Virol* **74**: 2531–2537.

Malik KT, Even J & Karpas A (1988) Molecular cloning and complete nucleotide sequence of an adult T cell leukaemia virus/human T cell leukaemia virus type I (ATLV/HTLV-I) isolate of Caribbean origin: relationship to other members of the ATLV/HTLV-I subgroup. *J Gen Virol* **69**: 1695–1710.

Maloney EM, Biggar RJ, Neel JV et al (1992) Endemic human T-cell lymphotropic virus type-II infection among isolated Brazilian Amerindians. *J Infect Dis* **166**: 100–107.

Mauclère P, Gessain A, Garcia-Calleja JM et al (1993) HTLV-II in African prostitutes from Cameroon. *AIDS* **7**: 1394–1395.

Mauclère P, Mahieux R, Garcia-Calleja JM et al (1995) A new HTLV type II subtype A isolate in an HIV type 1-infected prostitute from Cameroon, Central Africa. *AIDS Res Hum Retroviruses* **11**: 989–993.

Mauclère P, Le Hesran JY, Mahieux R et al (1997a) Demographic, ethnic, and geographic differences between human T cell lymphotropic virus (HTLV) type-I seropositive carriers and persons with HTLV-I *gag*-indeterminate western blots in Central Africa. *J Infect Dis* **176**: 505–509.

Mauclère P, Froment A, Ruffié A et al (1997b) HTLV-I and HTLV-II are both endemic among Bakola Pygmies of Cameroon. Abstract ED59. *VIIIth International Conference on Human Retrovirology: HTLV*, 9–13 June 1997, Rio, Brazil.

Meyerhans A, Cheynier R, Albert J et al (1989) Temporal fluctuation in HIV quasispecies in vivo are not reflected by sequential HIV isolations. *Cell* **58**: 901–910.

Miura T, Fukunaga T, Igarashi T et al (1994) Phylogenetic subtypes of human T-lymphotropic virus type I and their relations to the anthropological background. *Proc Natl Acad Sci USA* **91**: 1124–1127.

Miyoshi I, Kubonishi I, Yoshimoto S et al (1981) Type C virus particles in a cord T-cell line derived by cocultivating normal human cord leukocytes and human leukaemic T cells. *Nature* **294**: 770–771.

Miyoshi I, Yoshimoto S, Fujishita M et al (1982) Natural adult T-cell leukemia virus infection in Japanese monkeys. *Lancet* ii: 658.

Moynet D, Cosnefroy JY, Bedjabaga I, Roelants G, Georges-Courbot MC & Guillemain B (1995) Identification of new genetic subtypes of human T cell leukemia virus type I in Gabon from encoding sequence of surface envelope glycoprotein. *AIDS Res Hum Retroviruses* 11: 1407–1411.

Mueller N (1991) The epidemiology of HTLV-I infection. *Cancer Causes Control* 2: 37–52.

Mukhopadhyaya R & Sadaie MR (1993) Nucleotide sequence analysis of HTLV-I isolated from cerebrospinal fluid of a patient with TSP/HAM: comparison to other HTLV-I isolates. *AIDS Res Hum Retroviruses* 9: 109–114.

Murphy EL (1996) The clinical epidemiology of human T-lymphotropic virus type II (HTLV-II). *J AIDS Hum Retrovirol* 13: S215–S219.

Murphy EL, Mahieux R, de Thé G et al (1998) Molecular epidemiology of HTLV-II among US blood donors and intravenous drug users: an age-cohort effect for HTLV-II RFLP type A0. *Virology* 242: 425–434.

Neel JV, Biggar RJ & Sukernik RI (1994) Virologic and genetic studies relate Amerind origins to the indigenous people of the Mongolia/Manchuria/southeastern Siberia region. *Proc Natl Acad Sci USA* 91: 10737–10741.

Nei M, Takezaki N & Sitnikova T (1995) Assessing molecular phylogenies. *Science* 267: 253–254.

Nerurkar VR, Miller MA, Leon-Monzon ME et al (1992) Failure to isolate human T cell lymphotropic virus type I and to detect variant-specific genomic sequences by polymerase chain reaction in Melanesians with indeterminate western immunoblot. *J Gen Virol* 73: 1805–1810.

Nerurkar VR, Babu PG, Song KJ et al (1993a) Sequence analysis of human T cell lymphotropic virus type I strains from southern India: gene amplification and direct sequencing from whole blood blotted onto filter paper. *J Gen Virol* 74: 2799–2805.

Nerurkar VR, Song KJ, Saitou N, Melland RR & Yanagihara R (1993b) Interfamilial and intrafamilial genomic diversity and molecular phylogeny of human T-cell lymphotropic virus type I from Papua New Guinea and the Solomon Islands. *Virology* 196: 506–513.

Nerurkar VR, Song KJ, Bastian IB, Garin B, Franchini G & Yanagihara R (1994a) Genotyping of human T cell lymphotropic virus type I using Australo-Melanesian topotype-specific oligonucleotide primer-based polymerase chain reaction: insights into viral evolution and dissemination. *J Infect Dis* 170: 1353–1360.

Nerurkar VR, Song KJ, Melland RR & Yanagihara R (1994b) Genetic and phylogenetic analyses of human T-cell lymphotropic virus type I variants from Melanesians with and without spastic myelopathy. *Mol Neurobiol* 8: 155–173.

Nerurkar VR, Achiron A, Song KJ et al (1995) Human T-cell lymphotropic virus type I in Iranian-born Mashhadi Jews: genetic and phylogenetic evidence for common source of infection. *J Med Virol* 45: 361–366.

Niewiesk S, Daenke S, Parker CE et al (1994) The transactivator gene of human T-cell leukemia virus type I is more variable within and between healthy carriers than patients with tropical spastic paraparesis. *J Virol* 68: 6778–6781.

Niewiesk S, Daenke S, Parker CE et al (1995) Naturally occurring variants of human T-cell leukemia virus type I Tax protein impair its recognition by cytotoxic T lymphocytes and the transactivation function of Tax. *J Virol* 69: 2649–2653.

Oguma S, Imamura Y, Kusumoto Y et al (1992) Accelerated declining tendency of human T-cell leukemia virus type I carrier rates among younger blood donors in Kumamoto. *Jpn J Cancer Res* 52: 2620–2623.

Osame M, Usuku K, Izumo S et al (1986) HTLV-I associated myelopathy, a new clinical entity. *Lancet* i: 1031–1032.

Osame M, Matsumoto M, Usuku K et al (1987) Chronic progressive myelopathy associated with elevated antibodies to human T-lymphotropic virus type I and adult T-cell leukemia like cells. *Ann Neurol* 21: 117–122.

Ozden S, Coscoy L & Gonzalez-Dunia D (1996) HTLV-I transgenic models: an overview. *J AIDS Hum Retrovirol* **13:** S154–S161.

Paine E, Garcia J, Philpott TC, Shaw G & Ratner L (1991) Limited sequence variation in human T-lymphotropic virus type 1 isolates from North American and African patients [published erratum appears in *Virology* (1992) **188(1):** 414]. *Virology* **182:** 111–123.

Pardi D, Kaplan JE, Coligan JE, Folks TM & Lal RB (1993a) Identification and characterization of an extended Tax protein in human T-cell lymphotropic virus type II subtype b isolates. *J Virol* **67:** 7663–7667.

Pardi D, Switzer WM, Hadlock KG, Kaplan JE, Lal RB & Folks TM (1993b) Complete nucleotide sequence of an Amerindian human T-cell lymphotropic virus type II (HTLV-II) isolate: identification of a variant HTLV-II subtype b from a Guaymi Indian. *J Virol* **67:** 4659–4664.

Pecon-Slattery J, Franchini G & Gessain A (1999) Molecular phylogenetics and viral emergence: applications in the evolutionary history of HTLV/STLV. (in press).

Picard FJ, Coulthart MB, Oger J et al (1995) Human T-lymphotropic virus type 1 in coastal natives of British Columbia: phylogenetic affinities and possible origins. *J Virol* **69:** 7248–7256.

Picchio GR, Hayami M, Veronesi R & Yamashita M (1997) Molecular epidemiology of 5 Argentinean human T-cell leukemia type-1 (HTLV-I) strains present in asymptomatic and adult T-cell leukemia (ATL) HTLV-I seropositive patients. Abstract ME06. *VIIIth International Conference on Human Retrovirology: HTLV*, 9–13 June 1997, Rio, Brazil.

Poiesz BJ, Ruscetti FW, Reitz MS et al (1980) Detection and isolation of a type C retrovirus particles from fresh and cultured lymphocytes of a patient with cutaneous T-cell lymphoma. *Proc Natl Acad Sci USA* **77:** 7415–7419.

Poiesz BJ, Ruscetti FW, Reitz MS, Kalyanaraman VS & Gallo RC (1981) Isolation of a new type C retrovirus (HTLV) in primary uncultured cells of a patient with Sezary T-cell leukaemia. *Nature* **294:** 268–271.

Poiesz B, Sherman M, Saksena N et al (1993) The biology and epidemiology of the human T-cell lymphoma/leukemia viruses. In: Neu HC, Levy JA & Weiss RA (eds) *Frontiers of Infectious Diseases*, pp 189–205. Edinburgh: Churchill Livingstone.

Popovic M, Reitz MS, Sarngadharan MG et al (1982) The virus of Japanese adult T-cell leukemia is a member of the human T-cell leukaemia virus group. *Nature* **330:** 63–66.

Rassart E, Nelbach L & Jolicoeur P (1986) Cas-Br-E murine leukemia virus: sequencing of the paralytogenic region of its genome and derivation of specific probes to study its origin and the structure of its recombinant genomes in leukemic tissues. *J Virol* **60:** 910–919.

Ratner L, Josephs SF, Starcich B et al (1985) Nucleotide sequence analysis of a variant human T-cell leukemia virus (HTLV-Ib) provirus with a deletion in pX-I. *J Virol* **54:** 781–790.

Ratner L, Philpott T & Trowbridge DB (1991) Nucleotide sequence analysis of isolates of human T-lymphotropic virus type 1 of diverse geographical origins. *AIDS Res Hum Retroviruses* **7:** 923–941.

Renjifo B, Borrero I & Essex M (1995) *Tax* mutation associated with tropical spastic paraparesis/human T-cell leukemia virus type I-associated myelopathy. *J Virol* **69:** 2611–2616.

Saito M, Furukawa Y, Kubota R et al (1995) Frequent mutation in pX region of HTLV-1 is observed in HAM/TSP patients, but is not specifically associated with the central nervous system lesions. *J Neurovirol* **1:** 286–294.

Saito M, Furukawa Y, Kubota R et al (1996) Mutation rates in LTR of HTLV-I in HAM/TSP patients and the carriers are similarly high to *tax/rex* coding sequence. *J Neurovirol* **2:** 330–335.

Saitou N & Nei M (1987) The neighbor-joining method: a new method for reconstructing phylogenetic trees. *Mol Biol Evol* **4:** 406–425.

Saksena NK, Sherman MP, Yanagihara R, Dube DK & Poiesz BJ (1992) LTR sequence and phylogenetic analyses of a newly discovered variant of HTLV-I isolated from the Hagahai of Papua New Guinea. *Virology* **189:** 1–9.

Saksena NK, Herve V, Sherman MP et al (1993) Sequence and phylogenetic analyses of a new STLV-I from a naturally infected tantalus monkey from Central Africa. *Virology* **192**: 312–320.

Saksena NK, Herve V, Durand JP et al (1994) Seroepidemiologic, molecular, and phylogenetic analyses of simian T-cell leukemia viruses (STLV-I) from various naturally infected monkey species from central and western Africa. *Virology* **198**: 297–310.

Salemi M, Cattaneo E, Casoli C & Bertazzoni U (1995) Identification of IIa and IIb molecular subtypes of human T-cell lymphotropic virus type II among Italian injecting drug users. *J AIDS Hum Retrovirol* **8**: 516–520.

Salemi M, Vandamme AM, Guano F et al (1996) Complete nucleotide sequence of the Italian human T-cell lymphotropic virus type II isolate Gu and phylogenetic identification of a possible origin of South European epidemics. *J Gen Virol* **77**: 1193–1201.

Salemi M, Vandamme AM, Gradozzi C et al (1998a) Evolutionary rate and genetic heterogeneity of human T-cell lymphotropic virus type II (HTLV-II) using isolates from European injecting drug users. *J Mol Evol* **(46)**: 602–611.

Salemi A, Van Dooren S, Audenaert E et al (1998b) Two new human T-lymphotropic Virus Type I Phylogenetic subtypes in seroindeterminates, a Mbuti Pygmy and a Gabones, have closest relatives among African STLV-I strains. *J Virol* **(246)**: 277–287.

Sandler SG, Fang CT & Williams AE (1991) Human T-cell lymphotropic virus type I and II in transfusion medecine. *Transfus Med Rev* **5**: 93–107.

Schätzl H, Tschikobava M, Rose D et al (1993) The Sukhumi primate monkey model for viral lymphomogenesis: high incidence of lymphomas with presence of STLV-I and EBV-like virus. *Leukemia* **7 (suppl. 2)**: S86–S92.

Schulz TF, Calabro ML, Hoad JG, Carrington CV, Matutes E & Catovsky D (1991) HTLV-1 envelope sequences from Brazil, the Caribbean, and Romania: clustering of sequences according to geographic origin and variability in an antibody epitope. *Virology* **184**: 483–491.

Seiki M, Hattori S & Yoshida M (1982) Human adult T-cell leukemia virus: molecular cloning of the provirus DNA and the unique terminal structure. *Proc Natl Acad Sci USA* **79**: 6899–6902.

Seiki M, Hattori S, Hirayama Y & Yoshida M (1983) Human adult T-cell leukemia virus: complete nucleotide sequence of the provirus genome integrated in leukemia cell DNA. *Proc Natl Acad Sci USA* **80**: 3618–3622.

Sherman MP, Saksena NK, Dube DK, Yanagihara R & Poiesz BJ (1992) Evolutionary insights on the origin of human T-cell lymphoma/leukemia virus type I (HTLV-I) derived from sequence analysis of a new HTLV-I variant from Papua New Guinea. *J Virol* **66**: 2556–2563.

Sherman MP, Dube S, Spicer TP et al (1993) Sequence analysis of an immunogenic and neutralizing domain of the human T-cell lymphoma/leukemia virus type I gp46 surface membrane protein among various primate T-cell lymphoma/leukemia virus isolates including those from a patient with both HTLV-I-associated myelopathy and adult T-cell leukemia. *Cancer Res* **53**: 6067–6073.

Shimotohno K, Takahashi Y, Shimizu N et al (1985) Complete nucleotide sequence of an infectious clone of human T-cell leukemia virus type II: an open reading frame for the protease gene. *Proc Natl Acad Sci USA* **82**: 3101–3105.

Shirabe S, Nakamura T, Tsujihata M, Nagataki S, Seiki M & Yoshida M (1990) Retrovirus from human T-cell leukemia virus type I-associated myelopathy is the same strain as a prototype human T-cell leukemia virus type I. *Arch Neurol* **47**: 1258–1260.

Song KJ, Nerurkar VR, Saitou N et al (1994) Genetic analysis and molecular phylogeny of simian T-cell lymphotropic virus type I: evidence for independent virus evolution in Asia and Africa. *Virology* **199**: 56–66.

Song KJ, Nerurkar VR, Pereira-Cortez AJ et al (1995) Sequence and phylogenetic analyses of human T cell lymphotropic virus type 1 from a Brazilian woman with adult T cell leukemia: comparison with virus strains from South America and the Caribbean basin. *Am J Trop Med Hyg* **52**: 101–108.

Switzer WM, Owen SM, Pieniazek DA et al (1995a). Molecular analysis of human T-cell lymphotropic virus type II from Wayuu Indians of Colombia demonstrates two subtypes of HTLV-IIb. *Virus Genes* **10**: 153–162.

Switzer WM, Pieniazek D, Swanson P et al (1995b). Phylogenetic relationship and geographic distribution of multiple human T-cell lymphotropic virus type II subtypes. *J Virol* **69**: 621–632.

Switzer WM, Black FL, Pieniazek D, Biggar RJ, Lal RB & Heneine W (1996) Endemicity and phylogeny of the human T cell lymphotropic virus type II subtype A from the Kayapo Indians of Brazil: evidence for limited regional dissemination. *AIDS Res Hum Retroviruses* **12**: 635–640.

Szurek PF, Yuen PH, Jerzy R & Wong PK (1988) Identification of point mutations in the envelope gene of Moloney murine leukemia virus TB temperature-sensitive paralytogenic mutant ts1: molecular determinants for neurovirulence. *J Virol* **62**: 357–360.

Tajima K & Ito SI (1990) Prospective studies of HTLV-I and associated diseases in Japan. *Hum Retrovirol HTLV* pp. 267–279.

Tajima K, Inoue M, Takezaki T, Ito M & Ito SI (1994) Ethnoepidemiology of ATL in Japan with special reference to the Mongoloid dispersal. In: Takatsuki K (ed.) *Adult T-cell Leukaemia*, pp 91–112. Oxford: Oxford University Press.

Takahashi K, Takezaki T, Oki T et al (1991) Inhibitory effect of maternal antibody on mother-to-child transmission of human T-lymphotropic virus type I. The Mother-to-Child Transmission Study Group. *Int J Cancer* **49**: 673–677.

Takahashi H, Zhu SW, Ijichi S, Vahlne A, Suzuki H & Hall WW (1993) Nucleotide sequence analysis of human T cell leukemia virus, type II (HTLV-II) isolates. *AIDS Res Hum Retroviruses* **9**: 721–732.

Takatsuki H, Uchiayama T, Sagawa K & Yodoi J (1977) Adult T-cell leukemia in Japan. In: Seno S, Takaku F & Irino S (eds) *Topics in Hematology*, pp 73–77. Amsterdam: Excerpta Medica.

Takezaki T, Tajima K, Komoda H & Imai J (1995) Incidence of human T lymphotropic virus type I seroconversion after age 40 among Japanese residents in an area where the virus is endemic. *J Infect Dis* **171**: 559–565.

Tsujimoto A, Teruuchi T, Imamura J, Shimotohno K, Miyoshi I & Miwa M (1988) Nucleotide sequence analysis of a provirus derived from HTLV-1-associated myelopathy (HAM). *Mol Biol Med* **50**: 481–492.

Tuppin P, Gessain A, Kazanji M et al (1996) Evidence in Gabon for an intrafamilial clustering with mother-to-child and sexual transmission of a new molecular variant of human T-lymphotropic virus type-II subtype B. *J Med Virol* **48**: 22–32.

Ureta-Vidal A, Gessain A, Yoshida M et al (1994a) Molecular epidemiology of HTLV type I in Japan: evidence for two distinct ancestral lineages with a particular geographical distribution. *AIDS Res Hum Retroviruses* **10**: 1557–1566.

Ureta-Vidal A, Gessain A, Yoshida M et al (1994b) Phylogenetic classification of human T cell leukaemia/lymphoma virus type I genotypes in five major molecular and geographical subtypes. *J Gen Virol* **75**: 3655–3666.

Vallejo A, Benito A, Varela JM et al (1994) Human T-cell leukemia virus-I/II infection in Equatorial Guinea. *J AIDS* **8**: 1501–1503.

Vallejo A, Soriano V, Gutierrez M, Gomez C & Gonzalez-Lahoz J (1995) HTLV-I and HTLV-II in Africans. *Genitourin Med* **71**: 53–59.

Vallejo A, Ferrante P, Soriano V et al (1996) Nucleotide sequence and restriction fragment length polymorphism analysis of human T-cell lymphotropic virus type II (HTLV-II) in southern Europe. Evidence for the HTLV-IIa and HTLV-IIb subtypes. *J AIDS Hum Retrovirol* **13**: 384–391.

Van Brussel M, Goubau P, Rousseau R, Desmyter J & Vandamme AM (1996) The genomic structure of a new simian T-lymphotropic virus, STLV-PH969, differs from that of human T-lymphotropic virus types I and II. *J Gen Virol* **77**: 347–358.

Van Brussel M, Goubau P, Rousseau R, Desmyter J & Vandamme AM (1997) Complete nucleotide sequence of the new simian T-Lymphotropic virus, STLV-PH969 from a Hamadryas baboon, and unusual features of its long terminal repeat. *J Virol* **71**: 5464–5472.

Van Brussel M, Salemi M, Liu H-F et al (1998) The Simian T-Lymphotropic virus STLV-PP1664 from *Pan paniscus* is distinctly related to HTLV-2 but differs in genomic organization. *J Virol* **(243)**: 366–379.

Vandamme AM, Liu HF, Goubau P & Desmyter J (1994) Primate T-lymphotropic virus type I LTR sequence variation and its phylogenetic analysis: compatibility with an African origin of PTLV-I. *Virology* **202**: 212–223.

Vandamme AM, Liu HF, Van Brussel M, De Meurichy W, Desmyter J & Goubau P (1996) The presence of a divergent T-lymphotropic virus in a wild-caught pygmy chimpanzee (*Pan paniscus*) supports an African origin for the human T-lymphotropic/simian T-lymphotropic group of viruses. *J Gen Virol* **77**: 1089–1099.

Vandamme A-M, Van Laethem K, Liu HF et al (1997a) Use of generic PCR assay detecting human T-lymphotropic virus (HTLV) types I, II and divergent simian strains in the evaluation of individuals with indeterminate HTLV serologies. *J Med Virol* **52**: 1–7.

Vandamme A-M, Salemi M, Desmyter J (1998b) The simian origins of the pathogenic human T-cell lymphotropic virus type I. *Trend micro* **6(12)**: 477.

Vandamme AM, Van Brussel M, Liu HF et al (1997b) A new subtype of human T-lymphotropic virus type II, HTLV-IId, in Zairian Bambuti Efe Pygmies and its implications for the origin of HTLV-II. Abstract ME11. *VIIIth International Conference on Human Retrovirology: HTLV*, 9–13 June 1997, Rio, Brazil.

Vandamme AM, Salemi M, Van Brussel M et al (1998) African origin of human T-lymphotropic virus type 2 (HTLV-2) supported by a potential new HTLV-2d subtype in Congolese Bambuti Efe Pygmies. *J Virol* **72**: 4327–4340.

Van Dooren S, Gotuzzo E, Salemi M et al (1998) Evidence for a post-Columbian introduction of human T-cell lymphotropic virus in Latin America. *J Virol* **(79)**: 2695–2708

Voevodin A & Gessain A (1997) Common origin of human T-lymphotropic virus type-I from Iran, Kuwait, Israel, and La Réunion Island. *J Med Virol* **51**: 77–82.

Voevodin A, al-Mufti S, Farah S, Khan R & Miura T (1995) Molecular characterization of human T-lymphotropic virus, type 1 (HTLV-1) found in Kuwait: close similarity with HTLV-1 isolates originating from Mashhad, Iran. *AIDS Res Hum Retroviruses* **11**: 1255–1259.

Voevodin A, Miura T, Samilchuk E & Schatzl H (1996a) Phylogenetic characterization of simian T lymphotropic virus type I (STLV-I) from the Ethiopian sacred baboon (*Papio hamadryas*). *AIDS Res Hum Retroviruses* **12**: 255–258.

Voevodin A, Samilchuk E, Schätzl H, Boeri E & Franchini G (1996b) Interspecies transmission of macaque simian T-cell leukemia/lymphoma virus type 1 in baboons resulted in an outbreak of malignant lymphoma. *J Virol* **70**: 1633–1639.

Voevodin AF, Johnson BK, Samilchuk EI et al (1997a) Phylogenetic analysis of simian T lymphotropic virus type 1 (STLV-1) in common chimpanzees (*Pan troglodytes*): evidence for interspecies transmission of the virus between chimpanzees and humans in Central Africa. *Virology* **238**: 212–220.

Voevodin A, Samilchuk E, Allan J, Rogers J & Broussard S (1997b). Simian T-lymphotropic virus type 1 (STLV-1) infection in wild yellow baboons (*Papio hamadryas cynocephalus*) from Mikumi National Park, Tanzania. *Virology* **228**: 350–359.

Wain-Hobson S (1992) Human immunodeficiency virus type 1 quasispecies in vivo and ex vivo. *Curr Top Microbiol Immunol* **176**: 181–193.

Watanabe T, Seiki M, Tsujimoto H, Miyoshi I, Hayami M & Yoshida M (1985) Sequence homology of the simian retrovirus genome with human T-cell leukemia virus type I. *Virology* **144**: 59–65.

Watanabe T, Seiki M, Hirayama Y & Yoshida M (1986) Human T-cell leukemia virus type I is a member of the African subtype of simian viruses (STLV). *Virology* **148**: 385–388.

Wattel E, Vartanian JP, Pannetier C & Wain-Hobson S (1995) Clonal expansion of human T-cell leukemia virus type I-infected cells in asymptomatic and symptomatic carriers without malignancy. *J Virol* **69**: 2863–2868.

Wattel E, Cavrois M, Gessain A & Wain-Hobson S (1996) Clonal expression of infected cells: a way of life for HTLV-I. *J AIDS Hum Retrovirol* **13**: S92–S99.

Xu X, Kang SH, Heidenreich O, Brown DA & Nerengerg MI (1995) HTLV-I associated myelopathy/tropical spastic paraparesis isolates bear mutations in the HTLV-I LTR R region which prevents binding of CREB and ATF-2. *J AIDS* **10**: 217.

Yamashita M, Senyuta NB, Paylish OA et al (1995a) Prevalence and phylogenetic characterization of HTLV-I isolates from far Eastern Russia. *Mol Biol* **29**: 697–703.

Yamashita M, Takehisa J, Miura T et al (1995b) Presence of the widespread subtype of HTLV-I in South Africa. *AIDS Res Hum Retroviruses* **11**: 645–647.

Yamashita M, Ido E, Miura T & Hayami M (1996) Molecular epidemiology of HTLV-I in the world. *J AIDS Hum Retrovirol* **13**: S124–S131.

Yamashita M, Ibuki K, Agus L et al (1997) Phylogenetic study of Asian STLV-I including those of orangutans. Abstract ME04. *VIIIth International Conference on Human Retrovirology: HTLV*, 9–13 June 1997, Rio, Brazil.

Yanagihara R (1994) Geographic-specific genotypes or topotypes of human T-cell lymphotropic virus type I as markers for early and recent migrations of human populations. *Adv Virus Res* **43**: 147–186.

Yanagihara R, Nerurkar VR, Garruto RM et al (1991a) Characterization of a variant of human T-lymphotropic virus type I isolated from a healthy member of a remote, recently contacted group in Papua New Guinea. *Proc Natl Acad Sci USA* **88**: 1446–1450.

Yanagihara R, Nerukar VR & Ajdukiewicz AB (1991b) Comparison between strains of human T lymphotropic virus type I isolated from inhabitants of the Solomon Islands and Papua New Guinea. *J Infect Dis* **164**: 443–449.

Yoshida M, Miyoshi I & Hinuma Y (1982) Isolation and characterization of retrovirus from cell lines of human adult T-cell leukemia and its implication in the disease. *Proc Natl Acad Sci USA* **79**: 2031–2035.

Zella D, Mori L, Sala M et al (1990) HTLV-II infection in Italian drug abusers. *Lancet* **336**: 575–576.

Zella D, Cavicchini A, Salemi M et al (1993) Molecular characterization of two isolates of human T cell leukaemia virus type II from Italian drug abusers and comparison of genome structure with other isolates. *J Gen Virol* **74**: 437–444.

Chapter 17

HUMAN HERPESVIRUS 6

Mark K. Williams and Paolo Lusso

INTRODUCTION

Human herpesvirus 6 (HHV-6) was isolated for the first time in 1985 by Salahuddin and colleagues at the National Cancer Institute in Bethesda, Maryland, while searching for the aetiological agent of the B cell lymphomas associated with the acquired immune deficiency syndrome (AIDS) (Salahuddin et al, 1986). The putative association with B lymphoproliferative disorders, which was not confirmed by subsequent epidemiological studies, together with a preliminary phenotypic characterization of infected cells, led to the hypothesis of an elective B lymphocyte tropism of HHV-6. Hence, the original designation 'human B-lymphotropic virus' (HBLV), which was later abandoned when it became evident that the primary target cells of HHV-6 are T lymphocytes, both in vitro (Lusso et al, 1988) and in vivo (Takahashi et al, 1989). Four years after the discovery of HHV-6, a second human herpesvirus with a predominant T lymphocyte tropism, namely HHV-7, was identified (Frenkel et al, 1990b; Berneman et al, 1992). These viruses share a close genetic, biological and immunological relatedness. However, there is an important difference in their cellular host range, because HHV-7 appears to target CD4+ T cells selectively, as it utilizes the CD4 glycoprotein as a critical membrane receptor (Lusso et al, 1994a), whereas HHV-6 can infect a broad range of immune and non-immune cell types. In spite of the fact that a lymphocyte tropism is a typical feature of the γ-herpetovirinae, HHV-6 and HHV-7 have been classified as β-herpetovirinae based on the degree of genetic homology with human cytomegalovirus (hCMV). In fact, both viruses differ from the other known lymphotropic herpesviruses, e.g. Epstein–Barr virus (EBV or HHV-4) and the simian *Herpesvirus saimiri* and *Herpesvirus ateles*, because of their inability to induce direct cellular transformation in vitro (Salahuddin et al, 1986; Lusso et al, 1988; Frenkel et al, 1990b; Berneman et al, 1992). Based on these peculiar characteristics, HHV-6 and HHV-7 may eventually be classified into a novel subfamily among the *Herpesviridae*.

Since the initial report of the discovery of HHV-6, the history of this virus has been animated by heated controversies, and for a long time HHV-6 has remained 'a virus in search of a disease'. In spite of the unquestionable association with subitum exanthema, established by Yamanishi et al (1988), the difficulties in defining the aetiological role of HHV-6 continue even today; similar complications are encountered also with other herpesviruses, owing to the fact that they are ubiquitous agents, induce latent infection and persist in the host indefinitely. There is still considerable confusion about the appropriate methods for the clinical diagnosis of HHV-6 infection, as some of the most

HIV and the New Viruses Second Edition
ISBN 0-12-200741-7

commonly employed diagnostic tests, e.g. titration of serum IgG and cellular DNA amplification by the polymerase chain reaction (PCR), cannot distinguish between active and latent infection and thus are generally useless, except during primary infection. Another debated issue has been the classification of the different viral isolates: with the characterization of an increasing number of viral strains collected throughout the world, it has become evident that significant biological and genetic heterogeneity exists and that the different isolates segregate into two well-defined subgroups, referred to as HHV-6 A and B (Ablashi et al, 1991; Aubin et al, 1991; Schirmer et al, 1991), leading some authors to propose their classification as different viruses. Last but not least, there is still uncertainty about the efficacy of antiviral compounds against HHV-6. Along with all the controversies, however, the study of HHV-6 has generated a remarkable interest, undoubtedly stimulated by the recognition of its unique interactions with the cells of the immune system and, more recently, by the accumulating evidence of a pathogenetic role in vivo. It was largely by virtue of this continuous interest that our knowledge of the genetic, biological, and epidemiological features of HHV-6 has progressed rapidly in a relatively short period.

STRUCTURE AND GENETIC CHARACTERISTICS

On a morphological basis, HHV-6 is a typical herpesvirus, indistinguishable from other members of the family (Biberfeld et al, 1987) (Figure 17.1). The diameter of the mature particle is 200 nm. Its structure includes an internal core formed by the capsid, consisting of 162 capsomers arranged according to the typical icosadeltahedral symmetry, which encloses a double-stranded DNA genome of approximately 160 000 base pairs

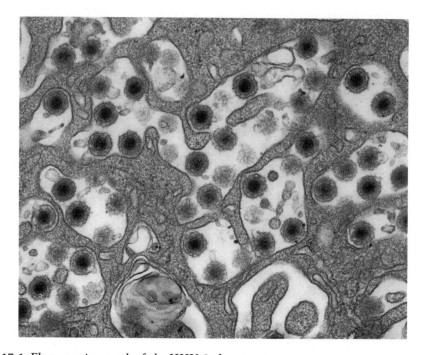

Figure 17.1 *Electron micrograph of the HHV-6 ultrastructure*

(bp). The nucleocapsid is surrounded by an amorphous tegument and, externally, by a lipid-containing envelope with spikes projecting from its surface (Figure 17.1). The virus replicates by the rolling-circle mechanism, with the formation of head-to-tail concatamers in infected nuclei (Martin et al, 1991). A lytic phase origin of DNA replication has been identified upstream of the major DNA-binding protein gene (Dewhurst et al, 1993a, 1994).

The genome of HHV-6 is large and complex, containing more than 100 genes. The complete nucleotide sequence of strain U1102 (Downing et al, 1987), which is closely related to GS prototype of subgroup A (Salahuddin et al, 1986), has been obtained (Gompels et al, 1995), whereas to date no complete sequence of strains belonging to subgroup B has been made available. The complete sequence of HHV-7 has also been recently reported (Nicholas, 1996). The overall G + C content of HHV-6A is 43%, one of the lowest amongst the *Herpesviridae*. Human herpesvirus 6 shares with HHV-7 the highest degree of nucleotide sequence homology, while both HHV-6 and HHV-7 are more distantly related to hCMV (Efstathiou et al, 1988; Frenkel et al, 1990b; Lawrence et al, 1990; Neipel et al, 1991; Berneman et al, 1992; Gompels et al, 1995). On average, the genetic distance between HHV-6 and HHV-7, based on amino acid identity within the conserved herpesvirus gene products, ranges between 25% and 59% (Nicholas, 1996). The genomes of HHV-6, HHV-7 and hCMV are roughly colinear. More than 65% of the HHV-6 genes have a recognized counterpart in hCMV, with many of the remaining genes having no recognized counterpart in other herpesviruses (Gompels et al, 1995). The genome of HHV-6 consists of a single long unique (UL) element of approximately 140 000 bp, rich in A + T, flanked by two identical direct repeat (DR) segments, rich in G + C, of approximately 8000 bp each. The DR, which contain essential elements for the viral DNA packaging and replication, as well as possibly for the maintenance of the latency state (Lindquester and Pellett, 1991; Martin et al, 1991; Thomson et al, 1994a; Gompels and Macaulay, 1995), include a tandem repetitive sequence (GGGTTA) that is also present in the genome of Marek's disease virus (Kishi et al, 1988), as well as in the telomers of human chromosomes. The herpesvirus cleavage/packaging homologues pac-1 and pac-2 are also located at the termini of the left and right DR elements, respectively (Thomson et al, 1991). Pac-1 and pac-2 allow the formation of the cleavage/packaging signal upon genome circularization or concatamerization. Restriction endonuclease polymorphism has been documented amongst different HHV-6 isolates (Josephs et al, 1988a; Jarrett et al, 1989; Kikuta et al, 1989); as discussed below, two major viral subgroups (A and B) have been defined. The overall genetic divergence between the two subgroups is approximately 5%, ranging from 2% to 25% according to the loci analysed (Aubin et al, 1991; Gompels et al, 1993, 1995).

The analysis of the complete nucleotide sequence of HHV-6 (Gompels et al, 1995) has led to the identification of five major families of genes (hCMV US22 homologues, DR1/6, U4/5, G protein-coupled receptor-like and immunoglobulin-like families). Several genes of HHV-6 have been characterized, mostly based on genetic homology or colinearity with other human herpesviruses, particularly with hCMV. A conserved HHV-6 homologue has been found for all seven genes believed to be necessary and sufficient for the replication of HSV-1, i.e. the transactivator U27, the DNA polymerase (U38), a single-stranded DNA-binding protein (U41), the helicase/primase complex (U43, U74, U77) and the origin binding protein (U73). Additional genes which are critical for the replication of hCMV also have homologues in HHV-6 (Gompels et al, 1995). It is noteworthy that HHV-6, like hCMV, lacks a thymidine kinase, but possesses a gene (U69) homologous to the ganciclovir kinase of hCMV (Gompels et al, 1995). Among the genes that are unique to HHV-6, one has a striking homology to the *rep* gene of adeno-associated virus type 2 (AAV-2), a defective parvovirus dependent on adenovirus for its replication (Thomson et al, 1991). The Rep protein of AAV-2 possesses site-specific ATP-dependent endonuclease and helicase activities and blocks cellular transformation by papillomaviruses. The Rep-like protein of HHV-6 was shown to be functional, as it mediates AAV-2 DNA replication (Thomson et al, 1994b) and inhibits cellular transformation by H-*ras* (Araujo et al, 1995); moreover, contrasting effects have been reported on HIV expression, with either increase (Thomson et al, 1994b) or suppression (Araujo et al, 1995) of the transcription driven by the long terminal repeat (LTR) of HIV-1. The function of the *rep*-like gene and its protein product in the life cycle of HHV-6 is currently unknown.

The genome of HHV-6, like that of other herpesviruses, has coding regions on both DNA strands and contains only limited non-coding segments. Nevertheless, post-transcriptional modifications are thought to occur for a number of HHV-6 transcripts. Differential messenger RNA (mRNA) splicing has been documented for the gene encoding the glycoproteins gp82–gp105, with 12 exons (for a total of 2.5 kb) spanning 20 kb of genomic sequence (Pfeiffer et al, 1995). Additionally, the *rep*-like gene (Braun et al, 1997), the G protein-coupled receptor homologue U12 (Gompels et al, 1995), and the CC chemokine homologue U83 (M. K. Williams and P. Lusso, unpublished observations) are expressed as differentially spliced transcripts.

VIRAL SUBGROUPS OR VARIANTS

Two major subgroups of HHV-6 have been identified (A and B) (Ablashi et al, 1991; Aubin et al, 1991; Schirmer et al, 1991). The two subgroups, also referred to as 'variants', differ genetically (restriction endonuclease polymorphism), immunologically (monoclonal antibody reactivity) and biologically (cellular tropism) (Table 17.1). It has also been suggested that HHV-6 A and B may have a different epidemiological distribution in the human population and, possibly, different disease associations, although systematic studies of the epidemiology of HHV-6 A are still wanting. The most rapid and efficient methods currently available for subgroup identification are based either on the reactivity of specific monoclonal antibodies with productively infected cells, or on the restriction patterns of specific regions of the viral genome. After amplification by PCR or by virus propagation in vitro, a different pattern is observed for each of the two subgroups upon digestion with specific restriction enzymes (e.g. *Bgl*II, *Sal*I, *Hind*III, *Taq*1), the most notable heterogeneity being observed upon digestion with *Sal*I. Restriction analysis has also permitted the definition of at least one further subdivision within subgroup B (B1 and B2) (Di Luca et al, 1992).

The prototype of HHV-6 A is strain GS, originally isolated from a young man with acute T cell lymphoblastic leukaemia (Salahuddin et al, 1986). Albeit less frequently detected and isolated, subgroup A has been thoroughly characterized biologically. To date, no definitive links with human disease have been established, but the fact that isolates of the A type were derived predominantly from immunocompromised patients has led many investigators to suggest a role in immunodeficiency conditions,

Table 17.1 *Major biological features of the two HHV-6 subgroups*

	Subgroup A (GS-like)	Subgroup B (Z29-like)
Cellular tropism		
Primary blood cells:		
CD4+ T cells	+	+
CD8+ T cells	+	+/−
NK cells	+	−
γδ T cells	+	−
Cell lines:		
HSB-2	+	−
Molt-3	−	+
Sup-T1	+	+
Jurkat	+	+
EBV + LCL	+	−
Cytopathic effect	+	+
Cell immortalization in vitro	−	−
Downregulation of CD3	+	+
Upregulation of CD4	+	+/−
HIV-LTR transactivation	+	+

including AIDS. This concept is strengthened by the broad 'immunotropism' demonstrated by these isolates (see below).

The HHV-6 subgroup B, prototyped by strain Z29, was first identified by Lopez and colleagues at the Centers for Disease Control (CDC) in Atlanta Georgia (Lopez et al, 1988). It is responsible for most of the cases of primary infection in childhood (Dewhurst et al, 1993b) and has been firmly associated with the aetiology of *exanthema subitum* (Yamanishi et al, 1988), a usually benign febrile disorder of infants, also referred to as roseola infantum or sixth disease. The prevalence of type B isolates in the general population is almost universal.

BIOLOGICAL FEATURES

Like other herpesviruses, HHV-6 establishes latent infection in vivo and thereby persists in the host indefinitely after primary infection. No well-characterized models of latent infection in vitro are currently available, but persistent non-productive infection has been documented in cultured macrophages (Kondo et al, 1991), in papillomavirus-immortalized cervical epithelial cells (Chen et al, 1994b) and, more recently, in an EBV-negative Burkitt's lymphoma cell line (Bandobashi et al, 1997). Although circulating monocytes and epithelial cells of the bronchial and salivary glands have been suggested as possible reservoirs (Krueger et al, 1990; Kondo et al, 1991), there is still uncertainty regarding the exact sites of viral persistence and latency in vivo. In vitro, HHV-6 behaves as a highly cytopathic virus. Exposure of activated primary human mononuclear cells to HHV-6 results in dramatic cytomorphological changes after 3–7 days. The cells enlarge and lose their structured, blastic shape, eventually becoming evenly rounded and refractile. After their appearance, these enlarged cells survive in culture for 3–4 days. Unlike EBV and the monkey T lymphotropic herpesviruses (e.g. *Herpesvirus saimiri* and *ateles*), HHV-6 does not directly cause immortalization of its target cells, at least in vitro. The mechanisms underlying the dramatic cytopathicity of HHV-6 have not been definitively elucidated. Evidence has been provided that infection of primary CD4+ T cells is associated with widespread induction of apoptosis (programmed cell death) both in vitro and in vivo (Inoue et al, 1997; Yasukawa et al, 1998); nevertheless, membrane fusion resulting in the formation of giant multinucleated cells also occurs, and has been exploited for the establishment of fusion inhibition tests in vitro (P. Lusso, unpublished observations).

The picture emerging from the study of the cellular tropism of HHV-6, particularly for strains of the A variant, is that of a broadly 'immunotropic' herpesvirus which may dramatically affect, directly or indirectly, both the cellular and humoral arms of the immune system. This concept is consistent with a possible immunosuppressive role of HHV-6 infection in vivo (see below). There is now consensus that HHV-6 (both A and B) has a primary tropism for CD4+ T cells, both in vitro and in vivo (Lusso et al, 1988; Takahashi et al, 1989). When activated mononuclear cells of different tissue origin (e.g. peripheral blood, umbilical cord blood, thymus, lymph node, tonsil, bone marrow) are exposed to HHV-6, the vast majority of the infected cells display the phenotype of activated CD4+ T cells (CD2+, CD4+, CD5+, CD7+, CD26+, CD38+, CD71+). Although both major viral subgroups (A and B) infect CD4+ T cells preferentially, there are some important differences in their cellular tropism. For example, the two subgroups differ in their capacity to infect established human T cell lines and B lymphoblastoid cell lines (LCL) (Table 17.1). Moreover, HHV-6 A has the unique ability to infect productively and cytopathically different types of cytotoxic effector cells, such as CD8+ T lymphocytes (Lusso et al, 1991a,b), natural killer (NK) cells (Lusso et al, 1993), and γδ T lymphocytes (Lusso et al, 1995). Because such cells are involved in the mechanisms of antiviral defense in vivo, this strategy may allow HHV-6 to counteract the protective immune surveillance of the host and thereby establish persistent infection. By contrast, HHV-6 subgroup B seems to have a more restricted cellular tropism, and in particular to be unable to infect cytotoxic effector cells (P. Lusso, unpublished observations).

Besides T cells and NK cells, HHV-6 can also infect other cell types, including B cells, mononuclear phagocytes and cells of neural, muscular, fibroblastic and epithelial origin (reviewed by Lusso and Gallo, 1995a). It is remarkable that B lymphoid cells can be productively infected in vitro only after their immortalization with EBV (Ablashi et al, 1988), suggesting that EBV infection may induce the expression of a functional receptor for HHV-6. However, a Burkitt's lymphoma cell line with an immature B cell phenotype, which is persistently and non-productively infected with HHV-6 in the absence of EBV, has been reported (Bandobashi et al, 1997). Other types of interaction between HHV-6 and EBV have also been documented (Flamand et al, 1993). The cells of the mononuclear phagocytic system represent another target that can be infected and killed by HHV-6 (Kondo et al, 1991; Carrigan, 1992), although the infection is typically non-productive and the cytopathic effect, when occurring, is most probably indirect. Consistent with the documented 'neurotropism' of HHV-6 in vivo (see below), primary neural cells can also be productively infected in vitro. For example, infected astrocytes demonstrate cytopathic effects, and the progeny virus is able to reinfect both primary T cells and astrocytes (He et al, 1996). Finally, several immortalized human cell lines can be infected by HHV-6, including lines of T cell (Ablashi et al, 1988; Cermelli et al, 1997), epithelial (Chen et al, 1994b), fibroblastic (Luka et al, 1990; Robert et al, 1996), monocytoid (Arena et al, 1997), immature haematopoietic (Furlini et al, 1996), megakaryoblastoid and neuroblastoma (Ablashi et al, 1988) and hepatocytic (Inagi et al, 1996) origin. Although the relevance of these observations to the cellular tropism of HHV-6 in vivo remains unknown, these models may be useful for studies in vitro.

Little information is currently available on the surface membrane receptors of HHV-6. The fact that activation signals are necessary to induce susceptibility to HHV-6 infection in T cells (Lusso et al, 1989b; Frenkel et al, 1990a) indicates that the receptor, or at least one component of the HHV-6 receptor complex, is a T cell activation antigen. Indeed, as inferred from the study of other herpesviruses, the HHV-6 receptor is likely to encompass different molecules forming a multimolecular complex expressed on the cellular surface membrane under specific conditions of activation. The differences in cellular tropism documented between A and B strains of HHV-6 suggest that the composition of the receptor complex may be different for the two viral subgroups. Evidence has been reported that the receptor for HHV-6 is not the CD4 glycoprotein (Lusso et al, 1989b), which instead serves as a major receptor for HIV (Dalgleish et al, 1984; Klatzmann et al, 1984) and HHV-7 (Lusso et al, 1994b). In fact, HHV-7 and HIV-1 compete for CD4 occupancy and reciprocally interfere for infection of both CD4+ T cells (Lusso et al, 1994b) and mononuclear phagocytes (Crowley et al, 1996).

Complex interactions have been described between HHV-6 and its host cells, particularly related to the ability of HHV-6 to activate or suppress the expression of cellular genes at the transcriptional level. Terminally infected T cells fail to express the T cell receptor (TCR) complex (the TCR $\alpha\beta$ heterodimer associated with the different CD3 chains) on their surface membrane (Lusso et al, 1988), as a consequence of the ability of HHV-6 to downregulate transcriptionally the expression of several CD3 chains in the course of its lytic infection (Lusso et al, 1991b). This effect is induced by both A and B HHV-6 strains (P. Lusso, unpublished observation). Because of the critical role played by the TCR complex in T cell activation, downregulation of CD3 may have an immunosuppressive effect. Another unexpected phenotypic feature observed in infected T lymphocytes is that a variable proportion of them coexpresses both CD4 and CD8 (Lusso et al, 1988). This phenomenon is related to the unique ability of HHV-6 to activate transcriptionally the expression of CD4 in cells that physiologically do not express it, such as mature CD8+ T cells (Lusso et al, 1991a). This effect seems to be mediated by early gene products of HHV-6, as indicated by experiments with the viral DNA polymerase inhibitor phosphonoformic acid (PFA) (Lusso et al, 1991a); direct activation of the CD4 promoter has been suggested (L. Flamand et al, 1998). Similar observations were subsequently made in NK cells and $\gamma\delta$ T cells (Lusso et al, 1993, 1995). Owing to the inefficient growth of subgroup B isolates in CD4-negative cells, de novo CD4 induction was hitherto documented only with HHV-6 A. Nonetheless, increased levels of CD4 expression were observed, upon infection with different HHV-6 B strains, in Jurkat, a CD4+^{low} neoplastic T cell line (P. Lusso, unpublished observation). Among the other virus–host interactions documented, HHV-6 infection has been shown to induce the release of inflammatory cytokines, such as interferon-α (IFN-α) (Kikuta et al, 1990), tumour necrosis factor α (TNF-α), interleukin (IL) 1β (Flamand et al, 1991), IL-8 (Inagi et al, 1996) and IL-15 (Flamand et al, 1996), as well as the expression of the G protein-coupled receptor EBI-1, which is typically induced by EBV infection (Hasegawa et al, 1994).

Interestingly, HHV-6 was found to possess genes with homology to the G protein-coupled receptors for chemokines (U12 and U51), as well as a unique gene with homology to CC chemokines (U83). Although the exact function of their gene products remains to be determined, it can be speculated that the virus may exploit such molecular mimicry mechanisms to generate decoys for the immune system and thereby evade the immunological control of the host.

ANIMAL MODELS OF HHV-6 INFECTION

Several animal species have been tested both in vitro and in vivo for their susceptibility to infection by HHV-6. The virus seems to have a very restricted host species range, as it was shown to infect only non-human primates, besides humans. The prototype of subgroup A, GS, can replicate in chimpanzees (*Pan troglodytes*) in vitro (Lusso et al, 1990), in pig-tailed macaques (*Macaca nemestrina*) both in vitro (Lusso et al, 1994a) and in vivo (P. Lusso et al, unpublished observations), and in cynomolgus macaques (*Macaca fascicularis*) in vivo (P. Lusso et al, unpublished observations), but not in African green monkeys (*Cercopithecus aethiops*) and New World primates such as tamarins and marmosets (Lusso et al, 1990). In contrast, a strain belonging to subgroup B, i.e. HST, was shown to infect cynomolgus macaques and African green monkeys both in vitro and in vivo (Yalcin et al, 1992). Thus, not only the cellular tropism but also the species tropism of the two major HHV-6 variants appear to be different.

In addition to monkeys, another potentially interesting study model for HHV-6 infection is the heterochimeric severe combined immunodeficient (SCID) mouse repopulated with tissues of human origin. In the SCID-hu thy/liv model, both HHV-6 A and HHV-6 B were found to replicate efficiently and to induce characteristic pathological consequences (A. Gobbi et al, unpublished observations). The increasing utilization of this and other animal model systems will certainly be instrumental in fostering our understanding of the pathophysiology of HHV-6 infection.

TRANSMISSION, DIAGNOSIS AND EPIDEMIOLOGY

Infection by HHV-6 subgroup B is widespread in the human population and is acquired almost universally within the first 2 years of life (reviewed by Lusso and Gallo, 1995a). The infection is believed to be transmitted via oropharyngeal secretions, as suggested by the repeated isolation and detection of the virus in saliva or salivary glands using different techniques (Pietroboni et al, 1988; Fox et al, 1990a; Gopal et al, 1990; Harnett et al, 1990; Cone et al, 1993a; Lyall and Cubie, 1995). In contrast, HHV-6 was not found in breast milk (Dunne and Jevon, 1993; Kusuhara et al, 1997), suggesting that breast-feeding is not an important route of transmission. Transmission of HHV-6 to infants through the genital tract of their mothers at delivery seems to occur rarely, if at all (Maeda et al, 1997); in one case, intrauterine transmission has been suggested (Aubin et al, 1992).

At present, information regarding the epidemiology and routes of transmission of HHV-6 A is still limited and inconclusive. Whereas HHV-6 B has been found to be virtually ubiquitous, limited studies have been performed to determine the prevalence of HHV-6 A and the rate of A/B co-infection. In a recent study HHV-6 B was detected by PCR in oropharyngeal secretions and peripheral blood mononuclear cells (PBMCs) of the vast majority (> 90%) of children, whereas HHV-6 A was found in less than 3% of cases (Hall et al, 1998). However, a higher prevalence of subtype A (14–17%) was seen in the cerebrospinal fluid (CSF). Interestingly, dual infection by HHV-6 A and B was documented in some cases, with HHV-6 A in CSF and HHV-6 B in PBMCs (Hall et al, 1998). Other studies have confirmed the low frequency of infection by HHV-6 A in saliva (Aberle et al, 1996; Tanaka-Taya et al, 1996), although in one of them (Aberle et al, 1996) a relatively high prevalence (16%) was documented in the PBMCs of healthy adult individuals. These studies underscore the need to analyse different tissues in order to assess the prevalence of HHV-6 A relative to HHV-6 B, as the viral reservoirs for the two variants may be different. The use of reliable diagnostic methods and the analysis of a broad spectrum of tissue sources will help elucidate more precisely the relative prevalence of HHV-6 strains A and B in the human population.

The question of the diagnosis of HHV-6 infection is complex and has generated considerable confusion. To understand the reasons for this complexity, one has to consider the typical features of the HHV-6 life cycle in the host. The virus enters the body during primary infection, which is followed by a variable period of latency, and it is probably never cleared from the organism, persisting for the entire lifetime in selected anatomical sites. Subsequently, either reactivation or exogenous reinfection may occur, particularly in concomitance with episodes of immune dysregulation or deficiency; these can be followed, again, by variable periods of latency. Unfortunately, as a consequence of these characteristics of the viral life cycle, the diagnostic value of the classical seroepidemiological tools such as IgG antibody testing, as well as of the most popular molecular diagnostic techniques such as DNA PCR on blood cells, is greatly reduced. Indeed, these methods are unable to distinguish between latent (clinically silent) and active (clinically relevant) infection, and their inappropriate use or interpretation may lead to erroneous conclusions about the pathogenic role of HHV-6. In particular, serological IgG antibody testing for HHV-6 has several important shortcomings:

- the inability to discriminate between active and pregressed (latent) infection;
- the generalized IgG seropositivity in the adult population (Briggs et al, 1988; Saxinger et al, 1988), which is due to the ubiquitous distribution of the infection (at least by subgroup B);
- the persistence of maternal IgG antibodies, which may lead to confusing results even in early childhood;
- the frequent fluctuations of the IgG titres, in the absence of any appreciable clinical manifestations;
- the reported immunological cross-reactivity of HHV-6 with other herpesviruses (e.g. hCMV, HHV-7) (Irving et al, 1988; Larcher et al, 1988; Chou and Scott, 1990; Berneman et al, 1992; Adler et al, 1993; Osman et al, 1997);
- the inability of the current tests to identify the two major HHV-6 subgroups (A vs. B).

In spite of these shortcomings, however, IgG antibody serology has been widely used, especially in the early days of HHV-6 research, and the results of these studies have led in several instances to definitive, yet totally unsubstantiated, conclusions.

A summary of the specificities of different diagnostic tests available for HHV-6 is given in Table 17.2. In contrast to IgG, anti-HHV-6 IgM antibodies may be a useful, albeit inconsistent, marker of active infection (Fox et al, 1990b; Suga et al, 1992; Secchiero et al, 1995a). The main limitations of IgM antibody testing are the low sensitivity of the current tests and the putative cross-reactivity with other herpesviruses (Osman et al, 1997). Virus isolation methods, though more consistently positive during episodes of active virus replication, are unable to distinguish formally between active and latent infection, as they require preactivation of the cells in vitro. Even less accurate is qualitative PCR analysis of DNA extracted from blood cells or tissues. In fact, this type of DNA PCR has almost the same lack of specificity as IgG antibody testing, with the additional complication of the wide variations in the sensitivity of the test performed in different laboratories. As an example, in five studies performed in Western countries, the proportion of HHV-6 DNA-positive cases in peripheral blood leukocytes among unselected healthy individuals ranged from 5% to 95% (Cone et al, 1993b; Cuende et al, 1994; Di Luca et al, 1994; Rajcani et al, 1994; Wilborn et al, 1994). While not allowing a direct diagnosis of active infection, quantitative methods, such as quantitative PCR (Secchiero et al, 1995b; G. Locatelli et al, unpublished observations), in situ hybridization on DNA and Southern blot hybridization, may be more informative, as they provide – with different degrees of accuracy – an estimate of the viral load in vivo, which often correlates with the degree of virus replication. Particularly in tissues involved by inflammatory or neoplastic disorders, the occasional presence of infected cells may be totally irrelevant to pathogenesis: in such cases, quantitative techniques may be helpful for the establishment of an aetiologic link. However, even quantitative methods may sometimes be misleading (as discussed by Carrigan, 1995). Thus, the diagnosis of active HHV-6 infection must rely on the use of direct markers of virus replication in vivo.

To date, one HHV-6 antigen capture test has been reported, but its suitability for use in clinical diagnosis remains uncertain (Marsh et al, 1996). The most accurate and practical test currently available for the diagnosis of active HHV-6 infection is the detection of HHV-6 DNA by PCR in serum, plasma or other body fluids (Secchiero et al, 1995b). Using this method, HHV-6 replication in vivo has been documented both in children with primary infection (*exanthema subitum*) and in immunocompromised patients

Table 17.2 Diagnostic accuracy of different tests of HHV-6 infection

	Primary infection	Latency	Reactivation/reinfection
IgG serology	+*	+	+
IgM serology	+	–	+/–
Virus isolation	+	+/–	+/–
Antigen capture	+	–	+
Qualitative PCR on cellular DNA	+	+	+
Quantitative PCR on cellular DNA	++	+	++
Cell-free DNA-PCR (plasma)	+	–	+
RT-PCR on cellular RNA	+	–	+
Immunohistochemistry	+	–	+
In situ hybridization (DNA)	++	+	++
In situ hybridization (RNA)	+	–	+

*Positive during the convalescence period, after a variable seronegative window.

(Secchiero et al, 1995a). A good correlation was observed in the same study between serum PCR and IgM antibody reactivity. Other suitable methods for the demonstration of active HHV-6 infection are reverse transcriptase (RT) PCR, immunohistochemistry (Carrigan et al, 1991; Knox and Carrigan, 1994) and in situ hybridization on RNA, the latter two offering the additional bonus of an exact identification of the infected cell lineage. In conclusion, the diagnosis of HHV-6 infection is a delicate process that requires the use of markers of active virus replication in vivo and, usually, the assessment of multiple parameters simultaneously. Clinical studies based on low-specificity markers or lacking a critical evaluation of the diagnostic potential of the techniques used may lead to confusing and sometimes erroneous conclusions on the pathogenic role of HHV-6.

CLINICAL MANIFESTATIONS OF HHV-6 INFECTION

There have been numerous claims of possible associations of HHV-6 with human diseases. Most of these claims have not been adequately substantiated, and some have been contradicted by subsequent studies. In many instances, these inconsistencies are related, at least in part, to the inappropriate choice or interpretation of the diagnostic techniques employed to document HHV-6 infection.

HHV-6 infection in children

Primary infection by HHV-6 B in early childhood is firmly associated with *exanthema subitum* (Yamanishi et al, 1988). The disease is manifested by high-grade fever, typically lasting for 3 days, followed by the appearance of a characteristic skin eruption (roseola). However, it must be emphasized that the rash is often absent and primary infection may be manifested only by an acute, non-specific, febrile illness (Suga et al, 1989; Pruksananonda et al, 1992). Although the febrile disease is usually self-limiting and does not require specific therapeutic interventions, HHV-6 has been reported as a major cause of visits to paediatric emergency departments (Breese-Hall et al, 1994), in a high proportion of cases because of the high fever, sometimes associated with seizures, or neurological complications, such as meningitis and meningo-encephalitis (Ishiguro et al, 1990; Huang et al, 1991; Asano et al, 1992; Lakeman and Whitley, 1995; McCullers et al, 1995; Yanagihara et al, 1995; Jee et al, 1998). The neurological manifestations seem to be related to the direct invasion of the central nervous system by HHV-6 (Kondo et al, 1993; McCullers et al, 1995), a finding that is consistent with the increasing evidence that HHV-6 may remain latent in the nervous tissue (Liedtke et al, 1995a; Luppi et al, 1995; Tang et al, 1997). In one case, an acute disseminated demyelination was aetiologically associated with HHV-6 infection of the brain (Kamei et al, 1997). However, in a case–control study of 86 patients, no association between primary HHV-6 infection and an increased risk of seizures was observed (Hukin et al, 1998). Other abnormalities that may be associated with *exanthema subitum* are granulocytopenia and liver dysfunction (Takikawa et al, 1992). It is not clear as yet whether all the cases of *exanthema subitum* are due to HHV-6 B: in fact, there has been one report of a causal link between HHV-6 A and a clinical picture similar to that of *exanthema subitum* in a child less than 2 months of age (Hidaka et al, 1997); moreover, primary HHV-7 infection has been implicated as another cause of first or second episodes of *exanthema subitum*, although its incidence seems to be slightly delayed compared with that of HHV-6 (Hidaka et al, 1994; Tanaka et al, 1994; Asano et al, 1995; Torigoe et al, 1995; Huang et al, 1997; Suga et al, 1998).

Several disorders have been sporadically associated with HHV-6 infection in children, including fulminant and chronic hepatitis (Asano et al, 1990; Mendel et al, 1995; Tajiri et al, 1997), idiopathic thrombocytopenic purpura (Yoshikawa et al, 1993; Kitamura et al, 1994; Saijo et al, 1995), Rosai-Dorfman disease (Levine et al, 1992), fatal haemophagocytic syndrome (Huang et al, 1990; Portolani et al, 1997), transient erythroblastopenia of childhood (Penchansky and Jordan, 1997), Griscelli syndrome (Wagner et al, 1997) and fatal multiple organ failure (Prezioso et al, 1992). Active co-infection by HHV-6 and varicella-zoster virus can be life-threatening, as in the case of an infant with disseminated dual infections in multiple organs, including the brain (Ueda et al, 1996).

HHV-6 infection in immunocompetent adults

After primary infection, HHV-6 infection has been associated with several different disorders in immunocompetent adults, albeit in some instances only by anecdotal evidence. They include EBV-negative infectious mononucleosis (Steeper et al, 1990; Akashi et al, 1993), persistent lymphadenopathy (Niederman et al, 1988), fulminant hepatitis (Sobue et al, 1991), autoimmune disorders (Krueger et al, 1991) and chronic fatigue syndrome (Buchwald et al, 1992; Patnaik et al, 1995). Moreover, several reports have suggested a link with neurological disorders, including fulminant demyelinating encephalomyelitis (Novoa et al, 1997), chronic myelopathy (Mackenzie et al, 1995), meningoencephalitis (Ikusaka et al, 1997), subacute leukoencephalitis (Carrigan et al, 1996) and multiple sclerosis (see below). Despite intriguing evidence obtained in individual patients, the exact significance of these proposed associations remains to be defined in large epidemiological studies.

HHV-6 in multiple sclerosis

As in other putative autoimmune diseases, researchers have been striving for decades to discover the cause of multiple sclerosis (MS). A long list of factors has been proposed, including viral, genetic and environmental causes (Johnson, 1994; Boccaccio and Steinman, 1996; Brod et al, 1996; Carton, 1996; McDonnell and Hawkins, 1996; Collins, 1997; Sadovnick, 1997). In particular, diverse viral agents have been implicated as the aetiologic agents responsible for triggering the autoimmune responses. Although the cause of MS is far from certain, the potential role of HHV-6 is at present the object of animated debate (Challoner et al, 1995; Gilden et al, 1996; Carrigan and Knox, 1997; Soldan et al, 1997; Kimberlin, 1998).

The first association between HHV-6 infection and MS was suggested in a study conducted by representational difference analysis (RDA), a molecular subtraction technique, which identified HHV-6 as a 'different' element between the DNA extracted from MS brains and that extracted from normal brains (Challoner et al, 1995). Using a nested PCR assay and immunohistochemistry, the same study documented the presence and the active expression of HHV-6 in the brain tissue of more than 75% of patients with MS. Surprisingly, however, HHV-6 DNA and antigen expression were also detected in a similar percentage of non-MS brains, including brains of people who died from non-neurological causes. An apparently MS-specific expression of viral antigens was detected in the nucleus of oligodendrocytes around the plaques that constitute the hall-mark of the disease (Challoner et al, 1995).

In a subsequent study, the serum levels of IgM against an HHV-6 early antigen were found to be significantly higher in patients with the relapsing–remitting form of MS, compared to those with the chronic progressive form or with non-MS controls, although the number of patients studied was small (Soldan et al, 1997). Furthermore, by serum PCR analysis, 15 of 50 (30%) MS patients (14 with relapsing–remitting MS) and none of the controls demonstrated positivity for cell-free HHV-6 DNA. The CSF was not analysed for the presence of HHV-6 DNA. Surprisingly, no attempt was made to correlate the IgM titres and the serum PCR positivity. Another study provided con-flicting results, with the detection of cell-free HHV-6 DNA in the CSF of only 4 of 36 MS patients, whereas both MS and control patients invariably demonstrated the pres-ence of HHV-6 DNA in the cellular fraction of CSF (Liedtke et al, 1995a). In contrast, Merelli et al (1997) detected HHV-6 DNA in four of eight non-MS brains, but in none of five MS brains. Similarly, Martin et al (1997) found no evidence of HHV-6 infection in 115 CSF samples from patients with MS, optic neuritis and other neurological dis-orders. Unfortunately, many of these studies, though suggestive, were conducted on small numbers of patients and without adequate controls; moreover, the methodology used was often limited to non-quantitative DNA PCR.

In a controlled study conducted in Sweden, the efficacy in MS of the broad-spectrum antiherpes drug aciclovir was investigated. To test the hypothesis that herpesvirus infection may be involved in the pathogenesis of MS, aciclovir was administered to patients with the relapsing–remitting form (Lycke et al, 1996). Clinically defined exacer-bations of MS were encountered 34% less frequently in the group receiving aciclovir, but the difference with the placebo group was not statistically significant. Furthermore, in patients with a disease duration longer than 2 years, the previously established rate of exacerbation decreased following the initiation of therapy. Nevertheless, neurological deterioration was unaffected during aciclovir treatment. It is documented that aciclovir, while efficacious against certain herpesviruses, is relatively ineffective against HHV-6 in vitro. However, its efficacy may be greater in vivo than in vitro, as observed for other antiherpesvirus agents (see below).

Despite the increasing number of clinical reports, the role of HHV-6 in MS remains controversial. Carefully controlled longitudinal studies investigating the activity of HHV-6 infection in MS patients will be fundamental in determining its implication in the aetiology of this disorder.

HHV-6 in neoplasia

Like other human herpesviruses, HHV-6 has been proposed as a possible aetiological factor in neoplastic disorders. The virus or its genetic sequences have been identified in Hodgkin's disease (HD) (Clark et al, 1990; Luppi et al, 1993a; Di Luca et al, 1994), in non-Hodgkin's lymphoma (NHL) (Jarrett et al, 1988; Josephs et al, 1988b; Fillet et al, 1995), in cerebral lymphoma (Liedtke et al, 1995a), in neuroglial tumours (Luppi et al,

1995), in angioimmunoblastic lymphadenopathy with dysproteinaemia (AILD) (Luppi et al, 1993a; Daibata et al, 1997), in Langerhans cell hysticocytosis (Leahy et al, 1993), in Kaposi's sarcoma (Bovenzi et al, 1993; Kempf et al, 1997), in oral carcinoma (Yadav et al, 1994, 1997) and in cervical carcinoma (Chen et al, 1994a). Unfortunately, most of these studies were conducted without sufficient and appropriate controls, and were limited to qualitative DNA PCR analyses. Indeed, without a precise identification of the cell type involved by HHV-6 infection (i.e. the neoplastic or the infiltrating cell populations) and in the absence of an accurate quantitative evaluation of the viral load, it is impossible to draw any conclusion about the possible role played by HHV-6 in these neoplastic disorders. That HHV-6 may have a tumorigenic role is nevertheless indicated by studies in vitro on the transforming potential of isolated HHV-6 genes transfected into murine fibroblasts (Razzaque, 1990; Kashanchi et al, 1997) or human keratinocytes (Razzaque et al, 1993). Moreover, evidence has been provided that HHV-6 can integrate into the host cell's chromosomes (Luppi et al, 1993b) and accelerate the tumorigenesis of human papillomavirus-immortalized human cervical cells in immunodeficient mice (Chen et al, 1994b). Further investigation will be essential to elucidate the relationship of HHV-6 with human cancer.

HHV-6 as an opportunistic agent in immunocompromised patients

To date, the strongest aetiological associations of post-primary HHV-6 infection have been documented in immunocompromised individuals, particularly in patients who received bone marrow or organ transplantation and in HIV-infected individuals (reviewed by Lusso and Gallo, 1995a). An increasing body of evidence suggests that HHV-6 may act in these settings as an opportunistic agent causing severe, sometimes life-threatening, organ infections; moreover, its inherent immunosuppressive potential may in itself aggravate the immunodeficiency, as suggested by clinical and experimental studies in HIV-infected people (see below). The evidence suggesting a possible aetiopathogenic role of HHV-6 is based on the direct detection of viral antigen expression or viral DNA in diseased tissues, and on the temporal association between signs of disease and serum IgM antibody positivity or virus isolation. The major clinical manifestations associated with HHV-6 infection in immunocompromised hosts include interstitial pneumonitis (Carrigan et al, 1991; Cone et al, 1993b; Knox and Carrigan, 1994), encephalitis (Drobyski et al, 1994; Knox and Carrigan, 1995; Knox et al, 1995a), retinitis (Qavi et al, 1992; Reux et al, 1992) and bone marrow or organ graft failure (Drobyski et al, 1993; Carrigan and Knox, 1994; Appleton et al, 1995; Wang et al, 1996; Chan et al, 1997; Singh et al, 1997; Lautenschlager et al, 1998). Consistent with the latter observations, studies in vitro have demonstrated a suppressive activity of HHV-6 on the maturation and growth of normal bone marrow precursor cells committed toward different lineages (Knox and Carrigan, 1992). However, in one study, no delay in engraftment was found in the majority of patients who received autologous or allogeneic bone marrow transplantation, despite evidence of active HHV-6 infection (Kadakia et al, 1996). In a longitudinal study performed by serum PCR and IgM antibody testing in three bone marrow transplanted patients, fluctuating HHV-6 reactivation/reinfection episodes were documented, in concomitance with fever and respiratory or neurological signs (Secchiero et al, 1995a). Most of the viral isolates obtained from the circulating leukocytes or bone marrow of immunosuppressed patients with HHV-6-associated clinical manifestations were found to belong to subgroup B (Carrigan and Knox, 1994; Drobyski et al, 1994), whereas the strains detected in serum from post-transplantation and AIDS patients were mostly of the A type (Secchiero et al, 1995a), suggesting that co-infection by HHV-6 A and B may occur in immunocompromised hosts.

Despite all these intriguing observations, there is still uncertainty regarding the role of HHV-6 as a bona fide opportunistic agent in immunodeficient patients. Carefully controlled, longitudinal studies will be essential to address this important question. Nevertheless, clinical practice may sometimes impose critical decisions, in cases of life-threatening infection of suspected HHV-6 aetiology, in the absence of any other recognized causes. The apparent similarity between the profile of susceptibility to antiviral agents of HHV-6 and hCMV, with relative resistance to aciclovir and sensitivity to ganciclovir and foscarnet, may in such cases direct the therapeutic choice. Moreover, in the follow-up of bone marrow and solid organ transplant recipients, appropriate prophylactic measures for preventing HHV-6-related complications should be taken into account.

HHV-6 in AIDS

The first suggestion that HHV-6 could play a role in the pathogenesis or the disease progression of human immunodeficiency virus (HIV) infection dates back to 1988, after the demonstration of the preferential tropism and cytopathic effect of HHV-6 for CD4+ T lymphocytes, the T cell subset that is directly targeted by HIV and selectively depleted in AIDS (Lusso et al, 1988). Subsequently, a series of intriguing experimental observations on the interactions between HHV-6 and HIV have corroborated this hypothesis. Like other DNA viruses, HHV-6 (both A and B) transactivates the LTR of HIV (Ensoli et al, 1989; Horvat et al, 1989; Lusso et al, 1989a; Demarchi et al, 1996), a phenomenon that is likely to have biological significance because HHV-6, unlike other DNA viruses, is able productively to co-infect individual CD4+ T cells together with HIV (Lusso et al, 1989a) and thus a direct interaction between the two viruses may take place. In turn, the Tat protein of HIV-1 was shown to enhance the replication of HHV-6 (Sieczkowski et al, 1995). These observations may explain, at least in part, the dramatic acceleration of HIV-1 expression and the cytopathic effect observed in primary mononuclear cell cultures co-infected by HHV-6 and HIV-1 (Lusso et al, 1989a). Although some authors have subsequently reported a suppressive effect of HHV-6 on HIV-1 replication (Carrigan et al, 1990; Levy et al, 1990), possibly related to the high multiplicity of infection used for HHV-6, there is consensus that co-infection leads to accelerated cell death. As mentioned above, HHV-6 is a powerful inducer of the release of cytokines, such as TNF-α and IL-1β (Flamand et al, 1991), which can enhance the replication of HIV in vitro. However, the most striking and unique interaction between HHV-6 and HIV is the HIV receptor regulation by HHV-6. As mentioned earlier, infection with HHV-6 positively regulates the expression of CD4, the major receptor for HIV (Dalgleish et al, 1984; Klatzmann et al, 1984): not only is CD4 upregulated in cells that already express it at low levels (e.g. the Jurkat T cell line), but it is even induced de novo, by transcriptional activation mechanisms, in cells that physiologically do not express it, such as CD8+ T cells (Lusso et al, 1991a), NK cells (Lusso et al, 1993) and $\gamma\delta$ T cells (Lusso et al, 1995). Importantly, the induced CD4 receptor is functional, as it can mediate productive infection of these cells with HIV. Similarly, previous infection with HHV-6 A renders haematopoietic precursor cell lines susceptible to HIV-1 superinfection (Furlini et al, 1996). Thus, HHV-6 may significantly expand the range of cells susceptible to HIV. Positive interactions have also been documented between HHV-6 and simian immunodeficiency virus (SIV), in co-infected macaque T lymphocyte cultures (Lusso et al, 1994b). Together, the above observations indicate that HHV-6 is an excellent candidate for a role as a cofactor in the course of HIV infection: it is an 'immunotropic' herpesvirus that may cause direct damage to the immune system and, by direct or indirect mechanisms, significantly boost the replication and/or the spread of HIV in co-infected individuals.

Although the possible role of HHV-6 as a cofactor in AIDS is still controversial, several lines of clinical evidence are consistent with this concept. First, HHV-6 may act as an immunosuppressive agent in its own right, as suggested by the findings in an HIV-negative infant with progressive immunodeficiency, thymic atrophy and severe T lymphocytopenia associated with widespread HHV-6 infection (Knox et al, 1995b). Besides the frequent isolation of HHV-6, particularly subgroup A, from HIV-infected individuals (reviewed by Lusso and Gallo, 1995b), there have been reports of high rates of detection of HHV-6 by PCR in circulating leukocytes and in lymph nodes of HIV-infected patients (Buchbinder et al, 1988; Dolcetti et al, 1994, 1996; Knox and Carrigan, 1996). A significant correlation was observed between the viral load in vivo and the number of circulating CD4+ T cells, with a decline of the HHV-6 copy number in blood cells in parallel with the depletion of CD4+ T cells (Fairfax et al, 1994; Dolcetti et al, 1996). Although this phenomenon may be simply a consequence of the decreased number of target cells for HHV-6 infection, it may also indicate an aetiological association between HHV-6 and the disappearance of CD4+ T cells. Active replication of HHV-6 was documented by serum PCR in patients with early symptomatic HIV infection (AIDS-related complex) and full-blown AIDS, but not in asymptomatic HIV-seropositive individuals (Secchiero et al, 1995a), and was found to correlate with a loss of circulating CD4+ cells (Iuliano et al, 1997). In both studies, the A variant of HHV-6 was detected in serum. Two studies on post-mortem tissues, one conducted by PCR (Corbellino et al, 1993) and the other by immunohistochemistry (Knox and Carrigan, 1994), suggested that HHV-6 infection is active and disseminated in terminal AIDS patients. In another report, a higher viral load of HHV-6 was documented in AIDS tissues compared with controls (Clark et al, 1996). Unfortunately, these necropsy studies could not elucidate the time of acquisition/reactivation of HHV-6 and thereby distinguish between a primary (pathogenetic) and a secondary (epiphenomenal) event. An analysis of bioptic material derived from live patients, however, confirmed the presence of actively replicating HHV-6 A in 100% of the lymph nodes studied (eight with follicular hyperplasia and two with follicular involution) from HIV-infected patients (Knox and Carrigan, 1996). Carefully controlled longitudinal studies, utilizing markers of active HHV-6 infection, are needed to establish whether a temporal association exists between HHV-6 reactivation/reinfection and the progression of the immunodeficiency of AIDS. Another approach that could help in elucidating the role played by HHV-6 in AIDS is the experimental co-infection with HHV-6 and HIV (or SIV) in susceptible animal models: the evaluation of virological, immunological and clinical parameters in these animals, compared with singly infected controls, will help to define the influence of HHV-6 in the natural history of HIV- or SIV-induced AIDS. Finally, as discussed more in detail in the next section, important information may come from the implementation of combination therapeutic trials in vivo, employing selective anti-HHV-6 compounds combined with antiretroviral drugs.

ANTIVIRAL AGENTS ACTIVE AGAINST HHV-6

The progressive unravelling of the pathogenic role of HHV-6 in humans increasingly requires the development and testing of effective therapeutic measures against this virus. Several antiviral agents have been tested for their efficacy on HHV-6 infection in cell culture systems in vitro. Conversely, apart from sporadic attempts, no systematic trial has been performed in vivo on well-documented HHV-6 infection. Table 17.3 summarizes the most important in vitro studies conducted thus far. With few exceptions, most studies demonstrated that aciclovir exerts no inhibitory effect on the replication of HHV-6, whereas there is still debate about the efficacy of ganciclovir. Some investigators have reported a very narrow therapeutic window for this compound (Streicher et al, 1988; Akesson-Johansson et al, 1990); other groups have observed significant inhibition of HHV-6 replication, with a good selectivity index (Russler et al, 1989; Burns and Sandford, 1990; Agut et al, 1991). Whether such discrepancies reflect a different sensitivity of the viral strains used (not only their subgroup classification, but also the fact that laboratory-passaged isolates may have an increased drug resistance) or different experimental conditions remains to be established. It has to be underlined that HHV-6, like hCMV, lacks a thymidine kinase, although both viruses encode another nucleoside kinase. Mutations of this kinase have been shown to confer resistance to ganciclovir to hCMV (Littler et al, 1992).

Table 17.3 *Susceptibility of HHV-6 to antiviral agents*

Compound	Viral subgroup	Selectivity index ID_{50}/TD_{50}	Reference
Aciclovir	A	<5	Streicher et al (1988)
	B	<5	Kikuta et al (1989)
	B	<5	Russler et al (1989)
	B	<5	Di Luca et al (1990)
	A	<10	Akesson-Johansson et al (1990)
	A,B	<10	Agut et al (1991)
	A	<5	Reymen et al (1995)
Ganciclovir	A	<5	Streicher et al (1988)
	B	>50	Russler et al (1989)
	B	>10	Burns and Sandford (1990)
	A	8	Akesson-Johansson et al (1990)
	A,B	>50	Agut et al (1991)
PFA	A	>20	Streicher et al (1988)
	A	31	Akesson-Johansson et al (1990)
	A,B	>100	Agut et al (1991)
	A,B	>100	P. Lusso et al (unpublished)
	A	125	Reymen et al (1995)
H2G	A	47	Akesson-Johansson et al (1990)
PMEDAP	A	18	Reymen et al (1995)
CyA, CyG	A	>40	P. Lusso et al (unpublished)

The compounds with the best therapeutic window against HHV-6, together with a novel guanosine analogue, H2G (Akesson-Johansson et al, 1990), and selected acyclic nucleoside phosphonates (Reymen et al, 1995), are the pyrophosphate analogues phosphonoformic acid (PFA or foscarnet) (Streicher et al, 1988) and phosphonoacetic acid (Di Luca et al, 1990). Foscarnet (Öberg, 1989) is approved for clinical use and is of proven efficacy against drug-resistant HSV and hCMV infections in patients with AIDS. Interestingly, foscarnet has a documented ability to induce an increase in the survival time in patients with AIDS (Studies of Ocular Complications of AIDS Research Group, 1992). It is tempting to hypothesize that the beneficial effects of foscarnet in AIDS may be due, at least in part, to the direct inhibition of HHV-6 replication, and thus that HHV-6 infection has a significant effect on the mortality of HIV-infected patients. However, this hypothesis is not supported by any direct experimental or clinical evidence, and it must be considered that foscarnet is not selectively active on herpesviruses, but also exerts a direct antiretroviral effect (Öberg, 1989). The concept that herpesviruses may act as cofactors in AIDS is also corroborated by the results of different clinical trials employing zidovudine (azidothymidine AZT) combined with aciclovir (reviewed by Griffiths, 1995). In this respect, it should be emphasized again that discrepancies may exist between the results observed in cell culture and the actual effectiveness of aciclovir in vivo (Griffiths, 1995). For example, the frequency of HHV-6 activity in bone marrow transplant recipients monitored by routinely culturing recipient PBMCs was reduced during periods of aciclovir treatment (Kadakia et al, 1996; Wang et al, 1996). Ideally, future trials in HIV-infected patients should include compounds selectively active against HHV-6 in combination with antiretroviral agents. This approach, paralleled by careful monitoring of markers of HHV-6 replication in vivo, will be extremely valuable for the definition of the relative role played by HHV-6 in AIDS.

Besides chemotherapy, cytokines should also be considered as potential therapeutic agents against HHV-6. For example, it has been reported that IFN-α (Kikuta et al, 1990) and IL-2 (Roffman and Frenkel, 1990) exert an inhibitory effect on infection by HHV-6 in cell culture systems. It is unclear, at present, whether their antiviral action is direct or indirect (i.e. mediated by the activation of antiviral effector cells in vitro). The use of these immunomodulating agents may combine the virus-suppressing effect with a positive action on the immune system.

CONCLUSION

Since the discovery of HHV-6, our knowledge of the biology, genetics, epidemiology and clinical manifestations of this unique agent has grown rapidly. Human herpesvirus 6 is the first example of an 'immunotropic' human herpesvirus, and its fascinating interactions with the immune system have led to the hypothesis that it may be involved in the pathogenesis of immunodeficiency states. After some years of uncertainty, sensitive and specific diagnostic methods have been defined, which permit the distinction between active and latent HHV-6 infection, and HHV-6 is emerging as an important human pathogen which may cause severe opportunistic infections in immunocompromised patients and may act as a cofactor in AIDS.

Conversely, its suggested involvement in neoplastic disorders and in multiple sclerosis needs to be further evaluated. Future research will likely be focused on a better understanding of the fine mechanisms of HHV-6 cytopathogenicity, as well as of the complex interactions between the virus and its host cells, both during the lytic cycle, and during latency. Finally, the progressive unraveling of the pathogeneic role of HHV-6 will undoubtedly boost the efforts aimed at identifying effective therapeutic agents for the control of HHV-6 infection *in vivo*.

REFERENCES

Aberle SW, Mandl CW, Kunz C & Popow-Kraupp T (1996) Presence of human herpesvirus 6 variants A and B in saliva and peripheral blood mononuclear cells of healthy adults. *J Clin Microbiol* **34**: 3223–3225.

Ablashi DV, Josephs SF, Buchbinder A et al (1988) Human T-lymphotropic virus (human herpesvirus-6). *J Virol Methods* **21**: 65–88.

Ablashi DV, Balachandran N, Josephs SF et al (1991) Genomic polymorphism, growth properties and immunologic variations in human herpesvirus-6 isolates. *Virology* **184**: 545–552.

Adler SP, McVoy M, Chou S, Hempfling S, Yamanishi K & Britt W (1993) Antibodies induced by a primary cytomegalovirus infection react with human herpesvirus 6 proteins. *J Infect Dis* **168**: 1119–1126.

Agut H, Aubin JT & Huraux JM (1991) Homogeneous susceptibility of distinct human herpesvirus 6 strains to antivirals in vitro. *J Infect Dis* **163**: 1382–1383.

Akashi K, Eizuru Y, Sumiyoshi Y et al (1993) Severe infectious mononucleosis-like syndrome and primary human herpesvirus 6 infection in an adult. *N Engl J Med* **329**: 168–171.

Akesson-Johansson A, Harmenberg J, Wahren B & Linde A (1990) Inhibition of human herpesvirus 6 replication by 9-[4-hydroxy-2-(hydroxymethyl)butyl]guanine (2HM-HBG) and other antiviral compounds. *Antimicrob Agents Chemother* **34**: 2417–2419.

Appleton AL, Sviland L, Peiris JS et al (1995) Human herpes virus-6 infection in marrow graft recipients: role in pathogenesis of graft-versus-host disease. Newcastle upon Tyne Bone Marrow Transport Group. *Bone Marrow Transplant* **16**: 777–782.

Araujo JC, Doniger J, Kashianchi F, Hermonat P L, Thompson J & Rosenthal LJ (1995) Human herpesvirus 6A *ts* suppresses both transformation by H-*ras* and transcription by the H-*ras* and human immunodeficiency virus type 1 promoters. *J Virol* **69**: 4933–4940.

Arena A, Liberto MC, Capozza AB & Foca A (1997) Productive HHV-6 infection in differentiated U937 cells: role of TNF alpha in regulation of HHV-6. *New Microbiol* **20**: 13–20.

Asano Y, Yoshikawa T, Suga S, Yazaki T, Kondo Y & Yamanishi K (1990) Fatal fulminant hepatitis in an infant with human herpesvirus-6 infection. *Lancet* **335**: 862–863.

Asano Y, Yoshikawa T, Kajita Y et al (1992) Fatal encephalitis/encephalopathy in primary human herpesvirus-6 infection. *Arch Dis Child* **67**: 1484–1485.

Asano Y, Suga S, Yoshikawa T, Yazaki T & Uchikawa T (1995) Clinical features and viral excretion in an infant with primary human herpesvirus 7 infection. *Pediatrics* **95**: 187–190.

Aubin JT, Collandre H, Candotti D et al (1991) Several groups among human herpesvirus 6 strains can be distinguished by southern blotting and polymerase chain reaction. *J Clin Microbiol* **29**: 367–372.

Aubin JT, Poirel L, Agut H et al (1992) Intrauterine transmission of human herpesvirus 6. *Lancet* **340**: 482–483.

Bandobashi K, Daibata M, Kamioka M et al (1997) Human herpesvirus 6 (HHV-6)-positive Burkitt's lymphoma: establishment of a novel cell line infected with HHV-6. *Blood* **90**: 1200–1207.

Berneman ZN, Ablashi DV, Li G et al (1992) Human herpesvirus 7 is a T-lymphotropic virus and is related to, but significantly different from, human herpesvirus 6 and human cytomegalovirus. *Proc Natl Acad Sci USA* **89**: 10552–10556.

Biberfeld P, Kramarsky B, Salahuddin SZ & Gallo RC (1987) Ultrastructural characterization of a new human B lymphotropic DNA virus. *J Natl Canc Inst* **79**: 933–941.

Boccaccio GL & Steinman L (1996) Multiple sclerosis: from a myelin point of view. *J Neurosci Res* **45**: 647–654.

Bovenzi P, Mirandola P, Secchiero P, Strumia R, Cassai E & Di Luca D (1993) Human herpesvirus 6 (variant A) in Kaposi's sarcoma. *Lancet* **341**: 1288–1289.

Braun DK, Dominguez G & Pellett PE (1997) Human herpesvirus 6. *Clin Microbiol Rev* **10**: 521–567.

Breese-Hall C, Long CE, Schnabel K et al (1994) Human herpesvirus 6 infection in children. *N Engl J Med* **331**: 432–438.

Briggs M, Fox J & Tedder RS (1988) Age prevalence of antibody to human herpesvirus 6. *Lancet* **i**: 1058–1059.

Brod SA, Lindsey JW & Wolinsky JS (1996) Multiple sclerosis: clinical presentation, diagnosis and treatment [published erratum appears in *Am Fam Physician* (1997) **55(2)**: 448]. *Am Fam Physician* **54**: 1301–1306, 1309–1311.

Buchbinder A, Josephs SF, Ablashi DV, Salahuddin SZ & Gallo RC (1988) Polymerase chain reaction amplification and in situ hybridization for the detection of human B-lymphotropic virus. *J Virol Methods* **21**: 133–140.

Buchwald D, Cheney PR, Peterson DL et al (1992) A chronic illness characterized by fatigue, neurologic and immunologic disorders, and active human herpesvirus type 6 infection. *Ann Intern Med* **116**: 103–113.

Burns WH & Sandford GR (1990) Susceptibility of human herpesvirus 6 to antivirals in vitro. *J Infect Dis* **162**: 634–637.

Carrigan DR (1992) Human herpesvirus-6 and bone marrow transplantation. In: Ablashi DV, Krueger GRF & Salahuddin SZ (eds) *Human Herpesvirus-6: Epidemiology, Molecular Biology and Clinical Pathology*, pp 281–302. Amsterdam: Elsevier.

Carrigan DR (1995) Human herpesvirus-6 and bone marrow transplantation. *Blood* **85**: 294–295.

Carrigan DR & Knox KK (1994) Human herpesvirus 6 (HHV-6) isolation from bone marrow: HHV-6-associated bone marrow suppression in bone marrow transplant patients. *Blood* **84**: 3307–3310.

Carrigan DR & Knox KK (1997) Human herpesvirus six and multiple sclerosis. *Mult Scler* **3**: 390–394.

Carrigan DR, Knox KK & Tapper MA (1990) Suppression of human immunodeficiency virus type 1 replication by human herpesvirus-6. *J Infect Dis* **162**: 844–851.

Carrigan DR, Drobyski WR, Russler SK, Tapper MA, Knox KK & Ash RC (1991) Interstitial pneumonitis associated with human herpesvirus-6 infection after marrow transplantation. *Lancet* **338**: 147–149.

Carrigan DR, Harrington D & Knox KK (1996) Subacute leukoencephalitis caused by CNS infection with human herpesvirus-6 manifesting as acute multiple sclerosis. *Neurology* **47**: 145–148.

Carton H (1996) Multiple sclerosis. A historical review with emphasis on the last 20 years. *Acta Neurol Belg* **96**: 224–227.

Cermelli C, Pietrosemoli P, Meacci M et al (1997) SupT-1: a cell system suitable for an efficient propagation of both HHV- 7 and HHV-6 variants A and B. *New Microbiol* **20**: 187–196.

Challoner PB, Smith KT, Parker JD et al (1995) Plaque-associated expression of human herpesvirus 6 in multiple sclerosis. *Proc Natl Acad Sci USA* **92**: 7440–7444 .

Chan PK, Peiris JS, Yuen KY et al (1997) Human herpesvirus-6 and human herpesvirus-7 infections in bone marrow transplant recipients. *J Med Virol* **53**: 295–305.

Chen M, Wang H, Berneman Z, Lusso P, Ablashi DV & DiPaolo JA (1994a) Detection of human herpesvirus-6 and human papillomavirus 16 in cervical carcinoma. *Am J Pathol* **145**: 1509–1516.

Chen M, Popescu N, Woodworth C et al (1994b) Human herpesvirus-6 infects cervical epithelial cells and transactivates human papillomavirus gene expression. *J Virol* **68:** 1173–1178.

Chou SW & Scott KM (1990) Rise in antibody to human herpesvirus 6 detected by enzyme immunoassay in transplant recipients with primary cytomegalovirus infection. *J Clin Microbiol* **28:** 851–854.

Clark DA, Alexander FE, McKinney PA et al (1990) The seroepidemiology of human herpesvirus-6 (HHV-6) from a case–control study of leukaemia and lymphoma. *Int J Cancer* **45:** 829–833.

Clark DA, Taled M, Kidd IM et al (1996) Quantification of human herpesvirus 6 in immunocompetent persons and post-mortem tissues from AIDS patients by PCR. *J Gen Virol* **77:** 2271–2275.

Collins JG (1997) Prevalence of selected chronic conditions: United States, 1990–1992. *Vital Health Stat* **10:** 1–89.

Cone RW, Huang ML, Ashley R & Corey L (1993a) Human herpesvirus 6 DNA in peripheral blood cells and saliva from immunocompetent individuals. *J Clin Microbiol* **31:** 1262–1267.

Cone RW, Hackman RC, Huang ML et al (1993b) Human herpesvirus 6 in lung tissue of transplanted patients. *N Engl J Med* **329:** 156–161.

Corbellino M, Lusso P, Gallo RC, Parravicini C, Galli M & Moroni M (1993) Disseminated human herpesvirus 6 infection in AIDS. *Lancet* **342:** 1242.

Crowley RW, Secchiero P, Zella D, Cara A, Gallo RC & Lusso P (1996) Interference between human herpesvirus 7 and HIV-1 in mononuclear phagocytes. *J Immunol* **156:** 2004–2008.

Cuende JI, Ruiz J, Civeira MP & Prieto J (1994) High prevalence of HHV-6 DNA in peripheral blood mononuclear cells of healthy individuals detected by nested PCR. *J Med Virol* **43:** 115–118.

Daibata M, Ido E, Murakami K et al (1997) Angioimmunoblastic lymphadenopathy with disseminated human herpesvirus 6 infection in a patient with acute myeloblastic leukemia. *Leukemia* **11:** 882–885.

Dalgleish AG, Beverley PCL, Clapham PR, Crawford DH, Greaves MF & Weiss RA (1984) The CD4 (T4) antigen is an essential component of the receptor for the AIDS retrovirus. *Nature* **312:** 763–766.

Demarchi F, Bovenzi P, Di Luca D & Giacca M (1996) Transcriptional activation of human immunodeficiency virus type 1 by herpesvirus infection: an in vivo footprinting study. *Intervirology* **39:** 236–241.

Dewhurst S, Dollard SC, Pellett PE & Dambaugh TR (1993a) Identification of a lytic-phase origin of DNA replication in human herpesvirus 6B strain Z29. *J Virol* **67:** 7680–7683.

Dewhurst S, McIntyre K, Schnabel K & Hall CB (1993b) Human herpesvirus 6 (HHV-6) variant B accounts for the majority of symptomatic primary HHV-6 infections in a population of US infants. *J Clin Microbiol* **31:** 416–418.

Dewhurst S, Krenitsky DM & Dykes C (1994) Human herpesvirus 6B origin: sequence diversity, requirement for two binding sites for origin-binding protein, and enhanced replication from origin multimers. *J Virol* **68:** 6799–6803.

Di Luca D, Katsafanas G, Schirmer EC, Balachandran N & Frenkel N (1990) The replication of viral and cellular DNA in human herpesvirus 6-infected cells. *Virology* **175:** 199–210.

Di Luca D, Mirandola P & Secchiero P (1992) Characterization of human herpesvirus 6 strains isolated from patients with exanthem subitum with or without cutaneous rash. *J Infect Dis* **166:** 689.

Di Luca D, Dolcetti R, Mirandola P et al (1994) E. Human herpesvirus 6. A survey of presence and variant distribution in normal peripheral lymphocytes and lymphoproliferative disorders. *J Infect Dis* **170:** 211–215.

Dolcetti R, Di Luca D, Mirandola P et al (1994) Frequent detection of human herpesvirus 6 DNA in HIV-associated lymphadenopathy. *Lancet* **344:** 543.

Dolcetti R, Di Luca D, Carbone A et al (1996) Human herpesvirus 6 in human immunodeficiency virus-infected individuals: association with early histologic phases of lymphadenopathy syndrome but not with malignant lymphoproliferative disorders. *J Med Virol* **48**: 344–353.

Downing RG, Sewankambo N, Serwadda D et al (1987) Isolation of human lymphotropic herpesviruses from Uganda. *Lancet* **ii**: 390.

Drobyski WR, Dunne WM, Burd EM et al (1993) Human herpesvirus-6 (HHV-6) infection in allogeneic bone marrow transplant recipients: evidence of a marrow-suppressive role for HHV-6 in vivo. *J Infect Dis* **167**: 735–739.

Drobyski WR, Knox KK, Majewski D & Carrigan DR (1994) Fatal encephalitis due to variant B human herpesvirus-6 infection in a bone marrow-transplant recipient. *N Engl J Med* **330**: 1356–1360.

Dunne WM & Jevon M (1993) Examination of human breast milk for evidence of human herpesvirus 6 by polymerase chain reaction. *J Infect Dis* **168**: 250.

Efstathiou S, Gompels UA, Craxton MA, Honess RW & Ward K (1988) DNA homology between a novel human herpesvirus (HHV-6) and human cytomegalovirus. *Lancet* **i**: 63–64.

Ensoli B, Lusso P, Schachter F et al (1989) Human herpesvirus-6 increases HIV-1 expression in co-infected T-cells via nuclear factors binding to the HIV-1 enhancer. *EMBO J* **8**: 3019–3028.

Fairfax MR, Schacker T, Cone RW, Collier AC & Corey L (1994) Human herpesvirus 6 DNA in blood cells of human immunodeficiency virus-infected men: correlation of high levels with high CD4 cell counts. *J Infect Dis* **169**: 1342–1345.

Fillet AM, Raphael M, Visse B, Audouin J, Poirel L & Agut H (1995) Controlled study of human herpesvirus 6 detection in acquired immunodeficiency syndrome-associated non-Hodgkin's lymphoma. *J Med Virol* **45**: 106–112.

Flamand L, Romerio M, Visse B et al (1998) CD4 promoter transactivation by human herpesvirus 6. *J Virol* **72**: 8797–8805.

Flamand L, Gosselin J, D'Addario M et al (1991) Human herpesvirus 6 induces interleukin 1β and tumor necrosis factor alpha, but not interleukin 6, in peripheral blood mononuclear cell cultures. *J Virol* **65**: 5105–5110.

Flamand L, Stefanescu I, Ablashi DV & Menezes J (1993) Activation of the Epstein–Barr virus replicative cycle by human herpesvirus 6. *J Virol* **67**: 6768–6777.

Flamand L, Stefanescu I & Menezes J (1996) Human herpesvirus-6 enhances natural killer cell cytotoxicity via IL-15. *J Clin Invest* **97**: 1373–1381.

Fox JD, Briggs M, Ward PA & Tedder RS (1990a) Human herpesvirus 6 in salivary glands. *Lancet* **ii**: 590–593.

Fox J, Ward P & Briggs M (1990b) Production of IgM antibody to HHV-6 in reactivation and primary infection. *Epidemiol Infect* **104**: 289–296.

Frenkel N, Schirmer EC, Wyatt LS et al (1990a) Isolation of a new herpesvirus from CD41 T cells. *Proc Natl Acad Sci USA* **87**: 748–752.

Frenkel N, Schirmer EC, Katsafanas G & June CH (1990b) T-cell activation is required for efficient replication of human herpesvirus 6. *J Virol* **64**: 4598–4602.

Furlini G, Vignoli M, Ramazzotti E, Re MC, Visani G & La P (1996) A concurrent human herpesvirus-6 infection renders two human hematopoietic progenitor (TF-1 and KG-1) cell lines susceptible to human immunodeficiency virus type-1. *Blood* **87**: 4737–4745.

Gilden DH, Devlin ME, Burgoon MP & Owens GP (1996) The search for virus in multiple sclerosis brain. *Mult Scler* **2**: 179–183.

Gompels UA & Macaulay H (1995) Characterization of human telomeric repeat sequences from human herpesvirus 6 and relationship to replication. *J Gen Virol* **76**: 451–458.

Gompels UA, Carrigan DR, Carss AL & Arno J (1993) Two groups of human herpesvirus 6 identified by sequence analyses of laboratory strains and variants from Hodgkin's lymphoma and bone marrow transplant patients. *J Gen Virol* **74**: 613–622.

Gompels UA, Nicholas J, Lawrence G et al (1995) The DNA sequence of human herpesvirus 6: structure, coding content, and genome evolution. *Virology* **209**: 29–51.

Gopal MR, Thompson BJ, Fox J, Tedder RS & Honess RW (1990) Detection by PCR of HHV-6 and EBV DNA in blood and oropharynx of healthy adults and HIV-seropositive. *Lancet* **i:** 1598–1599.

Griffiths PD (1995) A proposal that herpesviruses are co-factors of HIV disease. *Antiv Chem Chemother* **6(suppl. 1):** 17–21.

Hall CB, Caserta MT, Schnabel KC et al (1998) Persistence of human herpesvirus 6 according to site and variant: possible greater neurotropism of variant A. *Clin Infect Dis* **26:** 132–137.

Harnett GB, Farr TJ, Pietroboni GR & Bucens MR (1990) Frequent shedding of human herpesvirus 6 in saliva. *J Med Virol* **30:** 128–130.

Hasegawa H, Utsunomiya Y, Yasukawa M, Yanagisawa K & Fujita S (1994) Induction of G protein-coupled peptide receptor EBI 1 by human herpesvirus 6 and 7 infection in CD41 T cells. *J Virol* **68:** 5326–5329.

He J, McCarthy M, Zhou Y, Chandran B & Wood C (1996) Infection of primary human fetal astrocytes by human herpesvirus 6. *J Virol* **70:** 1296–1300.

Hidaka Y, Okada K, Kusuhara K, Miyazaki C, Tokugawa K & Ueda K (1994) Exanthem subitum and human herpesvirus 7 infection. *Pediatr Infect Dis J* **13:** 1010–1011.

Hidaka Y, Kusuhara K, Takabayashi A et al (1997) Symptomatic primary infection with human herpesvirus 6 variant A. *Clin Infect Dis* **24:** 1022–1023.

Horvat RT, Wood C & Balachandran N (1989) Transactivation of human immunodeficiency virus promoter by human herpesvirus 6. *J Virol* **63:** 970–973.

Huang LM, Lee CY, Lin KH et al (1990) Human herpesvirus-6 associated with fatal haemophagocytic syndrome. *Lancet* **336:** 60–61.

Huang LM, Lee CY, Lee PI, Chen JM & Wang PJ (1991) Meningitis caused by human herpesvirus-6. *Arch Dis Child* **66:** 1443–1444.

Huang LM, Lee CY, Liu MY & Lee PI (1997) Primary infections of human herpesvirus-7 and herpesvirus-6: a comparative, longitudinal study up to 6 years of age. *Acta Paediatr* **86:** 604–608.

Hukin J, Farrell K, MacWilliam LM et al (1998) Case–control study of primary human herpesvirus 6 infection in children with febrile seizures. *Pediatrics* **101:** E3.

Ikusaka M, Ota K, Honma Y, Shibata K, Uchiyama S & Iwata M (1997) Meningoencephalitis associated with human herpesvirus-6 in an adult (letter). *Intern Med* **36:** 157.

Inagi R, Guntapong R, Nakao M et al (1996) Human herpesvirus 6 induces IL-8 gene expression in human hepatoma cell line, Hep G2. *J Med Virol* **49:** 34–40.

Inoue Y, Yasukawa M & Fujita S (1997) Induction of T-cell apoptosis by human herpesvirus 6. *J Virol* **71:** 3751–3759.

Irving WL, Cunningham AL, Keogh A & Chapman JR (1988) Antibody to both human herpesvirus 6 and cytomegalovirus. *Lancet* **ii:** 630–631.

Ishiguro N, Yamada S, Takahashi T, Takahashi Y, Okuno T & Yamanishi K (1990) Meningoencephalitis associated with HHV-6 related exanthem subitum. *Acta Paediatr Scand* **79:** 987–989.

Iuliano R, Trovato R, Lico S et al (1997) Human herpesvirus-6 reactivation in a longitudinal study of two HIV-1 infected patients. *J Med Virol* **51:** 259–264.

Jarrett RF, Gledhill S, Qureshi F et al (1988) Identification of human herpesvirus 6-specific DNA sequences in two patients with non-Hodgkin's lymphoma. *Leukemia* **2:** 496–501.

Jarrett RF, Gallagher A, Gledhill S, Jones MD, Teo I & Griffin BE (1989) Variation in restriction map of HHV-6 genome. *Lancet* **i:** 448–449.

Jee SH, Long CE, Schnabel KC, Sehgal N, Epstein LG & Hall CB (1998) Risk of recurrent seizures after a primary human herpesvirus 6-induced febrile seizure. *Pediatr Infect Dis J* **17:** 43–48.

Johnson RT (1994) The virology of demyelinating diseases. *Ann Neurol* **36:** S54–60.

Josephs SF, Ablashi DV, Salahuddin SZ et al (1988a) Molecular studies of HHV-6. *J Virol Methods* **21:** 179–190.

Josephs SF, Buchbinder A, Streicher HZ et al (1988b). Detection of human B-lymphotropic virus (human herpesvirus 6) sequences in B cell lymphoma tissues of three patients. *Leukemia* **2:** 132.

Kadakia MP, Rybka WB, Stewart JA et al (1996) Human herpesvirus 6: infection and disease following autologous and allogeneic bone marrow transplantation. *Blood* **87:** 5341–5354.

Kamei A, Ichinohe S, Onuma R, Hiraga S & Fujiwara T (1997) Acute disseminated demyelination due to primary human herpesvirus-6 infection. *Eur J Pediatr* **156:** 709–712.

Kashanchi F, Araujo J, Doniger J et al (1997) Human herpesvirus 6 (HHV-6) ORF-1 transactivating gene exhibits malignant transforming activity and its protein binds to p53. *Oncogene* **14:** 359–367.

Kempf W, Adams V, Wey N et al (1997) CD681 cells of monocyte/macrophage lineage in the environment of AIDS-associated and classic-sporadic Kaposi sarcoma are singly or doubly infected with human herpesviruses 7 and 6B. *Proc Natl Acad Sci USA* **94:** 7600–7605.

Kikuta H, Lu H, Matsumoto S, Josephs SF & Gallo RC (1989) Polymorphism of human herpesvirus 6 DNA from five Japanese patients with exanthem subitum. *J Infect Dis* **160:** 550–551.

Kikuta H, Nakane A, Lu H, Taguchi Y, Minagawa T & Matsumoto S (1990) Interferon induction by human herpesvirus 6 in mononuclear cells. *J Infect Dis* **162:** 35.

Kimberlin DW (1998) Human herpesviruses 6 and 7: identification of newly recognized viral pathogens and their association with human disease. *Pediatr Infect Dis J* **17:** 59–67.

Kishi M, Harada H, Takahashi M et al (1988) A repeat sequence, GGGTTA, is shared by DNA of human herpesvirus 6 and Marek's disease virus. *J Virol* **62:** 4824–7.

Kitamura K, Ohta H, Ihara T et al (1994) Idiopathic thrombocytopenic purpura after human herpesvirus 6 infection. *Lancet* **344:** 830.

Klatzmann D, Champagne E, Chamaret S et al (1984) T-lymphocyte T4 molecule behaves as receptor for human retrovirus HTLV. *Nature* **312:** 767–770.

Knox KK & Carrigan DR (1992) In vitro suppression of bone marrow progenitor cell differentiation by human herpesvirus 6 infection. *J Infect Dis* **165:** 925–929.

Knox KK & Carrigan DR (1994) Disseminated and active infections by human herpesvirus 6 in patients with AIDS. *Lancet* **343:** 577–578.

Knox KK & Carrigan DR (1995) Active human herpesvirus (HHV-6) infection of the central nervous system in patients with AIDS. *J AIDS Hum Retrovirol* **9:** 69–73.

Knox KK & Carrigan DR (1996) Active HHV-6 infection in the lymph nodes of HIV-infected patients: in vitro evidence that HHV-6 can break HIV latency. *J AIDS Hum Retrovirol* **11:** 370–378.

Knox KK, Harrington DP & Carrigan DR (1995a) Fulminant human herpesvirus 6 encephalitis in a human immunodeficiency virus-infected infant. *J Med Virol* **45:** 288–292.

Knox KK, Pietryga D, Harrington DJ, Franciosi R & Carrigan DR (1995b) Progressive immunodeficiency and fatal pneumonitis associated with human herpesvirus 6 infection in an infant. *Clin Infect Dis* **20:** 406–413.

Kondo K, Kondo T, Okuno T, Takahashi M & Yamanishi K (1991) Latent human herpesvirus 6 infection of human monocytes/macrophages. *J Gen Virol* **72:** 1401–1408.

Kondo K, Nagafuji H, Hata A, Tomomori C & Yamanishi K (1993) Association of human herpesvirus 6 infection of the central nervous system with recurrence of febrile convulsions. *J Infect Dis* **167:** 1197–1200.

Krueger GRF, Wasserman K, de Clerck LS et al (1990) Latent human herpesvirus 6 in salivary and bronchial glands. *Lancet* **336:** 1255–1256.

Krueger GR, Sander C, Hoffmann A, Barth A, Koch B & Braun M (1991) Isolation of human herpesvirus-6 (HHV-6) from patients with collagen vascular diseases. *In Vivo* **5:** 217–225.

Kusuhara K, Takabayashi A, Ueda K et al (1997) Breast milk is not a significant source for early Epstein–Barr virus or human herpesvirus 6 infection in infants: a seroepidemiologic study in 2 endemic areas of human T-cell lymphotropic virus type I in Japan. *Microbiol Immunol* **41:** 309–312.

Lakeman FD & Whitley RJ (1995) Diagnosis of herpes simplex encephalitis: application of polymerase chain reaction to cerebrospinal fluid from brain-biopsied patients and correlation with disease. National Institute of Allergy and Infectious Diseases Collaborative Antiviral Study Group. *J Infect Dis* **171**: 857–863.

Larcher C, Huemer HP, Margreiter R & Dierich MP (1988) Serological crossreaction of human herpesvirus-6 with cytomegalovirus. *Lancet* **ii**: 963–964.

Lautenschlager I, Hockerstedt K, Linnavuori K & Taskinen E (1998) Human herpesvirus-6 infection after liver transplantation. *Clin Infect Dis* **26**: 702–707.

Lawrence GL, Chee M, Craxton MA, Gompels UA, Honess RW & Barrel BG (1990) Human herpesvirus 6 is closely related to human cytomegalovirus. *J Virol* **64**: 287–299.

Leahy MA, Krejci SM, Friednash M et al (1993) Human herpesvirus 6 is present in lesions of Langerhans cell histiocytosis. *J Invest Dermatol* **101**: 642–645.

Levine PH, Jahan N, Murari P, Manak M & Jaffe ES (1992) Detection of human herpesvirus 6 in tissues involved by sinus histiocytosis with massive lymphadenopathy (Rosai–Dorfman disease). *J Infect Dis* **166**: 291–295.

Levy JA, Landay A & Lennette ET (1990) Human herpesvirus 6 inhibits human immunodeficiency virus type 1 replication in cell culture. *J Clin Microbiol* **28**: 2362–2364.

Liedtke W, Trubner K & Schwechheimer K (1995a) On the role of human herpesvirus 6 in viral latency in nervous tissue and in cerebral lymphoma. *J Neurol Sci* **134**: 184–188.

Lindquester GJ & Pellett PE (1991) Properties of the human herpesvirus strain Z29 genome: G1C content, length and presence of variable-length directly repeated terminal sequence elements. *Virology* **182**: 102–110.

Littler E, Stuart AD & Chee MS (1992) Human cytomegalovirus UL97 open reading frame encodes a protein that phosphorylates the antiviral nucleoside analogue ganciclovir. *Nature* **358**: 160–162.

Lopez C, Pellett P, Stewart J et al (1988) Characteristics of human herpesvirus 6. *J Infect Dis* **157**: 1271–1273.

Luka J, Okano M & Thiele G (1990) Isolation of human herpesvirus-6 from clinical specimens using human fibroblast cultures. *J Clin Lab Anal* **4**: 483–486.

Luppi M, Marasca R, Barozzi P, Artusi T & Torelli G (1993a) Frequent detection of human herpesvirus-6 sequences by polymerase chain reaction in paraffin-embedded lymph nodes from patients with angioimmunoblastic lymphadenopathy and angioimmunoblastic lymphadenopathy-like lymphoma. *Leuk Res* **17**: 1003–1011.

Luppi M, Marasca R, Barozzi P et al (1993b) Three cases of human herpesvirus-6 latent infection: integration of viral genome in peripheral blood mononuclear cell DNA. *J Med Virol* **40**: 44–52.

Luppi M, Barozzi P, Maiorana A et al (1995) Human herpesvirus-6: a survey of presence and distribution of genomic sequences in normal brain and neuroglial tumors. *J Med Virol* **47**: 105–111.

Lusso P & Gallo RC (1995a) Human herpesvirus 6. *Baillière's Clin Haematol* **8**: 201–223.

Lusso P & Gallo RC (1995b) Human herpesvirus 6 in AIDS. *Immunol Today* **16**: 67–71.

Lusso P, Markham PD, Tschachler E et al (1988) In vitro cellular tropism of human B-lymphotropic virus (human herpesvirus 6). *J Exp Med* **167**: 1659–1670.

Lusso P, Ensoli B, Markham PD et al (1989a) Productive dual infection of human CD4+ T-lymphocytes by HIV-1 and HHV-6. *Nature* **337**: 368–370.

Lusso P, Gallo RC, DeRocco SE & Markham PD (1989b) CD4 is not the membrane receptor for HHV-6. *Lancet* **i**: 730.

Lusso P, Markham PD, DeRocco SE & Gallo RC (1990) In vitro susceptibility of T lymphocytes from chimpanzees (*Pan troglodytes*) to human herpesvirus 6 (HHV-6): a potential animal model to study the interaction between HHV-6 and human immunodeficiency virus type 1 in vivo. *J Virol* **64**: 2751–2758.

Lusso P, De Maria A, Malnati M et al (1991a) Induction of CD4 and susceptibility to HIV-1 infection in CD8+ human T lymphocytes by human herpesvirus 6. *Nature* **349**: 533–535.

Lusso P, Malnati M, De Maria A, De Rocco S, Markham PD & Gallo RC (1991b) Productive infection of CD41 and CD81 mature T-cell populations and clones by HHV-6: transcriptional down-regulation of CD3. *J Immunol* **147**: 2147–2153.

Lusso P, Malnati M, Garzino-Demo A, Crowley RW, Long EO & Gallo RC (1993) Infection of natural killer cells by human herpesvirus 6. *Nature* **362**: 458–462.

Lusso P, Secchiero P, Crowley RW, Garzino-Demo A, Berneman ZN & Gallo RC (1994a) CD4 is a critical component of the receptor for human herpesvirus 7: interference with human immunodeficiency virus. *Proc Natl Acad Sci USA* **91**: 3872–3876.

Lusso P, Secchiero P & Crowley RW (1994b) In vitro susceptibility of *Macaca nemestrina* to human herpesvirus-6: a potential animal model of coinfection with primate immunodeficiency viruses. *AIDS Res Hum Retroviruses* **10**: 181–187.

Lusso P, Garzino-Demo A, Crowley RW & Malnati MS (1995) Infection of T lymphocytes by human herpesvirus 6 (HHV-6): transcriptional induction of CD4 and susceptibility to HIV infection. *J Exp Med* **181**: 1303–1310.

Lyall EG & Cubie HA (1995) Human herpesvirus-6 DNA in the saliva of paediatric oncology patients and controls. *J Med Virol* **47**: 317–322.

Lycke J, Svennerholm B, Hjelmquist E et al (1996) Acyclovir treatment of relapsing–remitting multiple sclerosis. A randomized, placebo-controlled, double-blind study. *J Neurol* **243**: 214–224.

Mackenzie IR, Carrigan DR & Wiley CA (1995) Chronic myelopathy associated with human herpesvirus-6. *Neurology* **45**: 2015–2017.

Maeda T, Okuno T, Hayashi K et al (1997) Outcomes of infants whose mothers are positive for human herpesvirus-6 DNA within the genital tract in early gestation. *Acta Paediatr Jpn* **39**: 653–657.

Marsh S, Kaplan M, Asano Y et al (1996) Development and application of HHV-6 antigen capture assay for the detection of HHV-6 infections. *J Virol Methods* **61**: 103–112.

Martin C, Enbom M, Soderstrom M et al (1997) Absence of seven human herpesviruses, including HHV-6, by polymerase chain reaction in CSF and blood from patients with multiple sclerosis and optic neuritis. *Acta Neurol Scand* **95**: 280–283.

Martin ME, Thomson BJ, Honess RW et al (1991) The genome of human herpesvirus 6: maps of unit-length and concatameric genomes for nine restriction endonucleases. *J Gen Virol* **72**: 157–168.

McCullers JA, Lakeman FD & Whitley RJ (1995) Human herpesvirus 6 is associated with focal encephalitis. *Clin Infect Dis* **21**: 571–576.

McDonnell GV & Hawkins SA (1996) Primary progressive multiple sclerosis: a distinct syndrome? *Mult Scler* **2**: 137–141.

Mendel I, de Matteis M, Bertin C et al (1995) Fulminant hepatitis in neonates with human herpesvirus 6 infection. *Pediatr Infect Dis J* **14**: 993–997.

Merelli E, Bedin R, Sola P et al (1997) Human herpes virus 6 and human herpes virus 8 sequences in brains of multiple sclerosis patients, normal adults and children. *J Neurol* **244**: 450–454.

Neipel F, Ellinger K & Fleckenstein B (1991) The unique region of the human herpesvirus 6 genome is essentially colinear with the UL segment of human cytomegalovirus. *J Gen Virol* **72**: 2293–2297.

Nicholas J (1996) Determination and analysis of the complete nucleotide sequence of human herpesvirus-7. *J Virol* **70**: 5975–5989.

Niederman JC, Liu C-R, Kaplan MH & Brown NA (1988) Clinical and serological features of human herpesvirus-6 infection in three adults. *Lancet* **ii**: 817–819.

Novoa LJ, Nagra RM, Nakawatase T, Edwards-Lee T, Tourtellotte WW & Cornford ME (1997) Fulminant demyelinating encephalomyelitis associated with productive HHV-6 infection in an immunocompetent adult. *J Med Virol* **52**: 301–308.

Öberg B (1989) Antiviral effects of phosphonoformate (PFA, foscarnet sodium). *Pharmacol Ther* **40**: 213–285.

Osman HK, Peiris JS, Taylor CE, Karlberg JP & Madeley CR (1997) Correlation between the detection of viral DNA by the polymerase chain reaction in peripheral blood leukocytes and serological responses to human herpesvirus 6, human herpesvirus 7, and cytomegalovirus in renal allograft recipients. *J Med Virol* **53**: 288–294.

Patnaik M, Komaroff AL, Conley E, Ojo-Amaize EA & Peter JB (1995) Prevalence of IgM antibodies to human herpesvirus 6 early antigen (p41/38) in patients with chronic fatigue syndrome [published erratum appears in *J Infect Dis* (1995) **172(6)**: 1643]. *J Infect Dis* **172**: 1364–1367.

Penchansky L & Jordan JA (1997) Transient erythroblastopenia of childhood associated with human herpesvirus type 6, variant B. *Am J Clin Pathol* **108**: 127–132.

Pfeiffer B, Thomson B & Chandran B (1995) Identification and characterization of a cDNA derived from multiple splicing that encodes envelope glycoprotein gp105 of human herpesvirus 6. *J Virol* **69**: 3490–3500.

Pietroboni GR, Harnett GB, Bucens MR & Honess RW (1988) Antibody to human herpesvirus 6 in saliva. *Lancet* **i**: 1059.

Portolani M, Cermelli C, Meacci M et al (1997) Primary infection by HHV-6 variant B associated with a fatal case of hemophagocytic syndrome. *New Microbiol* **20**: 7–11.

Prezioso PJ, Cangiarella J, Lee M et al (1992) Fatal disseminated infection with human herpesvirus-6. *J Pediatr* **120**: 921–923.

Pruksananonda P, Hall CB, Insel RA et al (1992) Primary human herpesvirus 6 infection in young children. *N Engl J Med* **326**: 1445–1450.

Qavi HB, Green MT, SeGall GK, Lewis DE & Hollinger FB (1992) Demonstration of HIV-1 and HHV-6 in AIDS-associated retinitis. *Curr Eye Res* **11**: 315.

Rajcani J, Yanagihara R, Godec MS, Nagle JW, Kudelova M & Asher DM (1994) Low-incidence latent infection with variant B or roseola type human herpesvirus 6 in leukocytes of healthy adults. *Arch Virol* **134**: 357–368.

Razzaque A (1990) Oncogenic potential of human herpesvirus-6 DNA. *Oncogene* **5**: 1365–1370.

Razzaque A, Williams O, Wang J & Rhim JS (1993) Neoplastic transformation of immortalized human epidermal keratinocytes by two HHV-6 DNA clones. *Virology* **195**: 113–120.

Reux I, Fillet AM, Agut H, Katlama C, Hauw JJ & LeHoang P (1992) In situ detection of human herpesvirus-6 in AIDS-associated retinitis. *Am J Ophthalmol* **114**: 375–377.

Reymen D, Naesens L, Balzarini J, Holy A & De Clercq E (1995) Antiviral activity of selected nucleoside analogues against human herpes virus type 6. *Nucleos Nucleot* **14**: 567–570.

Robert C, Aubin JT, Visse B, Fillet AM, Huraux JM & Agut H (1996) Difference in permissiveness of human fibroblast cells to variants A and B of human herpesvirus-6. *Res Virol* **147**: 219–225.

Roffman E & Frenkel N (1990) Interleukin-2 inhibits the replication of human herpesvirus-6 in mature thymocytes. *Virology* **175**: 591–594.

Russler SK, Tapper MA & Carrigan DR (1989) Susceptibility of human herpesvirus 6 to acyclovir and ganciclovir. *Lancet* **ii**: 382.

Sadovnick AD (1997) Update on management and genetics of multiple sclerosis. *J Neural Transm Suppl* **50**: 167–172.

Saijo M, Saijo H, Yamamoto M et al (1995) Thrombocytopenic purpura associated with primary human herpesvirus 6 infection (letter). *Pediatr Infect Dis J* **14**: 405.

Salahuddin SZ, Ablashi DV, Markham PD et al (1986) Isolation of a new virus, HBLV, in patients with lymphoproliferative disorders. *Science* **234**: 596–601.

Saxinger C, Polesky N, Eby S et al (1988) Antibody reactivity with HBLV (HHV-6) in a US population. *J Virol Methods* **21**: 199–204.

Schirmer EC, Wyatt LS, Yamanishi K, Rodriguez WJ & Frenkel N (1991) Differentiation between two distinct classes of viruses now classified as human herpesvirus 6. *Proc Natl Acad Sci USA* **88**: 5922.

Secchiero P, Carrigan DR, Asano Y et al (1995a) Detection of human herpesvirus 6 in plasma of children with primary infection and immunosuppressed patients by polymerase chain reaction. *J Infect Dis* **171:** 273–280.

Secchiero P, Zella D, Crowley RW, Gallo RC & Lusso P (1995b) Quantitative polymerase chain reaction for human herpesvirus 6 and 7. *J Clin Microbiol* **33:** 2124–2130.

Sieczkowski L, Chandran B & Wood C (1995) The human immunodeficiency virus *tat* gene enhances replication of human herpesvirus-6. *Virology* **211:** 544–553.

Singh N, Carrigan DR, Gayowski T & Marino IR (1997) Human herpesvirus-6 infection in liver transplant recipients: documentation of pathogenicity. *Transplantation* **64:** 674–678.

Sobue R, Miyazaki H, Okamoto M et al (1991) Fulminant hepatitis in primary human herpesvirus-6 infection. *N Engl J Med* **324:** 1290.

Soldan SS, Berti R, Salem N et al (1997) Association of human herpes virus 6 (HHV-6) with multiple sclerosis: increased IgM response to HHV-6 early antigen and detection of serum HHV-6 DNA. *Nature Med* **3:** 1394–1397.

Steeper TA, Horwitz CA, Ablashi DV et al (1990) The spectrum of clinical and laboratory findings resulting from human herpesvirus-6 (HHV-6) in patients with mononucleosis-like illnesses not resulting from Epstein–Barr virus or cytomegalovirus. *Am J Clin Pathol* **93:** 776–783.

Streicher HL, Hung CL, Ablashi DV et al (1988) In vitro inhibition of human herpesvirus-6 by phosphonoformate. *J Virol Methods* **21:** 301–304.

Studies of Ocular Complications of AIDS Research Group (1992) Mortality in patients with the acquired immunodeficiency syndrome treated with either foscarnet or ganciclovir for cytomegalovirus retinitis. *N Engl J Med* **326:** 213.

Suga S, Yoshikawa T, Asano Y, Yazaki T & Hirata S (1989) Human herpesvirus 6 infection (exanthem subitum) without rash. *Pediatrics* **83:** 1003–1006.

Suga S, Yoshikawa T, Asano Y et al (1992) IgM neutralizing antibody response to human herpesvirus 6 in patients with exanthema subitum or organ transplantation. *Microbiol Immunol* **36:** 495–506.

Suga S, Suzuki K, Ihira M, Furukawa H, Yoshikawa T & Asano Y (1998) Clinical features of primary HHV-6 and HHV-7 infections. *Nippon Rinsho* 56, 203–207.

Tajiri H, Tanaka-Taya K, Ozaki Y, Okada S, Mushiake S & Yamanishi K (1997) Chronic hepatitis in an infant, in association with human herpesvirus-6 infection. *J Pediatr* **131:** 473–475.

Takahashi K, Sonoda S, Higashi K et al (1989) Predominant CD4 T-lymphocyte tropism of human herpesvirus 6–related virus. *J Virol* **63:** 3161–3163.

Takikawa T, Hayashibara H, Harada Y & Shiraki K (1992) Liver dysfunction, anaemia, and granulocytopenia after exanthema subitum. *Lancet* **340:** 1288–1289.

Tanaka K, Kondo T, Torigoe S, Okada S, Mukai T & Yamanishi K (1994) Human herpesvirus 7: another causal agent for roseola (exanthem subitum). *J Pediatr* **125:** 1–5.

Tanaka-Taya K, Kondo T, Mukai T et al (1996) Seroepidemiological study of human herpesvirus-6 and -7 in children of different ages and detection of these two viruses in throat swabs by polymerase chain reaction. *J Med Virol* **48:** 88–94.

Tang YW, Espy MJ, Persing DH & Smith TF (1997) Molecular evidence and clinical significance of herpesvirus coinfection in the central nervous system. *J Clin Microbiol* **35:** 2869–2872.

Thomson BJ, Efstathiou S & Honess RW (1991) Acquisition of the human adeno-associated virus type-2 *rep* gene by human herpesvirus type-6. *Nature* **351:** 78–80.

Thomson BJ, Dewhurst S & Gray D (1994a) Structure and heterogeneity of the a sequences of human herpesvirus 6 strain variants U1102 and Z29 and identification of human telomeric repeat sequences at the genomic termini. *J Virol* **68:** 3007–3014.

Thomson BJ, Weindler FW, Gray D, Schwaab V & Heilbronn R (1994b) Human herpesvirus 6 (HHV-6) is a helper virus for adeno-associated virus type 2 (AAV-2) and the AAV-2 *rep* gene homologue in HHV-6 can mediate AAV-2 DNA replication and regulate gene expression. *Virology* **204:** 304–311.

Torigoe S, Kumamoto T, Koide W, Taya K & Yamanishi K (1995) Clinical manifestations associated with human herpesvirus 7 infection. *Arch Dis Child* **72:** 518–519.

Ueda T, Miyake Y, Imoto K et al (1996) Distribution of human herpesvirus 6 and varicella-zoster virus in organs of a fatal case with exanthem subitum and varicella. *Acta Paediatr Jpn* **38:** 590–595.

Wagner M, Muller-Berghaus J, Schroeder R et al (1997) Human herpesvirus-6 (HHV-6)-associated necrotizing encephalitis in Griscelli's syndrome. *J Med Virol* **53:** 306–312.

Wang FZ, Dahl H, Linde A, Brytting M, Ehrnst A & Ljungman P (1996) Lymphotropic herpesviruses in allogeneic bone marrow transplantation. *Blood* **88:** 3615–3620.

Wilborn F, Schmidt CA, Zimmerman C, Brinkman V, Neipel F & Siegert W (1994) Detection of herpesvirus type 6 by polymerase chain reaction in blood donors: random tests and prospective longitudinal studies. *Br J Haematol* **88:** 187–192.

Yadav M, Chandrashekran A, Vasudevan DM & Ablashi DV (1994) Frequent detection of human herpesvirus 6 in oral carcinoma. *J Natl Cancer Inst* **86:** 1792–1794.

Yadav M, Arivananthan M, Chandrashekran A, Tan BS & Hashim BY (1997) Human herpesvirus-6 (HHV-6) DNA and virus-encoded antigen in oral lesions. *J Oral Pathol Med* **26:** 393–401.

Yalcin S, Mukai T, Kondo K et al (1992) Experimental infection of cynomolgus and African green monkeys with human herpesvirus 6. *J Gen Virol* **73:** 1673–1677.

Yamanishi K, Okuna T & Shiraki K (1988) Identification of human herpesvirus 6 as a causal agent for exanthem subitum. *Lancet* **i:** 1065.

Yanagihara K, Tanaka-Taya K, Itagaki Y et al (1995) Human herpesvirus 6 meningoencephalitis with sequelae. *Pediatr Infect Dis J* **14:** 240–242.

Yasukawa M, Inoue Y, Ohminami H, Terada K & Fujita S (1998) Apoptosis of CD41 T lymphocytes in human herpesvirus-6 infection. *J Gen Virol* **79:** 143–147.

Yoshikawa T, Asano Y & Kobayashi I (1993) Exacerbation of idiopathic thrombocytopenic purpura by primary human herpesvirus 6 infection. *Pediatr Infect Dis J* **12:** 409–410.

Chapter 18

KAPOSI'S SARCOMA AND HUMAN HERPESVIRUS 8

Simon J. Talbot and Denise Whitby

INTRODUCTION

In 1994 Chang et al described the identification of sequences from a new human herpesvirus in tumour tissue from AIDS-related Kaposi's sarcoma (KS) (Chang et al, 1994), using the technique of representational difference analysis (RDA) (Lisitsyn et al, 1993). This technique uses cycles of polymerase chain reaction (PCR) and subtractive hybridization to amplify and enrich low-abundance DNA sequences present in only one of two otherwise identical DNA populations, and is therefore ideal for determining small differences between complex genomes (i.e. the presence of viral DNA genomes). Viral sequences have now been detected, using PCR, in over 95% of all KS lesions, including those from non-HIV-infected patients (Boshoff et al, 1995a; Collandre et al, 1995; Dupin et al, 1995; Chang et al, 1996b). The entire genome of the new virus has now been cloned and sequenced and shown to be a gamma-2-herpesvirus with homology to the human gammaherpesvirus Epstein–Barr virus (EBV) and the monkey herpesvirus saimiri (HVS) (Figure 18.1) (Moore et al, 1996b; Russo et al, 1996; Neipel et al, 1997a). This new herpesvirus was initially given the descriptive name Kaposi's sarcoma-associated herpesvirus (KSHV), but is also known as human herpesvirus 8 (HHV-8), and this term will be used throughout this chapter.

KAPOSI'S SARCOMA

Kaposi's sarcoma was originally described by the Austro-Hungarian dermatologist Moritz Kaposi in 1872 (Kaposi, 1872). The tumour usually appears as brownish-purple lesions, often on the extremities, but may in more aggressive forms of the disease progress to involve organs such as the lungs, lymph nodes and gastrointestinal tract. The lesions contain many types of cell but are characterized histologically by the presence of elongated 'spindle' cells and irregular slit-like vascular spaces.

The origin of the spindle cells is disputed; immunohistochemistry of KS lesions has been used by many groups to demonstrate the expression of various cell markers on spindle cells. Several studies have demonstrated the expression of markers specific for lymphatic or vascular endothelial cells (Beckstead et al, 1985; Rutgers et al, 1986). Other studies have shown expression of markers for monocytes/macrophages (Uccini et al, 1994), dermal dendrocytes (Nickoloff and Griffiths, 1989) and smooth muscle cells (Thompson et al, 1991; Weich et al, 1991). Because of the conflicting data it has been

HIV and the New Viruses Second Edition
ISBN 0-12-200741-7

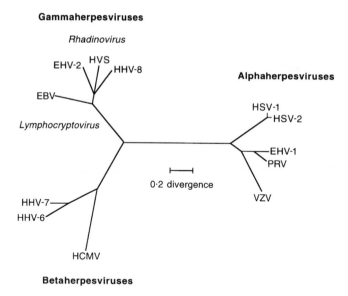

Figure 18.1 *Phylogenetic tree showing relation of HHV-8 to other herpesviruses*
The tree, derived by the neighbour joining method is shown in unrooted form with branch
lengths proportional to divergence between nodes binding each branch. HSV, herpes simplex
virus; EHV, equine herpesvirus; PRV, pseudorabies virus; VZV, varicella-zoster virus; HCMV,
human cytomegalovirus; HVS, *Herpesvirus saimiri*. Redrawn with permission from Moore et al
(1996b).

suggested that spindle cells in KS lesions may be derived from more than one cell type
(Roth, 1991) or from a mesenchymal progenitor cell (Weiss, 1996).

There is also controversy as to whether KS is a true malignant neoplasm. Several
aspects of KS indicate that it may be a reactive hyperplasia – a proliferative lesion which
grows as a result of a cascade of growth factors. The majority of cell lines derived from
KS lesions require growth factors for prolonged growth in vitro and when injected into
nude mice result in lesions that are of mouse rather than human origin. Two KS cell
lines have been described which are fully transformed and which give rise to human
tumours when injected into nude mice (Herndier et al, 1994a,b; Lunardi et al, 1995). An
important characteristic of neoplastic tumours is that they arise from clonal expansion
of a single cell, whereas reactive hyperplasia is polyclonal. Recent studies using an X
chromosome inactivation assay indicate that KS tumour cells are clonal in origin and
hence suggest that KS is a true neoplasm (Rabkin et al, 1997). Thus KS has features of
both a hyperplasia and a neoplasm, and may be a disease in which a hyperplastic lesion
may progress to become clonal, of which there are several examples, including EBV-
related post-transplant or AIDS-associated lymphoproliferative disease.

Four forms of KS have been described, all of which are histologically identical but are
distinguished by epidemiological factors and disease severity. Classic KS is characterized
by benign, indolent tumours, often limited to the extremities. It is rare in the USA and
northern Europe and occurs mostly in elderly men of Mediterranean, Middle Eastern
or Jewish descent (Wahman et al, 1991). African, or endemic, KS is common in parts of
sub-Saharan Africa. In adults it occurs more often in men, but unlike classic KS it is also
seen in children of both sexes. Kaposi's sarcoma is generally more aggressive in patients

from Africa, particularly in young children in whom lymph node involvement is common and the disease is rapidly fatal (Taylor et al, 1972; Ziegler and Katongole-Mbidde, 1996). Iatrogenic KS occurs in individuals who are immunosuppressed because of organ transplantation (Farge et al, 1994). The best-known form of KS is AIDS-related KS, which is often very aggressive with widely disseminated cutaneous lesions, visceral involvement and associated mortality. Indeed, it was the appearance of KS in young homosexual and bisexual men in the USA which was one of the first signs of the AIDS epidemic (Service, 1981).

Strong epidemiological evidence has accumulated since the 1980s suggesting that KS is caused by an infectious agent; KS is 300 times more common in AIDS patients than in other immunosuppressed groups and 20 000 times more common than in the general population. Importantly, KS is 10 times more common in homosexual or bisexual men with AIDS than in groups of HIV-infected individuals who have become infected by non-sexual routes, such as individuals with haemophilia, injecting drug users or blood transfusion recipients (Beral, 1991). The lesion also occurs in young homosexual men who are not HIV-infected (Friedman Kien et al, 1990). In Africa, AIDS-KS is seen in those who acquire HIV by heterosexual contact (Beral et al, 1990b). This suggests that any putative causative virus is likely to be sexually transmitted.

Despite strong evidence for the association of an infectious agent with KS, numerous studies had failed to show an association with any known virus. Cytomegalovirus (Giraldo et al, 1984), human papillomavirus 16 (HPV-16) (Huang et al, 1992), retrovirus-like particles (Rappersberger et al, 1990), human herpesvirus 6 (Kempf et al, 1995; Kempf and Adams, 1996), BK virus (Monini et al, 1996) *Mycoplasma penetrans* (Wang et al, 1993) and HIV itself (Ensoli et al, 1990, 1994; Barillari et al, 1992) have all been proposed as candidate KS agents. In each case, however, the evidence has been unconvincing and has not been corroborated in subsequent studies.

The role of HHV-8 in Kaposi's sarcoma

Since the discovery of sequences from this new human herpesvirus (Chang et al, 1994), significant evidence has accumulated to support an association between HHV-8 and KS. Using PCR techniques, several groups have shown HHV-8 to be present in the vast majority of KS tumour biopsies from all forms of KS (Boshoff et al, 1995a,b; Dupin et al, 1995; Schalling et al, 1995; Chang et al, 1996b) (Table 18.1). A study using in situ PCR demonstrated that HHV-8 DNA is found in KS spindle cells as well as in flat endothelial cells lining the slit-like vascular spaces of KS lesions (Boshoff et al, 1995a,b). This finding has been confirmed by studies using conventional in situ techniques (Li et al, 1996; Staskus et al, 1997).

It has been suggested, however, that HHV-8 may be ubiquitous in the general population, as is the case for the other human herpesviruses, and that it preferentially replicates in KS tissue without being the cause of the tumour. For this reason the prevalence of the virus in populations both affected and unaffected by KS is of crucial importance in establishing its role in KS.

Initial HHV-8 prevalence studies were based on the detection of viral DNA by PCR in blood, biopsies or other body fluids. Using nested PCR techniques, it was shown that 52% of HIV-infected patients with KS had detectable HHV-8 in the peripheral blood cells compared with only 11% of HIV-infected controls without KS. The virus was not

Table 18.1 *Detection of HHV-8 in Kaposi's sarcoma*
Summary of the prevalence of HHV-8 sequences in lesions from the different epidemiological forms of Kaposi's sarcoma (KS) compared with other malignancies or normal tissue.

Type of KS lesion	Number positive	Number tested	Percentage positive (%)
AIDS-KS	252	259	97
Classic KS	160	175	91
Iatrogenic KS	13	13	100
African endemic	71	80	89
HIV-negative homosexual men with KS	8	9	89
Control tissues	14	743	2

Data from Chang et al, 1994, 1996b; Ambroziak et al, 1995; Boshoff et al, 1995a; Dupin et al, 1995; Huang et al, 1995; Lebbe et al, 1995, 1997; Moore et al, 1995a; Schalling et al, 1995; Su et al, 1995; Buonaguro et al, 1996; Cathomas et al, 1996; Chuck et al, 1996; Corbellino et al, 1996; Dictor et al, 1996; Gaidano et al, 1996; Jin et al, 1996; Luppi et al, 1996a; Marchiolo et al, 1996; McDonagh et al, 1996; O'Neill et al, 1996.

detected in the peripheral blood of 134 blood donors or 26 cancer patients (Whitby et al, 1995). Additionally, HHV-8 was detected in the peripheral blood of 60% of patients with classic KS but not in age-matched controls from the same area (D. Whitby and A. Hatzakis, unpublished observations). These findings have been confirmed by several other groups (Ambroziak et al, 1995; Humphrey et al, 1996; Moore et al, 1996c). Some groups have reported that HHV-8 can be detected in around 10% of healthy individuals in parts of Italy (Bigoni et al, 1996; Viviano et al, 1997). Since the prevalence of KS is higher in Italy than in the UK or the USA, these findings are consistent with an association between HHV-8 and KS. We have also shown that detection of HHV-8 in asymptomatic HIV-positive individuals strongly predicts subsequent progression of those patients to development of KS (Whitby et al, 1995). These data are highly suggestive of a causal role for HHV-8 in KS. Studies to determine the cell types infected in vivo by HHV-8 have shown detection of HHV-8 DNA in B cells (Ambrosiak et al, 1995; Whitby et al, 1995), circulating 'spindle cells' in peripheral blood (Sirianni et al, 1997a) and CD8 cells (Harrington et al, 1996; Sirianni et al, 1997b).

Human herpesvirus 8 has been detected, infrequently in the saliva of AIDS-KS patients (Whitby et al, 1995; Luppi et al, 1996b; Koelle et al, 1997) and also in the bronchoalveolar lavage (BAL) fluid from over 80% of patients with pulmonary KS (Howard et al, 1995). There have been, as yet, no reports of detection of HHV-8 in saliva of classic KS patients or in symptom-free individuals infected with HHV-8; thus the role of saliva in the transmission of HHV-8 remains unclear. The virus has not been detected in the faeces of AIDS-KS patients (Whitby et al, 1995).

Conflicting data have emerged as to the prevalence of HHV-8 in semen. The detection of HHV-8 DNA in the vast majority of semen samples from AIDS-KS patients and also in a significant proportion of samples from healthy semen donors in the USA has been reported (Lin et al, 1995, since withdrawn). Another study reported the detection of HHV-8 in 90% of semen samples and prostate tissue from healthy individuals in Italy (Corbellino et al, 1996). Several others, however, have not detected HHV-8 DNA in any semen samples from healthy individuals in the USA or Italy (Tasaka et al, 1996; Howard et al, 1997). We have been able to detect HHV-8 in semen samples of 20% of AIDS-KS patients but not in 115 samples from healthy donors (Howard et al, 1997). Another report detected HHV-8 DNA in 13% of semen samples in Sicily, an area with very high rates of KS (Viviano et al, 1997). Taken together, these reports suggest that sexual transmission of HHV-8 via semen is likely to occur but that HHV-8 is not prevalent outside the known KS risk groups. An intriguing finding related to this topic was recently reported. A study using in situ hybridization to examine expression of transcripts shown to be specific for latent and lytic HHV-8 infection in KS and other tissues found frequent detection of latent HHV-8 genes in prostate tissue from elderly HIV-negative men or younger HIV-positive men. These same tissues were, however, negative by conventional HHV-8 PCR. The significance of these findings is unclear and they await independent confirmation.

There have been fewer studies examining the prevalence of HHV-8 in the female genital tract. Two small studies reported that HHV-8 was not detected in small numbers of cervical or vulval cancer specimens in the US (Tasaka et al, 1996) and Italy (Viviano et al, 1997). In a larger study of cervical smears from HIV-positive and negative women in the UK, it was found that HHV-8 could be detected in 3 out of 11 HHV-8-seropositive women of African origin, but not in 78 HHV-8-seronegative women of both African and non-African origin (D. Whitby et al, 1999). Such detection has important implications for both sexual and vertical transmission of HHV-8.

First-generation serological assays have been developed that can detect antibodies to latent or lytic HHV-8 antigens (Figure 18.2); see Latency-associated antigens encoded by HHV-8. Initial serosurveys have shown that these assays detect antibodies to HHV-8 in the sera of over 95% of classic KS patients and 80% of AIDS-KS patients (Gao et al, 1996; Kedes et al, 1996; Lennette et al, 1996; Miller et al, 1996; Simpson et al, 1996)

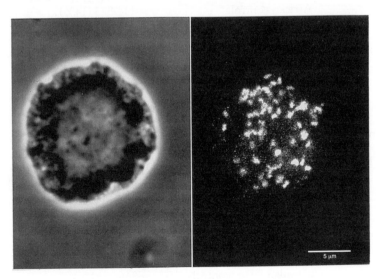

Figure 18.2 *Immunofluorescent analysis of a PEL cell line (BCP-1) latently infected with HHV-8 using Kaposi's sarcoma-positive antisera*
The left-hand panel shows a phase contrast image, and the right-hand panel shows immunofluorescent analysis of the same cell showing the characteristic discrete nuclear stippling pattern. Photomicrographs were taken with a BioRad MRC600 confocal imaging system in conjunction with a Nikon Fluorescence Optiphot ×60 plan-apo.

(Table 18.2). An immunofluorescence assay (IFA) which measures antibodies to lytically infected cells has been reported to detect antibodies in 20% of healthy US blood donors (Lennette et al, 1996). The reported seroprevalence in healthy blood donors measured by other assays, both to latent and lytic antigens, is much lower, 0–3% (Gao et al, 1996; Kedes et al, 1996; Miller et al, 1996; Simpson et al, 1996). Seroprevalence of antibodies to HHV-8 has been shown to be significantly higher in groups at risk for KS, such as HIV-infected homosexual men (Gao et al, 1996; Lennette et al, 1996; Simpson et al, 1996). It is unclear whether the discrepancy in the measured prevalence of HHV-8 in the general population is underestimated by the low sensitivity of most assays, or

Table 18.2 *Summary of prevalence of antibodies against lytic or latent HHV-8 antigens in KS risk groups versus the general population*
Butyrate and 12-0-tetradecanoylphorbol B-acetate (TPA) induce lytic replication of HHV-8 in the primary effusion lymphoma cell lines BC-1, BCBL-1 and BCP-1. The latent immuno-fluorescence assay (IFA) detects antibodies against LNA-1 (see Figure 18.2 and pp. 370–371).

Assay	Population	No. positive/total (percentage)	Reference
Assays to lytic antigens			
Butyrate-induced BC-1 western blot	US AIDS-KS	32/48 (67%)	Miller et al (1996)
	US HIV+ homosexual men	7/54 (13%)	
Butyrate-induced BC-1 IFA	US AIDS-KS	31/48 (65%)	Miller et al (1996)
	US HIV+ homosexual men	7/54 (13%)	
BCBL-1 TPA-induced IFA	US AIDS-KS, African KS	47/91 (52%)	Lennette et al (1996)
	US HIV+ homosexual men	28/28 (100%)	
	US HIV– controls	87/94 (93%)	
	African controls	112/825 (14%)	
		162/239 (68%)	
ORF65 ELISA/Western blot	US/UK AIDS-KS	46/57 (81%)	Simpson et al (1996)
	Uganda AIDS-KS	14/17 (82%)	
	Greek classic KS	17/18 (94%)	
	UK HIV+ homosexual men	5/16 (31%)	
	UK/US HIV– controls	10/382 (3%)	
	Ugandan controls	22/51 (43%)	
	Greek controls	3/26 (12%)	
Assays to latent antigens			
BCBL-1 latent IFA	US AIDS-KS	38/46 (83%)	Kedes et al (1996)
	US blood donors	2/141 (1%)	
	US STD clinic attenders	23/176 (13%)	
BCP-1 latent IFA	US/Italy/Uganda AIDS-KS	59/72 (82%)	Gao et al (1996)
	Classic/endemic KS	12/12 (100%)	
	US HIV+ homosexual men	12/40 (30%)	
	US blood donors	0/122 (0%)	
	Italian blood donors	4/107 (4%)	
	Ugandan HIV– controls	24/47 (51%)	
BCBL-1 latent IFA	US AIDS-KS	47/91 (52%)	Lennette et al (1996)
	African KS	28/28 (100%)	
	US HIV+ homosexual men	16/71 (23%)	
	US HIV– controls	0/825 (0%)	
	African controls	31/239 (13%)	
	Greek controls	3/26 (12%)	

whether the higher rate reported using lytic IFA reflects non-specific binding of cross-reactive antibodies in this system, although the authors tried to control for this (Lennette et al, 1996). Whether the background prevalence of HHV-8 is 2% or 20%, however, it is clearly much lower than in those with KS or at risk for KS, and therefore the seroepidemiology of HHV-8 reported so far supports a causal role for HHV-8 in KS.

Serological studies have also indicated that HHV-8 is sexually transmitted, since antibodies are more frequently detected in sexually transmitted disease (STD) clinic attendees (Kedes et al, 1996; Simpson et al, 1996). A recent study in a men's cohort has shown that HHV-8 infection is associated with number of homosexual partners and correlates with history of sexually transmitted disease and HIV infection (Martin et al, 1998). This study provides clear evidence that HHV-8 is sexually transmitted among homosexual men; however, the routes of transmission of HHV-8 may be different in homosexual men and in KS endemic regions. We have reported that in Italy, where the incidence of classic KS is significantly higher than in the UK or the USA (Geddes et al, 1994), the prevalence of antibodies to HHV-8 in blood donors is significantly higher than rates reported in the UK and the USA (Whitby et al, 1998a). The prevalence of antibodies is significantly higher in blood donors from the south of Italy, where the incidence of KS is highest, than in donors from the north. In addition, the geometric mean titre of anti-HHV-8 antibodies is highest in blood donors from the south, where the incidence of KS and the prevalence of HHV-8 is highest. A similar regional distribution of HHV-8 in Italy has been reported by others (Calabro et al, 1998), who also observed that seroprevalence of HHV-8 increased with age in Italian blood donors. Studies in Africa, where KS rates are high, have also shown that the prevalence of antibodies to HHV-8 is higher than in the UK or the USA (Gao et al, 1996; Simpson et al, 1996; Sitas et al, 1998) and that the prevalence of antibodies to HHV-8 increases steadily with age, even before puberty (Sitas et al, 1998; Whitby et al, 1998b). Such an age distribution of infection is not typical for an agent that is sexually transmitted. Martin et al (1998) did not observe any increase in prevalence of HHV-8 with age in their study population. A study of risk factors for HHV-8 in a London STD clinic has shown that, while HHV-8 is clearly associated with number of sexual partners, HIV infection and history of STD in homosexual men, this is not the case for heterosexuals where the only independent risk factor is country of birth (Whitby et al, 1998b). Thus the route and pattern of transmission of HHV-8 may be different in homosexual men in the USA and UK from that in KS endemic regions, such as Africa and the Mediterranean. Such differences also occur with other viruses such as hepatitis B virus which is transmitted horizontally in Africa but largely sexually or parenterally in the West. Serological evidence for mother-to-child transmission of HHV-8 in Africa has also been reported (Bourboulia et al, 1998; Whitby et al, 1998b). Early acquisition of HHV-8 in Africa is likely because KS occurs in African children (Zeigler and Katongole-Mbidde, 1996). The seroprevalence of HHV-8 is highest in populations with raised rates of KS, with the exception of West Africa, where the rate of KS is low but the prevalence of HHV-8 is high (Lennette et al, 1996; Ariyoshi et al, 1998). Interestingly, in The Gambia, when KS does occur it is mostly associated with HIV-1 rather than HIV-2 infection (Ariyoshi et al, 1998). The differences in KS rates between East and West Africa, which both have a high seroprevalence of HHV-8, affords a unique opportunity to study viral, host and environmental cofactors in the pathogenesis of KS.

HHV-8 and other diseases

Human herpesvirus 8 is found consistently in a rare form of AIDS-associated B cell lymphoma, primary effusion lymphoma (PEL), also known as body cavity-based lymphoma (BCBL), which is associated with KS (Chang et al, 1994; Cesarman et al, 1995; Ansari et al, 1996). It has also been reported by several groups to be present in Castleman's disease (CD), a lymphoproliferative disorder also frequently associated with KS (Dupin et al, 1995; Soulier et al, 1995; Luppi et al, 1996b). Detection of HHV-8 DNA has been reported in squamous cell skin carcinoma biopsies from four HIV-uninfected immunosuppressed transplant recipients (Rady et al, 1995). However, other groups have not been able to repeat these findings in larger series of skin cancer patients (Boshoff et al, 1996; Uthman et al, 1996).

Human herpesvirus 8 has been reported to be present in the bone marrow dendritic cells of multiple myeloma patients and in 25% of patients with monoclonal gammopathy of undetermined significance (MGUS), which is a precursor to myeloma (Rettig et al, 1997). The cytokine interleukin 6 (IL-6) is known both to stimulate myeloma growth and to prevent apoptosis of malignant plasma cells by paracrine mechanisms. Human herpesvirus 8 encodes a homologue of human IL-6 (Moore et al, 1996a; Neipel et al, 1997b) which has been shown to be transcribed in the myeloma bone marrow dendritic cells (Rettig et al, 1997). It was postulated that HHV-8 may be required for the transformation from MGUS to myeloma and may perpetuate the growth of malignant plasma cells by secretion of viral IL-6 (vIL-6) from infected bone marrow dendritic cells (Rettig et al, 1997). However, we and others have been unable to confirm an association between HHV-8 and multiple myeloma or MGUS (Mackenzie et al, 1997; Marcelin et al, 1997; Masood et al, 1997; Parravacini et al, 1997; Whitby et al, 1997; Tarte et al, 1998). Another report suggested an association between HHV-8 and sarcoidosis, based on PCR detection of HHV-8 in tissues from sarcoidosis patients from southern Italy (Di Alberti et al, 1997). The authors failed to point out that the patients in this study were from an area of high HHV-8 prevalence, and may not have controlled adequately for PCR contamination. We could not detect antibodies to HHV-8 in sera from UK sarcoidosis patients and others have also refuted the findings of this paper (Moore, 1998; Regamey et al, 1998).

THE HHV-8 GENOME

A preliminary analysis of the HHV-8 genome based on the sequence of 20·7 kb derived from an AIDS-KS genomic DNA library (Moore et al, 1996b) indicated that HHV-8 belongs to the rhadinovirus subgroup of gammaherpesviruses (see Figure 18.1). Rhadinoviruses share a common genomic structure. The linear double-stranded DNA (approximately 165 kb) has a central segment of 110–130 kb with a low GC content (L-DNA) that is flanked by long terminal repetitive DNA with high GC content (H-DNA) (Bornkamm et al, 1976). The complete nucleotide sequence of HHV-8 DNA has now been determined from a PEL cell line (BC-1) (Russo et al, 1996) and from a KS biopsy (Neipel et al, 1997a). Sequence analysis revealed that HHV-8 has the characteristic genome structure of rhadinoviruses with several 801-nucleotide H-DNA terminal repeats (84·5% GC) and 140·5 kb L-DNA (53·5% GC). The HHV-8 L-DNA long unique region (LUR) contains 81 open reading frames (ORFs), of which 66 have sequence similarity and are in colinear orientation with respect to the L-DNA genes of the oncogenic gammaherpesviruses, Epstein–Barr virus and herpesvirus saimiri (Figure 18.3) (Albrecht et al, 1992; Russo et al, 1996; Neipel et al, 1997c). A well-characterized New World-primate tumour virus (Fleckenstein and Desrosiers, 1982), HVS is not associated with any disease in its natural host, the squirrel monkey, but causes fulminant lymphoproliferation in other New World primates and has been shown to transform human T lymphocytes in vitro (Table 18.3).

Rose et al (1997) have provided DNA sequence evidence for two new herpesviruses closely related to HHV-8 from simian retroperitoneal fibromatosis tissues of two macaque species, *Macaca nemestrina* and *M. mulatta*. Retroperitoneal fibromatosis (RF) is a vascular fibroproliferative neoplasm which has many morphological and histological similarities to human KS. Like epidemic KS in AIDS patients, RF is highly associated with an immunodeficiency syndrome, simian acquired immunodeficiency syndrome (SAIDS), caused by a retrovirus infection, simian retrovirus type 2 (SRV-2). Previous experimental transmission studies and epidemiological data suggested that RF also has an infectious aetiology. These data suggest that the putative macaque gamma-2-herpesviruses are associated with KS-like neoplasms in different primate species (Rose et al, 1997) (Table 18.3). A second family of gamma-2-herpesviruses infecting rhesus monkeys has been reported, which does not appear to be related to disease and is not as closely related to HHV-8 (Desrosiers et al, 1997).

Human herpesvirus 8 has numerous open reading frames with striking homology to known cellular genes, which have probably been captured from the host cell during viral evolution. These include genes that code for proteins that regulate the cell cycle and proteins that interfere with the immune system (Table 18.4). HHV-8 lacks genes with homology to those implicated in EBV-induced cell immortalization and oncogenesis (*EBNA-1* and *-2*, and *LMP-1* and *-2*) and two genes implicated in HVS transformation of T lymphocytes (*StpC* and *Tip*) (Russo et al, 1996; Neipel et al, 1997b). However, HHV-8 encodes other genes whose products could play a role in cellular transformation and tumour induction. These include a viral cyclin D homologue (v-cyc) (Cesarman et al, 1996; Chang et al, 1996a; Godden-Kent et al, 1997; Li et al, 1997),

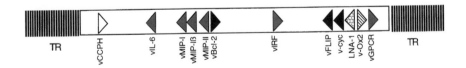

Figure 18.3 *Genomic organization of HHV-8 and* Herpesvirus saimiri
The rhadinovirus subgroup of gammaherpesviruses are characterized by a long unique region of DNA (LUR) between terminal repeats (TR). The equivalent positions of cell-homologous genes in the HHV-8 and HVS genomes are indicated by arrows. Core gene blocks conserved between herpesviruses lie between the unique cell-homologous genes. Genes involved in signalling or signal transduction mechanism are shown by grey arrows: GPCR (G protein-coupled receptor); IRF (interferon regulatory factor); vIL-6, vMIP-I, vMIP-II, vMIP-III. Genes involved in cell proliferation or apoptosis are shown by black arrows: v-cyc (D-type cyclin homologue); FLIP (FLICE inhibitory protein); Bcl-2. Genes interacting with the host immune response are shown by open arrows: CCPH (complement-controlled protein homologue); cellular adhesion molecule, Ox-2 homologue; LNA-1, latent nuclear antigen (ORF73); CD59.

Table 18.3 *Gammaherpesviruses and disease*

Gammaherpesvirus	Host	Associated tumours
Human herpesvirus 8 (HHV-8)	Human	Kaposi's sarcoma Primary effusion lymphoma Castleman's disease
Epstein–Barr virus (EBV)	Human	Burkitt's lymphoma Nasopharyngeal carcinoma Hodgkin's disease Post-transplant lymphoproliferations
Herpesvirus saimiri (HVS)	New World primates	T cell lymphoma Lymphosarcomas
Retroperitoneal fibrosis herpesvirus (RFHV)	Old World primates	Retroperitoneal fibrosis
Rhesus monkey rhadinovirus (RRV)	Old World primates	None
Murine herpesvirus 68	Mouse	Lymphoproliferation

Table 18.4 *Homology of HHV-8-encoded proteins with cellular genes*

Host cell homologue	HHV-8-encoded protein	Possible function
D-type cyclin	v-cyc	Inactivation of pRB Promotes G1 to S phase transition
IL-8 GPCR	vGPCR	Cellular growth signal
Interferon regulatory factor	vIRF	Inhibits p21 and MHC class I expression
CC chemokines	vMIP-I, vMIP-II, vMIP-1	Chemoattraction Angiogenesis
IL-6	vIL-6	Growth factor for KS cells
Bcl-2 family protein	vBcl-2	Inhibition of apoptosis
FLICE inhibitory protein	vFLIP	Inhibition of CD95L and TNF-induced apoptosis
N-CAM family protein	vOx-2	Cellular adhesion molecule
CD21/CR2 complement binding protein	Orf4	Escape from host immune response

a Bcl-2 homologue (Cheng et al, 1997; Sarid et al, 1997), a G protein-coupled receptor (GPCR) homologue (Cesarman et al, 1996; Arvanitakis et al, 1997; Guo et al, 1997) and *K1*, a gene which is at an equivalent position to HVS STP (Lee et al, 1998a). Additional genes have been identified that encode products that may also be important in pathogenesis, including a viral FLICE inhibitor (vFLIP) (Bertin et al, 1997; Thome et al, 1997), an interferon regulatory factor (vIRF) (Gao et al, 1997), an IL-6 homologue (vIL-6)

(Moore et al, 1996a; Neipel et al, 1997a) and three chemokines, macrophage inhibitory protein homologues (vMIP-I, vMIP-II and vMIP-1) (Moore et al, 1996b; Neipel et al, 1997c).

HHV-8 genes involved in cellular proliferation

Several oncogenic viruses interact with their host cells by providing an additional growth stimulus, which extends the proliferative capacity of the virus. Cell cycle control can be deregulated by viral oncogenes by interference with receptor-mediated signal transduction pathways and the function of nuclear cell cycle regulatory proteins. Several DNA viruses, for example, encode proteins that specifically target and inhibit both the retinoblastoma protein (pRB) and the p53 tumour suppressor pathways involved in cell cycle regulation and apoptosis respectively (Jansen-Durr, 1996). One such pathway by which pRB induces growth arrest is the repression of E2F-bound promoters. It was recently reported that the histone deacetylase HDAC1 forms a complex with pRB to repress E2F function. Viral oncogenic proteins such as human papillomavirus E7 disrupt the interaction between pRB and HDAC1 (Brehm et al, 1998; Magnaghi-Jaulin et al, 1998). It is as yet unknown whether any HHV-8 or EBV proteins also target this complex.

Viral proteins that target growth-suppressive signalling pathways controlling cell cycle progression in untransformed cells promote progression into the S phase of the cell cycle (DNA synthesis), which is probably necessary for efficient replication of the viral genome (Jansen-Durr, 1996). A side-effect of deregulated progression through the cell cycle may be unchecked cellular proliferation and ultimately transformation.

Cell proliferation and cell cycle progression are partly regulated by cellular cyclins, which are components of cellular kinases (Peters, 1994; Sherr, 1995). Expression of cyclins D and E in tissue culture cells accelerates transit through the checkpoint in the G1 phase of the cell cycle into S phase (Resnitzky et al, 1994; Sherr, 1994; Wang et al, 1994). The aberrant expression of cellular D type cyclins is strongly implicated in various human cancers (Bodrug et al, 1994; Hunter and Pines, 1994; Lovec et al, 1994; Peters, 1994; Bates and Peters, 1995; Sherr, 1995; Jacks and Weinberg, 1996). A hallmark of cyclins is their ability to associate with and stimulate cellular kinase subunits (cyclin-dependent kinases, CDK) (Morgan, 1995). When active, certain of these kinases phosphorylate and inactivate cell cycle checkpoint molecules and thereby facilitate progression of cells through the cell cycle. The HHV-8-encoded cyclin gene (v-cyc; ORF72) has 27% sequence identity with cellular D type cyclins and, when ectopically expressed in cells, promotes phosphorylation of the retinoblastoma protein (pRB) (primarily through activation of CDK6) and can overcome a cell cycle checkpoint imposed by this tumour suppresser (Chang et al, 1996a, Godden-Kent et al, 1997; Li et al, 1997). These data suggest that the HHV-8-encoded cyclin protein may act to stimulate cellular CDKs and, in ways analogous to cell cycle-regulating cellular cyclins, promote cell cycle progression.

The closely related rhadinovirus HVS also encodes a v-cyclin (Jung et al, 1994) which has been shown to phosphorylate pRB, also through the activation of CDK6, but there is no cyclin homologue encoded by EBV. However, EBV *EBNA-2* and *EBNA-LP* genes have been shown to induce the expression of cellular D type cyclins in infected lymphocytes (Sinclair et al, 1994). These two human herpesviruses, EBV and HHV-8, have therefore evolved different mechanisms for modifying the same signal transduction pathway.

Human herpesvirus 8 encodes a G protein-coupled receptor (vGPCR; ORF75) that is homologous to a GPCR encoded by HVS and to human interleukin 8 (IL-8) receptors (Cesarman et al, 1996; Arvanitakis et al, 1997; Guo et al, 1997). Although vGPCR has an amino acid sequence that is homologus to IL-8 receptors, binds IL-8 with high affinity, and activates the same signal transduction cascade, HHV-8 GPCR is constitutively active (i.e. it does not require activation by an agonist) and couples to a wider variety of G proteins than do IL-8 receptors (Arvanitakis et al, 1997). Constitutive signalling occurs with mutated cellular GPCRs in many human diseases and in tumours (Julius et al, 1989; Coughlin, 1994; Milano et al, 1994; Alblas et al, 1996). Ectopic expression of vGPCR in rat fibroblasts in tissue culture leads to enhanced proliferation, supporting a role for this receptor in tumorigenesis (Arvanitakis et al, 1997). There are two potential mechanisms whereby vGPCR could act in the development of HHV-8-associated diseases: directly with constitutive signalling of vGPCR leading to altered growth and neoplastic transformation; and indirectly, whereby signalling by vGPCR induces the activation of inflammatory and angiogenic growth factors potentially involved in the pathogenesis of KS (Arvanitakis et al, 1997; Bais et al, 1998; Boshoff, 1998).

The *K1* gene of HHV-8 occupies an equivalent position in the HHV-8 genome to HVS *STP*, a gene that is required for transformation of cells by that virus, though it has no sequence homology to *STP*. The *K1* gene has been shown to transform rat fibroblasts (Lee et al, 1998a) and recombinant HVS containing *K1* instead of *STP* was shown to immortalize primary T lymphocytes of common marmosets in culture and induce lymphoma in a common marmoset (Lee et al, 1998a). K1 is a transmembrane protein (Lagunoff and Ganem, 1997) with a distinctive immunoreceptor tyrosine-based activation signalling motif (ITAM) in its cytoplasmic tail (Lee et al, 1998b; Zong et al, 1998), which is involved in signal transduction via the SH2 domain of syk (Lee et al, 1998b). The *K1* gene is also the most variable gene in the HHV-8 genome, which may differ by up to 30%. Four distinct groups and 10 clades have been reported (Zong et al, 1998).

Host cell responses to viral infection include shutting down the cell cycle (see above), induction of apoptosis (Hinshaw et al, 1994; Vaux et al, 1994; Shen and Shenk, 1995, 1996; Dittmer and Mocarski, 1997) (see Apoptosis), and enhancement of immune responses through upregulation of major histocompatibility complex (MHC) antigens (Biron, 1994; Bot et al, 1996). Many viruses have evolved specific proteins that can defeat these host defences (Gooding, 1992, 1994; Jagus and Gray, 1994; Morris et al, 1995). Herpesviruses such as cytomegalovirus and herpes simplex virus 1 (HSV-1) inhibit cell-mediated antiviral immunity by interfering with the expression of MHC antigens (Browne et al, 1990; Beersma et al, 1993; Wiertz et al, 1996). HHV-8 also encodes a 449 amino acid protein with homology to the interferon regulatory factor family of proteins (vIRF; ORF K9), and has been shown to inhibit interferon-γ (IFN-γ) signal transduction (Gao et al, 1997). Interferon has various cellular effects including the inhibition of cellular proliferation (Kirchhoff et al, 1993, 1995; Vaughan et al, 1995), induction of apoptosis (Tanaka et al, 1994; Tamura et al, 1996) and upregulation of MHC antigens (Sen and Ransohoff, 1993; Kamijo et al, 1994; Kimura et al, 1994; Martin et al, 1994). It is possible that vIRF may block binding of cellular IRF1 to IFN-stimulated response elements (ISRE) in responsive promoter sites, thereby inhibiting induction of IFN-inducible genes such as MHC class I antigen and the cyclin-dependent kinase inhibitor p21$^{WAF/CIP1}$ (Gao et al, 1997). Inhibition of the expression of these cellular proteins may contribute to viral escape from immune surveillance or cell cycle shutdown which is initiated as an antiviral response. A consequence of this survival mechanism may be cellular transformation, although it seems likely that HHV-8 induced tumorigenesis probably involves several oncogenes (e.g. v-cyc, vGPCR).

Apoptosis

Host cells typically respond to viral infection by the induction of apoptosis. Deregulation of the host cell cycle mechanisms (discussed above) by viral proteins leads to the upregulation of the tumour suppressor p53, which then activates genes encoding for apoptosis-mediating proteins such as Bax, Bik, and other proteins of this family (Sato et al, 1994; Boyd et al, 1995). The p53 protein may also upregulate expression of death receptors such as CD95 (Fas, Apo-1) and its ligand CD95L (FasL) which respond to signals from cytotoxic T lymphocytes or via an autocrine mechanism induced by soluble CD95L produced by the infected cell, by signalling to the cell to apoptose (Thome et al, 1997). Similar to EBV and HVS, HHV-8 encodes a gene (ORF16) with homology to cellular Bcl-2 (Cheng et al, 1997; Sarid et al, 1997). The heterodimerization of cellular Bcl-2 with Bax is probably important in preventing Bax-mediated apoptosis (Sato et al, 1994). The homology between HHV-8 ORF16 and members of the Bcl-2 family suggests that it prolongs the life span of persistently infected cells. Deregulated Bcl-2 expression has been shown to occur in human cancers, such as follicular lymphoma (Cleary et al, 1986; Tsujimoto et al, 1987; Korsmeyer, 1992), and therefore ectopic expression of vBcl-2 may contribute to tumorigenesis through its antiapoptotic effect.

A new family of viral apoptosis inhibitors (viral FLICE inhibitory proteins, vFLIPs) has been identified which interfere with apoptosis signalled through death receptors. Several gammaherpesviruses (including HHV-8; ORF K13) encode vFLIPs, as well as the oncogenic human molluscipoxvirus (Bertin et al, 1997; Thome et al, 1997). The vFLIPs contain two death-effector domains which interact with the adapter protein FADD (Fas-associating protein with death domain) (Boldin et al, 1995; Chinnaiyan et al, 1996), and inhibit the recruitment and activation of the protease FLICE (FADD-like interleukin 1β-converting enzyme) (Boldin et al, 1996; Muzio et al, 1996) by the CD95 death receptor (Nagata, 1997). Recruitment of FLICE by the CD95 receptor in response to the binding of its ligand (CD95L) leads to the assembly of a receptor-associated death-inducing signalling complex (DISC) (Kischkel et al, 1995). The DISC-associated FLICE subsequently initiates proteolytic activation of other ICE protease family members, which in turn leads to apoptosis (Boldin et al, 1996; Muzio et al, 1996). Cells expressing vFLIP are protected against apoptosis induced by CD95 or by TNF-R1 (Bertin et al, 1997; Thome et al, 1997). The HVS FLIP is detected late during the lytic replication cycle, suggesting that protection of virus-infected cells from death receptor-induced apoptosis may lead to higher virus production and contribute to the persistence and oncogenicity of FLIP-encoding viruses. It has recently been shown that the lack of responsiveness to CD95L-mediated apoptosis due to downregulation of CD95 may contribute to the development of certain tumours (e.g. melanomas and hepatomas) (Hahne et al, 1996; Strand et al, 1996). Therefore, it is possible that vFLIPs may not only promote viral spread and persistence, but also contribute to the transforming capacity of herpesviruses such as HHV-8.

The gammaherpesviruses which encode a vFLIP also encode Bcl-2 homologues. The antiapoptotic Bcl-2 family members block cell death induced by growth factor depriva-tion, gamma-irradiation and cytotoxic drugs (Yang and Korsmeyer, 1996; Huang et al, 1997). However, these proteins have a less potent effect on CD95-mediated apoptosis of lymphoid cell lines, in contrast to the vFLIPs. Some viruses may therefore take advan-tage of two complementary antiapoptotic functions provided by a Bcl-2 homologue and by a vFLIP.

Chemokines and cytokines

Human interleukin 6 (hIL-6) has long been suspected to be involved in the pathogene-sis of KS. The expression of hIL-6 is raised in KS lesions, and cultured KS cells have been reported to respond to recombinant hIL-6 with increased growth (Miles, 1994; Scala et al, 1994). Human herpesvirus 8 encodes a homologue of hIL-6, viral IL-6 (vIL-6), which has 24·7% amino acid identity, significant conservation of amino acid sequence in the receptor-binding domain, and has been shown to substitute for hIL-6 in an in vivo functional assay (inhibition of apoptosis in the mouse hybridoma cell line B9) (Moore et al, 1996a; Neipel et al, 1997b). Viral IL-6 has been shown to activate the JAK/STAT signalling pathway via the gp130 subunit of the IL-6 receptor, and does not require the alpha subunit of this receptor (Molden et al, 1997). Although vIL-6 has only been shown to be expressed in a limited number of KS lesions (Moore et al, 1996a), a possible role for vIL-6 in KS pathogenesis is supported by the fact that KS spindle cells express the high-affinity IL-6 receptor in vivo (Miles et al, 1990). Human IL-6 has also been shown to be a growth factor for two other HHV-8-associated diseases, primary effusion lymphomas (PEL) and multicentric Castleman's disease.

Two virus proteins similar to two human macrophage inflammatory protein (MIP-1α) chemokines, and a third protein similar to both MIP-1 and macrophage chemoattrac-tant protein (MCP) chemokines, have been identified in the HHV-8 genome (ORFs K4, K6 and K4·1) (Moore et al, 1996a; Neipel et al, 1997b). They show some 30–40% amino acid identity to the equivalent cellular CC chemokines. The MIP/RANTES family of CC chemokines have been shown to block entry and replication of non-syncytium-inducing (NSI) HIV-1 strains by binding to the CCR5 chemokine receptor, which also serves as an HIV-1 coreceptor on monocytes (Cocchi et al, 1995; Deng et al, 1996; Dragic et al, 1996). In a similar functional assay, vMIP-I was shown to inhibit the entry of HIV-1 strains SF162 and M23 into CCR5-expressing cells, suggesting that vMIP-I is able to bind CCR5 and may contribute to interactions between HHV-8 and HIV-1 (Moore et al, 1996a). Viral MIP-I and vMIP-II have since been shown to be able to bind to both CC and CXC receptors (Boshoff et al, 1997; Kledal et al, 1997). The viral chemokines have been shown to induce angiogenesis, unlike the human chemokines MIP-1α or RANTES, and thus may also be involved directly in the pathogenesis of KS (Boshoff et al, 1997).

Latency-associated antigens encoded by HHV-8

Cell lines latently infected with HHV-8 have been established from PEL (Cesarman et al, 1995; Renne et al, 1996). Patients' sera have been tested in an indirect immunofluores-cence assay (IFA) against such PEL cells (Gao et al, 1996; Kedes et al, 1996; Lennette et al, 1996; Simpson et al, 1996), with positive sera showing a distinct nuclear localizing stippling pattern (see Figure 18.2). The nuclear antigens being detected in such assays are formally analogous to the EBNAs of EBV (Kedes et al, 1996), which show similar nuclear localizing patterns with IFA. We and others have employed this IFA to detect antibodies to latently HHV-8-infected PEL cells in sera of nearly 95% of patients with KS (classical and AIDS-associated), in up to 50% of Ugandans without HIV, in 30% of HIV-positive homosexual men without KS, and at very low prevalence in HIV-positive patients with haemophilia and in healthy blood donors in the UK or USA (Gao et al, 1996; Kedes et al, 1996; Lennette et al, 1996; Simpson et al, 1996). Although such studies indicate that antibodies against latent HHV-8 proteins are frequently present in patients with or at risk of KS, this first-generation serological assay probably underestimates the true prevalence of HHV-8 in the general population (Lennette et al, 1996; Rickinson, 1996).

The identification and cloning of the latent nuclear antigen (LNA-1) recognized by KS patients' sera in PEL cell lines have been described (Kellam et al, 1997; Kedes et al, 1997; Rainbow et al, 1997). These data showed that HHV-8 ORF73 encodes the latently controlled immunoreactive nuclear antigen which is analogous to the EBV nuclear antigens (EBNAs). The LNA-1 protein has many features reminiscent of several EBNAs; it is a very hydrophilic (38% charged residues) and proline-rich protein, with an extensive repetitive domain and a leucine zipper motif. Potential nuclear localization signals consisting of runs of basic amino acids also occur in the C-terminal domain of ORF73. Epstein–Barr virus establishes a stable, latent infection of primary B lymphocytes (Klein, 1994). At least 11 EBV genes are expressed in latent infection (*EBER 1* and *2*, *EBNAs 1–6*, and *LMP 1, 2A, 2B*), although only a subset of these are expressed in most EBV latently infected B cell lines (Liebowitz and Kieff, 1993). The EBNA and LMP proteins are required for B cell growth transformation (Klein, 1994). Each of the EBNAs has been shown by immune microscopy to give a distinctive nuclear staining pattern (Liebowitz and Kieff, 1993) similar to that observed with BCP-1 IFA or ORF73 transfected into 293 cells (Kellam et al, 1997).

It is not known whether LNA-1, by analogy to EBV latent nuclear proteins, plays a role in cellular transformation. Future work will establish whether LNA-1 transactivates other viral or cellular proteins and also whether variations in the repetitive sequence of ORF73 are suitable for subtyping of HHV-8.

Molecular mimicry of human genes

Molecular mimicry, the piracy and incorporation of host cell genes into the viral genome, is a newly recognized characteristic of some DNA viruses, particularly herpesviruses and poxviruses (Moore et al, 1996b). As discussed above, HHV-8 encodes homologues of human cyclin D (ORF72), an IL-8-like GPCR (ORF74), three chemokine homologues, vMIP-I, vMIP-II and vMIP-I (ORFs K6, K4, K4·1), a homologue of IL-6, a gene with homology to interferon regulatory factors (vIRF; ORF K9), and a Bcl-2 homologue (ORF16). In addition, HHV-8 also encodes genes similar to the complement-binding proteins CD21/CR2 (ORF4), and a neural cell adhesion molecule (N-CAM)like adhesion protein (ORF K14) (Russo et al, 1996). It has been noted that there is a significant correlation between the cell-homologous genes encoded by HHV-8 and the cellular genes induced by EBV (the closest known relative of HHV-8 in humans;

see Figure 18.1) (Moore et al, 1996b; Neipel et al, 1997a; Russo et al, 1996). The EBV EBNA and LMP proteins (homologous genes are absent in HHV-8) are largely responsible for the induction of hIL-6 (Tanner et al, 1996), cyclin D (Sinclair et al, 1994), the IL-8-like receptor EBI1 (Birkenbach et al, 1993; Burgstahler et al, 1995), cellular Bcl-2 (Henderson et al, 1991), adhesion molecules, and the complement controlling protein CR-2 (Larcher et al, 1995). Therefore, it is apparent that HHV-8 and EBV have developed different strategies to attain the same objective: to overcome cell cycle arrest, apoptosis and activation of cellular immunity, which are typical host cell responses to viral infection. It has been proposed that these convergent strategies utilized by both HHV-8 and EBV contribute to the malignant transformation of B cells in humans.

IN VITRO CULTURE OF HHV-8

Although HHV-8 is present in all KS lesions, cell cultures and cell lines derived from KS lesions lose detectable HHV-8 after a few passages. The reasons for this are unclear at present. It is possible that the HHV-8-infected spindle cells do not grow in culture, or that the culture systems employed result in HHV-8 being lost from cultured cells. In contrast, several cell lines have been established from PEL cells, which have been shown to be latently infected with a high number of HHV-8 genome copies per cell (Cesarman et al, 1995; Renne et al, 1996). Lytic replication of HHV-8 can be induced in these cell lines using phorbol esters. Primary effusion lymphoma cell lines have been invaluable tools in studying latent and lytic infection in vitro and have also formed the basis of most serological assays (see above).

   Isolation and propagation of HHV-8 in the human embryonal kidney cell line 293 has been described (Foreman et al, 1997), but others have failed to repeat these findings (Renne et al, 1998). Culture in vitro of HHV-8 in primary dermal microvascular endothelial cells (Panyutich et al, 1998) and primary bone marrow endothelial cells (Flore et al, 1998) has also been described, and this may prove a more fruitful development.

CONCLUSION

Seroepidemiology and PCR studies strongly suggest that HHV-8 is the cause of Kaposi's sarcoma and primary effusion lymphoma, and is also associated with multicentric Castleman's disease. Although HHV-8 has not yet been shown to be a transforming virus in vitro, the virus encodes several proto-oncogenes capable of the deregulation of cell cycle control (v-cyc, vGPCR, vIRF, K1), inhibition of apoptosis (vBcl-2, vFLIP) and control of growth differentiation (vIL-6, vMIP-I, vMIP-II, vMIP-I), which could contribute to tumour formation.

REFERENCES

Alblas J, Van Etten I & Moolenaar WH (1996) Truncated, desensitization-defective neurokinin receptors mediate sustained MAP kinase activation, cell growth and transformation by a Ras-independent mechanism. *EMBO J* **15:** 3351–3360.

Albrecht JC, Nicholas J, Biller D et al (1992) Primary structure of the herpesvirus saimiri genome. *J Virol* **66:** 5047–5058.

Ambroziak JA, Blackbourn DJ, Herndier BG et al (1995) Herpes-like sequences in HIV-infected and uninfected Kaposi's sarcoma patients. *Science* **268:** 582–583.

Ansari MQ, Dawson DB, Nador R et al (1996) Primary body cavity-based AIDS-related lymphomas. *Am J Clin Pathol* **105:** 221–229.

Ariyoshi K, van der Loeff M, Cook P et al (1998) Kaposi's sarcoma in the Gambia, West Africa is less frequent in human immunodeficiency virus type 2 than in human immunodeficiency virus type 1 infection despite a high prevalence of human herpesvirus 8. *J Hum Virol* **1:** 192–199.

Arvanitakis L, GerasRaaka E, Varma A, Gershengorn MC & Cesarman E (1997) Human herpesvirus KSHV encodes a constitutively active G-protein-coupled receptor linked to cell proliferation. *Nature* **385:** 347–349.

Bais C, Santomasso B, Coso O et al (1998) Kaposi's sarcoma-associated herpesvirus (KSHV/HHV-8) G protein-coupled receptor is a viral oncogene and angiogenesis activator. *Nature* **391:** 86–89.

Barillari G, Buonaguro L, Fiorelli V et al (1992) Effects of cytokines from activated immune cells on vascular cell growth and HIV-1 gene expression. Implications for AIDS–Kaposi's sarcoma pathogenesis. *J Immunol* **149:** 3727–3734.

Bates S & Peters G (1995) Cyclin D1 as a cellular proto-oncogene. *Semin Cancer Biol* **6:** 73–82.

Beckstead JH, Wood GS & Fletcher V (1985) Evidence for the origin of Kaposi's sarcoma from lymphatic endothelium. *Am J Pathol* **119:** 294–300.

Beersma MF, Bijlmakers MJ & Ploegh HL (1993) Human cytomegalovirus down-regulates HLA class I expression by reducing the stability of class I H chains. *J Immunol* **151:** 4455–4464.

Beral V (1991) Epidemiology of Kaposi's sarcoma. In: Beral V, Jaffe HW & Weiss RA (eds) *Cancer, HIV and AIDS*, pp. 5–22. New York: Cold Spring Harbor Laboratory Press.

Beral V, Peterman TA, Berkelman RL & Jaffe HW (1990b) Kaposi's sarcoma among persons with AIDS: a sexually transmitted infection? *Lancet* **335:** 123–128.

Bertin J, Armstrong RC, Ottilie S et al (1997) Death effector domain-containing herpesvirus and poxvirus proteins inhibit both Fas- and TNFR1-induced apoptosis. *Proc Natl Acad Sci USA* **94:** 1172–1176.

Bigoni B, Dolcetti R, de Lellis L et al (1996) Human herpesvirus 8 is present in the lymphoid system of healthy persons and can reactivate in the course of AIDS. *J Infect Dis* **173:** 542–549.

Birkenbach M, Josefsen K, Yalamanchili R, Lenoir G & Kieff, E (1993) Epstein–Barr virus-induced genes: first lymphocyte-specific G protein-coupled peptide receptors. *J Virol* **67:** 2209–2220.

Biron CA (1994) Cytokines in the generation of immune responses to, and resolution of, virus infection. *Curr Opin Immunol* **6:** 530–8.

Bodrug SE, Warner BJ, Bath ML, Lindeman GJ, Harris AW & Adams JM (1994) Cyclin D1 transgene impedes lymphocyte maturation and collaborates in lymphomagenesis with the *myc* gene. *EMBO J* **13:** 2124–2130.

Boldin MP, Varfolomeev EE, Pancer Z, Mett IL, Camonis JH & Wallach D (1995) A novel protein that interacts with the death domain of Fas/APO1 contains a sequence motif related to the death domain. *J Biol Chem* **270:** 7795–7798.

Boldin MP, Goncharov TM, Goltsev YV & Wallach D (1996) Involvement of MACH, a novel MORT1/FADD-interacting protease, in Fas/APO-1- and TNF receptor-induced cell death. *Cell* **85:** 803–815.

Bornkamm GW, Delius H, Fleckenstein B, Werner FJ & Mulder C (1976) Structure of herpesvirus saimiri genomes: arrangement of heavy and light sequences in the M genome. *J Virol* **19:** 154–161.

Boshoff C, Schulz TF, Kennedy MM et al (1995a) Kaposi's sarcoma-associated herpesvirus infects endothelial and spindle cells. *Nature Med* **1:** 1274–1278.

Boshoff C, Whitby D, Hatziioannou T et al (1995b) Kaposi's-sarcoma-associated herpesvirus in HIV-negative Kaposi's sarcoma. *Lancet* **345:** 1043–1044.

Boshoff C, Talbot S, Kennedy M, O'Leary J, Schulz T & Chang Y (1996) HHV8 and skin cancers in immunosuppressed patients (letter) [published erratum *Lancet* (1996) 348(9020):138]. *Lancet* **347:** 338–339.

Boshoff C, Endo Y, Collins PD et al (1997) Angiogenic and HIV inhibitory functions of KSHV-encoded chemokines. *Science* **278:** 290–293.

Boshoff C (1998) Kaposi's sarcoma. Coupling herpesvirus to angiogenesis *Nature* **391(6662):** 86–89.

Bot A, Reichlin A, Isobe H et al (1996) Cellular mechanisms involved in protection and recovery from influenza virus infection in immunodeficient mice. *J Virol* **70:** 5668–5672.

Bourboulia D, Whitby D, Boshoff C et al (1998) Serological evidence for mother to child transmission of Kaposi's sarcoma-associated herpesvirus in healthy South African children. *JAMA* **280(1):** 31–33.

Boyd JM, Gallo GJ, Elangovan B et al (1995) Bik, a novel death-inducing protein shares a distinct sequence motif with Bcl-2 family proteins and interacts with viral and cellular survival-promoting proteins. *Oncogene* **11:** 1921–1928.

Brehm A, Miska EA, McCance DJ, Reid JL, Bannister AJ & Kourzarides T (1998) Retinoblastoma protein recruits histone deacetylase to repress transcription. *Nature* **391:** 597–601.

Browne H, Smith G, Beck S & Minson T (1990) A complex between the MHC class I homologue encoded by human cytomegalovirus and beta 2 microglobulin. *Nature* **347:** 770–772.

Buonaguro FM, Tornesello ML, Beth-Giraldo E et al (1996) Herpesvirus-like DNA sequences detected in endemic, classic, iatrogenic and epidemic Kaposi's sarcoma biopsies. *Int J Cancer* **65:** 25–28.

Burgstahler R, Kempkes B, Steube K & Lipp M (1995) Expression of the chemokine receptor BLR2/EBI1 is specifically transactivated by Epstein–Barr virus nuclear antigen 2. *Biochem Biophys Res Commun* **215:** 737–743.

Calabro L, Sheldon J, Favero A et al (1998) Seroprevalence of Kaposi's sarcoma-associated herpesvirus/human herpesvirus 8 in several regions of Italy. *J Hum Virol* **1:** 207–213.

Cathomas G, McGandy CE, Terracciano LM, Itin PH, de Rosa G & Gudat F (1996) Detection of herpesvirus-like DNA by nested PCR on archival skin biopsy specimens of various forms of Kaposi's sarcoma. *J Clin Pathol* **49:** 631–633.

Cesarman E, Moore PS, Rao PH, Inghirami G, Knowles DM & Chang Y (1995) In vitro establishment and characterisation of two acquired immunodeficiency syndrome-related lymphoma cell lines (BC-1 and BC-2) containing Kaposi's sarcoma-associated herpesvirus-like (KSHV) DNA sequences. *Blood* **86:** 2708–2714.

Cesarman E, Nador RG, Bai F et al (1996) Kaposi's sarcoma-associated herpesvirus contains G protein-coupled receptor and cyclin D homologs which are expressed in Kaposi's sarcoma and malignant lymphoma. *J Virol* **70:** 8218–8223.

Chang Y, Cesarman E, Pessin MS et al (1994) Identification of herpesvirus-like DNA sequences in AIDS-associated Kaposi's sarcoma. *Science* **266:** 1865–1869.

Chang Y, Moore PS, Talbot SJ et al (1996a) Cyclin encoded by KS herpesvirus (letter). *Nature* **382:** 410.

Chang Y, Ziegler J, Wabinga H et al (1996b) Kaposi's sarcoma-associated herpesvirus and Kaposi's sarcoma in Africa. *Arch Intern Med* **156:** 202–204.

Cheng EHY, Nicholas J, Bellows DS et al (1997) A Bcl-2 homolog encoded by Kaposi sarcoma-associated virus, human herpesvirus 8, inhibits apoptosis but does not heterodimerize with Bax or Bak. *Proc Natl Acad Sci USA* **94:** 690–694.

Chinnaiyan AM, Tepper CG, Seldin MF et al (1996) FADD/MORT1 is a common mediator of CD95 (Fas/APO-1) and tumor necrosis factor receptor-induced apoptosis. *J Biol Chem* **271:** 4961–4965.

Chuck S, Grant RM, Katongole-Mbidde E, Conant M & Ganem D (1996) Frequent presence of a novel herpesvirus genome in lesions of HIV negative Kaposi's sarcoma. *J Infect Dis* **173:** 248–251.

Cleary ML, Smith SD & Sklar J (1986) Cloning and structural analysis of cDNAs for bcl-2 and a hybrid bcl-2/immunoglobulin transcript resulting from the t(14;18) translocation. *Cell* **47:** 19–28.

Cocchi F, DeVico AL, GarzinoDemo A, Arya SK, Gallo RC, & Lusso P (1995) Identification of RANTES, MIP-1 alpha and MIP beta as the major HIV-suppressive factors produced by CD8+ T cells. *Science* **270:** 1811–1815.

Collandre H, Ferris S, Grau O, Montagnier L & Blanchard A (1995) Kaposi's sarcoma and new herpesvirus [1]. *Lancet* **345:** 1043.

Corbellino M, Bestetti G, Poirel L et al (1996) Is human herpesvirus type 8 fairly prevalent among healthy subjects in Italy? *J Infect Dis* **174:** 668–670.

Coughlin SR (1994) Expanding horizons for receptors coupled to G proteins: diversity and disease. *Curr Opin Cell Biol* **6:** 191–197.

Deng H, Liu R, Ellmeier W et al (1996) Identification of a major co-receptor for primary isolates of HIV-1. *Nature* **381:** 661–666.

Desrosiers RC, Sasseville VG, Czajak SC et al (1997) A herpesvirus of rhesus monkeys related to the human Kaposi's sarcoma-associated herpesvirus. *J Virol* **71:** 9764–9769.

Di Alberti L, Piattelli A, Arteses L et al (1997) Human herpesvirus 8 variants in sarcoid tissues. *Lancet* **350:** 1655–1661.

Dictor M, Rambech E, Way DM & Bendsoe N (1996) HHV-8 DNA in Kaposi's sarcoma lesions, AIDS Kaposi's sarcoma cell lines, endothelial Kaposi's sarcoma simulators and the skin of immunosuppressed patients. *Am J Pathol* **148:** 2009–2016.

Dittmer D & Mocarski ES (1997) Human cytomegalovirus infection inhibits G1/S transition. *J Virol* **71:** 1629–1634.

Dragic T, Litwin V, Allaway GP et al (1996) HIV-1 entry into CD4+ cells is mediated by the chemokine receptor CC-CKR-5. *Nature* **381:** 667–673.

Dupin N, Gorin I, Deleuze J, Agut H, Huraux JM & Escande JP (1995) Herpes-like DNA sequences, AIDS-related tumors, and Castleman's disease (letter). *N Engl J Med* **333:** 798–799.

Ensoli B, Barillari G, Salahuddin SZ, Gallo RC & Wong SF (1990) Tat protein of HIV-1 stimulates growth of cells derived from Kaposi's sarcoma lesions of AIDS patients. *Nature* **345:** 84–86.

Ensoli B, Gendelman R, Markham P et al (1994) Synergy between basic fibroblast growth factor and HIV-1 Tat protein in induction of Kaposi's sarcoma *Nature* **371:** 674–680.

Farge D, Herve R, Mikol J et al (1994) Simultaneous progressive multifocal leukoencephalopathy, Epstein–Barr virus (EBV) latent infection and cerebral parenchymal infiltration during chronic lymphocytic leukemia. *Leukemia* **8:** 318–321.

Flore O, Shahin R, Scott E et al (1998) Transformation of primary human endothelial cells by Kaposi's sarcoma-associated herpesvirus. *Nature* **394:** 588–591.

Fleckenstein B & Desrosiers RC (1982) *Herpesvirus saimiri* and *herpesvirus ateles*. In: Roizman B (ed.) *The Herpesviruses*, pp 253–331. New York: Plenum.

Foreman KE, Friborg J, Kong WP et al (1997) Propagation of a human herpesvirus from AIDS-associated Kaposi's sarcoma. *N Engl J Med* **336:** 163–171.

Friedman Kien AE, Saltzman BR, Cao YZ et al (1990) Kaposi's sarcoma in HIV-negative homosexual men (letter). *Lancet* **335:** 168–169.

Gaidano G, Pastore C, Gloghini A et al (1996) Distribution of HHV-8 sequences throughout the spectrum of AIDS-related neoplasia. *AIDS* **10:** 941–949.

Gao SJ, Kingsley L, Li M et al (1996) KSHV antibodies among Americans, Italians and Ugandans with and without Kaposi's sarcoma. *Nature Med* **2:** 925–928.

Gao SJ, Boshoff C, Weiss RA, Chang Y & Moore PS (1997) KSHV vIRF is an oncogene that inhibits the interferon signalling pathway. *Oncogene* **15**: 1979–1985.

Geddes M, Franchesci S, Barchielli A et al (1994) Kaposi's sarcoma in Italy before and after the AIDS epidemic. *Br J Cancer* **69**: 333–336

Giraldo G, Beth E & Kyalwazi SK (1984) Role of cytomegalovirus in Kaposi's sarcoma *IARC Sci Publ* **63**: 583–606.

Godden-Kent D, Talbot SJ, Boshoff C et al (1997) The cyclin encoded by Kaposi's sarcoma-associated herpesvirus (KSHV) stimulates cdk6 to phosphorylate the retinoblastoma protein and histone H1. *J Virol* **71**: 4193–4198.

Gooding LR (1992) Virus proteins that counteract host immune defenses. *Cell* **71**: 5–7.

Gooding LR (1994) Regulation of TNF-mediated cell death and inflammation by human adeno-viruses. *Infect Agents Dis* **3**: 106–115.

Guo HG, Browning P, Nicholas J et al (1997) Characterisation of a chemokine receptor-related gene in human herpesvirus 8 and its expression in Kaposi's sarcoma. *Virology* **228**: 371–378.

Hahne M, Rimoldi D, Schroter M et al (1996) Melanoma cell expression of Fas(Apo-1/CD95) ligand: implications for tumor immune escape. *Science* **274**: 1363–1366.

Harrington WJ, Bagsara O, Sosa CE et al (1996) Human herpesvirus type 8 DNA sequences in cell free plasma and mononuclear cells of Kaposi's sarcoma patients. *J Infect Dis* **174**: 1101–1105.

Henderson BE, Ross RK & Pike MC (1991) Toward the primary prevention of cancer. *Science* **254**: 1131–1138.

Herndier BG, Kaplan LD & McGrath MS (1994a) Pathogenesis of AIDS lymphomas. *AIDS* **8**: 1025–1049.

Herndier BG, Werner A, Arnstein P et al (1994b) Characterization of a human Kaposi's sarcoma cell line that induces angiogenic tumors in animals. *AIDS* **8**: 575–581.

Hinshaw VS, Olsen CW, Dybdahl Sissoko N & Evans D (1994) Apoptosis: a mechanism of cell killing by influenza A and B viruses. *J Virol* **68**: 3667–3673.

Howard M, Brink N, Miller R & Tedder R (1995) Association of human herpesvirus with pulmonary Kaposi's sarcoma. *Lancet* **346**: 712.

Howard MR, Whitby D, Bahadur G et al (1997) Detection of human herpesvirus 8 DNA in semen from HIV-infected individuals but not healthy semen donors. *AIDS* **11**: F15–F19.

Huang YQ, Li JJ, Rush MG et al (1992) HPV-16-related DNA sequences in Kaposi's sarcoma. *Lancet* **339**: 515–518.

Huang YQ, Li JJ, Kaplan MH et al (1995) Human herpesvirus – like nucleic acid in various forms of Kaposi's sarcoma. *Lancet* **345(8952)**: 759–761.

Huang DCS, Cory S & Strasser A (1997) Bcl-2, Bcl-x(L) and adenovirus protein E1B19kD are functionally equivalent in their ability to inhibit cell death. *Oncogene* **14**: 405–414.

Humphrey RW, O'Brien TR, Newcomb FM et al (1996) Kaposi's sarcoma (KS)-associated herpesvirus-like DNA sequences in peripheral blood mononuclear cells: association with KS and persistence in patients receiving anti-herpesvirus drugs. *Blood* **88**: 297–301.

Hunter T & Pines J (1994) Cyclins and cancer. II: Cyclin D and CDK inhibitors come of age. *Cell* **79**: 573–582.

Jacks T & Weinberg RA (1996) Cell-cycle control and its watchman. *Nature* **381**: 643–644.

Jagus R & Gray MM (1994) Proteins that interact with PKR. *Biochimie* **76**: 779–791.

Jansen-Durr P (1996) How viral oncogenes make the cell cycle. *Trends Genet* **12**: 270–275.

Jin YT, Tsai ST, Yan JJ et al (1996) Detection of KSHV-like DNA sequence in vascular lesions. *Clin Pathol* **105**: 360–363.

Julius D, Livelli TJ, Jessell TM & Axel R (1989) Ectopic expression of the serotonin 1c receptor and the triggering of malignant transformation. *Science* **244**: 1057–1062.

Jung JU, Stager M & Desrosiers RC (1994) Virus-encoded cyclin. *Mol Cell Biol* **14**: 7235–7244.

Kamijo R, Harada H, Matsuyama T et al (1994) Requirement for transcription factor IRF-1 in NO synthase induction in macrophages. *Science* **263**: 1612–1615.

Kaposi M (1872) Idiopathisches multiples Pigmentsarcom der Haut. *Arch Dermatol Syph* **4:** 265–273.

Kedes DH, Operskalski E, Busch M, Kohn R, Flood J & Ganem D (1996) The seroepidemiology of human herpesvirus 8 (Kaposi's sarcoma-associated herpesvirus): distribution of infection in KS risk groups and evidence for sexual transmission. *Nature Med* **2:** 918–924.

Kedes DH, Lagunoff M, Renne R & Ganem D (1997) Identification of the gene encoding the major latency-associated nuclear antigen of the Kaposi's sarcoma-associated herpesvirus. *J Clin Invest* **100:** 2606–2612.

Kellam P, Boshoff C, Whitby D, Matthews S, Weiss RA & Talbot SJ (1997) Identification of a major latent nuclear antigen (LNA-1) in the human herpesvirus 8 (HHV-8) genome. *J Hum Virol* **1:** 19–29.

Kempf W & Adams V (1996) Viruses in the pathogenesis of Kaposi's sarcoma – a review. *Biochem Mol Med* **58:** 1–12.

Kempf W, Adams V, Pfaltz M et al (1995) Human herpesvirus type 6 and cytomegalovirus in AIDS-associated Kaposi's sarcoma: no evidence for an etiological association. *Hum Pathol* **26:** 914–919.

Kimura T, Nakayama K, Penninger J et al (1994) Involvement of the IRF-1 transcription factor in antiviral responses to interferons. *Science* **264:** 1921–1924.

Kirchhoff S, Schaper F & Hauser H (1993) Interferon regulatory factor 1 (IRF-1) mediates cell growth inhibition by transactivation of downstream target genes. *Nucl Acids Res* **21:** 2881–2889.

Kirchhoff S, Koromilas AE, Schaper F, Grashoff M, Sonenberg N & Hauser H (1995) IRF-1 induced cell growth inhibition and interferon induction requires the activity of the protein kinase PKR. *Oncogene* **11:** 439–445.

Kischkel FC, Hellbardt S, Behrmann I et al (1995) Cytotoxicity-dependent APO-1 (Fas/CD95)-associated proteins form a death-inducing signaling complex (DISC) with the receptor. *EMBO J* **14:** 5579–5588.

Kledal TN, Rosenkilde MM, Coulin F et al (1997) A broad spectrum chemokine antagonist encoded by Kaposi's sarcoma-associated herpesvirus. *Science* **277:** 1656–1659.

Klein G (1994) Epstein–Barr virus strategy in normal and neoplastic B cells. *Cell* **77:** 791–793.

Koelle DM, Huang ML, Chandran B, Vieira J, Piepkorn M & Corey L (1997) Frequent detection of Kaposi's sarcoma-associated herpesvirus (human herpesvirus 8) DNA in saliva of human immunodeficiency virus – infected men: clinical and immunologic correlates. *J Infect Dis* **176(1):** 94–102.

Korsmeyer SJ (1992) Bcl-2 initiates a new category of oncogenes: regulators of cell death. *Blood* **80:** 879–886.

Lagunoff M & Ganem D (1997) The structure and coding organisation of the genomic termini of Kaposi's sarcoma-associated herpesvirus (human herpesvirus 8). *Virology* **236:** 147–154.

Larcher C, Kempkes B, Kremmer E et al (1995) Expression of Epstein–Barr virus nuclear antigen-2 (EBNA2) induces CD21/CR2 on B and T cell lines and shedding of soluble CD21. *Eur J Immunol* **25:** 1713–1719.

Lebbe C, de Cremoux P, Rybojad M et al (1995) Kaposi's sarcoma and new herpesvirus. *Lancet* **345:** 1180.

Lebbe C, Agbalika F, de Cremoux P et al (1997) Detection of HHV-8 and HTLV-1 sequences in Kaposi's sarcoma. *Arch Dermatol* **133:** 25–30.

Lee H, Veazy R, Williams K et al (1998a) Deregulation of cell growth by the *K1* gene of Kaposi's sarcoma-associated herpesvirus. *Nature Med* **4:** 435–440.

Lee H, Guo J, Li M, et al (1998b) Identification of an immunoreceptor tyrosine-based activation motif of K1 transforming protein of Kaposi's sarcoma-associated herpesvirus. *Mol Cell Biol* **18(19):** 5219–5228.

Lennette ET, Blackbourn DJ & Levy JA (1996) Antibodies to human herpesvirus type 8 in the general population and in Kaposi's sarcoma patients. *Lancet* **348:** 858–861.

Li JJ, Huang YQ, Cockerell CJ & Friedman Kien AE (1996) Localization of human herpes-like virus type 8 in vascular endothelial cells and perivascular spindle-shaped cells of Kaposi's sarcoma lesions by in situ hybridization. *Am J Pathol* **148:** 1741–1748.

Li M, Lee H, Yoon DW et al (1997) Kaposi's sarcoma-associated herpesvirus encodes a functional cyclin. *J Virol* **71:** 1984–1991.

Liebowitz D & Kieff E (1993) Epstein–Barr virus. In: Roizman B, Whitley RJ, Lopez C (eds) *The Human Herpesviruses*, pp 107–172. New York: Raven.

Lin JC, Lin SC, Mar EC et al (1995) Is Kaposi's-sarcoma-associated herpesvirus detectable in semen of HIV-infected homosexual men? *Lancet* **346:** 1601–1602.

Lisitsyn N, Lisitsyn N & Wigler M (1993) Cloning the differences between two complex genomes. *Science* **259:** 946–951.

Lovec H, Grzeschiczek A, Kowalski MB & Moroy T (1994) Cyclin D1/bcl-1 co-operates with myc genes in the generation of B-cell lymphoma in transgenic mice. *EMBO J* **13:** 3487–3495.

Lunardi IY, Bryant JL, Zeman RA et al (1995) Tumorigenesis and metastasis of neoplastic Kaposi's sarcoma cell line in immunodeficient mice blocked by a human pregnancy hormone [published erratum *Nature* (1995) **376(6539):** 447]. *Nature* **375:** 64–68.

Luppi M, Barozzi P, Maiorana A et al (1996a) Human herpesvirus-8 DNA sequences in human immunodeficiency virus-negative angioimmunoblastic lymphadenopathy and benign lymphadenopathy with giant germinal center hyperplasia and increased vascularity. *Blood* **87:** 3903–3909.

Luppi M, Barozzi P, Maiorana A et al (1996b) Frequency and distribution of herpesvirus-like DNA sequences (KSHV) in different stages of classic Kaposi's sarcoma and in normal tissues from an Italian population. *Int J Cancer* **66:** 427–431.

Mackenzie J, Sheldon J, Morgan G, Cook G, Schulz TF & Jarret RF (1997) HHV-8 and multiple myeloma in the UK. *Lancet* **350:** 1144–1145.

Magnaghi-Jaulin L, Groisman R, Naguibneva I et al (1998) Retinoblastoma protein represses transcription by recruiting a histone deacetylase. *Nature* **391:** 601–604.

Marcelin AG, Dupin N, Bourscary D et al (1997) HHV-8 and multiple myeloma in France. *Lancet* **350:** 1144.

Marchiolo CC, Love JL, Abbott LZ et al (1996) Prevalence of HHV-8 DNA sequences in several patient populations. *J Clin Microbiol* **34:** 2635–2638.

Martin E, Nathan C & Xie QW (1994) Role of interferon regulatory factor 1 in induction of nitric oxide synthase. *J Exp Med* **180:** 977–984.

Martin JN, Ganem DE, Osmond DH, Page-Shafer KA, Macrae D & Kedes DH (1998) Sexual transmission and the natural history of human herpesvirus 8 infection. *N Engl J Med* **338:** 948–954.

Masood R, Zheng T, Tupule A et al (1997) Kaposi's sarcoma-associated herpesvirus infection and multiple myeloma. *Science* **278:** 1969.

McDonagh DP, Liu J, Gaffey MJ, Layfield LJ, Azumi N & Traweek TS (1996) Detection of KSHV-like DNA sequences in angiosarcoma. *Am J Pathol* **149:** 1363–1368.

Milano CA, Allen LF, Rockman HA et al (1994) Enhanced myocardial function in transgenic mice overexpressing the beta2-adrenergic recetor. *Science* **264:** 582–586.

Miles SA (1994) Pathogenesis of HIV-related Kaposi's sarcoma. *Curr Opin Oncol* **6:** 497–502.

Miles SA, Rezai AR, Salazar Gonzalez JF et al (1990) AIDS Kaposi's sarcoma-derived cells produce and respond to interleukin 6. *Proc Natl Acad Sci USA* **87:** 4068–4072.

Miller G, Rigsby MO, Heston L et al (1996) Antibodies to butyrate-inducible antigens of Kaposi's sarcoma-associated herpesvirus in patients with HIV-1 infection. *N Engl J Med* **334:** 1292–1297.

Molden J, Chang Y, You Y, Moore PS & Goldsmith MA (1997) A Kaposi's sarcoma-associated herpesvirus-encoded cytokine homolog (vIL-6) activates signaling through the shared gp 130 receptor subunit. *J Biol Chem* **272:** 19625–19631.

Monini P, Rotola A, de Lellis L et al (1996) Latent BK virus infection and Kaposi's sarcoma pathogenesis. *Int J Cancer* **66:** 717–722.

Moore PS, Chang Y (1995) Detection of herpesvirus-like DNA sequences in Kaposi's sarcoma in patients with and without HIV infection. *N Engl J Med* **332(18):** 1181–1185.

Moore PS (1998) Human herpesvirus 8 variants. *Lancet* **351:** 679–680.

Moore PS, Boshoff C, Weiss RA & Chang Y (1996a) Molecular mimicry of human cytokine and cytokine response pathway genes by KSHV. *Science* **274:** 1739–1744.

Moore PS, Gao SJ, Dominguez G et al (1996b) Primary characterization of a herpesvirus agent associated with Kaposi's sarcoma. *J Virol* **70:** 549–558.

Moore PS, Kingsley LA, Holmberg SD et al (1996c) Kaposi's sarcoma-associated herpesvirus infection prior to onset of Kaposi's sarcoma. *AIDS* **10:** 175–180.

Morgan DO (1995) Principles of CDK regulation. *Nature* **374:** 131–134.

Morris JD, Eddleston AL & Crook T (1995) Viral infection and cancer. *Lancet* **346:** 754–758.

Muzio M, Chinnaiyan AM, Kischkel FC et al (1996) FLICE, a novel FADD-homologous ICE/CED-3-like protease, is recruited to the CD95 (Fas/APO-1) death-inducing signaling complex. *Cell* **85:** 817–827.

Nagata S (1997) Apoptosis by death factor. *Cell* **88:** 355–365.

Neipel F, Albrecht JC & Fleckenstein B (1997a) Cell-homologous genes in the Kaposi's sarcoma-associated rhadinovirus human herpesvirus 8: determinants of its pathogenicity? *J Virol* **71:** 4187–4192.

Neipel F, Albrecht JC, Ensser A et al (1997b) Human herpesvirus 8 encodes a homolog of interleukin-6. *J Virol* **71:** 839–842.

Neipel F, Albrecht JC, Ensser A et al (1997c) Primary structure of the Kaposi's sarcoma associated human herpesvirus 8. Genbank accession no. U93872.

Nickoloff BJ & Griffiths CE (1989) Factor XIIIa-expressing dermal dendrocytes in AIDS-associated cutaneous Kaposi's sarcomas. *Science* **243:** 1736–1737.

O'Neill E, Henson TH, Ghorbani AJ, Land MA, Webber B, L & Garcia JV (1996) Herpes virus-like sequences are specifically found in Kaposi's sarcoma lesions. *J Clin Pathol* **49:** 306–308.

Panyutich EA, Said JW & Miles SA (1998) Infection of primary dermal microvascular endothelial cells by Kaposi's sarcoma-associated herpesvirus. *AIDS* **12:** 467–472.

Parravacini C, Lauri L, Baldini L et al (1997) Kaposi's sarcoma-associated herpesvirus infection and multiple myeloma. *Science* **278:** 1969.

Peters G (1994) The D-type cyclins and their role in tumorigenesis. *J Cell Sci Suppl* **18:** 89–96.

Rabkin CS, Janz S, Lash A et al (1997) Monoclonal origin of multicentric Kaposi's sarcoma lesions. *N Engl J Med* **336:** 988–993.

Rady PL, Yen A, Rollefson JL et al (1995) Herpesvirus-like DNA sequences in non-Kaposi's sarcoma skin lesions of transplant patients. *Lancet* **345:** 1339–1340.

Rainbow L, Platt GM, Simpson GR et al (1997) The 222- to 234-kilodalton latent nuclear protein (LNA) of Kaposi's sarcoma-associated herpesvirus (human herpesvirus 8) is encoded by orf73 and is a component of the latency-associated nuclear antigen. *J Virol* **71:** 5915–5921.

Rappersberger K, Tschachler E, Zonzits E et al (1990) Endemic Kaposi's sarcoma in human immunodeficiency virus type 1-seronegative persons: demonstration of retrovirus-like particles in cutaneous lesions. *J Invest Dermatol* **95:** 371–381.

Regamey N, Erb P, Tamm M & Cathomas G (1998) Human herpesvirus 8 variants. *Lancet* **351:** 680.

Renne R, Zhong W, Herndier B et al (1996) Lytic growth of Kaposi's sarcoma-associated herpesvirus (human herpesvirus 8) in culture. *Nature Med* **2:** 342–346.

Renne R, Blackbourn D, Whitby D, Levy J & Ganem D (1998) Transmission of Kaposi's sarcoma-associated herpes virus (KSHV/HHV-8) is limited in cultured cells *J Virol* (Unpublished results).

Resnitzky D, Gossen M, Bujard H & Reed SI (1994) Acceleration of the G1/S phase transition by expression of cyclins D1 and E with an inducible system. *Mol Cell Biol* **14:** 1669–1679.

Rettig MB, Ma, HJ, Vescio RA et al (1997) Kaposi's sarcoma associated herpesvirus infection of bone marrow dendritic cells from multiple myeloma. *Science* **276:** 1851–1854.

Rickinson AB (1996) Changing seroepidemiology of HHV-8. *Lancet* **348:** 1110–1111.

Rose TM, Strand KB, Schultz ER et al (1997) Identification of two homologs of the Kaposi's sarcoma-associated herpesvirus (human herpesvirus 8) in retroperitoneal fibromatosis of different macaque species. *J Virol* **71:** 4138–4144.

Roth WK (1991) HIV-associated Kaposi's sarcoma: new developments in epidemiology and molecular pathology. *J Cancer Res Clin Oncol* **117:** 186–191.

Russo JJ, Bohenzky RA, Chien MC et al (1996) Nucleotide sequence of the Kaposi sarcoma-associated herpesvirus (HHV8). *Proc Natl Acad Sci USA* **93:** 14862–14867.

Rutgers JL, Wieczorek R, Bonetti F et al (1986) The expression of endothelial cell surface antigens by AIDS-associated Kaposi's sarcoma. Evidence for a vascular endothelial cell origin. *Am J Pathol* **122:** 493–499.

Sarid R, Sato T, Bohenzky RA, Russo JJ & Chang Y (1997) Kaposi's sarcoma-associated herpesvirus encodes a functional bcl-2 homologue. *Nature Med* **3:** 293–298.

Sato T, Hanada M, Bodrug S et al (1994) Interactions among members of the Bcl-2 protein family analysed with a yeast two-hybrid system. *Proc Natl Acad Sc USA* **91:** 9238–9242.

Scala G, Ruocco MR, Ambrosino C et al (1994) The expression of the interleukin 6 gene is induced by the human immunodeficiency virus 1 TAT protein. *J Exp Med* **179:** 961–971.

Schalling M, Ekman M, Kaaya EE, Linde A & Biberfeld P (1995) A role for a new herpes virus (KSHV) in different forms of Kaposi's sarcoma. *Nature Med* **1:** 705–706.

Sen GC & Ransohoff RM (1993) Interferon-induced antiviral actions and their regulation. *Adv Virus Res* **42:** 57–102.

Service PH (1981) Kaposi's sarcoma and *Pneumocystis* pneumonia among homosexual men in New York city and California. *MMWR Morb Mortal Wkly Rep* **30:** 305–308.

Shen Y & Shenk TE (1995) Viruses and apoptosis. *Curr Opin Genet Dev* **5:** 105–111.

Shen Y & Shenk T (1996) Relief of p53-mediated transcriptional repression by the adenovirus E1B 19-kDa protein or the cellular Bcl-2 protein. *Proc Natl Acad Sci USA* **91:** 8940–8944.

Sherr CJ (1994) G1 phase progression: cycling on cue. *Cell* **79:** 551–555.

Sherr CJ (1995) D-type cyclins. *Trends Biochem Sci* **20:** 187–190.

Simpson GR, Schulz TF, Whitby D et al (1996) Prevalence of Kaposi's sarcoma associated herpesvirus infection measured by antibodies to recombinant capsid protein and latent immunofluorescence antigen. *Lancet* **348:** 1133–1138.

Sinclair AJ, Palmero I, Peters G & Farrell PJ (1994) EBNA-2 and EBNA-LP cooperate to cause G0 to G1 transition during immortalization of resting human B lymphocytes by Epstein–Barr virus. *EMBO J* **13:** 3321–3328.

Sirianni MC, Uccini S, Angeloni A, Faggioni A, Cottoni F & Ensoli B (1997a) Circulating spindle cells: correlation with human herpesvirus-8 (HHV-8) infection and Kaposi's sarcoma (letter). *Lancet* **349:** 255.

Sirianni MC, Vicenzi L, Topino S et al (1997b) Human herpesvirus 8 DNA sequences in CD8+ T cells. *J Infect Dis* **176:** 541.

Sitas F, Carrara H, Beral V et al (1998) Association between HHV-8, HIV, cancer and selected risk factors in South Africa. Abstract 1. Second National AIDS Malignancy Conference, Bethesda, 1998. *J AIDS* **17:** 12.

Soulier J, Grollet L, Oksenhendler E et al (1995) Kaposi's sarcoma-associated herpesvirus-like DNA sequences in multicentric Castleman's disease. *Blood* **86:** 1276–1280.

Staskus KA, Zhong W, Gebhard K et al (1997) Kaposi's sarcoma-associated herpesvirus gene expression in endothelial (spindle) tumor cells. *J Virol* **71:** 715–719.

Strand S, Hofmann WJ, Hug H et al (1996) Lymphocyte apoptosis induced by CD95 (APO-1/Fas) ligand-expressing tumor cells – a mechanism of immune evasion? *Nature Med* **2:** 1361–1366.

Su J, Hsu YS, Chang YC & Wang IW (1995) Herpesvirus-like DNA sequences in Kaposi's sarcoma from AIDS and non-AIDS patients in Taiwan. *Lancet* **435:** 722–723.

Tamura T, Ueda S, Yoshida M, Matsuzaki M, Mohri H & Okubo T (1996) Interferon-gamma induces Ice gene expression and enhances cellular susceptibility to apoptosis in the U937 leukemia cell line. *Biochem Biophys Res Commun* **229**: 21–26.

Tanaka N, Ishihara M, Kitagawa M et al (1994) Cellular commitment to oncogene-induced transformation or apoptosis is dependent on the transcription factor IRF-1. *Cell* **77**: 829–839.

Tanner JE, Alfieri C, Chatila TA & DiazMitoma F (1996) Induction of interleukin-6 after stimulation of human B-cell CD21 by Epstein–Barr virus glycoproteins gp350 and gp220. *J Virol* **70**: 570–575.

Tarte K, Olsen SJ, Yang LZ et al (1998) Clinical grade functional dendritic cells from patients with multiple myeloma are not infected with Kaposi's sarcoma-associated herpesvirus. *Blood* **91**: 1852–1857.

Tasaka T, Said JW & Koeffler HP (1996) Absence of HHV-8 in prostate and semen (letter). *N Engl J Med* **335**: 1237–1238.

Taylor JF, Smith PG, Bull D & Pike MC (1972) Kaposi's sarcoma in Uganda: geographic and ethnic distribution. *Br J Cancer* **26**: 483–497.

Thome M, Schneider P, Hofmann K et al (1997) Viral FLICE-inhibitory proteins (FLIPs) prevent apoptosis induced by death receptors. *Nature* **386**: 517–521.

Thompson EW, Nakamura S, Shima TB et al (1991) Supernatants of acquired immunodeficiency syndrome-related Kaposi's sarcoma cells induce endothelial cell chemotaxis and invasiveness. *Cancer Res* **51**: 2670–2676.

Tsujimoto Y, Bashir MM, Givol I et al (1987) DNA rearrangements in human follicular lymphoma can involve the 5′ or the 3′ region of the bcl-2 gene. *Proc Natl Acad Sci USA* **84**: 1329–1331.

Uccini S, Ruco LP, Monardo F et al (1994) Co-expression of endothelial cell and macrophage antigens in Kaposi's sarcoma cells. *J Pathol* **173**: 23–31.

Uthman A, Brna C, Weninger W & Tschachler E (1996) No HHV8 in non-Kaposi's sarcoma mucocutaneous lesions from immunodeficient HIV-positive patients (letter). *Lancet* **347**: 1700–1701.

Vaughan PS, Aziz F, van Wijnen AJ et al (1995) Activation of a cell-cycle-regulated histone gene by the oncogenic transcription factor IRF-2. *Nature* **377**: 362–365.

Vaux DL, Haecker G & Strasser A (1994) An evolutionary perspective on apoptosis. *Cell* **76**: 777–779.

Viviano E, Vitale F, Ajello F et al (1997) Human herpesvirus type 8 DNA sequences in biological samples of HIV-positive and negative individuals in Sicily. *AIDS* **11**: 607–612.

Wahman A, Melnick SL, Rhame FS & Potter JD (1991) The epidemiology of classic, African, and immunosuppressed Kaposi's sarcoma. *Epidemiol Rev* **13**: 178–199.

Wang RYH, Shih JWK, Weiss SH et al (1993) *Mycoplasma penetrans* infection in male homosexuals with AIDS: high seropervalence and association with Kaposi's sarcoma. *Clin Infect Dis* **17**: 724–729.

Wang TC, Cardiff RD, Zukerberg L, Lees E, Arnold A & Schmidt EV (1994) Mammary hyperplasia and carcinoma in MMTV-cyclin D1 transgenic mice. *Nature* **369**: 669–671.

Weich HA, Salahuddin SZ, Gill P et al (1991) AIDS-associated Kaposi's sarcoma-derived cells in long term culture express and synthesize smooth muscle alpha-actin. *Am J Pathol* **139(6)**: 1251–1258.

Weiss RA (1996) Human herpesvirus 8 in lymphoma and Kaposi's sarcoma: now the virus can be propagated. *Nature Med* **2**: 277–278.

Whitby D, Howard MR, Tenant Flowers M et al (1995) Detection of Kaposi sarcoma associated herpesvirus in peripheral blood of HIV-infected individuals and progression to Kaposi's sarcoma. *Lancet* **346**: 799–802.

Whitby D, Boshoff C, Luppi M & Torelli G (1997) Kaposi's sarcoma associated herpesvirus infection and multiple myeloma. *Science* **278**: 1971–1972.

Whitby D, Luppi M, Barozzi P, Boshoff C, Weiss RA & Torreli G (1998a) Human herpesvirus 8 seroprevalence in blood donors and lymphoma patients from different regions of Italy. *J Natl Cancer Inst* **90:** 395–397.

Whitby D, Smith N, Ariyoshi K et al (1998b) Abstract 3. Second National AIDS Malignancy Conference, Bethesda, 1998. *J AIDS* **17:** A12.

Whitby D, Smith NA, Matthews et al (1999) Human herpesvirus 8: Seroepidemiology among women and detection in the genital tract of seropositive women. *J Infect Dis* **179(1):** 234–236.

Wiertz EJ, Jones TR, Sun L, Bogyo M, Geuze HJ & Ploegh HL (1996) The human cytomegalovirus US11 gene product dislocates MHC class I heavy chains from the endoplasmic reticulum to the cytosol. *Cell* **84:** 769–779.

Yang E & Korsmeyer SJ (1996) Molecular thanatopsis: a discourse on the BCL2 family and cell death. *Blood* **88:** 386–401.

Ziegler JL & Katongole-Mbidde E (1996) Kaposi's sarcoma in childhood: an analysis of 100 cases from Uganda and relationship to HIV infection. *Int J Cancer* **65:** 200–203.

Zong JC, Poole L, Cuifo D et al (1998) Identification of four major subgroups and up to ten distinct clades of KSHV based on the analysis of the hypervariable of K1 membrane receptor protein. Abstract 63. Second National AIDS Malignancy Conference, Bethesda, 1998. *J AIDS* **17:** A27.

Chapter 19

PATHOGENESIS AND CELL BIOLOGY OF KAPOSI'S SARCOMA

B. Ensoli, P. Monini and C. Sgadari

INTRODUCTION

Kaposi's sarcoma (KS) is a neoplasm of vascular origin, classified clinically and epidemiologically into four groups: 'classic' or 'Mediterranean' (CKS); 'endemic' or 'African' (AKS); 'post-transplant' (PKS); and 'AIDS-associated' (AIDS-KS). It generally arises with multiple lesions in the form of purple maculae, papulae or plaques which are often located on the lower extremities. Lesions can progress toward a nodular or tumoral form and can coalesce to form large cutaneous tumours which ulcerate. Classic KS has an indolent course and affects elderly men of Mediterranean and Eastern European origin (Rothman, 1962; Safai and Good, 1981; Franceschi and Geddes, 1995; Geddes et al, 1995). Endemic KS is frequent in subequatorial Africa and is characterized by a more aggressive course, with involvement of visceral organs, infiltrative cutaneous lesions and a high morbidity rate (Slavin et al, 1969; Taylor et al, 1971, 1972). A form with massive involvement of the lymph nodes is often seen in African children and young adults and has a rapid and fatal course (Taylor et al, 1971). Post-transplant KS occurs most frequently in kidney allograft recipients treated with cyclosporin and it usually develops after few months from the beginning of the immunosuppressive therapy (Harwood et al, 1979; Penn, 1979; Civati et al, 1988; Trattner et al, 1993). The AIDS-associated form is the most frequent tumour of human immunodeficiency virus type 1 (HIV-1) infected individuals, particularly homosexual and bisexual men, and develops 300 times more frequently than in individuals with primary immunodeficiencies but without HIV infection (Beral et al, 1990). It is usually aggressive, disseminated and fatal, with involvement of visceral organs and lymph nodes (Friedman-Kien, 1981; Haverkos and Drotman, 1985; Safai et al, 1985).

Although these forms of KS have different clinical courses, they share the same histopathological characteristics: neoangiogenesis, oedema, infiltration of lympho-mononuclear cells and growth of spindle-shaped cells (KS spindle cells, KSC considered to be the tumour cells of the lesions. This suggests that the development of the different forms of KS is mediated by the same mechanisms and aetiological agents. In this regard, the newly discovered human herpesvirus 8 (HHV-8) which has been found in all forms of KS and in high-risk individuals may play a key role in disease development, as suggested by epidemiological studies (Chang et al, 1994; Ambroziak et al, 1995; Boshoff et al, 1995; De Lellis et al, 1995; Huang et al, 1995; Moore and Chang, 1995;

HIV and the New Viruses Second Edition
ISBN 0-12-200741-7

Schalling et al, 1995; Whitby et al, 1995; Buonaguro et al, 1996; Gao et al, 1996; Kedes et al, 1996; Lennette et al, 1996; Moore et al, 1996).

Many observations suggest that, at least in its early stages, KS is not a true sarcoma. For example, all forms of KS can start as multiple lesions that appear simultaneously at different sites of the body with a symmetrical or dermatome distribution and without the typical histopathological features of metastatic lesions from sarcoma or angiosarcoma (Brooks, 1986). There is no evidence of aneuploidy in most KS lesions analysed although, as discussed later, rare high-grade lesions may have aneuploid cells and a clonal origin (Saikevych et al, 1988; Dictor et al, 1991; Rabkin et al, 1995, 1997). Patients with PKS can show spontaneous lesion regression and even complete remission after the withdrawal or reformulation of immunosuppressive therapy (Klepp et al, 1978; Harwood et al, 1979; Hoshaw and Schwartz, 1980; Zisbrod et al, 1980; Akhtar et al, 1984; Real et al, 1986; Bencini et al, 1993). Spontaneous regressions have been documented also in AIDS-KS patients (Janier et al, 1985; Real and Krown, 1985) and, more recently, upon treatment with HIV-1 protease inhibitors (Conant et al, 1997; Rizzieri et al, 1997).

Evidence indicates that KS onset is associated with a disturbance of the immune system leading to activation. Patients with all forms of KS and KS risk groups show activation of CD8+ T cells and increased expression of T helper 1 (Th1) type cytokines with a high level of production of interferon gamma (IFN-γ) (Fiorelli et al, 1998; Sirianni et al, 1998). The sarcoma itself starts as a granulation-like tissue rich in inflammatory cells consisting of lymphocytes and monocyte/macrophages. These infiltrating cells produce the same inflammatory cytokines (IFN-γ, interleukins 1 and 2, tumour necrosis factor, etc.) produced by the peripheral blood mononuclear cells (PBMCs) of KS patients (Fiorelli et al, 1998; Sirianni et al, 1998). These phenomena may be triggered or enhanced by infection with HHV-8. In turn, these inflammatory cytokines may trigger KS lesion formation through the activation of endothelial cells (EC) and the induction of cytokines that mediate KSC growth, angiogenesis and oedema (Barillari et al, 1992; Ensoli et al, 1994a; Fiorelli et al, 1995, 1998). In this context, the Tat protein of HIV-1 may act as a progression factor in AIDS-KS by molecular mimicry of extracellular matrix (ECM) proteins (Ensoli et al, 1994a) and by enhancing inflammatory cytokine production (Chang et al, 1995).

Together, these observations suggest that, at least in the early stages, KS may be a reactive inflammatory-angiogenic process in response to or enhanced by HHV-8 infection, triggered by the action of inflammatory cytokines and, for AIDS-KS, further promoted by the HIV-1 Tat protein. Other observations, however, suggest the possibly malignant nature of late nodular KS. Although cultured KSCs are not immortal or transformed, two KS cell lines tumorigenic in severe combined immunodeficient (SCID) mice have now been isolated (Siegal et al, 1990; Lunardi-Iskandar et al, 1995). In addition, although KSCs are typically diploid, aneuploid cells have been detected in some high-grade lesions (Saikevych et al, 1988; Dictor et al, 1991). As shown in a recent study, AIDS-KS lesions can have a high rate of microsatellite instability, although this trait is lacking in HIV-negative patients with CKS (Bedi et al, 1995). Evidence of clonality of KSCs has also been detected by analysis of the methylation patterns of the androgen receptor gene in some advanced lesions from women infected with HIV (Rabkin et al, 1997), although others have found polyclonality of the lesions with a similar approach (Gill et al, 1997).

PREDISPOSING FACTORS

The immune system seems to play a major role in KS pathogenesis. Clinical, epidemiological and laboratory studies suggest that immunoactivation rather than immunodeficiency is a predisposing factor to KS development (Maurice et al, 1982; Levy and Ziegler, 1983; Ensoli et al, 1989, 1991; Beckstead, 1992; Roth et al, 1992; Ensoli and Gallo, 1994). All patients with KS or at high risk for KS have signs of immunoactivation and KS can arise in the absence of immunodeficiency.

Homosexual men and HIV-1-infected individuals, the groups at the highest risk of developing KS, show increased levels of inflammatory cytokines such as interleukin (IL) 1, IL-6, tumour necrosis factor alpha (TNFα), IFN-γ, and of soluble intracellular adhesion molecule 1 (ICAM-1), soluble CD8 and neopterin, even prior to HIV-1 infection or KS development (Lepe-Zanuga et al, 1987; Lahdevirta et al, 1988; Hober et al, 1989; Caruso et al, 1990; Honda et al, 1990; Fan et al, 1993; Kalinkovich et al, 1993; DePaoli et al, 1994; Schlessinger et al, 1994; Fiorelli et al, 1995). The lifestyle of homosexual men may result in chronic antigenic and allogeneic stimulation and infection by multiple agents, preceding or accompanying HIV-1 infection. In homosexual men, a milder form of KS can arise prior to HIV-1 infection (Maurice et al, 1982; Friedman-Kien et al, 1990). African subjects are also immunoactivated, probably owing to frequent exposure to different infections (Master et al, 1970; Kestens et al, 1985; Rizzardini et al, 1996). Elderly individuals may present an oligoclonal expansion of CD8+ T cells producing IL- I and TNFα (Levin et al, 1992; Fagiolo et al, 1993; Fagnoni et al, 1996). In transplanted patients, KS can regress after withdrawal of immunosuppressive therapy (Klepp et al, 1978; Harwood et al, 1979; Hoshaw and Schwartz, 1980; Zisbrod et al, 1980; Akhtar et al, 1984; Real et al, 1986; Bencini et al, 1993). However, despite therapeutic immunosuppression, in these patients allogeneic stimulation may result in the emergence of local foci of stimulated immune cells. In addition, although the increased incidence of KS in organ transplant recipients is mostly associated with the use of cyclosporin, KS has also been described in patients undergoing glucocorticoid therapy (Real et al, 1986; Schulhafer et al, 1987; Gill et al, 1989; Trattnier et al, 1993). It is noteworthy that glucocorticoids have a direct proliferative effect on KSCs (Guo and Antakly, 1995; Cai et al, 1997). These spindle cells express high levels of glucocorticoid receptors that are upregulated by inflammatory cytokines (Guo et al, 1996). Moreover, glucocorticoids synergize with oncostatin M (OSM), a major growth factor for KSC (Miles et al, 1992; Nair et al, 1992; Guo et al, 1996) (see below). Finally, allogeneic stimulation and opportunistic infections in transplant recipients can result in reactivation of latent herpesviruses, particularly cytomegalovirus (Söderberg-Nauclér et al, 1997). By analogy, reactivation of HHV-8 may also occur and contribute to the emergence of KS (see below).

Consistent with these findings, PBMCs from AIDS-KS and CKS patients produce high levels of Th1 type cytokines, particularly IFN-γ, resulting from CD8+ T cell activation (Fiorelli et al, 1998; Sirianni et al, 1998). The importance of T cell activation and inflammatory cytokine production in KS pathogenesis is also suggested by clinical reports of KS progression or onset in patients treated with IFN-γ, TNFα or IL-2 (Aboulafia et al, 1989; Krigel et al, 1989; Albrecht et al, 1994). Disease progression is also reported in the course of opportunistic infections, naturally associated with inflammatory cytokine production (Mitsuyasu, 1993).

A systemic increase of inflammatory cytokines may be responsible for several features

Table 19.1 *Effects of inflammatory cytokines in KS*

Systemic effects on:

Circulating KSC progenitors: proliferation, differentiation into macrophages and endothelial macrophages
Vessels: activation with adhesion and extravasation of inflammatory cells
HIV-1: induction of gene expression/replication, production/release of HIV-1 Tat protein
HHV-8: activation, increase of viral load in circulating B cells and monocytes

---

Local (KS lesion) effects on:

Inflammatory cells: recruitment, activation and differentiation into macrophages, endothelial macrophages and dendritic cells
TIL: maintenance of phenotype and survival
E-KSC and M-KSC: differentiation from precursors (EC, circulating KSC progenitors) and proliferation
E-KSC: enhancement of cytokine production and angiogenic activity
EC activation with:
    induction of a spindle morphology
    downregulation of VWF:Ag (due to its release)
    upregulation of adhesion molecules and activation marker expression
    expression of bFGF, VEGF, IL-6, IL-8, MCP-1, GM-CSF, IL-1, PDGF-A and induction of bFGF release
    responsiveness to the angiogenic effects of the HIV-1 Tat protein
    acquisition of an angiogenic phenotype in vivo (KS-like lesions in nude mice)
Angiogenesis: induction of KS-like lesions after injection in nude mice

---

See text for details. E-KSC, M-KSC, endothelial and macrophagic KSC; TIL; tumour-infiltrating lymphocyte; EC; endothelial cell.

described in KS patients such as (1) the expansion of circulating KSC progenitors and their differentiation into macrophages and endothelial macrophages (see below); (2) the vessel activation with increased levels of von Willebrand factor (VWF):Ag serum levels (Cotran and Pober, 1988; Pober and Cotran, 1990; Fiorelli et al, 1995) and the adhesion and extravasation of inflammatory cells into tissues (MacPhail et al, 1996; Zietz et al, 1996); (3) the activation of HHV-8 replication (see below); and (4) the activation of HIV-1 gene expression and replication with further production of inflammatory cytokines and increase of extracellular HIV-1 Tat protein, which acts as a progression factor in AIDS-KS (Table 19.1).

DEVELOPMENT OF KS LESIONS

Inflammatory cell infiltration

The inflammatory cell infiltrate seen in the very early stage of KS lesions or in the perilesional tissue generally precedes the appearance of the typical KSCs and is associated with the earliest histological changes of KS. In the initial stages the inflammatory cell infiltrate, activated proliferating endothelial cells and neoangiogenesis, which are prominent histological traits associated with the appearance of KSCs, may not be seen. With time, the KSCs become the predominant cell type and the lesions acquire a more monomorphic aspect resembling a fibrosarcoma, although neoangiogenesis is always present (McNutt et al, 1983; Dorfman and Path, 1984; Ruszczak et al, 1987a,b; Ensoli et al, 1991, 1994; Stürzl et al, 1992).

Immunohistochemical studies indicate that the inflammatory cell infiltrate of KS consists of CD8+ and CD4+ T cells, dendritic cells, and monocytes/macrophages expressing CD4, CD14, CD68, CD45, Pulmonary Alveolar macrophage-1 (PAM-1), and often presenting a spindle morphology and a subendothelial localization. The CD8+ T cells are always predominant with respect to CD4+ T lymphocytes. In contrast, B cells expressing CD19, CD20 or CD30 are few or absent (Nickoloff and Griffiths, 1989; Regezi et al, 1993; Tabata et al, 1993; Uccini et al, 1994, 1997; MacPhail et al, 1996; Fiorelli et al, 1998; Sirianni et al, 1998). The analysis of tumours infiltrating lymphocytes (TILs) isolated immediately after biopsy and tissue mincing or after short-term culture shows that these cells consist of T cell receptor (TCR) α/β+CD8+ and TCRα/β+CD4+ (Sirianni et al, 1998). In addition, TCRα/β+CD8−CD4− cells which usually represent a low fraction (2%) of normal lymphoid cells (Kusnoki et al, 1992) are present in high percentage (25%) in TILs (Sirianni et al, 1998). The analysis of the cytokine profile of TILs by immunohistochemistry and in the supernatants of cultured macrophages from lesional skin specimens shows that both AIDS-KS and CKS patients have a predominant Thl-type cytokine profile with high level of expression of IFN-γ and low or absent IL-4 (Fiorelli et al, 1998; Sirianni et al, 1998). Interestingly, TCRα/β+CD8−CD4− cells have been associated with autoimmune disorders and are able to produce IFN-γ (Shivakumar et al, 1989; Wirt et al, 1989; Porcelli et al, 1992).

The events that recruit the early inflammatory cell infiltrate are most probably associated with the condition of immunoactivation and inflammatory cytokine production, which, as discussed above, are characteristic of KS patients and individuals at risk for KS. Increased levels of inflammatory cytokines lead to vessel activation, adhesion of monocytes–macrophages and T lymphocytes and their extravasation into tissues. This has been shown in individuals infected with HIV-1 (Zietz et al, 1996). In addition, IFN-γ production can be detected in yet uninvolved tissues from KS patients (Fiorelli et al, 1998). This cytokine has been shown to play a major role in leukocyte recruitment into the skin, suggesting that the upregulation of IFN-γ seen in these individuals may be essential to amplify the very early tissue infiltration of inflammatory cells (Colditz and Watson, 1992). The early appearance of activated proliferating endothelial cells and the initial angiogenesis seen at onset of KS can be mediated by inflammatory cytokines produced by these infiltrating inflammatory cells; these cytokines, in fact, induce the expression of angiogenic factors. In later stages, KSCs expressing adhesion molecules and releasing chemokines may become important mediators of leukocyte recruitment.

Origin of KS spindle cells present in lesions and in blood

The origin of KS spindle cells has been debated for a long time. The most recent studies show that lesional KSCs consist of an heterogeneous cell population dominated by activated endothelial cells (positive for VWF:Ag, PAL-E, ULEX, VE-cadherin, CD34, CD36, CD31, ICAM-1, vascular cell adhesion molecule-1 (VCAM-1), endothelial leukocyte adhesion molecule-1 (ELAM-1), CD40, DR) mixed with cells of macrophagic origin (positive for CD14, CD68, CD36, CD4, CD45, PAM-1, DR, ICAM-1) (Nadji et al, 1981; Kraffert et al, 1991; Huang et al, 1993a; Regezi et al, 1993; Gendelman et al,

1994; Uccini et al, 1994, 1997; Fiorelli et al, 1995, 1998; Kaaya et al, 1995; MacPhail et al, 1996; Sirianni et al, 1998) (Table 19.2). Long-term cultures of both endothelial (E-KSC) and macrophagic (M-KSC) KSCs have been established from KS lesions by utilizing the same inflammatory cytokines expressed in the lesions but with modification of IFN-γ and IL-2 content, which at higher concentrations preferentially support the growth of M-KSCs (Nakamura et al, 1988; Barillari et al, 1992; Sirianni et al, 1998) (see Tables 19.1, 19.3). These cultured cells have the same immunophenotype as the E-KSCs and M-KSCs found in the lesions (Regezi et al, 1993; Gendelman et al, 1994; Fiorelli et al, 1995, 1998; MacPhail et al, 1996; Sirianni et al, 1998). Some of the KSCs of the lesions express markers of macrophages but are also positive for VE-cadherin, a marker of vascular endothelial cells (Uccini et al, 1994, 1997) (Table 19.2). In normal tissues, coexpression of these markers is detected only in sinus-lining cells of the spleen, in lymphatic sinus cells of the palatine tonsils, and in the 'endothelial macrophages' of the lymph nodes (Byckley et al, 1985; Baroni et al, 1987; Lampugnani et al, 1992; Uccini et al, 1994, 1997).

The circulating KSC progenitors expanded in the blood of KS patients have a similar phenotype. These cells express markers of monocyte/macrophages, but express also VE-cadherin, whereas CD34 and VWF:Ag are low or undetectable (Browning et al, 1994; Sirianni et al, 1997a; Uccini et al, 1997; Colombini et al, 1998) (Table 19.2). KSC progenitors are increased in all forms of KS and in individuals at risk of KS, and can be identified in the adherent cell fraction of cultured PBMCs from these individuals (Browning et al, 1994; Sirianni et al, 1997a; Uccini et al, 1997; Colombini et al, 1998). In addition, the appearance of this circulating KSC population in culture is enhanced by the same inflammatory cytokine increase as in KS (Browning et al, 1994; Colombini et al, 1998) (Table 19.1). As discussed below, circulating KSC progenitors are infected by HHV-8 and may play a role in virus recruitment into tissues (Sirianni et al, 1997a; Colombini et al, 1998).

These data support the concept that lesional KSCs and circulating KSCs from the blood of KS patients are related cell populations. Since they are increased in KS high-risk groups (Browning et al, 1994; Colombini et al, 1998) and their appearance can be enhanced by the same inflammatory cytokines increased in KS and KS risk groups, this suggests that circulating KSCs may be the cell progenitors of lesional KSCs, possibly contributing to the multifocal nature of KS. Further, the increase of this cell population may represent a prognostic marker for KS development and progression (Browning et al, 1994; Sirianni et al, 1997a; Colombini et al, 1998).

Local inflammatory cytokines trigger KS lesion development

The hypothesis that KS behaves as a cytokine-mediated disease is supported by the observation that the inflammatory cytokines increased in KS patients or in high-risk individuals and found at increased levels in the lesions from all forms of KS support the growth of KSCs. A variety of these cytokines is expressed in lesions from all forms of KS. Infiltrating lymphocytes and monocytes/macrophages produce IFN-γ, TNF, IL-1, IL-6, granulocyte–monocyte colony stimulating factor (GM-CSF) and others (Oxholm et al, 1989; Miles et al, 1990; Stürzl et al, 1995, 1997a; Fiorelli et al, 1998; Sirianni et al, 1998). Interleukins 1 and 6 and GM-CSF are also produced by E-KSCs and activated

Table 19.2. *Spindle cells present in KS lesions or cultured from lesions or from the peripheral blood show the phenotype of endothelial or macrophagic cells, and express activation markers*

Marker	Specificity	E-KSC	M-KSC	Circulating KSC progenitors
VWF:Ag	Vascular endothelium	$-/+$*	$-$	$-$†
CD34	Vascular endothelium and haematopoietic cell progenitors	$+$	$-$	$-$†
VE-cadherin	Vascular endothelium, endothelial macrophages	$+$	n.d.	$+$
CD31	Macrophages, endothelial cells	$+$	$+$	$+$
CD14	Monocytes–macrophages	$-$	$+$	$+$
CD68	Tissue macrophages	$-$	$+$	$+$
CD36	Macrophages, capillary endothelium	$+$	$+$	n.d.
HLA-DR	Macrophages, activated endothelial cells, others	$+$	$+$	$+$
PAM-1	Macrophages	$-$	$+$	$+$
CD45	Leukocytes	$-$	$+$	$+$
CD4	T cells, monocytes–macrophages	$-$	$+$	$+$‡
ICAM-1	Macrophages, activated endothelial cells, others	$+$	$+$	$+$
VCAM-1	Activated endothelial cells, others	$+$	n.d.	n.d.
ELAM-1	Activated endothelial cells	$+$	$-$	n.d.
CD40	Vascular endothelium	$+$	n.d.	n.d.
$\alpha_5\beta_1$ and $\alpha_v\beta_3$	Activated endothelial cells, others	$+$	n.d.	n.d.

Endothelial and macrophagic KSC (E-KSC and M-KSC, respectively) derived from lesions, KSC derived from blood after 6–7 days of culture and frozen sections of AIDS-KS and CKS lesions were stained by immunohistochemistry for the indicated markers (Regezi et al, 1993; Ensoli et al, 1994a; Fiorelli et al, 1995, 1998; MacPhail et al, 1996; Pammer et al, 1996; Colombini et al, 1998; Sirianni et al, 1998). Cultured cells were grown with inflammatory cytokines with modification in some cytokine content, as described in the text and elsewhere (Fiorelli et al, 1995, 1998; Sirianni et al, 1998).
*Positive in early stage KS, lost in late stages. †In a previous report few weakly positive cells have been detected (Browning et al, 1994). ‡No expression has been detected in one report (Browning et al, 1994). Key: +, positive expression; −, negative expression; n.d., not done.

endothelial cells (Ensoli et al, 1989; Miles et al, 1990; Yang et al, 1994a; Stürzl et al, 1995; Faris et al, 1998).

These inflammatory cytokines induce endothelial cells to acquire the E-KSC phenotype (Nakamura et al, 1988; Salahudin et al, 1988; Ensoli et al, 1989) (Table 19.3). Specifically, they induce in normal endothelial cells:

1. The spindle morphology (Barillari et al, 1992); a modulation of the expression of several surface markers, including a downregulation of VWF:Ag expression due to its release (Fiorelli et al, 1995, 1998), as observed for cultured E-KSCs and in late KS lesions; activation of ELAM-1, ICAM-1, VCAM-1, and HLA-DR expression (Yang et al, 1994b; Fiorelli et al, 1995).
2. The expression of integrin receptors ($\alpha v\beta3$, $\alpha5\beta1$) at levels similar to those detected in E-KSCs (Barillari et al, 1993; Fiorelli et al, 1995); as discussed below, these receptors mediate the effects of extracellular HIV-1 Tat protein (Ensoli et al, 1990; Barillari et al, 1992, 1993; Fiorelli et al, 1995).
3. The production of angiogenic factors such as basic fibroblastic growth factor (bFGF), vascular endothelial growth factor (VEGF), IL-8, platelet-derived growth factor (PDGF) and other cytokines and chemokines with effects on cell recruitment, proliferation and angiogenesis. In addition, the inoculation of these inflammatory cytokines or cytokine-activated endothelial cells into nude mice induces vascular lesions of mouse cell origin closely resembling early KS (Fiorelli et al, 1995; Samaniego et al, 1997) and indistinguishable from the lesions induced by inoculation of cultured KSCs (Salahuddin et al, 1988; Ensoli et al, 1994a,b) (see Figure 19.1C).

We interpret these results as indicating that KSCs represent an activated functional state of endothelial cells. This hypothesis is in agreement with histological findings indicating that the very early stage of KS is characterized by the presence of activated, proliferating endothelial cells (McNutt et al, 1983; Dorfman and Path, 1984; Ruszczak et al, 1987a,b; Ensoli et al, 1991).

Table 19.3 *Local and systemic expression of inflammatory cytokines and activity in KS*

Factor	Expression in lesions	Increase of serum levels	Possible role in KS pathogenesis
IL-1α,β	+	+	Induction of E-KSC proliferation (mediated by bFGF)
			Activation and induction of a spindle morphology in endothelial cells
			Cell recruitment and differentiation
IL-6	+	+	Role in KS pathogenesis still debated
TNFα,β	+	+	Induction of E-KSC proliferation
			Activation and induction of a spindle morphology in endothelial cells
			Cell recruitment and differentiation
			KS progression after systemic inoculation
IFN-γ	+	+	Activation and induction of a spindle morphology in endothelial cells
			Cell recruitment and differentiation
			Induction of Tat-responsiveness
			KS progression after systemic inoculation
			In vitro growth of lesional M-KSC
OSM	?	n.d.	Induction of E-KSC proliferation

Inflammatory cytokines are mostly produced by infiltrating inflammatory cells (CD8 T cells and monocytes–macrophages) and some others by KSCs (IL-1, IL-6). These cytokines are also increased in high-risk patients and KS patients, and synergize in inducing both systemic and tissue-localized effects (see Table 19.1). E-KSC, endothelial KSC; OSM, oncostatin M; n.d., not done; +, increased; ?, conflicting data.

Inflammatory cytokines may also play a role in disease progression (Tables 19.1 and 19.3). By using conditioned media (TCM) derived from retrovirus-infected CD4+ T cells or from activated primary T cells or PBMCs, rich in IFN-γ, TNFα, IL-1, IL-2, IL-6 and OSM, it is possible to establish and maintain in long-term culture E-KSCs from KS lesions (Nakamura et al, 1988; Barillari et al, 1992). These inflammatory cytokines are increased in KS lesions and stimulate growth of KSCs. They include TNF, IL- I (Ensoli et al, 1989; Barillari et al, 1992) and OSM, which has been found to be a strong growth factor for E-KSCs (Miles et al, 1992; Nair et al, 1992; Cai et al, 1994; Faris et al, 1996). The effect of inflammatory cytokines on KSC growth is mediated by a synergistic stimulation of bFGF production and release (Samaniego et al, 1995, 1997; Fiorelli et al, 1995, 1998; Wijelath et al, 1997; Faris et al, 1998). In turn, bFGF acts as an autocrine KSC growth factor (Ensoli et al, 1989, 1994a,b; Samaniego et al, 1995; Faris et al, 1998) (Table 19.4). Further, the treatment of E-KSCs with TCM, or the combined inflammatory cytokines contained in it, increases not only the release of angiogenic cytokines but also the proliferation of endothelial cells induced by KSCs and the ability of cultured KSCs to induce KS-like lesions after inoculation in nude mice (Ensoli et al, 1989; Samaniego et al, 1995). Inflammatory cytokines also support the establishment of M-KSCs from the lesions (see Tables 19.1 and 19.3). However, this cell type requires slight modifications in culture conditions for growth (Sirianni et al, 1998).

Inflammatory cytokines are also able to maintain in culture KS-derived TILs with the same phenotype as those found in situ in KS lesions, whereas in the absence of TCM these cells undergo apoptosis and disappear rapidly (Sirianni et al, 1998) (Table 19.1).

Interferon gamma appears to be the major mediator of the changes induced by inflammatory cytokines on KSCs and normal endothelial cells, although the other cytokines,

Table 19.4 *Expression and activity of growth and angiogenic factors in KS*

Factor	Expression* in vivo	Expression* in vitro	Possible role in KS pathogenesis
bFGF	+	+	Potent inducer of E-KSC proliferation and angiogenesis Synergy with HIV-1 Tat protein and VEGF in induction of angiogenesis and oedema
VEGF	+	+	Synergy with bFGF in inducing angiogenesis and oedema
IL-8	+	+	Chemotaxis of endothelial cells, KSCs and other cells of KS lesions
PDGF-A	+	+	Induction of KSC proliferation and angiogenesis
PDGF-B	+	?	Induction of KSC proliferation
SF/HGF	+	+	May induce KSC proliferation and contribute to angiogenesis
aFGF	+	+	May contribute to KSC growth and angiogenesis (more data required)
MCP-1	+	+	Chemoattractive for monocytes
GM-CSF	+	+	Induction of monocyte differentiation to macrophages and dendritic cells
PAF	n.d.	+	May contribute to angiogenesis (more data required)
TGFβ	+	+	Induction of KSC proliferation (more data required)

*Expression in primary lesions and in cultured KSCs.
E-KSC, endothelial KSC; n.d., not done; +, increased; ?, conflicting data.

particularly IL-1 and TNF, contribute to these effects in a synergistic fashion (Montesano et al, 1985; Fiorelli et al, 1995, 1998; Samaniego et al, 1995, 1997) (see Table 19.3). In addition, IFN-γ upregulates CD40 expression in cultured E-KSCs (Pammer et al, 1996). CD40 is also highly expressed by KSCs of AIDS-KS and CKS lesions and by vascular endothelial cells in areas within and adjacent the tumour (Pammer et al, 1996). This and the HLA-DR expression in KS lesions indicate that IFN-γ is active on KSCs and endothelial cells present in the lesions. By signalling through CD40, IFN-γ may also be able to prevent apoptosis through the induction of the bcl-2 proto-oncogene (Liu et al, 1991) which is expressed in lesional KSCs and endothelial cells during progression of all forms of KS (Bohan-Morris et al, 1996).

Although inflammatory cytokines induce endothelial cells to acquire the phenotype of E-KSCs, a few differences still exist. These include the lack of production of VEGF (Samaniego et al, 1998) and the lack of growth response to RGD peptides in inflammatory cytokine-activated endothelial cells (Barillari et al, 1998). Endothelial KSCs, in contrast, produce VEGF, its expression being increased by inflammatory cytokines (Weindel et al, 1992; Cornali et al, 1996; Samaniego et al, 1998), and proliferate in response to RGD peptides, suggesting alterations in their integrin pathway (Barillari et al, 1998). Moreover, in contrast to inflammatory cytokine-activated endothelial cells, E-KSCs do not proliferate in response to VEGF, perhaps because of the low expression of flt-1, one of the VEGF receptors (Samaniego et al, 1998). These data suggest that E-KSCs have acquired a 'transdifferentiated' phenotype, although they are not transformed or tumorigenic in SCID mice.

Taken together, these results indicate that local production of inflammatory cytokines is capable of triggering a cascade of events leading to lesion formation and to maintenance and progression of KS.

INFLAMMATORY CYTOKINES UPREGULATE HHV-8 AND HIV INFECTION

Human herpesvirus 8 has been identified in KS lesions and shown to be present in all epidemiological forms of KS (Chang et al, 1994; Ambroziak et al, 1995; Boshoff et al, 1995; De Lellis et al, 1995; Huang et al, 1995; Moore and Chang, 1995; Schalling et al, 1995; Albini et al, 1996b; Buonaguro et al, 1996). It has been postulated that HHV-8 is the causal agent of KS, and recent studies indicate that HHV-8 infection can be predictive of KS development (Whitby et al, 1995; Gao et al, 1996; Moore et al, 1996). This virus is particularly prevalent in geographical areas with a high incidence of KS, where the virus can also be present in patients without KS and in normal blood donors (Bigoni et al, 1996; Lennette et al, 1996; Monini et al, 1996a; Blackbourn et al, 1997; Viviano et al, 1997; Rezza et al, 1998), but with a lower viral load or prevalence than in KS patients (Corbellino et al, 1996; Decker et al, 1996; Gao et al, 1996; Kedes et al, 1996; Rezza et al, 1998).

The virus is detected in circulating B cells (Ambroziak et al, 1995; Mesri et al, 1995; Harrington et al, 1996; Blackbourn et al, 1997), but recent data indicate that it is also present in monocytes–macrophages from peripheral blood and lesions (Blasig et al, 1997; Orenstein et al, 1997; Sirianni et al, 1998; Colombini et al, 1998) and, more rarely (in advanced KS), in T cells (Harrington et al, 1996; Sirianni et al, 1997b; Colombini et al, 1998). It is also detected in the circulating KSC progenitors (Sirianni et al, 1997a; Colombini et al, 1998). As B cells are few or absent in KS lesions, the data suggest that circulating monocytes and KSC progenitors may play a role in the recruitment of the virus into tissues.

At the lesional level, HHV-8 is present in endothelial cells and KSCs mostly in a latent form (Boshoff et al, 1995; Li et al, 1996; Rainbow et al, 1997; Staskus et al, 1997; Stürzl et al, 1997b), whereas mononuclear cells including monocytes–macrophages are lytically infected (Blasig et al, 1997; Orenstein et al, 1997) and may support virus production and spread to other cell types. The recruitment of lytically infected monocytes into KS tissues has also been demonstrated by in situ hybridization studies (Blasig et al, 1997). In addition, the virus is lost upon culture of E-KSCs from the lesions (Lebbé et al, 1995; Dictor et al, 1996; Monini et al, 1996b), but it is maintained in the M-KSC cultures derived from the lesions (Sirianni et al, 1998).

More recent data indicate that the same inflammatory cytokines found increased in KS lesions can maintain and rescue viral infection in PBMCs from KS patients and KS risk groups by activating viral lytic replication in B cells and monocytes–macrophages (Colombini et al, 1998) and allowing viral transmission to other cells (see Table 19.1). In addition, increased IFN-γ and DR activation can be found in early lesions prior to HHV-8 detection (Fiorelli et al, 1998), suggesting that inflammatory cytokines are at least partially responsible for the higher viral load and spread in KS patients and in individuals at high risk of developing KS.

Finally, inflammatory cytokines activate HIV-1 transcription, replication and production of Tat in infected cells (Chang et al, 1995). Specifically, TNF, IL-1, GM-CSF and other inflammatory cytokines activate viral gene expression and, in combination, exert synergistic effects on viral replication (see Table 19.1) (Barillari et al, 1992). At the same time, by increasing the HIV-1 promoter activity, these cytokines synergize with Tat in activating and enhancing viral gene expression and replication (Chang et al, 1995) and further production of Tat. Tat, in turn, activates inflammatory cytokine production in infected cells (Buonaguro et al, 1994; Chang et al, 1995), suggesting that a vicious cycle may be established in which HIV uses both viral and host factors to increase infectivity and transmission. As Tat may represent a progression factor for AIDS-KS (see below), the effect of inflammatory cytokines on its expression may potentiate its action in KS.

Although more studies are needed to understand the role of viruses in KS development, these results suggest that the virus–host interplay is modulated by the same inflammatory cytokines inducing cell recruitment, endothelial cell activation, angiogenesis and KS cell growth.

THE ROLE OF ANGIOGENIC, CHEMOTACTIC AND GROWTH FACTORS

The molecular and biochemical characterization of the factors produced in vitro and in primary lesions by KSCs indicates that these cells constitutively express a variety of angiogenic molecules, growth factors, cytokines and chemokines (Table 19.4). The capability of E-KSCs to induce angiogenesis in the chorioallantoic membrane assay and KS-like lesions upon inoculation in nude mice provided the first experimental evidence that angiogenic factors are involved in KS lesion formation (Salahuddin et al, 1988; Ensoli et al, 1989; Ensoli and Gallo, 1994). These E-KSC-induced lesions are of mouse cell origin, regress within 2 weeks and are mediated by specific angiogenic factors produced by the cells (Figure 19.1B). Among these factors, bFGF appears to play a major role in KS lesion development.

Basic fibroblastic growth factor is very highly expressed by lesional KSCs in all forms of KS. High levels of bFGF are also released in the extracellular fluid by cultured KSCs and inflammatory cytokine-activated endothelial cells, and in absence of cell death or permeability changes (Ensoli et al, 1989, 1994a,b; Samaniego et al, 1995, 1997). This bFGF induces the migration, invasion and growth of endothelial cells and KSCs in vitro and in vivo (Ensoli et al, 1989, 1994a,b) (Table 19.4). When inoculated in nude mice, bFGF induces vascular lesions very similar to those induced by KSCs (Ensoli et al, 1994a) (Figure 19.1C). Inhibition studies with specific neutralizing antibodies or antisense oligodeoxynucleotide directed against bFGF mRNA demonstrated that bFGF is required for KS-like lesion formation induced by E-KSCs in mice (Ensoli et al, 1994b). Moreover, as mentioned above, inflammatory cytokines induce bFGF production and release in both E-KSCs and normal endothelial cells (Samaniego et al, 1995, 1997; Faris et al, 1998; Fiorelli et al, 1998) and induce KS-like lesions in nude mice that are mediated by bFGF and are indistinguishable from those induced by E-KSCs (Barillari et al, 1998). Interferon gamma is also essential for this effect, although TNF and IL-1 contribute and synergize with IFN-γ to induce production and release of bFGF and the angiogenic phenotype of endothelial cells (Fiorelli et al, 1995, 1998; Samaniego et al, 1995, 1997; Barillari et al, 1998).

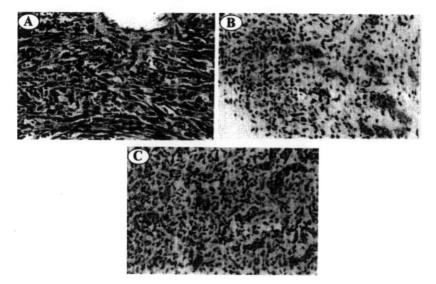

Figure 19.1 *Histology of a human KS lesion and KS-like lesions induced by inoculation of nude mice with KSCs or bFGF*
Kaposi's sarcoma lesions (A) are characterized by intense angiogenesis, proliferation of spindle-shaped cells, inflammatory cell infiltrate and oedema. Similar lesions are induced by subcutaneous inoculation of nude mice with cultured E-KSCs (B) or with bFGF (C). These lesions are of mouse cell origin, are transient and regress in 2 weeks. However, the inflammatory cell infiltrate and oedema are less prominent when bFGF is inoculated alone, suggesting that other factors are involved in lesion formation (haematoxylin–eosin staining).

Although bFGF appears to be key to KS lesion formation, the inflammatory cell infiltrate and oedema are less prominent in bFGF-induced lesions than in E-KSC-induced lesions (Figure 19.1C). This observation suggested earlier that other factors are involved in KS lesion formation.

Vascular endothelial growth factor is another angiogenic factor expressed in KS lesions and by E-KSCs as the two secreted forms, VEGF121 and VEGF165. Production of VEGF is induced by inflammatory cytokines and other cytokines found in KS lesions such as PDGF-B (Cornali et al, 1996; Samaniego et al, 1998). It synergizes with bFGF in inducing not only endothelial cell proliferation and angiogenesis in vitro and in vivo, but also oedema, as demonstrated by injecting the two cytokines alone or in combination in guinea pigs (Cornali et al, 1996; Samaniego et al, 1998) (Table 19.4). Endothelial KSCs express in vitro both VEGF receptors (KDR/FLK-1 and flt-1); however, flt-1 is at a much lower level, suggesting an explanation for the lack of stimulation of E-KSC growth (Cornali et al, 1996; Samaniego et al, 1998). Nevertheless, VEGF seems to mediate the growth of the two KS tumour cell lines established from KS (Masood et al, 1997).

Other molecules are found in KS and they may participate in lesion formation (Table 19.4). Scatter factor/hepatocyte growth factor (SF/HGF) induces KSC morphology in endothelial cells and KSCs proliferation. It is expressed in KSCs together with its cognate receptor, the c-met protein (Naidu et al, 1994; Maier et al, 1996).

Platelet-derived growth factor-B is also mitogenic for KSCs. It is expressed in vivo in cells infiltrating KS lesions, while its receptor is expressed on KSCs, suggesting that it may activate KSC proliferation in a paracrine fashion (Stürzl et al, 1992, 1995).

Interleukin 8, a chemokine active on a wide range of immune cells and with angiogenic properties, is expressed in KS lesions in vivo, in cultured KSCs and in inflammatory cytokine-activated endothelial cells (Sciacca et al, 1994). However, this cytokine seems to activate the migration rather than the proliferation of KSCs and endothelial cells (B. Ensoli, unpublished data).

Interleukin 1 is also expressed in cultured KSCs (Ensoli et al, 1989) and in vivo in KS lesions (Stürzl et al, 1995, 1997a). It is mitogenic for KSCs, but its effect seems to be mediated by the induction of bFGF (Ensoli et al, 1994b; Samaniego et al, 1995, 1997; Faris et al, 1998). In addition, IL-1 promotes leukocyte recruitment and synergizes with TNF and IFN-γ to induce endothelial cell activation and the acquisition of the KSC phenotype (Fiorelli et al, 1995, 1998).

Interleukin 6 and GM-CSF, produced by inflammatory infiltrating cells, KSCs and activated endothelial cells, are weak or controversial inducers of KSC growth (Bussolino et al, 1989; Ensoli et al, 1989; Miles et al, 1990; Yang et al, 1994a; Murakami-Mori et al, 1995). However, they may contribute to the leukocyte recruitment and differentiation (Mantovani et al, 1997).

Another chemoattractant detected in KSCs in vivo and in vitro, monocyte chemotactic protein 1 (MCP-1) (Sciacca et al, 1994; Mantovani et al, 1997), may amplify monocyte recruitment into KS lesions (Sciacca et al, 1994; Stürzl et al, 1997a).

Several other molecules are expressed in KSCs, including Acidic FGF (Ensoli et al, 1989; Werner et al, 1989; Corbeil et al, 1991; Li et al, 1993), FGF-6 (Li et al, 1993; Roth, 1993), platelet-activating factor (PAF) (Sciacca et al, 1994; Bussolino et al, 1995), PDGF-(Ensoli, et al, 1989; Roth et al, 1989; Werner et al, 1990; Stürzl et al, 1995) and transforming growth factor β (TGFβ) (Ensoli et al, 1989; Roth, 1993; Williams et al, 1995), but their role in KS lesion formation remains to be determined (Table 19.4).

KAPOSI'S SARCOMA AND AIDS

All the data reported herein do not explain why AIDS-associated Kaposi's sarcoma is more common and has a more aggressive course than other clinical forms. The disease is rare in people infected with HIV-2. Furthermore, AIDS patients are 300 times more likely to develop KS than individuals with primary immunodeficiency (Beral et al, 1990).

Transgenic mice bearing the HIV-1 *tat* gene develop KS-like lesions or, depending on the levels of transgene expression, tumours of various cell origin (Vogel et al, 1988; Corallini et al, 1993). Further, KS-like lesions occur mostly in male mice, just as KS occurs predominantly in men. This and other studies indicate that the Tat protein of HIV-1 may be responsible for the aggressive nature of AIDS-KS.

Tat is a viral protein produced early during the infection of T cells by HIV-1; it trans-activates viral gene expression and is essential for optimal virus replication (Arya et al, 1985; Fisher et al, 1986; Ensoli et al, 1990, 1993; Chang et al, 1995) (Table 19.5). During acute infection of T cells by HIV-1, Tat is also released into the extracellular fluid in the absence of cell death and via a leaderless secretion pathway as a biologically active protein (Ensoli et al, 1990, 1993; Chang et al, 1997). Extracellular Tat is able to induce some levels of viral replication in neighbouring cells (Ensoli et al, 1993) and can exert activating/proliferating effects on KSCs and on cells of the vascular system. Extracellular Tat is capable of inducing the growth, migration and invasion of E-KSCs (Ensoli et al, 1990, 1993, 1994a; Barillari et al, 1992; Albini et al, 1995) and of inflammatory cytokine-activated endothelial cells (Barillari et al, 1992, 1993; Albini et al, 1995; Fiorelli et al, 1995) (Table 19.5). Tat also induces endothelial cells to differentiate in tube-like structures on a matrigel substrate and to express the 72 kDa collagenase IV, known to be associated with angiogenesis and tumour growth (Barillari et al, 1993; Ensoli et al, 1994a; Albini et al, 1995). When Tat is immobilized on culture plates, it induces adhesion of both normal vascular cells and KSCs (Barillari et al, 1993). As mentioned above, the activities of Tat on normal endothelial cells require the previous exposure of cells to the inflammatory cytokines increased in KS patients, and IFN-γ is essential for this effect (Barillari et al, 1992, 1993; Fiorelli et al, 1995, 1998) (see Tables 19.3 and 19.5). Inflammatory cytokines induce the Tat cell responsiveness by promoting both the induction of $\alpha_5\beta_1$ and $\alpha_v\beta_3$ integrins, which function as the receptors for Tat, and the induction of bFGF production (Barillari et al, 1993, 1997, 1998; Ensoli et al, 1994a; Fiorelli et al, 1995, 1998; Samaniego et al, 1995, 1997; Faris et al, 1998). In turn, bFGF is capable of inducing the same integrins (Strömblad and Cheresh, 1996) and, as discussed below, is the final mediator of the cell growth and in vivo angiogenic effects induced by the viral protein (Ensoli et al, 1994a; Barillari et al, 1997). The integrins $\alpha_v\beta_3$ and $\alpha_5\beta_1$ and bFGF are constitutively expressed by E-KSCs which are able to respond to Tat in the absence of other stimuli (Barillari et al, 1993, 1997, 1998).

Consistent with these data, inoculation of Tat alone in nude mice does not lead to angiogenesis. However, when Tat is inoculated in the presence of suboptimal (non-lesion-forming) amounts of bFGF or heparin, it greatly enhances bFGF-mediated KS-like lesion formation in terms of both number of mice developing lesions and intensity of the histological alterations, including angiogenesis and KSC growth (Albini et al, 1994; Ensoli et al, 1994a). Synergistic angiogenic effects in vivo are also observed by inoculating mice with a combination of inflammatory cytokines and Tat, since the cytokines induce both integrins and bFGF expression (Barillari et al, 1998; Fiorelli et al, 1998) (Table 19.5).

Tat angiogenic activities are mediated by two different domains: the basic region, a heparin-binding domain, and the RGD region, which mediates binding to the $\alpha_v\beta_3$ and $\alpha_5\beta_1$ integrins. The heparin-binding basic sequence of Tat competes with bFGF for binding to heparan sulfate proteoglycans of the cell surface and ECM (Chang et al, 1997). By this competitive effect Tat releases ECM-bound bFGF and maintains it in a soluble form available for growth induction (Barillari et al, 1997, 1998). At the same time, the RGD region of Tat binds the $\alpha_v\beta_3$ and $\alpha_5\beta_1$ integrins (Barillari et al, 1993), induces the phosphorylation of the focal adhesion kinase p125 FAK (B. Ensoli, unpublished data) and promotes cell adhesion, migration and invasion (Barillari et al, 1993, 1997, 1998). However, bFGF released by the Tat basic region represents the final mediator of Tat-induced cell growth, whereas Tat-induced cell adhesion is required and increases the growth response to bFGF (Ensoli et al, 1994a). These data may explain why Tat synergizes with bFGF but not with VEGF in inducing angiogenesis in vivo (Barillari et al, 1997). This, in fact, is due to the ability of Tat to bind to the bFGF-induced integrins $\alpha_v\beta_3$ and $\alpha_5\beta_1$ but not to $\alpha_v\beta_5$ which is induced by VEGF and is involved in its angiogenic pathway (Strömblad and Cheresh, 1996; Barillari et al, 1997, 1998).

Altogether, the data suggest that Tat exerts its angiogenic activity via a molecular mimicry of ECM proteins such as fibronectin and vitronectin, two adhesion proteins required for angiogenesis. In addition, the heparin-binding activity of Tat leads to the retrieval of angiogenic factors present in the lesions.

Extracellular Tat is detectable in AIDS-KS lesions (Ensoli et al, 1994a). In addition, lesional endothelial cells and KSCs express both bFGF and $\alpha_v\beta_3$ and $\alpha_5\beta_1$-Tat receptors, and extracellular Tat costains with these receptors on KSCs and activated vessels (Ensoli et al, 1994a), suggesting that the mechanisms described here are operative in vivo and that Tat may explain the higher frequency and aggressiveness of KS in the setting of HIV-1 infection.

Table 19.5 *Effects of the HIV-1 Tat protein in the course of KS*

System/targets	*Extracellular Tat protein activity*
Activated EC	Induces proliferation, migration, invasion, adhesion, differentiation and increases cytokine/chemokine production
E-KSC	Induces proliferation, migration, invasion, adhesion Increases cytokine/chemokine production, ELAM-1, VCAM-1 and ICAM-1 activation
Monocytes–macrophages	Induces adhesion and extravasation
In vivo (mice)	Synergy with bFGF or inflammatory cytokines in inducing angiogenesis and KS-like lesions in mice
HHV-8-infected cells	Increases viral load
	Tat protein activity in HIV-1-infected or tat-transfected cells
Bcl-2	Increases expression in PBMCs
HIV-1-infected cells	Enhances viral replication, activates inflammatory cytokine production and TGFβ production

See text for details. EC, endothelial cell. E-KSC, endothelial KSC.

Other data further support a role of Tat in AIDS-KS. Tat basic region can bind KDR-1 (Albini et al, 1996b), one of the VEGF receptors expressed in KS lesions in vivo (Brown et al, 1996). However, KS tumour cell lines but not E-KSCs proliferate with VEGF, suggesting that this cytokine is more likely to have a role in transformed–progressed KS.

Tat has also been shown to promote the adhesion of monocytes–macrophages to vessels with production of collagenases and vascular damage, and to increase their migration and extravasation into tissues (Lafrenie et al, 1996) (Table 19.5).

The other mechanism by which Tat may contribute to AIDS-KS pathogenesis is through the activation of cellular gene expression, especially cytokine genes involved in KS pathogenesis (Chang et al, 1995). For example, Tat activates TNFα and β (Buonaguro et al, 1994), IL-6 (Scala et al, 1994) and other genes in HIV-1-infected cells. As an extracellular protein, Tat can induce TGFβ production in monocytes–macrophages (Zauli et al, 1992), ELAM- I expression in endothelial cells (Hofman et al, 1993) and VCAM-1, ICAM-1, MCP-1 and IL-6 in E-KSCs (Domingo et al, 1998). Other data suggest that Tat can increase HHV-8 viral load (Harrington et al, 1997) and Bcl-2 expression (Zauli et al, 1995) (Table 19.5).

The presence of detectable extracellular Tat in sera from AIDS patients (Westendorp et al, 1995) and in AIDS-KS lesions (Ensoli et al, 1994a) supports the hypothesis of its role as a progression factor in AIDS-KS.

ONCOGENE EXPRESSION IN KS

Only a few oncogenes have been found to be expressed in KS. Among these are *bcl-2*, *p53*, *c-myc*, *ras* and *int-2*.

Recent data indicate that Bcl-2 is expressed in endothelial cells and KSCs of the lesions from all KS clinical forms and that its expression increases with lesion stage, reaching maximal levels in nodular lesions (Bohan-Morris et al, 1996; Montmayeur et al, 1997). Bcl-2 is also induced during angiogenesis (Strömblad and Cheresh, 1996) upon ligation of $\alpha_v\beta_3$ integrin, suggesting that its expression is associated with cell proliferation and angiogenesis. The reasons for the induction of Bcl-2 are under study; however, preliminary results suggest that the same inflammatory cytokines and angiogenic factors present in KS lesions upregulate Bcl-2 expression in endothelial cells and KSCs (Sgadari et al, 1998). As mentioned above, IFN-γ can contribute to Bcl-2 upregulation by inducing the expression of CD40 (Pammer et al, 1996). Moreover, HIV-1 Tat protein can also induce Bcl-2 expression (Zauli et al, 1995) (Table 19.5).

The role of Bcl-2 in KS is proved by the high remission rate obtained in KS patients with paclitaxel (Taxol) as a single agent in phase I clinical trials (Saville et al, 1995). This compound is known to inhibit Bcl-2 function (Haldar et al, 1995) by phosphorylating the protein on a serine residue. Moreover, recent data indicate that paclitaxel can inhibit E-KSC growth and KS-like lesion formation in nude mice (Sgadari et al, 1998). Thus, Bcl-2 expression coupled with cell growth stimuli may divert cells from apoptosis towards continued cell proliferation and this may represent a step toward the lesion transformation and monoclonality that has been observed in some nodular KS lesions.

Heterozygous *p53* mutations have been detected in KS tissues (Scincariello et al, 1994) and p53 has also been detected by immunohistochemistry in late-stage KS lesions but not in early lesions; however, very few cells (about 1%) of the lesions express detectable p53 (Bergman et al, 1996; Montmayeur et al, 1997), so its role in KS pathogenesis has still to be determined.

Expression of c-myc is upregulated by PDGF-B in cultured E-KSCs and downregulated by IFN-Con1, a derivative of IFN-α, which is cytostatic on E-KSCs and has KS therapeutic effects (Köster et al, 1996). Moreover, c-myc-specific antisense oligodeoxynucleotides inhibit specifically the proliferation and the migration of E-KSCs in vitro. In addition, in vivo c-myc expression in KSCs increases in late nodular KS lesions as compared to early lesions (Köster et al, 1996). This indicates that c-myc regulates KSC proliferation and migration and may have a key role in disease progression (Köster et al, 1996).

A significant overexpression of Ras protein has been observed in CKS, without correlation with disease stage (Spandidos et al, 1990). Ongoing studies suggest that ras activation may represent the pathway toward which many stimuli convert. Int-2 (FGF-3) mRNA and protein have also been detected in KS (Huang et al, 1993b). However, the significance of these findings is yet undetermined and requires further study.

The data reported above support the hypothesis that the expression of Bcl-2, p53 and c-myc in late nodular KS may be involved in KS progression and suggest that in later stages of development KS may transform from a reactive process to a true sarcoma (Figure 19.2).

CONCLUSION

The data described here suggest that early Kaposi's sarcoma is a reversible, reactive process that in its later stages can progress to a true sarcoma (Costa and Rabson, 1983; Brooks, 1986). The 'multicentric' nature of KS without evidence of real metastasis (Costa and Rabson, 1983), the symmetrical dermatome distribution of the lesions (Brooks, 1986), the lack of aneuploidy (Fukunaga and Silverberg, 1990; Kaaya et al, 1992), the spontaneous regression observed in PKS (KIepp et al, 1978; Harwood et al, 1979; Hoshaw and Schwartz, 1980; Zisbrod et al, 1980; Akhtar et al, 1984; Bencini et al, 1993) and AIDS-KS (Real and Krown, 1985; Janier et al, 1985), the regression of AIDS-KS lesions upon therapy with HIV-1 protease inhibitors (Conant et al, 1997) and the induction of reversible KS-like lesions in mice inoculated with KS spindle cells (Salahuddin et al, 1988; Ensoli et al, 1989, Ensoli and Gallo, 1994) all support the concept that early KS is not a true sarcoma, but an inflammatory–angiogenic lesion triggered by inflammatory cytokines and infection with HHV-8.

We propose that the immunoactivation observed in individuals at risk of developing KS leads to an increased production of inflammatory cytokines which, in turn, induce the production of angiogenic molecules, growth factors and cytokines/chemokines that mediate lesion formation (Figure 19.2). Early lesions can be indistinguishable from granulation tissue, which also contains spindle-shaped cells, while typical crops of tumour cells are absent. Inflammatory cytokines, which seem to have a pivotal role in early KS pathogenesis, induce endothelial cells to acquire the KSC phenotype and trigger leukocyte recruitment and differentiation of monocytes into macrophages, endothelial macrophages and dendritic cells. Inflammatory cytokines also promote KSC proliferation and angiogenesis by inducing angiogenic factors and increase replication of

HHV-8 and HIV-1. The continuous stimulation of reactive KSCs by inflammatory cytokines, growth factors, HHV-8 and HIV-1 Tat protein may result in the transformation of a reactive KS lesion to a real sarcoma (Figure 19.2). This hypothesis is supported by the increased expression of oncogenes such as *bcl-2* and *p53* in late-stage KS lesions, and by the observation of clonality in some nodular KS lesions. Moreover, two cell lines have been established from late-stage KS patients which, in contrast to the E-KSCs and M-KSCs usually cultured from KS lesions and described so far, present karyotype abnormalities and tumorigenicity in SCID mice. This suggests that in some cases a clone of KS cells may transform, and that at this stage immunodeficiency may play an important

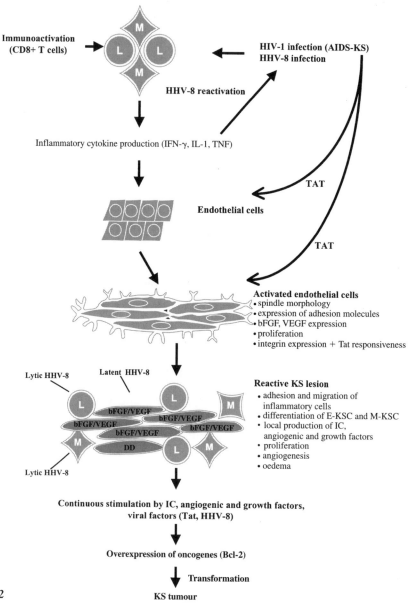

Figure 19.2

Figure 19.2 *Key events in KS pathogenesis*
Immunoactivation increases the systemic levels of inflammatory cytokines (IC), including (IFN-γ, interleukin 1 (IL-1) and tumour necrosis factor (TNF). This causes endothelial cell activation and increased expression of adhesion molecules. These events support extravasation of inflammatory cells including CD8+ T cells and monocytic-macrophagic cells into the tissues and secretion of inflammatory cytokines. Local inflammatory cytokines cause phenotypic transformation of endothelial cells and monocytic/macrophagic cells to E-KSCs and M-KSCs, respectively. Alternatively or in addition to this event, inflammatory cytokines induce the expansion of circulating KSC precursors (of monocytic origin) which may also be recruited into the tissues and differentiate to macrophagic/endothelial KSCs (endothelial macrophages). Factors secreted by the KSCs and inflammatory cells amplify these events and stimulate KSC proliferation (bFGF, PDGF), angiogenesis (bFGF, VEGF, SF/HGF, PDGF-B), oedema and further recruitment of T cells, monocytes and other immune cells (MCP-1, IL-8). At the same time, infected monocytes–macrophages and circulating KSC progenitors recruit HHV-8 into tissues. In AIDS-KS infected lymphocytes and monocytes also recruit HIV into tissue. Local inflammatory cytokines upregulate HHV-8 and HIV infection and increase viral load, creating a vicious cycle of virus–host interactions that amplify these events. In individuals infected with HIV-1, the Tat protein released systemically or in loco by HIV-infected cells binds KSCs and activated endothelial cells, enhances KSC growth, angiogenesis and Bcl-2 expression, and further upregulates inflammatory cytokine production. In the course of this mutual stimulation among the different cell types, tumour transformation may occur over time, as indicated by the increased expression of proto-oncogenes (e.g. *bcl-2*) in advancing KS lesions. In the presence of immunodeficiency, and particularly in HIV-1-infected individuals, this may cause the transformation of a reactive and potentially reversible early KS lesion to a true sarcoma. L, lymphocytes; M, monocyte–macrophages; DD, dermal dendritic cells. The darkest shaded cells represent KSCs (modified from Ensoli and Gallo, 1994).

role. Progression of KS into a real sarcoma – more common in AIDS-KS patients, particularly from Africa – may indeed require a profound immunodeficiency.

The complex interaction among all the factors involved in KS pathogenesis is not yet completely understood and requires further study, particularly concerning the role of HHV-8. However, the recognition of two stages in the course of the disease is important for therapeutic intervention. Early-stage KS occurs in the absence of immunodeficiency, is mediated and supported by cytokines, and can regress. In contrast, late-stage KS may be growth-independent, it is associated with immunodeficiency, does not regress and is often resistant to conventional therapies. A pathogenetic therapy targeting specific factors and the monitoring of markers associated with KS development or progression, such as activation markers, levels of circulating KSCs, Bcl-2 expression, HHV-8 viral load and the grade of the immunodeficiency, may have prognostic value and address disease treatment.

REFERENCES

Aboulafia D, Miles SA, Saks SR et al (1989) Intravenous recombinant tumor necrosis factor in the treatment of AIDS-related Kaposi's sarcoma. *J AIDS* **2:** 54–58.

Akhtar M, Bunar H, Ashraf-Ali M et al (1984) Kaposi's sarcoma in renal transplant recipients. Ultrastructural and immunoperoxidase study of four cases. *Cancer* **53:** 258.

Albini A, Fontanini G, Masiello L et al (1994) Angiogenic potential in vivo by Kaposi's sarcoma cell-free supernatants and HIV-1 Tat product: inhibition of KS-like lesions by tissue inhibitor of metalloproteinase-2. *AIDS* **8:** 1237–1244.

Albini A, Barillari G, Benelli R et al (1995) Angiogenic properties of human immunodeficiency virus type I Tat protein. Proc Natl Acad Sci USA 92: 4838–4842.

Albini A, Aluigi MG, Benelli R et al (1996a) Oncogenesis in HIV-infection: KSHV and Kaposi's sarcoma. Int J Oncol 9: 5–8.

Albini A, Soldi R, Giunciuglio D et al (1996b) The angiogenesis induced by HIV- I Tat protein is mediated by the Flk/l/KDR receptor on vascular endothelial cells. Nature Med 2: 1371–1375.

Albrecht H, Stellbrink H-J, Gross G et al (1994) Treatment of atypical leishmaniasis with interferon γ resulting in progression of Kaposi's sarcoma in an AIDS patient. Clin Invest 72: 1041–1047.

Ambroziak JA, Blackbourn DJ, Herndier BG et al (1995) Herpes-like sequences in HIV-infected and uninfected Kaposi's sarcoma patients. Science 268: 582–583.

Arya SK, Guo C, Josephs SF et al (1985) Transactivator gene of human T-lymphotropic virus type III (HTLV-III). Science 229: 69–73.

Barillari G, Buonaguro L, Fiorelli V et al (1992) Effects of cytokines from activated immune cells on vascular cell growth and HIV- I gene expression: implications for AIDS–Kaposi's sarcoma pathogenesis. J Immunol 149: 3727–3734.

Barillari G, Gendelman R, Gallo RC et al (1993) The Tat protein of human immunodeficiency virus type 1, a growth factor for AIDS Kaposi's sarcoma and cytokine-activated vascular cells, induces adhesion of the same cell types by using integrin receptors recognizing the RGD amino acid sequence. Proc Natl Acad Sci USA 90: 7941–7945.

Barillari G, Fiorelli V, Gendelman R et al (1997) HIV-1 Tat protein enhances angiogenesis and Kaposi's sarcoma (KS) development triggered by inflammatory cytokines (IC) or bFGF by engaging the αvβ3 integrin. J AIDS Hum Retrovirol 14: A33.

Barillari G, Gendelman R, Fiorelli V et al (1998) HIV-1 Tat protein promotes angiogenesis triggered with inflammatory cytokines by retrieving heparin-bound basic fibroblast growth factor and by engaging the αvβ3 integrin (submitted for publication).

Baroni CD, Vitolo D, Remotti D et al (1987) Immunohistochemical heterogeneity of macrophage subpopulations in human lymphoid tissues. Histopathology 11: 1029–1042.

Beckstead JH (1992) Oral presentation of Kaposi's sarcoma in a patient without severe immunodeficiency. Arch Pathol Lab Med 116: 543–545.

Bedi GC, Westra WH, Farzadegan H et al (1995) Microsatellite instability in primary neoplasms from HIV+ patients. Nature Med 1: 65–68.

Bencini PL, Montagnino G, Tarantino A et al (1993) Kaposi's sarcoma in kidney transplant recipient. Arch Dermatol 129: 248–250.

Beral V, Peterman TA, Berkelman RL et al (1990) Kaposi's sarcoma among persons with AIDS: a sexually transmitted infection? Lancet 335: 123–128.

Bergman R, Ramon M, Kilim S et al (1996) An immunohistochemical study of p53 protein expression in classical Kaposi's sarcoma. Am J Dermatol 18: 367–370.

Bigoni B, Dolcetti R, De Lellis L et al (1996) Human herpesvirus 8 is present in the lymphoid system of healthy persons and can reactivate in the course of AIDS. J Infect Dis 173: 542–549.

Blackbourn DJ, Ambroziak J, Lennette E et al (1997) Infectious human herpesvirus 8 in a healthy North American blood donor. Lancet 349: 609–611.

Blasig C, Zietz C, Haar B et al (1997) Monocytes in Kaposi's sarcoma lesions are productively infected by human herpesvirus-8. J Virol 10: 7963–7968.

Bohan-Morris C, Gendelman R, Marrogi AJ et al (1996) Immunohistochemical detection of bcl-2 in AIDS-associated and classical Kaposi's sarcoma. Am J Pathol 148: 1055–1063.

Boshoff C, Acholz T, Kennedy MM et al (1995) Kaposi's sarcoma-associated herpesvirus infects endothelial and spindle cells. Nature Med 1: 1274–1278.

Brooks JJ (1986) Kaposi's sarcoma: a reversible hyperplasia. Lancet ii: 1309–1311.

Brown LF, Tognazzi K, Dvorak HF et al (1996) Strong expression of kinase insert domain-containing receptor, a vascular permeability factor/vascular endothelial growth factor receptor in AIDS-associated Kaposi's sarcoma and cutaneous angiosarcoma. Am J Pathol 148: 1065–1074.

Browning PJ, Sechler JMG, Kaplan M et al (1994) Identification and culture of Kaposi's sarcoma-like spindle cells from the peripheral blood of HIV-1-infected individuals and normal controls. *Blood* **84:** 2711–2720.

Buonaguro L, Buonaguro FM, Giraldo G et al (1994) The human immunodeficiency virus type 1 Tat protein transactivates tumor necrosis factor beta gene expression through a TAR-like structure. *J Virol* **68:** 2677–2682.

Buonaguro FM, Tornesello ML, Beth-Giraldo E et al (1996) Herpesvirus-like DNA sequences detected in endemic, classic, iatrogenic and epidemic Kaposi's sarcoma (KS) biopsies. *Int J Cancer* **65:** 25–28.

Bussolino F, Wang JM & De Filippi P (1989) Granulocyte and granulocyte–macrophages colony stimulating factors induce human endothelial cells to migrate and proliferate. *Nature* **337:** 471–473.

Bussolino F, Arese M, Montrucchio G et al (1995) Platelet activating factor produced in vitro by Kaposi's sarcoma cells induces and sustains in vivo angiogenesis. *J Clin Invest* **96:** 940–952.

Byckley PJ, Dickson SA & Walker WS (1985) Human splenic sinusoidal lining cells express antigens associated with monocytes, macrophages, endothelial cells and T lymphocytes. *J Immunol* **134:** 2310–2315.

Cai BJ, Gill PS, Masood P et al (1994) Oncostatin-M is an autocrine growth factor in Kaposi's sarcoma. *Am J Pathol* **145:** 74–79.

Cai BJ, Zheng T, Lotz M et al (1997) Glucocorticoids induce Kaposi's sarcoma cell proliferation through the regulation of transforming growth factor-β. *Blood* **89:** 1491–1500.

Caruso A, Gonzales R, Stellini R et al (1990) Interferon-γ marks activated T lymphocytes in AIDS patients. *AIDS Res Hum Retroviruses* **6:** 899–940.

Chang Y, Cesarman E, Pessin MS et al (1994) Identification of herpesvirus-like DNA sequences in AIDS-associated Kaposi's sarcoma. *Science* **266:** 1865–1869.

Chang HC, Gallo RC & Ensoli B (1995) Regulation of cellular gene expression and function by the human immunodeficiency virus type 1 Tat protein. *J Biomed Sci* **2:** 189–202.

Chang HC, Samaniego F, Nair BC et al (1997) HIV-1 Tat protein exits from cells via leaderless secretory pathway and binds to extracellular matrix-associated heparan sulfate proteoglycans through its basic region. *AIDS* **11:** 1421–1431.

Civati G, Busnach G, Brando B et al (1988) Occurrence of Kaposi's sarcoma in renal transplant recipients treated with cyclosporine. *Transplant Proc* **20:** 924.

Colditz IG & Watson DL (1992) The effect of cytokines and chemotactic agonists on the migration of T lymphocytes into skin. *Immunology* **76:** 272.

Colombini S, Monini P, Stürzl M et al (1998) Activation and persistence of human herpesvirus-8 infection in B cells and monocyte–macrophages by inflammatory cytokines increased in Kaposi's sarcoma. (submitted for publication).

Conant MA, Opp KM, Poretz D et al (1997) Reduction of Kaposi's sarcoma lesions following treatment of AIDS with ritonavir. *AIDS* **11:** 1300–1301.

Corallini A, Altavilla G, Pozzi L et al (1993) Systemic expression of HIV-1 *tat* gene in transgenic mice induces endothelial proliferation and tumors of different histotypes. *Cancer Res* **53:** 5569–5575.

Corbeil J, Evans LA, Vasak E et al (1991) Culture and properties of cells derived from Kaposi sarcoma. *J Immunol* **146:** 2972–2976.

Corbellino M, Poirel L, Bestetti G et al (1996) Restricted tissue distribution of extralesional Kaposi's sarcoma-associated herpesvirus-like DNA sequences in AIDS patients with Kaposi's sarcoma. *AIDS Res Human Retroviruses* **12:** 651–657.

Cornali E, Zietz C, Benelli R et al (1996) Vascular endothelial growth factor regulates angiogenesis and vascular permeability in Kaposi's sarcoma. *Am J Pathol* **149:** 1851–1869.

Costa J & Rabson AS (1983) Generalized Kaposi's sarcoma is not a neoplasm. *Lancet* **i:** 58.

Cotran RS & Pober JS (1988) Endothelial activation: its role in inflammatory and immune reaction. In: Simionescu N & Simionescu M (eds) *Endothelial Cell Biology*, pp. 335–347, New York: Plenum.

Decker LL, Shankar P, Khan G et al (1996) The Kaposi's sarcoma-associated herpesvirus (KSHV) is present as an intact latent genome in KS tissue but replicates in the peripheral blood mononuclear cells of KS patients. *J Exp Med* **184:** 283–288.

De Lellis L, Fabris M, Cassai E et al (1995) Herpesvirus-like DNA sequences in non-AIDS Kaposi's sarcoma. *J Infect Dis* **172:** 1605–1607.

DePaoli P, Caffau C, D'Andrea M et al (1994) Serum levels of intercellular adhesion molecule 1 in patients with HIV-1 related Kaposi's sarcoma. *J AIDS* **7:** 695–699.

Dictor M, Ferno M & Baldetorp B (1991) Flow cytometric DNA content in Kaposi's sarcoma by histologic stage: comparison with angiosarcoma. *Anal Quant Cytol Histol* **13:** 201–208.

Dictor M, Rambech E, Way D et al (1996) Human herpesvirus 8 (Kaposi's sarcoma-associated herpesvirus) DNA in Kaposi's sarcoma lesions, AIDS Kaposi's sarcoma cell lines, endothelial Kaposi's sarcoma simulators, and the skin of immunosuppressed patients. *Am J Pathol* **148:** 2009–2016.

Domingo Kelly G, Ensoli B, Gunthel CJ et al (1998) Induction of inflammatory response genes by purified Tat in Kaposi's sarcoma cells. (submitted for publication).

Dorfman RF & Path FRC (1984) The histogenesis of Kaposi's sarcoma. *Lymphology* **17:** 76–77.

Ensoli B & Gallo RC (1994) Growth factors in AIDS-associated Kaposi's sarcoma: cytokines and HIV- I Tat protein. In: DeVita VT, Hellman S & Rosenberg SA (eds) *Cancer – Principles and Practice of Oncology*, pp 1–12. Philadelphia: Lippincott.

Ensoli B & Gallo RC (1995) AIDS-associated Kaposi's sarcoma: a new perspective of its pathogenesis and treatment. *Proc Assoc Am Physicians* **107:** 8–18.

Ensoli B, Nakamura S, Salahuddin SZ et al (1989) AIDS-Kaposi's sarcoma-derived cells express cytokines with autocrine and paracrine growth effects. *Science* **243:** 223–226.

Ensoli B, Barillari G, Salahuddin SZ et al (1990) The Tat protein of HIV-1 stimulates growth of cells derived from Kaposi's sarcoma lesions of AIDS patients. *Nature* **345:** 84–86.

Ensoli B, Barillari G & Gallo RC (1991) Pathogenesis of AIDS-related Kaposi's sarcoma. *Hematol Oncol Clin North Am* **5:** 281–295.

Ensoli B, Buonaguro L, Barillari G et al (1993) Release, uptake, and effects of extracellular HIV-1 Tat protein on cell growth and viral transactivation. *J Virol* **67:** 277–287.

Ensoli B, Gendelman R, Markham P et al (1994a) Synergy between basic fibroblast growth factor and human immunodeficiency virus type I Tat protein in induction of Kaposi's sarcoma. *Nature* **371:** 674–680.

Ensoli B, Markham P, Kao V et al (1994b) Block of AIDS-Kaposi's sarcoma (KS) cell growth, angiogenesis and lesion formation in nude mice by antisense oligonucleotides targeting basic fibroblast growth factor: a novel strategy for the therapy of KS. *J Clin Invest* **94:** 1736–1746.

Fagiolo U, Cossarizza A, Scala E et al (1993) Increased cytokine production in mononuclear cells of healthy elderly people. *Eur J Immunol* **23:** 2375–2378.

Fagnoni FF, Vescovini R, Mazzola M et al (1996) Expansion of cytotoxic CD8$^+$ CD28$^-$ T cells in healthy aging people, including centenarians. *Immunology* **88:** 501.

Fan J, Bass HZ & Fahev JL (1993) Elevated IFNγ and decreased IL-2 gene expression are associated with HIV-1 infection. *J Immunol* **151:** 5031–5040.

Faris M, Ensoli B, Stahl N et al (1996) Differential activation of the ERK, Jun Kinase and Janus Kinase-Stat pathways by oncostatin M and basic fibroblast growth factor in AIDS-derived Kaposi's sarcoma cells. *AIDS* **10:** 369–378.

Faris M, Ensoli B, Kokot N et al (1998) Inflammatory cytokines induce the expression of bFGF isoforms required for growth of Kaposi's sarcoma and endothelial cells through the activation of AP-1 response elements of the bFGF promoter. *AIDS* **12:** 19–27.

Fiorelli V, Markham P, Gendelman R et al (1995) Cytokines from activated T cells induce normal endothelial cells to acquire the phenotypic and functional features of AIDS-Kaposi's sarcoma spindle cells. *J Clin Invest* **95:** 1723–1734.

Fiorelli V, Gendelman R, Sirianni MC et al (1998) γ-Interferon produced by CD8+ T cells infiltrating Kaposi's sarcoma induces spindle cells with angiogenic phenotype and synergy with HIV-1 Tat protein: an immune response to HHV-8 infection? *Blood* **91:** 956–967.

Fisher AG, Feinberg MB, Josephs SE et al (1986) The trans-activator gene of HTLV-III is essential for virus replication. *Nature* **320:** 367–371.

Franceschi S & Geddes M (1995) Epidemiology of classic Kaposi's sarcoma, with special reference to Mediterranean population. *Tumori* **81:** 308–314.

Friedman-Kien AE (1981) Disseminated Kaposi's sarcoma syndrome in young homosexual men. *J Am Acad Dermatol* **5:** 468–471.

Friedman-Kien AE, Saltzman BR, Cao Y et al (1990) Kaposi's sarcoma in HIV-negative homosexual men. *Lancet* **i:** 168–169.

Fukunaga M & Silverberg SG (1990) Kaposi's sarcoma in patients with acquired immune deficiency syndrome. A flow cytometric DNA analysis of 26 lesions in 21 patients. *Cancer* **66:** 758–764.

Gao S, Kingsley L, Hoover DR et al (1996) Seroconversion to antibodies against Kaposi's sarcoma-associated herpesvirus-related latent nuclear antigens before the development of Kaposi's sarcoma. *N Engl J Med* **335:** 233–241.

Geddes M, Franceschi M, Balzi D et al (1995) Birthplace and classic Kaposi's sarcoma in Italy. *J Natl Cancer Inst* **87:** 1015–1017.

Gendelman R, Fiorelli V, Kao V et al (1994) Spindle cells from both AIDS-associated and classical Kaposi's sarcoma (KS) lesions are of endothelial cell origin and present an activated phenotype. *AIDS Res Hum Retroviruses* **10:** S101.

Gill PS, Loureiro C, Bernstein-Singer M et al (1989) Clinical effects of glucocorticoids on Kaposi's sarcoma related to the acquired immunodeficiency syndrome (AIDS). *Ann Intern Med* **110:** 937–940.

Gill PS, Tsai Y, Rao AP et al (1997) Clonality in Kaposi's sarcoma. *N Engl J Med* **337:** 570–571.

Guo WX & Antakly T (1995) AIDS-related Kaposi's sarcoma: evidence for direct stimulatory effect of glucocorticoid on cell proliferation. *Am J Pathol* **146:** 727–734, 1995.

Guo WX, Antakly T, Cadotte M et al (1996) Expression and cytokine regulation of glucocorticoid receptors in Kaposi's sarcoma. *Am J Pathol* **148:** 1999–2008.

Haldar S, Jena N & Croce CM (1995) Inactivation of bcl-2 by phosphorylation. *Proc Natl Acad Sci USA* **92:** 4507–4511.

Harrington MJ, Bagasra 0, Sosa CE et al (1996) Human herpesvirus type 8 DNA sequences in cell-free plasma and mononuclear cells of Kaposi's sarcoma patients. *J Infect Dis* **174:** 1101–1105.

Harrington W, Sieczkowski L, Sosa C et al (1997) Activation of HHV-8 by HIV-1 Tat. *Lancet* **349:** 774.

Harwood AR, Osoba SD, Hofstadler SL et al (1979) Kaposi's sarcoma in recipients of renal transplants. *Am J Med* **67:** 759.

Haverkos HW & Drotman DP (1985) Prevalence of Kaposi's sarcoma among patients with AIDS. *N Engl J Med* **312:** 1518.

Hober D, Haque A, Wattre P et al (1989) Production of tumor necrosis factorα (TNFα) and interleukin 1 (IL-1) in patients with AIDS. Enhanced level of TNFα is related to higher cytotoxic activity. *Clin Exp Immunol* **78:** 329–333.

Hofman FM, Wright AD, Dohadwala MM et al (1993) Exogenous Tat protein activates human endothelial cells. *Blood* **82:** 2774–2780.

Honda M, Kitamura K, Mizutani Y et al (1990) Quantitative analysis of serum IL-6 and its correlation with increased levels of serum IL2R in HIV-1 -induced disease. *J Immunol* **145:** 4059–4064.

Hoshaw R & Schwartz R (1980) Kaposi's sarcoma after immunosuppressive therapy: prednisone. *Arch Dermatol* **116:** 1280.

Huang YQ, Friedman-Kien AE, Li JJ et al (1993a) Cultured Kaposi's sarcoma cell lines express factor XIIIa, CD14, and VCAM-1, but not factor VIII or ELAM-1. *Arch Dermatol* **129:** 1291–1296.

Huang YQ, Li JJ, Moscatelli D et al (1993b) Expression of the *int-2* oncogene in Kaposi's sarcoma lesions. *J Clin Invest* **91:** 1191–1197.

Huang YQ, Li JJ, Kaplan MH et al (1995) Human herpesvirus-like nucleic acid in various forms of Kaposi's sarcoma. *Lancet* **345:** 759–762.

Janier M, Vignon MD & Cottenot F (1985) Spontaneously healing Kaposi's sarcoma in AIDS. *N Engl J Med* **312:** 1638–1639.

Kaaya EE, Parravicini C, Sundelin B et al (1992) Spindle cell ploidy and proliferation in endemic and epidemic African Kaposi's sarcoma. *Eur J Cancer* **11:** 1890–1894.

Kaaya EE, Parravicini C, Ordonez C et al (1995) Heterogeneity of spindle cells in Kaposi's sarcoma: comparison of cells in lesions and culture. *J AIDS Hum Retroviruses* **10:** 295–305.

Kalinkovich A, Livshits G, Engelmann H et al (1993) Soluble tumor necrosis factor receptors (sTNF-R) and HIV infection: correlation to CD8+ lymphocytes. *Clin Exp Immunol* **93:** 350–355.

Kedes DH, Operskalski E, Busch M et al (1996) The seroepiderniology of human herpesvirus-8 (Kaposi's sarcoma-associated herpesvirus): distribution of infection in KS risk groups and evidence for sexual transmission. *Nature Med* **2:** 918–924.

Kestens L, Melbye M, Biggar RJ et al (1985) Endemic African Kaposi's sarcoma is not associated with immunodeficiency. *Int J Cancer* **36:** 49–54.

Klepp O, Dahl O & Stenwig J (1978) Association of Kaposi's sarcoma and prior immunosuppressive therapy. A five-year material of Kaposi's sarcoma in Norway. *Cancer* **42:** 2626.

Köster R, Blatt LM, Streubert M et al (1996) Consensus-interferon and platelet-derived growth factor adversely regulate proliferation and migration of Kaposi's sarcoma cells by control of c-myc expression. *Am J Pathol* **149:** 1871–1885.

Kraffert C, Planus L & Penneys NS (1991) Kaposi's sarcoma: further immunohistologic evidence of a vascular endothelial origin. *Arch Dermatol* **127:** 1734–1735.

Krigel RL, Padavic-Shaller KA, Rudolph AR et al (1989) Exacerbation of epidemic Kaposi's sarcoma with a combination of interleukin-2 and gamma interferon: results of a phase 2 study. *J Biol Response Mod* **8:** 359–365.

Kusnoki Y, Hirai Y, Kyoizumi S et al (1992) Evidence for in vivo clonal proliferation of unique population of blood CD4⁻/CD8⁻ T cells bearing T-cell receptor α and β chains in two normal men. *Blood* **79:** 2965.

Lafrenie RM, Wahl LM, Epstein JS et al (1996) HIV-1 Tat protein promotes chemotaxis and invasive behavior by monocytes. *J Immunol* **157:** 974–977.

Lahdevirta J, Maury CPJ, Teppo AM et al (1988) Elevated levels of circulating cachectin/tumor necrosis factor in patients with acquired immunodeficiency syndrome. *Am J Med* **85:** 289–291.

Lampugnani MG, Resnati M, Raitieri M et al (1992) A novel endothelial-specific membrane protein is a marker of cell–cell contacts. *J Cell Biol* **118:** 1511–1522.

Lebbé C, de Crémoux P, Rybojad M et al (1995) Kaposi's sarcoma and new herpesvirus. *Lancet* **345:** 1180.

Lennette ET, Blackbourn DJ & Levy JA (1996) Antibodies to human herpesvirus type-8 in the general population and in Kaposi's sarcoma patients. *Lancet* **348:** 858–861.

Lepe-Zanuga U, Mansell TWA & Hersh EM (1987) Idiopathic production of interleukin-1 in acquired immunodeficiency syndrome. *J Clin Microbiol* **5:** 1695–1700.

Levin MJ, Murray M, Rotbart HA et al (1992) Immune response of elderly individual to live attenuated varicella vaccine. *J Infect Dis* **166:** 253–259.

Levy JA & Ziegler JL (1983) Acquired immunodeficiency syndrome is an opportunistic infection and Kaposi's sarcoma results from secondary immune stimulation. *Lancet* **ii:** 78–81.

Li JJ, Huang YQ, Moscatelli D et al (1993) Expression of fibroblast growth factors and their receptors in acquired immunodeficiency syndrome associated Kaposi's sarcoma tissue and derived cells. *Cancer* **72:** 2253–2259.

Li JJ, Huang YQ, Cockerell CJ et al (1996) Localization of human herpes-like virus type 8 in vascular endothelial cells and perivascular spindle-shaped cells of Kaposi's sarcoma lesions by in situ hybridization. *Am J Pathol* **148:** 1741–1748.

Liu YJ, Mason DY, Johnson GD et al (1991) Germinal center cells express bcl-2 protein after activation by signals which prevent their entry into apoptosis. *Eur J Immunol* **21:** 1905–1910.

Lunardi-Iskandar Y, Gill P, Lam V et al (1995) A neoplastic cell line (KS Y-1) from Kaposi's sarcoma (KS): evidence that late stage KS can be a true malignancy. *J Natl Cancer Inst* **87:** 974–981.

MacPhail LA, Dekker NP & Regezi JA (1996) Macrophages and vascular adhesion molecules in oral Kaposi's sarcoma. *J Cutan Pathol* **23:** 464–472.

Maier JA, Mariotti M, Albini A et al (1996) Over-expression of hepatocyte growth factor in human Kaposi's sarcoma. *Int J Cancer* **65:** 168–172.

Mantovani A, Bussolino F & Introna M (1997) Cytokine regulation of endothelial cell function: from molecular level to the bedside. *Immunol Today* **18:** 231–240.

Masood R, Cai J, Zheng T et al (1997) Vascular endothelial growth factor/vascular permeability factor is an autocrine growth for AIDS-Kaposi's sarcoma. *Proc Natl Acad Sci USA* **94:** 979–984.

Master SP, Taylor JF, Kyalwazi SK et al (1970) Immunologic studies in Kaposi's sarcoma in Uganda. *BMJ* **1:** 600–602.

Maurice PD, Smith NP & Pinching AJ (1982) Kaposi's sarcoma with benign course in a homosexual. *Lancet* **i:** 571.

McNutt NS, Fletcher V & Conant MA (1983) Early lesions of Kaposi's sarcoma in homosexual men. An ultrastructural comparison with other vascular proliferations in skin. *Am J Pathol* **111:** 62–77.

Mesri EA, Cesarman E, Arvanitakis L et al (1995) Human herpesvirus-8/Kaposi's sarcoma-associated herpesvirus is a new transmissible virus that infects B cells. *J Exp Med* **183:** 2385–2390.

Miles SA, Rezai AR, Salazar-Gonzalez JF et al (1990) AIDS Kaposi sarcoma-derived cells produce and respond to interleukin 6. *Proc Natl Acad Sci USA* **87:** 4068–4072.

Miles SA, Martinez-Maza O, Rezai A et al (1992) Oncostatin M as a potent mitogen for AIDS-KS-derived cells. *Science* **255:** 1432–1434.

Mitsuyasu RT (1993) Clinical aspects of AIDS-related Kaposi's sarcoma. *Curr Opin Oncol* **5:** 835–844.

Monini P, De Lellis L, Fabris M et al (1996a) Kaposi's sarcoma-associated herpesvirus DNA sequences in prostate tissue and human sperm. *N Engl J Med* **334:** 1168–1172.

Monini P, Rotola A, De Lellis L et al (1996b) Latent BK virus infection and Kaposi's sarcoma pathogenesis. *Int J Cancer* **66:** 717–722.

Montesano R, Orci L & Vassalli P (1985) Human endothelial cell cultures: phenotypic modulation by leukocyte interleukins. *J Cell Physiol* **122:** 424–434.

Montmayeur F, Krajewski S, Bejui-Thivolet F et al (1997) In vivo patterns of apoptosis-regulating protein expression in Kaposi's sarcoma. *Appl Immunohistochem* **5:** 104–110.

Moore PS & Chang Y (1995) Detection of herpesvirus-like DNA sequences in Kaposi's sarcoma in patients with and those without HIV infection. *N Engl J Med* **332:** 1181–1185.

Moore PS, Kingsley L, Holmberg SD et al (1996) Kaposi's sarcoma-associated herpesvirus infection prior to onset of Kaposi's sarcoma. *AIDS* **10:** 175–180.

Murakami-Mori K, Taga T, Kishimoto T et al (1995) AIDS-associated Kaposi's sarcoma (KS) cell express oncostatin M (OM)-specific receptor but not leukemia inhibitory factor/OM receptor or interleukin-6 receptor. *J Clin Invest* **96:** 1319–1327.

Nadji M, Morales AR, Zigler-Weissman J et al (1981) Kaposi's sarcoma: immunohistologic evidence for an endothelial origin. *Arch Pathol Lab Med* **105:** 274–275.

Naidu YM, Rosen EM, Zitnick R et al (1994) Role of scatter factor in the pathogenesis of AIDS-related Kaposi's sarcoma. *Proc Natl Acad Sci USA* **91**: 5281–5285.

Nair BC, DeVico AL, Nakamura S et al (1992) Identification of a major growth factor for AIDS-Kaposi's sarcoma cells as oncostatin M. *Science* **255**: 1430–1432.

Nakamura S, Salahuddin SZ, Biberfeld P et al (1988) Kaposi's sarcoma cells: long-term culture with growth factor from retrovirus-infected CD4+ T cells. *Science* **242**: 427–430.

Nickoloff BJ & Griffiths CRM (1989) Factor XIIIa-expressing dermal dendrocytes in AIDS-associated cutaneous Kaposi's sarcoma. *Science* **243**: 1736–1737.

Orenstein JM, Alkan S, Blauvelt A et al (1997) Visualization of human herpesvirus type 8 in Kaposi's sarcoma by light and transmission electron microscopy. *AIDS* **11**: 35–45.

Oxholm A, Oxholm P, Permin H et al (1989) Epidermal tumour necrosis factor (alpha) and interleukin 6-like activities in AIDS-related Kaposi's sarcoma. *Acta Pathol Micro Scand* **97**: 533–538.

Pammer J, Plettenberg A, Weninger W et al (1996) CD40 antigen is expressed by endothelial cells and tumor cells in Kaposi's sarcoma. *Am J Pathol* **148**: 1387–1396.

Penn I (1979) Kaposi's sarcoma in organ transplant recipients: report of 20 cases. *Transplantation* **27**: 8–11.

Pober JS & Cotran RS (1990) Cytokines and endothelial cell biology. *Phys Rev* **70**: 427–451.

Porcelli S, Morita CA & Brenner MB (1992) CD1β restricts the response of human CD4⁻CD8⁻ T lymphocytes to a microbial agent. *Nature* **360**: 593.

Rabkin CS, Bedi G, Musaba E et al (1995) AIDS-related Kaposi's sarcoma is a monoclonal neoplasm. *Clin Cancer Res* **1**: 257–260.

Rabkin CS, Janz S, Lash A et al (1997) Monoclonal origin of multicentric Kaposi's sarcoma lesions. *N Engl J Med* **336**: 988–993.

Rainbow L, Platt GM, Simpson GR et al (1997) The 222- to 223-kilodalton latent nuclear protein (LNA) of Kaposi's sarcoma-associated herpesvirus (human herpesvirus 8) is encoded by orf 73 and is a component of the latency-associated nuclear antigen. *J Virol* **71**: 5915–5921.

Real FX & Krown SE (1985) Spontaneous regression of Kaposi's sarcoma in patients with AIDS. *N Engl J Med* **313**: 1659.

Real FX, Krown, SE & Kroziner B (1986) Steroid-related development of Kaposi's sarcoma in a homosexual man with Burkitt's lymphoma. *Am J Med* **80**: 119.

Regezi SA, MacPhail LA, Daniels TE et al (1993) Human immunodeficiency virus-associated oral Kaposi's sarcoma: a heterogeneous cell population dominated by spindle-shaped endothelial cells. *Am J Pathol* **43**: 240–249.

Rezza G, Lennette ET, Giuliani M et al (1998) Prevalence and determinants of anti-lytic and anti-latent antibodies to HHV-8 among Italian individuals at risk of sexually and parenterally transmitted infections. *Int J Cancer* **77**: 361–365.

Rizzardini G, Piconi S, Ruzzante S et al (1996) Immunological activation markers in the serum of African and European HIV-seropositive and seronegative individuals. *AIDS* **10**: 1535–1542.

Rizzieri DA, Liu J, Traweek ST et al (1997) Clearance of HHV-8 from peripheral blood mononuclear cells with a protease inhibitor. *Lancet* **349**: 775–776.

Roth WK (1993) TGF-β and FGF-like growth factors involved in the pathogenesis of AIDS-associated Kaposi's sarcoma. *Res Virol* **144**: 105–109.

Roth WK, Werner S, Schirren CU et al (1989) Depletion of PDGF from serum inhibits growth of AIDS-related and sporadic Kaposi's sarcoma cells in culture. *Oncogene* **4**: 483–487.

Roth WK, Brandstetter H & Stürzl M (1992) Cellular and molecular features of HIV-associated Kaposi's sarcoma. *AIDS* **6**: 895–913.

Rothman S (1962) Some clinic aspects of Kaposi's sarcoma in the European and American populations. *Acta Unio Internationalis Contra Cancrum* **18**: 364.

Ruszczak Z, Mayer-Da Silva A & Orfanos CE (1987a) Angioproliferative changes in clinically noninvolved, perilesional skin in AIDS-associated Kaposi's sarcoma. *Dermatologica* **175**: 270–279.

Ruszczak Z, Mayer-Da Silva A & Orfanos CE (1987b) Kaposi's sarcoma in AIDS. *Am J Dermatopathol* **9**: 388–398.

Safai B & Good RA (1981) Kaposi's sarcoma, a review and recent development. *CA Cancer J Clin* **31**: 2–12.

Safai B, Johnson KG, Myskowski PL et al (1985) The natural history of Kaposi's sarcoma in the acquired immunodeficiency syndrome. *Ann Intern Med* **103**: 744–750.

Saikevych IA, Mayer M, White RL et al (1988).Cytogenetic study of Kaposi's sarcoma associated with acquired immunodeficiency syndrome. *Arch Pathol Lab Med* **112**: 825–828.

Salahuddin SZ, Nakamura S, Biberfeld P et al (1988) Angiogenic properties of Kaposi's sarcoma-derived cells after long-term culture in vitro. *Science* **242**: 430–433.

Samaniego F, Markham P, Gallo RC et al (1995) Inflammatory cytokines induce AIDS-Kaposi's sarcoma-derived spindle cells to produce and release basic fibroblast growth factor and enhance Kaposi's sarcoma-like lesion formation in nude mice. *J Immunol* **154**: 3582–3592.

Samaniego F, Markham PD, Gendelman R et al (1997) Inflammatory cytokines induce endothelial cells to produce and release basic fibroblast growth factor and to promote Kaposi's sarcoma-like lesions in nude mice. *J Immunol* **158**: 1887–1894.

Samaniego F, Markham PD, Gendelman R et al (1998) Vascular endothelial growth factor and basic endothelial growth factor present in Kaposi's sarcoma (KS) are induced by inflammatory cytokines and synergize to induce vascular permeability and KS lesion development. *Am J Pathol* **152**: 1433–1443.

Saville MW, Lietzau J, Pluda JM et al (1995) Treatment of HIV-associated Kaposi's sarcoma with paclitaxel. *Lancet* **346**: 26–28.

Scala G, Ruocco MR, Ambrosino C et al (1994) The expression of the interleukin 6 gene is induced by the human immunodeficiency virus I Tat protein. *J Exp Med* **179**: 961–971.

Schalling M, Ekman M, Kaaya EE et al (1995) A role for a new herpesvirus (KSHV) in different forms of Kaposi's sarcoma. *Nature Med* **1**: 707–708.

Schlessinger MYN, Chu FN, Badamchian M et al (1994) A distinctive form of soluble CD8 is secreted by stimulated CD8+ cells in HIV-1-infected and high risk individuals. *Clin Immunol Immunopathol* **73**: 252–260.

Schulhafer EP, Grossman ME, Fagin G et al (1987) Steroid-induced Kaposi's sarcoma in a patient with pre-AIDS. *Am J Med* **82**: 313–317.

Sciacca FL, Stürzl M, Bussolino F et al (1994) Expression of adhesion molecules, platelet-activating factor, and chemokines by Kaposi's sarcoma cells. *J Immunol* **153**: 4816–4825.

Scincariello F, Dolan MJ, Nedelcu I et al (1994) Occurrence of human papillomavirus and *p53* gene mutations in Kaposi's sarcoma. *J Virol* **203**: 153–157.

Sgadari C, Toschi E, Palladino C et al (1998) Block of experimental Kaposi's sarcoma by Taxol is due to apoptosis following inhibition of bcl-2 expression. (submitted for publication)..

Shivakumar S, Tsokos GC & Datta SK (1989) T cell receptor α/β expressing double negative (CD4–/CD8–) and CD4+ T helper cells in humans augment the production of pathogenic anti-DNA autoantibodies associated with lupus nephritis. *J Immunol* **143**: 103.

Siegal B, Levinston-Kriss S, Schiffer A et al (1990) Kaposi's sarcoma in immunosuppression: possibly the result of dual viral infection. *Cancer* **65**: 492–498.

Sirianni MC, Uccini S, Angeloni A et al (1997a) Circulating spindle cells: correlation with human herpesvirus-8 (HHV-8) infection and Kaposi's sarcoma. *Lancet* **349**: 225.

Sirianni MC, Vincenzi L, Topino S et al (1997b) Human herpesvirus 8 DNA sequences in CD8+ T cells. *J Infect Dis* **176**: 541.

Sirianni MC, Vincenzi L, Fiorelli V et al (1998) γ-Interferon production in peripheral blood mononuclear cells (PBMC) and tumour infiltrating lymphocytes from Kaposi's sarcoma patients: correlation with the presence of human herpesvirus-8 in PBMC and lesional macrophages. *Blood* **91**: 968–976.

Slavin G, Cameron HM & Singh H (1969) Kaposi's sarcoma in mainland Tanzania: a report of 117 cases. *Br J Cancer* **23**: 349–357.

Söderberg-Nauclér C, Fish KN & Nelson JA (1997) Reactivation of latent human cytomegalovirus by allogeneic stimulation of blood cells from healthy donors. *Cell* **91**: 119–126.

Spandidos DA, Kaloterakis A, Yiagnisis M et al (1990) Ras, c-myc and c-erbB-2 oncoprotein expression in non-AIDS Mediterranean Kaposi's sarcoma. *Anticancer Res* **10**: 1619–1626.

Staskus KA, Zhong W, Gebhard K et al (1997) Kaposi's sarcoma-associated herpesvirus gene expression in endothelial (spindle) tumor cells. *J Virol* **71**: 715–719.

Strömblad S & Cheresh DA (1996) Cell adhesion and angiogenesis. *Cell Biol* **6**: 462–468.

Stürzl M, Brandstetter H & Roth WK (1992) Kaposi's sarcoma: a review of gene expression and ultrastructure of KS spindle cells in vivo. *AIDS Res Hum Retroviruses* **8**: 1753–1763.

Stürzl M, Brandstetter H, Zietz C et al (1995) Identification of interleukin-1 and platelet-derived growth factor-B as major mitogens for the spindle cells of Kaposi's sarcoma: a combined in vitro and in vivo analysis. *Oncogene* **10**: 2007–2016.

Stürzl M, Cornali E, Gahlemann C et al (1997a) Expression of growth factors, cytokines, and chemokines in Kaposi's sarcoma in HIV and cytokines. In: Guneounou M (ed.) *Research in HIV and Cytokines*, pp 349–357. Paris, les Edition, INSERM.

Stürzl M, Blasig C, Schreier A et al (1997b) Expression of HHV-8 latency-associated T0.7 RNA in spindle cells and endothelial cells of AIDS-associated, classical and African Kaposi's sarcoma (KS). *Int J Cancer* **72**: 68–71.

Tabata M, Langford A, Becker J et al (1993) Distribution of immunocompetent cells in oral Kaposi's sarcoma (AIDS). *Oral Oncol* **29B**: 209–213.

Taylor JF, Templeton AC, Vogel CL et al (1971) Kaposi's sarcoma in Uganda: a clinicopathological study. *Int J Cancer* **8**: 122.

Taylor JF, Smith PG, Bull D et al (1972) Kaposi's sarcoma in Uganda: geographic and ethnic distribution. *Br J Cancer* **6**: 483–497.

Trattner A, Hodak E, David M et al (1993) The appearance of Kaposi's sarcoma during corticosteroid therapy. *Cancer* **72**: 1779–1783.

Uccini S, Ruco LP, Monardo F et al (1994) Co-expression of endothelial cell and macrophage antigens in Kaposi's sarcoma cells. *J Pathol* **173**: 23–31.

Uccini S, Sirianni MC, Vincenzi L et al (1997) Kaposi's sarcoma (KS) cells express the macrophage associated antigen mannose receptor and develop in peripheral blood cultures of KS patients. *Am J Pathol* **150**: 929–938.

Viviano E, Vitale F, Ajello F et al (1997) Human herpesvirus type 8 DNA sequences in biological samples of HIV-positive and negative individuals in Sicily. *AIDS* **11**: 607–612.

Vogel J, Hinrichs SH, Reynolds RK, Luciw PA & Jay G (1988) The HIV tat gene induces dermal lesions resembling Kaposi's sarcoma in transgenic mice. *Nature* **335**: 601–661.

Weindel K, Marmé D & Weich HA (1992) AIDS associated Kaposi's sarcoma cells in culture express vascular endothelial growth factor. *Biochem Biophys Res Commun* **183**: 1167–1174.

Werner S, Hofschneider PH, Stürzl M et al (1989) Cytochemical and molecular properties of simian virus 40 transformed Kaposi's sarcoma-derived cells: evidence for the secretion of a member of the fibroblast growth factor family. *J Cell Physiol* **141**: 490–502.

Werner S, Hofschneider PH, Heldin CH et al (1990) Cultured Kaposi's sarcoma-derived cells express functional PDGF-A type and B-type receptors. *Exp Cell Res* **187**: 98–103.

Westendorp MO, Frank R, Ochsenbauer C et al (1995) Sensitization of T cells to CD95-mediated apoptosis by HIV-1 Tat and gp120. *Nature* **275**: 497–500.

Whitby D, Howard MR, Tenant-Flowers M et al (1995) Detection of Kaposi's sarcoma associated herpesvirus in peripheral blood of HIV-infected individuals and progression to Kaposi's sarcoma. *Lancet* **346**: 799–803.

Wijelath ES, Carisen B, Cole T et al (1997) Oncostatin M induces basic fibroblast growth factor expression in endothelial cells and promotes endothelial cell proliferation, migration and spindle morphology. *J Cell Sci* **100**: 871–879.

Williams AO, Ward JM, Li JF et al (1995) Immunohistochemical localization of transforming growth factor-β 1 in Kaposi's sarcoma. *Hum Pathol* **26**: 469–473.

Wirt DP, Brooks EG, Vaidya S et al (1989) Novel T-lymphocyte population in combined immunodeficiency with features of graft-versus-host disease. *N Engl J Med* **321:** 370.

Yang J, Hagan M & Offerman MK (1994a) Induction of IL-6 gene expression in Kaposi's sarcoma cells. *J Immunol* **152:** 943–955.

Yang J, Xu Y, Zhu C et al (1994b) Regulation of adhesion molecular expression in Kaposi's sarcoma cells. *J Immunol* **152:** 361–373.

Zauli G, Davis BR, Re MC et al (1992) Tat protein stimulates production of transforming growth factor-β 1 by marrow macrophages: a potential mechanism for human immunodeficiency virus-1-induced hematopoietic suppression. *Blood* **80:** 3036–3043.

Zauli G, Gibellini D, Caputo A et al (1995) The human immunodeficiency virus type 1 Tat protein upregulates *bcl-2* gene expression in Jurkat T-cell lines and primary peripheral blood mononuclear cells. *Blood* **86:** 3823–3834.

Zietz C, Hotz B, Stürzl M et al (1996) Aortic endothelium in HIV-1 infection: chronic injury, activation and increased leukocyte adherence. *Am J Pathol* **149:** 1887–1898.

Zisbrod Z, Haymov M, Schanzer H et al (1980) Kaposi's sarcoma after kidney transplantation. Report of complete remission of cutaneous and visceral involvement. *Transplantation* **30:** 383.

Chapter 20

HEPATITIS B AND DELTA VIRUSES

William F. Carman and Christian Trautwein

HEPATITIS B

Much progress has been made in all spheres of our understanding of hepatitis B virus (HBV) and its associated diseases. Here, we do not reiterate the past (which can be found in many standard textbooks) but look to recent findings which we believe pave the way to future research into this medically important virus.

Genome organization

The genome consists of a circular DNA of 3·2 kb in length, the smallest genome of any DNA virus infecting humans. The whole genome codes for protein through four open reading frames which are partly overlapping and generate at least seven viral gene products. They are under the transcriptional control of four promoter elements (pre-S_1, S, C and X promoter) and two enhancer elements (enhancer I and II) (Figure 20.1), which are embedded within the open reading frames. The pre-S promoter is located upstream of the first ATG of the S gene and regulates transcription of the 2·4 kb pre-S mRNA which encodes the large and middle envelope proteins of the virus (Figure 20.2). The S promoter, situated between the first and second ATG of the S gene, controls transcription of the 2·1 kb S mRNA. The C and X promoters are directly located upstream of their appropriate genes. The C promoter controls transcription of the pregenomic and the precore mRNAs which encode the core protein, the hepatitis B e antigen (HBeAg) protein and DNA polymerase. The pregenomic RNA is essential for the viral life cycle as it serves as the RNA template for DNA replication. The core protein is assembled into the capsid; the precore protein is processed intracellularly to HBeAg, which is secreted (Figure 20.2). Note that HBeAg is not a breakdown product of HBcAg. All mRNAs transcribed from the HBV genome use the same polyadenylation site.

The structure of HBV particles

The various particles of HBV are well described in standard texts. Recent developments include the three-dimensional structure of core particles and further elucidation of the antigenic structure of hepatitis B surface antigen (HBsAg).

HIV and the New Viruses Second Edition
ISBN 0-12-200741-7

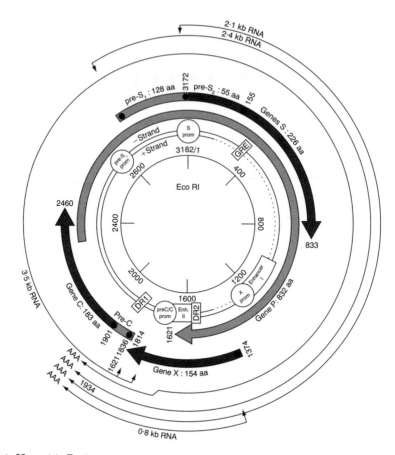

Figure 20.1 *Hepatitis B virus genome*
Schematic representation of the HBV genome. The numbered circle depicts the 3182 kb HBV genome. The transcribed length of the mRNAs and the translated proteins are shown. On the partial double-stranded genome the control elements are shown, the pre-S$_1$, S, X and pre-C/C promoter and the two enhancer elements I and II which overlap with either the X or the pre-C/C promoter.

Core particles

Cryoelectron microscopy (Bottcher et al, 1997) has revealed that there are both T=3 and T=4 structures where T = triangulation number. The basic building block is a hepatitis B core antigen (HBcAg) dimer, which interacts with adjacent dimers by means of a pair of α helices. There are spikes at the five fold axes, which are the sites of the major conformational B cell epitope. Core interacts on its outer surface with pre-S$_1$ and S envelope sequences (see below). On its inner surface, it has both RNA and DNA binding domains which can order the encapsidated nucleic acid.

HBsAg antigenic structure

The major hydrophilic region (MHR) of HBsAg is from amino acid (aa) 100 to 160. There is much indirect evidence for a two-loop structure (reviewed in Berting et al, 1995; Wallace and Carman, 1997), defined by disulphide bridging, from aa 124 to 147. Peptides that encompass this 28 aa stretch bind the majority of anti-HBs stimulated by vaccination; at least 50% binds to the sequence between aa 139 and 147 (Brown et al, 1984).

Circularizing the peptides aa 124–137 and 139–147 leads to greater antigenicity compared with the linear peptides. There are a number of highly conserved cysteines in HBsAg; their effect on assembly is reviewed in Prange et al (1995). Mutagenesis experiments (Ashton-Rickardt and Murray, 1989; Mangold and Streeck, 1993; Mangold et al, 1995), as well as natural antigenic variants, point to the complex interleaving of a number of epitopes within the MHR and also indicate that the cysteine at aa 149 is more critical than that at aa 147. Use of a phage display library (Chen et al, 1996) has led to the description of a potentially novel cysteine web, consisting of all cysteines in the HBsAg MHR, which sits on the virion membrane. Based on this, the two-loop model has been modified, so that there are now two loops projecting from the web, from aa 107 to aa 138, and from aa 139 to aa 147 or 149 (Figure 20.3). In addition, there is a tight loop from aa 121 to aa 124 (Qiu et al, 1996); insertions here (Yamamoto et al, 1994; Carman et al, 1995a; Hou et al, 1995) can lead to global disruption of the epitope structure of the MHR. The region aa 154–160 is on the outer surface, so forms at least one anti-HBs epitope (Chen et al, 1996). Some epitopes are formed from amino acids on both large loops, so they must be spatially close.

Knowledge of the structure is important for the analysis of antigenic variants which have been found in a number of clinical situations. The finding that insertions can occur in the loop between aa 121 and aa 124 (Yamamoto et al, 1994; Carman et al, 1995a; Hou et al, 1995) also indicates that this region is not functionally critical and is probably not involved in receptor binding.

There are two other envelope proteins: large (L; containing pre-S$_1$, pre-S$_2$ and S), and middle (M), containing pre-S$_2$ and S (see Figure 20.2). These are found in different proportions in virions and subviral particles. They are not only antigenic in themselves, but provide T cell help for responses against core and HBsAg proteins (see below). They are thus particularly potent immunogens and useful in vaccines.

Genotypes and subtypes

There are six genotypes, four fundamental antigenic subtypes and five other sub-sub-types. The genotypes are based on an 8% nucleotide divergence (Okamoto et al, 1988;

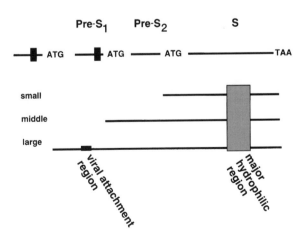

Figure 20.2A

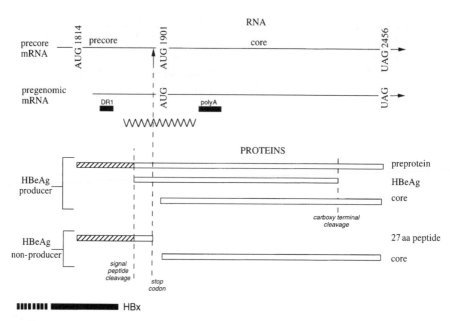

Figure 20.2B

Figure 20.2 *Molecular biology of S and C genes*
(A) There are three regions on this gene: pre-S₁, pre-S₂ and S. Each begins with its own translational start codon, ATG, but all end in a common translational stop codon, TAA. The three resulting proteins, S (also known as HBsAg), M and L are indicated. Note that the common carboxy terminus contains the major hydrophilic region sequence, in which are found the neutralizing epitopes on S protein. The promoters for transcription of the mRNAs which translate the three proteins are also shown (upright black rectangles). The putative viral attachment site within pre-S₁ is shown; S also contains a region that binds to hepatocytes. Thus, there may be two sites; perhaps both are found on L protein rather than on L and S proteins. The hydrophilic region is shown in greater detail in Figure 20.3. If M protein translation is inhibited by loss of the ATG, the pre-S₂ region is still translated because it is included within L protein.
(B) The two mRNAs driven from the basal core promoter are shown. Within the precore message are two translational start codons (AUG); translation from the first leads to the preprotein that, after loss of the signal sequence (hatched bar, cleavage at dashed line) and the carboxy terminal nucleophilic sequence (dashed line), leads to HBeAg. HBeAg is secreted into the serum. Translation from the second AUG, but from the pregenomic message, leads to core protein (HBcAg), which is the building block of core particles and therefore virions. Both HBeAg and HBcAg translation end at the common translational stop signal, UAG. Within the gene sequence is direct repeat (DR) 1, necessary for reverse transcription, the encapsidation signal (zigzag line) (see also Figure 20.4) and a polyadenylation tract (polyA), the common point for termination of all virus mRNAs. In an HBeAg non-producing strain, there is often a premature stop codon at the end of the precore region (dashed line leading to arrow; A₁₈₉₆) which leads to a short peptide (the function, if any, is unknown). However, HBcAg is still produced.

Norder et al, 1992a,b) and tend to be distributed geographically. Recent phylogenetic data indicate that genotype F is the most ancient (P. Bollyky et al, 1997, unpublished observations), suggesting the possibility that the virus seems to have spread out from the New World to the Old World around the time of European conquest, then diverged

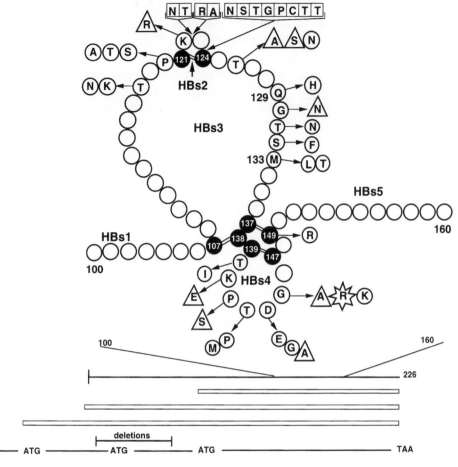

Figure 20.3 *Proposed two-loop model of major hydrophilic region of HBsAg and variants seen in vaccinees*

The cysteine web model of the major hydrophilic region (MHR) of HBsAg (aa 101–160). A web of cysteines is found on the surface of the viral envelope with the two loops directed into the exterior of the virus. Each circle signifies an amino acid. Each black numbered circle is a cysteine molecule. Single-letter codes for amino acids are used. Numbers indicate amino acids of HBsAg. This is an enlargement of the hydrophilic region shown in Figure 20.2a. There is a highly conformational structure; antigenicity is dependent upon it. One possible set of disulphide bridges is shown here (double lines) between amino acids 107 and 138, aa 139 and 147, aa 137 and 149, and aa 121 and 124. HBs1–5 are proposed antigenic regions; note that there are many epitopes formed from more than one region and that HBs2 (aa 121–124) and HBs4 (aa 139–147) are close spatially and constitute the neutralizing epitope cluster. The other regions are important antigenically, mostly in relation to the sensitivity of diagnostic assays. The variants shown are those published as associated with vaccine failure at the time of writing. The arrows indicate mutations; those in circles represent a single case, those in triangles more than one case. The starburst indicates the most common mutation observed. Squares denote an insertion.

into genotypes; it therefore may not have migrated in parallel with humans but infected established populations. The subtypes, which are antigenically defined, are based on amino acid substitutions in the MHR. Classically, aa 122 and 160 are critical, but variants at other positions, such as aa 159 and aa 126/127 allow full antigenic expression

(Norder et al, 1992a,b). It seems that genotypes and subtypes have evolved in ethnic backgrounds probably as the result of selection (possibly antigenic) pressure. The clinical importance of a variant of HBsAg may be different depending upon the subtype backbone in which it is found.

There are no reliable data to indicate that genotypes or subtypes have any replication advantage over one another or different pathogenic potential, although there are hints that subtype axx may lead more often to long-term carriage (Okamoto et al, 198X?) and genotype D may be associated with fulminant hepatitis more commonly than predicted from its prevalence in chronic carriers (Yasmin, 1997).

Immunology

CTL responses rarely occur in chronic hepatitis
During acute hepatitis, there is a strong, polyclonal and multispecific response to all HBV proteins. However, during chronic hepatitis, very little cytotoxic T lymphocyte (CTL) activity can be found (Penna et al, 1991; Bertoletti et al, 1993; Missale et al, 1993). When it is present, it is weak and monospecific. The CTL responses in chronic carriers can be similar to those in acute clearance when anti-HBe seroconversion occurs either naturally or after interferon therapy (Rehermann et al, 1996a). However, it may be that the CTLs are merely 'mopping up' the resulting antigenic load from destroyed hepatocytes, but are not the primary effectors. As in HIV infection, CTL antagonism has been shown in two patients who had evident CTL responses during chronic carriage (Bertoletti et al, 1994); they had variant sequences of the well-described HLA-A2 CTL epitope in HBcAg. These variants led to either poor binding of the peptide to the class I groove or inhibition of the response to the standard peptide. We do not believe these results are relevant to the majority of chronic carriers; in fact, variants of this epitope are found in similar numbers in HLA-A2 positive and negative carriers, implying the lack of a selective immune pressure (Ferrari et al, 1995).

This raises the issue of what causes hepatocyte death during exacerbations of disease. There is precious little evidence of a direct cytopathic effect of HBV, so immune reactivity is presumed to be the cause of the inexorable progression to end-stage disease seen in a large proportion of cases with active replication. The mediators are unknown, but perhaps antibodies are involved (see below).

Correlation of Th responses with viral load and exacerbations of hepatitis
Seroconversion to antibody to HBeAg (anti-HBe) usually leads to clinical remission and lowering of viral load. As the immune response is believed to be critical in viral clearance and the accompanying raised levels of aminotransferases, it makes sense that a surge in T cell proliferation (Tsai et al, 1992) and increased IgM anti-HBc should accompany these serological changes. This is also seen during flares of hepatitis during both HBeAg and anti-HBe positive phases. Interestingly, an increase in the blood HBV-DNA level often precedes such immunological activity. As this has also been seen in trials of therapeutic vaccination (W. Carman et al, 1997, unpublished observations), it raises the possibility that the immune response in some way switches on viral replication. It is entirely unknown why hepatitis activity is sporadic and intermittent in some patients.

CTLs can clear infection without killing hepatocytes

Transgenic mouse models of HBV replication have given us fresh insights into patho-genesis (reviewed by Chisari, 1996). However, some cynicism is justified concerning the relevance of such models to human infection. These models have indicated, for exam-ple, that both L and X proteins (see below), if overproduced, can lead to liver tumours; however, others have failed to show this.

Most interestingly, CTLs have been shown to downregulate HBV replication in hepa-tocytes through tumour necrosis factor alpha (TNFα) and interferon gamma (IFN-γ) at a post-transcriptional level (Guidotti et al, 1994; Tsui et al, 1995). Antibody infusions against these effectors have proved the specificity of this effect. This is clearly a poten-tially important mechanism of viral clearance, although it is at variance with the com-monly observed hepatitis with inflammatory infiltrates associated with immune clearance. It does help to explain how clearance can occur without massive hepatitis and death.

The function of serum HBeAg and excess HBsAg may be to induce tolerance

HBeAg is translated from the first ATG of the precore mRNA (see Figure 20.2). The amino terminus is a signal peptide which leads to secretion into blood. It is not neces-sary for virus replication or survival; in fact, it may even downregulate virus replication (Guidotti et al, 1996a,b). Why should a virus with a highly compact genome produce a secreted protein which is not part of the virion? Similarly, why expend cellular energy on production of excess HBsAg which is then secreted as DNA free particles into the blood? One explanation is that they in some way modulate the immune response to sustain chronic infection. In mice, HBeAg can cross the placenta and induce a state of tolerance in the fetus (Milich et al, 1990). This is in keeping with chronic infection being almost universal in unvaccinated infants born to HBeAg-positive mothers. The toler-ance is eventually lost in the mice, paralleling the appearance of immune activity in early human adulthood. Little work has been done on HBsAg and tolerance, but the long period of HBsAg positivity observed after seroconversion to anti-HBe would fit with this hypothesis.

Genetically determined immune non-responsiveness can be overcome with additional antigens

Strains of mice with specific H2 types (the equivalent of human HLA) that do not respond to HBsAg have been well characterized. Both pre-S$_1$ and pre-S$_2$ antigens are able to overcome this anergy and allow production of high anti-HBs levels (Milich et al, 1986). Similarly, circumvention of CTL non-response has been observed in H2 restricted mice with CTL responses by using DNA vaccines, probably due to an antigen processing pathway other than the classical HLA class I route (Schirmbeck et al, 1995). As discussed below, this is relevant both to decreasing vaccine non-responsiveness and increasing the efficacy of DNA vaccines as therapeutic agents.

Genetic factors are linked to chronicity

Both HLA-DR1301 and HLA-DR1302 (both class II antigens) are linked to clearance of HBV during the acute infection (Thursz et al, 1995). Although the association is not per-fect and does not explain all causes, it provides clues to the pathogenesis. However, the clear links with age of acquisition and viral sequence variability to chronicity are more difficult to explain by a host genetic factor.

The anti-HBe-positive phase of active infection is associated with selection in HBcAg B cell epitopes

Within geographically or ethnically defined confines, most sequences from HBeAg-positive persons are relatively conserved, irrespective of disease activity (Carman et al, 1995b). However, after seroconversion to anti-HBe and selection of a precore variant which switches off HBeAg, such as A_{1896} (see below), amino acid substitutions collect within HBcAg. It is interesting that seroconversion to anti-HBe is not sufficient: the production of HBeAg has to cease. These steps occur in a rush soon after precore mutants become dominant and statistically are concentrated within B cell epitopes in those with ongoing disease (Carman et al, 1997a). In cross-sectional studies of patients with active hepatitis, variants are also found within T helper (Th) epitopes (Carman et al, 1995b), but in longitudinal studies they tend to collect only in those who go on to remission. Unfortunately, no study providing formal evidence of immune escape has been published, so this conclusion is inferential at present.

Th cell responses against HBcAg and HBeAg are separate and balanced

The antigens HBcAg and HBeAg are generated from the same genetic material although HBeAg has an additional 10 amino acids at the amino terminus and around 35 amino acids fewer at the carboxyl terminus (Figure 20.2). This leads to HBcAg becoming particulate, but HBeAg remaining as a monomer which is secreted into the blood. Therefore, they share what are considered to be the major Th epitopes from aa 1–20 and aa 50–85 (Ferrari et al, 1995). Yet, it has now been shown that HBcAg stimulates predominantly a Th1 response, leading to activation of CTLs, hepatitis and immune clearance, whereas HBeAg causes proliferation of Th2 cells (Milich et al, 1997). The Th2 cells predominantly cause secretion of cytokines involved in moderation of the inflammatory process. This implies that the balance helps to prevent severe hepatitis which might otherwise lead to the death of the host. This is in keeping with some observations that anti-HBe-positive viraemia associated with the precore variant is linked to severe disease, but it is not in keeping with similar observations in those with clinical remission and quiescent, mild hepatitis (Okamoto et al, 1990; Tur-Kaspa et al, 1992; Carman et al, 1995b).

Pathogenesis

Acute hepatitis

Acute hepatitis with viral clearance is probably due to CTL-mediated killing or apoptosis of infected hepatocytes. Although the transgenic mouse model of HBV replication indicates that some hepatocytes can be cleared of HBV without being lysed or undergoing apoptosis, the high levels of aminotransferases seen during viral clearance in vivo are in keeping with the former mechanisms. Whether natural killer (NK) cell activity (perhaps mediated by antibodies; antibody-dependent cell-mediated cytoxicity) is involved is not known. Even after complete clinical recovery, some virus can be found peripherally for years which can restimulate the CTLs (Rehermann et al, 1996b). It is likely that HBV is never fully cleared from hepatocytes, which explains how viral reactivation with hepatitis can occur during and after immunosuppression. An alternative hiding place is in white blood cells although there is not universal agreement on the relevance of this site to HBV biology.

Anti-HBe seroconversion in chronic carriers is not always beneficial
The antigen HBeAg is lost in at least 5% of chronic carriers per year. For much of the HBeAg-positive phase, the antigen and its antibody are in complexes (Maruyama et al, 1993); as HBeAg is lost, so anti-HBe becomes predominant. There are, broadly, two outcomes after this event: clinical remission (with low-level viraemia), or worsening disease (with high-level viraemia). The latter is commonly seen in Far Eastern and Mediterranean populations. In both clinical scenarios, variants of the precore region which switch off HBeAg production emerge (Okamoto et al, 1990). This can come about by selection of a stop codon (Figure 20.2), disruption of a start codon, or by insertions or deletions (and so a frameshift). The most common is a G to A substitution at position 1896 (A_{1896}) which leads to a stop codon (Brunetto et al, 1989; Carman et al, 1989; Tong et al, 1990). However, as these variants are common in both clinical outcomes, they do not explain the clinical picture. Also, the sequence requirements of the epsilon RNA encapsidation signal (Tong et al, 1992; Lok et al, 1994) do not allow A_{1896} emergence in some genotypes, yet severe disease occurs (Lindh et al, 1996) (Figure 20.4). As translation of the precore region is not necessary for viral survival, loss of HBeAg, which is a potent antigen, probably represents a form of immune escape. Probably, precore variants are a manifestation of a basic immunological inadequacy (an inability to clear the virus completely) in certain individuals, and thus immune escape from an ineffective response. It thus seems that there is a subpopulation of carriers who are unable to suppress virus replication even though they can induce seroconversion: it is these who go on to viraemic, anti-HBe-positive disease. The description of this group has put to rest (at last) the widely held contention that HBeAg is the most sensitive or useful marker of active viral replication.

Is chronic carriage due to host or viral factors?
Chronic carriage in adults appears to be primarily due to host factors. HLA class II alleles have been associated with viral clearance (Thursz et al, 1995), implying that failure to present or respond to viral peptides is fundamental. However, as only 5% of adults fail to clear HBV during acute infection, it is unlikely to be as clear-cut as this. It is as likely that other genetically linked factors, such as mannose binding protein (Thomas et al, 1996) or TNFα alleles, are involved. Viral factors have not been shown unequivocally to be critical, but there are some hints. For example, patients who are infected ab initio with HBeAg non-producer strains (precore variants) seldom go on to develop chronicity, but usually get acute disease (Raimondo et al, 1993; Mphahlele et al, 1997), even sometimes fulminant hepatitis. This is probably because HBeAg induces tolerance of anti-HBc/e Th cells; if HBeAg is not present, this cannot occur. However, this clearly does not explain the large number of infections due to HBeAg-producing strains in adults which lead to acute hepatitis and viral clearance. Although it has not been formally proved, HBsAg is also believed to have an immunomodulatory effect. Finally, HBV can downregulate the synthesis or action of IFN in vitro (Twu et al, 1988), but the significance of this in vivo is unclear.

Once chronic carriage is established, the virus undergoes a series of mutations, usually beginning with a precore variant leading to the loss of HBeAg. This occurs around the time of anti-HBe seroconversion, and it is hypothesized that anti-HBe itself selects viruses that cannot produce this antigen, possibly via NK cells. Immediately following this process, multiple amino acid substitutions are selected in B cell epitopes of HBcAg, in greater amounts in those who have ongoing disease (Ehata et al, 1992; Akarca and Lok, 1995; Carman et al, 1997a), perhaps leading to viral escape. Escape has not been formally shown in vitro, but the association with such epitopes is statistically significant (see Tables 20.1a,b). Interestingly, in those who go into clinical remission after loss of HBeAg, variants are preferentially selected in Th cell epitopes. Chronic carriers with ongoing hepatitis often also have deletions and mutations in the pre-S regions of the

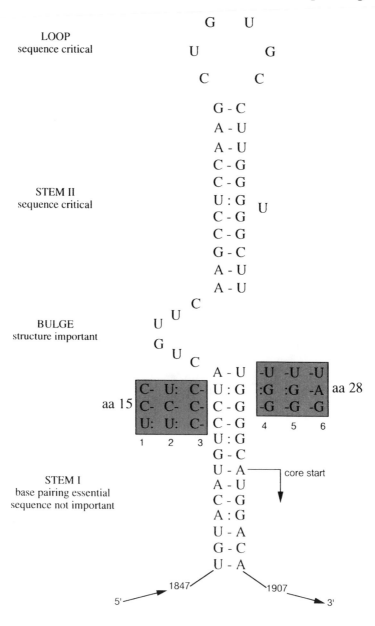

Figure 20.4

genome (Santantonio et al, 1992), also proposed to be a mechanism of viral escape. At some time during this process, variants in the X protein and S protein can be selected. These last may be attempts by HBV to downregulate viral replication and to escape the evolving anti-HBs responses, respectively. For much of the period of chronic carriage, HBsAg/anti-HBs are in a balanced state; eventually, it is thought that the production (or immunomodulatory effect) of HBsAg is overwhelmed and full seroconversion occurs. This viral evolutionary cascade represents sequential attempts to survive the immune response; the downside is a gradual diminution of replication competence and full infectability. It is likely that the majority of transmission occurs during acute infection or the HBeAg-positive phase of chronic carriage. Thus, although there is substantial microevolution (within individuals), little of this shows up in the macroevolutionary picture (within populations) because many of these viruses are either not transmissible or they are easily outcompeted by their accompanying wild-type viruses.

Some 90% of infants with HBeAg-positive mothers become infected, the majority of them chronically. The reason for this is most probably that the immunomodulatory effect of HBeAg is more dominant in an immature fetal immune system. In mouse models, HBeAg crosses the placenta; when infection occurs at, or shortly after, birth (infection is thought to be uncommon in utero), HBcAg in hepatocytes is not recognized and chronic infection ensues. HBeAg preferentially stimulates Th2 cells and downregulates Th1 responses. Thus, there is a dominance of antibody responses but little CTL activity. The different HBcAg and HBeAg structures in some way dictate the Th response, although they share peptide epitopes. Why the immunomodulatory effect is lost over the years, as shown in mice, but also observed in vivo in humans, is not known.

What is the relevance of A_{1896}?

Much has been written about this single nucleotide substitution. It was initially found in anti-HBe-positive persons with severe disease (Carman et al, 1989; Brunetto et al, 1989), and was believed to be directly causal. The following is now clear (reviewed by Carman, 1996; Miyakawa et al, 1997). It is seen in persons with and without active hepatitis, it is present in a low proportion of total sequences during the HBeAg-positive phase, and it becomes dominant around or after the time of seroconversion. It is only seen in certain genotypes (Tong et al, 1992; Lok et al, 1994) because of structural

Figure 20.4 *Variability and stability of the encapsidation signal of HBV*
This is an RNA sequence which folds up into a secondary structure with the four marked regions. Base pairing in stem 1 has been shown to be crucial for stability. Instability will lead to breakdown of the structure and an inability of the viral (pregenomic) mRNA to be taken up (encapsidated) into core particles. If this fails, then new virions will not be produced. The effect of the common precore mutant, A_{1896} (G to A at nt 1896), is shown in the context of a number of possible triplets at amino acid 15 opposite on the stem. CCU and CCC code for proline; UCC and UCU code for serine. A dash between two bases indicates a strong pairing and a colon a weak pairing. In the standard sequence shown, note that there are weak pairings in two positions between or close to these two triplets. By selecting A_{1896} (6), this is strengthened. However, if a serine is encoded by UCC (1), there are still two weak pairings, and clearly a third selection of A_{1896} (6) would be unacceptable. Thus, a stop codon is never seen with serine at aa 15. For the same reasons, serine is never encoded by UCU (2) because there would be three weak pairings before there was a chance to select A_{1896} (6) in order to strengthen the stem – this virus would not survive long enough. Genotype A of HBV has CCC (3). This is already stable and selection of A_{1896} (6) would be unlikely to occur.

constraints of the RNA encapsidation signal (epsilon; see below). There are some data that patients with A_{1896} have a better outcome than those without (Lindh et al, 1996) but this is probably related to the infecting genotype; these findings need to be confirmed. The A_{1896} substitution does not per se influence response to interferon (Aikawa et al, 1995; Fattovich et al, 1995; Lampertico et al, 1995); that these are people who have been unable to clear the virus naturally, even after seroconversion to anti-HBe, must surely influence the response to imune-stimulating antiviral agents. Some persons who are anti-HBe-positive are infectious; whether A_{1896} is directly relevant or merely a marker for the serological picture is unclear. Although HBeAg has been shown to inhibit replication, HBV DNA levels in A_{1896}-infected persons are no higher than those with G_{1896}. Finally, there are data showing that A_{1896}-containing strains can be transmitted independently. A case report (Mphahlele et al, 1998) describes transmission of a pure A_{1896} strain leading to acute hepatitis and clearance in an adult. Transmission to infants usually leads to acute hepatitis (Raimondo et al, 1993). Phylogenetic data show that these strains cluster together into multiple lineages within genotypes (P. Bollyky et al, unpublished observations). Considering that anti-HBe-positive patients have become the most common new presentation to hepatitis clinics, it may be that this strain is becoming dominant.

Is fulminant hepatitis due to viral or host factors?

Fulminant hepatitis B (FHB) is a form of acute hepatitis, so it is expected that a link with HLA class II alleles will be found. Also consistent with a role for host factors is the observation that the isolate from a patient with FHB had an identical nucleotide sequence to that observed in the patient with minimal hepatitis who was the source of the infection (Karayiannis et al, 1995). There have been outbreaks of HBV infection with only a minority of FHB cases, implying either that a host factor is involved or that the contact has a mixture of strains.

What is more difficult to reconcile with a simple host factor is the finding of outbreaks of FHB disease and the preferential transmission of FHB from anti-HBe carriers; for example, there are two cases of a man transmitting this disease to two sexual partners (Fagan et al, 1986) and there have been several outbreaks of FHB from a single source patient (Oren et al, 1989; Tanaka et al, 1995). Transmission of an A_{1896} strain from an FHB case to chimpanzees resulted in more severe hepatitis than would be expected (Ogata et al, 1993), although this was not accompanied by a higher than average level of viraemia. The finding of A_{1896}-containing strains in transmission pairs has lent support to the role of this variant. The pathogenesis of FHB in neonates, and the observation that transmission of A_{1896} strains to children seldom gives rise to chronic carriage (Raimondo et al, 1993), may be partially explained by the absence of HBeAg as an immunomodulator. Thus, there is strong epidemiological evidence that viral factors are also involved in FHB, rather than it being solely dependent on a strong immune response to a 'normal' viral strain. Although A_{1896} is often found, this does not necessarily mean that it is the variant per se which is causing the FHB. In fact, a number of studies do not find a convincing association with A_{1896}, even in neonatal disease. Further, in acute infection with recovery, the HBeAg-negative variant may emerge (Carman et al, 1991). Either there are multiple strains with A_{1896} (with different outcomes) or a host factor is crucial.

Complete genome sequencing of two epidemiologically unrelated FHB strains (Ogata et al, 1993; Hasegawa et al, 1994), both with A_{1896}, revealed widespread substitutions, concentrated in the *cis*-acting elements. In particular, these were clustered in two areas within the basal core promoter (BCP)/enhancer II region (Figure 20.5), and correlated with binding regions of transcription factors. Clustering of substitutions within this gene has been confirmed in other studies (Sato et al, 1995). Increased transcriptional efficiency may lead to very high levels of HBV DNA being produced, which may result (in certain individuals) in a massive immune response, and therefore more severe disease. In vitro experiments showed that an A_{1896} genome did not have higher replication efficiency. However, a genome from an FHB patient with A_{1896}, T_{1762} and A_{1764} (within the BCP) gave rise to very high levels of secreted HBV DNA; the major effect was on encapsidation (Baumert et al, 1996). Variability at nucleotides 1766 and 1768 seems to be important in this process, implying that there is another region outside epsilon involved in encapsidation, perhaps indirectly. We have completed a detailed phylogenetic analysis which includes 20 new epidemiologically unrelated FHB sequences, revealing clusters of fulminant strains each with distinct mutational patterns (Bollyky, 1997). In keeping with the finding that contacts and index cases have highly related sequences is that contact sequences are found within the clusters. Variants are again clustered in the *cis*-acting regions and HBx proteins. No unique variant was specifically associated with this disease in all cases, but the combination ('motifs') of nucleotide variants in the enhancer/promoter regions plus amino acid substitution in HBx was almost uniquely associated with this disease. Variants out of their usual genotype context were also found to be important. A_{1896} strains with such motifs were found only in fulminant isolates, explaining why A_{1896} is common, but FHB is rare. There was also statistical evidence of a heightened replication rate of the fulminant genomes. Supporting the hypothesis that these strains are relevant clinically, the transcription efficiency in vitro of the BCP from A_{1896}-associated fulminant strains is two to three times greater than that from non-A_{1896} fulminants and A_{1896} non-fulminant sequences (Yasmin, 1997). Finally, oligonucleotides containing the variants failed to bind to some nuclear transcription factors, implying that a central pathogenetic mechanism in FHB is loss of inhibition of transcription.

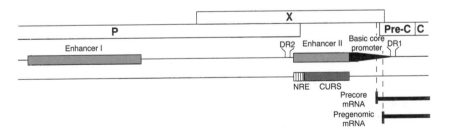

Figure 20.5 *Outline molecular structure upstream of precore initiation site*
The X gene encodes for HBx; P is the gene coding for reverse transcriptase. Within the P gene is enhancer I. Within the X gene is enhancer II and a number of other nucleotide sequences which have an effect on transcription (and thus replication, as there is an intermediate RNA stage). CURS, core upstream regulatory sequence; NRE, negative regulatory element. The start sites of the precore (which translates to HBeAg; Figure 20.2) and the pregenomic mRNA (which is reverse transcribed to DNA and is also translated to core and polymerase proteins; Figure 20.2) are shown. Note that any mutations found within the X gene may act either on the function of HBx protein or transcription of mRNA. Direct repeats DR1 and DR2 are important for virus replication (see text and Figure 20.8).

Studies of chronic mother and fulminant infant pairs show selection of one or more strains from the mother's mixed population. This helps to explain how direct sequencing has observed identical sequences in both carrier and fulminant patient and how the disease can be so different after transmission. It seems credible that, in the milieu of a fresh liver, selection of the more efficiently replicating strain may be easier than in a liver that is adapted to chronic infection. The fulminant-defining mutations must appear gradually (microevolution) on the background of a 'fulminant cluster sequence' in a chronically infected host, as a virus that kills its host is, by definition, a dead end (thus there is no macroevolution).

Transcriptional control

Hepatotropism of HBV is controlled on different levels. Besides specific binding, it has become evident that hepatocyte-specific gene transcription is also essential. The transcriptional activity of the four promoter elements of HBV is controlled by the two enhancers I and II.

Organization of the enhancer I

The enhancer I element (Shaul et al, 1985; Tognoni et al, 1985) is located between nucleotides 970 and 1240 of the HBV genome (Figure 20.6). It regulates the activity of the promoters of the S, X and C genes (Bulla and Siddiqui, 1988; Antonucci and Rutter, 1989; Guo et al, 1991; Hu and Siddiqui 1991; Loperz-Cabrera et al, 1991; Zhang et al, 1992). Although enhancer I is controlled by hepatocyte-specific transcription factors (Shaul et al, 1985; Jameel and Siddiqui, 1986; Honigwachs et al, 1989), it retains some transcriptional activity in dedifferentiated hepatocytes and in non-hepatocyte cell types (Yen, 1993; Kosovsky et al, 1996). This activity in different cell types points to its complex regulation by hepatocyte-specific and ubiquitous transcription factors (Schaller et al, 1991; Yen, 1993; Kosovsky et al, 1996). In cells of non-hepatocyte origin the enhancer I seems mainly regulated by the ubiquitous factors, while in hepatocytes there appears to be cooperative interaction between transcriptionally active proteins that are specifically expressed in hepatocytes and those that are found in many cell types.

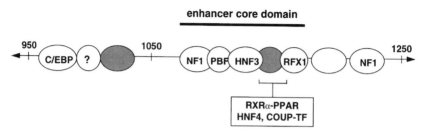

Figure 20.6 *Organization of enhancer I*
The known DNA-binding motifs of the enhancer I region are shown between nucleotides 950 and 1250 of the hepatitis B virus genome. The region between nucleotides 1080 and 1165 is indicated as the enhancer core domain. The grey circle is a binding site for C/EBP, NF-1, CREB/ATF2 and AP1. The circle with thick stripes is a binding motif for C/EBP, HNF-1 and OCT2. The round circle with a question mark represents a site with unknown binding factor.

Within enhancer I, a central, highly active region which is located between nucleotides 1080 and 1165 has been called the enhancer core domain (Figure 20.6) (Trujillo et al, 1991). At least four motifs that bind nuclear factors are located here (Ben-Levy et al, 1989; Dikstein et al, 1990; Truiillo et al, 1991; Garcia et al, 1993; Chen et al, 1994; Huan et al, 1995; Ori and Shaul, 1995; Kosovosky et al, 1996). The motif located nearest the 5' terminal in the core domain binds nuclear factor 1 (NF-1) and the region nearest 3' is functionally important for the interaction with RF-1. In between are located two motifs which are controlled mainly by hepatocyte-specific transcription factors. The 5'-located motif of this inner domain binds members of the nuclear receptor family, for example HNF-4 which is liver-enriched (Garcia et al, 1993; Huan et al, 1995). The last motif in the inner domain is an HNF-3 binding site; of the several isoforms of HNF-3, HNF3b is probably the one that preferentially binds to this motif (Kosovsky et al, 1996).

Organization of the X promoter

The X promoter is located between enhancer I and the ATG of the X gene. It partly overlaps with the enhancer I and therefore some of the elements are essential for both elements. The X mRNA is 0·8 kb in length. The X promoter itself contains two major elements; the one closest to the ATG of the X gene binds NF-1, and the other element between this and the enhancer I binds C/EBP, NF-1, CREB/ATF2 and AP1 (Maguire et al, 1991). As the activity of CREB and AP1 family members may be modulated by intracellular signalling cascades, this is a potential target of extracellular stimuli by which X gene transcription can be modified. Besides the cellular factors important for X gene transcription, it has been shown that the HBx itself can enhance the activity of the X promoter by binding to a 20 bp element located between nucleotides 1095 and 1120 in the HBV genome (Takada et al, 1996).

Organization of the core promoter and enhancer II

The core promoter is located between nucleotides 1620 and 1785 and controls the transcription of both the pre-C and the pregenomic mRNAs. A region of the core promoter between nucleotides 1687 and 1774 also functions as enhancer II (Yee, 1989). It seems especially important for upregulating the S promoter (Zhou and Yen, 1990), and it is probably critical for the preferential expression of the S promoter in hepatocytes. In contrast, the enhancer II only minimally activates the pre-S promoter (leading to L and M proteins) (Zhou and Yen, 1990).

The factors HNF-3, HNF-4 and SP1 are especially important for regulating core promoter and enhancer II activity (Figure 20.7) (Guo et al, 1993; Zhang et al, 1993; Johnson et al, 1995; Li et al, 1995; Raney et al, 1997). The second HNF-4 site is also a binding site for other members of the nuclear receptor family such as retinoid X receptor and peroxisome proliferator-activated receptor (Raney et al, 1997). Besides these positive regulatory elements, three negative regulatory elements were described which are located 5'. They also downregulate the core promoter/enhancer II in non-hepatic cells, which indicates that hepatocyte-specific expression of HBV is controlled by negative and positive transcriptional mechanisms. The factors that bind to the 5' negative control region have not been identified yet.

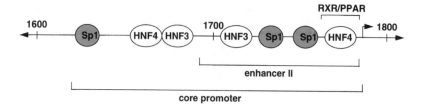

Figure 20.7 *Organization of enhancer II/core promoter*
The region between 1600 and 1800 of the HBV genome is shown. The core promoter is located between nucleotides 1620 and 1785 and the enhancer II region between nucleotides 1687 and 1774. The different DNA-binding regions are shown in circles.

Organization of the S promoters

The pre-S and the S promoter control transcription of the 2·4 and 2·1 kb mRNAs respectively. The pre-S promoter, nucleotides 2710–2800, is thus located 5′ of the ATG of the L protein. The S promoter, which is TATA-less, is found in the coding region of the pre-S region between nucleotides 3045 and 3180. For the pre-S promoter, two motifs have been described between nucleotides −90 and −76, and nucleotides −58 and −50 from the start site of transcription. The first, HNF-1 (−90 to −76), mainly controls the activity of the pre-S promoter and contributes to its liver specificity. Oct-1 binds to the second motif (−58 to −50). Although it is not liver-specific, together with HNF-1 it contributes to the hepatocyte-specific activity of the pre-S promoter (Zhou and Yen, 1991). The ultimate reason for high S promoter activity in hepatocytes is unresolved. Two factors control its activity: Sp1 and NF-Y (Zhou et al, 1991; Raney et al, 1992; Lu et al, 1996; C. T. Bock and C. Trautwein, unpublished observations). Sp1 binds to three elements in the S promoter and NF-Y to the CCAAT box located between nucleotides 3105 and 3110 of the HBV genome. This box regulates transcription of both the pre-S and the S promoter: it enhances S promoter activity, but it downregulates transcription from the pre-S promoter (Lu et al, 1995).

Post-transcriptional regulation

Several factors bind to the viral RNAs that have been transcribed from the HBV genome. Between nucleotides 1151 and 1684 is the post-transcriptional regulatory element (PRE), which has two different subelements PREa and PREb (Donello et al, 1996). The PRE region is most likely to be involved in mediating the nucleocytoplasmic export of unspliced HBV-RNA. A 30 kDa and a 45 kDa RNA-binding protein bind to this region (Huang et al, 1996). As discussed elsewhere, TNFα and IFN-α/β can abolish HBV replication by a non-cytopathic mechanism (Guidotti et al, 1996); inhibition of HBV replication is associated with the binding of a 26 kDa protein to the PRE. At the same time, binding of the 30 kDa and 45 kDa proteins to the PRE is impaired and no more viral RNA is detected (F. V. Chisari, 1997, unpublished observations). These results indicate that stability of the viral RNAs could be controlled by RNA-binding proteins.

The 170 nucleotides at the 3′ end of the pregenomic RNA are essential, functioning in polyadenylation, translation, RNA encapsidation and DNA synthesis. A 65 kDa RNA-binding protein binds to two RNA motifs in this region and first results indicate that these proteins may be involved in regulating HBV replication (Perri and Ganem, 1996).

Replication strategy

Hepatitis B virus replicates via an intermediate RNA step which requires a reverse transcriptase mechanism. This form of replication closely resembles that of the *Retroviridae*. It was first described by Summers and Mason (1982) for the duck hepatitis B virus (DHBV). Essentially four different steps (reviewed by Nassal and Schaller, 1996) can be distinguished during HBV replication (Figure 20.8).

1. After the uptake into hepatocytes the viral genome reaches the nucleus and the partially double-stranded DNA is converted into a 3200 bp covalently closed circular DNA (ccDNA).
2. The ccDNA serves as a template which transcribes the viral RNA pregenome and the different viral RNAs.
3. The viral RNAs are translated into proteins. The viral nucleocapsid packs the RNA pregenome and the viral polymerase which synthesize the DNA minus strand.
4. In the last step the viral polymerase synthesizes part of the plus strand, virus formation occurs at the endoplasmic reticulum and the virus is released from the cell.

The most spectacular step during HBV replication is the intermediate RNA step. The RNA pregenome is terminally redundant and thus around 200 bp longer than the viral genome. It contains the DR2 (nucleotide direct repeat) sequence near the polyadenylated 3′ end but the DR1 sequence and U5 homologous sequence at both ends. This redundancy is required to serve as a template for viral replication. A specific sequence at the 5′ region of the RNA pregenome (e) is the essential signal for nucleocapsid packaging (Knauss and Nassal, 1993; Pollack and Ganem, 1993; Chen et al, 1994; Wang et al,

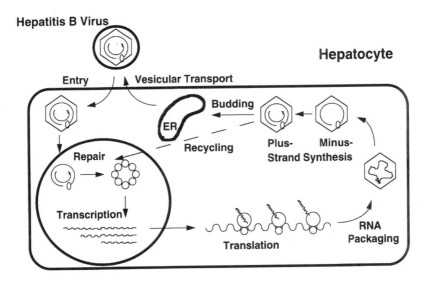

Figure 20.8 *HBV replication*
The different stages of HBV replication. After uptake only the viral genome reaches the nucleus and the genome is converted into closed circular DNA. The viral RNAs are transcribed and translated into the viral proteins. The RNA genome is packed into the nucleocapsid and the DNA minus strand is synthesized by the viral polymerase. Finally the plus strand is formed by the polymerase, the virus is formed in the endoplasmic reticulum and the virus leaves the hepatocyte.

1994). A primer protein of yet unidentified origin (viral or cellular) is located at the 3′ end of the pregenomic RNA at the DR1. This region is recognized by the DNA polymerase which starts synthesis of the DNA minus strand. At the same time as DNA synthesis proceeds, the pregenomic RNA is degraded through the RNase H-like activity of the DNA polymerase up to an approximately 20 bp oligonucleotide at the 5′ end of the RNA pregenome (Lien et al, 1986; Seeger and Maragos, 1989; Radziwill et al, 1990). The oligonucleotide which contains the redundant DR1 locus hybridizes with the partially complementary sequence at the DR2 locus of the minus strand (Lien et al, 1986; Seeger et al, 1986). After this priming step the partial synthesis of the plus strand starts (Seeger and Maragos, 1989). Termination of plus strand synthesis varies, and the mechanism that determines the length of the plus strand is not understood. The DNA synthesis occurring in the nucleocapsid is not directly coordinated with the assembly of the virus and its secretion. The final step in the maturation of the virus occurs in the endoplasmic reticulum where the surface protein and thus nucleocapsid is packed into the envelope before the virus leaves the cell (Ostapchuk et al, 1994; Sheu and Lo, 1994).

Hepatitis B virus X protein and its role in hepatocarcinogenesis

The X gene (see Figure 20·5) codes for a 154 amino acid protein (HBx), which is not essential for replication in vitro (Blum et al, 1992). However, it seems to be required for HBV replication in vivo. This protein is a transcriptional activator without a DNA-binding domain and several studies indicate that it is involved in hepatocellular carcinogenesis induced by chronic HBV infection. Hepatocellular carcinoma (HCC) is only associated with *Hepadnaviridae* infection when the genome codes for HBx (there is no HBx in the DHBV genome and no HCC). Several groups have created transgenic lines overexpressing HBx. In only one transgenic strain did HCC develop (Lee et al, 1990; Kim et al, 1991; Perfumo et al, 1992), so HBx alone is a weak oncogene. More probably HBx is a cofactor and at least one 'second hit' seems essential for HCC development.

Recent studies have therefore concentrated on four different topics: the activation of intracellular signalling pathways, the regulation of cell growth/cell cycle progression, the interaction of HBx with p53 and the general impact of HBx on gene transcription.

Activation of intracellular pathways by HBx
The NF-kB family of transcription factors are normally trapped in the cytoplasm by binding to I-kB. This complex formation prevents nuclear translocation. After phosphorylation of I-kB, the protein degrades, allowing nuclear translocation; NF-kB then binds to the target sequence in the promoter of different genes, leading to gene activation (for review, see Barnes and Karin, 1997). The crucial step in the activation of NF-kB is thus the disruption of the complex with I-kB. Activation of NF-kB is common during cellular events like inflammation, stress or during cell growth, so X gene expression could interact with this signal transduction pathway. The protein HBx can induce nuclear translocation of NF-kB and its subsequent nuclear DNA binding. However, it was shown that NF-kB activation does not involve the Ras/Raf, the protein kinase C pathway, oxygen radicals or tyrosine kinases (Lucito and Schneider, 1992; Chirillo et al, 1996; Su and Schneider, 1996).

Another family of transcription factors that is controlled by post-transcriptional mechanisms are the AP1 proteins. In particular, c-jun phosphorylation in its N-terminal part at serine 63 and 73 has been found to increase c-jun-dependent transcription and stabilizes c-jun protein expression because it blocks ubiquitin degradation (Musti et al, 1997). Expression of c-jun and c-jun-dependent gene transcription – as shown for NF-kB – are induced by several pathophysiological stimuli (for review, see Karin, 1994). The essential step in the activation of c-jun is the activation of jun kinase (JNK) which leads to the phosphorylation of the protein. Earlier results showed that intracellular X protein expression leads to higher c-jun protein expression and DNA binding (Natoli et al, 1994). Meanwhile Benn and co-workers showed that X protein can lead directly to JNK activation. Additionally they showed that by cotransfecting dominant-negative mitogen-activated protein (MEK) kinases which are located upstream of JNK its activation could be blocked (Benn et al, 1996). Therefore HBx seems to activate JNK through the classical signalling cascade which is also induced by TNF, ultraviolet radiation or stress.

The protein, HBx may also be a kinase which can directly phosphorylate different intracellular targets. Dopheide and Azad (1996) reported that recombinant HBx is a potent AMP kinase which phosphorylates adenosine monophosphate to adenosine diphosphate.

Activation of transcription by HBx

The activation of signalling pathways by HBx is linked to its cytoplasmic localization. By truncating the nuclear localization signal, HBx loses its ability to translocate into the nucleus and to act as a transactivator (Doria et al, 1995). Once in the nucleus, HBx can transactivate various viral and cellular promoter or enhancer elements through a region between amino acids 58 and 140 (Kumar et al, 1996). It does not bind directly with a specific DNA element, but via specific transactivation domains of transcription factors (Feitelson et al, 1993; Haviv et al, 1995). Via the enhanced gene transcription, HBx can also modulate the DNA-binding specificity of the transcription factor itself (Maguire et al, 1991). By binding to CREB or ATF-2, HBx confers DNA binding of both CREB family members to an element in the HBV enhancer. In contrast, CREB or ATF-2 itself cannot bind the element in the HBV enhancer I (Maguire et al, 1991).

Besides interacting with different acidic transactivation domains, HBx can also bind to different components of the basal RNA polymerase machinery, for example TF-IIH, TF-IIB and TBP. This interaction does not require any other TBP-associated factors (TAFs), suggesting that HBx itself may act as a coactivator of transcription (Qadri et al, 1995, 1996; Wang et al, 1995a; Haviv et al, 1996). By interacting with acidic transactivation domains (Haviv et al, 1995) and by binding members of the basal machinery, HBx links both components to each other (Figure 20.9), explaining how HBx can activate gene transcription of a variety of different promoters (Maguire et al, 1991; Natoli et al, 1994; Haviv et al, 1995, 1996; Truant et al, 1995).

Conservation by *Hepadnaviridae* of the X gene is probably due to their reliance on transcription initiation for both replication and differential gene expression. On the other hand, as multiple mRNAs are transcribed from at least four different promoters, a specific strategy may be required to increase transcription complex assembly. A virus-specific coactivator, like HBx, may accelerate viral replication, so may be indispensable for viral survival.

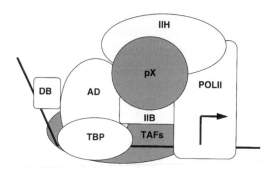

Figure 20.9 *Interaction of HBx with the basal machinery*
Molecular model for transcription regulation by the X protein (pX). The X protein interacts with several components of cellular transcription machinery – TF-IIH, TF-IIB and the RNA polymerase (POLII). The activator protein is dissected in the DNA-binding domain (DB) and activation domain (AD) which interacts directly with pX.

Interaction of the X protein with the *p53* suppressor gene

The *p53* tumour suppressor gene has been called the 'guardian of the genome' as it is involved in DNA repair and leads to apoptosis when cells undergo oncogenic transformation (for review, see Ko and Prives, 1996). p53 is a transcriptional activator with an acidic transactivation domain which interacts directly with HBx (Feinston et al, 1993). In certain regions of Asia, HCC in chronic HBV carriers is associated with mutations in *p53* (Hsu et al, 1991), for example at codon 249, which leads to a lack of DNA binding. This mutation is apparent in HCC only when the host has chronic HBV infection and also high aflatoxin intoxication (Hsu et al, 1991).

Interaction between p53 and HBx is known to have two effects: control of HBV-dependent gene transcription, and apoptosis. High expression of wild-type p53 inhibits HBV replication in cell culture (Lee et al, 1995); this can be relieved by coexpressing HBx. Overexpression of wild-type p53 directly blocks the pregenomic/core promoter (Lee et al, 1995). While HBx enhances transcription from its own promoter, wild-type p53 blocks this effect. Wild-type p53 and HBx also have an antagonistic effect on a 20 bp promoter element in the X promoter (Takada et al, 1996). In contrast, mutant p53, which is unable to bind DNA, can no longer confer this negative regulation. As only small amounts of wild-type p53 are expressed in vivo, these results await confirmation of their importance in real life. As one strategy for elimination of invading viruses is apoptosis, and wild-type p53 may be essential for triggering apoptotic signals, the role of HBx may be to bind p53 and prevent apoptosis of the infected cells. Expression of HBx inhibits p53-dependent target gene transcription – for example p21/WAF1 or Bax, which are involved in the apoptotic pathways. It also inhibits interactions with the TF-IIH transcription–nucleotide excision repair complex (Wang et al, 1995a) which is also involved in p53-dependent apoptosis.

The role of HBx for cell cycle progression and growth control

Chronic HBV infection results in loss of liver tissue and an increase in hepatocyte proliferation. Such entry into the cell cycle is a controlled process; the G_0 to S transition is crucial to prevent oncogenic transformation. Several growth factors are involved in expression of receptors on the membrane of hepatocytes (for review, see Michalopoulos and DeFrances, 1997); for example, the insulin-like growth factor receptor 1 and the transforming growth factor β_1. HBx increases transcription of the insulin-like growth factor receptor 1 gene and thus enhances the mitogenic effect induced by insulin-like growth in human hepatoma cells (Kim et al, 1996). In contrast, HBx inhibits the growth inhibitory effect of TGF-β (Oshikawa et al, 1996). Both results indicate that HBx may alter normal growth control of hepatocytes during regenerative processes in the liver towards a more proliferative state.

Even more dramatic is the effect of HBx on cell cycle progression in vitro. Independently two different groups showed that overexpression of HBx leads to cell cycle progression. HBx triggers quiescent mouse fibroblasts into the cell cycle without leading to oncogenic transformation (Koike et al, 1994). It also stimulates cell cycle progression by shortening the emergence of cells from quiescence (G_0) into S phase so that transit time through the checkpoints of the cell cycle is decreased (Benn and Schneider, 1995).

Animal and in vitro models of HBV replication

One of the major difficulties in assessing the true relevance in vivo of this wealth of molecular data and immunological observations is the lack of realistic, faithful models.

Cells are transfectable, but not truly infectable

Human cells can be transfected by full-length HBV genomes, but the virions produced cannot infect other cells in the culture. The raw replication efficiency of natural variant or mutagenized genomes can be compared, but this does not take account of small differences in efficiency, which can be uncovered only in a system that allows passage. Most of the HBV antigens are expressed after transfection. Even primary hepatocytes are very difficult to infect; whether this is due to their rapid loss of full differentiation, leading to lack of the appropriate receptor, or possibly to the lack of an as yet undefined cofactor which is not found in isolated hepatocyte preparations, remains to be seen. Addition of polyethylene glycol and dimethyl sulphoxide (Rumin et al, 1996) increases infection efficiency, but passage still does not clearly occur. Recently, continuous cultures of primary hepatocyte lines have been established using cocktails of hypothalamic and pituitary extracts; they appear to be infectable with HBV (Vitulli et al, 1998). Intriguingly, rat hepatocyte lines which have been permanently transfected with annexin V, a human and liver specific surface molecule (reviewed by De Meyer et al, 1997), are infectable in vitro. The efficiency of this system is also not fully established. There are many other candidate receptors of HBV. Some bind to pre-S_1 and others to HBsAg (such as annexin V). It may well be that, as for HIV, there are two receptors, both of which are critical, one binding to each envelope protein.

Even chimpanzees are not the perfect model

Chimpanzees are endangered and there is a general moral aversion to their use in human experimentation. Also, their response to infection with HBV is not identical to that of humans and little is known about their immune system. There are animal hepadnaviruses, such as those infecting duck (and other birds), woodchuck, tree squirrel and others. All of these are useful models, but have one or another drawback, for example no chronicity, no acute disease, no cancer, etc. Also, the human HBV cannot infect any of them.

Transgenic mice have been created with single HBV open reading frames (Mancini et al, 1996) or the whole genome (reviewed by Chisari, 1996; Koziel and Liang, 1997). The former have largely been used to investigate antigenic effects or tumorigenesis. Replicative transgenics have most expression in the liver and have been very useful for studying immune responses (see above).

Immunization

What is available?

There are two widely used vaccines. The plasma-derived version consists of particles harvested from chronic carriers, from which all infectious activity is removed by a number of sequential steps. The second is the only recombinant vaccine on the market for humans and is made in yeast. The plasma-derived vaccine is used widely in the Far East and in Africa. It is not available in Europe and the USA owing to safety concerns. The plasma vaccine tends to be cheaper, but the price of both vaccines is negotiated usually on a country-wide basis. Both vaccines are equally efficacious, although there are subtle advantages of the plasma vaccine for those on renal dialysis and other immunosuppressed groups. The overall constitution of both vaccines is similar: both consist of particles made of HBsAg. However, there are small quantities of L and M envelope proteins in the natural product, whereas the recombinant version is derived from a single gene, so only S protein is present. It is possible that the small amount of pre-S_1 (in L protein) and pre-S_2 (in M and L) provides T cell help for anti-HBs production.

Do they work?

Seroconversion rates vary according to age, genetic factors and sex. Overall, some 85% of persons will seroconvert after three doses; this increases to 90–95% after a fourth dose. Whether the three doses are given at 0, 1 and 6 months or 0, 1 and 2 months does not greatly affect this rate. A low dose (especially for some plasma-derived versions) and intradermal route of administration can lower the rate of seroconversion substantially. Any non-responder after three doses should receive a fourth dose.

The implementation of mass campaigns has led to reductions in carriage rates. In China, the HBsAg positivity rate in vaccinated children is around 2%, but 10–15% in control groups. New cases of HCC in Taiwanese children young enough to have received vaccination are almost non-existent (Chang et al, 1997). In The Gambia (Whittle et al, 1995), the long-term follow-up of children shows that, despite rapid falls in anti-HBs levels, protection was very good: vaccine efficacy for prevention of carriage was almost 95%.

Do we need to boost?

There is much debate about the level of anti-HBs which correlates with protection from infection. The subject is reviewed by Lemon and Thomas (1997). There are no good data to support the popular contention that 100 mIU ml^{-1} are required. Old studies indicated that persons who had more than 10 mIU ml^{-1} were protected against disease even if their level declined below this; in such cases, infection can occasionally occur (manifested by the appearance of anti-HBc), but this is asymptomatic and not accompanied by subsequent carriage. In those who once had more than 10 mIU ml^{-1}, B cells expressing anti-HBs can still be found in peripheral blood once the titre declines below this

Table 20.1a *Statistical analysis of HBc substitutions in patients who seroconverted from HBeAg to anti-HBe and went into clinical remission*

Two samples were taken from patients who who seroconverted from HBeAg to anti-HBe and went into clinical remission. The amino acid substitutions between the two samples were allocated to regions known to be antigenically recognized as detailed in the table. The odds ratio and 95% confidence intervals represent the likelihood of amino acid substitutions occurring in the epitope in comparison with the remaining HBc sequence. Probability (P) values are calculated using the χ^2 test or Fisher's exact method. T/B, Th or B cell epitope.

Epitope	T/B	Odds ratio	95% confidence interval	P value
1–20	T	0·81	0·13–3·74	0·99
50–69	T	5·64	2·12–14·85	0·00045
74–83	B	0	–	–
76–89	B	0	–	–
107–118	B	1·42	0·02–6·69	0·65
128–135	B	1·04	0·02–7·06	0·99
130–138	B	0·92	0·02–6·18	0·99

Table 20.1b *Statistical analysis of HBc substitutions in patients with active hepatitis who were followed during the anti-HBe-positive phase of their disease*

Two samples were taken from patients who were continuously anti-HBe positive with severe disease. The amino acid substitutions between the two samples were allocated to regions known to be antigenically recognized as detailed in the table. The odds ratio and 95% confidence intervals represent the likelihood of amino acid substitutions occurring in the epitope in comparison with the remaining HBc sequence. Probability (P) values are calculated using the χ^2 test or Fisher's exact method. T/B, Th or B cell epitope.

Epitope	T/B	Odds ratio	95% confidence interval	P value
1–20	T	0·86	0·88–1·52	0·59
50–69	T	0·55	0·28–1·08	0·06
74–83	B	2·63	1·36–4·74	0·00067
76–89	B	1·21	0·59–2·43	0·56
107–118	B	1·31	0·61–2·71	0·45
128–135	B	2·15	0·94–4·37	0·03
130–138	B	2·56	1·35–4·80	0·0014

threshold; the implication is that an anamestic response will occur on exposure to HBV. The protocol used in the West of Scotland for health care workers is given in Table 20.2. A low anti-HBs level is, however, an important factor in the selection of escape mutants after vaccination (see below).

The issue of a protective cut-off has been further muddied by the variable results of commercial anti-HBs assays. Some assays read consistently (and reliably) higher than others; which are right is impossible to tell, because of the lack of a 'gold standard'. Clearly, using an assay that reads higher gives an impression of better protection. At the lower end of the dynamic range (around 10–100 mIU ml^{-1}), assays give similar results; so as long as the protective threshold is set at 10 mIU ml^{-1}, this interassay variability should not be of major concern.

Table 20.2 *Follow-up of hepatitis B vaccine programme reporting grid*

Titre (IU l^{-1})	Recommendation
After course of three doses	
<10	Non-responder to hepatitis B immunization
	Recommend a second course of vaccine is completed and check antibody titre
10–99	Poor response to hepatitis B immunization
	Booster dose required now and check in 2–3 months
100 or more	Protective response to hepatitis B immunization
	Give final booster dose in 5 years' time
	No further antibody check is required
After fourth dose (i.e. booster)	
<10	Non-responder to hepatitis B immunization
	This patient remains susceptible to HBV infection and requires HBIG in the event of a significant needlestick or similar injury
10–49	Adequate immunity following hepatitis B immunization
(tested in two assays)	Boost in 5 years
	No further antibody check required
50–99	Adequate immunity following hepatitis B immunization
	Boost in 5 years
	No further antibody check required
100 or more	As above

Whom should we vaccinate?

There are two vaccination policies in current use. The first is universal childhood vaccination, which is in force in some 100 countries including much of Europe, Asia and the USA. The UK is a notable exception. The second policy, which has failed, is selective immunization of the classical high-risk groups. If eradication is our goal, then universal vaccination worldwide is the only realistic and sensible option.

Will escape variants of HBsAg become dominant?

In most studies of vaccinated children born to HBeAg-positive mothers, some 15% of children develop evidence of infection (they become anti-HBc-positive). In 1988, several vaccinated children were reported to have also been HBsAg-positive, although often transiently (Zanetti et al, 1988). The isolate from one child had 145R, an arginine at aa 145 of HBsAg (Carman et al, 1990), whereas the mother had 145G. This finding has been confirmed in children from Singapore (Oon et al, 1995), Japan (Fujii et al, 1992; Okamoto et al, 1992; Hino et al, 1995) and the USA (Nainan et al, 1997). The study in the USA showed that, overall, 0·8% of vaccinated children born to HBeAg-positive mothers were infected with 145R; the other 4% with breakthrough were infected with 145G. Other variants also appear (see Figure 20.3) but, unlike 145R, often rapidly revert to the strain seen in the mother, indicating that 145R is not only replication competent, as shown in chimpanzees (Ogata et al, 1997), but its altered antigenicity gives it a significant survival advantage over 145G. Other variants are often single examples or not enough information has been published to interpret their significance fully. It is of interest that de novo selection in the infant as well as selection from a maternal quasispecies can occur. Vaccination at birth is an ideal situation for selection of escape variants: the virus is already within the new host while the immune response is building up. A parallel situation is seen in liver transplant recipients who receive high-titre anti-HBs preparations to prevent infection of the graft (Cariani et al, 1995; Hawkins et al, 1996; Carman et al, 1996a,b). By far the most common variant observed to arise in this situation is G145R. It is not often seen in the pretransplant sample, indicating that it has arisen de novo. A minority proportion of variant strains can be seen in the mothers of variant-infected infants, but the alternative scenario, that they arise de novo and cannot be found in the mother is also true. So far, variants have not been seen in children exposed to carriers (family and other close contacts) after vaccination. Much of the anti-HBs response is directed towards the second loop of the MHR, so it is not surprising (in retrospect) that variants in this region arise after such highly focused immune pressure. There is no doubt that the failure of the passive/active immunization protocol to control infection with this variant of HBV is the result of altered antigenicity. It is interesting to speculate whether the emergence of this variant in vaccinees is the result of infection in a host with a humoral immune response in the absence of a cytotoxic T cell response. In acute infection, when variants do not become dominant, the humoral response is always accompanied by a cytotoxic T cell response. It is not known whether there is any specific host factor that predisposes individuals to selection of G145R. This virus is replication competent as it can infect chimpanzees. Phylogenetic analyses (J. Wilson et al, unpublished observations) indicate that the examples of this variant described to date are not closely related, so they can arise around the world from almost any background. One caveat to this is that selection of G145R in some strains leads to a translational stop codon in the overlapping polymerase protein, so such viruses would not be viable.

Most of the vaccine-associated cases of 145R have occurred in children with a low level of anti-HBs and the chimpanzee experiments indicate that high titres protect against infection with this variant (Ogata et al, 1994). It remains to be seen, however, whether the lower levels of immunity often seen in immunocompromised and older vaccinated patients will be sufficient to prevent infection with this variant. It should be kept in mind that, when antibody titres in vaccinees are low, infection does occur with 145G, so it seems probable that in poorly immunized patients susceptibility to HBV variants may be even greater.

Some epitopes present normally on 145G particles are also present on 145R, so it is possible that high-titre vaccine-induced responses may be capable of generating protection against both. Decisions will need to be made on whether there is a need to improve the immunogenicity and broaden the specificity of the response induced by the current vaccine, in order to generate a more effective immune response which is capable of neutralizing both strains with the same efficiency.

There are a number of crucial questions that need to be addressed before the importance of this situation can be assessed. First is the current prevalence. A few examples of naturally occurring 145R strains have been reported (Yamamoto et al, 1994; Carman et al, 1995a; Kidd-Ljunggren et al, 1995). On the other hand, screening studies in Papua New Guinea, Sardinia and South Africa, based on differential results in two HBsAg assays in unvaccinated populations (Carman et al, 1997b), failed to detect it. The second is whether there will be secondary spread of this replication-competent, infectious virus in unvaccinated and vaccinated subjects. Unfortunately, this cannot be determined until the current generation of infected children reaches sexual maturity. Only then will we know whether it is sexually or vertically transmitted. However, mathematical modelling

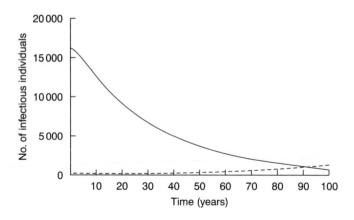

Figure 20.10 *Predicting the emergence of HBsAg variants after vaccination*
The solid line represents the wild-type and the dashed line the vaccine escape mutant strain.

(Wilson et al, 1998) shows that it is likely to become the dominant viral strain, even in the epidemiological setting of horizontal transmission (Figure 20.10). A notable feature is the time predicted for the emergence of the mutants. This is because, although highly contagious, HBV is not very infectious compared with measles or influenza. The spread of HBV is dependent on chronic infection generating new infections over a long period, so the incidence of new infections is low and the mutant takes decades to spread through the population. However, just because the change in HBV type is slow does not mean the mutants are unimportant; if they are allowed to attain high prevalence, it will also take several decades to eradicate them, even with cross-reactive vaccines, because of the chronic nature of the infection. It is highly relevant that, of the vaccine escape variants so far described, 145R is the most persistent and is not lost in favour of 145G over time. This model does not take account of vertical transmission (which is the most common setting of emergence), nor of a baseline prevalence in the unvaccinated population (which has been clearly shown). Thus, these predictions fall at the latest point of a range of time periods. Third, even if it is transmitted or emerges in vaccinated persons, will it cause disease? The 145R strain definitely causes chronic hepatitis; in the first reported vaccinated case the person was HBeAg positive and had chronic hepatitis.

What can be done to prevent this situation, or is it inevitable (as modelling suggests)? Additional epitopes from other regions such as HBc or pre-S could be added to the current vaccine. These provide Th help to the anti-HBs response and have minor neutralizing epitopes. The S gene-containing DNA vaccines (Michel et al, 1995) may lead to very high anti-HBs titres and may circumvent HLA-mediated non-responsiveness, but suffer from the same problem as current HBsAg vaccines of highly focused immune responses that do not mimic the natural immune response to virus infection. The simplest and quickest solution, while the situation is addressed epidemiologically over the next 10–20 years, is to use a hybrid vaccine that contains both (or mixed) 145G and 145R HBsAg particles. Unfortunately, pre-S_2/S vaccines (see below) do not appear to be the answer.

Do we need third-generation vaccines?

Because of the 5% (at least) failure rate of HBsAg-only vaccines after three doses and the need to give additional hyperimmunoglobulin (HBIG) to infants of HBeAg-positive mothers at birth (as well as vaccine), there has been a move to generate vaccines with additional antigenic components. This has taken the route of adding pre-S_2, or pre-S_1 plus pre-S_2. These antigens, commonly made in mammalian cells, can lead to neutralizing antibody responses but also provide T cell help to anti-HBs responses. Recent studies show conclusively that the seroconversion rates are higher and the mean anti-HBs titre is also more impressive (Hourvitz et al, 1996; Zuckerman et al, 1997). Of subjects who failed to respond to multiple doses of HBsAg-only vaccines, 64% developed protective levels of anti-HBs after a single dose of these improved vaccines. Whether the vaccines will be used for universal vaccination or will be employed in niche markets such as health care workers, renal dialysis patients and infants born to HBeAg-positive mothers (to avoid the use of HBIG, which is expensive) remains to be seen. Whether such third-generation vaccines will be able to prevent the emergence of escape variants will probably not be known for many years, but the preliminary data are not encouraging.

The use of vaccines to treat carriers

Therapeutic vaccination shows significant promise in the treatment of chronic HBV infection. A study using a pre-S_2-S vaccine looked promising (Pol et al, 1998). We have used a pre-S/S vaccine (Hepagene; Medeva plc, UK) to stimulate seroconversion to anti-HBe and loss of HBV-DNA in 8 of 22 carriers; although this is a similar rate to interferon (see above), it is promising that these particular patients were not believed to be good candidates for standard therapies (Carman et al, 1998). Only eight doses over 8 months were required rather than thrice-weekly injections (for interferon), with side-effects, over 6 months. A parallel approach has been to combine HBsAg with anti-HBs (HBsAg in excess) attached to solid matrices or to aluminium hydroxide (Wen et al, 1995). The proposed mechanism is that the anti-HBs enhances uptake by APC and thus presentation of antigenic epitopes derived from HBsAg. A third approach has been to generate a peptide that contains the HLA-A2 class I HBcAg epitope, linked to a Th cell epitope (tetanus toxoid) and lipid tail (supposedly to enhance peptide uptake). Naïve volunteers all generated CTL responses (Vitiello et al, 1995). HBV-DNA vaccines (containing the S or C, or both, genes) are clearly an attractive approach to stimulating immune responses. They result in anti-HBs, Th and CTL responses in animal models; significantly, they appear able to bypass the HLA processing pathway, as non-responder mice (which do not generate CTLs to HBsAg) make anti-HBs CTLs if given a DNA vaccine. This is a distinct advantage over the HLA-restricted peptide approach mentioned above.

Therapy

Acute infection

There is no specific therapy for treating acute HBV infection. When liver function is deteriorating, patients have to be transferred to hospital. Several criteria exist to determine the need for liver transplantation (Bernuau et al, 1986; O'Grady et al, 1989, 1993) but most important is the experience of the clinician and the dynamics of the disease.

Therapy of chronic HBV infection

The treatment of chronic HBV infection has been reviewed by Hoofnagle and diBisceglie (1997).

Interferon

Interferon alpha belongs to a family of naturally occurring proteins with several properties. It is used for therapy of viral infections because of its antiviral and immunomodulatory effects. At present, in most countries, IFN-α is the only approved therapy for treating chronic HBV infection. For the treatment of chronic HBV infection, usually between 5 million units per day and 5–10 million units three times per week are administered subcutaneously (Perillo et al, 1994; diBisceglie, 1995; Fried, 1996). At lower doses, side-effects are less severe and the response rate does not vary from higher doses.

There are contraindications: decompensated liver cirrhosis (Child grade C), hypersplenism (thrombocytopenia $<75 \times 10^9$ l^{-1} and leukopenia $< 1.5 \times 10^9$ l^{-1}), concurrent autoimmune diseases, psychiatric disorders (depression) and drug abuse. After the start of therapy, patients should be closely monitored. Every week during the first month patients should be examined, questioned for side-effects, and blood aminotransferases and blood counts should be controlled. After the first month the frequency of visits depends on the severity of side-effects and the degree of liver disease (Perillo et al, 1994; diBisceglie, 1995a,b; Fried, 1996).

The most frequent side-effects during IFN-α therapy are flu-like symptoms (fatigue, myalgia, arthralgia); bone marrow suppression; psychiatric disorders (depression and suicide); and induction of autoimmune diseases (e.g. dysfunction of the thyroid, autoantibodies). If side-effects do occur, either treatment must be stopped or the interferon dose can be reduced (Fattovich et al, 1996; Fried, 1996).

The response to IFN-α treatment is defined as loss of HBV-DNA, loss of HBeAg (with or without the appearance of anti-HBe) and normalization or a marked decrease in serum aminotransferases. Several studies and meta-analysis showed that the response rate of chronic HBV patients to IFN-α treatment is 30–40%. After cessation of therapy there is normally no relapse of HBV. During long-term follow-up around 70% of the patients who responded to interferon treatment also lost HBsAg. Thus most of the patients who initially respond to interferon treatment will have a complete resolution of HBV infection (Bisceglie, 1995a,b; Niederau et al, 1996).

Several markers in children and adults (Ruiz-Moreno et al, 1995) with chronic HBV infection are found to be associated with response. Positive predictive markers are low levels of HBV-DNA, high aminotransferase levels, female sex, and short duration between infection and start of therapy. In contrast, negative predictive markers before treatment are low aminotransferase levels, high levels of HBV-DNA, perinatal infection, immunosuppression and co-infection with other viruses (HIV, HBV, HCV) (Perrillo and Mason, 1994; diBisceglie, 1995a,b; Fried, 1996). Aminotransferases have a predictive value for HBV therapy because an increase in their level reflects the active interaction of the immune system with the virus. Therefore, during periods of active interactions between virus and the immune system – when aminotransferase levels are high – interferon further triggers immune-mediated elimination of HBV.

What happens to extrahepatic syndromes after interferon treatment? One study showed that patients suffering from HBV-related glomerulonephritis who responded to IFN-α therapy had a recovery of their renal function. In contract, in patients who did not respond to this treatment, further deterioration of renal function was observed (Conjeevaram et al, 1995). Should patients with more progressive liver disease be treated with interferon? The rate of side-effects is increased and (especially in patients with Child C cirrhosis) a dramatic deterioration of liver function is frequently observed, necessitating liver transplantation. In patients with cirrhosis, therapy should be performed in a transplant centre and patients monitored frequently (Perrillo et al, 1995; Perrillo, 1995).

Nucleoside analogues
The virus replicates via an intermediate RNA step requiring reverse transcriptase. Two nucleoside analogues have been shown to block HBV-DNA replication in vivo: lamivudine and penciclovir (Krüger et al, 1994; Dienstag et al, 1995). The oral form of penciclovir is famciclovir, which is normally used in clinical trials. The modes of action of these two nucleoside analogues differ. Lamivudine is a deoxycytidine analogue, while penciclovir/famciclovir is a deoxyguanosine analogue.

Lamivudine acts as a chain terminator, so can inhibit both DNA- and RNA-dependent activities of the HBV-DNA polymerase. In contrast to lamivudine, penciclovir/famciclovir is not an obligate chain inhibitor but it has an additional mechanism of action which is not shared by lamivudine: inhibition of the priming step of reverse transcription.

Lamivudine, given in different doses over a 12-week period, resulted in rapid and complete inhibition of HBV-DNA (Dienstag et al, 1995) (Figure 20.11). Twenty per cent of the patients remained negative for HBV-DNA 6 months after cessation of treatment. Preliminary results obtained for penciclovir/famciclovir also showed inhibition of viral replication, but compared with lamivudine inhibition of HBV replication seems less rapid. The end-point of therapy with nucleoside analogues has to be defined, as these agents cannot eliminate the ccDNA of HBV, which has a long half-life. Another important issue is the occurrence of mutations in the polymerase gene, which confer resistance. The mutations found during either lamivudine or penciclovir/famciclovir

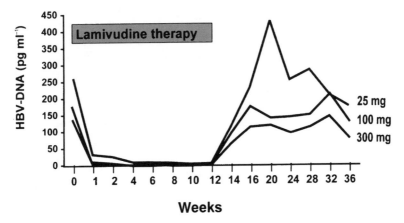

Figure 20.11 *Effect of lamivudine on HBV-DNA levels*
Changes in HBV-DNA levels of chronic HBV carriers receiving lamivudine for a 12-week period. Three different doses of lamivudine were given: 25 mg, 100 mg or 300 mg per day. Ten or eleven patients were treated in each group. The average HBV-DNA levels of each subgroup are shown. From Dienstag et al (1995).

treatment differ (Aye et al, 1996; Ling et al, 1996; Tipples et al, 1996), so combination or sequential treatment may offer advantages. It has been estimated that about a third of persons treated long-term with lamivudine will relapse owing to mutations in the active site of the polymerase.

Future therapies
First clinical trials are being performed with therapeutic vaccines which induce a cyto-toxic T cell response (see above). Amongst these are pre-S protein vaccines or naked DNA (Bohm et al, 1996; Davies et al, 1996). The latter is particularly interesting as it led to complete inhibition of HBsAg production in transgenic mice by a post-transcriptional mechanism.

Strategies that interfere with viral replication are the administration of antisense oligonucleotides or ribozymes. Both approaches interfere specifically with viral RNA and lead to inhibition of HBV replication (Offensperger et al, 1994). However, they cannot eliminate intracellular viral ccDNA, and clinical trials have not been performed.

Liver transplantation
Fulminant or chronic hepatitis B infection can lead to the need for liver transplantation. Fulminant hepatitis B infection leads to acute deterioration of liver function, the need for organ replacement becoming evident in hours or days. Chronic hepatitis B infection leads either to cirrhosis, HCC or both (Beasley et al, 1981). The time point for liver transplantation in these conditions is less defined. The decision for liver transplantation is mainly dependent on the clinical situation and the availability of organs.

In patients suffering from chronic HBV requiring liver transplantation, a specific problem is reinfection of the donor liver with the HBV (Todo et al, 1991). After reinfection, the course of the disease progresses more rapidly towards liver cirrhosis, often requiring retransplantation (Todo et al, 1991; Shah et al, 1992). In these cases the histological picture of fibrosing cholestatic hepatitis (FCH) can be found (Benner et al, 1990; Davies et al, 1991). The hepatocytes of patients with FCH stain prominently for HBsAg (Lau et al, 1992), and results from a transgenic mouse model revealed that high amounts of L protein are cytotoxic (Chisari et al, 1987). After liver transplantation, therefore, mechanisms may be activated leading to retention of S protein in the endoplasmic reticulum; this is cytotoxic and leads to the histological picture of FCH. A mutation in the CCAAT box of the S promoter inverses the ratio of large to small S protein, which increases cellular retention of S protein and may further impair liver function in these patients (Trautwein et al, 1996). However, the pathogenesis of FCH remains unclear.

Prevention of hepatitis B reinfection after liver transplantation mainly entails anti-HBs (Lauchert et al, 1987). Anti-HBs titres of more than 200 mIU ml^{-1} have to be maintained for at least 1 year after transplantation. Prophylactic anti-HBs reduces the risk of reinfection and prolongs survival (Figure 20.12) (Lauchert et al, 1987; Müller et al, 1991; Samuel et al, 1991, 1993). However, during anti-HBs prophylaxis some patients suffer reinfection; frequently mutations in the coding region of the S gene are observed, especially in the MHR. These mutations reduce the binding of anti-HBs and therefore are thought to be the result of an escape mechanism of the virus to circumvent anti-HBs prophylaxis (Carman et al, 1996). Interferon alpha after liver transplantation is not recommended (Neuhaus et al, 1991; Lucy, 1994).

Lamivudine and famciclovir appear useful to prevent reinfection after liver transplan-

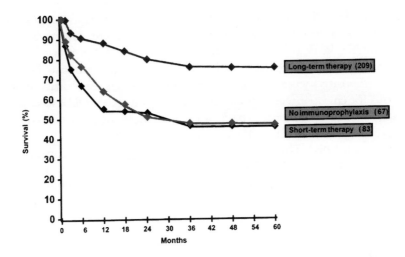

Figure 20.12 *Anti-HBs prophylaxis and HBV reinfection after liver transplantation*
Actuarial survival after liver transplantation due to chronic hepatitis B infection, dependent on
the duration of anti-HBs prophylaxis. After liver transplantation patients with chronic HBV
infection were assigned to three groups: one received no anti-HBs immunoprophylaxis; the
others received either short-term (less than 6 months) or long-term (longer than 6 months)
therapy. Patients receiving long-term immunoprophylaxis had a significantly better survival rate
after liver transplantation. From Samuel et al (1993).

tation (Böker et al, 1994; Krüger et al, 1994; Grellier et al, 1996; Perillo et al, 1996). As
both drugs reduce or stop HBV replication, they presumably improve the outcome of
patients. However, patients on lamivudine or famciclovir select mutations in the polymerase gene, rendering the virus resistant against further therapy (Ling et al, 1996;
Tipples et al, 1996).

Diagnostics

Hepatitis B is not diagnosed clinically, nor can we predict with any precision that a unit
of donated blood is HBV-free by looking at it, even with a cynical eye. Therefore we are
reliant upon highly sensitive, specific and cheap commercial diagnostic assays. Hepatitis
B virus is unique in virology in that the primary diagnostic tool is the detection of antigen (HBsAg) in the blood. A wide range of techniques, ranging from electron
microscopy to radioimmunoassay, is available, but in general enzyme-linked immunoassay (ELISA) is employed. Each manufacturer employs different anti-HBs antibodies,
monoclonal and/or polyclonal, on the solid phase. Serum is then added and a detector
anti-HBs completes the sandwich. It is therefore critical that the antibodies chosen are
able to detect HBsAg whatever the antigenicity and whatever the serum concentration.
This has become a major issue with the discovery of variants of HBsAg that affect antigenicity (see above). This subject, which is far from resolved, has been the subject of
recent reviews (e.g. Wallace and Carman, 1997). All the diagnostic companies are
actively assessing their kits, with a view to improvement. What is most critical is knowledge of the prevalence of such HBsAg variants and whether they are geographically
related.

The other major issue is the move away from HBeAg as a marker of infectivity and the introduction of HBV DNA measurement. The commercial assays are useful, but it is a pity that there is so little comparability between them; there is not even a reliable conversion factor (Kapke et al, 1997). Measurement of HBV DNA is used mainly as a predictor of response before interferon therapy and to define response. Polymerase chain reaction (PCR) positivity is a difficult area because sera that are positive are not necessarily infectious for chimpanzees (Kaneko et al, 1989). Yet, HBV DNA assays are not sensitive enough to exclude a person as being infectious; note the report of anti-HBe-positive carrier surgeons, who were HBV DNA-negative but PCR-positive, infecting their patients (Heptonstall et al, 1997). The solution is an assay with an intermediate sensitivity.

Finally, it is well known that persons who are HBsAg-negative and anti-HBc-positive may yet be PCR-positive and infectious after blood transfusion (Hoofnagle et al, 1978). These cases, plus the occasional person who has no markers of HBV infection (including anti-HBc) but is PCR-positive, are redefining our view of 'chronic carriage'.

Conclusion

Much has been learnt about HBV since the 1980s and there is clearly much still to be learnt. This virus has provided the lead in achieving an understanding of natural genetic variants of viruses and their interaction with the host; it is still one of the paradigms of how to diagnose viral infections and, unusual amongst viruses, can be treated with an increasing number of agents. Its molecular biology and interaction with the cell is being dissected at a frantic rate. The issue is whether we will really understand HBV before a decision is made to destroy the last stocks after it has been eliminated by vaccines, either the current or modified ones.

DELTA VIRUS

The subject of hepatitis D (or delta) virus (HDV) has been covered recently by Casey (1996), Monjardino (1996), Beard et al (1997), Casey et al (1997), Hadziyannis (1997a,b) and Taylor et al (1997). A short update follows.

Biology

Hepatitis D virus comprises a 1700 nucleotide single-strand circular RNA that forms a rod-like structure owing to two-thirds of its genome being complementary. These features resemble those of viroids (plant pathogens). Viroids have much smaller genomes because they have no open reading frame (ORF). Nucleotides 1608 to 1669 are probably an RNA-RNA promoter containing two bulge regions in a stem-loop structure, encompassing a GC-rich motif (Beard et al, 1996). In HDV, the single ORF encodes hepatitis delta antigen (HDAg), which is absolutely necessary for replication. Its exact function is unknown: it has no polymerase activity, so may, for example, act as a transcription factor. It certainly binds to delta RNA. There are two forms of HDAg, which differ only by the presence of an amber stop codon. RNA editing at adenosine 1012, within this stop codon, in the antigenomic RNA allows synthesis of both from a single open reading frame. The editing is thought to be catalysed by a cellular adenosine deaminase (Polson et al, 1998). The one sequence is thus derived from the other during the intracellular life cycle. The small protein promotes replication, possibly by binding via a leucine repeat region in a central domain (Chang et al, 1993). The large form, which has an additional 19 aa residues at the C terminus, is highly replication-inhibitory, but, with HBsAg (which provides the coat of the viroid), promotes packaging of the genome, possibly mediated by a proline- and glycine-rich region at the C terminus (Lazinski and Taylor, 1993). Woodchucks can also become infected with HDV, indicating that wood-chuck hepatitis virus (WHV) as well as HBV envelope proteins can provide helper func-tions. There is some evidence that woodchucks can become infected de novo without the help of WHV (Netter et al, 1994). However, WHV is necessary for ongoing infec-tion. By the yeast two-hybrid system, a cellular factor that binds to HDAg yet has 24% homology to HDAg has been identified (Brazas and Ganem, 1996). Overproduction of this homologue inhibits HDV replication in competent cells, via downregulation of mRNA synthesis. It was proposed that this cellular homologue was the origin of HDAg, acquired during infection of a cell by a viroid. The proposed method of replication is as a rolling circle, with the genomic and antigenomic strands acting as template in turn for production of multimeric full-length copies which are self-cleaved by ribozyme activity. This cleavage has variable efficiency depending upon the cell type, implying that a cellular factor is required for full activity. Both HDAgs (particularly the large form) enhance, but are not required for, the cleavage (Jeng et al, 1996). HDAg could not restore the ribozyme activity of mutant HDV RNAs which had lost the ribozyme function. This enhancement does not require HDV RNA replication but does need the RNA-binding activity of HDAg. This suggests that HDAg can regulate the cleavage and liga-tion of HDV RNA during the HDV life cycle.

Prevalence and genotypes

Hepatitis D virus is the only virus in the satellite family known to infect animal species. Recently, the world map of prevalence of HDV has been modified, with decreasing prevalence in certain areas and some new foci of HDV endemicity (such as Albania, Okinawa, China and India). Worldwide, some 5% of HBV carriers are co-infected with HDV. There are three genotypes, with type II being found predominantly in the Far East and type III in South America. There is 30–40% variability amongst the genotypes and about 10–15% within the same genotype (Casey and Gerin, 1998). The vast majority of Italian patients with chronic hepatitis, for example, had genotype I, without any sub-types being identified; however, clustering of some sequences suggested correlations with geography and transmission route (Niro et al, 1997). These Italian sequences were more diverse than those from east Asia and North America, suggesting introduction into Italy earlier and/or from multiple sources compared with those areas. In one study (Zhang et al, 1996), phylogenetic analyses revealed three clusters of type I, each of them corresponding to certain geographical regions: the 'western' group from western Europe and the USA; the 'eastern' group from Moldavia, Bulgaria, mainland China and Taiwan; and the 'African–Middle East' group from Ethiopia, Somalia, Jordan and Lebanon. African isolates were interrelated and formed a novel group within genotype I.

Clinical matters

Cirrhosis is common in chronic HDV (about two-thirds of cases). Although genotype II superinfection is associated with fulminant hepatitis in Taiwan (Wu et al, 1995), it is difficult to distinguish host from viral factors. The role of HDV was investigated as the possible cause of severe acute hepatitis in military personnel at four different jungle outposts during 1992–1993 (Casey et al, 1996); 64% were infected exclusively with genotype III. The data suggested focal sources in the jungle environment. Within genotypes, genetic variations are not clearly associated with different disease patterns. Hepatotropic viruses like hepatitis C virus and HDV may cause autoimmune phenomena (Manns, 1997) which are similar to those in idiopathic autoimmune hepatitis and linked to Liver Kidney Microsomal (LKM) antibodies. Superinfection with HDV does not accelerate the development of hepatocellular carcinoma (Huo et al, 1996).

Diagnosis is by detection of HDV antigens, antibodies or RNA in serum. Although dot-blot hybridization is useful, the sensitivity of the polymerase chain reaction (PCR) is superior (Jardi et al, 1995); further, PCR correlated with intrahepatic HDAg detection. The IgM antibody to HDV represents a valid surrogate marker of liver damage; it also provides an excellent predictor of impending resolution of chronic disease, whether spontaneous or IFN-induced (Borghesio et al, 1998). For example, IgM antibody disappeared in each patient who responded to IFN therapy with the persistent normalization of aminotransferases and with the clearance of serum HBsAg and HDV-RNA. The IgM reactivity did not decline either in the 45 non-responders or in those who experienced a relapse. Following transplantation IgM rapidly disappears. The only available treatment for chronic delta hepatitis is IFN-α (Farci et al, 1994). With high doses (9 MU thrice weekly for a year), about half of those treated have a histological remission with normal aminotransferase levels; long-term virological remission is rare as most patients relapse after stopping treatment.

Analysis of peripheral blood mononuclear cell proliferation in response to recombinant large HDAg (Nisini et al, 1997) revealed specific response in 8 of 30 patients (27%). All responders had signs of inactive disease; none of the patients with active disease and none of the control subjects showed any significant proliferation. Peptides of aa 50–65 and aa 106–121 were presented to specific T cells in association with multiple class II molecules. All the CD4+ T cell clones assayed were able to produce high levels of interferon gamma and belonged either to T helper 1 (Th1) or Th0 subsets; some of them were cytotoxic in a specific assay. Thus, detection of a specific T cell response to HDAg in the peripheral blood is related to a decrease of disease activity.

REFERENCES

Aikawa T, Kanai K, Kako M et al (1995) Interferon-α2a for chronic hepatitis B with e antigen or antibody: comparable antiviral effects on wild-type virus and precore mutant. *J Viral Hep* **2:** 243–250.

Akarca US & Lok ASF (1995) Naturally occurring core-gene-defective hepatitis B viruses. *J Gen Virol* **76:** 1821–1826.

Antonucci TK & Rutter WJ (1989) Hepatitis B virus (HBV) promoters are regulated by the HBV enhancer in a tissue specific manner. *J Virol* **63:** 579–583.

Ashton-Rickardt PG & Murray K (1989) Mutants of the hepatitis B virus surface antigen that define some antigenically essential residues in the immunodominant alpha region. *J Med Virol* **29:** 196–203.

Aye TT, Bartholomeusz AJ, Shaw T et al (1996) Hepatitis B virus polymerase mutations during famciclovir therapy in patients following liver transplantation. *Hepatology* **24** (suppl.): 285A.

Barnes PJ & Karin M (1997) Nuclear factor-kB – a pivotal transcription factor in chronic inflammatory diseases. *N Engl J Med* **336:** 1066–1071.

Baumert TF, Rogers SA, Hasagawa K & Liang TJ (1996) Two core promoter mutations identified in a hepatitis B virus strain associated with fulminant hepatitis result in enhanced viral replication. *J Clin Invest* **98:** 2268–2276.

Beard MR, Gowans EJ & MacNaughton TB (1996) Identification and characterization of a hepatitis delta virus RNA transcriptional promoter. *J Virol* **70:** 4986–4995.

Beard MR, Gowans EJ & MacNaughton TB (1997) A transcriptional promoter for antigenomic-sense HDV RNA. In: Rizzetto M, Purcell RH, Gerin JL & Verme G (eds) *Viral Hepatitis and Liver Disease*, pp 298–301. Turin: Minerva.

Beasley PR, Lin CC, Hwang LY & Chien CS (1981) Hepatocellular carcinoma and hepatitis B virus a prospective study of 22 707 men in Taiwan. *Lancet* **ii:** 1129–1133.

Ben-Levy R, Faktor O, Berger I & Siddiqui A (1989) Cellular factors that interact with the hepatitis B virus enhancer. *Mol Cell Biol* **9:** 1804–1809.

Benn J & Schneider RJ (1994) Hepatitis B virus HBx protein deregulates cell cycle checkpoint controls. *Proc Natl Acad Sci USA* **92:** 11215–11219.

Benn J, Su F, Doria M & Schneider RJ (1996) Hepatitis B virus HBx protein induces transcription factor AP-1 by activation of extracellular signal-regulated and c-Jun N-terminal mitogen-activated protein kinases. *J Virol* **70:** 4978–4985.

Benner KG, Lee RG, Keeffe EB, Lopez RR, Sasaki AW & Pinson CW (1990) Fibrosing cytolytic liver failure secondary to recurrent hepatitis B after liver transplantation. *Gastroenterology* **103:** 1307–1312.

Bernuau J, Rueff B & Benhamou JP (1986) Fulminant and subfulminant liver failure: definition and causes. *Sem Liver Dis* **6:** 97–106.

Berting A, Hahnen J, Kröger M & Gerlich WH (1995) Computer-aided studies on the spatial structure of the small hepatitis B surface protein. *Intervirology* **38:** 8–15.

Bertoletti A, Chisari FV, Penna A et al (1993) Definition of a minimal optimal cytotoxic T-cell epitope within the hepatitis B virus nucleocapsid protein. *J Virol* **67:** 2376–2380.

Bertoletti A, Sette A, Chisari FV et al (1994) Natural variants of cytotoxic epitopes are T-cell receptor antagonists for antiviral cytotoxic T cells. *Nature* **369:** 407–410.

diBisceglie AM (1995a) Chronic hepatitis B. *Postgrad Med* **98:** 99–106.

diBisceglie AM (1995b) Long-term outcome of interferon-alpha therapy for chronic hepatitis B. *J Hepatol* **22(suppl. 1):** 65–70.

Blum HE, Zhang ZS, Galun E et al (1992) Hepatitis B virus X protein is not central to viral life cycle in vitro. *J Virol* **66:** 1223–1227.

Bohm W, Kuhrober A, Paier T, Mertens T, Reimann J & Schirmbeck R (1996) DNA vector constructs that prime hepatitis B surface antigen-specific cytotoxic T lymphocyte and antibody reponses in mice after intramuscular injection. *J Immunol Methods* **193:** 29–40.

Böker KH, Ringe B, Krüger M, Pichlmayr R & Manns MP (1994) Prostaglandin E plus famciclovir – a new concept for the treatment of severe hepatitis B after liver transplantation. *Transplantation* **57:** 1706–1708.

Bollyky P (1997) PhD Thesis, University of Oxford.

Borghesio E, Rosina F, Smedile A et al (1998) Serum immunoglobulin M antibody to hepatitis D as a surrogate marker of hepatitis D in interferon-treated patients and in patients who underwent liver transplantation. *Hepatology* **27:** 873–876.

Bottcher B, Wynne SA & Crowther RA (1997) Determination of the fold of the core protein of hepatitis B virus by electron cryomicroscopy. *Nature* **386(6620):** 88–91.

Brazas R & Ganem D (1996) A cellular homolog of hepatitis delta antigen: implications for viral replication and evolution. *Science* **274**: 90–94.

Brown SE, Howard CR, Zuckerman AJ & Steward MW (1984) Affinity of antibody responses in man to hepatitis B vaccine determined with synthetic peptides. *Lancet* **ii**: 184–187.

Brunetto MR, Stemmler M, Schodel F et al (1989) Identification of HBV variants which cannot produce precore-derived HBeAg and may be responsible for severe hepatitis. *Ital J Gastroenterol* **21**: 151–154.

Bulla GA & Siddiqui A (1988) The hepatitis B virus enhancer modulates transcription of the hepatitis B virus surface antigen gene from an internal location. *J Virol* **62**: 1437–1441.

Cariani E, Ravaggi A, Tanzi E et al (1995) Emergence of hepatitis B virus S gene mutant in a liver transplant recipient. *J Med Virol* **47**: 410–415.

Carman WF, Jacyna MR, Hadziyannis S et al (1989) Mutation preventing formation of e antigen in patients with chronic HBV infection. *Lancet* **ii**: 588–591.

Carman WF, Zanetti AR, Karayiannis P et al (1990) Vaccine-induced escape mutant of hepatitis B virus. *Lancet* **336**: 325–329.

Carman WF, Hadziyannis S, Karayiannis P et al (1991) Association of the precore variant of HBV with acute and fulminant hepatitis B infection. In: Hollinger, Lemon & Margolis (eds) *Viral Hepatitis and Liver Disease*, pp 216–219. Baltimore: Williams & Wilkins.

Carman WF, Korula J, Wallace L, MacPhee R, Mimms L & Decker R (1995a) Fulminant reactivation of hepatitis B due to envelope protein mutant of HBV that escaped detection by monoclonal HBsAg ELISA. *Lancet* **345**: 1406–1407.

Carman WF, Thursz M, Hadziyannis S et al (1995b) HBeAg negative chronic active hepatitis: HBV core mutations occur predominantly in known antigenic determinants. *J Viral Hep* **2**: 77–84.

Carman WF, Trautwein C, Colman K et al (1996a) Hepatitis B virus envelope variation under polyclonal antibody pressure after liver transplantation. *Hepatology* **24**: 489–493.

Carman WF, Trautwein C, van Deursen FJ et al (1996b) Hepatitis B virus envelope variation after transplantation with and without hepatitis B immune globulin prophylaxis. *Hepatology* **24**(3): 489–493.

Carman WF, Boner WF, Fattovich G et al (1997a) HBV core protein mutations are concentrated in B cell epitopes in progressive disease and T helper cell epitopes in clinical remission. *J Infect Dis* **175**: 1093–1100.

Carman WF, van Deursen FJ, Mimms LT et al (1997b) The prevalence of surface antigen variants of hepatitis B virus in Papua New Guinea, South Africa and Sardinia. *Hepatology* **1997**(26): 1658–1666.

Carman WF, Tucker T, Song E, Hawarden D, Kirsch R, Lobidel D, Maloney E, Williams E (1998). Efficacy of a third generation pre-S1/pre-S2 containing HBV vaccine. *Progress in Clinical Virology IV*. Hamburg (Germany). Abstract No. 111.

Casey JL (1996) Hepatitis delta virus. Genetics and pathogenesis. *Clin Lab Med* 451–464.

Casey JL & Gerin JL (1998) Genotype-specific complementation of hepatitis delta virus RNA replication by hepatitis delta antigen. *J Virol* **72**: 2806–2814.

Casey JL, Niro GA, Engle RE et al (1996) Hepatitis B virus (HBV)/hepatitis D virus (HDV) coinfection in outbreaks of acute hepatitis in the Peruvian Amazon basin: the roles of HDV genotype II and HBV genotype F. *J Infect Dis* **174**: 920–926.

Casey JL, Polson AG, Bass BL & Gerin JL (1997) Molecular biology of HDV: analysis of RNA editing and genotype variations. In: Rizzetto M, Purcell RH, Gerin JL & Verme G (eds) *Viral Hepatitis and Liver Disease*, pp 290–294. Turin: Minerva.

Chang MF, Sun CY, Chen CJ & Chang SC (1993) Functional motifs of delta antigen essential for RNA binding and replication of hepatitis delta virus. *J Virol* **67**: 2529–2536.

Chang MH, Chen CJ, Lai MS et al (1997) Universal hepatitis B vaccination in Taiwan and the incidence of hepatocellular carcinoma in children. *N Engl J Med* **336**(26): 1855–1859.

Chen M & Ou JH (1995) Cell type-dependent regulation of the activity of the negative regulatory element of the hepatitis B virus core promoter. *Virology* **214**: 198–206.

Chen M, Hieng S, Qian X, Costa R & Ou JH (1994) Regulation of hepatitis B virus ENI enhancer activity by hepatocyte-enriched transcription factor HNF3. *Virology* **205**: 127–132.

Chen Y-CJ, Delbrook K, Dealwis C, Mimms L, Mushahwar IK & Mandecki W (1996) Discontinuous epitopes of hepatitis B surface antigen derived from a filamentous phage peptide library. *Proc Natl Acad Sci USA* **93**: 1997–2001.

Chirillo P, Falco M, Puri PL et al (1996) Hepatitis B virus pX activates NF-kappa B dependent transcription through a Raf-independent pathway. *J Virol* **70**: 641–646.

Chisari FV (1996) Hepatitis B virus transgenic mice: model of viral immunobiology and pathogenesis. *Curr Top Microbiol Immunol* **206**: 149–173.

Chisari FV, Filippi P, Buras J et al (1987) Structural and pathological effects of synthesis of hepatitis B virus large envelope polypeptide in transgenic mice. *Proc Natl Acad Sci USA* **84**: 6909–6913.

Conjeevaram HS, Hoofnagle JH, Austin HA, Park Y, Fried MW & di Biscegli AM (1995) Long-term outcome of hepatitis B virus-related glomerulonephritis after therapy with interferon alpha. *Gastroenterology* **109**: 540–546.

Davies SE, Portmann BC, O'Grady JG et al (1991) Hepatic histologic findings after transplantation for chronic hepatitis B virus infection, including a unique pattern of fibrosing cholestatic hepatitis. *Hepatology* **13**: 150–157.

De Meyer S, Gong ZJ, Suwandhi W, van Pelt J, Soumillion A & Yap SH (1997) Organ and species specificity of hepatitis B virus (HBV) infection: a review of literature with a special reference to preferential attachment of HBV to human hepatocytes. *J Viral Hep* **4**: 145–153.

Dienstag JL, Perrillo RP, Schiff ER et al (1995) A preliminary trial of lamivudine for chronic hepatitis B infection. *N Engl J Med* **333**: 1657–1661.

Dikstein R, Faktor O & Shaul Y (1990) Hierarchic and cooperative binding of the rat liver nuclear protein C/EBP at the hepatitis B virus enhancer. *Mol Cell Biol* **10**: 4427–4440.

Donello JE, Beeche AA, Smith GJ, Lucero GR & Hope TJ (1996) The hepatitis B virus posttranscriptional regulatory element is composed of two subelements. *J Virol* **93**: 4589–4594.

Dopheide TA & Azad AA (1996) The hepatitis B virus X protein is a potent AMP kinase. *J Gen Virol* **77**: 173–176.

Doria M, Klein N, Lucito R & Schneider RJ (1995) The hepatitis B virus HBx protein is a dual specificity cytoplasmic activator of Ras and nuclear activator of transcription. *EMBO J* **19**: 4747–4757.

Ehata T, Omata M, Yokosuka O, Hosodo K & Ohto M (1992) Variations in codons 84–101 in the core nucleotide sequence correlate with hepatocellular injury in chronic hepatitis B virus infection. *J Clin Invest* **89**: 332–338.

Fagan E, Smith P, Davison F & Williams R (1986) Fulminant hepatitis B in successive female sexual partners of two anti-HBe positive males. *Lancet* **ii**: 538–540.

Farci P, Mandas A, Coiana A et al (1994) Treatment of chronic hepatitis D with interferon alfa-2a. *N Engl J Med* **330**: 88–94.

Fattovich G, Colman K, Thursz M et al (1995) Pre-core/core gene variation and interferon therapy. *Hepatology* **22**: 1355–1362.

Fattovich G, Giustina G, Favarato S et al (1996) A survey of adverse events in 11241 patients with chronic viral hepatitis treated with alfa inteferon. *J Hepatol* **24**: 38–47.

Feitelson MA, Zhu M, Duan LX & London WT (1993) Hepatitis B x antigen and p53 are associated in vitro and in liver tissues from patients with primary hepatocellular carcinoma. *Oncogene* **8**: 1109–1117.

Ferrari C, Bertoletti A, Fiaccadori F & Chisari FV (1995) Is antigenic variability a strategy adopted by hepatitis B virus to escape cytotoxic T-lymphocyte surveillance? *Virology* **6**: 1–8.

Fried MW (1996) Therapy of chronic viral hepatitis. *Med Clin North Am* **80**: 957–972.

Fujii H, Moriyama K, Sakamoto N et al (1992) Gly 145 to Arg substitution in HBs antigen of immune escape mutant of hepatitis B virus. *Biochem Biophys Res Commun* **184:** 1152–1157.

Garcia AD, Ostapchuck P & Hearing P (1993) Functional interaction of nuclear factors EF-C, HNF-4, and RXRa with hepatitis B virus enhancer I. *J Virol* **67:** 3940–3950.

Grellier L, Mutimer D, Ahmed M et al (1996) Lamivudine prophylaxis against reinfection in liver transplantation for hepatitis B cirrhosis. *Lancet* **348:** 1212–1215.

Guidotti LG, Ando K, Hobbs MV et al (1994) Cytotoxic T lymphocytes inhibit hepatitis B virus gene expression by a noncytolytic mechanism in transgenic mice. *Proc Natl Acad Sci USA* **91:** 3764–3768.

Guidotti LG, Borrow P, Hobbs MV et al (1996a) Viral cross talk: intracellular inactivation of the hepatitis B virus during an unrelated viral infection of the liver. *Proc Natl Acad Sci USA* **93:** 4589–4594.

Guidotti LG, Matzke B, Pasquinelli C, Shoenberger JM, Rogler CE & Chisari FV (1996b) The hepatitis B virus (HBV) precore protein inhibits HBV replication in transgenic mice. *J Virol* **70:** 7056–7061.

Guo W, Bell KD & Ou JH (1991) Characterization of the hepatitis B virus EnhI enhancer and X promoter complex. *J Virol* **65:** 6686–6692.

Guo WM, Chen M, Yen TSB & Ou JH (1993) Hepatocyte specific expression of the hepatitis B virus core promoter depends on both positive and negative regulation. *Mol Cell Biol* **13:** 443–448.

Hadziyannis SJ (1997a) Hepatitis delta: an overview. In: Rizzetto M, Purcell RH, Gerin JL & Verme G (eds) *Viral Hepatitis and Liver Disease*, pp 283–289. Turin: Minerva.

Hadziyannis SJ (1997b) Review: hepatitis delta. *J Gastroenterol Hepatol* **12:** 289–298.

Hasegawa K, Huang J, Rogers SA, Blum HE, Liang TJ. Enhanced replication of a hepatitis B virus mutant associated with an epidemic of fulminant hepatitis. *J Virol* **68:** 1651–1659.

Haviv I, Vaizel D & Shaul Y (1995) The X protein of hepatitis B virus coactivates potent activation domains. *Mol Cell Biol* **15:** 1079–1085.

Haviv I, Vaizel D & Shaul Y (1996) pX, the HBV-encoded coactivator, interacts with components of the transcription machinery and stimulates transcription in a TAF-independent manner. *EMBO J* **15:** 3413–3420.

Hawkins A, Gilson R, Gilbert N et al (1996) Hepatitis B virus surface mutations associated with infection after liver transplantation. *J Hepatol* **24**(1): 8–14.

Heptonstall J, Barnes J, Burton E, et al (1997) Transmission of hepatitis B to patients from four infected surgeons without hepatitis B e antigen. *N Eng J Med* **336**(3): 178–184.

Hino K, Okuda M, Hashimoto O, Ishiko H & Okita K (1995) Glycine-to-arginine substitution at codon 145 of HBsAg in two infants born to hepatitis B e antigen-positive carrier. *Dig Dis Sci* **40:** 566–570.

Honigwachs J, Faktor O, Dikstein R, Shaul Y & Laub O (1989) Liver-specific expression is determined by the combined action of the core gene promoter and the enhancer. *J Virol* **63:** 919–924.

Hoofnagle AJJ & diBisceglie AM (1997) The treatment of chronic viral hepatitis. *N Engl J Med* **336**(5): 347–355.

Hoofnagle JH, Seeff LB, Bales ZB & Zimmerman HJ (1978) Type B hepatitis after transfusion with blood containing antibody to hepatitis B core antigen. *N Engl J Med* **298:** 1379–1383.

Hou JL, Karayiannis P, Thomas HC, Luo KX, Liang C & Thomas HC (1995) A unique insertion in the S gene of surface antigen negative HBV in Chinese carriers. *Hepatology* **21:** 273–278.

Hourvitz A, Mosseri R, Solomon A et al (1996) Reactogenicity and immunogenicity of a new recombinant hepatitis B vaccine containing PreS antigens: a preliminary report. *J Viral Hep* **3:** 37–42.

Hsu IC, Metcalf RA, Sun T, Welsh JA, Wang NJ & Harris CC (1991) Mutational hotspot in the *p53* gene in human hepatocellular carcinomas. *Nature* **350:** 427–428.

Hu KQ & Siddiqui A (1991) Regulation of the hepatitis B virus gene expression by the enhancer element I. *Virology* **181:** 721–726.

Huan B, Kosovsky MJ & Siddiqui A (1995) Retinoid X receptor a transactivates the hepatitis B virus enhancer I element by forming a heterodimer complex with the peroxisome proliferator-activated receptor. *J Virol* **69:** 547–551.

Huang ZM, Zang WQ & Yen TSB (1996) Cellular proteins that bind to the hepatitis B virus post-transcriptional regulatory element. *Virology* **217:** 573–581.

Huo TI, Wu JC, Lai CR, Lu CL, Sheng WY & Lee SD (1996) Comparison of clinico-pathological features in hepatitis B virus-associated hepatocellular carcinoma with or without hepatitis D virus superinfection. *J Hepatol* **25:** 439–444.

Jameel S & Siddiqui A (1986) The human hepatitis B virus enhancer requires transactivation cellular factor(s) for activity. *Mol Cell Biol* **6:** 710–715.

Jardi R, Buti M, Cotrina M et al (1995) Determination of hepatitis delta virus RNA by polymerase chain reaction in acute and chronic delta infection. *Hepatology* **21:** 25–29.

Jeng KS, Su PY & Lai MM (1996) Hepatitis delta antigens enhance the ribozyme activities of hepatitis delta virus RNA in vivo. *J Virol* **70:** 4205–4209.

Johnson JL, Raney AK & McLachlan A (1995) Characterization of a functional hepatocyte nuclear factor 3 binding site in the hepatitis B virus nucleocapsid promoter. *Virology* **208:** 147–158.

Kaneko S, Miller RH, Feinstone SM et al (1989) Detection of serum hepatitis B virus DNA in patients with chronic hepatitis using the polymerase chain reaction assay. *Proc Natl Acad Sci USA* **86:** 312–316.

Kapke GF, Watson G, Sheffler S, Hunt D & Frederick C (1997) Comparison of the Chiron Quantiplex branched DNA (bDNA) assay and the Abbott Genostics solution hybridization assay for quantification of hepatitis B viral DNA. *J Viral Hep* **4:** 67–75.

Karayiannis P, Alexopoulou A, Hadziyannis S et al (1995) Fulminant hepatitis associated with hepatitis B virus antigen-negative infection: importance of host factors. *Hepatology* **22:** 1628–1634.

Karin M (1994) Signal transduction from the cell surface to the nucleus through the phosphorylation of the transcription factors. *Curr Opin Cell Biol* **6:** 415–424.

Kidd-Ljunggren K, Ekdahl K, Öberg M, Kurathong S, Lolekha S (1995) Hepatitis B virus strains in Thailand: Genomic variants in chronic carriers. *Journal of Medical Virology* **47:** 454–461.

Kim C, Koike K, Saito I, Miyamura T & Jay G (1991) Hepatitis B virus HBx gene induces liver cancer in transgenic mice. *Nature* **351:** 317–320.

Kim SO, Park JG & Lee YI (1996) Increased expression of the insulin-like growth factor I (IGF-I) receptor gene in hepatocellular carcinoma cell lines: implications of IGF-I receptor gene activation by hepatitis B virus X gene product. *Cancer Res* **56:** 3831–3836.

Ko LJ & Prives C (1996) p53: puzzle and paradigm. *Genet Dev* **10:** 1054–1072.

Kohno H, Inoue T, Tsuda F, Okamoto H & Akahane Y (1996) Mutations in the envelope gene of hepatitis B virus variants co-occurring with antibody to surface antigen in sera from patients with chronic hepatitis B. *J Gen Virol* **77:** 1825–1831.

Koike K, Moriya K, Yotsuyanagi H, Lino S & Kurokawa K (1994) Induction of cell cycle progression by hepatitis B virus HBx gene expression in quiescent mouse fibroblasts. *J Clin Invest* **94:** 44–49.

Kosovsky MJ, Huan B & Siddiqui A (1996) Purification and properties of rat liver nuclear proteins that interact with the hepatitis B virus enhancer I. *J Biol Chem* **71:** 21859–21869.

Koziel MJ & Liang TJ (1997) DNA vaccines and viral hepatitis: are we going around in circles? *Gastroenterology* **112:** 1410–1414.

Krüger M, Tillmann HL, Trautwein C et al (1994) Treatment of hepatitis B virus reinfection after liver transplantation with famciclovir. *Hepatology* **20:** 130A.

Kumar V, Jayasuryan N & Kumar R (1996) A truncated mutant (residues 58–140) of the hepatitis B virus X protein retains transactivation function. *Proc Natl Acad Sci USA* **93**: 5647–5652.

Lampertico P, Manzin A, Rumi MG et al (1995) Hepatitis B virus precore mutants in HBeAg carriers with chronic hepatitis treated with interferon. *J Viral Hep* **2**: 251–256.

Lau J, Bain VG, Davies SE et al (1992) High level expression of hepatitis B viral antigens in fibrosing cholestatic hepatitis. *Gastroenterology* **102**: 956–962.

Lauchert W, Muller R & Pichlmayr R (1987) Long-term immunoprophylaxis of hepatitis B virus reinfection in recipients of human liver allografts. *Transplant Proc* **19**: 4051–4053.

Lazinski DW & Taylor JM (1993) Relating structure to function in the hepatitis delta virus antigen. *J Virol* **67**: 2672–2680.

Lee H, Lee YH, Huh YS, Moon H & Yun Y (1995) X-gene product antagonizes the p53-mediated inhibition of hepatitis B virus replication through regulation of the pregenomic/core promoter. *J Biol Chem* **270**: 31405–31412.

Lee TH, Finegold MJ, Shen RF, DeMayo JL, Woo SLC & Butel JS (1990) Hepatitis B virus transactivator X protein is not tumorigenic in transgenic mice. *J Virol* **64**: 5939–5947.

Lemon SM & Thomas DL (1997) Vaccines to prevent viral hepatitis. *N Engl J Med* **336(3)**: 196–204.

Li M, Xie Y, Wu X, Kong Y & Wang Y (1995) HNF3 binds and activates the second enhancer, ENII, of hepatitis B virus. *Virology* **214**: 371–378.

Lien JM, Aldrich CE & Mason WS (1986) Evidence that a capped oligoribonucleotide is the primer for duck hepatitis B virus plusstrand DNA synthesis. *J Virol* **57**: 229–236.

Lindh M, Horal P, Dhillon AP, Furuta Y & Norkrans G (1996) Hepatitis B virus carriers without precore mutations in hepatitis B e antigen-negative stage show more severe liver damage. *Hepatology* **24**: 494–501.

Ling R, Mutimer D, Ahmed M et al (1996) Selection of mutations in the hepatitis B virus polymerase during therapy of transplant recipients with lamivudine. *Hepatology* **24**: 711–713.

Lok ASF, Akarca U & Greene S (1994) Mutations in the pre-core region of the hepatitis B virus serve to enhance the stability of the secondary structure of the pre-genome encapsidation signal. *Proc Natl Acad Sci USA* **91**: 4077–4081.

Lopez-Cabrera M, Letovsky J, Hu KQ & Siddiqui A (1991) Transcriptional factor C/EBP binds to and transactivates the enhancer II element of the hepatitis B virus. *Virology* **183**: 825–829.

Lu C-C & Yen TSB (1996) Activation of the hepatitis B virus S promoter by transcription factor NF-Y via a CCAAT element. *Virology* **225**: 387–394.

Lu CC, Chen M, Ou JH & Yen TSB (1995) Key role of a CCAAT element in regulating hepatitis B virus surface protein expression. *Virology* **206**: 1155–1158.

Lucy MR (1994) Hepatitis B infection and liver transplantation: the art of the possible. *Hepatology* **19**: 245–247.

Lucito R & Schneider RJ (1992) Hepatitis B virus X protein activates transcription factor NF-kB without a requirement for protein kinase C. *J Virol* **66**: 983–991.

Maguire HF, Hoeffler JP & Siddiqui A (1991) HBV X protein alters the DNA binding specificity of CREB and ATF-2 by protein–protein interactions. *Science* **252**: 842–844.

Mancini M, Hadchouel M, Davis HL, Whalen RG, Tiollais P & Michel M-L (1996) DNA-mediated immunization in a transgenic mouse model of the hepatitis B surface antigen chronic carrier state. *Proc Natl Acad Sci USA* **93**: 12496–12501.

Mangold CMT & Streeck RE (1993) Mutational analysis of the cysteine residues in the hepatitis B virus small envelope protein. *J Virol* **67**: 4588–4597.

Mangold CMT, Unckell F, Werr M & Streeck RE (1995) Secretion and antigenicity of hepatitis B virus small envelope proteins lacking cysteines in the major antigenic region. *Virology* **211**: 535–543.

Manns MP (1997) Hepatotropic viruses and autoimmunity 1997. *J Viral Hep* **4**: 7–10.

Maruyama T, Lino S, Koike K, Yasuda K & Milich DR (1993) Serology of acute exacerbation in chronic hepatitis B virus infection. *Gastroenterology* **105**: 1141–1151.

Michalopoulos GK & DeFrances MC (1997) Liver regeneration. *Science* **276**: 60–66.

Michel ML, Davis HL, Schleef M, Mancini M, Tiollais P & Whalen RG (1995) DNA-mediated immunization to the hepatitis B surface antigen in mice: aspects of the humoral response mimic hepatitis B viral infection in humans. *Proc Natl Acad Sci USA* **92**: 5307–5311.

Milich DR, McLachlan A, Chisari FV, Kent SBH & Thornton GB (1986) Immune response to the pre-S1 region of the hepatitis B surface antigen [HBsAg]: a pre-S1 specific T cell response can bypass nonresponsiveness to the pre-S2 and S regions of HBsAg. *J Immunol* **137**: 315–322.

Milich DR, Jones JE, Hughes JL, Price J, Raney AK & McLachlan A (1990) Is a function of the secreted hepatitis B e antigen to induce immunologic tolerance in utero? *Proc Natl Acad Sci USA* **87**: 6599–6603.

Milich DR, Schodel F, Hughes JL, Jones JE & Peterson DL (1997) The hepatitis B virus core and e antigens elicit different Th cell subsets: antigen structure can affect Th cell phenotype. *J Virol* **71**: 2192–2201.

Missale J, Redeker A, Person A et al (1993) HLA A31 and HLA AW68 restricted cytotoxic T-cell responses to a single hepatitis B virus nucleocapsid epitope during acute viral hepatitis. *J Exp Med* **177**: 751–762.

Miyakawa Y, Okamoto H & Mayumi M (1997) The molecular basis of hepatitis B e antigen (HBeAg)-negative infections. *J Viral Hep* **4**: 1–8.

Monjardino J (1996) Replication of hepatitis delta virus. *J Viral Hep* **3**: 163–166.

Mphahlele MJ, Shattock AG, Bone, W, Quinn J, McCormic PA, Carman WF (1997) Transmission of a homogeneous HBV population of A_{1896} containing strains leading to mild resolving acute hepatitis and seroconversion to anti-HBE in an adult. *Hepatology* **26**: 743–746.

Mphahlele MJ, Shattock AG, Boner W, Quinn J, McCormick PA & Carman WF (1997) Transmission of a homogeneous HBV population of A_{1896} containing strains leading to mild resolving acute hepatitis and seroconversion to anti-HBe in an adult. *Hepatology* (in press).

Müller R, Gubernatis G, Farle M et al (1991) Liver transplantation in HBs antigen (HBsAg) carriers. Prevention of hepatitis B virus (HBV) recurrence by passive immunization. *J Hepatol* **13**: 90–96.

Musti AM, Treier M & Bohmann D (1997) Reduced ubiquitin-dependent degradation of c-Jun after phosphorylation by MAP kinases. *Science* **275**: 400–402.

Nainan OV, Stevens CE & Margolis HS (1996) Hepatitis B virus (HBV) antibody resistant mutants: frequency and significance. *Ninth Triennial International Symposium on Viral Hepatitis and Liver Disease*, Rome. Abstr. **98**: p 29.

Nainan OV, Stevens CE, Taylow RE, Margolis HS (1997) Hepatitis B virus (HBV) antibody resistant mutants among mothers and infants with chronic HBV infection. Viral Hepatitis and Liver Disease. Eds: Rizzetto M, Purcell RH, Gerin JL and Verme G. *Edizioni Minerva Medica*; Turin 132–134.

Nassal M & Schaller H (1996) Hepatitis B virus replication – an update. *J Viral Hep* **3**: 217–226.

Natoli G, Avantaggiati ML, Chirillo P et al (1994) Induction of the DNA-binding activity of c-Jun/c-Fos heterodimers by the hepatitis B virus transactivator pX. *Mol Cell Biol* **14**: 989–998.

Netter HJ, Gerin JL, Tennant BC & Taylor JM (1994) Apparent helper-independent infection of woodchucks by hepatitis delta virus and subsequent rescue with woodchuck hepatitis virus. *J Virol* **68**: 5344–5350.

Neuhaus P, Steffen R, Blumhardt G et al (1991) Experience with immunoprophylaxis and interferon therapy after liver transplantation in HBsAG positive patients. *Transplant Proc* **23**: 1522–1524.

Niederau C, Heintges T, Lange S et al (1996) Long-term follow-up of HBeAg-positive patients treated with interferon alfa for chronic hepatitis B. *N Engl J Med* **334**: 1422–1427.

Niro GA, Smedile A, Andriulli A, Rizzetto M, Gerin JL & Casey JL (1997) The predominance of hepatitis delta virus genotype I among chronically infected Italian patients. *Hepatology* **25**: 728–734.

Nisini R, Paroli M, Accapezzato D et al (1997) Human CD4+ T-cell response to hepatitis delta virus: identification of multiple epitopes and characterization of T-helper cytokine profiles. *J Virol* **71:** 2241–2251.

Norder H, Couroucé A-M & Magnius LO (1992a) Molecular basis of hepatitis B virus serotype variations within the four major subtypes. *J Gen Virol* **73:** 3141–3145.

Norder H, Hammas B, Löfdahl S, Couroucé AM & Magnius LO (1992b) Comparison of the amino acid sequences of nine different serotypes of hepatitis B surface antigen and genomic classification of the corresponding hepatitis B virus strains. *J Gen Virol* **73:** 1201–1208.

Offensperger WB, Blum HE & Gerok W (1994) Molecular therapeutic strategies in hepatitis B virus infection. *Clin Invest* **72:** 737–741.

Ogata N, Miller RH, Ishak KG & Purcell RH (1993) The complete nucleotide sequence of a pre-core mutant of hepatitis B virus implicated in fulminant hepatitis and its biological character-ization in chimpanzees. *Virology* **194:** 263–276.

Ogata N, Miller RH, Ishak KG, Zanetti AR & Purcell RH (1994) Genetic and biological charac-terization of two hepatitis B virus variants: a precore mutant implicated in fulminant hepatitis and a surface mutant resistant to immunoprophylaxis. In: Nishioka K, Suzuki H, Mishiro S, Oda T (eds) *Viral Hepatology Liver Disease*, pp 238–242.

Ogata N, Zanetti AR, Yu M, Miller RH & Purcell RH (1997) Infectivity and pathogenicity in chimpanzees of a surface gene mutant of hepatitis B virus that emerged in a vaccinated infant. *J Infect Dis* **175(3):** 511–523.

O'Grady JG, Alexander GJM, Hayllar KM & Williams R (1989) Early indicators of prognosis in fulminant hepatic failure. *Gastroenterology* **97:** 439–445.

O'Grady JG, Schalm SW & Williams R (1993) Acute liver failure: redefining the syndrome. *Lancet* **342:** 273–275.

Okamoto H, Tsuda F, Sakugawa H et al (1988) Typing hepatitis B virus by homology in nucleotide sequence: comparison of surface antigen subtypes. *J Gen Virol* **69:** 2575–2583.

Okamoto H, Yotsumoto S, Akahane Y et al (1990) Hepatitis B viruses with pre-core region defects prevail in persistently infected hosts along with seroconversion to the antibody against e antigen. *J Virol* **64:** 1298–1303.

Okamoto H, Yano K, Nozaki Y et al (1992) Mutations within the s gene of hepatitis-B virus trans-mitted from mothers to babies immunized with hepatitis-B immune globulin and vaccine. *Pediatr Res* **32:** 264–268.

Oon CJ, Lim GK, Ye Z et al (1995) Molecular epidemiology of hepatitis B virus vaccine variants in Singapore. *Vaccine* **13:** 699–702.

Oren I, Hershow RC, Ben-Porath E et al (1989) A common-source outbreak of fulminant hepati-tis B in a hospital. *Ann Intern Med* **110:** 691–698.

Ori A & Shaul Y (1995) Hepatitis B virus enhancer binds and is activated by the hepatocyte nuclear factor 3. *Virology* **207:** 98–106.

Oshikawa O, Tamura S, Kawata S et al (1996) The effect of hepatitis B virus X gene expression on response to growth inhibition by transforming growth factor-beta 1. *Biochem Biophys Res Commun* **222:** 770–773.

Ostapchuk P, Hearing P & Ganem DA (1994) A dramatic shift in the transmembrane topology of a viral envelope glycoprotein accompanies hepatitis B viral morphogenesis. *EMBO J* **13:** 1048–1057.

Penna A, Chisari FV, Bertoletti A et al (1991) Cytotoxic T lymphocytes recognize an HLA-A2 restricted epitope within the hepatitis B virus nucleocapsid antigen. *J Exp Med* **174(6):** 1565–1570.

Perfumo S, Amicone L, Colloca S, Giorgio M, Pozzi L & Tripodi M (1992) Recognition effi-ciency of the hepatitis B virus polyadenylation signal is tissue specific in transgenic mice. *J Virol* **66:** 6819–6823.

Perrillo RP (1995) Chronic hepatitis B: problem patients. *J Hepatol* **22** (suppl. 1): 45–48.

Perrillo RP & Mason AL (1994) Therapy for hepatitis B virus infection. *Gastroenterol Clin North Am* **23:** 581–601.

Perrillo R, Tamburro C, Regenstein F et al (1995) Low-dose, tirable interferon alfa in decompensated liver disease caused by chronic infection with hepatitis B virus. *Gastroenterology* **109**: 908–916.

Perillo R, Rakela J, Martin P et al (1996) Lamivudine for hepatitis B after liver transplantation (OLT). *Hepatology* **24** (suppl.): 182A.

Pol S, Driss F, Carnot F, Michel ML, Bethelot P & Brechot C (1993) Efficacité d'une immunothérapie par vaccination contre le virus de l'hépatite B sur la multiplication virale B. *C R Acad Sci* **316**: 688–691.

Pollack JR & Ganem D (1993) A stem-loop structure directs hepatitis B virus genomic product: domain structure and Rnase H activity. *J Virol* **61**: 1384–1390.

Polson AG, Ley HL, Bass BL & Casey JL (1998) Hepatitis delta virus RNA editing is highly specific for the amber/W site and is suppressed by hepatitis delta antigen. *Mol Cell Biol* **18**: 1919–1926.

Prange R, Mangold CMT, Hilfrich R & Streeck RE (1995) Mutational analysis of HBsAg assembly. *Intervirology* **38**: 16–23.

Qadri I, Maguire HF & Siddiqui A (1995) Hepatitis B virus transactivator protein X interacts with the TATA-binding protein. *Proc Natl Acad Sci USA* **92**: 1003–1007.

Qadri I, Conaway JW, Conaway RC, Schaak J & Siddiqui A (1996) Hepatitis B virus transactivator protein, HBx associates with the components of TFIIH and stimulates the DNA helicase activity of TFIIH. *Proc Natl Acad Sci USA* **93**: 10578–10583.

Qiu X, Schroeder P & Bridon D (1996) Identification and characterization of a C(K/R)TC motif as a common epitope present in all subtypes of a hepatitis B surface antigen. *J Immunol* **156**: 3350–3356.

Radzwill G, Tucker W & Schaller H (1990) Mutational analysis of the hepatitis B virus P gene product: domain structures and Rnase H activity *J Virol* **61**: 1384–1390.

Raimondo G, Tanzi E, Brancatelli S et al (1993) Is the course of perinatal hepatitis B virus infection influenced by genetic heterogeneity of the virus? *J Med Virol* **40**: 87–90.

Raney AK, Le HB & McLachlan A (1992) Regulation of transcription from the hepatitis B virus major surface antigen promoter by the Sp1 transcription factor. *J Virol* **66**: 6912–6921.

Raney AK, Johnson JL, Palmer CNA & McLachlan A (1997) Members of the nuclear receptor superfamily regulate transcription from the hepatitis B virus nucleocapsid promoter. *J Virol* **71**: 1058–1071.

Rehermann B, Lau D, Hoofnagle JH & Chisari FV (1996a) Cytotoxic T lymphocyte responsiveness after resolution of chronic hepatitis B virus infection. *J Clin Invest* **97**: 1654–1665.

Rehermann B, Ferrari C, Pasquinelli C & Chisari FV (1996b) The hepatitis B virus persists for decades after patients' recovery from acute viral hepatitis despite active maintenance of a cytotoxic T-lymphocyte response. *Nature Med* **2**: 1104–1108.

Ruiz-Moreno M, Camps T, Jimenez J et al (1995) Factors of response to interferon therapy in children with chronic hepatitis B. *J Hepatology* **22**: 540–544.

Rumin S, Gripon P, Le Seyec J, Corral-Debrinski M & Guguen-Guillouzo C (1996) Long-term productive episomal hepatitis B virus replication in primary cultures of adult human hepatocytes infected in vitro. *J Viral Hep* **3**: 227–238.

Samuel D, Bismuth A, Mathieu D et al (1991) Passive immunoprophylaxis after liver transplantation in HBsAg-positive patients. *Lancet* **337**: 813–815.

Samuel D, Muller R, Alexander G et al (1993) Liver transplantation in European patients with the hepatitis B surface antigen. *N Engl J Med* **329**: 1842–1847.

Santantonio T, Jung MC, Schneider R et al (1992) Hepatitis-B virus genomes that cannot synthesise pre-S2 proteins occur frequently and as dominant virus populations in chronic carriers in Italy. *Virology* **188**: 948–952.

Sato S, Suzuki K, Akahane Y et al (1995) Hepatitis B virus strains with mutations in the core promoter in patients with fulminant hepatitis. *Ann Intern Med* **122**: 241–248.

Schirmbeck R, Bohm W, Ando K, Chisari FV & Reimann J (1995) *J Virol* **69(10):** 5929–5934.

Seegar C & Maragos J (1989) Molecular analysis of the function of direct repeats and a polypurine tract for plus-strand DNA priming in wood chuck hepatitas virus *J. Virol* **63:** 1907–1925.

Shah G, Demetris AJ, Gavaler JS et al (1992) Incidence, prevalence, and clinical course of hepatitis B following liver transplantation. *Gastroenterology* **103:** 323–329.

Shaul Y, Rutter WJ & Laub O (1985) A human hepatitis B virus enhance element. *EMBO J* **4:** 427–430.

Sheu S & Lo SJ (1994) Biogenesis of the hepatitis B viral middle (M) surface protein in a human hepatoma cell line: demonstration of an alternative secretion pathway. *J Gen Virol* **75:** 3031–3039.

Su F & Schneider RJ (1996) Hepatitis B virus HBx protein activates transcription factor NF-kB by acting on multiple cytoplasmic inhibitors of rel-related proteins. *J Virol* **70:** 4558–4566.

Su H & Yee JK (1992) Regulation of hepatitis B virus gene expression by its two enhancers. *Proc Natl Acad Sci USA* **89:** 2708–2712.

Summers J & Mason WS (1982) Replication of the genome of a hepatitis B-like virus by reverse transcription of an RNA intermediate. *Cell* **29:** 403–415.

Takada S, Kaneniwa N, Tsuchida N & Koike K (1996) Hepatitis B virus x gene expression is activated by x protein but repressed by p53 tumor suppressor gene product in the transient expression system. *Virology* **216:** 80–89.

Tanaka S, Yoshiba M, Iino S, Fukuda M, Nakao H, Tsuda F, Okamoto H, Miyakawa Y, Mayumi M (1995) A common-source outbreak of fulminant hepatitis B in hemodialysis patients induced by pre core mutant. *Kidney International* **48:** 1972–1978.

Taylor JM, Wu T-T & Bichko V (1997) Variations on the HDV RNA genome. In: Rizzetto M, Purcell RH, Gerin JL & Verme G (eds) *Viral Hepatitis and Liver Disease*, pp 295–297. Turin: Minerva.

Thomas HC, Foster GR, Sumiya M et al (1996) Mutation of gene for mannose-binding protein associated with chronic hepatitis B viral infection. *Lancet* **348:** 1417–1419.

Thursz M, Kwiatkowski D, Allsopp CEM, Greenwood BM, Thomas HC & Hill AVS (1995) Association between an MHC class II allele and clearance of hepatitis B virus in the Gambia. *N Engl J Med* **332:** 1065–1069.

Tipples GA, Ma MM, Fischer KP, Bain VG, Knetman MM & Tyrrell DLJ (1996) Mutation in HBV RNA-dependent DNA polymerase confers resistance to lamivudine in vivo. *Hepatology* **24:** 714–717.

Todo S, Demetris AJ, Van Thiel D, Teperman L, Fung JJ & Sarzl TE (1991) Orthotopic liver transplantation for patients with hepatitis B virus-related liver disease. *Hepatology* **13:** 619–626.

Tognoni A, Cattaneo R, Serfling E & Schaffner W (1985) A novel expression selection approach allows precise mapping of the hepatitis B virus enhancer. *Nucl Acids Res* **13:** 7457–7472.

Tong S, Li J, Vitvitski L & Trepo C (1990) Active hepatitis B virus replication in the presence of anti-HBe is associated with viral variants containing an inactive pre-C region. *Virology* **176:** 596–603.

Tong SP, Li JS, Vitvitski L & Trèpo C (1992) Replication capacities of natural and artificial precore stop codon mutants of hepatitis B virus: relevance of pregenome encapsidation signal. *Virology* **191:** 237–245.

Trautwein C, Schrem H, Tillmann H, Kubicka S & Manns MP (1996) Functional relevance of mutations in the hepatitis B virus pre-S genome before and after liver transplantation. *Hepatology* **24:** 482–488.

Trautwein C, Bock CT, Tillmann H & Manns MP (19—) A hepatitis B (HBV) preS mutation causes intracellular retention of viral proteins and is associated with fibrosing cholestatic hepatitis. *Hepatology?*

Truant R, Antunovic J, Greenblatt J, Prives C & Cromlish JA (1995) Direct interaction of the hepatitis B virus HBx protein with p53 leads to inhibition by HBx of p53 response element-directed transactivation. *J Virol* **69:** 1851–1859.

Trujillo MA, Letovsky J, Maguire HF, Lopez-Cabrera M & Siddiqui A (1991) Functional analysis of a liver-specific enhancer of the hepatitis B virus. *Proc Natl Acad Sci USA* **88:** 3797–3801.

Tsai SL, Chen PJ, Lai MY et al (1992) Acute exacerbations of chronic type B hepatitis are accompanied by increased T cell responses to hepatitis B core and e antigens. *J Clin Invest* **89:** 87–96.

Tsiquaye K, Sutton D, Manns M & Boyd MR (1993) Pharmacokinetics and antiviral activities of penciclovir in Pekin ducks chronically infected with duck hepatitis B virus. *Proceedings of the IAAC*, New Orleans, p 1594.

Tsui LV, Guidotti LG, Ishikawa T & Chisari FV (1995) Posttranscriptional clearance of hepatitis B virus RNA by cytotoxic T lymphocyte-activated hepatocytes. *Proc Natl Acad Sci USA* **92:** 12398–12402.

Tur-Kaspa R, Klein A & Aharonson S (1992) Hepatitis B virus precore mutants are identical in carriers from various ethnic origins and are associated with a range of liver disease severity. *Hepatology* **16:** 1338–1342.

Twu JS, Lee CH, Lin PM & Schloemer RH (1988) Hepatitis B virus suppresses expression of human β-interferon. *Proc Natl Acad Sci USA* **85:** 252–256.

Vitiello A, Ishioka G, Grey HM et al (1995) Development of a lipopeptide-based therapeutic vaccine to treat chronic HBV infection. *J Clin Invest* **95:** 341–349.

Vitulli D, Pirisi M, Rebus S, Dornan E, Boner W, Curcio F & Carman WF (1998) Long term infection with HBV of primary cultured hepatocytes. *Progress in Clinical Virology IV*. Hamburg (Germany) Abstr. 110.

Wallace L & Carman WF (1997) Surface gene variation of HBV: scientific and medical relevance. *Viral Hep Rev* **3(1):** 5–16.

Wang GH, Zoulim EH, Leber EH et al (1994) Role of RNA in enzymatic activity of the reverse transcriptase of hepatitis B virus. *J Virol* **68:** 8437–8442.

Wang HD, Yuh CH, Dang CV & Johnson DL (1995a) The hepatitis B virus X protein increases the cellular level of TATA-binding protein, which mediates transactivation of RNA polymerase III genes. *Mol Cell Biol* **15:** 6720–6728.

Whittle HC, Maine N, Pilkington J et al (1995) Long-term efficacy of continuing hepatitis B vaccination in infancy in two Gambian villages. *Lancet* **345:** 1089–1092.

Wu JC, Choo KB, Chen CM, Chen TZ, Huo TL & Lee SD (1995) Genotyping of hepatitis D virus by restriction-fragment length polymorphism and relation to outcome of hepatitis D. *Lancet* **346:** 939–941.

Yamamoto K, Horita M, Tsuda F et al (1994) Naturally occurring escape mutants of hepatitis B virus with various mutations in the S gene in carriers seropositive for antibody to hepatitis B surface antigen. *J Virol* **68:** 2671–2676.

Yasmin M (1997) Molecular biology of fulminant hepatitis B viruses. *Thesis of University of Glasgow*.

Yee JK (1989) A liver-specific enhancer in the core promoter region of human hepatitis B virus. *Science* **246:** 658–661.

Yen TSB (1993) Regulation of the hepatitis B virus expression. *Semin Virol* **4:** 33–42.

Zanetti AR, Tanzi E, Manzillo G et al (1988) Hepatitis B variant in Europe (letter). *Lancet* **ii:** 1132–1133.

Zhang P, Raney AK & McLachlan A (1992) Characterization of the hepatitis B virus X- and nucleocapsid gene transcriptional regulatory elements. *Virology* **191:** 31–41.

Zhang P, Raney AK & McLachlan A (1993) Characterization of functional Sp1 transcription factor binding sites in the hepatitis B virus nucleocapsid promoter. *J Virol* **67:** 1472–14881.

Zhang YY, Tsega E & Hansson BG (1996) Phylogenetic analysis of hepatitis D viruses indicating a new genotype I subgroup among African isolates. *J Clin Microbiol* **34:** 3023–3030.

Zhou DX & Yen TSB (1990) Differential regulation of the hepatitis B viral surface gene promoters by a second viral enhancer. *J Biol Chem* **265:** 20731–20734.

Zhou DX & Yen TSB (1991a) The hepatitis B virus S promoter comprises a CCAAT motif and two initiation regions. *J Biol Chem* **266:** 23416–23421.

Zhou DX & Yen TSB (1991b) The ubiquitous transcription factor Oct-1 and the liver-specific factor HNF-1 are both required to activate transcription of a hepatitis B virus promoter. *Mol Cell Biol* **11:** 1353–1359.

Zuckerman JN, Sabin C, Craig FM, Williams A & Zuckerman AJ (1997) Immune response to a new hepatitis B vaccine in healthcare workers who had not responded to standard vaccine: randomised double blind dose–response study. *BMJ* **314:** 329–333.

Chapter 21

HEPATITIS C AND G VIRUSES – OLD OR NEW?

Peter Simmonds and Donald B. Smith

DISCOVERY OF A NOVEL AGENT CAUSING HEPATITIS

The discovery and characterization of hepatitis C virus (HCV) in 1989 was the culmination of a large research endeavour to find the causes of post-transfusion hepatitis. In the 1970s, several groups produced evidence for an infectious agent, distinct from the hepatitis A and hepatitis B viruses (HAV, HBV), that caused chronic hepatitis and was frequently transmitted by blood and blood products (Prince et al, 1974; Alter et al, 1975; Feinstone et al, 1975). Clinically, the so-called non-A, non-B hepatitis (NANBH) was a mild disease, which caused jaundice in less than half of cases, and was frequently asymptomatic. However, in its clinical features it differed in several respects from acute HAV and HBV infection, having a different mean incubation period from HAV and HBV, and a high rate of chronicity following acute infection (at least 50% of cases, compared with around 5% for adult HBV infection). While HCV infection is almost invariably asymptomatic in the first 10–20 years, a significant proportion of cases eventually progress to cirrhosis and hepatocellular carcinoma (HCC). Because of this extremely slow rate of disease progression, the full impact of past silent epidemics of HCV in the 1950s and 1960s is only now becoming clinically apparent. For example, in many Western countries, HCV is now the most common identifiable cause of HCC, and the most frequent indication for liver transplantation.

Progress towards identifying the cause of NANBH was hampered by difficulties in culturing the agent in cell or organ culture; the eventual characterization of the infectious agent of NANBH was first achieved only after the transmission of the agent to chimpanzees. Using this experimental model, it was found that the agent could pass through an 80 nm filter and was inactivated by lipid solvents such as chloroform; together these data suggested a small enveloped virus. Passaging and titration of the NANBH agent in chimpanzees provided high-titre stocks of virus for the eventual cloning and sequence analysis of the infectious agent (Bradley et al, 1979; Bradley and Maynard, 1986). Nucleic acid (both DNA and RNA) was extracted following ultracentrifugation of a large volume of a plasma sample with a high infectivity titre in chimpanzees (approximately 10^6 infectious units per ml). The nucleic acid was reverse transcribed and the resulting DNA fragments cloned into the λ-gt11 bacteriophage, so that they were expressed as fusion proteins during replication of the bacteriophage in *Escherichia coli*. Immunoscreening of large numbers of bacteriophage plaques with antisera from

HIV and the New Viruses Second Edition
ISBN 0-12-200741-7

patients with NANBH led to the eventual identification of an immunoreactive clone (designated 5-1-1) that appeared to be specific for the NANBH agent, HCV (Choo et al, 1989). This clone encoded a protein that was consistently recognized by sera from patients with NANBH, but not by sera from control individuals, or from patients with hepatitis of a different aetiology (Kuo et al, 1989).

The discovery of the 5-1-1 clone led to the assembly of a nearly complete nucleotide sequence of the NANBH agent (Choo et al, 1991). Knowledge of the length, nucleic acid composition and genome organization of HCV has since provided the basis for its proposed classification in the *Flaviviridae* family and clues about its likely method of replication. From this sequence it was possible to design primers for amplification of the HCV genome by the polymerase chain reaction (PCR), and to manufacture recombinant proteins for use in assays for antibody to HCV. This has enabled effective screening of blood donors for HCV infection to be adopted worldwide.

Taxonomy of HCV

The organization of the HCV genome shares several features with other positive-stranded RNA viruses. Most striking is the existence of a single continuous open reading frame that occupies almost the entire genome (Figure 21.1). By analogy with other positive-stranded RNA viruses, this would be cleaved into individual proteins that are enzymes necessary for virus replication or are structural components of the virion. In overall genome organization and presumed method of replication, HCV is most similar to members of the *Flaviviridae*. The roles of the different proteins encoded by HCV have been inferred by comparison with homologues in other flaviviruses; the NS5B protein of HCV contains the GDD (glycine-aspartate-aspartate) motif associated with the active site of virus RNA-dependent RNA polymerases (RdRp) of positive-stranded RNA viruses (Miller and Purcell, 1990; Koonin, 1991) (Figure 21.2). There is also amino acid sequence similarity in part of NS3 to helicase polypeptides of other RNA viruses (Miller and Purcell, 1990).

Comparison with other positive-stranded RNA virus groups whose coding capacity is contained within a single open reading frame provides several clues about the replication of HCV. The *Flaviviridae* (e.g. dengue fever virus, yellow fever virus) are enveloped viruses resembling HCV in that their genomes code for structural proteins at the 5' end and non-structural proteins at the 3' end. However, there are differences between viruses in genome size, the number of proteins produced, the mechanism by which the polyprotein is cleaved, and in the detailed mechanism of genome replication. These differences are consistent with an extremely distant evolutionary relationship (Zanotto et al, 1996).

A more fundamental aspect of genome organization that differs between members of the *Flaviviridae* is the structure of the 5' and 3' untranslated regions (UTRs). These parts of the genome are involved in replication, and in the initiation of translation of the virus genome by cellular ribosomes. There is evidence that both pestiviruses and HCV have highly structured 5' UTRs, in which internal base pairing produces a complex set of stem-loop structures that are thought to interact with various host cell and virus proteins during replication and translation (Tsukiyama Kohara et al, 1992; Brown et al, 1992; Wang et al, 1994; Poole et al, 1995; Smith et al, 1995). In both viruses, there is evidence for internal initiation of translation, in which binding to the host cell ribosome

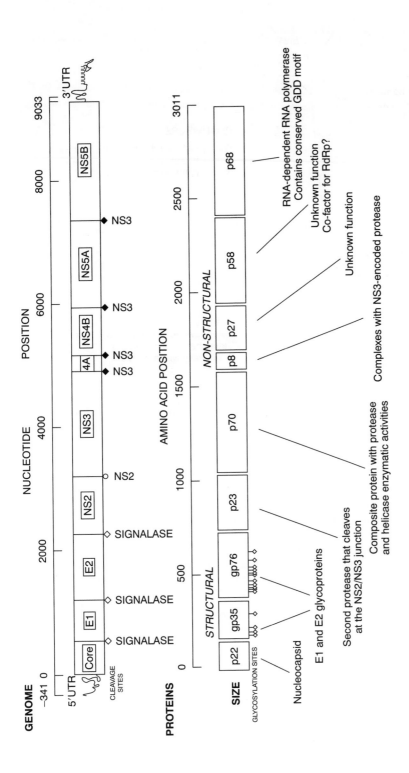

Figure 21.1 Genomic organization and gene products of HCV. Translation is initiated through ribosomal binding to an internal ribosome entry element in the 5′ untranslated region (5′ UTR), from which a >3000 amino acid polyprotein is translated. Cellular proteases (signalase in the lumen of the endoplasmic reticulum) and viral proteases (NS2 and NS3) subsequently cleave the protein to release structural components of the virion (core, E1 and E2) and non-structural proteins (NS2-NS5B) with enzymatic activities required for virus replication.

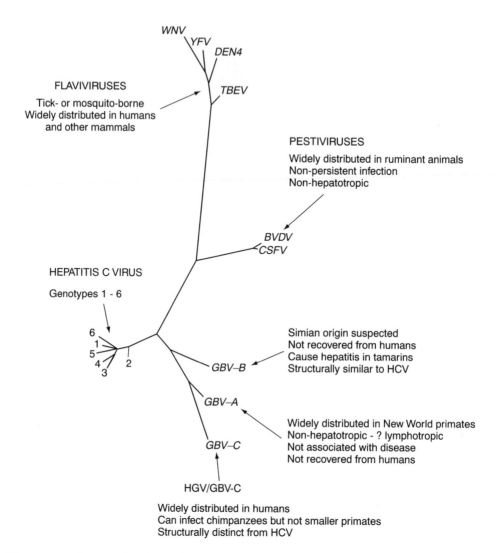

Figure 21.2 Genetic relatedness of the Flaviviridae
Phylogenetic analysis of a 100 amino acid sequence surrounding the highly conserved GDD motif in the RdRp (NS5B) of HCV and related flaviviruses. The relatedness between viruses is proportional to their separation in the tree. See Zanotto et al (1997) for the origin of sequences, and for the alignment used.

directs translation to an internal methionine (AUG) codon. This contrasts strongly with translation of flavivirus genomes which act much like cellular mRNAs, for which ribosomal binding occurs at the capped 5′ end of the RNA, followed by scanning of the sequence in the 5′ to 3′ direction with translation commencing at the first AUG codon.

Sequence heterogeneity of HCV

Variants of HCV can be classified into at least six major genotypes, each of which comprises a number of subtypes. Phylogenetic analysis of nucleotide sequences of either structural genes, such as the envelope glycoprotein, E1 (Bukh et al, 1993), non-structural genes such as NS5B, the RNA polymerase (Simmonds et al, 1993), or of the complete genome (Chamberlain et al, 1997) reveals a total of six genetically distinct groups, approximately equally divergent from each other (Figure 21.3).

The six genotypes differ from one another by approximately 30% at the nucleotide level over the complete genome of 9400 bases. The genome codes for proteins with a combined size of 3008 to 3037 amino acids, differing by up to 31% between genotypes. The E1 protein is the most divergent (49% sequence difference between genotypes 1 and 2) while the region encoding the core protein is the most conserved (approximately 10% sequence difference). For the E1, E2 and non-structural proteins, this degree of variability produces substantial antigenic differences between genotypes. As the E1 and E2 envelope glycoproteins are the likely targets of neutralizing antibody, it is possible that the immune response elicited upon infection with one genotype will not be protective against infection with another genotype, analogous to the lack of cross-neutralization between serotypes of other RNA viruses such as dengue and polio. Genotypes of HCV also differ biologically from one another, with type 1 variants being associated with a greater likelihood of causing HCC, while types 1 and 4 show a poorer response to interferon treatment than types 2 or 3 (reviewed by Bukh, 1995; Miyakawa et al, 1995; Simmonds, 1997). The genetic determinants of these differences remain uncertain, although there is empirical evidence that the amino terminus of NS5A influences response to interferon treatment (Enomoto et al, 1995, 1996), even though the function of this protein remains largely unknown.

DISCOVERY OF HEPATITIS G VIRUS

Since the discovery and characterization of HCV, screening of blood has proved highly effective at preventing post-transfusion hepatitis. The few residual cases that have occurred may be referred to as non A-E hepatitis, although these lack the commonality of symptoms and signs that characterized NANBH, and do not immediately suggest a

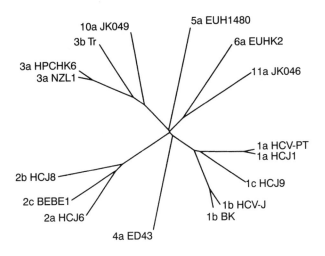

Figure 21.3A

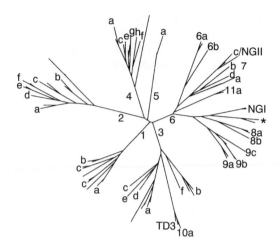

Figure 21.3B

Figure 21.3 *Phylogenetic analysis of HCV*
(A) Phylogenetic analysis of complete genomic sequences of HCV (Chamberlain et al, 1997), showing six main genetic groups, each of which may contain a number of more closely related yet distinct subsidiary groupings. Sequences are labelled according the consensus nomenclature, in which the six main groupings are described as genotypes 1–6, while the groupings within clades are described as subtypes a, b, c, etc. (Simmonds et al, 1994).
(B) Phylogenetic analysis of sequences from NS5B (positions 7975–8196 numbered according to Choo et al, 1991) showing equivalent relationships to those observed upon comparison of complete genome sequences (A). Comparison of this extended data set indicates the existence of a much greater range of subtypes within five of the six clades, most of which have restricted geographical distributions in Central and West Africa or southeast Asia. Isolates previously labelled as types 7, 8 and 9 belong to clade 6.

common, undetected infectious agent that causes hepatitis. These clinical observations have not deterred investigators from seeking further viruses, and recently there was the coincidence of two research groups simultaneously but independently announcing the discovery of the same virus (Erker et al, 1996; Leary et al, 1996b; Linnen et al, 1996). Equally remarkable is the subsequent evidence that this novel virus does not actually cause hepatitis in humans, or, as far as has been determined, any other identifiable disease.

One group used an expression library to identify a novel RNA virus, described as hepatitis G virus (HGV), in an individual with chronic hepatitis (Linnen et al, 1996). The virus was shown to resemble HCV, in having a positive-sense RNA genome coding for a continuous open reading frame occupying most of the genome, with evidence for an internal ribosome entry site (IRES) at its 5′ end (Simons et al, 1996) (Figure 21.4). Remarkably, the immunoreactive clone originally identified coded for a peptide that was translated from the genome in an antisense direction and, as far as can be currently determined, would not be expressed in vivo. In a separate series of investigations, two novel infectious agents (GBV-A, GBV-B) were discovered in a tamarin used in the final passage of an infectious agent recovered from a surgeon (G.B.) who suffered an acute hepatitis of unknown aetiology in 1963 (Prince et al, 1974; Simons et al, 1995). However, it has been subsequently shown that neither virus is likely to infect humans, and both probably originated from the marmosets and tamarins used in the passaging experiments. Indeed, a series of viruses related to GBV-A has subsequently been found to be endemic in a variety of New World primate species (Leary et al, 1996a; Bukh and Apgar, 1997). The genomes of GBV-A and GBV-B are arranged similarly to HCV and HGV, and it was the application of PCR using primers to target a region of NS3 that is highly conserved between GBV-A, GBV-B and HCV that led to the discovery of GBV-C in individuals from The Gambia (Leary et al, 1996b). This agent is similar to HGV. As a final irony, testing of current and stored samples from the original surgeon (G.B.) by PCR and antibody assays for HGV/GBV-C provided no evidence for either past or current infection with the virus; some have suggested that the original episode of hepatitis resulted from acute infection with HAV.

The degree of relatedness between HCV, HGV/GBV-C, GBV-A and GBV-B with other flaviviruses can be illustrated by phylogenetic analysis of the polypeptide region surrounding the GDD motif (see Figure 21.2). Using such information and differences in genome organization, the International Committee on the Taxonomy of Viruses proposes to divide the *Flaviviridae* into three genera, the flaviviruses, the pestiviruses and the hepaciviruses, the latter group consisting of HCV, with perhaps GBV-A and related viruses forming several subgenera within it (Robertson et al, 1999).

EPIDEMIOLOGY AND GEOGRAPHICAL DISTRIBUTION OF VIRUS GENOTYPES

HCV

Hepatitis C virus is found at varying frequencies in populations throughout the world. The main route of transmission of HCV is through parenteral exposure such as blood transfusion (hence the frequent occurrence of post-transfusion NANBH), and the use of non-inactivated blood products such as clotting factor concentrates, which were responsible for the high frequency of infection in haemophiliac patients. Since the development of serological screening assays, it has become apparent that HCV infection is also frequent in injecting drug abusers, with this risk group currently accounting for the majority of infected individuals identified in blood donor screening. However, in older populations past transfusion or medical treatment with inadequately sterilized equipment may be the predominant risk factor. It has been suggested that previous large-scale immunization campaigns such as those for yellow fever virus or parenteral treatment for bilharzia in Egypt and elsewhere in Africa in the 1950s and 1960s may have led to epidemic spread of HCV infection in particular populations.

There is scant evidence for other non-parenteral routes of transmission of this virus. For example, sexual transmission appears to be inefficient at best, occurring considerably less frequently than transmission of other viruses such as hepatitis B virus (HBV) or human immunodeficiency virus (HIV). Similarly the frequency of mother-to-child transmission is low (around 5%), and those cases that do occur may result from unrecognized parenteral exposure. Furthermore, unlike flaviviruses such as yellow fever and dengue viruses, there is no evidence for vector-borne transmission of HCV. If efficient transmission of HCV is confined to parenteral routes, it is somewhat difficult to understand how HCV became widespread in the past and how it achieved its current high level of genetic diversity. Blood transfusion, medical treatment with unsterilized needles and needle sharing became widespread only in the twentieth century, yet the existing variants of HCV are likely to have diverged from a common ancestor much further in the past.

Some clues about the past spread of HCV are provided by the current geographical and risk-group distribution of different genotypes. In Western countries and in Japan and other countries in the Far East, the number of HCV variants infecting various risk groups is relatively restricted (data reviewed by Bukh, 1995; Simmonds, 1995; Stuyver et al, 1996). For example, exhaustive analysis of hepatitis C patients and blood donors in Japan has provided evidence that infection is almost exclusively restricted to types 1b, 2a and 2b. Similarly, types 1a, 1b, 2b, 2c and 3a account for almost all infections in Europe and the USA, although the relative frequencies of these genotypes varies between countries. In Northern and Western Europe, types 1a and 3a are the predominant genotypes, while type 1b (and in Mediterranean countries, type 2c) is the most frequently detected genotype in Southern and Eastern Europe. Revealingly, the genotype distribution in Europe varies with age and therefore with the inferred duration of infection (Tisminetzky et al, 1994; Nousbaum et al, 1995; Pawlotsky et al, 1995; Simmonds et al, 1996). Types 1a and 3a are found predominantly amongst a younger age group than those infected with types 1b and 2, suggesting a series of separate transmission episodes perhaps associated with different risk behaviours. Analysis of the degree of diversity accumulated between epidemiologically unlinked variants within each of the genotypes found in Western countries has proved to be a particularly useful method to estimate their times of origin (Smith et al, 1997b), which can be used to substantiate other inferences concerned with their past and current transmission (see below).

The relatively restricted range of genotypes found in Western countries contrasts strongly with other geographical areas, where a markedly different pattern of diversity is found. For example, amongst individuals from Congo (Zaïre), Gabon and other Central African countries, HCV infection is largely confined to variants of the genotype 4 clade, although an extreme diversity of different subtypes of type 4 is observed. For example, amongst 16 variants sequenced from these countries in either E1 or NS5B, a total of 12 different subtypes was found (Stuyver et al, 1994). Similarly, there is evidence for extreme subtype diversity of other genotypes of HCV elsewhere in Africa and southeast Asia; subtypes of type 1 and 2 are particularly numerous in West Africa (Mellor et al, 1995; Ruggieri et al, 1996; Stuyver et al, 1996), while multiple and highly divergent subtypes of types 3 and 6 are found in India, throughout southeast Asia and probably in southern China (Apichartpiyakul et al, 1994; Tokita et al, 1994a,b, 1995, 1996; Mellor et al, 1995, 1996). The coexistence of multiple subtypes of a particular genotype suggests their long-term presence within human communities occupying particular geographical regions, although it tells us little about how the virus populations were maintained in such populations. Clearly an estimate of the time of divergence of such subtypes would be of value in understanding the history and spread of HCV in historic and possibly prehistoric times.

HGV/GBV-C

Although HGV/GBV-C is the most closely related human virus to HCV, current knowledge of the epidemiology and genetic diversity of the two viruses provides a series of marked – and frequently baffling – contrasts. Infection with HGV/GBV-C has been found in every human population sampled to date, with frequencies of viraemia ranging from 1% to 30% in the general population, and evidence for an even higher rate of past, resolved infection from the detection of specific antibodies to the envelope glycoprotein, E2. Unlike HCV, higher frequencies of active and past infection with HGV/GBV-C are associated with the number of sexual partners (Stark et al, 1996; Fiordalisi et al, 1997; Kao et al, 1997; Wu et al, 1997). Consistent with these studies, a frequency of viraemia of 26% amongst prostitutes and homosexual men has been found, as well as evidence for a total exposure (past and current infection) of 45% (Scallan et al, 1998). This increased prevalence contrasts with frequencies of viraemia and antibody of 3% and 2% in control groups and 2·3% viraemia in the local blood donor population (Jarvis et al, 1996). Also in contrast to HCV is the current evidence for a high frequency of mother-to-child transmission (Feucht et al, 1996; Moaven et al, 1996; Fischler et al, 1997). Finally, HGV/GBV-C may also be transmitted parenterally, with evidence for high frequencies of past or current infection in drug users, haemophiliac patients and multitransfused individuals.

Despite the existence of 'natural' routes of transmission and its worldwide distribution, HGV/GBV-C shows remarkably little sequence variability. A maximum of 13% nucleotide sequence divergence and 4% amino acid sequence divergence has been found upon comparison of variants from Western countries, Central and West Africa, Southeast Asia and the Far East, consistently lower than that found between the major genotypes or subtypes of HCV. Comparison of complete genome sequences of HGV/GBV-C provides evidence for genetic grouping of some variants (Figure 21.5A). However, for reasons that are currently unclear, comparison of even relatively long subgenomic regions does not reproduce the phylogenetic relationships observed with complete genomic sequences (Smith et al, 1997a). The only exception is a specific region of the 5' UTR which, in contrast to HCV, shows a degree of sequence diversity comparable to that of the rest of the genome. Comparison of sequences obtained from this latter region reveals marked differences in geographical distribution of the three main genetic groups of HGV/GBV-C (Figure 21.5B).

Despite the progress made in identifying routes of transmission of HGV, very little is currently understood concerning the past spread of the virus, and how it has come to have a global distribution in the absence of a substantial degree of sequence diversity. As is the case for HCV, even an approximate indication of the rate change of HGV/GBV-C over time would provide valuable information on the likely time scale for transmission of HGV/GBV-C in the past, and on the underlying reasons for the current distribution of genetic groups in different geographical regions.

RATE OF SEQUENCE CHANGE OF HCV AND HGV/GBV-C

The methods and underlying assumptions required to estimate times of divergence by comparison of nucleotide sequences of extant species underlie much current discussion in molecular evolution. Fundamental to any such attempt is the development of a 'molecular clock' that indicates the degree of nucleotide (or amino acid) sequence change over a specified time interval for a particular organism. In the cases of animal and plant evolution, the fossil record combined with the analysis of nucleotide sequence infor-

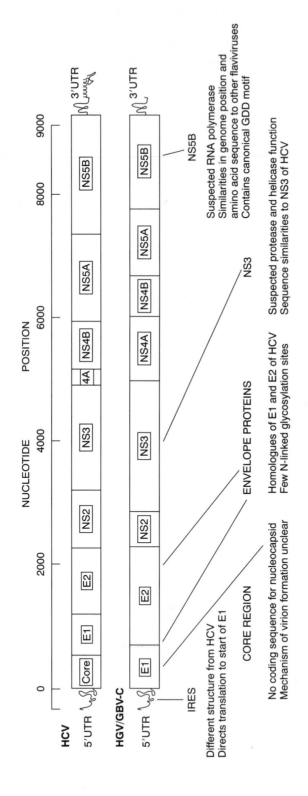

Figure 21.4 *Comparison of the organization of the genomes of HCV and HGV/GBV-C*

Apart from the differences in the predicted sizes of the structural and non-structural proteins, HGV/GBV-C (in common with GBV-A) shows no homologue of the core protein of HCV (and of GBV-B). It therefore remains unclear how virion assembly can occur. Other differences include a distinct secondary structure of the IRES of HGV/GBV-C that is more similar to those of picornaviruses than of HCV (Simons et al, 1996), and a virtual absence of potential *N*-linked (or O-linked) glycosylation sites in the putative E1 and E2 proteins of HGV/GBV-C, which presumably leads to a distinct virion structure and physicochemical properties from HCV. The 3′ UTR lacks the polyU tract found in most isolates of HCV, and does not contain the highly base-paired terminal RNA sequence recently described for HCV.

mation from present-day species allows such a calibration, albeit one subject to continuing refinements, major adjustments and ongoing disagreement. In the absence of a fossil record, as is the case for bacteria and viruses, reconstruction of their evolutionary history is even more problematic. This is particularly true for viruses where different families have replication strategies associated with markedly different frequencies of mutation. Viruses with RNA genomes, such as HCV and HGV/GBV-C, are replicated by a virally encoded RNA polymerase that transcribes an RNA copy from an RNA template without proofreading, and which is therefore considerably less accurate than the DNA polymerases of bacteria and eukaryotes. Most commentators have tended to ascribe the diversity observed amongst current RNA virus populations to this feature of RNA virus replication, but many anomalies arise from this simple explanation. Other important influences on the diversity observed within a virus population include its transmission history and the relative frequencies of the descendant viruses, while the observed rate of sequence change over time is strongly influenced by the degree and nature of constraints at different nucleotide positions in the genome. These factors are analysed below as another means of understanding the current epidemiology of HCV and HGV/GBV-C.

Overall rate of nucleotide substitution for HCV

For hepatitis C virus, numerous longitudinal studies have been made of virus sequence diversity, although not all of these provide useful measures of the rate of sequence change. For example, sequence analysis of more than half the virus genome during

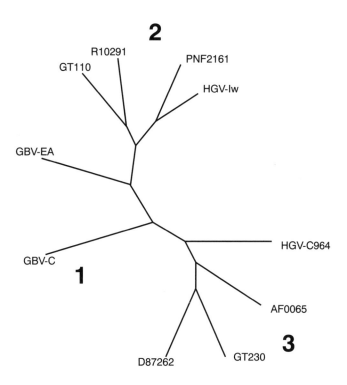

Figure 21.5A

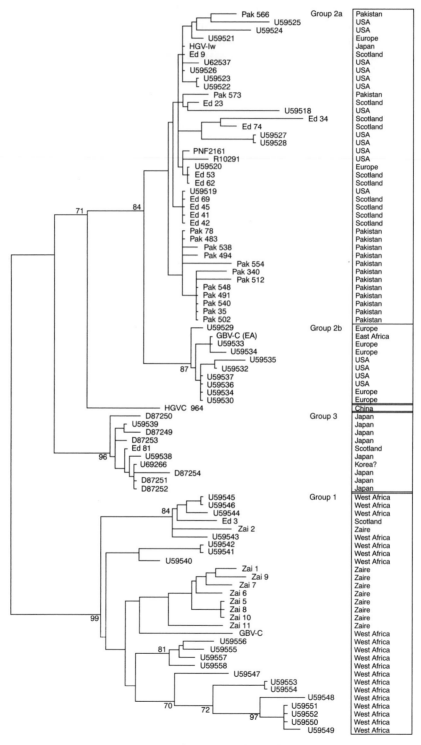

is equivalent to a distance of 0.02

Figure 21.5B

acute infection and from the same chronically infected individual 13 years later gave a rate of nucleotide substitution of $1\cdot9 \times 10^{-3}$ per site per year (Ogata et al, 1991). Two other studies of large or complete genome segments in chronically infected chimpanzees gave slightly lower figures of $1\cdot4 \times 10^{-3}$ (Okamoto et al, 1992) and $0\cdot9 \times 10^{-3}$ (Abe et al, 1992). Extrapolating the average of these rates of substitution ($1\cdot4 \times 10^{-3}$) to the divergence observed between different genotypes of HCV gives estimated times of origin of 115–240 years ago for virus subtypes and 170–370 years ago for virus types. If correction is made for the occurrence of multiple substitutions at some nucleotide positions, these times increase to 135–286 years ago for virus subtypes and 230–480 years ago for virus types.

However, despite the similar rates of substitution observed in these three different studies, their accuracy can be doubted because of uncertainty in each case about the relationship between the virus sequences compared. For example, if infection was initiated with many virus particles it is possible that a virus sequence detected some years later was present as a minor, undetected variant in the infective source, and that little sequence evolution had in fact occurred. For example, in one of these studies (Okamoto et al, 1992), a third as much variation was observed between different clones of virus sequences at the beginning of infection as occurred over time, while just as much variation was observed amongst different clones from the second sample.

Variation in the rate of nucleotide substitution across the genome

Another difficulty in interpreting these estimates of the overall rate of substitution is that this rate varies across the virus genome. During chronic infection, very few substitutions occur in the 5′ UTR or core genes relative to the two envelope genes and the NS2 gene (Ogata et al, 1991; Okamoto et al, 1992), while nucleotide substitution is especially frequent in a small region at the NH_2 terminus of the E2 gene called the hypervariable region. These inequalities in the frequency of substitution mean that variation in the hypervariable region may become saturated within two decades of evolution from a common source (McAllister et al 1998; Casino et al, 1999).

Similar inequalities between different regions of the HCV genome are observed when comparisons are made between sequences of the same virus subtype, or between different virus types. For example, viruses belonging to different virus types, and which are more than 30% divergent over the whole genome, may nevertheless be almost identical in the 5′ UTR (Mellor et al, 1996), while synonymous substitutions are suppressed in the core (Ina et al, 1994) and NS5B genes (Smith and Simmonds, 1997). The suppression

Figure 21.5 *Phylogenetic analysis of HGV/GBV-C and geographical distribution of variants*
(A) Phylogenetic analysis of complete genomic sequences of HGV/GBV-C with the main genetic groups indicated as 1, 2 and 3. An unrooted tree was generated using the program DNAML of Phylip (Felsenstein, 1993). The sequences compared were GBV-C (West Africa, U36380), HGVC964 (China, U75356), AF0065 (China, AF006500), GT230 (Japan, D90601), D87262 (Japan), GBV-EA (East Africa, U63715), GT110 (Japan, D90600), R10291 (USA, U45966), PNF2161 (USA, U44402) and HGV-Iw (Japan, D87255).
(B) Geographical distribution of genetic groups 1–3 of HGV/GBV-C identified by comparison of the region −366 to −235 in the 5′ UTR (numbered as in Smith et al, 1997a).

of nucleotide substitution in all of these regions is probably due to constraints imposed by functionally important RNA secondary structures (Brown et al, 1992; Smith et al, 1995; Smith and Simmonds, 1997). These inequalities in the rate of sequence substitution mean that the average rate of substitution across the entire virus genome over short periods is difficult to extrapolate to longer periods, and suggest that more accurate estimates will be obtained from the analysis of subgenomic regions.

Rate of substitution in subgenomic regions

Several studies have estimated the rate of substitution in defined regions of the virus genome. For example, a study of two individuals infected by the same HCV-contaminated batch of anti-D immunoglobulin gave rates of 1×10^{-3} and 5.2×10^{-3} for the E1 gene (Hohne et al, 1994) compared with 2.1×10^{-3}, 2.4 and 10^{-3} and 1×10^{-3} for the same region in the studies described above. For part of NS3, rates of between 1.1×10^{-3} and 2.1×10^{-3} were observed for sequences from blood transfusion recipients and contemporary samples from their implicated donors (Cuypers et al, 1991), compared with an overall rate for NS3 of 1.3×10^{-3} or 1.5×10^{-3}. These rates of substitution are equivalent to times of origin of virus subtypes of 60–300 years ago (E1) or 110–220 years ago (NS3), and for virus types 100–536 years ago (E1) and 180–350 years ago (NS3). Another way of improving the accuracy of estimates of the rate of substitution is to reduce sampling bias by measuring the rate of substitution in multiple recipients of the same infectious source. For example, amongst 26 individuals infected with HCV by the same source of contaminated anti-D immunoglobulin (Power et al, 1994, 1995), the average rate of nucleotide substitution in part of the NS5B gene over 17 years of infection was 0.4×10^{-3} per site per year, while for the E1 gene for 23 different anti-D recipients the rate was slightly higher at 0.7×10^{-3} substitutions per site per year (Smith et al, 1997b). One explanation for the low rate of change observed in this study is that sequence diversity of the infectious source was relatively low, so that all of the divergence observed between different individuals developed during chronic infection rather than deriving from pre-existing variation. Using these rates of substitution and correcting for the occurrence of multiple substitutions gives estimated times of origin of 260–280 years ago for virus subtypes and 400–570 years ago for virus types.

Biases in nucleotide substitution

Even within subgenomic regions several types of bias in the pattern of nucleotide substitution can be distinguished.

First, the ratio of evolutionary distances at non-synonymous sites (resulting in amino acid replacements) to those at synonymous sites is less than 0.24 in the longitudinal studies described above, and less than 0.1 in comparisons between epidemiological unrelated subtype 1b sequences (Smith and Simmonds, 1997). Since a ratio of about 3 would be expected in the absence of selection, this implies the existence of negative selection against those amino acid replacements that affect the function of virus proteins and that the rate of non-synonymous substitution observed in the short term may not be sustained over longer periods.

A second type of bias is that transition substitutions (A↔G or C↔U) exceed transversions (A or G↔C or U) by four-to five-fold in sequential samples (Ogata et al, 1991;

Okamoto et al, 1992) or between viruses of different genotypes (Tanaka et al, 1993). An even greater bias (8- to 16-fold) towards transitions is observed if only synonymous positions or the third position of codons are considered (Smith and Simmonds, 1997).

Taking account of these biases, analysis of synonymous sites in the core gene gives relatively recent times of origin for virus subtypes (100–200 years ago) and virus types (200–400 years ago) (Ina et al, 1994), although these estimates may be affected by the suppression of synonymous substitutions in this region. Similar analysis of the E1 and NS5B genes reveals that substitutions are saturated in many comparisons between different HCV subtypes and types (Figure 21.6). Estimated times of origin are therefore minimums and are pushed back to more than 320 years ago (subtypes) and more than 450–600 years ago (types) respectively, and could be much earlier (Smith et al, 1997b).

An additional type of bias, not taken into account in these estimates, is that G and C nucleotides are over-represented at the third position of codons where most substitutions are synonymous (Smith and Simmonds, 1997). There are also biases in dinucleotide frequencies and for the presence of termination codons in the +1 and −1 reading frames (Rima and McFerran, 1997). These biases imply that the rate of synonymous substitution may differ at a single nucleotide position depending on the particular nucleotide present or upon the flanking nucleotides.

In summary, several independent estimates of the time of divergence of HCV genotypes have been obtained based on the rate of substitution observed in longitudinal studies. Divergence of subtype 1b can be dated to about 70 years ago, divergence between subtypes of type 1 to 200–300 years ago, and divergence of HCV types at more

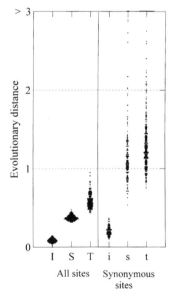

Figure 21.6 *Evolutionary distances for the E1 gene between HCV genotypes*
Evolutionary distances between E1 gene sequences of HCV isolates (I and i), subtypes (S and s) and types (T and t) were calculated for all nucleotide sites using the Kimura two-parameter method (Kumar et al, 1993). Distances at synonymous sites were calculated using the program dists1 (Ina, 1995) with an alpha/beta ratio of 16 (Smith and Simmonds, 1997). Sequence comparisons with a value of 1 mean that on average one substitution has occurred at each nucleotide position. Distances greater than 3 are assigned a value of 3.

than 450–600 years ago. Correction for biases in the pattern of substitution in the HCV genome tends to push estimates back to earlier times of origin, or to indicate that substitutions are becoming saturated so that distances become incalculable.

Rate of nucleotide substitution of HGV/GBV-C

Much less information is available concerning the rate of substitution during persistent infection with HGV/GBV-C. High rates of substitution have been observed for the NS3 helicase region in virus from individuals before and after interferon treatment ($2-20 \times 10^{-3}$ per nucleotide per year) (Berg et al, 1996), but lower rates ($1-3 \times 10^{-3}$) have been documented for the NS3 gene in eight patients infected for 2–16 years (Masuko et al, 1996) and for 2500 nucleotides of NS4/NS5 ($2 \cdot 7 \times 10^{-3}$) obtained from an individual chronically infected for $2 \cdot 5$ years (Khudyakov et al, 1997). A lower rate ($3 \cdot 9 \times 10^{-4}$) was observed from a longitudinal study of complete virus genomes from an individual chronically infected for $8 \cdot 4$ years (Nakao et al, 1997).

In this latter study only 5 out of the 29 substitutions in the coding region produced amino acid changes, giving a non-synonymous to synonymous site ratio of $0 \cdot 07$. Even lower ratios are observed in comparisons between the complete genome sequences of viruses from the 'Asian', 'African' and 'European' groupings (ratio $0 \cdot 03$) or within these groupings ($0 \cdot 02$). This strong bias against non-synonymous substitutions implies that there is selection against coding changes in the proteins encoded by HGV/GBV-C, although the biological reasons for this are unclear (Erker et al, 1996; Muerhoff et al, 1997; Smith et al, 1997a). If only synonymous sites are considered, the annual rate of substitution of HGV/GBV-C in studies in which substantial parts of the virus genome have been sequenced (Khudyakov et al, 1997; Nakao et al, 1997) is $1 \cdot 29 \times 10^{-3}$ or $5 \cdot 7 \times 10^{-3}$, and simple extrapolation of this implies that the diversity observed between the most divergent of the variants currently described would be produced after about 120 years or 500 years.

However, other information is difficult to understand given this time scale. The geographical distribution of different HGV/GBV-C phylogenetic groupings is still being defined, but already it is clear that distinct variants of HGV/GBV-C are found in Africa, Asia, and Europe/North America (Muerhoff et al, 1996, 1997; Smith et al, 1997a). Viruses within these groupings have been documented from both urban and remote rural communities, and this is consistent with a long-term endemic infection. In addition, the recent finding that HGV/GBV-C variants from South America are more closely related to variants previously found only in the Far East (GonzalezPerez et al, 1997) is consistent with this lineage having arisen before the colonization of the Americas via the Bering Strait some 10 000–20 000 years ago. This possibility is also suggested by the absence of pathological effects of infection by HGV/GBV-C and the recent identification in chimpanzees of a virus intermediate between GBV-A of New World monkeys and HGV IGBV-C of humans (Adams et al, 1998; Birkenmeyer et al, 1998). Distinct variants of GBV-A are confined to different New World primate species and are non-pathogenic (Leary et al, 1996a; Bukh and Apgar, 1997), suggesting that these variants may have co-speciated with their hosts many millions of years ago. Intriguingly, these variants differ at fewer than 40% of nucleotide positions in the 5′ UTR and NS3 regions (Bukh and Apgar, 1997) or 32% over the entire genome (Leary et al, 1997), only two to three times greater than the maximum divergence amongst currently described variants of HGV/GBV-C (10% for the 5' UTR, 20% for NS3 and 13% for entire coding regions). If the long-term evolution of both viruses is subject to similar constraints, this is consistent with divergence of HGV/GBV-C variants early in the history of *Homo sapiens*.

COMPARISONS BETWEEN HCV AND HGV/GBV-C

A surprising finding of these comparisons is the extent to which HCV and HGV/GBV-C differ in the characteristics of their sequence evolution. Although both viruses have similar overall rates of substitution ($0.4–3 \times 10^{-3}$ for HGV/GBV-C and $0.4–5 \times 10^{-3}$ for HCV) or rates at synonymous sites ($1.29–5.7 \times 10^{-3}$ for HGV/GBV-C compared with $1.1–3.5 \times 10^{-3}$ for HCV), the bias against non-synonymous substitution is stronger for HGV/GBV-C. For example, nucleotide sequence divergence between subtypes of HCV ranges from 20% to 25%, somewhat greater than between the most divergent variants of HGV/GBV-C (13%). However, while subtypes of HCV differ from each by around 13–18% at the amino acid level, the maximum divergence between variants of HGV/GBV-C is 4%. Why the coding sequence of HGV/GBV-C should be under a much greater degree of constraint than HCV is mysterious, given the likely functional equivalence at the level of viral replication and assembly of their encoded viral proteins.

One possible explanation is that the HGV/GBV-C genome contains extensive RNA secondary structures that coincide with local supplession of synonymous substitutions and would be expected to also limit substitution at non-synonymous sites (Simmonds & Smith, unpublished results).

This difference may partly explain why variants of HCV can be identified by phylogenetic analysis of subgenomic regions as short as 222 nucleotides (Simmonds et al, 1993) but analysis of HGV/GBV-C coding regions as large as 2000 nucleotides sometimes fails to group sequences belonging to different geographical groupings (Pickering et al, 1997; Smith et al, 1997a; Viazov et al, 1997). In addition, although infection with HGV/GBV-C can be persistent in a proportion of individuals, it apparently lacks a region corresponding to the hypervariable region at the N terminus of the E2 envelope protein of HCV. Variation of this region during HCV infection has been proposed to allow the virus to escape immune-mediated neutralization and so lead to the establishment of chronic infections (Weiner et al, 1992; Kato et al, 1993), but persistence of HGV/GBV-C obviously cannot be dependent on this type of interaction. Perhaps more germane is the observation that both viruses generate relatively weak immune responses against a limited range of virus-encoded antigens.

Another contrasting feature of HCV and HGV/GBV-C is that despite their similar genome organization the pattern of variability between the 5' UTR and coding regions differs. The 5' UTR of HCV is the most strongly conserved part of the virus genome with only 2.5–11% sequence divergence (mean 5.4%) between different virus types, compared with more than 30% for the coding region of the genome (Tokita et al, 1996). Comparisons between the three currently described groupings of HGV/GBV-C variants yield figures of 12–13% divergence over the coding region but divergence of the 5' UTR is similar at about 10%. Complex RNA secondary structures have been proposed for the 5' UTRs of both viruses (Brown et al, 1992; Smith et al, 1995; Simons et al, 1996) along with conserved sequence motifs, and are thought to be involved in the initiation of protein translation at internal entry sites (Tsukiyama Kohara et al, 1992; Simons et al, 1996) and in the replication of the virus genome. These functional properties presumably impose constraints that limit nucleotide substitutions within the 5' UTR to particular nucleotides or to covariant substitutions between base-paired nucleotides (Smith et al, 1995, 1997a; Simons et al, 1996). There is no obvious reason why these constraints should be any less severe for HGV/GBV-C and, if the annual rate of substitution in the 5' UTR is assumed to be similar to that for HCV, the greater divergence observed amongst HGV/GBV-C variants is consistent with an earlier time of origin than for the major genotypes of HCV.

Hence, comparison of the extent of divergence of 5' UTR sequences implies that HGV/GBV-C is older than HCV, while the rate of change and extent of divergence at synonymous sites suggests that HCV is older. Analysis of non-synonymous sites would imply the emergence of HGV/GBV-C within the last hundred years, but this is likely to be misleading because of the existence of strong constraints against coding changes.

CONCLUSION

Attempts to reconstruct the history of RNA viruses are at an early stage, and we are just beginning to appreciate the complex and interacting constraints that influence the long-term evolution of viruses. While extremely high rates of change have been documented in many RNA viruses over periods of years to decades, several lines of evidence suggest that these rates are not sustained over centuries and millennia. While there is no fossil record available to calibrate times of divergence, indirect evidence from the distribution of virus variants in contemporary human populations suggests that viruses such as HCV and HGV/GBV-C existed at the time of early human migrations some tens of thousands of years ago. Even earlier times of divergence of virus variants are implied by analogy with related viruses that have species-specific variants in New World primates.

For HCV this conflict between short-term and long-term rates of change is explained by the saturation of substitutions at synonymous sites in comparisons between virus types, so that times of divergence cannot be accurately estimated. Although variability is less extreme at non-synonymous sites, selection against changes in protein function provides an unquantified restriction on the accumulation of substitutions. Similarly, untranslated regions of the virus genome are subject to a variety of constraints imposed by the presence of RNA secondary structures and sequence motifs involved in virus replication and the initiation of transcription. Evolutionary distances between variants of HGV/GBV-C at synonymous sites are substantial, ranging up to about 0·7, but it is possible that this actually represents saturation because of (unidentified) constraints on substitution at these sites. For example, detailed analysis of the HCV genome revealed that synonymous substitutions were suppressed in the core and NS5B genes, probably because of constraints imposed by RNA secondary structures, and there is evidence for more extensive RNA structures in the HGV/GBV-C genome (Simmonds & Smith, unpublished results).

Although we are far from being able to date the time of origin of HCV or HGV/GBV-C, it is clear that these viruses are much older than is our experience of them. Both viruses have only been recognized since the 1970s because of changes in medical practice and in the sensitivity of methods for their detection. There is evidence for epidemics of HCV amongst recipients of blood products or amongst injecting drug users in the last few decades, but the origin of this pathogenic virus is much earlier. For HGV/GBV-C, there is little evidence for disease associated with infection, and it is probable that this virus has infected humans for a significant part of our history.

ACKNOWLEDGMENTS

P.S. is a Darwin Research Fellow; D.B.S. was supported by a career development grant from The Wellcome Trust.

REFERENCES

Abe K, Inchauspe G & Fujisawa K (1992) Genomic characterisation and mutation rate of hepatitis C virus isolated from a patient who contracted hepatitis during an epidemic on non-A, non-B hepatitis in Japan. *J Gen Virol* **73**: 2725–2729.

Adams NJ, Prescott LE, Jarvis LM et al (1998) Detection of a novel flavivirus related to hepatitis G virus/GB virus C in chimpanzees. *J Gen Virol* **79**: 1871–1877.

Alter HJ, Holland PV, Morrow AG, Purcell RH, Feinstone SM & Moritsugu Y (1975) Clinical and serological analysis of transfusion-associated hepatitis. *Lancet* **ii**: 838–841.

Apichartpiyakul C, Chittivudikarn C, Miyajima H, Homma M & Hotta H (1994) Analysis of hepatitis C virus isolates among healthy blood donors and drug addicts in Chiang Mai, Thailand. *J Clin Microbiol* **32**: 2276–2279.

Berg T, Dirla U, Naumann U et al (1996) Responsiveness to interferon alpha treatment in patients with chronic hepatitis C coinfected with hepatitis G virus. *J Hepatol* **25**: 763–768.

Birkenmeyer LG, Desal SM, Muerhoff AS et al (1988) Isolation of a GB virus-related genome from a chimpanzee. *J Med Virol* **56**: 44–51.

Bradley DW & Maynard JE (1986) Etiology and natural history of post-transfusion and enterically transmitted non-A, non-B hepatitis. *Semin Liver Dis* **6**: 56–66.

Bradley DW, Cook EH, Maynard JE et al (1979) Experimental infection of chimpanzees with antihemophilic (factor VIII) materials: recovery of virus-like particles associated with non-A, non-B hepatitis. *J Med Virol* **3**: 253–269.

Brown EA, Zhang HC, Ping LH & Lemon SM (1992) Secondary structure of the 5' nontranslated regions of hepatitis C virus and pestivirus genomic RNAs. *Nucl Acids Res* **20**: 5041–5045.

Bukh J (1995) Genetic heterogeneity of hepatitis C virus: quasispecies and genotypes. *Semin Liver Dis* **15**: 41–63.

Bukh J & Apgar CL (1997) Five new or recently discovered (GBV-A) virus species are indigenous to New World monkeys and may constitute a separate genus of the *Flaviviridae*. *Virology* **229**: 429–436.

Bukh J, Purcell RH & Miller RH (1993). At least 12 genotypes of hepatitis C virus predicted by sequence analysis of the putative E1 gene of isolates collected worldwide. *Proc Natl Acad Sci USA* **90**: 8234–8238.

Casino C, McAllister J, Davidson F et al (1999) Variation of hepatitis C virus following serial transmission: convergent evolution of envelope protein genes and rearrangements of the hypervariable region. *J Gen Virol* **80**: (in press).

Chamberlain RW, Adams NJ, Taylor LA, Simmonds P & Elliott RM (1997) The complete coding sequence of hepatitis C virus genotype 5a, the predominant genotype in South Africa. *Biochem Biophys Res Commun* **236**: 44–49.

Choo QL, Kuo G, Weiner AJ, Overby LR, Bradley DW & Houghton M (1989) Isolation of a cDNA derived from a blood-borne non-A, non-B hepatitis genome. *Science* **244**: 359–362.

Choo QL, Richman KH, Han JH et al (1991) Genetic organization and diversity of the hepatitis C virus. *Proc Natl Acad Sci USA* **88**: 2451–2455.

Cuypers HTM, Winkel IN, Van der Poel CL et al (1991) Analysis of genomic variability of hepatitis C virus. *J Hepatol* **13**: S15–S19.

Enomoto N, Sakuma I, Asahina Y et al (1995) Comparison of full-length sequences of interferon-sensitive and resistant hepatitis C virus 1b – sensitivity to interferon is conferred by amino acid substitutions in the NS5a region. *J Clin Invest* **96**: 224–230.

Enomoto N, Sakuma I, Asahina Y et al (1996) Mutations in the nonstructural protein 5A gene and response to interferon in patients with chronic hepatitis C virus 1b infection. *N Engl J Med* **334**: 77–81.

Erker JC, Simons JN, Muerhoff AS et al (1996) Molecular cloning and characterization of a GB virus C isolate from a patient with non-A-E hepatitis. *J Gen Virol* **77**: 2713–2720.

Feinstone SM, Kapikian AZ & Purcell RH (1975) Transfusion-associated hepatitis not due to viral hepatitis A or B. *N Engl J Med* **292**: 767–770.

Felsenstein J (1993) In: *PHYLIP Inference Package*, version 3·5. Seattle: Department of Genetics, University of Washington.

Feucht HH, Zollner B, Polywka S & Laufs R (1996) Vertical transmission of hepatitis G. *Lancet* **347**: 615–616.

Fiordalisi G, Bettinardi A, Zanella I et al (1997) Parenteral and sexual transmission of GB virus C and hepatitis C virus among human immunodeficient virus-positive patients. *J Infect Dis* **175**: 1025–1026.

Fischler B, Lara C, Chen M, Sonnerborg A, Nemeth A & Sallberg M (1997) Genetic evidence for mother-to-infant transmission of hepatitis G virus. *J Infect Dis* **176**: 281–285.

Gonzalez-Perez MA, Norder H, Bergstrom A, Lopez E, Visona KA & Magnius LO (1997) High prevalence of GB virus C strains genetically related to strains with Asian origin in Nicaraguan hemophiliacs. *J Med Virol* **52**: 149–155.

Gutierrez RA, Dawson GJ, Knigge MF et al (1997) Seroprevalence of GB virus C and persistence of RNA and antibody. *J Med Virol* **53**: 167–173.

Hohne M, Schreier E & Roggendorf M (1994) Sequence variability in the env-coding region of hepatitis C virus isolated from patients infected during a single source outbreak. *Arch Virol* **137**: 25–34.

Ina Y (1995) New methods for estimating the numbers of synonymous and nonsynonymous substitutions. *Mol Evol* **40**: 190–226.

Ina Y, Mizokami M, Ohba K & Gojobori T (1994) Reduction of synonymous substitutions in the core protein gene of hepatitis C virus. *J Mol Evol* **38**: 50–56.

Jarvis LM, Davidson F, Hanley JP, Yap PL, Ludlam CA & Simmonds P (1996) Infection with hepatitis G virus among recipients of plasma products. *Lancet* **348**: 1352–1355.

Kao JH, Chen W, Chen PJ, Lai MY, Lin RY & Chen DS (1997) GB virus-C/hepatitis G virus infection in prostitutes: possible role of sexual transmission. *J Med Virol* **52**: 381–384.

Kato N, Sekiya H, Ootsuyama Y et al (1993) Humoral immune response to hypervariable region-1 of the putative envelope glycoprotein (gp70) of hepatitis C virus. *J Virol* **67**: 3923–3930.

Khudyakov YE, Cong ME, Bonafonte MT et al (1997) Sequence variation within a nonstructural region of hepatitis G virus genome. *J Virol* **71**: 6875–6880.

Koonin EV (1991) The phylogeny of RNA-dependent RNA polymerases of positive-strand RNA viruses. *J Gen Virol* **72**: 2197–2206.

Kumar S, Tamura K & Nei M (1993) MEGA: *Molecular Evolutionary Genetics Analysis*, version 1·0. Pennsylvania State University.

Kuo G, Choo QL, Alter HJ et al (1989) An assay for circulating antibodies to a major etiologic virus of human non-A, non-B hepatitis. *Science* **244**: 362–364.

Leary TP, Desai SM, Yamaguchi J et al (1996a) Species-specific variants of GB virus A in captive monkeys. *J Virol* **70**: 9028–9030.

Leary TP, Muerhoff AS, Simons JN et al (1996b) Sequence and genomic organization of GBV-C: a novel member of the flaviviridae associated with human non-A-E hepatitis. *J Med Virol* **48:** 60–67.

Leary TP, Desai SM, Erker JC & Mushahwar IK (1997) The sequence and genomic organization of a GB virus A variant isolated from captive tamarins. *J Gen Virol* **78:** 2307–2313.

Linnen J, Wages J, Zhangkeck ZY et al (1996) Molecular cloning and disease association of hepatitis G virus: a transfusion-transmissible agent. *Science* **271:** 505–508.

Masuko K, Mitsui T, Iwano K et al (1996) Infection with hepatitis GB virus C in patients on maintenance haemodialysis. *N Engl J Med* **334:** 1485–1490.

McAllister J, Casino C, Davidson F et al (1998) Long-term evolution of the hypervariable region of hepatitis C virus in a common-source-infected cohort. *J Virol* **72:** 4893–4905.

Mellor J, Holmes EC, Jarvis LM et al (1995) Investigation of the pattern of hepatitis C virus sequence diversity in different geographical regions: implications for virus classification. *J Gen Virol* **76:** 2493–2507.

Mellor J, Walsh EA, Prescott LE et al (1996) Survey of type 6 group variants of hepatitis C virus in southeast Asia by using a core-based genotyping assay. *J Clin Microbiol* **34:** 417–423.

Miller RH & Purcell RH (1990) Hepatitis C virus shares amino acid sequence similarity with pestiviruses and flaviviruses as well as members of two plant virus supergroups. *Proc Natl Acad Sci USA* **87:** 2057–2061.

Missale G, Cariani E, Lamonaca V et al (1997) Effects of interferon treatment on the antiviral T-cell response in hepatitis C virus genotype 1b- and genotype 2c-infected patients. *Hepatology* **26:** 792–797.

Miyakawa Y, Okamoto H & Mayumi M (1995) Classifying hepatitis C virus genotypes. *Mol Med Today* **1:** 20–25.

Moaven LD, Tennakoon PS, Bowden DS & Locarnini SA (1996) Mother-to-baby transmission of hepatitis G virus. *Med J Aust* **165:** 84–85.

Muerhoff AS, Simons JN, Leary TP et al (1996) Sequence heterogeneity within the 5'-terminal region of the hepatitis GB virus C genome and evidence for genotypes. *J Hepatol* **25:** 379–384.

Muerhoff AS, Smith DB, Leary TP, Erker JC, Desai SM & Mushahwar IK (1997) Identification of GB virus C variants by phylogenetic analysis of 5'-untranslated and coding region sequences. *J Virol* **71:** 6501–6508.

Nakao H, Okamoto H, Fukuda M et al (1997) Mutation rate of GB virus C hepatitis G virus over the entire genome and in subgenomic regions. *Virology* **233:** 43–50.

Nousbaum JB, Pol S, Nalpas B et al (1995) Hepatitis C virus type 1b (II) infection in France and Italy. *Ann Intern Med* **122:** 161.

Ogata N, Alter HJ, Miller RH & Purcell RH (1991) Nucelotide sequence and mutation rate of the H strain of hepatitis C virus. *Proc Natl Acad Sci USA* **88:** 3392–3396.

Okamoto H, Kojima M, Okada SI et al (1992) Genetic drift of hepatitis C virus during an 8·2 year infection in a chimpanzee: variability and stability. *Virology* **190:** 894–899.

Pawlotsky JM, Tsakiris L, Roudotthoraval F et al (1995) Relationship between hepatitis C virus genotypes and sources of infection in patients with chronic hepatitis C. *J Infect Dis* **171:** 1607–1610.

Pickering JM, Thomas HC & Karayiannis P (1997) Genetic diversity between hepatitis G virus isolates: analysis of nucleotide variation in the NS-3 and putative 'core' peptide genes. *J Gen Virol* **78:** 53–60.

Poole TL, Wang CY, Popp RA, Potgieter LND, Siddiqui A & Marc S (1995) Pestivirus translation initiation occurs by internal ribosome entry. *Virology* **206:** 750–754.

Power JP, Lawlor E, Davidson F et al (1994) Hepatitis C viraemia in recipients of Irish intravenous anti-D immunoglobulin. *Lancet* **344:** 1166–1167.

Power JP, Lawlor E, Davidson F, Holmes EC, Yap PL & Simmonds P (1995) Molecular epidemiology of an outbreak of infection with hepatitis C virus in recipients of anti-D immunoglobulin. *Lancet* **345:** 1211–1213.

Prince AM, Brotman B, Grady GF, Kuhns WJ, Hazzi C, Levine RW & Millian SJ (1974) Long incubation post-transfusion hepatitis with evidence of exposure to hepatitis B virus. *Lancet* ii: 241–246.

Rima BK & McFerran NV (1997) Dinucleotide and stop codon frequencies in single-stranded RNA viruses. *J Gen Virol* **78**: 2859–2870.

Robertson B, Myers G, Howard C et al (1999) Classification, nomenclature and database development for hepatitis C virus (HCV) and related viruses: proposals for standardization. *Arch of Virol* (in press).

Ruggieri A, Argentini C, Kouruma F et al (1996) Heterogeneity of hepatitis C virus genotype 2 variants in West Central Africa (Guinea Conakry). *J Gen Virol* **77**: 2073–2076.

Scallan MF, Clutterbuck D, Jarvis LM, Scott GR & Simmonds P (1998) Sexual transmission of GB virus-C/hepatitis G virus. *J Med Virol* **55**: 203–208.

Simmonds P (1995) Variability of hepatitis C virus. *Hepatology* **21**: 570–583.

Simmonds P (1997) Clinical relevance of hepatitis C virus genotypes. *Gut* **40**: 291–293.

Simmonds P, Holmes EC, Cha TA et al (1993) Classification of hepatitis C virus into six major genotypes and a series of subtypes by phylogenetic analysis of the NS-5 region. *J Gen Virol* **74**: 2391–2399.

Simmonds P, Alberti A, Alter HJ et al (1994) A proposed system for the nomenclature of hepatitis C viral genotypes. *Hepatology* **19**: 1321–1324.

Simmonds P, Mellor J, Craxi A et al (1996) Epidemiological, clinical and therapeutic associations of hepatitis C types in western European patients. *J Hepatol* **24**: 517–524.

Simons JN, Pilot-Matias TJ, Leary TP et al (1995) Identification of two flavivirus-like genomes in the GB hepatitis agent. *Proc Natl Acad Sci USA* **92**: 3401–3405.

Simons JN, Desai SM, Schultz DE, Lemon SM & Mushahwar IK (1996) Translation initation in GB viruses A and C: evidence for internal ribosome entry and implications for genome organisation. *J Virol* **70**: 6126–6135.

Smith DB & Simmonds P (1997) Characteristics of nucleotide substitution in the hepatitis C virus genome: constraints on sequence change in coding regions at both ends of the genome. *J Mol Evol* **45**: 238–246.

Smith DB, Mellor J, Jarvis LM et al (1995) Variation of the hepatitis C virus 5′ non-coding region: implications for secondary structure, virus detection and typing. *J Gen Virol* **76**: 1749–1761.

Smith DB, Cuceanu N, Davidson F et al (1997a) Discrimination of hepatitis G virus/GBV-C geographical variants by analysis of the 5′ non-coding region. *J Gen Virol* **78**: 1533–1542.

Smith DB, Pathirana S, Davidson F et al (1997b) The origin of hepatitis C virus genotypes. *J Gen Virol* **78**: 321–328.

Stark K, Bienzle U, Hess G, Engel AM, Hegenscheid B & Schluter W (1996) Detection of the hepatitis G virus genome among injecting drug users, homosexual and bisexual men, and blood donors. *J Infect Dis* **174**: 1320–1323.

Stuyver L, Vanarnheim W, Wyseur A et al (1994) Classification of hepatitis C viruses based on phylogenetic analysis of the envelope 1 and nonstructural 5b regions and identification of five additional subtypes. *Proc Natl Assoc Sci USA* **91**: 10134–10138.

Stuyver L, Wyseur A, Vanarnhem W, Hernandez F & Maertens G (1996) Second-generation line probe assay for hepatitis C virus genotyping. *J Clin Microbiol* **34**: 2259–2266.

Tanaka T, Kato N, Hijikata M & Shimotohno K (1993) Base transitions and base transversions seen in mutations among various types of the hepatitis C viral genome. *FEBS Lett* **315**: 201–203.

Tisminetzky SG, Gerotto M, Pontisso P et al (1994) Genotypes of hepatitis C virus in Italian patients with chronic hepatitis C. *Int Hepatol Commun* **2**: 105–112.

Tokita H, Okamoto H, Tsuda F et al (1994a) Hepatitis C virus variants from Vietnam are classifiable into the seventh, eighth, and ninth major genetic groups. *Proc Natl Acad Sci USA* **91**: 11022–11026.

Tokita H, Shrestha SM, Okamoto H et al (1994b) Hepatitis C virus variants from Nepal with novel genotypes and their classification into the third major group. *J Gen Virol* **75:** 931–936.

Tokita H, Okamoto H, Luengrojanakul P et al (1995) Hepatitis C virus variants from Thailand classifiable into five novel genotypes in the sixth (6b), seventh (7c, 7d) and ninth (9b, 9c) major genetic groups. *J Gen Virol* **76:** 2329–2335.

Tokita H, Okamoto H, Iizuka H et al (1996) Hepatitis C virus variants from Jakarta, Indonesia classifiable into novel genotypes in the second (2e and 2f), tenth (10a) and eleventh (11a) genetic groups. *J Gen Virol* **77:** 293–301.

Tsukiyama Kohara K, Iizuka N, Kohara M & Nomoto A (1992) Internal ribosome entry site within hepatitis C virus RNA. *J Virol* **66:** 1476–1483.

Viazov S, Riffelmann M, Khoudyakov Y, Fields H, Varenholz C & Roggendorf M (1997) Genetic heterogeneity of hepatitis G virus isolates from different parts of the world. *J Gen Virol* **78:** 577–581.

Wang CY, Sarnow P & Siddiqui A (1994) A conserved helical element is essential for internal initiation of translation of hepatitis C virus RNA. *J Virol* **68:** 7301–7307.

Weiner AJ, Geysen HM, Christopherson C et al (1992) Evidence for immune selection of hepatitis C virus (HCV) putative envelope glycoprotein variants: potential role in chronic HCV infections. *Proc Natl Acad Sci USA* **89:** 3468–3472.

Wu JC, Sheng WY, Huang YH & Lee SD (1997) Prevalence and risk factor analysis of GBV-C/HGV infection in prostitutes. *J Med Virol* **52:** 83–85.

Zanotti PMD, Gibbs MJ, Gould EA, Holmes EC (1996) A re-evaluation of the higher taxonomy of viruses based on RNA polymerases. *J Virol* **70:** 6083–6096.

Chapter 22

ENDOGENOUS RETROVIRUSES

Clive Patience

INTRODUCTION

The genomes of all eukaryotes studied to date, including humans, contain elements called endogenous retroviruses (ERVs), so named because of their sequence similarity to infectious retroviruses. As shown in Figure 22.1, in comparison with other human retroelements, a full-length human ERV shows the same basic genome organization as an exogenous retrovirus, possessing regions with sequence similarity to the long terminal repeats (LTRs) and major open reading frames (ORFs). A comparison of the life cycles of exogenous and endogenous retroviruses is shown in Figure 22.2.

Two major theories exist regarding the mechanism by which ERVs may have evolved. Firstly, it has been suggested that ERVs may have developed as a result of reverse transcription and retrotransposition of ancestral retroelements (Temin, 1980). The presence

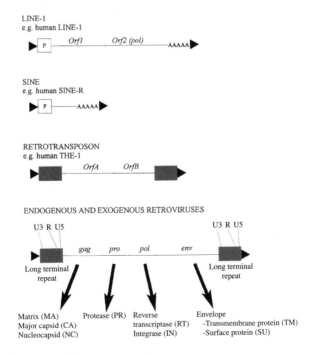

Figure 22.1 *General structure of human retroelements*
Open reading frames are indicated and where appropriate their protein products are listed.

HIV and the New Viruses Second Edition
ISBN 0-12-200741-7

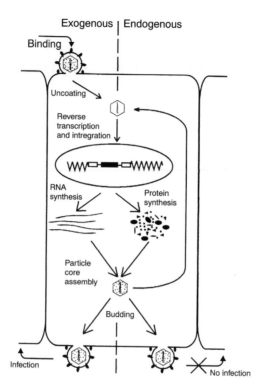

Figure 22.2 *Comparison of exogenous and endogenous retrovirus life cycles*
In addition to the cycles indicated, on rare occasions RNA transcripts can reintegrate into the genome without previously being associated with intracellular virus particles.

of retroelements among all eukaryotic kingdoms and the high degree of amino acid homology among all known reverse transcriptases adds support for this hypothesis. Alternatively, ERVs may have evolved as endogenized variants of exogenous viruses. Support for this theory can be taken from certain retroviruses, such as Jaagsiekte retrovirus and mouse mammary tumour virus (MMTV), which are found as both endogenous and exogenous agents in their host species (Coffin, 1982; Palmarini et al, 1996). It may be that these viruses represent viral agents that have yet to become fully endogenized. As ERVs form part of the germline DNA they will be inherited in a mendelian manner. Although generally considered to be stable mendelian traits of the host genome, in mice new ERV integrations have been reported (Szabo et al, 1993), suggesting that the acquisition of new sequences may still be an ongoing process, albeit a very slow one.

Isolation of the infectious retroviruses, murine leukaemia virus (MLV) and Rous-associated virus (RAV-0), from uninfected murine and avian cells respectively, was taken as the first proof that viruses could form part of the normal genome of cells (Aaronson et al, 1969; Vogt and Friis, 1971). Although the early studies into ERV which were performed in avian and murine systems were restricted to elements conferring a characteristic phenotype on their host, with the advances in hybridization and polymerase chain reaction (PCR) technologies we are now able to investigate elements that do not confer phenotypic traits. Also, as our ability to detect repetitive elements in the genome developed in advance of the techniques necessary to investigate their activities, the suggestion arose that the sequences were just 'junk' DNA capable of replication within the genome but serving no function. While this analysis may be partly true, it is now clear that retroelements, including ERV, play an important role in the biology of their host.

HUMAN ENDOGENOUS RETROVIRUSES

It has been known since the 1970s that the human germline DNA contains multiple ERV sequences (Benveniste and Todaro, 1974). Most human endogenous retrovirus (HERV) genomes appear to have been acquired at least 10 million years ago and possibly up to 100 million years ago (Shih et al, 1991) and may constitute up to 0·1% of our germline DNA (Callahan, 1988; Brack-Werner et al, 1989). Typically HERV families contain between 1 and 100 members, though some families can possess up to approximately 1000 copies (Wilkinson et al, 1993). HERV genomes are transmitted as part of the host's normal genetic material; consequently, the selective pressure to maintain functional ORFs no longer applies, resulting in the vast majority of HERV loci acquiring mutations and deletions which disrupt one or more of their ORFs. Some HERVs have, however, retained the ability to express proteins and in some cases produce virus particles.

Over evolutionary time periods ERV have been selected that are non-infectious and non-pathogenic for their host. However, some ERV have retained their ability to code for virus that can infect the cells of other species, a phenomenon called xenotropism. For example, xenotropic retroviruses have been described in mice that cannot replicate in mouse cells on account of receptor incompatibility, but can propagate to high titres in human cells in culture (Weiss, 1993). Also, chicken and pig ERVs rarely replicate in their own species but readily infect cultured cells other species (Coffin, 1982; Patience et al, 1997). Likewise, a cat ERV replicates in human cells, as does one from baboons, although neither replicates in its own host species.

Detection and classification of HERV

Several experimental approaches have been used to identify HERV. For example, the screening of human genomic libraries with probes derived from non-human endogenous and exogenous retroviruses or with oligonucleotides designed to detect the 3′ terminal sequence of tRNAs has proved successful, as has PCR amplification using degenerate primers, and analysis of flanking regions of cellular genes (reviewed by Wilkinson et al, 1994).

At present there is no universally accepted method for HERV classification. Sequence similarity to mammalian retroviruses has been used to classify HERV, with class I families having *pol* sequence similarity to the mammalian C type viruses, and class II families sharing *pol* sequence similarity with the mammalian B type, D type and avian C type retroviruses. A further widely used system of classification is based on the identity of the tRNA which has homology to the minus strand primer binding site located immediately downstream of the 5′ LTR, for example HERV-K elements utilizing a lysine tRNA.

Some of the HERV families that have been characterized in detail are shown in Table 22.1. The list is by no means complete, as many HERV *pol* sequences have been identi-

Table 22.1 *Summary of known HERV families*

Family	Class	Primer binding site	Approximate copy number	References
HERV-E	I	Glutamic acid	30–50	Repaske et al (1985)
ERV-1	I	Not determined	1–15	Bonner et al (1982)
HERV-R (ERV-3)	I	Arginine	1–10	O'Connell and Cohen (1984)
RRHERV-I	I	Not determined	10	Kannan et al (1991)
HERV-I	I	Isoleucine	3–25	Maeda (1985)
S7	I	Not determined	1–20	Leib-Mösch et al (1986)
HRES-I	I	Histidine?	1	Perl et al (1989)
HERP-P	I	Proline	10–20	Harada et al (1987)
ERV-9	I	Arginine	40	La Mantia et al (1991)
HERV-H	I	Histidine	1000	Mager and Henthorn (1984)
HERV-K	II	Lysine	30–50	Ono et al (1986)

fied that cannot be attributed to these families (Bangham et al, 1988; Shih et al, 1989; Wichman and Van Den Bussche, 1992; Medstrand and Blomberg, 1993; Cordonnier et al, 1995; Kabat et al, 1996; Widegren et al, 1996).

The following discussion does not attempt to summarize the current state of knowledge of all HERV families; rather, the diversity of HERVs, the roles that these elements play in the normal function of the host genome, and their pathogenic potential are described. Examples from non-human hosts are included where appropriate. For clarity, primer binding site nomenclature of the elements is used, with reference to other designations where informative.

HERV-E (4.1)

HERV-E elements, of which there are 35–50 copies per genome, entered the primate lineage at least 25 million years ago (Steele et al, 1984). Analysis of the prototype proviral clone (4.1) revealed a 8·8 kb full-length HERV with regions related to *gag*, *pol*, and *env* bounded by 490 bp LTRs (Repaske et al, 1985). The three coding regions all contained stop codons rendering the element incapable of coding for infectious virus, although a near full-length *env* ORF was present (Rabson et al, 1983; Gattoni-Celli et al, 1986). In addition to the full-length elements, solitary HERV-E LTRs have also been identified in the genome (Steele et al, 1984).

The HERV-E elements are probably active not through protein expression, but by controlling the expression of host genes via regulatory sequences in their LTRs. For example, HERV-E elements provide tissue-specific promoter activity to the salivary amylase and pleiotrophin genes (see Effects on adjacent genes). However, the presence of HERV-E Env proteins in alveolar macrophages has been described in patients with interstitial lung disease (Tamura et al, 1997) and the precise role of the Env expression in these patients warrants further investigation.

HERV-R (ERV-3)

HERV-R is a single copy element present on chromosome 7 which has a 9·9 kb genome including two LTRs together with *gag* and *pol* genes which are disrupted by multiple stop codons (O'Connell and Cohen, 1984) and an *env* gene that is free of stop codons except for one that truncates the TM region just upstream of the hydrophobic transmembrane motif (Cohen et al, 1985). The maintenance of the *env* ORF for at least 30 million years in the genomes of Old World primates and apes is intriguing and implies that selective pressure is maintaining the ORF, presumably because the protein supplies an essential or at least advantageous function to the host (Shih et al, 1991).

Transcripts of HERV-R *env* have been detected in many tissues, primarily in the placenta (Cohen et al, 1985; Kato et al, 1987; Larsson et al, 1994, 1997; Andersson et al, 1996). Placental expression of HERV-R is localized specifically to the fused syncytiotrophoblast sheet which forms the layer of cells between mother and fetus (Boyd et al, 1993; Venables et al, 1995). In view of the extraordinarily high level of expression of HERV-R envelope – up to 0·1% of the total cell protein – it has been speculated that this protein may augment the immunosuppressive role of the placenta (Venables et al, 1995). In this regard, it has been noted that HERV-R *env* has a region which has high similarity with the p15E motif of the TM protein of MLV which is known to have an immunosuppressive role in vivo (Boyd et al, 1993; Haraguchi et al, 1995; Venables et al, 1995). However, individuals have recently been identified who can express only severely truncated HERV-R Env proteins (de Parseval and Heidmann, 1998). The ability of these individuals to reproduce normally precludes an essential function in the placenta.

Although not essential for reproduction, several alternative activities for HERV-R have been suggested. The HERV-R proteins may play a role in the creation of the fused syncytiotrophoblast layer from cytotrophoblast cells, as retroviral Env proteins have fusogenic properties (Harris, 1991). Also, owing to the lack of expression in choriocarcinoma cell lines, HERV-R has been reported as having tumour suppressive properties (Kato et al, 1990). However, although the observation is correct, the interpretation is probably not, as HERV-R is expressed in syncytiotrophoblast cells and not the chorionic cells of the placenta. It has been noted that mothers of children with congenital heart block exhibit increased levels of anti-HERV-R antibodies, and the possible role of HERV-R expression in this condition warrants further investigation (Li et al, 1996).

With effectively only a single protein to investigate, it is likely that clarification of the activity of HERV-R Env protein will soon be forthcoming. It should be noted, however, that under low stringency hybridization conditions sequences related to HERV-R can be detected (O'Connell and Cohen, 1984). It may be that the human genome contains a small family of elements related to HERV-R and it will be interesting to identify the members of this family and to determine whether their *env* ORF is also conserved.

HERV-H (RTVL-H)

The HERV-H family was initially thought to comprise approximately 800–900 elements of approximately 5·8 kb in length and containing *gag* and *pol* regions flanked by 400–450 bp LTRs. It has become apparent, however, that these elements are actually the deleted derivatives of full-length HERVs which are present in much lower numbers, with only 100 or fewer copies per genome (Hirose et al, 1993; Wilkinson et al, 1993). The presence in addition to these elements of approximately a thousand solitary HERV-H LTRs makes this family by far the most abundant HERV identified to date. For the most part HERV-H elements appear to be fixed in the genome, although analysis of one locus in 70 individuals did detect two siblings in whom an element had recombined to form a solitary LTR by deletion of the intervening sequences (Mager and Goodchild, 1989).

Expression of HERV-H has been detected in a variety of cell lines including germ cell tumours (Hirose et al, 1993; Wilkinson et al, 1993), and in normal tissues such as the placenta (Johansen et al, 1989) and blood mononuclear cells (Medstrand et al, 1992). Further to the above examples, the packaging of RTVL-H sequences into the vimons produced by the human teratocarcinoma cell line GH has been demonstrated (Patience et al, 1998). Whether it is the particular RTVL-H sequences expressed in this cell line or whether it is the HERV-K-derived particles produced by this cell line which are responsible for the packaging remains to be determined.

HERV-K (HML-2)

HERV-K elements were originally identified due to their cross-hybridization with MMTV *pol* probes (Callahan et al, 1982; Westley and May, 1984; Deen and Sweet, 1986; Ono, 1986) and can be detected in the genomes of apes and Old World monkeys (Ono et al, 1986; Leib-Mösch et al, 1993). Also, a family of HERV-K-related sequences has been detected in New World monkeys (Simpson et al, 1996). In the human genome the HERV-K family comprises 30–50 proviral sequences (Ono, 1986) and approximately 10 000 solitary LTRs (Ono, 1986; Leib-Mösch et al, 1993).

HERV-K elements can be divided into two families depending on the structure of their genome. Type I HERV-K elements posses a characteristic 292 bp deletion between the *pol* and *env* genes and have been described as expressing exclusively unspliced RNA transcripts (Löwer et al, 1995) although spliced transcripts have also been reported (Etkind et al, 1997). Type II elements are full length, approximately 9·5 kb, and lack the 292 bp deletion. RNA transcripts from these elements are spliced to subgenomic *env* and two smaller RNAs (Löwer et al, 1993a,b, 1995). A 1·8 kb doubly spliced transcript encompasses most of the 292 bp segment and encodes for a small ORF, designated cORF (Löwer et al, 1993a,b, 1995). This region encodes a 12 kDa protein which has strong structural similarities to lentivirus Rev protein and, like Rev, accumulates in the nucleolus (Löwer et al, 1995). Functional activity of the protein has yet to be demonstrated.

HERV-K transcripts have been detected in many cell lines including breast carcinoma and teratocarcinoma cell lines (Ono et al, 1987; Löwer et al, 1993a,b; Patience et al, 1998), and in normal tissues such as the placenta, colonic mucosa, peripheral blood mononuclear cells (PBMCs) and leukocytes (Medstrand and Blomberg, 1993; Simpson et al, 1996; Tönjes et al, 1996). Control of HERV-K transcription in these cells may be regulated through the methylation of CpG dinucleotides (Gotzinger et al, 1996). Other HERV sequences related to HERV-K have been identified and have been divided into six families, termed HML-1 through 6 (Medstrand and Blomberg, 1993). Sequence similarity in comparison with HERV-K ranges from greater than 90% down to only 55–60%. Expression of the HML families in mammary and placental tissues and PBMCs has been reported, but the significance of the expression in these cells has yet to be determined (Andersson et al, 1996; Yin et al, 1997).

Typically the coding regions of HERV-K elements are far less disrupted by mutations than other HERV families and protein synthesis has been observed for all the main retroviral genes. The HERV-K Gag precursors are cleaved into major core, matrix and nucleocapsid components (Boller et al, 1993; Müeller-Lantzsch et al, 1993; Sauter et al, 1995), presumably by HERV-K protease, as functional activity has been demonstrated for this enzyme (Schommer et al, 1996). Also, functional activity of HERV-K dUTPase has recently been demonstrated (Harris et al, 1997). Within *pol*, functional activity of integrase (Kitamura et al, 1996) but not RT protein has been demonstrated, although virus particles that package HERV-K sequences do carry readily detectable levels of active RT enzyme (Patience et al, 1996; Simpson et al, 1996; Tönjes et al, 1996). Functional activity of the HERV-K envelope has yet to be proved. When expressed in insect cells, full-length Env precursor can be formed, but it is not properly cleaved, nor is it presented on the surface of cells or secreted into the medium (Tonjes et al, 1997).

Particles of HERV-K have been detected in breast carcinoma and teratocarcinoma cell lines (Bronson et al, 1979; Kurth et al, 1980; Patience et al, 1996). The HERV particles produced in the placenta have been shown to package HERV-K transcripts and are of similar morphology to the particles associated with the cell lines (Simpson et al, 1996).

There is increasing evidence for raised HERV-K protein expression in some human tumours. Antibodies to HERV-K, present only at low levels in normal donor blood, can be detected at markedly increased levels in patients with seminomas (Sauter et al, 1995). Furthermore, HERV-K *gag* expression has been identified in seminoma tumour biopsies (Herbst et al, 1996). Recently in situ hybridization studies for HERV-K *gag* and *env* have been performed upon a wide range of tumours including testicular and ovarian germ cell tumours (GCTs), GCT precursor lesions, and gestational trophoblastic disease (Herbst et al, 1996; Boller et al, 1997). Expression was detected in all GCTs and their precursor testicular lesions, with the exception of teratomas, and spermocytic seminomas. The production of antibodies is in itself interesting since HERV proteins can be regarded as self-antigens which under normal circumstances should induce immunological tolerance. The production of antibodies suggests either a dysfunction of the immune system or transcriptional activation of additional HERV loci capable of encoding immunogenic proteins. Reflecting such a dysregulation is the reported involvement of HERV-K in insulin-dependent diabetes mellitus (Conrad et al, 1997). It has been proposed that a MHC class II-dependent superantigen expressed from a HERV-K element may be responsible for the eventual destruction of the pancreatic beta cells. If this observation is correct it will be the first example of HERV causing disease.

IMPLICATIONS OF ERV

Effects on genome structure

The integration pattern of retroviruses is not entirely random, with accessible, often transcriptionally active, regions of the host genome being favoured. Consequently, HERV are frequently found in association with other retroelements and often in clusters, although some clustering may be a result of DNA amplification rather than multiple HERV integrations.

On occasions the host genome can acquire pseudogenes owing to reverse transcription and integration of mRNAs. Although rare, examples of such reverse flow of genetic material have been detected in several organisms (Wilkinson et al, 1994). These retroviruses may contribute to the formation of pseudogenes by supplying the RT activity necessary for the process and may also supply promoter activities which drive expression of the newly formed gene.

Being dispersed throughout the genome, HERVs can serve as foci for recombination and duplication events. HERV have been identified in the gene clusters coding for several proteins including amylase (see Effects on adjacent genes) (Samuelson et al, 1990), haptoglobin (Maeda, 1985), gammaglobulin (Fitch et al, 1991) and, most recently, the C2 and C4 components of the complement system. A retroposon derived from a HERV-K element is responsible for the restriction fragment length polymorphisms seen in the C2 component locus (Zhu et al, 1992), while for C4 the integration of a HERV-K element, prior to the separation of primate and human species, is responsible for the length polymorphisms seen with this gene (Dangel et al, 1994; Tassabehji et al, 1994; Chu et al, 1995). HERV-K loci have been identified in close proximity to several other host genes, including the breast and ovarian cancer susceptibility gene BCRA1 on chromosome 17 (Jones et al, 1994) and the glucose-6-phosphate dehydrogenase locus on the X chromosome (Sedlacek et al, 1993). However, the association of HERV with these genes may merely be a consequence of the sheer number of elements present in the normal human genome, as an interaction between the HERV and the proximal gene remains to be shown.

Effects on adjacent genes

HERV elements can affect adjacent gene expression as their LTRs contain transcriptional regulators. Probably the best-characterized example of control of gene expression by HERV is that of the salivary amylase genes where the insertion of an HERV-E element 5′ to a duplicated pancreatic amylase gene provides a parotid-specific enhancer element leading to tissue-specific expression of the gene (Samuelson et al, 1990; Ting et al, 1992). The integration of an HERV-E element 5′ to the gene encoding the human growth factor pleiotrophin has also been detected (Schulte et al, 1996). The integration generates an additional promoter which possesses trophoblast-specific activity. Fused HERV–pleiotrophin transcripts have been detected in normal trophoblast cultures and trophoblast-derived choriocarcinomas, but not in other cell lineages such as teratocarcinomas. It has been suggested that the fused gene product may be responsible for the invasive growth phenotype observed with normal human trophoblast, as depletion of the RNA prevented the growth, invasion and angiogenesis of human choriocarcinomas in mice.

Further examples whereby HERV-derived sequences affect normal host gene expression include the genes encoding cytochrome c_1 (Suzuki et al, 1990) and the phospholipase A_2 gene (Feuchter-Murthy et al, 1993). In addition to regulating gene expression via promoter sequences, examples have been identified in which HERV elements supply the polyadenylation signal for cellular sequences (Tomita et al, 1990; Goodchild et al, 1992).

HERV protein expression

The contribution of HERV proteins towards the biology of the host remains largely unknown as the majority of the proteins are currently poorly characterized. Even so, there is now increasing evidence that HERV proteins can be expressed and show biological activity. The *env* expression of HERV-R in the syncytiotrophoblast of the placenta is the best-documented case of ORF expression. Also in the placenta, expression of a protein related to the Gag proteins of ERV1 and baboon endogenous virus has been reported. In addition to the placenta, this protein has also been detected in ova, benign and malignant tumours, and choriocarcinoma cell lines. Although its function is unknown, expression of HERV Gag proteins in the placenta could benefit the host by protecting against infection with exogenous viruses, a possibility discussed in the section on Prevention of infection.

Retrovirus-like particles in human tissues

Particles resembling retroviruses are produced in quantity by the syncytiotrophoblast cell layer of humans, as well as of Old World and New World monkeys (Kalter et al, 1973). The particles are of D type morphology resembling HERV-K human teratocarcinoma-derived virus (HTDV) particles, and in humans package HERV-K-like sequences and an RT enzyme with a preference for Mg^{2+} (Nelson et al, 1978; Simpson et al, 1996). To date, attempts to culture the particles have not been successful; however, it may be that cell lines from species very distant to humans may be required to support the growth of the virus.

The wide distribution and abundance of the HERV-derived particles suggest a possible role in the function of the placenta. The conservation of particle production rather than just protein expression over evolutionary periods may serve as an indication that particle production or at least Gag expression is important. Alternatively it may be that Env protein expression is important but requires the protein to be presented in the context of virus particles. In the placenta HERV Gag and Env expression could serve to protect against infection by exogenous viruses (see Prevention of infection); also, HERV Env could be involved in the formation of the fused syncytiotrophoblast layer, or may act as an immunosuppressive agent. Whether this Env protein is derived from HERV-K elements or from other families, such as HERV-R, remains to be determined. It may be that HERV-K particles are providing a vehicle by which HERV-R Env proteins can be mobilized on virus particles.

Retrovirus-like particles have also been detected in teratocarcinoma cell lines and in the mammary carcinoma cell line T47D (Bronson et al, 1979; Kurth et al, 1980; Patience et al, 1996, 1998). The human teratocarcinoma-derived virus (HTDV) particles closely resemble those in the T47D cell line and the placenta, and are coded for by HERV-K elements. Although reports of similar retroviral particles in human milk and breast tissue, and the sequence similarity of HERV-K to MMTV, led to extensive searches for a human equivalent to MMTV, the investigations have not been encouraging.

Pathogenic consequences of HERV

A pathogenic effect of HERV has yet to be demonstrated but, by analogy with infectious viruses, the potential does exist.

Insertional mutagenesis by retrovirus-like elements has been shown to affect the expression of several genes in mice, either knocking out or increasing their expression (Jenkins et al, 1981; Stoye et al, 1988; Adachi et al, 1993). Murine ERVs play a central role in tumour induction in some strains of mice inbred for genetic predisposition to develop tumours owing to mutagenesis of proto-oncogenes (Stoye and Coffin, 1985; Favor and Morawetz, 1992). In humans a similar effect due to HERV has yet to be detected, although insertion of retroelements has been described. The insertion of LINE-1 retroelements has been detected in the genes encoding factor VIII in patients with haemophilia A (Kazazian et al, 1988) and in the *myc* and adenomatous polyposis coli genes of patients with breast and colon cancers respectively (Morse et al, 1988; Miki et al, 1992). Insertion of Alu retroelements has been detected in the factor IX gene of a patient with haemophilia B (Vidaud et al, 1993) and in the NF-1 gene of a patient with neurofibromatosis (Wallace et al, 1991). Although, with the above examples, it seems reasonable to assume that HERV could play a role in the development of at least some cancers, the extremely low rate of HERV transposition events argues against HERV being a major tumorigenic factor.

Human ERVs have the potential to act as foci for recombination events causing either duplication or deletion of regions of the genome. Examples with pathogenic consequences are again lacking for HERV, but have been reported for Alu retroelements affecting the genes coding for the low-density lipoprotein receptor, adenosine deaminase, and the dystrophin locus (Hobbs et al, 1986; Markert et al, 1988; Pizzuti et al, 1992).

The possible role of HERV in autoimmune disease has received much attention owing to the familial nature of several of the conditions (reviewed by Nakagawa and Harrison, 1996). To date the evidence for HERV involvement remains circumstantial (Ranki et al, 1992; Talal et al, 1992). However, the significance of ERV in autoimmune disease has been better characterized in murine systems, with ERVs being the targets of autoimmune responses in models of lupus (Izui et al, 1979), vasculitis (Miyazawa et al, 1987) and diabetes (Leiter and Kuff, 1984). Ultimately, to prove that an HERV antigen is responsible for an autoimmune lesion, specific expression of the antigen is required in conjunction with the exclusion of a role of other known autoimmune antigens. The presence of antibodies in normal human sera which cross-react with conserved retroviral antigens provides an additional complication. Mechanisms by which retroviruses, including HERV, could induce autoimmune responses are numerous; some are listed in Table 22.2.

Table 22.2 *Possible mechanisms for induction of autoimmunity by HERV*

Mechanism of induction	References
Molecular mimicry	Oldstone (1987)
Anti-idiotype network	Plotz (1983)
Mitogen or superantigen expression	Pahwa et al (1986)
	Marrack et al (1991)
	Chirmule et al (1994)
Immunosuppression	Krieg et al (1989)
	Lindeskog et al (1993)
HERV activation	Hynes et al (1981)
	Majors and Varmus (1983)

INTERACTION OF ENDOGENOUS AND EXOGENOUS VIRUSES

Using animal models, it has been shown that ERVs play a decisive role in the development and outcome of infection with some exogenous retroviruses. Although examples have not yet been described for humans, the mechanisms could be very similar and are briefly described on page 495.

Prevention of infection

Receptor interference by envelope glycoproteins is an obvious mechanism by which ERVs can inhibit the replication of an exogenous virus. This mechanism was originally reported in chickens (Payne et al, 1971) and has since been described in mice (Buller et al, 1987, 1988, 1990; Gardner et al, 1991) and cats (McDougall et al, 1994). The *Fv-4* or *Akvr1* gene of mice controls susceptibility to infection by ecotropic, but not amphotropic, murine leukaemia viruses (MLVs), and encodes a glycoprotein closely related to ecotropic MLV gp70 (Gardner et al, 1991; Inaguma et al, 1992; Limjoco et al, 1993). In addition to the *Fv-4* gene mice have a further locus, *Rmcf*, which encodes an endogenous Env product protecting against infection by exogenous recombinant mink cell focus-forming (MCF) viruses (Weiss, 1993).

Although interference often occurs by simple competition process at the plasma membrane, other sites of action are possible. For example, the binding of envelope glycoproteins to receptors in the endoplasmic reticulum of an infected cell can interfere with the normal translocation processes, resulting in the absence of receptor on the cell surface (Delwart and Panganiban, 1989). A further gene in mice, termed *Fv-4*, which acts at a stage after virus entry but before provirus formation, confers resistance to Friend virus-induced disease and has recently been cloned (Best et al, 1996). Sequence studies revealed that *Fv-1* was derived from the CA region of a HERV-L-related *gag* gene (Cordonnier et al, 1995; Benit et al, 1997).

Recombination and complementation

Genetic recombination of retroviruses is important in the pathogenesis of several retroviral diseases, and is classically illustrated by the development of leukaemia following MCF infection in certain mouse strains. In these mice, the transforming virus arises from recombination between endogenous viruses, namely ecotropic and MCF-like sequences (Kozak and Ruscetti, 1992). Further examples of recombination in vivo affecting disease progression are shown in Table 22.3.

Complementation, or phenotypic mixing, refers to the formation of virions containing proteins not encoded by the encapsidated genomes. Complementation is usually confined to envelope glycoproteins between distantly related viruses and frequently expands the host range of the virus (Linial and Blair, 1984). Such complementation has been shown to occur between exogenous and endogenous viruses including human immunodeficiency virus (HIV), human T cell leukaemia/lymphoma virus (HTLV) and MLVs (Canivet et al, 1990; Lusso et al, 1990; Spector et al, 1990; Landau et al, 1991).

Table 22.3 *Diseases influenced by retrovirus recombination*

Recombinant virus	Species	Disease	References
Polytropic murine leukaemia virus	Mouse	Leukaemia	Kozak and Ruscetti (1992)
Exogenous and endogenous feline leukaemia virus	Cat	Leukaemia	Pandey et al (1992) Sheets et al (1992, 1993)
Mouse mammary tumour virus	Mouse	Mammary tumour	Golovkina et al (1992)
Murine leukaemia virus	Mouse	Murine AIDS	Hugin et al (1991) Gayama et al (1995)

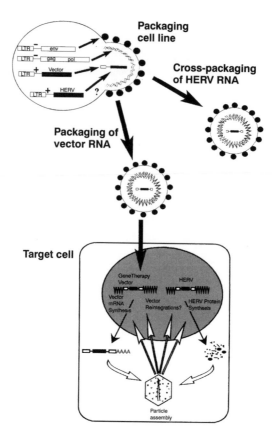

Figure 22.3 *Possible mechanism of gene therapy vector mobilization by HERV*
Coexpression of retroviral vector and HERV within a cell facilitates packaging of vector transcripts into HERV-derived virus particles leading to the possible mobilization of the vector.

Whether HERVs can phenotypically mix or recombine with human exogenous retroviruses is not known, but by analogy with the above observations the possibility cannot be dismissed.

A field of genetic research that relies on retroviral complementation is that of gene therapy. Replication-defective retroviral vectors can be used to deliver therapeutic genes into a host cell, where they are expressed in a stable manner owing to maintenance of the retroviral vector as part of the host genomic DNA. Human packaging cell lines have been developed to supply in *trans* all the structural constituents and enzymes required for assembly of a virus particle capable of a single round of infection. It will be important to examine fully the interaction of HERV with this system (Figure 22.3) as the interaction of murine ERV with murine packaging systems has been shown to have potentially pathogenic consequences (Donahue et al, 1992). Firstly, the packaging cell lines should nearly exclusively encapsidate retroviral vector transcripts and not related RNAs such as HERV. This will reduce the probability of recombination events and the production of replication-competent retroviruses. Secondly, therapeutic vectors used for gene therapy should be incapable of replication once transduced into target cells, because further rounds of replication carry with them a risk of insertional mutagenesis and recombination. Furthermore, if HERV Gag proteins could cross-package vector

RNA transcripts further replicative cycles might occur. Initial studies with these systems have proved encouraging (Patience et al, 1998), although further investigation is still required.

CONCLUSION

Although it may seem somewhat strange that a book dealing with new viruses should contain a chapter describing viruses that have been part of eukaryotic organisms for millions of years, their inclusion is apt. It is only recently that genetic techniques have been developed to allow us to begin to understand the role that these ancient elements play in the normal function of the human genome. While free passage through the germline DNA of the host is undoubtedly beneficial to the virus, the implications to the host are far more complex. Clearly endogenous retroviruses play a significant role in the structure and function of the human genome and of the genes within, and have even on occasions been utilized for a beneficial effect. The cost, however, may be the occasional pathogenic consequences of viral activation and expression.

It is clear that the field of HERV research still has much information to yield. A serious obstacle facing this research, however, is our inability to filter the interesting loci encoding biological activities from the defective or silent. It is only with continued investigation that we will fully appreciate the role that HERVs have played in our genetic heritage, and are still playing in the normal function of our genome.

REFERENCES

Aaronson SA, Hartley JW & Todaro GJ (1969) Mouse leukemia virus: 'spontaneous' release by mouse embryo cells after long-term in vitro cultivation. *Proc Natl Acad Sci USA* **64**: 87–94.

Adachi M, Watanabe-Fukunaga R & Nagata S (1993) Aberrant transcription caused by the insertion of an early transposable element in an intron of the Fas antigen gene of lpr mice. *Proc Natl Acad Sci USA* **90**: 1756–1760.

Andersson ML, Medstrand P, Yin H & Blomberg J (1996) Differential expression of human endogenous retroviral sequences similar to mouse mammary tumor virus in normal peripheral blood mononuclear cells. *AIDS Res Hum Retroviruses* **12**: 833–840.

Bangham CR, Daenke S, Phillips RE, Cruickshank JK & Bell JI (1988) Enzymatic amplification of exogenous and endogenous retroviral sequences from DNA of patients with tropical spastic paraparesis. *EMBO J* **7**: 4179–4184.

Benit L, De Parseval N, Casella JF, Callebaut I, Cordonnier A & Heidmann T (1997) Cloning of a new murine endogenous retrovirus, MuERV-L, with strong similarity to the human HERV-L element and with *gag* coding sequence closely related to the *Fv1* restriction gene. *J Virol* **71**: 5652–5657.

Benveniste RE & Todaro GJ (1974) Evolution of type C viral genes: I. Nucleic acid from baboon type C virus as a measure of divergence among primate species. *Proc Natl Acad Sci USA* **71**: 4315–4318.

Best S, Le Tissier P, Towers G & Stoye JP (1996) Positional cloning of the mouse retrovirus restriction gene *Fv1*. *Nature* **382**: 826–829.

Boller K, Konig H, Sauter M et al (1993) Evidence that HERV-K is the endogenous retrovirus sequence that codes for the human teratocarcinoma-derived retrovirus HTDV. *Virology* **196**: 349–353.

Boller K, Janssen O, Schuldes H, Tönjes RR & Kurth R (1997) Characterization of the antibody response specific for the human endogenous retrovirus HTDV/HERV-K. *J Virol* **71**: 4581–4588.

Bonner TI, O'Connell C, Cohen M, (1982) Cloned endogenous retroviral sequences from Human DNA. *Proc Natl Sci USA* **79**: 4709–4713.

Boyd MT, Bax CM, Bax BE, Bloxam DL & Weiss RA (1993) The human endogenous retrovirus ERV-3 is upregulated in differentiating placental trophoblast cells. *Virology* **196**: 905–909.

Brack-Werner R, Sander I, Leib-Mösch C, Ohlmann M, Erfle V & Werner T (1989) Human endogenous retrovirus-like sequences. In: Neth R (ed.) *Modern Trends in Human Leukemia VIII*. Berlin: Springer.

Bronson DL, Fraley EE, Fogh J & Kalter SS (1979) Induction of retrovirus particles in human testicular tumor (Tera-1) cell cultures: an electron microscopic study. *J Natl Cancer Inst* **63**: 337–339.

Buller RS, Ahmed A & Portis JL (1987) Identification of two forms of an endogenous murine retroviral *env* gene linked to the *Rmcf* locus. *J Virol* **61**: 29–34.

Buller RS, Sitbon M & Portis JL (1988) The endogenous mink cell focus-forming (MCF) gp70 linked to the *Rmcf* gene restricts MCF virus replication in vivo and provides partial resistance to erythroleukemia induced by Friend murine leukemia virus. *J Exp Med* **167**: 1535–1546.

Buller RS, Wehrly K, Portis JL & Chesebro B (1990) Host genes conferring resistance to a central nervous system disease induced by a polytropic recombinant Friend murine retrovirus. *J Virol* **64**: 493–498.

Callahan R (1988) Two families of endogenous retroviral genomes. In: Lambert M, McDonald J & Weinstein I (eds) Banbury Reports 30: *Eukaryotic Transposable Elements as Mutagenic Agents*, pp 91–100. New York: Cold Spring Harbor.

Callahan R, Drohan W, Tronick S & Schlom J (1982) Detection and cloning of human DNA sequences related to the mouse mammary tumor virus genome. *Proc Natl Acad Sci USA* **79**: 5503–5507.

Canivet M, Hoffman AD, Hardy D, Sernatinger J & Levy JA (1990) Replication of HIV-1 in a wide variety of animal cells following phenotypic mixing with murine retroviruses. *Virology* **178**: 543–551.

Chirmule N, Oyaizu N, Saxinger C & Pahwa S (1994) Nef protein of HIV-1 has B-cell stimulatory activity. *AIDS* **8**: 733–734.

Chu X, Rittner C & Schneider PM (1995) Length polymorphism of the human complement component C4 gene is due to an ancient retroviral integration. *Exp Clin Immunogenet* **12**: 74–81.

Coffin J (1982) Endogenous viruses. In: Weiss R, Teich N, Varmus H & Coffin J (eds) *RNA Tumor Viruses*, pp 1109–1204. New York: Cold Spring Harbor.

Cohen M, Powers M, O'Connell C & Kato N (1985) The nucleotide sequence of the *env* gene from the human provirus ERV3 and isolation and characterization of an ERV3-specific cDNA. *Virology* **147**: 449–458.

Conrad B, Weissmahr RN, Boni J, Arcari R, Schupbach J & Mach B (1997) A human endogenous retroviral superantigen as candidate autoimmune gene in type I diabetes. *Cell* **90**: 303–313.

Cordonnier A, Casella JF & Heidmann T (1995) Isolation of novel human endogenous retrovirus-like elements with foamy virus-related *pol* sequence. *J Virol* **69**: 5890–5897.

Dangel AW, Mendoza AR, Baker BJ et al (1994) The dichotomous size variation of human complement C4 genes is mediated by a novel family of endogenous retroviruses, which also establishes species-specific genomic patterns among Old World primates. *Immunogenetics* **40**: 425–436.

Deen KC & Sweet RW (1986) Murine mammary tumor virus *pol*-related sequences in human DNA: characterization and sequence comparison with the complete murine mammary tumor virus *pol* gene. *J Virol* **57**: 422–432.

Delwart EL & Panganiban AT (1989) Role of reticuloendotheliosis virus envelope glycoprotein in superinfection interference. *J Virol* **63:** 273–280.

De Parseval D & Heidmann T (1998) Physiological knockout of the envelope gene of the single-copy ERV-3 human endogenous retrovirus in a fraction of the Caucasian population. *J Virol* **72:** 3442–3445.

Donahue RE, Kessler SW, Bodine D et al (1992) Helper virus induced T cell lymphoma in non-human primates after retroviral mediated gene transfer. *J Exp Med* **176:** 1125–1135.

Etkind PR, Lumb K, Du J & Racevskis J (1997) Type 1 HERV-K genome is spliced into sub-genomic transcripts in the human breast tumor cell line T47D. *Virology* **234:** 304–308.

Favor J & Morawetz C (1992) Insertional mutations in mammals and mammalian cells. *Mutat Res* **284:** 53–74.

Feuchter-Murthy AE, Freeman JD & Mager DL (1993) Splicing of a human endogenous retrovirus to a novel phospholipase A_2 related gene. *Nucl Acids Res* **21:** 135–143.

Fitch DH, Bailey WJ, Tagle DA et al (1991) Duplication of the gamma-globin gene mediated by L1 long interspersed repetitive elements in an early ancestor of simian primates. *Proc Natl Acad Sci USA* 88, 7396–7400.

Gardner MB, Kozak CA & O'Brien SJ (1991) The Lake Casitas wild mouse: evolving genetic resistance to retroviral disease. *Trends Genet* **7:** 22–27.

Gattoni-Celli S, Kirsch K, Kalled S & Isselbacher KJ (1986) Expression of type C-related endogenous retroviral sequences in human colon tumors and colon cancer cell lines. *Proc Natl Acad Sci USA* 83, 6127–6131.

Gayama S, Vaupel BA & Kanagawa O (1995) Sequence heterogeneity of murine acquired immunodeficiency syndrome virus: the role of endogenous virus. *Int Immunol* **7:** 861–868.

Golovkina TV, Chervonsky A, Dudley JP & Ross SR (1992) Transgenic mouse mammary tumor virus superantigen expression prevents viral infection. *Cell* **69:** 637–645.

Goodchild NL, Wilkinson DA & Mager DL (1992) A human endogenous long terminal repeat provides a polyadenylation signal to a novel, alternatively spliced transcript in normal placenta. *Gene* **121:** 287–294.

Gotzinger N, Sauter M, Roemer K & Mueller-Lantzsch N (1996) Regulation of human endogenous retrovirus-K Gag expression in teratocarcinoma cell lines and human tumours. *J Gen Virol* **77:** 2983–2990.

Harada F, Tsukada N & Kato N (1987) Isolation of three kinds of human endogenous retrovirus-like sequences using tRNAP (Pro) as as probe. *Nucl Acids Res* **15:** 9153–9162.

Haraguchi S, Good RA & Day NK (1995) Immunosuppressive retroviral peptides: cAMP and cytokine patterns. *Immunol Today* **16:** 595–603.

Harris JR (1991) The evolution of placental mammals. *FEBS Lett* **295:** 3–4.

Harris JM, Haynes RH & McIntosh EM (1997) A consensus sequence for a functional human endogenous retrovirus K (HERV-K) dUTPase. *Biochem Cell Biol* **75:** 143–151.

Herbst H, Sauter M & Mueller-Lantzsch N (1996) Expression of human endogenous retrovirus K elements in germ cell and trophoblastic tumors. *Am J Pathol* **149:** 1727–1735.

Hirose Y, Takamatsu M & Harada F (1993) Presence of *env* genes in members of the RTVL-H family of human endogenous retrovirus-like elements. *Virology* **192:** 52–61.

Hobbs HH, Brown MS, Goldstein JL & Russell DW (1986) Deletion of exon encoding cysteine-rich repeat of low density lipoprotein receptor alters its binding specificity in a subject with familial hypercholesterolemia. *J Biol Chem* **261:** 13114–13120.

Hugin AW, Vacchio MS & Morse HC (1991) A virus-encoded 'superantigen' in a retrovirus-induced immunodeficiency syndrome of mice. *Science* **252:** 424–427.

Hynes NE, Kennedy N, Rahmsdorf U & Groner B (1981) Hormone-responsive expression of an endogenous proviral gene of mouse mammary tumor virus after molecular cloning and gene transfer into cultured cells. *Proc Natl Acad Sci USA* **78:** 2038–2042.

Inaguma Y, Yoshida T & Ikeda H (1992) Scheme for the generation of a truncated endogenous murine leukaemia virus, the *Fv-4* resistance gene. *J Gen Virol* **73**: 1925–1930.

Izui S, McConahey PJ, Theofilopoulos AN & Dixon FJ (1979) Association of circulating retroviral gp70–anti-gp70 immune complexes with murine systemic lupus erythematosus. *J Exp Med* **149**: 1099–1116.

Jenkins NA, Copeland NG, Taylor BA & Lee BK (1981) Dilute (d) coat colour mutation of DBA/2J mice is associated with the site of integration of an ecotropic MuLV genome. *Nature* **293**: 370–374.

Johansen T, Holm T & Bjorklid E (1989) Members of the RTVL-H family of human endogenous retrovirus-like elements are expressed in placenta. *Gene* **79**: 259–267.

Jones KA, Black DM, Brown MA et al (1994) The detailed characterisation of a 400 kb cosmid walk in the *BRCA1* region: identification and localisation of 10 genes including a dual-specificity phosphatase. *Hum Mol Genet* **3**: 1927–1934.

Kabat P, Tristem M, Opavsky R & Pastorek J (1996) Human endogenous retrovirus HC2 is a new member of the S71 retroviral subgroup with a full-length *pol* gene. *Virology* **226**: 83–94.

Kalter SS, Helmke RJ, Heberling RL et al (1973) Brief communication: C-type particles in normal human placentas. *J Natl Cancer Inst* **50**: 1081–1084.

Kannan P, Buettner R, Pratt DR et al (1991) Identification of a retinoic acid-inducible endogenous retroviral transcript in the human teratocarcinoma-derived cell line PA-1. *J Virol* **65**: 6343–6348.

Kato N, Pfeifer-Ohlsson S, Kato M et al (1987) Tissue-specific expression of human provirus ERV3 mRNA in human placenta: two of the three ERV3 mRNAs contain human cellular sequences. *J Virol* **61**: 2182–2191.

Kato N, Shimotohno K, Van Leeuwen D & Cohen M (1990) Human proviral mRNAs down regulated in choriocarcinoma encode a zinc finger protein related to Kruppel. *Mol Cell Biol* **10**: 4401–4405.

Kazazian HH, Wong C, Youssoufian H, et al (1988) Haemophilia A resulting from de novo insertion of L1 sequences represents a novel mechanism for mutation in man. *Nature* **332**: 164–166.

Kitamura Y, Ayukawa T, Ishikawa T, Kanda T & Yoshiike K (1996) Human endogenous retrovirus K10 encodes a functional integrase. *J Virol* **70**: 3302–3306.

Kozak CA & Ruscetti S (1992) Retroviruses in rodents. In: Levy JA (ed.) *The Retroviridae*, pp 405–482. New York: Plenum.

Krieg AM, Gause WC, Gourley MF & Steinberg AD (1989) A role for endogenous retroviral sequences in the regulation of lymphocyte activation. *J Immunol* **143**: 2448–2451.

Kurth R, Lower J, Lower R et al (1980) Onconavirus synthesis in human teratocarcinoma cultures and an increased antiviral reactivity in corresponding patients. In: *Cold Spring Harbor Conferences on Cell Proliferation*, pp 835–846. New York: Cold Spring Harbor.

Landau NR, Page KA & Littman DR (1991) Pseudotyping with human T-cell leukemia virus type I broadens the human immunodeficiency virus host range. *J Virol* **65**: 162–169.

La Mantia G, Maglione D, Pengue G et al (1991) Identification and characterization of novel human endogenous retroviral sequences preferentially expressed in undifferentiated embryonal carcinoma cells. *Nucl Acids Res* **19**: 1513–1520.

Larsson E, Andersson AC & Nilsson BO (1994) Expression of an endogenous retrovirus (ERV3 HERV-R) in human reproductive and embryonic tissues – evidence for a function for envelope gene products. *Ups J Med Sci* **99**: 113–120.

Larsson E, Venables P, Andersson AC et al (1997) Tissue and differentiation specific expression on the endogenous retrovirus ERV3 (HERV-R) in normal human tissues and during induced monocytic differentiation in the U-937 cell line. *Leukemia* **3**: 142–144.

Leib-Mösch C, Brack CR, Werner T et al (1986) Isolation of an SSAV-related endogenous sequence from human DNA. *Virology* **155**: 666–677.

Leib-Mösch C, Haltmeier M, Werner T et al (1993) Genomic distribution and transcription of solitary HERV-K LTRs. *Genomics* **18**: 261–269.

Leiter EH & Kuff EL (1984) Intracisternal type A particles in murine pancreatic B cells. Immunocytochemical demonstration of increased antigen (p73) in genetically diabetic mice. *Am J Pathol* **114**: 46–55.

Li JM, Fan WS, Horsfall AC et al (1996) The expression of human endogenous retrovirus-3 in fetal cardiac tissue and antibodies in congenital heart block. *Clin Exp Immunol* **104**: 388–393.

Limjoco TI, Dickie P, Ikeda H & Silver J (1993) Transgenic Fv-4 mice resistant to Friend virus. *J Virol* **67**: 4163–4168.

Lindeskog M, Medstrand P & Blomberg J (1993) Sequence variation of human endogenous retrovirus ERV9-related elements in an *env* region corresponding to an immunosuppressive peptide: transcription in normal and neoplastic cells. *J Virol* **67**: 1122–1126.

Linial M & Blair D (1984) Genetics of retroviruses. In: Weiss R, Teich N, Varmus H & Coffin J (eds) *RNA Tumor Viruses*, pp 649–784. New York: Cold Spring Harbor.

Löwer R, Boller K, Hasenmaier B et al (1993a) Identification of human endogenous retroviruses with complex mRNA expression and particle formation. *Proc Natl Acad Sci USA* **90**: 4480–4484.

Löwer R, Löwer J, Tondera-Koch C & Kurth R (1993b) A general method for the identification of transcribed retrovirus sequences (R-U5 PCR) reveals the expression of the human endogenous retrovirus loci HERV-H and HERV-K in teratocarcinoma cells. *Virology* **192**: 501–511.

Löwer R, Tönjes RR, Korbmacher C, Kurth R & Löwer J (1995) Identification of a Rev-related protein by analysis of spliced transcripts of the human endogenous retroviruses HTDV/HERV-K. *J Virol* **69**: 141–149.

Lusso P, Lori F & Gallo RC (1990) CD4-independent infection by human immunodeficiency virus type 1 after phenotypic mixing with human T-cell leukemia viruses. *J Virol* **64**: 6341–6344.

Maeda N (1985) Nucleotide sequence of the haptoglobin and haptoglobin-related gene pair. The haptoglobin-related gene contains a retrovirus-like element. *J Biol Chem* **260**: 6698–6709.

Mager DL & Goodchild NL (1989) Homologous recombination between the LTRs of a human retrovirus-like element causes a 5-kb deletion in two siblings. *Am J Hum Genet* **45**: 848–854.

Mager DL & Henthorn PS (1984) Identification of a retrovirus-like repetitive element in human DNA. *Proc Natl Acad Sci USA* **81**: 7510–7514.

Majors J & Varmus HE (1983) A small region of the mouse mammary tumor virus long terminal repeat confers glucocorticoid hormone regulation on a linked heterologous gene. *Proc Natl Acad Sci USA* **80**: 5866–5870.

Markert ML, Hutton JJ, Wiginton DA, States JC & Kaufman RE (1988) Adenosine deaminase (ADA) deficiency due to deletion of the ADA gene promoter and first exon by homologous recombination between two Alu elements. *J Clin Invest* **81**: 1323–1327.

Marrack P, Kushnir E & Kappler J (1991) A maternally inherited superantigen encoded by a mammary tumour virus. *Nature* **349**: 524–526.

McDougall AS, Terry A, Tzavaras T, et al (1994) Defective endogenous proviruses are expressed in feline lymphoid cells: evidence for a role in natural resistance to subgroup B feline leukemia viruses. *J Virol* **68**: 2151–2160.

Medstrand P & Blomberg J (1993) Characterization of novel reverse transcriptase encoding human endogenous retroviral sequences similar to type A and type B retroviruses: differential transcription in normal human tissues. *J Virol* **67**: 6778–6787.

Medstrand P, Lindeskog M & Blomberg J (1992) Expression of human endogenous retroviral sequences in peripheral blood mononuclear cells of healthy individuals. *J Gen Virol* **73**: 2463–2466.

Miki Y, Nishisho I, Horii A et al (1992) Disruption of the APC gene by a retrotransposal insertion of L1 sequence in a colon cancer. *Cancer Res* **52**: 643–645.

Miyazawa M, Nose M, Kawashima M & Kyogoku M (1987) Pathogenesis of arteritis of SL/Ni mice. Possible lytic effect of anti-gp70 antibodies on vascular smooth muscle cells. *J Exp Med* **166:** 890–908.

Morse B, Rotherg PG, South VJ, Spandorfer JM & Astrin SM (1988) Insertional mutagenesis of the *myc* locus by a LINE-1 sequence in a human breast carcinoma. *Nature* **333:** 87–90.

Mueller-Lantzsch N, Sauter M, Weiskircher A et al (1993) Human endogenous retroviral element K10 (HERV-K10) encodes a full-length Gag homologous 73-kDa protein and a functional protease. *AIDS Res Hum Retroviruses* **9:** 343–350.

Nakagawa K & Harrison LC (1996) The potential roles of endogenous retroviruses in autoimmunity. *Immunol Rev* **152:** 193–236.

Nelson J, Leong JA & Levy JA (1978) Normal human placentas contain RNA-directed DNA polymerase activity like that in viruses. *Proc Natl Acad Sci USA* **75:** 6263–6267.

O'Connell CD & Cohen M (1984) The long terminal repeat sequences of a novel human endogenous retrovirus. *Science* **226:** 1204–1206.

Oldstone MB (1987) Molecular mimicry and autoimmune disease. *Cell* **50:** 819–820.

Ono M (1986) Molecular cloning and long terminal repeat sequences of human endogenous retrovirus genes related to types A and B retrovirus genes. *J Virol* **58:** 937–944.

Ono M, Yasunaga T, Miyata T & Ushikubo H (1986) Nucleotide sequence of human endogenous retrovirus genome related to the mouse mammary tumor virus genome. *J Virol* **60:** 589–598.

Ono M, Kawakami M & Ushikubo H (1987) Stimulation of expression of the human endogenous retrovirus genome by female steroid hormones in human breast cancer cell line T47D. *J Virol* **61:** 2059–2062.

Pahwa S, Pahwa R, Good RA, Gallo RC & Saxinger C (1986) Stimulatory and inhibitory influences of human immunodeficiency virus on normal B lymphocytes. *Proc Natl Acad Sci USA* **83:** 9124–9128.

Palmarini M, Holland MJ, Cousens C, Dalziel RG & Sharp JM (1996) Jaagsiekte retrovirus establishes a disseminated infection of the lymphoid tissues of sheep affected by pulmonary adenomatosis. *J Gen Virol* **77:** 2991–2998.

Pandey R, Ghosh AK, Kumar DV, et al (1992) Recombination between feline leukemia virus subgroup B or C and endogenous *env* elements alters the in vitro biological activities of the viruses. *J Virol* **66:** 1.

Patience C, Simpson GR, Colletta AA, Welch HM, Weiss RA & Boyd MT (1996) Human endogenous retrovirus expression and reverse transcriptase activity in the T47D mammary carcinoma cell line. *J Virol* **70:** 2654–2657.

Patience C, Takeuchi Y & Weiss RA (1997) Infection of human cells by an endogenous retrovirus of pigs. *Nature Med* **3:** 282–286.

Patience C, Takeuchi Y, Cosset F-L & Weiss RA (1998) Packaging of endogenous retroviral sequences in retroviral vectors produced by murine and human packaging cells. *J Virol* **72:** 2671–2676.

Payne LN, Pani PK & Weiss RA (1971) A dominant epistatic gene which inhibits cellular susceptibility to RSV(RAV-O). *J Gen Virol* **13:** 455–462.

Perl A, Rosenblatt JD, Chen IS et al (1989) Detection and cloning of new HTLV-related endogenous sequences in man. *Nucl Acids Res* **17:** 6841–6854.

Pizzuti A, Pieretti M, Fenwick RG, Gibbs RA & Caskey CT (1992) A transposon-like element in the deletion-prone region of the dystrophin gene. *Genomics* **13:** 594–600.

Plotz PH (1983) Autoantibodies are anti-idiotype antibodies to antiviral antibodies. *Lancet* **ii:** 824–826.

Rabson AB, Steele PE, Garon CF & Martin MA (1983) mRNA transcripts related to full-length endogenous retroviral DNA in human cells. *Nature* **306:** 604–607.

Ranki A, Kurki P, Riepponen S & Stephansson E (1992) Antibodies to retroviral proteins in autoimmune connective tissue disease. Relation to clinical manifestations and ribonucleoprotein autoantibodies. *Arthritis Rheum* **35**: 1483–1491.

Repaske R, Steele PE, O'Neill RR, Rabson AB & Martin MA (1985) Nucleotide sequence of a full-length human endogenous retroviral segment. *J Virol* **54**: 764–772.

Samuelson LC, Wiebauer K, Snow CM & Meisler MH (1990) Retroviral and pseudogene insertion sites reveal the lineage of human salivary and pancreatic amylase genes from a single gene during primate evolution. *Mol Cell Biol* **10**: 2513–2520.

Sauter M, Schommer S, Kremmer E et al (1995) Human endogenous retrovirus K10: expression of Gag protein and detection of antibodies in patients with seminomas. *J Virol* **69**: 414–421.

Schommer S, Sauter M, Krausslich HG, Best B & Mueller-Lantzsch N (1996) Characterization of the human endogenous retrovirus K proteinase. *J Gen Virol* **77**: 375–379.

Schulte AM, Lai S, Kurtz A, et al (1996) Human trophoblast and choriocarcinoma expression of the growth factor pleiotrophin attributable to germ-line insertion of an endogenous retrovirus. *Proc Natl Acad Sci USA* **93**: 14759–14764.

Sedlacek Z, Korn B, Konecki DS et al (1993) Construction of a transcription map of a 300 kb region around the human G6PD locus by direct cDNA selection. *Hum Mol Genet* **2**: 1865–1869.

Sheets RL, Pandey R, Klement V, Grant CK & Roy-Burman P (1992) Biologically selected recombinants between feline leukemia virus (FeLV) subgroup A and an endogenous FeLV element. *Virology* **190**: 849–855.

Sheets RL, Pandey R, Jen WC & Roy-Burman P (1993) Recombinant feline leukemia virus genes detected in naturally occurring feline lymphosarcomas. *J Virol* **67**: 3118–3125.

Shih A, Misra R & Rush MG (1989) Detection of multiple, novel reverse transcriptase coding sequences in human nucleic acids: relation to primate retroviruses. *J Virol* **63**: 64–75.

Shih A, Coutavas EE & Rush MG (1991) Evolutionary implications of primate endogenous retroviruses. *Virology* **182**: 495–502.

Simpson GR, Patience C, Löwer R et al (1996) Endogenous D-type (HERV-K) related sequences are packaged into retroviral particles in the placenta and possess open reading frames for reverse transcriptase. *Virology* **222**: 451–456.

Spector DH, Wade E, Wright DA et al (1990) Human immunodeficiency virus pseudotypes with expanded cellular and species tropism. *J Virol* **64**: 2298–2308.

Steele PE, Rabson AB, Bryan T & Martin MA (1984) Distinctive termini characterize two families of human endogenous retroviral sequences. *Science* **225**: 943–947.

Stoye J & Coffin J (1985) Endogenous viruses. In: Weiss RA, Teich N, Varmus H & Coffin J (eds) *RNA Tumor Viruses*, pp 357–404. New York: Cold Spring Harbor.

Stoye JP, Fenner S, Greenoak GE, Moran C & Coffin JM (1988) Role of endogenous retroviruses as mutagens: the hairless mutation of mice. *Cell* **54**: 383–391.

Suzuki H, Hosokawa Y, Toda H, Nishikimi M & Ozawa T (1990) Common protein-binding sites in the 5'-flanking regions of human genes for cytochrome c_1 and ubiquinone-binding protein. *J Biol Chem* **265**: 8159–8163.

Szabo C, Kim YK & Mark WH (1993) The endogenous ecotropic murine retroviruses Emv-16 and Emv-17 are both capable of producing new proviral insertions in the mouse genome. *J Virol* **67**: 5704–5708.

Talal N, Flescher E & Dang H (1992) Are endogenous retroviruses involved in human autoimmune disease? *J Autoimmun* **5**: 61–66.

Tamura N, Iwase A, Suzuki K, Maruyama N & Kira S (1997) Alveolar macrophages produce the Env protein of a human endogenous retrovirus, HERV-E 4-1, in a subgroup of interstitial lung diseases. *Am J Respir Cell Mol Biol* **16**: 429–437.

Tassabehji M, Strachan T, Anderson M, Campbell RD, Collier S & Lako M (1994) Identification of a novel family of human endogenous retroviruses and characterization of one family member, HERV-K(C4), located in the complement C4 gene cluster. *Nucl Acids Res* **22**: 5211–5217.

Temin HM (1980) Origin of retroviruses from cellular moveable genetic elements. *Cell* **21:** 599–600.

Ting CN, Rosenberg MP, Snow CM, Samuelson LC & Meisler MH (1992) Endogenous retroviral sequences are required for tissue-specific expression of a human salivary amylase gene. *Genes Dev* **6:** 1457–1465.

Tomita N, Horii A, Doi S et al (1990) Transcription of human endogenous retroviral long terminal repeat (LTR) sequence in a lung cancer cell line. *Biochem Biophys Res Commun* **166:** 1–10.

Tönjes RR, Löwer R, Boller K et al (1996) HERV-K: the biologically most active human endogenous retrovirus family. *J AIDS Hum Retrovirol* **13:** S261–267.

Tönjes RR, Limbach C, Löwer R & Kurth R (1997) Expression of human endogenous retrovirus type K envelope glycoprotein in insect and mammalian cells. *J Virol* **71:** 2747–2756.

Venables PJ, Brookes SM, Griffiths D, Weiss RA & Boyd MT (1995) Abundance of an endogenous retroviral envelope protein in placental trophoblasts suggests a biological function. *Virology* **211:** 589–592.

Vidaud D, Vidaud M, Bahnak BR et al (1993) Haemophilia B due to a de novo insertion of a human-specific Alu subfamily member within the coding region of the factor IX gene. *Eur J Hum Genet* **1:** 30–36.

Vogt PK & Friis RR (1971) An avian leukosis virus related to RSV(O): properties and evidence for helper activity. *Virology* **43:** 223–234.

Wallace MR, Andersen LB, Saulino AM, et al (1991) A de novo Alu insertion results in neurofibromatosis type 1. *Nature* **353:** 864–866.

Weiss RA (1993) Cellular receptors and viral glycoproteins involved in retrovirus entry. In: Levy JA (ed.) *The Retroviridae*, pp 1–108. New York: Plenum.

Westley B & May FE (1984) The human genome contains multiple sequences of varying homology to mouse mammary tumour virus DNA. *Gene* **28:** 221–227.

Wichman HA & Van Den Bussche RA (1992) In search of retrotransposons: exploring the potential of the PCR. *Biotechniques* **13:** 258–265.

Widegren B, Kjellman C, Aminoff S, Sahlford LG & Sjogren HO (1996) The structure and phylogeny of a new family of human endogenous retroviruses. *J Gen Virol* **77:** 1631–1641.

Wilkinson DA, Goodchild NL, Saxton TM, Wood S & Mager DL (1993) Evidence for a functional subclass of the RTVL-H family of human endogenous retrovirus-like sequences. *J Virol* **67:** 2981–2989.

Wilkinson DA, Mager DL & Leong JA (1994) Endogenous human retroviruses. In: Levy JA (ed.) *The Retroviridae*, pp 465–535. New York: Plenum.

Yin H, Medstrand P, Andersson ML, et al (1997) Transcription of human endogenous retroviral sequences related to mouse mammary tumour virus in human breast and placenta: similar pattern in most malignant and non malignant breast tissues. *AIDS Res Hum Retroviruses* **13:** 507–516.

Zhu ZB, Hsieh SL, Bentley DR, Campbell RD & Volanakis JE (1992) A variable number of tandem repeats locus within the human complement C2 gene is associated with a retroposon derived from a human endogenous retrovirus. *J Exp Med* **175:** 1783–1787.

Chapter 23

NEW VIRUSES AND DISEASES OF THE HUMAN CENTRAL NERVOUS SYSTEM

Jeremy A. Garson and Jonathan R. Kerr

INTRODUCTION

This chapter describes the discovery and initial characterization of two novel viruses, multiple sclerosis-associated retrovirus (MSRV) and human Borna disease virus (BDV), which have been tentatively associated with multiple sclerosis and with affective disorders and schizophrenia, respectively. Multiple sclerosis (MS) is the most common central nervous system (CNS) disease of young adults in the UK, northern Europe, North America and Canada, with a prevalence of approximately 1 per 1000 individuals (Compston and Sadovnick, 1992). Major depressive disorder has a lifetime incidence of 15% (Kaplan and Sadock, 1996a), while bipolar affective disorder and schizophrenia have a lifetime incidence of approximately 1% (Kaplan and Sadock, 1996a,b). Although many of the findings discussed below are intriguing and highly suggestive, it should be borne in mind that research in both these fields is still at an early stage, and that further work is required before aetiological links can be firmly established.

MULTIPLE SCLEROSIS

Multiple sclerosis is characterized by focal demyelination of the CNS leading to progressive motor and sensory disability, generally over a period of many years. The disease typically starts in early adulthood and the course is usually one of relapses and remissions, although a chronic progressive evolution is well recognized in older patients. The major symptoms are of weakness, paralysis, incoordination, and sensory, visual and sphincter disturbances. Diagnosis is essentially based on clinical symptoms and signs; however, there are frequently abnormalities demonstrable on cerebrospinal fluid (CSF) analysis, recording of evoked potentials and magnetic resonance imaging (Poser et al, 1983). Postmortem histology reveals scattered plaques of demyelination in the white matter of the brain and spinal cord with preservation of axons. Old plaques are typically gliotic, while newer inflammatory lesions are characterized by oedema, disruption of the blood–brain barrier and extensive lymphoid cell infiltration.

HIV and the New Viruses Second Edition
ISBN 0-12-200741-7

Aetiological concepts

Although the precise aetiology of MS is unknown it is generally regarded as an immunopathological disorder, probably autoimmune, with putative environmental factor(s) initiating myelin destruction in genetically susceptible individuals (Martin et al, 1992). Several studies of twins have been performed in an effort to determine the relative importance of genetic and environmental factors and these have revealed a concordance rate of approximately 26% for monozygotic twins and about 2–3% for dizygotic twins (Compston and Sadovnick, 1992). These figures suggest that in addition to the genetic susceptibility (most probably polygenic) a substantial environmental effect is required. The prevalence of MS varies greatly with geographical location, generally being rare in the tropics and more common in temperate regions. Studies of migrants moving from low to high prevalence areas or vice versa suggest that whatever the nature of the environmental factor it is probably acquired before adolescence (Kurtzke, 1993). Many possible environmental factors have been proposed, including diet, climate and physical trauma, but the occurrence of apparent epidemics of MS in the Faroe Islands and elsewhere favours the possibility of an infectious agent (Kurtzke, 1993).

Viruses and multiple sclerosis

More than twenty viruses have been proposed as the possible cause of MS (Waksman, 1989; Dalgleish, 1997) including rubella virus, JC virus and several members of the *Paramyxoviridae* (e.g. paramyxovirus SV5, canine distemper virus, measles, mumps) and of the *Herpesviridae*, e.g. Marek's virus and the human herpesviruses herpes simplex virus (HSV), Epstein–Barr virus (EBV), varicella-zoster virus (VZV) and human herpesvirus 6 (HHV-6). However, in most cases the evidence remains inconclusive and the search for viral candidates continues unabated. In recent years attention has focused on retroviruses in particular. The reasons for this are, firstly, that retroviruses have been shown to cause degenerative neurological diseases (including demyelination) in a number of different species including mice, sheep, goats and non-human primates (Sigurdsson et al, 1957; Gardner, 1988) and, secondly, that the retroviruses human immunodeficiency virus (HIV-1) and human T cell leukaemia virus (HTLV-1) are both neurotropic. Indeed, the clinical similarities between some cases of HTLV-1-induced tropical spastic paraparesis (Gessain et al, 1985) and MS stimulated a number of studies in which evidence of HTLV-1 infection was sought in MS patients. Although two reports (Greenberg et al, 1989; Reddy et al, 1989) initially claimed to have detected HTLV-1 or HTLV-like sequences in patients with MS, the involvement of this retrovirus was not eventually confirmed (Ehrlich et al, 1991).

Isolation of a novel retrovirus from patients with MS

In 1989, Perron and colleagues in France isolated an unknown retrovirus (originally designated LM7) from the leptomeningeal cells of a patient with MS (Perron et al, 1989). They subsequently isolated LM7-like viruses from several other MS patients' monocytes, B lymphocytes and choroid plexus cells, and demonstrated a correlation between disease activity (i.e. relapse and exacerbation) and the production of virion-associated reverse transcriptase activity (Perron et al, 1991a,b, 1997a). As the LM7 virus had been

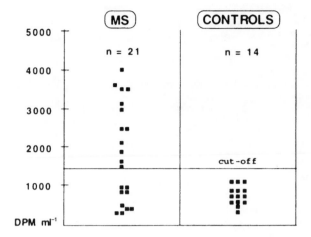

Figure 23.1 *Reverse transcriptase activity in multiple sclerosis patients*
Mean reverse transcriptase activity in monocyte culture supernatants (collected between day 3 and day 9) obtained from 21 patients with multiple sclerosis and from 14 control individuals. The cut-off is shown as a horizontal line. Adapted from Perron et al (1997a) with permission. Figure previously published in *Acta Neurologica Scandinavica.*

repeatedly isolated in vitro from MS patients but not from controls (Figure 23.1) (Perron et al, 1997a) it was renamed multiple sclerosis-associated retrovirus (MSRV). Similar findings were reported by Haahr's group in Denmark, who described retrovirus particles and associated reverse transcriptase activity in MS patient-derived lymphocyte cultures (Haahr et al, 1991, 1994).

Structural, biochemical and biological characteristics of MSRV
Electron microscopy of MSRV, concentrated by ultracentrifugation from tissue culture supernatant of the original MS patient-derived leptomeningeal cell line (Perron et al, 1989), revealed viral particles with a diameter of 100–120 nm. The location of the nucleoid in mature particles was markedly eccentric and the viral envelope bore numerous spike-like projections. Ultrathin sections of MSRV-infected cells showed the virions budding through the plasma membrane and also into cytoplasmic vacuoles (Figure 23.2). Morphologically the particles were unlike any previously described human retrovirus. Similar ultrastructural observations have subsequently been made on MSRV virions isolated from the choroid plexus cells and peripheral blood mononuclear cells (PBMCs) of a number of other patients with MS (Perron et al, 1991a,b).

The buoyant density of the MSRV virions detected by electron microscopy is approximately $1 \cdot 17$ g ml^{-1} as determined by sucrose density gradient ultracentrifugation (Perron et al, 1991a). Reverse transcriptase activity peaks in sucrose gradient fractions of the same density, suggesting that the enzyme in culture supernatants is indeed virion-associated. Optimum conditions for the assay of MSRV reverse transcriptase include use of a poly-Cm/oligo-dG$_{12-18}$ template and 20 mmol l^{-1} Mg^{2+} at pH $8 \cdot 2$. These conditions are distinct from those required by either HIV or HTLV. Antigenically MSRV also appears to be distinct from other human retroviruses and Western blotting of purified MSRV proteins using antibodies from the sera of MS patients reveals bands of 90, 65, 60, 50, 45 and 15 kDa (Perron et al, 1991a, 1992).

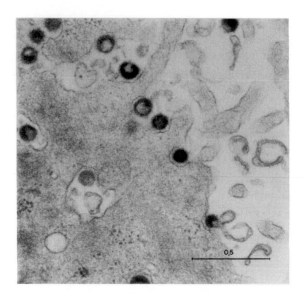

Figure 23.2 *Multiple sclerosis-associated retrovirus*
Electron micrograph of MSRV virions in a culture of human leptomeningeal cells from a patient with multiple sclerosis. A number of viral particles are seen along the cell membrane. The scale bar is 0·5 μm. Adapted from Perron et al (1989), with permission.

Transmission of MSRV to uninfected leptomeningeal and choroid plexus cells in vitro has been achieved following coculture with irradiated donor cells (Perron et al, 1992, 1997a). Serial passages of MSRV may be obtained by this means but the viral yield remains low, although a moderate increase in yield is seen in the presence of phorbol esters (Perron et al, 1989). Because MSRV infection was noted to induce interferon production in both leptomeningeal and choroid plexus cultures, attempts were made to enhance viral production by treating infected cultures with neutralizing anti-interferon antibodies. As predicted, the anti-interferon treatment resulted in an increased yield of MSRV (Perron et al, 1997a).

Although MSRV infection of primary leptomeningeal cell cultures does not produce any immediate cytopathic effects visible by light microscopy, the longevity of the cultures is reduced so that, unlike uninfected cells, they cannot be cultured beyond about 30 passages. After anti-interferon treatment, however, MSRV infection appears to result in the hyperproliferation of a subpopulation and the production of multinucleate giant cells, followed by recurrent synchronous waves of apoptosis, in both leptomeningeal and choroid plexus cultures. The possibility that MSRV may also possess oncogenic properties is suggested by the observation that the hyperproliferative subpopulation of infected, anti-interferon-treated leptomeningeal cells acquires the ability to produce neoplastic tumours when inoculated into nude mice. Significantly, anti-interferon-treated, *un*infected leptomeningeal cells fail to hyperproliferate or to produce tumours in nude mice (Perron et al, 1997a).

Interactions between MSRV and herpesviruses
The ability of herpesviruses to enhance the expression of retroviruses is well recognized and certain herpes simplex type 1 (HSV-1) immediate early gene products have been

shown to be responsible for the *trans*-activation of retroviral long terminal repeat (LTR) regulatory sequences (Mosca et al, 1987; Laurence, 1990). Superinfection with HSV-1 of MSRV-infected leptomeningeal cells was attempted in an effort to enhance MSRV expression, and this resulted in a dramatic but short-lived increase in MSRV yield prior to HSV-1-induced cell lysis. Transfection of MSRV-infected cells with plasmids expressing HSV-1 immediate early gene products ICP0 or ICP4 also resulted in enhanced MSRV expression. No enhancement was observed following transfection with plasmid expressing ICP27 or with the plasmid without insert (Perron et al, 1993). Thus it appears that herpesvirus immediate early gene products may enhance the expression of MSRV by *trans*-activation. This is of considerable interest in view of the long-standing and recently restated hypothesis (Soldan et al, 1997) that various herpesviruses (e.g. HSV-1, EBV, VZV, HHV-6) may be involved in the pathogenesis of MS (Dalgleish, 1997; Bergstrom et al, 1989). Indeed, Haahr and colleagues have proposed that EBV may act as a cofactor in MS by triggering retrovirus reactivation (Haahr et al, 1994); the demonstration by Challoner et al (1995) that human herpesvirus 6 is expressed in MS plaques is also relevant in this context.

Characterization of the MSRV genome

Despite numerous isolations of MSRV, initial attempts to characterize its genome using conventional molecular virological techniques were unsuccessful owing to inadequate viral yields obtainable in vitro. Success was eventually achieved by means of a 'panretrovirus' detection system based on the amplification, by polymerase chain reaction (PCR), of retroviral *pol* sequences with degenerate oligonucleotide primers (Tuke et al, 1997). The strategy is dependent upon the high degree of sequence conservation that occurs in a region of the *pol* gene shared by all known retroviruses (Mack and Sninsky, 1988).

The viral RNA was extracted from extracellular virions which had been purified by density gradient ultracentrifugation of supernatants from MSRV-infected leptomeningeal and choroid plexus cell cultures. Following reverse transcription, primed by random hexamers, the resulting cDNA was amplified using degenerate *pol* primers. Cloning and sequencing of the PCR product confirmed that the sequence was indeed *pol*-like, that it was novel and that it contained a potential open reading frame (Perron et al, 1997b). This *pol* fragment, designated MSRV-*cpol*, was detected only in those density gradient fractions containing the peak of reverse transcriptase activity, and only in supernatants derived from MSRV-infected, not from uninfected, cultures.

The sequence of the entire *pol* gene, and subsequently the sequence of almost all of the rest of the MSRV genome, was obtained from this initial MSRV-*cpol* fragment by sequence extension techniques (Perron et al, 1997b; H. Perron, personal communication). Comparative sequence analysis and hybridization experiments have shown that MSRV is related to the endogenous retroviral element ERV9 and more closely related to a family of MSRV-like endogenous sequences (Blond et al, 1999). It remains uncertain whether MSRV itself represents a replication-competent endogenous retrovirus capable of producing infectious virions, or an exogenous retrovirus with a closely related endogenous family – cf. mouse mammary tumour virus (MMTV) and Jaagseikte virus (Liegler and Blair, 1986; Bai et al, 1996). Phylogenetic analysis based on the amino acid sequence of the *pol* gene of MSRV suggests that it is related to the C type oncoviruses. This apparently contradicts earlier morphological observations which were more compatible with a B or D type oncovirus (Perron et al, 1989, 1992), although the *env* gene of MSRV appears to be consistent with a D type (H. Perron, personal communication). It is noteworthy in this context that D type to C type morphological conversion has been reported in association with even single amino acid substitutions (Rhee and Hunter, 1990).

A role for MSRV in the pathogenesis of MS?

Conceptual similarities exist with certain families of 'simple' retroviruses, exemplified by murine leukaemia virus (MuLV) and MMTV. Unlike most 'complex' retroviruses (e.g. HIV), these families of 'simple' retroviruses contain both exogenous and strongly homologous endogenous members. This category of agents is known to cause neurological disease (Portis, 1990) and to express 'endogenous superantigens' (Marrack et al, 1991; Rudge, 1991). Furthermore, interactions between the infectious retrovirus and the endogenous retroviral genetic background of the host can determine susceptibility or resistance to neurological disease (Gardner, 1990). It is tempting to speculate that similar interactions may be relevant in the pathogenesis of MS. Whether MSRV proves to be an exogenous retrovirus or a replication competent endogenous one, there are numerous possible mechanisms by which such an agent may be involved in the pathogenic process: antigenic mimicry (Perl and Banki, 1993; Banki et al, 1994; Wucherpfennig and Strominger, 1995); retrovirally encoded superantigens (recently implicated in diabetes) (Rudge, 1991; Conrad et al, 1997); direct neurocytotoxicity (Rasmussen et al, 1993); *trans*-activation (Perl and Banki, 1993; Rasmussen et al, 1993); and interactions with herpesviruses (Perron et al, 1993; Haahr et al, 1994; Challoner et al, 1995) can all be plausibly invoked. Nevertheless, the possibility that MSRV expression simply represents an epiphenomenon rather than a causative factor remains to be excluded.

Identification of an MSRV-associated gliotoxin

Recent observations on the toxic properties of MS patient-derived monocyte culture supernatant provide a tantalizing indication of a possible additional pathogenic mechanism. Ménard et al (1997a,b) tested monocyte culture supernatants from MS patients on rodent and human astrocyte and oligodendrocyte cultures. Intermediate filament network disorganization, leading to astrocyte and oligodendrocyte cell death, was observed with the supernatants obtained from patients with active MS but not from patients with other neurological diseases or from healthy controls. The cytotoxicity was dose dependent and correlated with the presence of both reverse transcriptase activity and MSRV-specific RNA in the monocyte culture supernatants. Partial purification and biochemical analysis of the gliotoxic factor suggests that it is a glycosylated protein with a molecular weight of approximately 17 kDa. The striking correlation ($r = 0.89$) in monocyte cultures between the level of gliocytotoxicity and the level of reverse transcriptase activity (accompanied by the presence of MSRV RNA) raises the possibility that the gliotoxin may be an MSRV-encoded protein or may be indirectly induced by MSRV. Further studies are in progress to address this possibility.

The 17 kDa gliotoxic protein has also been identified in the CSF of MS patients and has been shown to induce apoptotic cell death in astrocytes (including a human glial cell line) and oligodendrocytes, but not in fibroblasts, myoblasts, endothelial cells or neurones (Ménard et al, 1998). The cell type specificity of this potent gliotoxin is entirely compatible with its putative role in MS, since oligodendrocyte destruction would induce demyelination, and astrocyte damage would be expected to lead to impairment of the blood–brain barrier. A recently developed animal model confirms that both demyelination and disruption of the blood–brain barrier occur following a single intraventricular injection of the gliotoxin (Ménard et al, 1998).

Molecular epidemiological and seroepidemiological studies

As yet, relatively few epidemiological data exist on the prevalence of MSRV in MS patients and controls. Isolation of MSRV in tissue culture is both laborious and inefficient and therefore inappropriate for epidemiological studies. For this reason genome detection methods using PCR are currently the most suitable approach. Perron and colleagues detected MSRV RNA in the CSF of 5 of 10 patients with MS but in none of 10 patients with other neurological diseases (Perron et al, 1997b). In serum, virion-associated MSRV RNA has been detected in 9 of 17 MS patients but in only 3 of 41 controls (Garson et al, 1998). Intriguingly, in the latter study the prevalence in untreated MS patients was found to be significantly higher (100%) than the prevalence in MS patients treated with immunosuppressive drugs (27%).

Existing seroepidemiological data are equally limited. Antibodies recognizing MSRV proteins purified from tissue culture supernatant were detected by Western blotting in the serum of both of two MS patients but in neither of two controls (Perron et al, 1991a). Similarly, anti-MSRV antibodies were demonstrated by radioimmunoprecipitation assay in 12 of 20 MS sera (60%) and in 2 of 21 controls (Perron et al, 1992). Larger-scale epidemiological studies using tissue culture-derived MSRV proteins are not feasible owing to the poor viral yields obtainable in vitro, but pilot studies using recombinant MSRV proteins and synthetic peptides have given encouraging results with both serum and CSF (H. Perron, personal communication). More extensive seroepidemiological surveys using recombinant MSRV proteins and synthetic peptides are underway. It is hoped that these antibody-based techniques, in combination with PCR-based genome detection methods designed to exclude detection of confounding non-encapsidated homologous sequences, will eventually lead to a greater understanding of the relationship between MSRV and human disease (Garson et al, 1998).

AFFECTIVE DISORDERS AND SCHIZOPHRENIA

The two major diseases of affect (mood) are major depressive disorder and bipolar affective disorder. Major depressive disorder is a common condition with a lifetime incidence of 15% (Kaplan and Sadock, 1996a). Diagnosis is made clinically and the symptoms include depressed mood, loss of interest in life, weight loss, insomnia or hypersomnia, fatigue, psychomotor agitation, feelings of worthlessness and guilt, indecisiveness and suicidal ideation. Bipolar affective disorder is less common, with a lifetime incidence of about 1% (Kaplan and Sadock, 1996a). Diagnosis is clinical and symptoms alternate between those of depression, as described above, and those of mania. Manic symptoms include abnormal and persistent expansive mood, grandiosity, decreased need for sleep, flight of ideas, distractibility, increase in goal-directed activity, psychomotor agitation, and excessive involvement in pleasurable activities with a high potential for adverse consequences.

Schizophrenia is the most severe of the mental illnesses, with a lifetime incidence of 1–1·5% (Kaplan and Sadock, 1996b). Diagnosis is made clinically and symptoms include delusions, hallucinations, disorganized thought and speech, disorganized or catatonic behaviour, and affective components.

Aetiological concepts

Causative factors for major depressive disorder and bipolar affective disorder may be considered together and are divided into biological, psychosocial and genetic categories (Kaplan and Sadock, 1996a). However, this categorization is artificial in that the three may interact. First, biological factors: it is accepted that patients with mood disorders have abnormal regulation of biogenic amine neurotransmitters, principally noradrenaline (norepinephrine) and serotonin. Second, psychosocial factors: stressful life events, for example loss of a parent during childhood or the loss of a spouse, are known to precede episodes of major depression, and this is particularly true for first episodes. Third, genetic factors: family studies have shown that first-degree relatives of patients with mood disorders are at increased risk. This is most marked with bipolar affective disorder where first-degree relatives are 8–18 times more likely than controls to develop the condition (Kaplan and Sadock, 1996a). This is also supported by adoption studies. Twin studies have shown the concordance rate to be 33–90% and 50% for bipolar disorder and major depressive illness, respectively. Although genetic markers have been reported on various chromosomes, no association is consistent (Kaplan and Sadock, 1996a).

Schizophrenia probably includes a group of disorders with different causes and patients with various behavioural symptoms, natural histories and treatment responses. As with affective disorders, causative factors may be divided into biological, psychosocial and genetic factors (Kaplan and Sadock, 1996b). However, these may interact, and the stress-diathesis model postulates that a person may have a particular genetic predisposition, which when acted upon by a biological or psychosocial environmental stress allows the symptoms of schizophrenia to develop. Biological factors include an excess of dopaminergic activity, and possibly also abnormalities involving serotonin, noradrenaline (norepinephrine) and gamma-aminobutyric acid (GABA). In addition, infection, substance abuse and trauma are known predisposing factors. Psychosocial factors include low socioeconomic status and pathologically dysfunctional family relationships. Genetic predispositions include a family history of schizophrenia, and twin studies have shown a high concordance between monozygotic twins. Although genetic markers have been reported on various chromosomes, no association is consistent, and the literature may be summarized as indicating a potentially polygenic basis for schizophrenia (Kaplan and Sadock, 1996b).

Borna disease virus

Borna disease (BD), named after the town of Borna in Germany, was first described over 200 years ago as a fatal, disseminated, non-purulent meningoencephalomyelitis of sheep and horses. Borna disease virus (BDV) has recently been shown to be an enveloped, non-segmented, negative-strand RNA virus with a genome of 9 kb (Briese et al, 1994; Cubitt et al, 1994; de la Torre 1994; Schneemann et al, 1995). Six major open reading frames have been predicted from the genome sequence (de la Torre, 1994; Schneemann et al, 1995), and BDV has been classified as the prototype of the new family *Bornaviridae* within the order *Mononegavirales*, which also includes the families *Filoviridae*, *Paramyxoviridae* and *Rhabdoviridae*.

Animal infection

Although BD was originally described as a disease of horses, causing agitated, aggressive behaviour, rapidly progressing to paralysis, it has also been demonstrated in sheep, llamas, ostriches, cats and cattle (Rott and Becht, 1995). However, as a wide range of species have been experimentally infected, for example chickens, rabbits, tree shrews and rhesus macaques, it has been suggested that the host range may include all warm-blooded animals (Hatalski et al, 1997). The geographical extent of natural infection is not known, as this has been reported only in Europe, North America, Japan and Israel. Although the reservoir and mode of transmission are unknown, olfactory transmission has been suggested in naturally infected horses, as intranasal infection has been demonstrated followed by inflammation of the olfactory bulbs (Morales et al, 1988; Gosztonyi and Ludwig, 1995). However, demonstration of BDV RNA and proteins in PBMCs may indicate potential haematogenous transmission (Bode, 1995; Bode et al, 1995, 1996; Kishi et al, 1995a,b; Nakaya et al, 1996; Sauder et al, 1996; Sierra-Honigmann et al, 1996).

The Lewis rat has been selected as a rodent model as its disease resembles that in horses and sheep. Rats infected as adults show hyperactivity and exaggerated startle responses coinciding with the appearance of viral proteins in the brain (Narayan et al, 1983; Carbone et al, 1987); this is followed by dyskinesias and dystonias with distinct CNS changes, particularly in the dopamine system (Solbrig et al, 1995, 1996). Rats infected as neonates develop stunted growth, hyperactivity, learning difficulties and altered taste preferences, and fail to develop cellular immunity (Dittrich et al, 1989; Carbone et al, 1991). As experimental rodent infection leads to persistence with viral proteins in urine, faeces and saliva (Sierra-Honigmann et al, 1996), the rodent may be a natural reservoir and vector. However, natural BDV infection has not been reported in rodents. Although the pattern of infection differs between species, in all cases the virus has been found in areas of the brain that control basic emotions, such as the hippocampus.

Evidence implicating BDV in human neuropsychiatric disease

The broad host range and distribution of BDV, along with the similarity of the symptomatology in animals and human affective disorders, especially bipolar affective disorder, prompted a number studies designed to search for evidence of BDV infection in humans.

Serological evidence

The first study to link human BDV infection with neuropsychiatric illness was published by Rott et al (1985); this study examined 285 North American patients with mood disorders, 694 German psychiatric patients, and 200 normal, healthy controls. Indirect immunofluorescence assay (IFA) using a BDV-infected cell line showed serum reactivity in 4·2% of the North American patients, 0·58% of the German patients and in none of the controls. Serum reactivity was further characterized by Western blotting using BDV nucleoprotein (N) and phosphoprotein (P) from infected rabbit kidney cells (Fu et al, 1993). Anti-BDV-N antibodies were found in 53 of 138 (38%) patients with affective disorders and 19 of 117 (16%) controls. Anti-BDV-P antibodies were found in 16 of 138 (12%) patients and 5 of 117 (4%) controls. Anti-BDV-N+P antibodies were found in 9 of 138 (6·5%) patients and in only 1 of 117 (<1%) controls.

Bode et al (1993) studied 71 neuropsychiatric patients with various diagnoses (depression, paranoid psychosis, schizophrenia, anxiety, personality disorder) for serum reactivity using IFA. Anti-BDV antibodies were most frequently detected 17 days into the illness when seropositivity was seen in 37%, 25% and 6% of patients with major depression, paranoid psychosis and others, respectively. This represents a marked increase in seroprevalence from that found previously, and may be related to the fact that this study analysed anti-BDV at multiple time points in each patient. Fluorescence activated cell-sorting analysis (FACS) showed that of the 70 patients tested, more than 40% were BDV antigen carriers (Bode et al, 1994), approximately twice the number of patients who were previously shown to have anti-BDV antibodies (Bode et al, 1993).

Clinical and neurohistological similarities between BDV infection in neonatal rats and human schizophrenia prompted an investigation of serum immunoreactivity in 90 schizophrenic patients and 20 healthy controls using Western blotting based on BDV-N, P and matrix protein (M) from BDV-infected human neuroblastoma cells (Waltrip et al, 1995). Antibodies to a single protein were found in 29 of 90 (32%) patients and 4 of 20 (20%) controls. Antibodies to two or more proteins were found in 13 of 90 (14·4%) patients and in none of the controls. Antibodies to M protein were found in 12 of 90 (13·3%) patients and none in the controls. Antibodies to two or more proteins or M protein alone was associated with abnormal brain morphology detected by magnetic resonance imaging (MRI) and a clinical diagnosis of schizophrenic deficit syndrome, which is characterized by social withdrawal, neurological dysfunction and neuroanatomical abnormalities. Another independent study found an association between serum anti-BDV antibodies and cerebral atrophy in schizophrenia (Bechter et al, 1994).

Molecular evidence

Classical methods of virus purification are difficult; however, BDV RNA has been cloned from horse isolates using a molecular subtractive method (Lipkin et al, 1990; VandeWoude et al, 1990), and subsequently cloned from viral particles (Briese et al, 1994) and nuclear extracts of infected cells (Cubitt et al, 1994). Using reverse transcription polymerase chain reaction (RT-PCR) to detect BDV RNA in PBMCs, Bode et al (1995) detected BDV RNA in four of six psychiatric patients; Sauder et al (1996) detected BDV RNA of 13 of 26 (50%) neuropsychiatric patients (seven with schizophrenia and one with affective disorder); Kishi et al detected BDV RNA in 22 of 60 (37%) neuropsychiatric patients (Kishi et al 1995a) and 8 of 172 (5%) blood donors (Kishi et al, 1995b); Igata-Yi et al (1996) detected BDV RNA in 6 of 55 (11%) neuropsychiatric patients (five with schizophrenia and one with depression) but in none of 36 controls; and Nakaya et al (1996) detected BDV RNA in 3 of 25 (12%) chronic fatigue syndrome patients. Also, de la Torre et al (1996) detected BDV RNA at autopsy in the hippocampus of four of five patients with hippocampal sclerosis. In general, BDV sequences have been shown to be much less variable than other members of the *Mononegavirales*; however, the various studies of human BDV show differences in the degree of sequence conservation between isolates, which may reflect differences in RT-PCR methodologies employed (Sauder et al, 1996).

Response of bipolar affective disorder to antiviral therapy

Bode et al (1997) reported the successful antiviral treatment of a BDV-infected depressive patient with amantadine, a drug originally licensed to treat influenza A infections, and later parkinsonism. The patient was a 67-year-old woman who had suffered from bipolar affective disorder for the preceding 12 years and mild parkinsonism for the preceding 6 years. In May 1996, 3 months after onset of her last depressive episode, BDV nucleic acid and protein were detected in her blood. Daily oral amantadine treatment was started 9 days later, with 50–100 mg for 3 days, then 200 mg for 6 weeks, 100 mg for 14 weeks, and then 200 mg until the end of November 1996. From day 8 of treatment, depression dramatically improved, and this was maintained throughout follow-up. Circulating BDV antigen and RNA were eliminated 2 weeks and 6 weeks after the onset of treatment, respectively; clearance of these markers was maintained. The authors concluded that the patient's improvement, paralleled by BDV clearance, suggested that the drug's antidepressive action was the consequence of its anti-BDV activity.

In vitro, maximal inhibition of infection of young rabbit brain cells by a human strain of BDV was achieved with $0.1 \mu g \ ml^{-1}$ (10% of the dose required against influenza A virus infection), representing 25% of the blood level ($0.4 \mu g \ ml^{-1}$) following oral dosing with 200 mg per day; *meso*-inositol, a partially analogous structure, had no effect. The virus was also cleared from infected human oligodendroglial cells within 2 days (Bode et al, 1997). However, other groups have failed to demonstrate the anti-BDV activity of amantadine in vitro, and Lieb et al (1997) reported that the antidepressant effect of amantadine is well known, but that evidence suggests that this effect is mediated by binding to the N-methyl-D-aspartate (NMDA) receptor and not by anti-BDV activity.

Conclusion

While the above findings are exciting, the evidence linking BDV to human affective disorders is still speculative. Many of the findings have not been confirmed by independent research groups. It has also been claimed that some results achieved using RT-PCR may be due to laboratory contamination (Mestel, 1997). Even though BDV may infect humans with associated clinical consequences which are similar to those following animal infection, there are still unanswered questions. Reservoirs of BDV and the routes of transmission to humans are unknown. In addition, although BDV infection has been demonstrated in neuropsychiatric patients, clear links with distinct disease entities have not yet emerged. To address these issues, multicentre studies are underway using standardized methodologies.

ACKNOWLEDGMENTS

The authors are grateful to H. Perron, P. W. Tuke and W. Preiser for constructive criticism of the manuscript, and to bioMérieux SA and the Multiple Sclerosis Society of Great Britain and Northern Ireland for funding much of the work described in this chapter.

REFERENCES

Bai J, Zhu RY, Stedman K et al (1996) Unique long terminal repeat U3 sequences distinguish exogenous Jaagsiekte sheep retroviruses associated with ovine pulmonary carcinoma from endogenous loci in the sheep genome. *J Virol* **70**: 3159–3168.

Banki K, Colombo E, Sia F et al (1994) Oligodendrocyte-specific expression and autoantigenicity of transaldolase in multiple sclerosis. *J Exp Med* **180**: 1649–1663.

Bechter K, Bauer M, Estler HC et al (1994) Expanded nuclear magnetic resonance studies in Borna disease virus seropositive patients and control probands. *Nervenarzt* **65**: 169–174.

Bergstrom T, Andersen O & Vahlne A (1989) Isolation of herpesvirus type 1 during first attack of multiple sclerosis. *Ann Neurol* **26**: 283–285.

Blond J-L, Besème F, Duret L et al (1999) Molecular characterization and placental expression of HERV-W, a new human endogenous retrovirus family. *J Virol* **73**: (in press).

Bode L (1995) Human infections with Borna disease virus and potential pathologic implications. *Curr Top Microbiol Immunol* **190**: 103–130.

Bode L, Ferzst R & Czech G (1993) Borna disease virus infection and affective disorders in man. *Arch Virol* **7(suppl.)**: 159–167.

Bode L, Steinbach F & Ludwig H (1994) A novel marker for Borna disease virus infection. *Lancet* **343**: 297–298.

Bode L, Zimmermann W, Ferszt R et al (1995) Borna disease virus genome transcribed and expressed in psychiatric patients. *Nature Med* **1**: 232–236.

Bode L, Durrwald R, Rantam FA et al (1996) First isolates of infectious human Borna disease virus from patients with mood disorders. *Mol Psychiatry* **1**: 200–212.

Bode L, Dietrich DE, Stoyloff R et al (1997) Amantadine and Borna disease virus in vitro and in vivo in an infected patient with bipolar depression. *Lancet* **349**: 178–179.

Briese T, Schneemann A, Lewis AJ et al (1994) Genomic organization of Borna disease virus. *Proc Natl Acad Sci USA* **91**: 4362–4366.

Carbone K, Duchala C, Griffin J et al (1987) Pathogenesis of Borna disease in rats: evidence that intra-axonal spread is the major route for virus dissemination and the determination for disease incubation. *J Virol* **61**: 3431–3440.

Carbone K, Park S, Rubin S et al (1991) Borna disease: association with a maturation defect in the cellular immune response. *J Virol* **65**: 6154–6164.

Challoner PB, Smith KT, Parker JD et al (1995) Plaque-associated expression of human herpesvirus 6 in multiple sclerosis. *Proc Natl Acad Sci USA* **92**: 7440–7444.

Compston A & Sadovnick AD (1992) Epidemiology and genetics of multiple sclerosis. *Curr Opin Neurol Neurosurg* **5**: 175–181.

Conrad B, Weissmahr RN, Boni J et al (1997) A human endogenous retroviral superantigen as candidate autoimmune gene in type 1 diabetes. *Cell* **90**: 303–313.

Cubitt B, Oldstone C & de la Torre JC (1994) Sequence and genome organisation of Borna disease virus. *J Virol* **68**: 1382–1396.

Dalgleish AG (1997) Viruses and multiple sclerosis. *Acta Neurol Scand* **95 (suppl. 169)**: 8–15.

De la Torre JC (1994) Molecular biology of Borna disease virus: prototype of a new group of animal viruses. *J Virol* **68**: 7669–7675.

De la Torre JC, Gonzalez-Dunia D, Cubitt B et al (1996) Detection of Borna disease virus antigen and RNA in human autopsy brain samples from neuropsychiatric patients. *Virology* **223**: 272–282.

Dittrich W, Bode L, Ludwig H et al (1989) Learning deficiencies in Borna disease virus-infected but clinically healthy rats. *Biol Psychiatry* **26**: 818–828.

Ehrlich GD, Glaser JB, Bryz-Gornia V et al (1991) Multiple sclerosis, retroviruses, and PCR. *Neurology* **41**: 335–343.

Fu ZF, Amsterdam JD, Kao M et al (1993) Detection of Borna disease virus-reactive antibodies from patients with affective disorders by Western immunoblot technique. *J Affect Disord* **27:** 61–68.

Gardner MB (1988) Neurotropic retroviruses of wild mice and macaques. *Ann Neurol* **23:** 5201–5206.

Gardner MB (1990) Genetic resistance to a retroviral neurologic disease in wild mice. *Curr Top Microbiol Immunol* **160:** 3–10.

Garson JA, Tuke PW, Giraud P et al (1998) Detection of virion associated MSRV-RNA in the serum of patients with multiple sclerosis. *Lancet* **351:** 33.

Gessain A, Bairn F, Vernant JC et al (1985) Antibodies to human lymphotropic virus type I in patients with tropical spastic paraparesis. *Lancet* **ii:** 407–410.

Gosztonyi G & Ludwig H (1995) Borna disease – neuropathology and pathogenesis. *Curr Top Microbiol Immunol* **190:** 39–73.

Greenberg SJ, Ehrlich GD, Abbott MA et al (1989) Detection of sequences homologous to human retroviral DNA in multiple sclerosis by gene amplification. *Proc Natl Acad Sci USA* **86:** 2879–2882.

Haahr S, Sommerlund M, Moller-Larsen A et al (1991) Just another dubious virus in cells from a patient with multiple sclerosis? *Lancet* **337:** 863–864.

Haahr S, Sommerlund M, Christensen T et al (1994) A putative new retrovirus associated with multiple sclerosis and the possible involvement of Epstein–Barr virus in this disease. *Ann NY Acad Sci* **724:** 148–156.

Hatalski CG, Lewis AJ & Lipkin WI (1997) Borna disease. *Emerg Infect Dis* **3:** 129–135.

Igata-Yi R, Kazunari Y, Yoshiki K et al (1996) Borna disease virus and consumption of raw horse meat. *Nature Med* **2:** 948–949.

Kaplan HI & Sadock BJ (1996a) Mood disorders. In: Kaplan HI & Sadock BJ (eds) *Concise Textbook of Clinical Psychiatry*, pp 159–188. Baltimore: Williams & Wilkins.

Kaplan HI & Sadock BJ (1996b) Schizophrenia. In: Kaplan HI & Sadock BJ (eds) *Concise Textbook of Clinical Psychiatry*, pp 121–138. Baltimore: Williams & Wilkins.

Kishi M, Nakaya T, Nakamura Y et al (1995a) Demonstration of human Borna disease virus RNA in human peripheral blood mononuclear cells. *FEBS Lett* **3645:** 293–297.

Kishi M, Nakaya T, Nakamura Y et al (1995b) Prevalence of Borna disease virus RNA in peripheral blood mononuclear cells from blood donors. *Med Microbiol Immunol* **184:** 135–138.

Kurtzke JF (1993) Epidemiological evidence for multiple sclerosis as an infection. *Clin Microbiol Rev* **6:** 382–427.

Laurence J (1990) Molecular interactions among herpesviruses and human immunodeficiency viruses. *J Infect Dis* **162:** 338–346.

Lieb K, Hufert FT, Bechter K et al (1997) Depression, Borna disease and amantadine. *Lancet* **349:** 178–179.

Liegler TJ & Blair PB (1986) Direct detection of exogenous mouse mammary tumour virus sequences in lymphoid cells of BALB/cfC3H female mice. *J Virol* **59:** 159–162.

Lipkin WI, Travis G, Carbone K et al (1990) Isolation and characterisation of Borna disease agent cDNA clones. *Proc Natl Acad Sci USA* **87:** 4184–4188.

Mack DH & Sninsky JJ (1988) A sensitive method for the identification of uncharacterized viruses related to known virus groups: hepadnavirus model system. *Proc Natl Acad Sci USA* **85:** 6977–6981.

Marrack P, Kushnir E & Kappler J (1991) A maternally inherited superantigen encoded by a mammary tumor virus. *Nature* **349:** 524–526.

Martin R, McFarland HF & McFarlin DE (1992) Immunological aspects of demyelinating disease. *Annu Rev Immunol* **10:** 153–187.

Ménard A, Amouri R, Michel M et al (1997a) Gliotoxicity, reverse transcriptase activity and retroviral RNA in monocyte/macrophage culture supernatants from patients with multiple sclerosis. *FEBS Lett* **413:** 477–485.

Ménard A, Paranhos-Baccala G, Pelletier J et al (1997b) A cytotoxic factor for glial cells: a new avenue of research for multiple sclerosis? *Cell Mol Biol* **43**: 889–901.

Ménard A, Amouri R, Dobransky T et al (1998) A gliotoxic factor and multiple sclerosis. *J Neurol Sci* **154**: 209–221.

Mestel R (1997) Mind-altering bugs. *New Scientist* **2099**: 42–45.

Morales JA, Herzog S, Kompter C et al (1988) Axonal transport of Borna disease virus along olfactory pathways in spontaneously and experimentally infected rats. *Med Microbiol Immunol* **177**: 51–68.

Mosca JD, Bednarik DP, Raj NBK et al (1987) Herpes simplex virus type-1 can reactivate transcription of latent human immunodeficiency virus. *Nature* **325**: 67–70.

Nakaya T, Takahashi H, Nakamura Y et al (1996) Demonstration of Borna disease virus RNA in peripheral blood mononuclear cells derived from Japanese patients with chronic fatigue syndrome. *FEBS Lett* **378**: 145–149.

Narayan O, Herzog S, Frese K et al (1983) Behavioural disease in rats caused by immunopathological responses to persistent Borna virus in the brain. *Science* **220**: 1401–1403.

Perl A & Banki K (1993) Human endogenous retroviral elements and autoimmunity: data and concepts. *Trends Microbiol* **1**: 153–156.

Perron H, Geny C, Laurent A et al (1989) Leptomeningeal cell line from multiple sclerosis with reverse transcriptase activity and viral particles. *Res Virol* **140**: 551–561.

Perron H, Lalande B, Gratacap B et al (1991a) Isolation of retrovirus from patients with multiple sclerosis. *Lancet* **337**: 862–863.

Perron H, Geny C, Gratacap B et al (1991b) Isolations of an unknown retrovirus from CSF, blood and brain cells of patients with multiple sclerosis. In: Wietholter H et al (eds) *Current Concepts in Multiple Sclerosis*, pp 111–116. Amsterdam: Elsevier.

Perron H, Gratacap B, Lalande B et al (1992) *In vitro* transmission and antigenicity of a retrovirus isolated from a multiple sclerosis patient. *Res Virol* **143**: 337–350.

Perron H, Suh M, Lalande B et al (1993) Herpes simplex virus ICP0 and ICP4 immediate early proteins strongly enhance expression of a retrovirus harboured by a leptomeningeal cell line from a patient with multiple sclerosis. *J Gen Virol* **74**: 65–72.

Perron H, Firouzi R, Garson JA et al (1997a) Cell cultures and associated retrovirus in multiple sclerosis. *Acta Neurol Scand* **95 (suppl. 169)**: 22–31.

Perron H, Garson JA, Bedin F et al (1997b) Molecular identification of a novel retrovirus repeatedly isolated from patients with multiple sclerosis. *Proc Natl Acad Sci USA* **94**: 7583–7588.

Portis JL (1990) Wild mouse retrovirus: pathogenesis. *Curr Top Microbiol Immunol* **160**: 11–27.

Poser CM, Paty DW, Scheinberg L et al (1983) New diagnostic criteria for multiple sclerosis: guidelines for research protocols. *Ann Neurol* **13**: 227–231.

Rasmussen HB, Perron H & Clausen J (1993) Do endogenous retroviruses have etiological implications in inflammatory and degenerative nervous system diseases? *Acta Neurol Scand* **88**: 190–198.

Reddy EP, Sardberg-Wollheim M et al (1989) Amplification and molecular cloning of HTLV-1 sequences from DNA of multiple sclerosis patients. *Science* **243**: 529–533.

Rhee SS & Hunter E (1990) A single amino acid substitution within the matrix protein of a D-type retrovirus converts its morphogenesis to that of a C-type retrovirus. *Cell* **63**: 77–86.

Rott R & Becht H (1995) Natural and experimental Borna disease in animals. Borna disease. *Curr Top Microbiol Immunol* **190**: 17–30.

Rott R, Herzog S, Fleischer B et al (1985) Detection of serum antibodies to Borna disease virus in patients with psychiatric disorders. *Science* **228**: 755–756.

Rudge P (1991) Does a retrovirally encoded superantigen cause multiple sclerosis? *J Neurol Neurosurg Psychiatry* **54**: 853–855.

Sauder C, Muller A, Cubitt B et al (1996) Detection of Borna disease virus (BDV) antibodies and BDV RNA in psychiatric patients: evidence for high sequence conservation of human blood-derived BDV RNA. *J Virol* **70**: 7713–7724.

Schneemann A, Schneider PA, Lamb RA et al (1995) The remarkable coding strategy of Borna disease virus: a new member of the non-segmented negative strand RNA viruses. *Virology* **210:** 1–8.

Sierra-Honigmann AM, Rubin SA, Estafanous MG et al (1996) Borna disease virus in peripheral blood mononuclear and bone marrow cells of neonatally and chronically infected rats. *J Neuroimmunol* **45:** 31–36.

Sigurdsson B, Palsson PA & Grimsson H (1957) Visna, a demyelinating transmissible disease of sheep. *J Neuropathol Exp Neurol* **16:** 389–403.

Solbrig MV, Fallon JH & Lipkin WI (1995) Behavioural disturbances and pharmacology of Borna disease virus. *Curr Top Microbiol Immunol* **190:** 93–102.

Solbrig MV, Koob GF, Joyce JN et al (1996) A neural substrate of hyperactivity in Borna disease: changes in dopamine receptors. *Virology* **222:** 332–338.

Soldan SS, Berti R, Salem N et al (1997) Association of human herpes virus 6 (HHV-6) with multiple sclerosis: increased IgM response to HHV-6 early antigen and detection of serum HHV-6 DNA. *Nature Med* **3:** 1394–1397.

Tuke PW, Perron H, Bedin F et al (1997) Development of a 'pan-retrovirus' detection system for multiple sclerosis studies. *Acta Neurol Scand* **95** (suppl. 169): 16–21.

VandeWoude S, Richt J, Zink M et al (1990) A Borna virus cDNA encoding a protein recognized by antibodies in humans with behavioural diseases. *Science* **250:** 1276–1281.

Waksman BH (1989) Multiple sclerosis: relationship to a retrovirus? *Nature* **337:** 599.

Waltrip RW, Buchanan RW, Summerfelt A et al (1995) Borna disease virus and schizophrenia. *Psychiatry Res* **56:** 33–44.

Wucherpfennig KW & Strominger JL (1995) Molecular mimicry in T-cell mediated autoimmunity: viral peptides activate human T-cell clones specific for myelin basic protein. *Cell* **80:** 695–705.

Chapter 24

EMERGING VIRUSES:
Their Diseases and Identification

Paul Kellam

INTRODUCTION

Diseases caused by microbial infections have been present throughout human evolution. A large proportion are the result of virus infections. Since the 1970s, new human viruses have been identified which cause widely varying diseases. These viruses fall into the well-documented field of emerging infectious diseases. In fact, three distinct situations are included in this category: (a) resurgent or recurrent old diseases (usually caused by 'new' or mutated previously known agents); (b) well-known human diseases with epidemiological evidence of transmission, but with the discovery of a newly identified agent; or (c) diseases truly new to humans but caused by pre-existing zoonotic agents.

EMERGING VIRUS DISEASES

A commonly cited example of resurgent or recurrent disease is the yearly appearance of new antigenically different influenza viruses. These new variants are able to evoke disease in their host while causing the centuries-old symptoms of influenza. The variation that occurs in influenza viruses is of two kinds. The first, called antigenic drift, occurs as a result of accumulation of point mutations in the virus surface proteins, haemagglutinin (H) and neuraminidase (N). This can lead to viral escape from the immune response and the emergence of new influenza epidemics. The second kind of change, called antigenic shift, occurs when complete genome segments encoding the H or N genes of one virus variant are replaced by the corresponding segments from another variant. This process is known as reassortment. Both wildfowl and swine are endemically infected with influenza virus, and, periodically, avian and human influenza viruses co-infect pigs. It is therefore thought that pigs act as a mixing pot in the formation of pandemic influenza variants by allowing reassortment between avian, swine and human influenza viruses to occur. However, the 1918 'Spanish' pandemic strain of influenza was most likely derived solely from strains that infected humans and swine, not the avian subgroup (Taubenberger et al, 1997). This suggests that close observation of new pig influenza viruses could highlight the presence of potential new human pandemic variants.

HIV and the New Viruses Second Edition
ISBN 0-12-200741-7

New infectious diseases falling into the latter two categories – new disease or new agent – continue to be identified. Table 24.1 lists the major human viral diseases identified since 1977. This table also includes new human viruses for which a link with a specific human disease has not been conclusively established, for example hepatitis G virus (Simons et al, 1995a,b; Linnen et al, 1996), Borna disease virus (Lipkin et al, 1990; VandeWoude et al, 1990) and human herpesvirus 7 (Frenkel et al, 1990). Also included are subgenomic viral sequences identified in human tissue, namely human retrovirus 5 (HRV-5) (Griffiths et al, 1997) and multiple sclerosis-associated retrovirus (MSRV) (Perron et al, 1997), which have not yet yielded complete virus genomes.

Contributors to this book have described many viral diseases truly new in humans, but most probably caused by a zoonosis. Human immunodeficiency virus types 1 and 2 (HIV-1, HIV-2) have homology to Old World monkey virus, simian immunodeficiency virus (SIV), suggesting that a recent zoonosis resulted in the appearance of the new human disease of acquired immune deficiency syndrome (AIDS) (Allan et al, 1991; Myers et al, 1992). Another example of a new zoonotic infection, hantavirus pulmonary

Table 24.1 *Viruses associated with human disease identified since 1977*

Year	Agent	Disease	References
1977	Ebola virus	Ebola haemorrhagic fever	Johnson et al (1977)
1977	Hantaan virus	Haemorrhagic fever with renal syndrome	Lee et al (1978)
1980	Human T cell lymphotropic virus type 1 (HTLV-1)	T cell lymphoma/leukaemia	Poiesz et al (1980)
1982	Human T cell lymphotropic virus type 2 (HTLV-2)	Hairy cell leukaemia	Kalyanaraman et al (1981)
1983	Human papillomavirus types 16 and 18	Cervical cancer	Durst et al (1983)
1983	Human immunodeficiency virus type 1 (HIV-1)	AIDS	Barre Sinoussi et al (1983)
1986	Human immunodeficiency virus type 2 (HIV-2)	AIDS	Clavel et al (1986)
1988	Human herpesvirus 6 (HHV-6)	Roseola subitum	Yamanishi et al (1988)
1989	Hepatitis C virus (HCV)	Non-A, non-B hepatitis	Choo et al (1989)
1990	Borna disease virus	NAD	VandeWoude et al (1990) Lipkin et al (1990)
1990	Hepatitis E virus (HEV)	Acute faecal–oral hepatitis	Reyes et al (1990)
1990	Human herpesvirus 7 (HHV-7)	NAD	Frenkel et al (1990)
1991	Guantanto virus	Venezuelan haemorrhagic fever	Salas et al (1991)
1993	Hantavirus (SNV)	Hantavirus pulmonary syndrome (HPS)	Nichol et al (1993)
1994	Sabia virus	Brazilian haemorrhagic fever	Lisieux et al (1994)
1994	Human herpesvirus 8 (HHV-8)	Kaposi's sarcoma	Chang et al (1994)
1995	Hepatitis G virus (HGV, GBV-C)	NAD	Linnen et al (1996) Simons et al (1995a,b)
1997	Multiple sclerosis-associated retrovirus	Multiple sclerosis	Perron et al (1997)
1997	Human retrovirus 5 (HRV-5)	NAD	Griffiths et al (1997)

*NAD, no associated human disease.

syndrome (HPS), demonstrates how modern molecular virology can be instrumental in identifying new human pathogens. In May 1993 an outbreak of respiratory illness with a mortality rate in excess of 75% was reported in a border region of New Mexico, Arizona, Utah and Colorado in southwestern USA. Serology surveys initially failed to identify known agents associated with respiratory disease but did detect cross-reactive antibodies to a previously characterized hantavirus. This was thought unusual, as hantaviruses had not previously been associated with human disease in North America, nor had they been associated with a severe predominantly respiratory illness. Using degenerate PCR primers to conserved regions of the hantavirus genome led to the identification of a new hantavirus that differed from known hantaviruses by 30% at the nucleotide level (Nichol et al, 1993). The usual hosts for hantaviruses are rodents, and trapping of rodents in the Four Corners area revealed 33% of deer mice were seropositive for the new hantavirus. The outbreak of HPS was subsequently shown to be associated with a 10-fold increase in the deer mouse population in the Four Corners area. Four months after the HPS outbreak a culture system was established and the hantavirus responsible for HPS was finally designated Sin Nombre virus (SNV) (Elliott et al, 1994).

An example of a newly identified virus causing disease in humans with no currently known animal or insect host is provided by hepatitis C virus (HCV). Epidemiological evidence suggested that the causes of non-A, non-B hepatitis (NANBH) were of an infectious origin. However, conventional virological techniques had failed to identify the agent, even though evidence suggested NANBH was caused by a blood-borne, small enveloped virus, readily transmissible to chimpanzees. The hepatitis C virus was discovered by screening a cDNA expression library prepared from virus particles pelleted from a chimpanzee with a high NANBH virus titre. When the library was screened with serum from a patient with chronic NANBH one clone was identified from a library of one million clones (Choo et al, 1989). Identification of this clone led to the rapid characterization of the entire virus genome, and HCV was formally assigned to a new virus subfamily within the family *Flaviviridae*. Serology assays have now demonstrated that HCV is the major cause of NANBH throughout the world (Kuo et al, 1989; Linnen et al, 1996).

These examples show clearly how modern molecular biology techniques can be used to identify completely new viruses. These viruses can be associated with a new disease, or associated with a well-characterized disease present in humans for many years. In identifying an emerging virus, one is often presented with epidemiological data and clinical specimens that have no reactivity with diagnostic reagents available for known pathogens. The primary aim therefore is to identify any new infectious agent and build a body of data to support the existence of a causal link between organism and disease.

IDENTIFICATION OF EMERGING VIRUSES

Virus isolation and culture in vitro have long been the definitive method of new virus characterization and this is often the first method employed. However, in many viral diseases this approach may have several limitations. The main limitation is the need for a permissive host cell to propagate the new virus. Equally important is an effective detection method to show infection of the permissive cells. Classically, this was achieved through the observation of cytopathic effect or the detection of specific viral gene products. Isolation in cell culture will not work if the virus does not grow in vitro, or if no detectable cytopathic effect or viral gene product is produced. Indeed, in the case of HPS, discovery of a new hantavirus preceded the establishment of a tissue culture system for virus propagation (Elliott et al, 1994). For HCV, good culture systems still do not exist 10 years after the discovery of the virus, and keratinocytes permissive for human papillomavirus type16 replication are still not suitable for routine virus isolation 16 years after its discovery by molecular hybridization (Durst et al, 1983).

Electron microscopy (EM) of disease tissue and in vitro infected cell cultures is another method employed. It may reveal virus particles and common morphological features of a particular virus family, leading to clues for the further characterization of the virus. However, EM can suffer from artifactual structures present in the prepared material. In the late 1960s and throughout the 1970s, a number of groups reported the detection by EM of virus-like particles in human tumours (Dmochowski et al, 1968; Feller and Chopra, 1971; Seman et al, 1971; Denner and Dorfman, 1977). In contrast, Dalton (1975) demonstrated that many of these virus-like particles may simply be the result of breakdown products from normal cells. Also, the assumption that virus architecture is constant in a family of viruses is not always true, with viruses classified by EM requiring reclassification following molecular analysis of genome structure. One example of this is the *Arterivirus* genus of the coronavirus-like (CVL) superfamily. Equine anaemia virus (EAV – the prototype of the genus), lactate dehydrogenase-elevating virus (LDV), simian haemorrhagic fever virus (SHFV) and porcine reproductive and respiratory syndrome virus (PRRSV) are morphologically similar to viruses of the *Togaviridae*, and together with their genome size this resulted in their initial classification in the togavirus family. However, the arterivirus genome organization and replication strategy, along with the homology of proteins such as the RNA replicase, have since resulted in the reclassification of arteriviruses within the CVL superfamily. The final taxonomic fate of this superfamily is still to be decided (Snijder and Horzinek, 1993; Snijder et al, 1993). Electron microscopy can be successful when a completely new structure is observed, as with the observation of the worm-like virions of Marburg virus. This virus was later classified as the type species of the *Filovirus* family (Kiley et al, 1982).

Other standard techniques based on immunofluorescence assays use patient sera to detect viral antigens in infected pathological specimens. This was used successfully to detect SNV in hantavirus pulmonary syndrome. However, if the patient sera or clinical specimens do not react with known virus groups or virus-specific antibodies, this method provides little useful information. The tendency to use increasing specificity in routine diagnostic tests works against identification of related yet distinct virus strains.

A combination of immunological and molecular biological techniques has been used successfully to identify new human viruses. A general scheme for immunoscreening is outlined in Figure 24.1A. This relies on the production of high-quality cDNA expression libraries derived from infected tissue or, in the case of positive-sense RNA viruses, directly from virus present in plasma. Three hepatitis viruses have been identified using this technique. Hepatitis C virus was identified by Choo et al (1989) following the isolation of a virus-specific cDNA clone from NANB-infected chimpanzee plasma as described earlier. The same approach was successful in identifying hepatitis G virus (Linnen et al, 1996); this virus was found to be almost identical to the virus GBV-C, another human hepatitis-associated virus (Simons et al, 1995b), identified using degen-

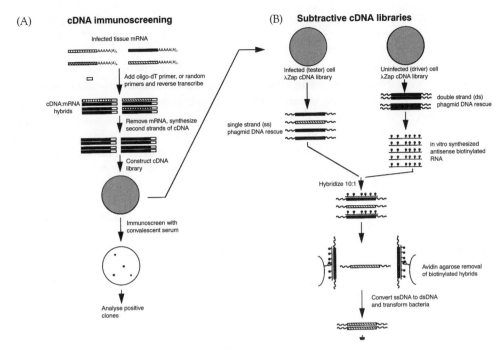

Figure 24.1 *The cDNA library approach to identifying new viruses*
Different cDNA library approaches to the identification of unculturable infectious agents. (A) Using cDNA immunoscreening. Modified from Gao and Moore (1996) with permission. (B) Using subtractive cDNA libraries, based on the method of Schweinfest et al (1990). Filled long boxes represent cDNA.

erate PCR primers. In addition hepatitis E virus (HEV)-specific cDNA clones were identified from a cDNA library derived from total RNA isolated from the bile juice of two experimentally infected rhesus monkeys (Uchida et al, 1992). An absolute requirement for cDNA immunoscreening is the availability of high antibody titre immune serum from patients or experimentally infected animals. However, these sera may also contain many antibodies that cross-react with human antigens expressed in the library, leading to false-positive clones being identified. This can be a particular problem for diseases in which autoantibodies are common. The requirement for a large representative cDNA library made in the absence of knowledge of the virus genome entails the use of randomly primed cDNA rather than conventional oligo-dT primed libraries. This was particularly relevant in the case of HCV where the viral positive-sense RNA genome contains a polyuridine rather than a polyadenine 3′ tail (Kolykhalov et al, 1996).

SEQUENCE-BASED METHODS OF IDENTIFYING NEW VIRUSES

The use of powerful molecular-based methods to analyse well-characterized biological specimens has created a new era of molecular identification of emerging viruses. The major advantage of molecular methodologies is the ability to look rapidly for new viruses, known viruses, or related but previously undetected members of established virus families. This was resoundingly successful for the discovery of genital papilloma-viruses (Durst et al, 1983).

With the widespread use of the polymerase chain reaction (PCR) and the availability of extensive sequence databases of virus genomes it is often possible to design PCR primers to conserved regions of virus genomes. It is then possible to survey samples with these primers to look for the presence of a given virus in new pathological speci-mens. These are very powerful identification tools but as such must be applied with care. The most common problems encountered are the nature of the assay conditions, where in addition to the general problems of PCR contamination, small modifications can dramatically alter the sensitivity of the PCR signal produced. Without appropriate controls, conditions can be accidentally contrived to amplify and detect many irrelevant DNA sequences from a disease. In addition, detection of virus genomes in disease tis-sue does not automatically produce a link between virus and disease (Fredericks and Relman, 1996). This has particularly relevance to PCR screening for known or new viruses. For example, a number of known infectious agents have been implicated in the pathogenesis of multiple sclerosis (MS) (Allen and Brankin, 1993; Kurtzke, 1993; Challoner et al, 1995; Perron et al, 1997), but none of these associations has been con-clusively proved to cause MS (Rice, 1992).

To look for new members of virus families, degenerate PCR primers can be designed to conserved regions of virus genomes at the nucleic acid level or, preferably, can be designed to conserved amino acid regions of virus proteins based on codon degeneracy for each amino acid. The latter cover a much wider variety of nucleic acid sequences and are therefore more divergent when used to look for new virus genomes. Many groups have published degenerate PCR primers to diverse virus families (Table 24.2). These have been used successfully to identify many new viruses ranging from human viruses such as GBV-C (hepatitis G virus) (Simons et al, 1995a), SNV (Nichol et al, 1993) and HRV-5 (Griffiths et al, 1997), to many animal viruses, pig endogenous retrovirus (PERV) (Patience et al, 1997), walleye dermal sarcoma virus (Zhang and Martineau, 1996) and a macaque gammaherpesvirus, retroperitoneal fibromatosis herpesvirus (RFHV Mn) (Rose et al, 1997). However, the lack of specific controls for all degenerate primers means standardization can only be achieved on existing members of virus families, thereby not guaranteeing detection of an unknown virus. Again, the problems of inap-propriate amplification conditions can lead to the production of many false positives or, in some cases, no amplification. It is therefore necessary to optimize carefully degen-erate PCR conditions on relevant controls. This usually requires optimization of reac-tion buffers and primer annealing temperatures, and often the use of nested or seminested PCR strategies. In addition, special amplification techniques such as 'Touch-down' PCR can often improve PCR amplification when using degenerate primers (Don et al, 1991; Zhang and Martineau, 1996).

The most recent array of molecular techniques to be adapted for use in new virus dis-covery are purely nucleic acid-based and make no assumptions about the nature of the viral agent present. These methods have evolved from one aim of molecular medicine, which is to identify differences at the nucleic acid level between disease-associated tis-sue and normal tissue. These methodologies have in common that one nucleic acid pop-ulation ('uninfected' or 'driver') is hybridized in excess with a second population ('infected' or 'tested') to remove common sequences, thereby enriching target sequences unique to the tester. The methods can be broadly divided into physical subtraction techniques and PCR-based kinetic enrichment techniques. Each has its own relative merits which are ultimately dependent on the nature of the nucleic acid sample to be analysed.

Physical subtraction techniques

Physical subtraction techniques are applicable only to detecting differences in mRNA expression between one cell type and another. Early uses of these techniques simply involved the solution hybridization of an excess of driver mRNA with cDNA made from tester mRNA. Common sequences present in both samples form cDNA:mRNA hybrids, leaving the unique sequences in the tester cDNA unhybridized. The double-stranded hybrids are removed by hydroxyapatite chromatography, exploiting the higher affinity of hydroxyapatite for double-stranded nucleic acids. The remaining sub-tracted cDNA is used to construct a subtracted cDNA library or, if labelled, as an enriched cDNA probe for library screening. This method was used in combination with immunoscreening to identify the virus associated with Borna disease (Lipkin et al, 1990; VandeWoude et al, 1990). Borna disease is an infectious neurological disease that occurs sporadically in horses and sheep in central Europe, and may also be associated with cer-tain human neuropsychiatric disorders (Bode et al, 1995).

Modern versions of this technique have been developed that use cDNA libraries con-structed in the cloning vector λZap (Stratagene, La Jolla, USA). The libraries are made from driver and tester cell lines or tissue samples (Figure 24.1B). This method has been used to isolate rare cDNAs (less than 0.01% abundance) from colon and hepatic carci-noma tissue (Schweinfest et al, 1990). The λZap vectors allow the rescue, directly from the library, of single-stranded DNA phagemids and DNA (ssDNA) or double-stranded DNA plasmids (dsDNA), all containing representative cDNAs. Initial methods made use of non-directionally cloned cDNA libraries (Schweinfest et al, 1990). However, newer vectors allowing the production of directionally cloned cDNAs can be used in a modified version of this technique (Figure 24.1B).

Both primary cDNA libraries are amplified and approximately 2 million plaque-form-ing units of the tester library are used to produce single-strand phage DNA. The same amount of driver library is used to produce double-stranded plasmid DNA. In vitro transcribed RNA incorporating dUTP-biotin is made from the driver dsDNA using a T3 promoter located in the vector. These biotinylated RNAs are complementary to the tester ssDNA and are subsequently hybridized to the tester ssDNA. The complemen-tary hybrids are removed using streptavidin beads. Multiple rounds of hybridization and subtraction can be performed, enriching for rare cDNAs in the tester population. Following subtraction the tester ssDNA is made double-stranded and transformed into *Escherichia coli*. This represents the enriched subtracted library which can then be analysed further.

PCR-based techniques

Representational difference analysis
Representational difference analysis (RDA) represented the first global approach to the analysis of differences between cellular genomes. Although originally developed to look for differences between tumour cell genomic DNA and normal cells (Lisitsyn et al,

Table 24.2 Degenerate PCR primers

Virus family	Gene	Primer name	Primary PCR primers (5' to 3')*	Nested/seminested PCR primers (5' to 3')	Reference
Retrovirus	RT	5'MOP-1 3'MOP-1	TGGAAAGTGYTRCCMCARGG GGMGGCCAGCAGSAKGTCATCCAYGTA		Shih et al (1989)
Retrovirus	RT	5'MOP-2 3'MOP-2 5'MOP-1 3'MOP-2	CCWTGGAATACTCCYRTWTT GTCKGAACCAATTWATATYYCC	TGGAAAGTGYTRCCMCARGG GGMGGCCAGCAGSAKGTCATC CAYGTA	Li et al (1996)
Retrovirus (A, B, D type)	RTing and protease	ABDPOL ABDPRO	TCCCCTTGGAATACTCCTGTTTTYGT CATTCCTGTGTGTAAAACTTTCCAYTG		Medstrand et al (1992)
Retrovirus (BLV, HTLV)	Integrase	110(+) 111(−)	CCCTACAATCCCACMAGCYTCRG RTGGTKATTTSCCATCKGGTYTT		Dube et al (1997)
Retrovirus (lentiviruses)	RT	LV1 LV2 LV3 DDMY	CCGGATCCDCAYCCNGSAGGAYTAMAA GGTCTAGAYRYARTTCATAACCCAKCCA	CCGGATCCGAYRTRGGKGAYGCMTA CCGGATCCRCRTCRTCCATRTA	Gelman et al (1992)
Herpesvirus	Pol[a]	DFA ILK KG1 TGV	GAYTTYGCNAGYYTNTAYCC TCCTGGACAAGCARNYSGCNMTNAA GTCTTGCTCACCAGNTCNACNCCYTT	TGTAACTCGGTGTAYGGNTTYACNG GNGT CACAGAGTCCGTRTCNCCRTADAT	VanDevanter et al (1996)
Herpesvirus	Pol[b]	IYG DFASA GDTD1B VYGA PCLNA KMLEA	GTGTTYGACTTYGCNAGYYTNTAYCC CGGCATGCGACAAACACGGAGTCNG TRTCNCCRTA	ACGTGCAACGCGGTGTAYGGNKTNA CNGG GTCGCCCTTGGCATCCTNCCNTGYC TNAA CAGGGCCGGAAGATGCTGGARACRT CNCARGC	Rose et al (1997)
Papillomavirus	L1[c]	GP17 GP18 GPR22	CGGGATCCGGNMGNGGNCARCCNY TNGG CGGGATCCAYNCCRTTRTTRTGNCCYTG	ARGAYGGNGAYATGRYNGAYAYNGG NTWYGG	Snijders et al (1991)
Coronavirus	S gene	55 56	GGAKAAGGTKAATGARTGYGT CCAKACVTACCAWGGCCAYTT		Tobler and Ackermann (1996)

Virus	Gene / region	Primer	Sequence	Reference
Hantavirus	G2[d]	+2548	GATATGAATGATTGYTTTGT	Nichol et al (1993)
		−2859	CCATCAGGGTCTYTCCA	
		+2590	TGTATAATTGGGACWGTATCTAA	
		−2751	GCAAAGTTACATTTYTTCCT	
		+2671	TTTAAGCAATGGTGYACTACWAC	
		−3108	CCATAACACATWGCAGC	
		+2770		
		−3012		
Morbillivirus	P gene[c]	UPPF	ATGTTTATGATCACAGCGGT	Barrett et al (1993)
		UPPR	ATTGGGTTGCACCACTTGTC	Shimizu et al (1994)
		MBV1	TATGCTGGGTGAAAGTAAGATCT	
		MBV3	GGATTGCTGAAATGATTTGTGAT	
		MBV2	AGAAAGAAATGTGCATTTGC	
			CCTGAACCCCATGCHCCATC	
Caliciviruses	orf1[f]	NVp110	ACDATYTCATCATCACCATA	Le Guyader et al (1996)
		NVp35	CTTGTTGGTTTGAGGCCATAT	
		NVp36	ATAAAAGTTGGCATGAACA	
		NVp69	GGCCTGCCATCTGGATTGCC	
		NI	GAATTCCATCGCCCACTGGCT	
			TTTGGCATTGAAACTATGTATCC	
Hepatitis C viruses[g]	NS3 helicase	5'	TYGCYACKGCKACCCCHCCKG	Simons et al (1995a)
		3'	TGCCMGCTYTCCCMCKGCC	
		3'	CRATRGTRAWRGTMGGGTCMAGG	
Flavivirus	NS5	FG1	TCAAGGAACTCCACACATGAGATGTACT	Fulop et al (1993)
		FG2	TGTATGCTGACACAGCAGGATGGGACAC	

*R = A or G; Y = C or T; M = A or C; S = C or G; W = A or T; K = G or T; D = G, T or A; H = A, C or T; V = A, C or G; N = all four nucleotides.

[a] 5' primers DFA and ILK are used in combination with 3' primer KG1.

[b] For pan-herpesvirus degenerate PCR, primary PCR is performed with primers DFSA and GDTD1B followed by seminested PCR with primers VYGA and GDTD1B. For gammaherpesvirus degenerate PCR, primary PCR is performed with primers VYGA and GDTD1B followed by two seminested PCRs with primers PCLNA and GDTD1B followed by primers KMLEA and GDTD1B.

[c] Oligonucleotide GPR22 is used as a radioactive probe to detect PCR products.

[d] Nested primer set +2548, −2859, +2590, −2751 detects Hantaan and Seoul serotype viruses and primer set +2671, −3108, +2770, −3012 detects Hantaan and Puumala

1993; Lisitsyn and Wigler, 1995), it has been adapted to identify differences in mRNA expression (Hubank and Schatz, 1994). The technique has been used successfully to identify new human viruses associated with Kaposi's sarcoma, namely human herpesvirus 8 (Chang et al, 1994) and the viruses GBV-A and B (Simons et al, 1995b), which are related to GBV-C/HGV (Simons et al, 1995a; Linnen et al, 1996), associated with viral hepatitis. More controversially, RDA has been used to identify human herpesvirus 6 sequences in plaques from patients with multiple sclerosis (Challoner et al, 1995) and cDNA clones from infectious Creutzfeldt–Jakob brain fractions that have no homology to any known database sequences (Dron and Manuelidis, 1996).

Representational difference analysis combines three elements: representation, subtractive enrichment and kinetic enrichment. The procedure is carried out in two stages. The first comprises the preparation of representations for driver and tester DNAs or cDNAs. Small restriction fragments derived from the starting nucleic acid are ligated to oligonucleotide adaptors and amplified by PCR. The second stage comprises the reiterative hybridization/selection steps (Figure 24.2A). Prior to hybridization, the oligonucleotide adaptors used for the initial representation PCR step are cleaved from both driver and tester amplicons and a new set of defined but different sequence adaptors is ligated onto the 5' ends of only the tester amplicons. After hybridization of tester and driver amplicons the mixture of molecules is treated with DNA polymerase. This adds a copy of the defined oligonucleotide to both 3' ends of only the self-annealed tester DNA fragments. The defined oligonucleotide adaptor/primer is then used during PCR of the mixture such that only the tester-annealed DNA fragments can participate in exponential amplification to yield a difference product. The cycle of cleavage of old adaptors and ligating new adaptors to the difference products, hybridizing with excess driver and PCR amplification is repeated two or three times and the final PCR products are cloned and analysed further.

Suppression subtraction hybridization

Suppression subtraction hybridization (SSH), based on similar principles to RDA, was described in June 1996 (Diatchenko et al, 1996). This technique is designed selectively to amplify differentially expressed cDNA fragments and simultaneously to suppress non-targeted DNA amplification. Like RDA, it can be used on a variety of nucleic acid targets (Figure 24.2B), and in both methods representations of both tester and driver DNAs are prepared by restriction enzyme digest. Tester DNA is then subdivided into two portions and each is ligated with a different oligonucleotide adaptor to the 5' end. However, in contrast to RDA, SSH involves two hybridization steps. In the first, an excess of denatured driver cDNA is added to each population of denatured tester cDNA. Owing to second-order reaction kinetics of hybridization, single-stranded molecules corresponding to high- and low-abundance sequences become normalized. Normalization occurs because the annealing process is faster for more abundant molecules and results in a proportion of the low-abundance tester cDNA remaining single-stranded. During the second hybridization the two primary hybrid samples are mixed together. Since the samples are not heat-denatured, only the remaining normalized and subtracted single-strand tester cDNAs are able to associate to form new hybrids. These hybrids have different oligonucleotide adaptor sequences at their 5' ends. Fresh denatured driver DNA is then added to enrich further for the differentially expressed sequences. Following hybridization and DNA polymerase end filling, the entire population is PCR amplified

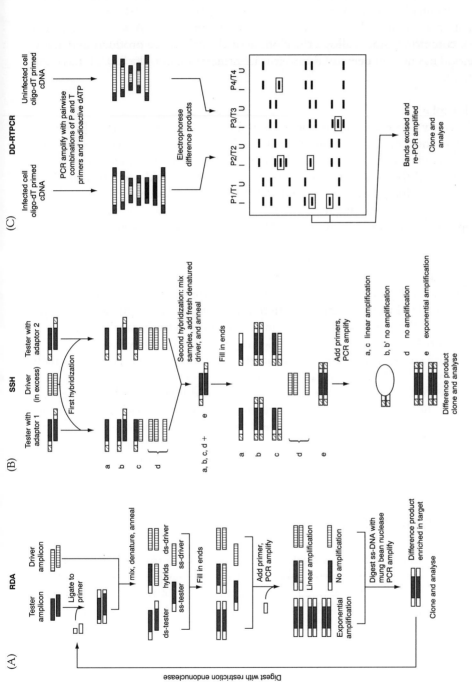

Figure 24.2 The PCR approach to identifying new viruses

Different sequence-based PCR methods used to identify unculturable infectious agents. (A) Representation difference analysis (RDA). Modified from Gao and Moore (1996) with permission. (B) Suppression subtraction hybridization (SSH). Modified from Diatchenko et al (1996) with permission. (C) Differential display RT PCR (DD-RTPCR). Long boxes represent DNA or cDNA, small boxes represent oligonucleotide primers. Note that in (B) two types of oligonucleotide adaptors are used, represented by different shading. Each set of adaptors has a common sequence shown by an open box corresponding to nested PCR primer sites. In (C), 'T primers' refer to composite oligo-dT primers and 'P primers' refer to composite 5' primers.

with oligonucleotides for both adaptors. Hybrids with the same adaptor at either end are suppressed from PCR amplification owing to the formation of panhandle structures between the terminally repeated primer sites (Diatchenko et al, 1996). Only hybrids with different adaptors at either end are exponentially amplified. A final nested PCR with internal common primers allows the cloning of the difference products and analysis. This method has not yet been used for virus identification, but like RDA it has enormous potential.

Differential display

Differential display reverse transcription PCR (DD-RTPCR), described by Liang and Pardee (1992), is only applicable to differences in mRNA expressed between cells and is essentially semirandomly primed PCR. This method is based on the assumption that every individual mRNA molecule can be reverse transcribed and amplified by PCR. Using the original principles described, DD-RTPCR has now been further developed and refined (Ayala et al, 1995) (Figure 24.2C).

Four composite 3′ primers containing a 10 base common sequence at the 5′ end followed by polyT and ending with a mixed base (A, C or G) plus a 3′ fixed base were designed. One such improved primer set is illustrated in Figure 24.3 (Ayala et al, 1995).

These primers are used to prime first strand cDNA synthesis of the driver and tester mRNA or total RNA. In addition, composite arbitrary primers containing the sequence CGTGAATTCG added to the 5′ end of different random 10-mer sequences were also designed. These are then used with the 3′ primers in a radioactively labelled PCR reaction involving initial low-stringency cycles followed by high-stringency cycles. The PCR products are resolved on either native or denaturing polyacrylamide gels. Differentially expressed or novel mRNAs are represented by specific bands in the tester tracks of the gels compared with the driver. The greater the number of different 5′ arbitrary primers used, the higher the chance of observing differences. However, as each new 5′ primer produces eight PCR reactions (four for tester and four for driver), numbers can become limiting. Any differentially displayed band is excised from the original gel, reamplified, cloned and studied further. Like SSH, DD-RTPCR has yet to result in the identification of a new virus.

NEW VIRUS DISCOVERY

The right combination of difference methods needs to be used for the variety of samples that are likely to arise in new virus discovery. If the biological samples available are cell-free, the choice of methods is limited. With no knowledge of the nature of the viral genome (i.e. DNA, positive-sense RNA or negative-sense RNA), only degenerate PCR, RDA or SSH methods are applicable. For infected cell lines or tissue samples, all methods are applicable. Again, if the nature of a new viral genome is not known, the most

$$5' \text{ CGGAATTCGGT}_{12} \text{ MA (M = A, C or G)}$$
$$\text{MC}$$
$$\text{MG}$$
$$\text{MT}$$

Figure 24.3 *Improved primer set for DD-RTPCR*

global method of analysis would be to study new mRNA species. The use of physical subtraction methods requires a relatively abundant supply of polyA+ mRNA which may limit the technique. However, since this method avoids PCR and thus PCR contamination, a problem in all of the PCR-based methods, it is still very useful. Another disadvantage of the physical subtraction method is the requirement for cDNA libraries, which are costly and time-consuming to make.

Both RDA and SSH have the disadvantages of requiring complex restriction digests, primer ligations, hybridization and PCR. Contamination with common laboratory DNA at the initial stages can lead to false positive clones. In addition, failure at any one stage can lead to no difference PCR products or false-positives which are impossible to control for when working with authentic samples. The RDA method is time-consuming compared with SSH, and SSH also requires less initial tester and driver nucleic acid than RDA. However, like DD-RTPCR, SSH has yet to be proved in the field of new virus discovery.

Differential display is very sensitive and prone to false positives, although no more so than other subtractive methods (Wan et al, 1996). The method also has the disadvantage of primer dependency for PCR amplification. Owing to the use of polyT primers DD-RTPCR does not allow detection of positive-sense virus genomes with alternative tails to polyA. However, it is possible to design and use alternative polynucleotide primers to cover such diverse virus groups. Overall, DD-RTPCR is quick and provides a qualitative result of the differences between two samples.

Following the identification of a candidate clone, further analysis to characterize whether the candidate is truly exogenous in origin is required. Northern and Southern blot analysis and/or PCR should reveal whether a clone is exogenous. Analysis of open reading frames should show whether the clone encodes for any known homologous protein and whether these are related to known viruses. However, as with many major advances in technology, sequenced-based approaches for new virus identification can lead to misleading conclusions. In particular, proving the link between a disease and an unculturable virus identified from nucleic acid sequences has stimulated much debate. This has led to the formation of revised guidelines for defining the causal relationship between a virus and a disease (Fredericks and Relman, 1996). The presence of viral sequences within disease tissue in the absence of other corroborating data is insufficient to prove that a new virus causes a particular disease. Detection of viruses may reflect either the presence of virus in surrounding cells and tissues, or the ability of a virus to replicate within the microenvironment of the disease tissue, rather than the virus causing the observed disease. In the example of hantavirus pulmonary syndrome, layers of evidence are compiled to reach the conclusion of disease causation. The opposite has been true for HGV. Following the virus discovery, the accumulation of more data now indicates that the virus is not the cause of hepatitis in humans but rather is a virus from a subgroup of individuals in whom it causes no apparent disease (Alter et al, 1997a,b).

The list of diseases with possible microbial aetiology is still large and new pathogens are likely to be identified for some of these by the molecular biological methods described. Specific diseases such as Kawasaki syndrome (Kawasaki et al, 1974), sarcoidosis (Newman et al, 1997) and multiple sclerosis (Allen and Brankin, 1993), as well as more general disease groups, such as neurodegenerative disorders, forms of arthritis, inflammatory bowel disease, autoimmune disease and certain cancers, may have an infectious origin. What may make new aetiological agents hard to define will be those agents that are part of a multifactorial origin of disease. Disease may result from multiple viral exposure, as is the case for dengue fever. Alternatively, viruses may act as triggers for a disease, or cause disease following long-term chronic asymptomatic infections. Clearly, molecular biological approaches should help unravel the causes of many preexisting diseases, while also enabling the rapid identification of any new, virulent, emerging infectious disease.

REFERENCES

Allan JS, Short M, Taylor ME et al (1991) Species-specific diversity among simian immunodeficiency viruses from African green monkeys. *J Virol* **65**: 2816–2828.

Allen I & Brankin B (1993) Pathogenesis of multiple sclerosis – the immune diathesis and the role of viruses. *J Neuropathol Exp Neurol* **52**: 95–105.

Alter HJ, Nakatsuji Y, Melpolder J et al (1997a) The incidence of transfusion-associated hepatitis G virus infection and its relation to liver disease. *N Engl J Med* **336**: 747–754.

Alter MJ, Gallagher M, Morris TT et al (1997b) Acute non-A-E hepatitis in the United States and the role of hepatitis G virus infection. Sentinel Counties Viral Hepatitis Study Team. *N Engl J Med* **336**: 741–746.

Ayala M, Balint RF, Fernandez-de-Cossio ME, Canaan-Haden L, Larrick JW & Gavilondo JV (1995) New primer strategy improves precision of differential display. *BioTechniques* **18**: 842–850.

Barre Sinoussi F, Chermann JC, Rey F et al (1983) Isolation of a T-lymphotropic retrovirus from a patient at risk for acquired immune deficiency syndrome (AIDS). *Science* **220**: 868–871.

Barrett T, Visser IK, Mamaev L, Goatley L, van Bressem MF & Osterhaust AD (1993) Dolphin and porpoise morbilliviruses are genetically distinct from phocine distemper virus. *Virology* **193**: 1010–1012.

Bode L, Zimmermann W, Ferszt R, Steinbach F & Ludwig H (1995) Borna disease virus genome transcribed and expressed in psychiatric patients. *Nature Med* **1**: 232–236.

Challoner PB, Smith KT, Parker JD et al (1995) Plaque-associated expression of human herpesvirus 6 in multiple sclerosis. *Proc Natl Acad Sci USA* **92**: 7440–7444.

Chang Y, Cesarman E, Pessin MS et al (1994) Identification of herpesvirus-like DNA sequences in AIDS-associated Kaposi's sarcoma. *Science* **266**: 1865–1869.

Choo QL, Kuo G, Weiner AJ, Overby LR, Bradley DW & Houghton M (1989) Isolation of a cDNA clone derived from a blood-borne non-A, non-B viral hepatitis genome. *Science* **244**: 359–362.

Clavel F, Guetard D, Brun Vezinet F et al (1986) Isolation of a new human retrovirus from West African patients with AIDS. *Science* **233**: 343–346.

Dalton AJ (1975) Microvesicles and vesicles of multivesicular bodies versus 'virus-like' particles. *J Natl Cancer Inst* **54**: 1137–1148.

Denner J & Dorfman NA (1977) Small virus-like particles in leukosis-like syndrome induced by certain antigens and immunostimulators. *Acta Biol Med Ger* **36**: 1451–1458.

Diatchenko L, Lau YF, Campbell AP et al (1996) Suppression subtractive hybridization: a method for generating differentially regulated or tissue-specific cDNA probes and libraries. *Proc Natl Acad Sci USA* **93**: 6025–6030.

Dmochowski L, Langford PL, Williams WC, Liebelt AG & Liebelt RA (1968) Electron microscopic and bioassay studies of milk from mice of high and low mammary-cancer and high and low leukemia strains. *J Natl Cancer Inst* **40**: 1339–1358.

Don RH, Cox PT, Wainwright BJ, Baker K & Mattick JS (1991) 'Touchdown' PCR to circumvent spurious priming during gene amplification. *Nucl Acids Res* **19**: 4008.

Dron M & Manuelidis L (1996) Visualization of viral candidate cDNAs in infectious brain fractions from Creutzfeldt–Jakob disease by representational difference analysis. *J Neurovirol* **2:** 240–248.

Dube S, Bachman S, Spicer T et al (1997) Degenerate and specific PCR assays for the detection of bovine leukaemia virus and primate T cell leukaemia/lymphoma virus *pol* DNA and RNA: phylogenetic comparisons of amplified sequences from cattle and primates from around the world. *J Gen Virol* **78:** 1389–1398.

Durst M, Gissmann L, Ikenberg H & zur Hausen H (1983) A papillomavirus DNA from a cervical carcinoma and its prevalence in cancer biopsy samples from different geographic regions. *Proc Natl Acad Sci USA* **80:** 3812–3815.

Elliott LH, Ksiazek TG, Rollin PE et al (1994) Isolation of the causative agent of hantavirus pulmonary syndrome. *Am J Trop Med Hyg* **51:** 102–108.

Feller WF & Chopra HC (1971) Virus-like particles in human milk. *Cancer* **28:** 1425–1430.

Fredericks DN & Relman DA (1996) Sequence-based identification of microbial pathogens: a reconsideration of Koch's postulates. *Clin Microbiol Rev* **9:** 18–33.

Frenkel N, Schirmer EC, Wyatt LS et al (1990) Isolation of a new herpesvirus from human CD4+ T cells [published erratum appears in *Proc Natl Acad Sci USA* (1990) **87(19):** 7797]. *Proc Natl Acad Sci USA* **87:** 748–752.

Fulop L, Barrett AD, Phillpotts R, Martin K, Leslie D & Titball RW (1993) Rapid identification of flaviviruses based on conserved NS5 gene sequences. *J Virol Methods* **44:** 179–188.

Gao SJ & Moore PS (1996) Molecular approaches to the identification of unculturable infectious agents. *Emerg Infect Dis* **2:** 159–167.

Gelman IH, Zhang J, Hailman E, Hanafusa H & Morse SS (1992) Identification and evaluation of new primer sets for the detection of lentivirus proviral DNA. *AIDS Res Hum Retroviruses* **8:** 1981–1989.

Griffiths DJ, Venables PJW, Weiss RA & Boyd MT (1997) A novel exogenous retrovirus sequence identified in humans. *J Virol* **71:** 2866–2872.

Hubank M & Schatz DG (1994) Identifying differences in mRNA expression by representational difference analysis of cDNA. *Nucl Acids Res* **22:** 5640–5648.

Johnson KM, Lange JV, Webb PA & Murphy FA (1977) Isolation and partial characterisation of a new virus causing acute haemorrhagic fever in Zaire. *Lancet* **i:** 569–571.

Kalyanaraman VS, Sarngadharan MG, Poiesz B, Ruscetti FW & Gallo RC (1981) Immunological properties of a type C retrovirus isolated from cultured human T-lymphoma cells and comparison to other mammalian retroviruses. *J Virol* **38:** 906–915.

Kawasaki T, Kosaki F, Okawa S, Shigematsu I & Yanagawa H (1974) A new infantile acute febrile mucocutaneous lymph node syndrome (MLNS) prevailing in Japan. *Pediatrics* **54:** 271–276.

Kiley MP, Bowen ET, Eddy GA et al (1982) *Filoviridae*: a taxonomic home for Marburg and Ebola viruses? *Intervirology* **18:** 24–32.

Kolykhalov AA, Feinstone SM & Rice CM (1996) Identification of a highly conserved sequence element at the 3' terminus of hepatitis C virus genome RNA. *J Virol* **70:** 3363–3371.

Kuo G, Choo QL, Alter HJ et al (1989) An assay for circulating antibodies to a major etiologic virus of human non-A, non-B hepatitis. *Science* **244:** 362–364.

Kurtzke JF (1993) Epidemiologic evidence for multiple sclerosis as an infection [published erratum appears in *Clin Microbiol Rev* (1994) **7(1):** 141]. *Clin Microbiol Rev* **6:** 382–427.

Lee HW, Lee PW & Johnson KM (1978) Isolation of the etiologic agent of Korean hemorrhagic fever. *J Infect Dis* **137:** 298–308.

Le Guyader F, Estes MK, Hardy ME et al (1996) Evaluation of a degenerate primer for the PCR detection of human caliciviruses. *Arch Virol* **141:** 2225–2235.

Li MD, Lemke TD, Bronson DL & Faras AJ (1996) Synthesis and analysis of a 640-bp *pol* region of novel human endogenous retroviral sequences and their evolutionary relationships. *Virology* **217:** 1–10.

Liang P & Pardee AB (1992) Differential display of eukaryotic messenger RNA by means of the polymerase chain reaction. *Science* **257:** 967–971.

Linnen J, Wages J, Zhang Keck ZY et al (1996) Molecular cloning and disease association of hepatitis G virus: a transfusion-transmissible agent. *Science* **271:** 505–508.

Lipkin WI, Travis GH, Carbone KM & Wilson MC (1990) Isolation and characterization of Borna disease agent cDNA clones. *Proc Natl Acad Sci USA* **87:** 4184–4188.

Lisieux T, Coimbra M, Nassar ES et al (1994) New arenavirus isolated in Brazil. *Lancet* **343:** 391–392.

Lisitsyn N & Wigler M (1995) Representational difference analysis in detection of genetic lesions in cancer. *Methods Enzymol* **254:** 291–304.

Lisitsyn N, Lisitsyn N & Wigler M (1993) Cloning the differences between two complex genomes. *Science* **259:** 946–951.

Medstrand P, Lindeskog M & Blomberg J (1992) Expression of human endogenous retroviral sequences in peripheral blood mononuclear cells of healthy individuals. *J Gen Virol* **73:** 2463–2466.

Myers G, MacInnes K & Korber B (1992) The emergence of simian/human immunodeficiency viruses. *AIDS Res Hum Retroviruses* **8:** 373–386.

Newman LS, Rose CS & Maier LA (1997) Sarcoidosis. *N Engl J Med* **336:** 1224–1234.

Nichol ST, Spiropoulou CF, Morzunov S et al (1993) Genetic identification of a hantavirus associated with an outbreak of acute respiratory illness. *Science* **262:** 914–917.

Patience C, Takeuchi Y & Weiss RA (1997) Infection of human cells by an endogenous retrovirus of pigs. *Nature Med* **3:** 282–286.

Perron H, Garson JA, Bedin F et al (1997) Molecular identification of a novel retrovirus repeatedly isolated from patients with multiple sclerosis. The Collaborative Research Group on Multiple Sclerosis. *Proc Natl Acad Sci USA* **94:** 7583–7588.

Poiesz BJ, Ruscetti FW, Gazdar AF, Bunn PA, Minna JD & Gallo RC (1980) Detection and isolation of type C retrovirus particles from fresh and cultured lymphocytes of a patient with cutaneous T-cell lymphoma. *Proc Natl Acad Sci USA* **77:** 7415–7419.

Reyes GR, Purdy MA, Kim JP et al (1990) Isolation of a cDNA from the virus responsible for enterically transmitted non-A, non-B hepatitis. *Science* **247:** 1335–1339.

Rice GP (1992) Virus-induced demyelination in man: models for multiple sclerosis. *Curr Opin Neurol Neurosurg* **5:** 188–194.

Rose TM, Strand KB, Schultz ER et al (1997) Identification of two homologs of the Kaposi's sarcoma-associated herpesvirus (human herpesvirus 8) in retroperitoneal fibromatosis of different macaque species. *J Virol* **71:** 4138–4144.

Salas R, de Manzione N, Tesh RB et al (1991) Venezuelan haemorrhagic fever. *Lancet* **338:** 1033–1036.

Schweinfest CW, Henderson KW, Gu JR et al (1990) Subtraction hybridization cDNA libraries from colon carcinoma and hepatic cancer. *Genet Anal Tech Appl* **7:** 64–70.

Seman G, Gallager HS, Lukeman JM & Dmochowski L (1971) Studies on the presence of particles resembling RNA virus particles in human breast tumors, pleural effusions, their tissue cultures, and milk. *Cancer* **28:** 1431–1442.

Shih A, Misra R & Rush MG (1989) Detection of multiple, novel reverse transcriptase coding sequences in human nucleic acids: relation to primate retroviruses. *J Virol* **63:** 64–75.

Shimizu H, Shimizu C & Burns JC (1994) Detection of novel RNA viruses: morbilliviruses as a model system. *Mol Cell Probes* **8:** 209–214.

Simons JN, Leary TP, Dawson GJ et al (1995a) Isolation of novel virus-like sequences associated with human hepatitis. *Nature Med* **1:** 564–569.

Simons JN, Pilot Matias TJ, Leary TP et al (1995b) Identification of two flavivirus-like genomes in the GB hepatitis agent. *Proc Natl Acad Sci USA* **92:** 3401–3405.

Snijder EJ & Horzinek MC (1993) Toroviruses: replication, evolution and comparison with other members of the coronavirus-like superfamily. *J Gen Virol* **74:** 2305–2316.

Snijder EJ, Horzinek MC & Spaan WJ (1993) The coronaviruslike superfamily. *Adv Exp Med Biol* **342:** 235–244.

Snijders PJF, Meijer CJLM & Walboomers JMM (1991) Degenerate primers based on highly conserved regions of amino acid sequence in papillomaviruses can be used in a generalized polymerase chain reaction to detect productive human papillomavirus infections. *J Gen Virol* **72:** 2781–2786.

Taubenberger JK, Reid AH, Krafft AE, Bijwaard KE & Fanning TG (1997) Initial genetic characterization of the 1918 'Spanish' influenza virus. *Science* **275:** 1793–1796.

Tobler K & Ackermann M (1996) [Identification and characterization of new and unknown coronaviruses using RT-PCR and degenerate primers]. *Schweiz Arch Tierheilkd* **138:** 80–86.

Uchida T, Suzuki K, Hayashi N et al (1992) Hepatitis E virus: cDNA cloning and expression. *Microbiol Immunol* **36:** 67–79.

VanDevanter DR, Warrener P, Bennett L et al (1996) Detection and analysis of diverse herpesviral species by consensus primer PCR. *J Clin Microbiol* **34:** 1666–1671.

VandeWoude S, Richt JA, Zink MC, Rott R, Narayan O & Clements JE (1990) A Borna virus cDNA encoding a protein recognized by antibodies in humans with behavioral diseases. *Science* **250:** 1278–1281.

Wan JS, Sharp SJ, Poirier GMC et al (1996) Cloning differentially expressed mRNAs. *Nature Biotech* **14:** 1685–1691.

Yamanishi K, Okuno T, Shiraki K et al (1988) Identification of human herpesvirus-6 as a causal agent for exanthem subitum. *Lancet* **i:** 1065–1067.

Zhang Z & Martineau D (1996) Single-tube heminested PCR coupled with 'touchdown' PCR for the analysis of the walleye dermal sarcoma virus *env* gene. *J Virol Methods* **60:** 29–37.

INDEX